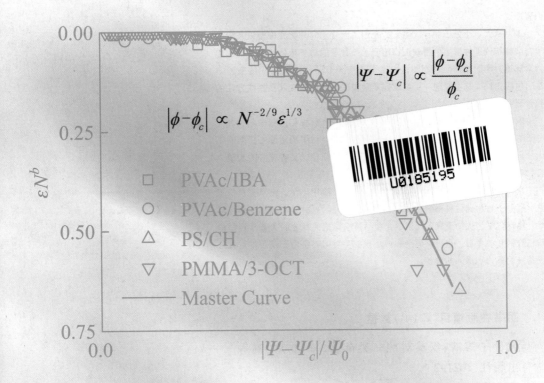

Macromolecular Solutions

大分子溶液

源自一次个人 40 年之旅的教科书
A Textbook from a Personal 40-Year Journey

吴奇 著

高等教育出版社·北京

内容提要

本书从大、小分子的异同出发,介绍了大分子的分类和来源以及大分子在结构和物理性质上的特殊性;阐明了大分子的链长依赖物理性质,包括溶解热力学、黏弹性和相图;枚举了研究大分子溶液时常用的实验方法,包括它们的基本原理、可测量物理量和数据处理方法以及这些方法的综合运用。

本书凝聚了作者近三十年教授"大分子导论"课程和近四十年研究大分子溶液的经验,从基本概念入门,循序渐进,介绍了历史演变,阐明了大分子溶液中的物理概念。本书既有详细的理论介绍,又有充分的实验内容,并结合了大分子溶液研究的最新成果。

本书为中英文对照,可作为高等学校化学、应用化学、化学工程与工艺、高分子科学、材料化学、物理学、应用物理学、材料物理学、环境科学、地球化学、生物工程、药剂学、食品科学等相关专业的大分子课程的双语教材和参考书,也可供硕士和博士研究生复习时参考,以获得对大分子溶液的整体和深入的理解。

图书在版编目(C I P)数据

大分子溶液:汉英对照 / 吴奇著. --北京:高等教育出版社,2021.2

ISBN 978 - 7 - 04 - 055346 - 8

Ⅰ.①大… Ⅱ.①吴… Ⅲ.①高分子溶液-高等学校-教材-汉、英 Ⅳ.①O631.4

中国版本图书馆 CIP 数据核字(2021)第 005507 号

DAFENZI RONGYE

策划编辑	曹 瑛	责任编辑	曹 瑛	封面设计	张 楠	版式设计	杜微言
插图绘制	于 博	责任校对	刘娟娟	责任印制	刘思涵		

出版发行	高等教育出版社	网　址	http://www.hep.edu.cn
社　址	北京市西城区德外大街 4 号		http://www.hep.com.cn
邮政编码	100120	网上订购	http://www.hepmall.com.cn
印　刷	佳兴达印刷(天津)有限公司		http://www.hepmall.com
开　本	787mm×1092mm　1/16		http://www.hepmall.cn
印　张	34.25		
字　数	780 千字	版　次	2021 年 2 月第 1 版
购书热线	010-58581118	印　次	2021 年 11 月第 2 次印刷
咨询电话	400-810-0598	定　价	70.00 元

前言

　　顾名思义,大分子是比那些通常的小分子或离子(如水、盐、糖、酒精、油等)大很多的分子,由若干个小分子通过化学键相接而成。那么大至何倍? 连接多少? 方可称作大分子呢? 粗略地讲,上百个小分子以化学键连接起来即为“大”分子。严格地说,大分子是具备了一些小分子欠缺的特殊物理和化学性质的长链分子。例如,大分子有一些链长依赖的物理性质,以及其在溶液中的形状无时不变,而小分子则无这些性质。

　　当大分子链长于缠结长度时,链间还可出现相互缠绕,导致大分子材料在常温、常压和通常的时间尺度上的黏弹性(同时既黏又弹)。小分子材料的性质则随着其环境(如温度和压强)变化呈现出非弹性(固体)即黏性(液体)。每种小分子有其确定的组成和结构,而合成的大分子链通常并非一致,有长有短、有大有小、组成不一、结构各异。

　　因历史原因,英文“polymers”(合成大分子)有不同的中文翻译,包括“高分子、聚合物或高聚物”。其原意只是“多聚体”,对应于“monomers”(单体)。然而,仅当大分子链长到足以相互缠结时,方可呈现一些小分子没有的物理性质。因此,基于这一大、小分子的本质区别,我们认为高度聚合的物质(简称“高聚物”)可能是“high polymer”最贴切的翻译。这可能也是为何在美国物理学会里与大分子相关的分会称作“High Polymer Physics Division”,而不是简单的“Polymer Physics Division”。如果将“高分子”一词理解成“高度聚合的或具有大摩尔质量的分子”,其翻译也勉强可接受。本书将采用“高聚物”代表高度聚合的物质。

　　整整一百年前,Hermann Staudinger 发表了第一篇有关大分子的科学论文[Berichte der Deutschen Chemischen Gesellschaft, 1920, 53:1073]。在过去的一百年中,人们从概念上到应用中逐步地认识到大、小分子在化学和物理性质上的异同性,涉及诸多学科,包括化学、物理、生物、食品、材料、医疗、工程以及其他领域。这些大、小分子的差异造就了高聚物科学,其源于高聚物的加工,逐渐产生了与加工紧密相关的高聚物化学和高聚物物理两个现代学科分支,故高聚物学科包括加工、合成和物理三个部分。

　　经过百年的飞速发展和工业化,合成大分子已深入人们的现代生活(衣、食、住、行、医……)中的每一个部分。可以毫不夸张地说,现今,从出生后睁开双眼的第一瞬间直至逝去时的最后一刻,无人能够使其视线偏离合成大分子材料。

　　大多数理工科毕业生,尤其是化学、化工、物理、材料等学科的毕业生,在一生中,或早或迟都会遇到制备、应用和处理大分子材料的有关问题。因此,在理工科的大学教纲中加入一门大分子导论的课程不仅是必要的,也是必不可少的。然而,在中国目前的高校课程体系中,除了与高聚物直接相关的专业或学系外,其他学科鲜有开设专门课程介绍大分子科学。

例如,在许多学校的食品专业和学系里,尽管其毕业生的工作主要涉及大分子(多糖、蛋白质、纤维素等),教学大纲并没有涉及大分子基本概念,大、小分子异同性及与食品加工密切相关的大分子物理的专门课程。由于对大分子的概念及其独特的化学和物理性质欠缺基本的了解,许多非高聚物专业的毕业生在其日后的工作中,每当遇到一个与大分子制备、加工或使用有关的问题时,经常不知所措。考虑到这一严重的问题及其深远影响,我决定将在过去四十年里研究大分子溶液的结果和我教授大分子科学的体会总结和综合成这本教科书《大分子溶液》,以飨读者。

本书强调基本概念,聚焦大、小分子的异同性,注重理论与实用的结合,尽可能地避免艰深的统计物理学和烦琐的数学推导,也不追求面面俱到和理论完整。本书可作为凝聚态物理、材料、化学、化工、食品、生物工程等专业本科高年级学生和研究生学习大分子(特别是高聚物)的入门教材。高聚物学系的师生们也可将它作为一本研究大分子溶液的参考书。

与传统大分子教科书中有关溶液的部分相比,除了枚举的许多实例,本书还汇聚了笔者过去四十年中在其工作或主导过的三个实验室里获得的主要研究结果,有以下特点:

尽可能地列出涉及的公式、引理、定律和方程的最早出处和发明人的全名。其目的有三:首先,先辈科学家们在历史上的贡献应被尊重,有些因历史悠久,早已淹没在文献浩卷之中,年轻的读者并不熟悉、甚至根本不知。其次,遵循古训"温故而知新",笔者希望青年读者可以感到科学研究的源远流长。试图了解和驾驭自然的我们,并不孤独,应该学会如何欣赏科学的进步和同行们的成就。再次,希望读者们在阅读和了解了一些过去的故事后,对科学先辈们有仰慕之意、对科学历史有敬畏之心、对自己的研究有自知之明。

同时,本书将大分子分成天然大分子和非天然(合成)大分子两大类。天然大分子包括核糖核酸(RNA)、脱氧核糖核酸(DNA)、蛋白质、多糖(淀粉和纤维素等)、几丁质和树脂等,和生命息息相关。多数时候强调的是它们的化学和生物功能,而不是它们的物理性质。非天然(合成)大分子(高聚物),涵盖了广泛使用的塑料、合成纤维、橡胶制品等种类繁多、数之不尽的材料,与日常生活和工业生产密不可分。以缠结长度为界,可将合成大分子再分成低聚物或高聚物。它们分别与小分子有相似和相异的物理性质。

此外,本书采用了两种与传统分类完全不同的、不重不漏的方法来归类不同的聚合反应。第一种分类方法基于一个聚合反应除了单体以外是否涉及其他小分子:a. 自由基、离子、配位、某些开环聚合等;b. 缩合、某些开环聚合等。第二种分类方法取决于每一条生长的大分子链是否仅有一个引发起点:a. 自由基、离子聚合、配位、某些开环聚合等;b. 缩合、某些开环聚合……实际使用时,可选用其中一种分类方法。这两种新的分类方法都避免了传统高分子教科书中现有分类方法互相交叠或涵盖不全的缺陷。

本书将英文"configuration"翻译成"组构"而不是常用的"构型"。主要有以下两点考虑:首先,对应的英文"configuration"一词源于拉丁文的"con"(together,一起)和"figurare"(to form,构成或形成)两个词根。采用的"组"和"构"分别很好地对应了二者。其次,"configuration"的中文翻译有多种,除了常用的"构型"以外,还包括"组态""构态"等。在小分子合成化学中,采用"构型"习以成俗,毫无疑义。这是因为每个小分子中的原子连接和空间分布方式之间存在一一对应的关系。然而,对一个给定的连接方式(组构或构型),大分子还仍可具有不同的结构;对给定的结构,一条大分子链还有不同的构象(每个单元在空间的分布形式)。因此,也许可以将"组构/构型"一词写成"组构结构/构型结构"。注意:此结构非彼结构。

尽管"组构/构型"和"构象/构形"二者明确地定义为"单体如何连接"和"单体如何在空间中分布","构型"较易和"构形"混淆。主要原因是中文的"型"和"形"读音一样、写法相似,但二者的内涵完全不同。此"型"非彼"形"也!"型"指一个物体所属的"分类或类型","形"则指一个物体的"形状、外形、形象",空间形状。

采用"构型"较易和"构象/构形"混淆。如果教师在讲课时没有每次都反复强调二者之别,或学生们将"构型"一旦误听或误解为"构形",就会混淆这两个完全不同的概念,陷入困境。在英文中,没有这一问题。也许,当初在小分子化学里将"configuration"翻译作"构型"后,已不能再将"conformation"译成"构形",因为二者极易混淆。于是不得不采用了日常生活中极少用的"构象"一词。这只是一个猜想而已。

在"linear polymers or chains"中,英文"linear"一词在中文翻译里对应着三个内涵和外延完全不同的词:"线型""线形"和"线性"。此处,"linear polymers or chains"指的是"单体被线性地连接在一起",为英文副词,而不是形容词。中文里的"线性"一词隐含了两个变量的相互依赖关系,与形状无关。知道一条链是如何合成(聚合类型)和单体连接的细节(如头尾还是尾头连接、等规还是无规连接等)并不重要。因此,本书仅用"线性(地)"描绘一条链是如何增长的;以及用"型"字描绘组构(同一类的单体连接方式),如线型链、支化型链、星型链等。

本书创新地将支化连接当作一个最基本的"根"组构,可按链上有、无多环结构分为两类。无多环:如果每一条支化子链都不再继续支化,就成了梳状组构。当一个梳状组构上所有的支化子链都连接在一个中心点上时,就成了星状组构。当星状组构中只有两条支化子链(两条臂)时,就成了线型组构。当线型链的两个末端相连时,就成了一个大的环状组构。有多环:支化链成为一个三维的空间网络。充分溶胀后成为一个凝胶,可看作一类介于液体和固体之间的特殊物质,也称湿物质。

笔者将与大分子链构象和平均构象的相关内容移至大分子溶液制备之后。这样,可以先回顾读者在物理化学中已学过的小分子溶解和相关的热力学内容。逻辑上,只有先将大分子样品溶解在一个溶剂中分开成一个一个单链,方可讨论有关单链的问题。另外,这一循序渐进的安排避免了先讨论艰深的统计物理学和数学计算,从而使数理基础薄弱的读者不至于产生畏惧心理。

与理想气体相比,真实气体分子具有一定的体积和相互作用。基于"分子达到体系内任意一点的概率",本书阐明了"硬核体积""相互作用"和"排除体积效应"三个相关但容易混淆的概念。在给定温度时,"硬核体积"减少了分子可达及的实际体积,增加了上述概率,导致压强增加。另一方面,"相互作用"会改变分子的运动速度,影响该概率,产生一个虚拟体积。"硬核体积"和"相互作用"产生的效应合而为一称作"排除体积效应"。

对中性分子,相互作用为"由色散引致的相互吸引",其减小上述概率,产生一个负内压效应,等效于减小了排除体积。注意,这只是一个效应,类似"离心力只是一个效应,不是一个真正的力"。一个真实分子的排斥体积总是存在,永远不会消失,更不会为负。相互吸引效应的大小决定了排斥体积效应为正、为零或为负。如果分子间的相互作用为排斥(如电解质或聚电解质),排斥体积效应就将永远为正。清楚地区分和明白三个概念的差别有助于领悟真实溶液中二阶维里系数、了解溶液性质如何随温度/压强变化。

无论是在讨论无规行走的理想链,还是推广后的高斯链时,本书都未引入相关的末端距分布函数,避免了传统教科书中对分布函数的积分,以减少读者的数学壁垒。在阐明链构象统计意义的基础上,本书先从实验的角度引入回转半径的概念、定义和物理意义,再指出依照定义算出一个运动物体或一条不停运动的链的回转半径的难处。在介绍方距和可测的回转半径之间的一个引理之前,本书详细地说明了如何计算物体内所有可能的任意两点之间直线距离的平方之后的加和,也称"方距加和"或简称"方距"。

对大分子链,不得不先考虑和统计平均给定链上两点后所有可能构象的方距以得到方均距。最终的结果称作"方均距加和",即"先对任意两个链段(i 和 j)之间在所有可能链构象中的方距作统计平均得构象方均距,再变化 i 和 j,对所有可能的两个链段的构象方均距求和"。"方均距加和"也简称"方均距"。由一个物体的方均距可得其方均回转半径。

传统教科书都停留在线型链的方距 $b^2|i-j|$,其中 b 为一个链段的长度。本书则给出了一般形式。在无环的任何链组构中,一条链上任意给定的两个链段之间,存在且只存在一条含有 x 个链段连接二者的线型路径,故二者之间的方均距为 b^2x。线型链仅是一个特例,$x=|i-j|$。本书以支化的梳状链为例,详细地阐明了如何计算一给定组构的大分子链构象的平均回转半径。希望读者可举一反三,求出其他组构链的平均回转半径。

本书还介绍了传统教科书避免的支化组构。即使给定了组成和组构,支化聚合反应还可生成具有完全不同连接细节的各种支化结构,包括不同的支化点数和不同的支化子链长度等。理论上,一个支化反应可产生从线型到超支化的各种结构。因此,除了"构象平均"外,还有另一个"结构平均"。利用一个简化的物理图像和标度理论代替繁杂的数学推导,本书阐述了如何对支化组构做两次统计平均,以得到支化链的平均回转半径分别与母链和子链的摩尔质量之间的一个双重标度关系。结果与基于艰深物理原理和复杂数学计算得到的完全吻合。在考虑了排除体积效应后,还介绍了上述标度关系如何随着溶剂性质变化。

在介绍每种常用的实验方法时,本书的一个特色是先给出基于物理直觉和图像的结论,再演示基于热力学原理的严格推导,有点类似"先讲解普通力学,再讨论理论力学"。对于每一种方法,本书主要关注的是其基本原理、其可测的物理参量,以及如何通过处理其实验数据寻获所需的物理量。特别地介绍了不同实验方法之间的内在联系。

因此,连用两个或多个实验方法,只需一个多分散的样品就可标定仪器。该部分内容独特,异于现存的其他大分子教科书。本书还有另一个特点:在进行必要的数学演绎时,尽量列出详细步骤,以帮助有较少数学训练的读者,尤其是化学、生物、食品科学等专业的学生们。因此,数学基础扎实的读者可能觉得啰唆和画蛇添足,敬请原谅。

自 20 世纪 60 年代起,分析型超速离心仪逐渐地淡出视线。在 20 世纪 90 年代,由于快电子、数字检测器、计算机及材料科学的飞速发展,其又慢慢地重新回到一些大分子研究实验室。现在，其转速更快、测量时间更短、采集数据更方便，以及处理信息更迅速。因此，本书详细地从理论和实验上介绍了超速离心技术。希望可以帮助读者了解这一研究大分子溶液和胶体粒子的有力方法。几乎每一本传统的大分子教科书都列出了不同平均摩尔质量的定义，但没有强调和阐明它们是如何源于不同的实验方法。本书努力避免这样一个错误，将每一种平均摩尔质量和其相关的实验方法连接。

英文"dynamics"和"kinetics"在中文中常常都被翻译成"动力学"，翻译无误！均指随时间的物理或化学变化。"kinetics"侧重"力和力矩"对物体运动的影响，在力学中也称为"kinematics"，在讨论化学反应速率时也常用；而"dynamics"则强调能量对变化的影响，具有更广的范畴，包括了"kinetics"。自20世纪20年代中叶，物理学家就逐渐地在教科书中采用"dynamics"一词。对应于"静态学"（研究平衡体系），将热能对大分子链整体和局部构象以及质心运动的动态影响称作"动态学"，而不是"动力学"可能更为恰当。

因此，本书将使用"动态学"。然而，在讨论大分子链从一个构象变成另一个构象的变化过程时，仍然使用"动力学"（kinetics）一词。另外，在讨论大分子相变、介观球相和胶体时，也使用"动力学稳定"描绘非平衡的亚稳态，以区分在平衡时的"动态稳定"。

讨论大分子动态学时，复杂和烦琐的理论推导有着其完美和优美的一面，但也让大部分实验工作者眩晕。撇开数学公式，仅从物理图像出发，理解链内不同简正模式的涨落振幅和特征弛豫时间并不复杂。对一条构象为无规线团的线型柔性大分子链，其链段相对于质心的运动千变万化。

如果不考虑各种相互作用，最慢的松弛时间应该和链运动一个自身尺寸的时间相近；内部运动的最大振幅（即整条链的转动/振动）应该接近但不会超过链本身的尺寸（平均末端距，R）；最小的涨落就是一个链段的长度（b），即两个相邻链段之间的相对运动。

另一方面，如果考虑流体力学相互作用的过度阻尼，最大的涨落幅度就应小于大分子链的平均末端距，而最慢的松弛时间应该比链移动一个自身尺寸的时间更慢。在本书中，链的各种简正模式（内部运动）的振幅和弛豫时间都被转换成可测的平均回转半径R_g和扩散一个R_g的距离所需的时间。本书还介绍一种更符合物理定义的、利用链的内部运动的相对贡献来确定Θ状态的全新实验方法。

本书还专门用一章介绍了近三十年来，两个关于大分子链在稀溶液中构象变化的最新实验结果和相关理论。其一，在一相区中，随着溶剂性质变差，链构象从无规线团蜷缩成均匀小球，中间经过一个新的"融化球"构象。该构象转变为单条大分子链上的一级相变，遵循大分子溶液一级相变的规律，即在链上形成若干个局部涨落和蜷缩的"核"。当其中一个核的尺寸超过临界值时，其将逐渐并吞较小的核，经过一个"粗粒化"过程，最终形成一个蜷缩球。其二，在拉伸流场中，当流量到达一临界值时，链构象发生一级突变，突然地从卷曲到伸展。讨论了临界流速与不同链组构之间的定量关系和相关理论。还展示了使用这些突变按照组构而不是尺寸分离大分子链的最新应用。

第十一章（大分子溶液中的介观球相）和第十二章（大分子胶体）原本可以并入第十章（大分子溶液的相变），考虑再三，还是将它们分为三章更好，以便读者更容易理解三者之间的差异。与小分子溶液相变类似，随着溶剂性质变差，大分子溶液也会分离成稀相（大分子在溶剂中的饱和溶液）和浓相（溶剂在大分子中的饱和溶液，即被溶剂溶胀的大分子)。与小分子溶液相变不同的是,大分子链在分相前后能够蜷缩。正是这一链内蜷缩和链间缔合之间的竞争造成了大分子聚集体特有的动力学稳定。

严格地说,介观球相可以归于大分子胶体。它们都是分相时形成的微浓相(液滴,或曰聚集体和粒子)。在这一亚稳态中,存在着一个稀相(一条条单链在溶剂中的饱和溶液)和微浓相之间的动态平衡。前提是微浓相的尺寸要足够的小、密度要接近分散介质(溶剂),故热能可以驱动它们在混合溶液中无规行走,在一定的时间区间内不会因沉降或上浮而出现宏观分层。依据是否需要额外的稳定剂以及其稳定是否主要为热力学控制,可粗略地将大分子溶液的介观球相从大分子胶体中分割出来,独成一章。

大分子溶液的介观球相还有不同的、新颖的称呼,包括"自组装"和"纳米粒子"等。然而,万变不离其宗!其和路径有关的亚稳态本质不变,其受动力学影响的稳定机理也不变。由于介观球相和大分子胶体依赖于其形成时的具体路径,所以不存在普适的物理规律。它们是运用而不是创造知识的两个典型例子。利用这一与路径有关的特性,从同一个体系出发,可形成不同的聚集体,产生不同的新材料,以解决日常生活、工业生产和医学中的问题。虽然此类研究没有知识上的贡献和真正的学术价值,但其具有一些应用价值。因此,该研究领域应聚焦在运用知识解决具体的实际问题。

感谢使用或采用此书作为教科书或参考书的教师和学生们。本书基于我对大分子溶液的理解以及我本人的教学和研究笔记,难免存在一些疏漏和错误,如蒙指出,不胜感激,以便再版时改进。

我还想借此机会由衷地感谢在我长达四十年求学和科研的旅途中曾经帮助过我的所有人。特别地感谢我的博士导师朱鹏年教授,是他引导我进入了大分子溶液研究的领域;与我亦师亦友的应琦琮教授,是她在我开始博士生学习时,传授给我很多实验技巧;我在德国巴斯夫公司总部工作时的导师和部门领导 Dieter Horn 博士,正是他为我提供了得以在工业界学习和研究三年的机会,以及我的组长 Eric Lueddeck 博士和他的夫人 Barbel Lueddeck。

感谢香港中文大学化学系,自一九九二年以来给了我一个教授"大分子导论"课程的讲台和研究大分子溶液的实验室。

我也想感谢我以前的老师、同事、合作者和学生们多年来对我的帮助、支持、鼓励、理解和疑义相与析。名单包括但不限于,中国科学技术大学的俞书勤教授、何天敬教授、马兴孝教授、朱清时教授,香港中文大学的黄乃正教授、欧阳植勤教授、谢作伟教授、夏克清教授、前校长高琨教授,中国科学院化学研究所的钱人元教授,南京大学的程镕时教授和游效曾教授,武汉大学的卓仁禧教授,复旦大学的江明教授,国家自然科学基金委员会的胡汉杰教授,美国芝加哥大学的何川教授和 Karl Freed 教授。

我也感谢以前的学生们,包括但不限于,吴加辉、周水琴、Muhammad Siddiq、汪晓辉、张玉宝、高均、彭树馥、魏涛、叶晓东、金帆、周科进、葛慧、何传新、龚湘君、李连伟、岳亚男、戴卓君、王建其、王艳景等人,以及曾经和现在的博士后和合作者:李梅、张广照、朱丹、刘世勇、刘光明、何卫东、梁好均等人,以及我的研究助理刘璐。

我还要感谢在家乡的一群青少年时的朋友,在我远离故乡的四十年间,他们逢年过节时总去看望我的父母;在我父母需要帮助时,总是伸出友谊之手。有他们,父母得到慰藉,我感到心安。

我更想谢谢所有家人多年来对我的倾心关爱、支持和信任。当年,尽管心中有着百般的不舍,但父母仍然为我默默地准备去美国的行装。后来,母亲又万里迢迢地来到美国帮助我们照顾年幼的孩子,使得我们可以专心地学习和工作。可是我却违背了古训:父母在,不远游! 在我留学美国时,我的祖母于 1988 年去世,我至今一直特别地遗憾没有机会给她老人家送终。

多年来,我唯一的妹妹和妹夫一直对年迈的父母精心照顾,使得作为游子的我可以远离家乡芜湖进行了一次长达四十年之久的个人求学和研究之旅。每念对家人永远无法弥补的所欠之情,我总是十分地愧疚和不安。仅将本书献给在我学习和研究的旅途中逝去的亲人们,愿他们在另一个世界里只有平安和幸福! 永无忧患和痛苦!

二〇二〇年二月十三日至二〇二〇年九月十三日作于"食草堂"

目录

Preface

As the name suggested, macromolecules are molecules much bigger than those common small molecules or ions (e. g., water, salt, sugar, alcohol, oil, etc.), made of many small molecules connected by chemical bonds. The questions are how large is "large" and how many is "many" to be called as macromolecules. Roughly speaking, hundred or more small molecules linked by chemical bonds are "big" molecules. Strictly speaking, macromolecules are long-chain molecules with some special physical and chemical properties that small molecules lack. For example, macromolecules have some chain-length dependent physical properties, and its shape in a solution constantly changes, while small molecules have no such properties.

When macromolecular chains are longer than the entanglement length, the intertwining between the chains can occur, leading to the viscoelasticity (viscous and elastic at the same time) of macromolecular materials at room temperature and normal pressure, as well as a usual time scale. The properties of a small molecular material are either elastic (solids) or viscous (liquids) as its environment (e. g., temperature or pressure) changes. Each kind of small molecules has its own defined composition and structure, while those synthesized macromolecular chains are normally not identical, longer or shorter, bigger or smaller, with different compositions and structures.

Due to historical reasons, " polymers " (synthetic macromolecules) have different translations in Chinese, including " high molecules, polymerized substances or highly polymerized substances ". Its literal meaning is only "multi-mers", corresponding to "monomer". However, only when the chains are sufficiently long that they can intertwine with each other, macromolecules can exhibit some physical properties that small molecules lack. Therefore, based on such an essential difference between macro-and small molecules, we believe that "high polymers" might be the most suitable translation of highly polymerized substances, simply called as " polymers ". This is why the division related to macromolecules in the American Physical Society is called " High Polymer Physics Division ", not simply " Polymer Physics Division ". If the term "high molecules" is correctly understood as "highly polymerized and entangled molecules" or "molecules with a higher molar mass", such a translation is also acceptable. This textbook will use "polymers" to represent "highly polymerized substances".

The first scientific paper related to macromolecules was exactly published one hundred years ago by Hermann Staudinger [Berichte der Deutschen Chemischen Gesellschaft, 1920, 53: 1073]. In the past one hundred years, people have gradually realized chemical and physical similarities and differences between small and macromolecules from concepts to applications, involving many subjects, including chemistry, physics, biology, food, materials and other fields. The differences between small and macromolecules created polymer science, which originated from polymer processing, gradually resulting in two modern disciplines: polymer chemistry and physics, so that polymer science includes three parts: processing, synthesis and physics.

After hundred years of rapid developments and industrialization, synthetic macromolecules have penetrated into every part of our modern life (clothing, food, housing, transportation, medicine …). It is no exaggeration to say that in nowadays, no one is able to deviate her/his sight from synthetic macromolecular materials, from the first moment of opening eyes after birth until the last breath before passing away.

Most of graduates from science and engineering, especially those from chemistry, chemical engineering, physics, materials and other disciplines, soon or later, will encounter problems related to the preparation, application and processing of macromolecular materials. Therefore, it is not only necessary, but also essential, to add an introduction course of macromolecules in the university syllabus of science and engineering. However, in the current system of high education in China, besides majors or departments directly related to polymers, few other disciplines offer a specialized course to introduce macromolecular science.

For example, in the food majors and departments of many schools, even jobs of their graduates mainly involve macromolecules (polysaccharides, proteins, cellulous and so on), there is no specialized courses in the syllabus, which involve basic concepts of macromolecules, differences between small and macro-molecules and macromolecular physics closely related to the food processing. Due to the lack of basic understanding about the concepts of macromolecules and their special chemical and physical properties, many graduates in non-polymer majors are often at a loss whenever they encounter a problem related to the preparation, processing or application of macromolecules in their daily works. Considering this serious problem and its far-reaching impact, I decided to summarize and combine our research results of studying macromolecular solutions in the last 40 years and my experience of teaching macromolecular science into this book "Macromolecular Solutions" for readers.

This book emphasizes basic concepts, focuses on similarities and differences between small and macro-molecules, concentrates on a combination of theories and applications, avoids difficult statistic physics and tedious mathematic derivation as much as possible, and also pursues no comprehensiveness and theoretical integrity. This book can be used as an introduction textbook of learning macromolecules (especially polymers) for senior undergraduate and postgraduate students in condensed matter physics, materials, chemistry, chemical engineering, food, bioengineering, and so on). Teachers and students in polymer departments can also use it as a reference book in studying macromolecular solutions.

In comparison with the solution part in traditional macromolecular textbooks, besides a list of many real examples, this textbook also gathers the main research results obtained by the author in three laboratories where he has worked or led over the past four decades, with the following characteristics.

This book lists all the formulas, lemmas, laws and equations involved in the earliest sources and the full names of the inventors as far as possible. It has three purposes. First, the contributions of scientific ancestors in history should be respected; some of them have been submerged in the vast volume of historical documents because of their long history, so that young readers are not familiar with it, or even don't know it at all. Second, following an ancient motto, "It is such a great pleasure to review the past knowledge and learn something new", the author hopes that young readers can feel the long history of scientific research. As members of human beings who are trying to understand and manipulate nature, we are not alone and should learn how to appreciate the progress of science and the achievements of their peers. Finally, hope that after reading and knowing some past stories, readers will admire scientific ancestors; awe scientific history; and correctly judge their own work.

Macromolecules are divided into two categories: natural and unnatural (synthetic). Natural macromolecules include ribonucleic acid (RNA), deoxyribonucleic acid (DNA), proteins, polysaccharides (starch and cellulose, etc.), chitin and resins, etc., closely related to life. Most of the time, the emphasis is on their chemical and biological functions, not their physical properties. Non-natural synthetic macromolecules (polymers), covering a very wide range of innumerous materials, such as widely used plastics, synthetic fibers, rubber products, etc., inseparable from daily lives and industrial productions. Taking the entanglement length as a boundary, synthetic macromolecules can be subdivided into oligomers and polymers, respectively. They respectively have similar and different physical properties with small molecules.

This textbook adopts two new ways to reclassifies different polymerizations, which are completely different from the traditional classifications, not only covering all polymerization, but also no overlapping. The first category is based on whether a polymerization involves other small molecules except monomers: a) radical, ionic, coordination, and certain ring-opening polymerizations, etc.; b) condensation, some ring-opening polymerization, etc. The second category depends on whether each growing macromolecular chain has only one initiation point: a) free radicals, ionic polymerization, coordination, some ring-opening polymerization, etc.; b) condensation, some ring-opening polymerization, etc. In the actual use, one of the classification methods can be selected. Logically, these two new classification methods avoid the overlap or incomplete coverage of existing classification methods in traditional polymer textbooks.

This book translates the English "configuration" into "organized structure" instead of the commonly used "structural form". There are two following considerations. First, the word "configuration" is derived from two Latin root words, "con" (together) and "figurare" (to form). The adoption of "organized" and "structure" corresponds them well, respectively. Second, there are different translations of "configuration" in Chinese, including "organization state, structural state, etc." besides the commonly used "structure form". In small molecular synthetic chemistry, it has become customary to use the "structure form", no ambiguity. This is because there is a one-to-one correspondence between how atoms are connected and spatially distributed in every small molecule. However, for a given way of connection (configuration), macromolecules can still have different structures. For a given structure, a macromolecular chain still has different conformations (spatial distributions of every unit). Therefore, it might be better to write the word "configuration" as "configurational structure". Note: this structure is not that structure.

Although the "organized structure/structural form" and "conformation/structural shape" are clearly defined as "how monomers are specifically connected" and "how monomers are spatially distributed", respectively, the "structural form" is easily mixed up with the "structural shape". The main reason for the confusion is that in Chinese the "form" and "shape" have the same pronunciation and a similar writing, but their connotations are completely different. This "form" is not that "shape"! The "form" implies to which "category and kind" an object belongs, and the "shape" refers to the "shape, appearance, and image" of an object, its spatial shape.

The adoption of "structural form" is easily mixed up with "structural imagine/ structural shape". If a teacher does not emphasize their differences each time repeatedly during a lecture, or if students mishear and misunderstand the "structural form" as the "structural shape", they will mix up these two completely different concepts, trapping in a difficult situation. In English, there is no such a problem. Perhaps, after the "structural form" was customarily translated as the "configuration" in small molecule chemistry, it is not possible to translate the "conformation" into the "structural shape" because they are extremely easily mixable. Therefore, in daily life, a rarely used term "structural imagine" has to be adopted. This is just a guess.

The English word "linear" in the "linear polymers or chains" corresponds to three words with completely different connotations and extensions in its Chinese translation: "linear form", "linear shape" and "linear relation". Here, the "linear polymers or chains" imply that the "monomers are linearly connected", an English adverb, not an adjective. In Chinese, the word "linear" implies the dependence of two variables, which has nothing to do with "shape". In physics, it is not important to know how a chain is synthesized (types of polymerization) and the details of the monomer connection (e.g., a head-to-tail or tail-to-head link, isotactic or atactic, etc.). Therefore, this textbook only uses "linearly" to describe how a chain grows; and uses the word "form" to describe the configuration (the same type of monomer connection), e. g., linear chains, branched chains, star chains, etc.

This book **creatively** regards the branched connection as the mostly fundamental "**root**" configuration, which is further divided into two groups, depending on whether there are many ring structures on a polymer chain. Without many rings, if each branching subchain no long continues to branch, it becomes a **comb configuration**. When all branching subchains on a comb chain are connected to a center point, It becomes a **star configuration**. When a star chain has only two branching subchains (two arms), it becomes a **linear configuration**. When the two ends of a linear chain are connected, it becomes a **ring configuration**. With many rings, the branched chain becomes a three dimensional **spatial network**. After sufficiently swollen, it becomes a gel, which is viewed as a special kind of matters between solids and liquids, also called wet matters.

This book moves the content related to the chain conformation and average conformation after the macromolecular solution preparation. In this way, the dissolution of small molecules and related thermodynamics, which readers had learned in physical chemistry, can be reviewed first. Logically, only after a sample of macromolecules is first dissolved in a solvent, separating into individual chains, one can discuss the issues of a single chain. In addition, this step-by-step arrangement avoids the discussion of difficult statistical physics and mathematical calculations first, making readers with weak mathematical background fearful.

In comparison with an ideal gas, molecules in a real gas have a certain volume and interaction. Based on "the probability of a molecule reaching any point in a system", this book clarifies three related but easily confused concepts of "the hardcore volume", "the interaction" and "the excluded volume effect". For a given temperature, "the hardcore volume" reduces the actual volume reachable by gas molecules, increasing the above probability, leading to an increase in pressure. On the other hand, "the interaction" can alternate the moving velocity of molecules, affecting the probability, generating a virtual volume. **A combination of "hardcore volume" and "interaction" is called "the excluded volume effect".**

For neutral molecules, the interaction is dispersion induced attraction, which reducing the above probability, generating **an effect** of "negative pressure", **equivalent to** a decrease of the excluded volume. Note: this is only **an effect**; similar to that "the centrifuge force is only an effect, not a real force". The hardcore volume of a real molecule always exists, let alone disappears or becomes negative. **The magnitude of the attraction effect determines whether the excluded volume effect is positive, zero or negative.** If the interaction among molecules is repulsive (electrolytes or polyelectrolytes), **the excluded volume effect will always be positive.** Clearly distinguishing and understanding differences among these three concepts will be helpful in comprehending the physical meaning of the second virial coefficient of real solutions and knowing how solution properties vary with temperature/pressure.

Whether it is discussing the random walking of an ideal chain or the extended Gaussian chain, this book introduce no relevant distribution function of the end-to-end distance, which avoids the integration of the distribution function in traditional textbooks, reducing mathematical barriers for readers. On the basis of clarifying the statistical significance of the chain conformation, this book first introduces the concept, definition and physical meaning of the radius of gyration from the experimental point of view, and then, state how difficult to calculate the radius of gyration of a moving object or an ever-moving chain according to its definition. Before introducing a lemma between the square distance and the measurable radius of gyration, this book illustrates how to calculate the sum of the square direct distance between all possible two points in an object in details, also called "the sum of square distance" or simply "the square distance".

For macromolecular chains, all possible conformation of the square distance of a given two points on the chain has to be considered and statistically averaged first to obtain the mean square distance. The result is called "the sum of mean square distances". Namely, first statistically average the square distance between any two segments (i and j) in all possible conformations to obtain the conformational average square distance, then change i and j to sum the conformational average square distances between all possible two chain segments. The sum of mean square distances is also simply called as "the mean square distances". The mean square radius of gyration of a subject can be obtained from its mean square distance.

Traditional textbooks stay at the square distance $b^2|i-j|$ of a linear chain, where b is the length of one segment. This book gives a general form. On any chain configurations with no ring, for any two given segments on a chain, there exists and only exists one linear path with x segments to connect them, so that the mean square distance between them is b^2x. The linear chain is just a special case, $x = |i-j|$. Using the comb chain as an example, This book explains how to calculate the average radius of gyration of a macromolecular chain in detail. Hope that readers can learn by analogy, and calculate the average radius of gyration of a chain with any configuration later.

This book also introduces the branched configuration that traditional textbooks avoid. Even for a given composition and configuration, a branching polymerization can still generate various branched structures with completely different connecting details, including different numbers of branching points and different subchain lengths. Theoretically, a branching reaction can produce various structures from linear to hyperbranched. Therefore, in addition to the "conformational average", there is another "structural average". Using a simplified physical picture and the scaling theory to replace the complicated mathematical derivation, this book explains how to do twice statistical average to obtain a double scaling relationship between the average radius of gyration of the branched chains and the molar masses of the parent and sub-chain, respectively.

The results are completely consistent with those obtained based on difficult physical principles and complex mathematical calculations. After taking into account the excluded volume effect, how the above scale relationships change with the solvent quality is introduced, too.

When introducing each commonly used experimental method, one feature of this book is to first present conclusions based on physical intuition and pictures, and then, show rigorous derivations based on thermodynamic principles, somewhat similar to "first explain general mechanics, and then, discuss theoretical mechanics". For each method, this book mainly focuses on its basic principle; its measurable physical parameters; and how to find the required physical quantity by processing its experimental data. The internal relationships among different experimental methods are specially introduced.

Therefore, using two or more experimental methods in conjunction, only one polydisperse sample is needed to calibrate the instruments. The content of this part is unique, different from other existing textbooks of macromolecules. This book has another feature. When performing necessary mathematical deductions, the detailed steps are listed as many as possible to help readers with less mathematic trainings, especially students in chemistry, biology, and food science. Therefore, readers with a solid foundation in mathematics may find it superfluous. The forgiveness is requested here.

Since the 1960s, analytical ultracentrifuge instruments have gradually faded out of sight. In the 1990s, due to the rapid development of fast electronics, digital detectors, computers, and materials science, it slowly returned to some macromolecular research laboratory. Now, it can rotate faster, shorten the measuring time, collect the data more conveniently, and process the information more quickly. Therefore, this book introduces ultracentrifugation in theoretical and experimental details. Hope to help readers to understand this powerful method for studying macromolecular solutions and colloidal particles. Almost every traditional textbook of macromolecules just lists definitions of different average molar masses, but does not emphasize and clarify how they are derived from different experimental methods. This book tries to avoid such a mistake, and link each kind of average molar mass to its related experimental methods.

The English "dynamics" and "kinetics" are often translated as "dynamics" in Chinese, no mistake in the translation! Both refer to physical or chemical changes over time. "Kinetics" focuses on the influence of "forces and moments" on the motion of an object, also called "kinematics" in mechanics, and also commonly used in discussing the chemical reaction rate; while "dynamics" emphasizes the influence of energy on the changes, having a broader scope, including kinetics. Since the mid-1920s, physicists have gradually adopted the term "dynamics" in textbooks. Corresponding to "statics" (study systems in equilibrium), it may be more appropriate to address dynamic influences of thermal energy on overall and local conformation of macromolecular chains as well as on the movement of the center of mass as "dynamics" instead of "kinetics".

Therefore, this book will use "dynamics". However, when discussing the changing process of macromolecular chains from one conformation to another, the term "kinetics" is still used. In addition, when discussing macromolecular phase transitions, mesoglobular phases and colloids, the "kinetic stability" is also used to describe the metastable non-equilibrium state to distinguish the "dynamic stability" in equilibrium.

When discussing the dynamics of macromolecules, the complex and tedious theoretical derivation has its perfect and beautiful side, but it also stuns most experimentalists. Leaving mathematical formulas on the side, only starting from the physical picture, it is not complicated to understand the intrachain fluctuation amplitude and characteristic relaxation time of different normal modes. For a linear flexible macromolecular chain with a random-coil conformation, the motion of its segments relative to the center of mass is ever changing.

If various interactions are ignored, the slowest relaxation time should be close to the time for the chain to move a distance of its own size. The maximum amplitude of the internal motions (i.e., the rotation/vibration of the entire chain) should be close to but not exceed the chain size itself (the average end-to-end distance, R); the smallest fluctuation is the length of a chain segment (b), i.e., the relative movement between two adjacent chain segments.

On the other hand, if the overdamping of the hydrodynamic interaction is considered, the maximum fluctuating amplitude should be less than the average end-to-end distance of a macromolecular chain, and the slowest relaxation time should be slower than the time for the chain to move its own size. In this book, the amplitude and relaxation time of various normal modes of the chain (the internal motions) are converted into the measurable average radius of gyration R_g and the time required to diffuse a distance of one R_g. This book also introduces a new and more physically defined experimental method of using the relative contributions of the internal motions of the chain to determine the Θ state.

This book also devoted a chapter to introduce the latest experimental results and related theories about the conformational changes of macromolecular chains in dilute solutions in the past three decades. First, in the one-phase region, as the solvent quality becomes poorer, the chain conformation collapses from a random coil into a uniform globule, undergoing a new "molten globule" conformation in the middle. This conformation transition is the first-order phase transition on a single macromolecular chain, following the law of the first-order phase transition of macromolecular solutions; namely, a number of local fluctuations and collapsed "nuclei" on the chain are formed. When the size of one of these "nuclei" is exceeds the critical value, it will gradually swallow smaller nuclei, via the "coarse-grain" process, finally forming a collapsed globule. Second, in an elongation flow field, when the flow rate reaches a critical value, the chain conformation undergoes a sharp first-order transition, suddenly from a coiled state to a stretched state. The quantitative relationship and related theories between the critical flow rate and different chain configurations are discussed. The latest applications of using these transitions to separate macromolecular chains according to their configurations instead of size are also demonstrated.

Chapter 11 (Mesoglobular Phase of Macromolecular Solution) and Chapter 12 (Macromolecular Colloids) could have been merged into Chapter 10 (Phase Transition of Macromolecular Solution), but it is better to divide them into three chapters, so that readers can understand their differences more easily. Similar to the phase transition of a solution of small molecules, a macromolecular solution also separates into a dilute phase (a saturated solution of macromolecules in solvent) and a concentrated phase (a saturated solution of solvent in macromolecules, i.e., macromolecules are swollen by solvent) as the solvent quality becomes poorer. Different from the phase transition of a solution of small molecules, each macromolecular chain is able to shrink before and after the phase separation. The competition between the intra-chain contraction and the inter-chain association results in the unique dynamic stabilization of macromolecular aggregates.

Strictly speaking, the mesoglobular phase can be attributed to the macromolecular colloids. They are the micro-concentrated phases (liquid droplets, or called aggregates or particles) in the metastable state formed during the phase transition. In this metastable state, in which there is a kinetic equilibrium between the dilute phase (a saturated solution of individual chains in solvent) and many micro-concentrated phases. The premise is that the size of these micro-concentrated phases must be sufficiently small and their densities should be close to the dispersion medium (solvent), so that the thermal energy can drive them to walk randomly in the solution mixture. In this way, there will be no macroscopic separation due to sedimentation or floating in a certain period. Depending on whether additional stabilizers are needed and whether their stability is mainly controlled by thermodynamics, the mesoglobular phase of macromolecular solutions can be separated from macromolecular colloids, forming an independent chapter.

There are also different and novel names for the mesoglobular phase of macromolecular solutions, including the "self-assembly" and "nanoparticles". However, these invariably changes touch no fundamentals! Its path-related metastable nature remains unchanged, and its stabilization influenced by kinetics remains unchanged, too. Since both the mesoglobular phase and the macromolecular colloids depend on each specific path on which they are formed, there exists no universal physical law. They are two typical examples of applying rather than creating knowledge. Using this path-related characteristic, starting from the same system, different aggregates can be formed, producing different new materials for solving problems in daily life, industrial production, and hospitals. Although this kind of researches has no intellectual contribution and no real academic value, it has some application values. Therefore, this research field should focus on applying knowledge to solve specific practical problems.

Thank teachers and students who used or adopted this book as reference or textbook. This book is based on my understanding of macromolecular solutions as well as my personal teaching and research notes. There are inevitably some omissions and errors. I would be very grateful if anyone points any mistakes in this book, so that the reprint will be improved.

I would like to take this opportunity to thank sincerely everyone who has helped me during my forty years journey of study and scientific research. Special thanks to my PhD supervisor, Professor Benjamin Chu, who guided me into the research field of macromolecular solutions; my teacher and friend, Professor Qicong Ying, who taught me many experimental skills when I was starting my PhD study; my mentor and departmental head, Dr. Dieter Horn, when I worked in the headquarter of BASF, Germany, who provided me with an opportunity to gain three years of industrial experience, and my group leader, Dr. Erik Lueddeck and his wife, Barbel Lueddeck.

Thank the Department of Chemistry, the Chinese University of Hong Kong, for giving me a podium to teach "Introduction to Macromolecules" and a laboratory for studying macromolecular solutions since 1992.

I would also like to thank my former teachers, colleagues, collaborators and students for their helps, supports, encouragements, understanding and discussion over the years. The list includes but not limited, Professors Shuqin Yu, Tianjin He, Xinxiao Ma, Qingshi Zhu from The University of Science and Technology of China; Professors Nai Ching Henry Wong, Steve Chik Fun Au-yeung, Zuowei Xie, Keqing Xia, former President Charles K. Kao from The Chinese University of Hong Kong; Professor Renyuan Qian from The Institute of Chemistry, Beijing, Professors Rongshi Cheng and Xiao zeng You from Nanjing University; Professor Renxi Zhuo from Wuhan University; Professor Ming Jiang from Fudan University; Professor Hanjie Hu from The National Natural Science Foundation; Professors Chuan He and Karl Freed from the University of Chicago.

My gratitude also goes to my former students, including but not limited to the following names, Ka Fai Woo, Shuiqin Zhou, Muhammad Siddiq, Xiaohui Wang, Yubao Zhang, Jun Gao, Shufu Peng, To Ngai, Xiaodong Ye, Fan Jin, Kejin Zhou, Hui Ge, Chuanxin He, Xiangjun Gong, Lianwei Li, Yanan Yue, Zhuojun Dai, Jianqi Wang, Yanjing Wang; as well as former and current postdoctoral fellows and collaborators, Mei Li, Guangzhao Zhang, Dan Zhu, Shiyong Liu, Guangming Liu, Weidong He, Haojun Liang, et al, and my research assistant Lu Liu.

I would also like to thank a group of my teenage friends in my hometown. During the 40 years I was away, they have always visited my parents during holidays and extended their hands of friendship to my parents whenever they needed helps. With them, my parents get comfort, and I get peace of mind.

I would also like to thank all my family members for their love, support and trust for me over the years. Back then, despite all reluctance in their hearts, my parents still silently prepared luggage for me to go to the United States. Later, my mother travelled thousands of miles to the United States to help us taking care of our young children so that we could focus on our studies and works. But I violated an old motto: when your parents are alive, do not travel far! When I was studying in the United States, my grandmother with whom I had great affection passed away in 1988. I have been very regretful until today for not being able to say the last goodbye to her.

Over the years, my only sister and brother-in-law have been taking care of my elderly parents, so that I as a wanderer can take a personal journey of study and research away from my hometown of Wuhu for forty years. Whenever I think about the debt that I can never make amends to my family, I feel deep guilty and disturbed. I wish to dedicate this book to my family members who have passed away during my study and research journey. May they only have peace and happiness in another world! Never feel worry and pain!

From February 13, 2020 to September 13, 2020, written in "Shi Cao Tang"

Contents

第一章　大分子的定义和来源

大分子通常定义为由成百上千个相同或不同的小分子(单体)通过化学键连接而成的分子。一个大分子上聚合了的单体总数被称为聚合度。大分子的摩尔质量(M)等于每条链的质量乘以 Avogadro 常数($N_{AV} = 6.02 \times 10^{23}\ mol^{-1}$)。小分子的尺寸约为一个纳米,而大分子通常则大了百倍左右。某些特殊大分子的链长则可达几十微米,包括制备超强纤维用的超高摩尔质量聚乙烯、作为标样的聚苯乙烯,以及三次采油时用的聚丙烯酰胺长链等人工合成大分子。脱氧核糖核酸(DNA)的长度甚至可达数米。随着链长(L)的增加,大分子链在较高浓度的溶液里或在熔融或固体状态下会互相纠结缠绕在一起,导致大分子构成的材料同时具有独特的、既黏又弹的物理性质,即黏弹性。每一个给定的大分子均有一个相应的缠结长度(L_e)或缠结聚合度($N_e \sim 10^2$)。

当链短于L_e或聚合度低于N_e时,增加浓度仅会导致每条链在空间里所占据的体积收缩,而不会引致它们的缠结。这样的"大分子"就和缺乏缠结的小分子一样,随着环境(如温度和压强)的变化,仅可展现出固体弹性或液体黏性,如右图所示。文献中,许多人在研究大分子物理性质时往往因忽略了缠结长度这一重要的概念而犯了一些本可避免的低级错误。还要注意:在文献和一些

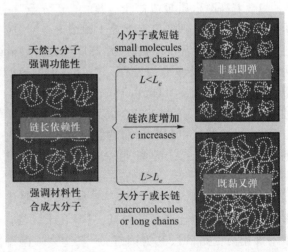

教科书中,一些由数十个单体(小分子)连接在一起短于缠结长度的链也被称为大分子。严格地说,它们应被称为寡聚链、寡聚分子或寡聚物。实际上,只有当链长或聚合度为缠结长度或聚合度的五至十倍时,由大分子组成的材料才可清楚地表现出小分子材料没有的黏弹性。

除了黏弹性以外,大分子与小分子还有许多其他差别。例如,大分子一般无法通过结晶的方法纯化,而只能利用溶解性对于链长的依赖性将不同长度的聚合物链分开,即逐渐地加入不良溶剂引致较长的链先沉淀出来,或加入良溶剂促使较短的链先溶解出来。大分子还有包括链组构(如线型和支化型)和链构象(如伸展和卷曲)等较为复杂的多级结

构。蛋白质链可折叠和自组装成特定的单链或多链高级结构。合成高聚物则有不同的多分散性,包括长度、构象、结构等;即一个样品中的大分子链有不同长度、构象、连接聚合单体的方式。纯的小分子则是均匀的,具有完全一致的结构和摩尔质量,与其制备方法和来源无关,如无论在何地和用何法获得纯净水分子均完全一样。

大分子可被分为天然生成和人工合成两类。前者为天然地存在于微生物、动物和植物内的大分子,通常由细胞合成,生成途径各异。它们还可进一步被细分为由多种单体组成的大分子,如细胞内的脱氧核糖核酸(DNA)、核糖核酸(RNA)和蛋白质;或由一种或少数几种单体在细胞里生成的长链,包括多糖(纤维素、淀粉、几丁质……)、天然橡胶、树脂等。

后者即为日常生活中的各种塑料制品、合成纤维和合成橡胶。区分塑料和橡胶的判据为其玻璃化温度(T_g,从弹性向黏性转变)是否高于室温。也许因为历史原因,合成大分子在中文中也常被称为聚合物、高分子或者高聚物,较为混乱。追宗求源,聚合反应中所用的那些起始小分子称作“单体”(monomer)。因此,英文“polymer”的直译应为“多聚体”。何为“多”? 何为“少”? 几十个还是几百个为多? 实际上,“多和少”永远都是相对的,取决于如何选择一个参照标准!

在大分子物理中,“多和少”的判据应为前述的“缠结长度或聚合度”。“聚合物”一词包括了聚合度小于缠结聚合度的寡聚物或曰“低度聚合物”(low polymer),简称“低聚物”。隐含着高摩尔质量或高聚合度的“高分子”一词则又重叠涵盖了一些天然大分子(如淀粉等)。因此,那些聚合度大于缠结聚合度、具有小分子缺乏的特殊物理性质(如黏弹性)的人工合成大分子应被称为“高度聚合物”(high polymer),简称“高聚物”。这也正是为何美国物理学会里专门设有一个 High Polymer Physics 而不是 Polymer Physics 分会。所以,“高聚物”是英文“high polymer”最贴切和正确的翻译。本书采用“高聚物”一词来表述链远长于缠结长度的人工合成大分子。

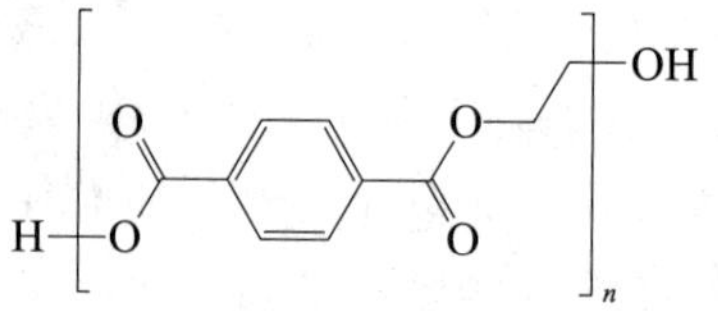

例如,由对苯二甲酸和乙二醇通过缩聚反应生成的聚对苯二甲酸乙二醇酯,也常简称为 PET,见右图,是目前最为广泛使用的热塑性人工合成树脂,也是排位第四的人工合成大分子材料。受限于缩聚反应机理,其聚合度通常较低,只有几十。当温度低于其熔融温度($\sim 250\ ^\circ\mathrm{C}$,随聚合度增加稍微变高)时,其为固态;一旦达到熔融温度,其立即成为液态,黏度较小,恍如由小分子组成的物质,如玻璃。因此,其常被用来制备注塑产品,如水杯和各种饮料的瓶子,而不可吹塑成很薄的膜类制品;在特殊的聚合条件下和采用相应的制备工艺,其聚合度也可高达几百,故其可被纺成合成纤维。

据不完全统计,化学专业的毕业生在其一生的职业生涯中有 70% 以上的机会从事与合成大分子材料有关的工作,包括它们的生产、研发和使用。所以了解和学习人工合成大分子的一些基本概念不仅是要求的,而且是必要的。另一方面,所有的生物体(包括我们自身)都是由天然大分子组成;大多数的食物也都是由天然大分子组成的,所以天然大分子的重要性已不言而喻。天然大分子的出现和使用都远远早于人工合成的高聚物。下面,先简要地介绍一些主要天然大分子的结构和它们在细胞中的生成过程。

在细胞内,两条互补的 DNA 链通常组成一个双螺旋结构,每条 DNA 链是一条线型长链,含有四种不同的脱氧核糖核苷酸通过 3',5'-磷酸二酯键聚合而成的,分别如下图(a)和(b)所示。在细胞分裂前,双链螺旋结构打开形成一对亲代 DNA 模板。在 DNA 聚合酶的催化作用下,由每条亲代模板可合成复制一条对应的子代 DNA 链。结果是,原本双螺旋中的两条 DNA 链被复制成四条 DNA 链。该复制过程不停地发生在所有的生物体内,是生物得以遗传的基础。

细胞分裂后,每个细胞仍然含有两条对应的组成双螺旋的 DNA 链。过去二十年里,分子生物学的飞速发展,特别是聚合酶链式反应(PCR),使得 DNA 链在细胞外的聚合和增殖成了一种廉价和简便的常规实验方法。细胞内,每条打开的 DNA 模板链中的一部分还可在 RNA 聚合酶的催化作用下,转录成一条编码 RNA 链(信使 RNA,mRNA)或非编码 RNA 链(输运 tRNA、核糖体 rRNA 等),其长度远小于其模板 DNA 链,如下图(c)所示。每种 RNA 的链长相同,分布单一。

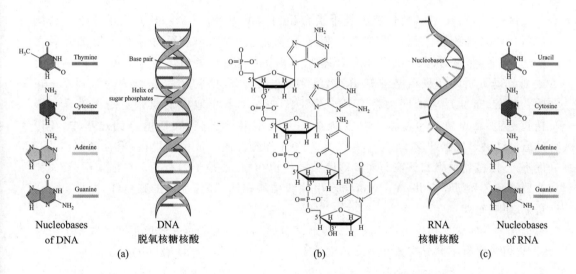

(a)　　　　　　　　　　　(b)　　　　　　　　　　　(c)

转录生成的信使 RNA 和带有氨基酸的转运 RNA 在核糖体内结合。信使 RNA 上每一组三个核苷酸被特别的核糖体结构解码和翻译成二十个氨基酸之一。将细胞内不同数目的不同氨基酸经过逐个地脱水缩聚反应生成一条蛋白质链。链上不同种类和不同数目的氨基酸以及它们的不同排列顺序造就了不同的蛋白质。理论上,细胞里二十种不同种类和不同数目的氨基酸的排列组合可以生成天文数字的不同蛋白质。然而,这不是一个随机的任意过程,而是一个精确、复杂、可调的"转录-翻译"的过程、受到 DNA 上遗传基因的严格制约。核苷酸序列的特异性决定了蛋白质氨基酸序列的特异性。细胞内蛋白质的种类远小于 DNA 上基因的数目。

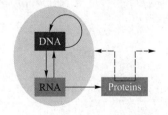

在真核细胞中,复制和转录的过程均发生在细胞核的里面,而转译则发生在细胞核和细胞膜之间的细胞液中,一些蛋白质被输运进细胞核,如上图所示。值得注意的是,该分子生物学中的"复制-转录-转译"过程是遗传信息得以保存的一个标准流程,与佛朗西斯·克里克于 1958 年提出的分子生物学中心法则不尽相同,即一连串不可逆转和逐步传送从核酸到蛋白质得遗传信息,The Central Dogma。有关的详细内容(分子组成、分子结构、细胞里的反应机理等),感兴趣的读者可参阅相关的细胞分子生物学教科书。

植物中,纤维素由许多微纤丝组成;每条微纤丝则含有多条互相平行的线型多糖长链,通过氢键作用形成结晶结构;每条多糖长链则含有成千上万个通过 $\beta(1\rightarrow 4)$ 糖苷链连接的 D-葡萄糖单元,如右图所示。纤维素是世界上最丰富的、细胞外的天然大分子。它是植物(也是许多海藻和细菌)细胞壁的一个基本和重要的结构组成。一些细菌通过分泌纤维素以形成生物膜。特别需要指出的是,在棉花中,纤维素的含量可高达 90%。

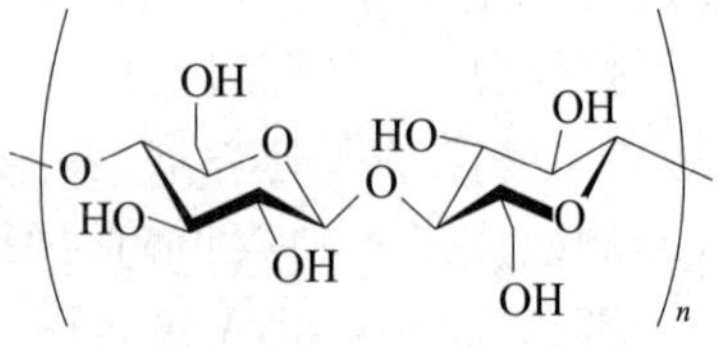

目前已知,线型多糖长链由纤维素合酶(CesA)催化生成。多个包括纤维素合酶的蛋白质在细胞膜上组成纤维素合酶复合体。每条微纤丝里线型多糖长链的数目可多达 36 条,取决于纤维素合酶复合体中含有多少 CesA 蛋白。纤维素在植物中的合成细节仍不明朗。仅粗略地知道,线型多糖长链合成的初始引物可能为谷甾醇-β-葡萄糖苷,催化反应的底物为 UDP-葡萄糖。CesA 不断地将葡萄糖基依次聚合到多糖长链上,由纤维素合酶复合体生成的多条线型多糖长链,在链增长的过程中,不断地通过结晶形成微纤丝。由于欠缺 DNA、RNA 和蛋白质合成中的模板,聚合多糖链的长度不一,随植物种类和环境而变。即使在同一植物中,也是长短有别。在细菌中,纤维素合酶被命名和缩写为 BesA。与植物产生的纤维素相比,细菌生成的则更纯、更长、更细、更强、更吸水。

各种各样的淀粉也主要由不同的聚合多糖链组成。它们可是长度不同的线型大分子链,也可为含有不同分支数和分支长度的支化大分子链,如下图(a)、(b)所示。

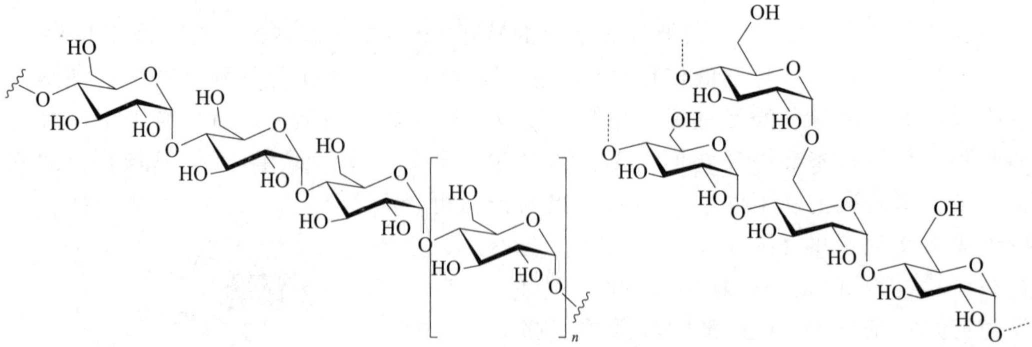

(a)

(b)

在一系列酶的协同作用下,其可在光合器官中的叶绿体内或非光合器官(如胚乳)中合成,二者之间,既有共性,也有不同之处。例如,在淀粉合成的最后阶段,淀粉合成酶(SS)、淀粉分支酶(SBE)和去分支酶(SDBE)是三个主要和关键的调控蛋白酶。在它们的协同催化作用下,聚合多糖链中直链和支化链的比例既取决于植物种子和种类(如小麦、玉米、土豆等),也可随天气和环境(如温度、降水、日照、土壤、肥料等)而变。详情略去,有兴趣的读者可从一些植物分子生物学教科书中获得有关的详细信息。

聚合多糖原料中线型链和支化链的比例还直接影响淀粉食物的口感(力学性能,即黏弹性,不是味道)。其中,直链与支化链的含量分别与淀粉食物的弹性和黏性正相关。对于给定的直链与支化链比例,支化链的长度也与弹性正相关。当然,食物的黏弹性还与加工中大分子取向、聚集和交联以及原料中蛋白质含量有关。特别需要指出的是,中国人对食物的口感有着痴迷地追求。这是为何中国人喜欢拉面以及各地的特色面条。另一方面,土豆淀粉中的支化链含量(~80%)与大米(75%~82%)和小麦(74%~81%)相似但高于玉米(~70%)。然而,由于其支化链较短,熟土豆偏软无弹性,故至今仍然无法成为中国人餐桌上的主食。因此,通过改性使其成为主食将直接关系到我国的粮食安全性。然而,对大分子化学和物理的深刻理解是修饰的前提。

几丁质具有和纤维素相似的结构,但其支化链中含有氮原子,可形成较强的链间氢键,使得其材料具有一些更好的力学性能。几丁质主要存在于节肢动物的外部骨骼和真菌的细胞壁中。由于其具有极好的生物相容性和生物降解性,常被用作生物医用材料,如手术缝合线。鉴于受到篇幅限制,本书略去有关几丁质和其他生物大分子的详细结构和生物合成细节。有兴趣的读者可参考相关的文献、专著和教科书。

另一方面,经过近八十年的飞速发展和工业化,聚合反应的技术飞速发展、日新月异。合成大分子(高聚物)已深入到人们现代生活(衣、食、住、行、医……)中的每一个角落,触手可及,造福人类。各种合成大分子材料价廉物美。可以毫不夸张地说,每一个来到今天这个世界的人,从出生时睁开双眼第一瞬间直至逝去时最后一刻,其视线都将始终无法偏离合成大分子材料。虽然,许多人开始关心和担忧合成塑料和其他各种合成大分子材料的环境污染问题,但我们今天已不可想象一个没有合成大分子的世界。例如,如果没有聚酯类人造纤维,我们该如何为地球上七十多亿人遮体避寒?

常见制备人工合成大分子的聚合方法包括自由基聚合、离子聚合、配位聚合、缩聚、开环聚合等。它们可依不同的方法分成不同的类别。任何一个好的分类方法应该最少包括以下两点:一、不同类别之间无重叠;二、各类别之和包含了所有的聚合反应类型,无遗缺。目前,高聚物化学教科书中的各种分类方法不尽人意,都没有满足以上的两条。

本书建议将不同的聚合方法重新分成与传统分类完全不同的、不重不漏的两类。第一种分类基于一个聚合反应除了单体以外是否涉及其他小分子:a. 自由基、离子、配位、某些开环聚合等;b. 缩合、某些开环聚合等。第二种分类取决于每一条生长的大分子链是否仅有一个引发起点:a. 自由基、离子聚合、配位、某些开环聚合等;b. 缩合、某些开环聚合等。实际使用时,可选用其中一种分类方法。逻辑上,这两种新的分类方法都避免了传统高分子教科书中现有分类方法的互相交叠或涵盖不全的缺陷。

在第一个类别中,当引发剂导致单体聚合开始后,单体一个接着一个地接到链的一个末端上。该类反应还可被进一步细分为活性和非活性聚合:在活性聚合中,链的增长速率低于引发速率,这样生成的高聚物链长度较均一,链的长度分布较窄;反之,在非活性聚合中,链的增长速率则远大于引发速率,所得高聚物的链长分布往往非常宽。同一个聚合反应得到大分子链长短不一。差别取决于反应条件和种类、单体纯度等。然而,每一种小分子的大小则为恒定,与其制备方法和来源完全无关。

运用这些聚合方法,合成化学家制备了成千上万种不同的高聚物。需要注意的是,大规模生产的种类大约只有二十种,包括聚烯烃、聚氨酯、聚苯乙烯、聚氯乙烯、聚碳酸酯、聚环氧乙烷等。关于各种聚合方法的反应条件、反应机理和工业流程,限于篇幅,恕不详述。有兴趣的读者可参考相关的教科书和工艺流程专著。

既然本章涉及了天然和合成大分子,进一步讨论"天然"和"非天然"将十分有趣。二者对立且统一,互相依赖,缺一不可。人类近代的文明几乎都建立在非天然的人造物质之上。在现今世界,人们享受着都市里的各种热闹,却又向往着荒野中的荒凉沉寂;享受着人造的精细物质,又感叹原始的粗犷之美;享受着现代农业提供的物廉价美食品,又抱怨化肥农药造成的不同的土壤和环境问题;享受着各种交通工具的便利舒适,又担心石油煤炭等能源在将来的枯竭……现实中,很少有人愿意长期住在荒野之地,回到刀耕火种的岁月,只是因为稀缺,向往而已!

在发展的同时保护环境是一个永久的课题,但不是将"天然"和"人造"对立起来!循环使用是保护环境的重要一环,强调的是节约资源、避免浪费,而不是将"天然"和"人造"对立起来!仅以合成大分子为例,生产可以完全生物降解的高聚物,并替代目前不可降解的塑料,在技术上已无明显的障碍,只是成本略高而已。所以,这只是一个市场和政策的问题。市场和政策又都取决于消费者(百姓)!有什么样的消费者,就有什么样的市场和政策。因此,人们也许不应该只是抱怨生产企业和各级政府。不幸的是,将一切错误和失败诿过于人已成为今天的现实,世界也许就是这样的荒谬。也许,人们应该想想是否需要重温"粒粒皆辛苦",避免浪费?是否需要在衣橱中塞满一百件衣服?是否需要每年更换一个新的手机?这些不是荒谬的问题。

小　　结

大分子由成百上千个相同或不同小分子(单体)通过共价化学键连接而成,有天然和人工合成之分。天然大分子通常由细胞生成,生成方式各不相同,千变万化。许多天然大分子在动物和植物细胞内具备确定的化学成分和特殊的三维空间结构;而合成大分子则由人为的聚合反应制备,即使在一样的聚合反应条件下,所得的合成大分子仍然长短不一。另一方面,对一条给定的没有链内特殊相互作用的合成大分子链,其仍有永远不停变化的三维空间形状。聚合反应的种类有数十种之多,可按不同的方式归类。

大分子和小分子组成的材料有着不同的物理性质。最明显的是,随着环境(温度、压强等条件)的变化,由小分子构成的材料不是弹性固体,就是黏性液体或气体,二者必居其一。而在大分子材料中,当链长超过缠结长度后,随着浓度的增加,长链会互相缠结,从而导致它们具有独特的物理性质,即可在一定的温度或时间范围内,它们可同时呈现出既黏又弹的性质,简称为"黏弹性"。实际上,仅当聚合物的链长为缠结长度五至十倍时,其方可呈现出明显的黏弹性。

严格地说,含有几十甚至一百个小分子单体、链长短于缠结长度的合成大分子应称为寡聚物或低聚物,而不是高聚物。它们组成的材料具有和小分子材料(如玻璃和水)类似的物理性质,非黏既弹,缺乏高聚物材料因链缠结特有的黏弹性。右图是一个小结。

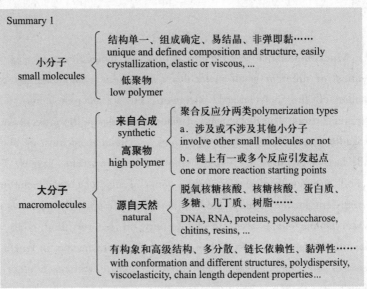

Summary 1

小分子 small molecules
- 结构单一、组成确定、易结晶、非弹即黏……
  unique and defined composition and structure, easily crystallization, elastic or viscous, ...
- 低聚物 low polymer

大分子 macromolecules
- 来自合成 synthetic
  高聚物 high polymer
  - 聚合反应分两类polymerization types
    - a. 涉及或不涉及其他小分子 involve other small molecules or not
    - b. 链上有一或多个反应引发起点 one or more reaction starting points
- 源自天然 natural
  - 脱氧核糖核酸、核糖核酸、蛋白质、多糖、几丁质、树脂…… DNA, RNA, proteins, polysaccharose, chitins, resins, ...
- 有构象和高级结构、多分散、链长依赖性、黏弹性…… with conformation and different structures, polydispersity, viscoelasticity, chain length dependent properties...

Chapter 1. Definition and Sources of Macromolecules

Macromolecules are usually defined as molecules made up of hundreds/thousands of identical or different small molecules (monomers) connected by chemical bonds. The total number of the polymerized monomers on a macromolecule is called **the degree of polymerization**. The molar mass of macromolecules is the mass of each chain multiplied by the Avogadro number ($N_{AV} = 6.02 \times 10^{23}$ mol^{-1}). Small molecules are about one nanometer in size, while large molecules are usually about a hundred times larger. The chain length of some special macromolecules can reach tens of micrometers, including ultrahigh molar mass polyethylene for the manufacture of super-strong fibers, polystyrene as a standard sample, and the polyacrylamide chains used in the tertiary oil recovery and other synthetic macromolecules. Deoxyribonucleic acid (DNA) can even reach several meters in length. As the chain length (L) increases, the macromolecular chains will entangle with each other in a higher concentrated solution or a molten or solid state, resulting in materials made of macromolecules simultaneously have a unique viscous and also an elastic physical property, i.e., viscoelasticity. Each given type of macromolecules has a corresponding **entanglement length** (L_e) or **entanglement degree of polymerization** ($N_e \sim 10^2$).

When the chain is shorter than L_e or the degree of polymerization is lower than N_e, increasing the concentration will only cause the spatial volume occupied by each chain to shrink, but will not lead to their entanglements. Such "macromolecules", just like small molecules with no entanglements, only display solid elasticity or liquid viscosity as the environment (such as temperature and pressure) changes, as shown on the right. In the literature, when studying the physical properties of macromolecules, many people often make some avoidable low-level mistakes by ignoring the important concept of the entanglement length. Also note: in literature and some

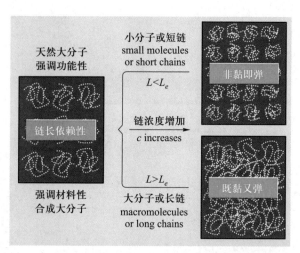

textbooks, some chains made of dozens connected monomers (small molecules) are also called macromolecules even they are shorter than the entanglement length. Strictly speaking, they should be called oligomeric chains or oligomers. In fact, only when the chain length or degree of polymerization is five to ten times the entanglement length or degree of polymerization, the material composed of macromolecules can clearly show the viscoelasticity, lacking in small molecular materials.

In addition tothe viscoelasticity, there are many other differences between macromolecules and small molecules. For example, macromolecules generally cannot be purified by crystallization. The polymer chains with different lengths can only be separated by using the chain length dependent solubility; namely, gradually adding a poor solvent to make longer chains to precipitate out first or adding a good solvent to promote the dissolution of short ones. Macromolecules also have complicated hierarchical structures, including chain configuration (such as linear and branched) and chain conformation (such as stretch and coiled). Protein chains can fold and self-assemble into specific single-chain or multi-chain higher order structures. Synthetic polymers have different polydispersity, including length, conformation, structure, etc.; namely, macromolecular chains in a sample have different lengths, conformations, and ways of connecting polymerized monomers. Pure small molecules are uniform with an identical structure and molar mass, independent of its preparation method and source, e.g., pure water molecules are identical no matter where and how they were obtained.

Macromolecules can be divided into two categories: naturally formed and artificially synthesized. The former are macromolecules that naturally exist in microorganisms, animals, and plants, usually synthesized by cells with different pathways. They can also be further subdivided into macromolecules composed of a variety of monomers, such as deoxyribonucleic acid (DNA), ribonucleic acid (RNA) and proteins in cells; or long chains made of one or a few monomers in cells, including polysaccharides (cellulose, starch, chitin, ...), natural rubber, resins, etc.

The latter are all kinds of plastic products, synthetic fibers and synthetic rubber in daily life. The criterion to distinguish plastics and rubber is whether its glass transition temperature (T_g, from elastic to viscous) is higher than room temperature. Maybe due to historical reasons, artificially synthesized macromolecules are often called " synthesized substances ", " high molecules" or "highly polymerized substances", more chaotic. Tracing the source, those starting small molecules used in polymerization are called "monomers". Therefore, the literal translation of "polymer" in English should be "multi-mer". What is "more"? What is "less"? Are dozens or hundreds more? In fact, "more and less" are always relative, depending on how to select a reference standard!

In macromolecular physics, the criterion of "more and less" should be the aforementioned "entanglement length or degree of polymerization". The term "polymer" includes "low polymers" whose degree of polymerization is less than that of the entanglement, or "oligomer" for short. The term "high polymer" with an implication of a high molar mass or a high degree of polymerization overlaps with some natural macromolecules (such as starch, etc.).

Therefore, those synthetic macromolecules with a degree of polymerization higher than the degree of entanglement and special physical properties (such as viscoelasticity) lacking in small molecules should be called "highly polymerized substance" (high polymer). This is why the American Physical Society has a "High Polymer Physics" instead of "Polymer Physics" Division. Therefore, "highly polymerized substance" is the most appropriate and correct translation of "High Polymer". This book uses the term "highly polymerized substance" to describe synthetic macromolecules whose chains are much longer than the entanglement length.

For example, polyethylene terephthalate, which is produced by polycondensation of terephthalic acid and ethylene glycol, is also often named as PET, as shown on the right, the most widespread thermoplastic synthetic resin used currently and also the fourth largest synthetic macromolecular material. Limited by the polycondensation reaction mechanism, the degree of polymerization is usually low, only a few dozen. When the temperature is lower than its melting temperature (~ 250 ℃, slightly higher as the degree of polymerization increases), it is a solid; as soon as reaching the melting temperature, it immediately becomes a liquid with a low viscosity, just like a substance composed of small molecules, such as glass. Therefore, it is often used to prepare injection-molded products, such as water cups and various beverage bottles, but cannot be blow molded into very thin-film products. Under special polymerization conditions and adopting corresponding preparation processes, the degree of polymerization can be as high as several hundred, so that it can be spun into synthetic fibers.

According to incomplete statistics, chemistry graduates have more than 70% chances to engage in works related to macromolecular materials, including their synthesis, research and developments, and applications during their career lives. Therefore, it is not only required, but also necessary to understand and learn some basic concepts of artificially synthetic macromolecules. On the other hand, all living organisms (including ourselves) are made of natural macromolecules; most of food is also composed of natural macromolecules, so that the importance of natural macromolecules is self-evident. The emergence and use of natural macromolecules are far earlier than synthetic polymers. The structures of some major natural macromolecules and their production processes in cells are briefly introduced as follows.

In the cell, two complementary DNA chains usually form a double helix structure. Each DNA strand is a linear long chain, consisting of four different deoxyribonucleotides polymerized by 3′,5′-phosphodiester bonds, respectively, shown in the figures (a) and (b) below. Before the cell division, the two-chain helical structure opens to form a pair of parental DNA templates. Under the catalytic action of DNA polymerase, each parent DNA acts as a template for making its complementary corresponding offspring DNA strand. As result, two DNA strands in the original double helix are duplicated into four DNA strands. This replication process occurs continuously in all organisms and is the basis for organisms to inherit.

After the cell division, each cell still contains two corresponding DNA strands that make up a double helix. In the past two decades, the rapid development of molecular biology, especially the polymerase chain reaction (PCR), has made the polymerization and proliferation of DNA strands outside the cell a cheap and convenient routine experimental method. In the cell, a part of each open DNA template strand can also be transcribed into a coding RNA strand (messenger RNA, mRNA) or non-coding RNA strand (transport tRNA, ribosomal rRNA, etc.) under the catalytic action of RNA polymerase.), its length is much smaller than its template DNA strand, as shown in the figure (c) below. Each kind of RNA has the same chain length, monodistributed.

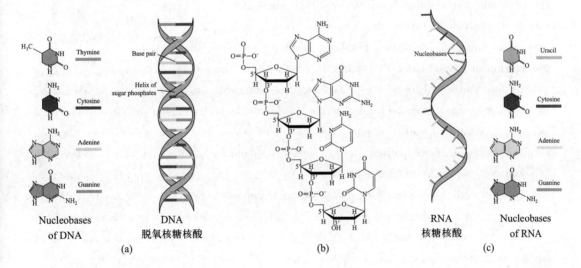

| Nucleobases of DNA | DNA 脱氧核糖核酸 | | RNA 核糖核酸 | Nucleobases of RNA |
| (a) | | (b) | | (c) |

The transcribed messenger RNA and transfer RNA with its carried amino acid are combined inside the ribosome. Each group of three nucleotides on a messenger RNA is decoded and translated into one of the twenty amino acids by the special structure of ribosome. These amino acids are dehydrated and polycondensed one by one to form a protein chain. Different types and numbers of amino acids as well as their different arrangements on a chain create different proteins. In theory, the permutation and combination of twenty different types and different numbers of amino acids in cells can generate astronomical different proteins. However,

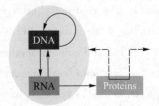

this is not a random arbitrary process, but a precise, complex, adjustable "transcription-translation" process, strictly constrained by genetic genes on DNA. The specificity of the nucleotide sequence determines the specificity of the amino acid sequence of the protein. The types of proteins in cells are much smaller than the number of genes on DNA.

In eukaryotic cells, the replication and transcription processes occur in the nucleus, while translation occurs in the cytosol between the nucleus and the cell membrane, and some of proteins are transported into the nucleus, as shown above. It is worth noting that the "replication-transcription-translation" process in molecular biology is a standard process for the preservation of genetic information, not completely in line with the central principle of molecular biology proposed by Francis Crick in 1958, i. e., a series of irreversible and gradual transmission of genetic information from nucleic acid to protein, The Central Dogma. For the detailed content (molecular composition, molecular structure, reaction mechanism in cells, etc.), interested readers can refer to related textbooks on cell and molecular biology.

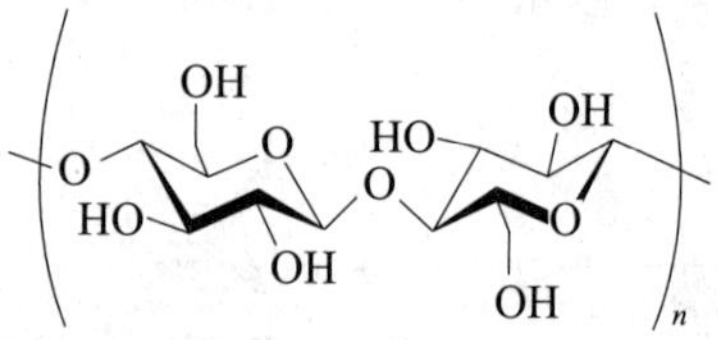

In plants, cellulose is composed of many micro-fibrils; each micro-fibril contains multiple parallel long linear polysaccharide chains, forming a crystalline structure through hydrogen bonding; each long polysaccharide chain contains thousands of The D-glucose unit connected by $\beta(1\rightarrow4)$, as shown on the right. Cellulose is the most abundant, extracellular natural macromolecule in the world. It is a basic and important structural component of the cell wall of plants (and also many seaweeds and bacteria). Some bacteria secrete cellulose to form biofilms. In particular, it should be pointed out that in cotton, the cellulose content can be as high as 90%.

It is currently known that long chains of linear polysaccharides are catalyzed by cellulose synthase (CesA). Multiple proteins including cellulose synthase form a cellulose synthase complex on the cell membrane. The number of linear polysaccharide chains in each micro-fibril can be as many as 36, depending on how much CesA protein is contained in the cellulose synthase complex. The details of the cellulose synthesis in plants remain unclear. It is only roughly known that the initial primer for synthesizing linear long polysaccharides chain may be sitosterol-β-glucoside, and the substrate for the catalytic reaction is UDP-glucose. CesA continuously polymerizes the glucose groups to the polysaccharide long chain in turn. The multiple linear polysaccharide chains generated by the cellulose synthase complex continuously crystallize to form micro-fibrils during the chain growth process. Due to the lack of templates in DNA, RNA, and protein synthesis, the length of polymerized polysaccharide chains varies with plant species and environment. Even in the same plant, there are differences in length. In bacteria, cellulose synthase is named and abbreviated as BesA. In comparison with the cellulose produced by plants, those made by bacteria is purer, longer, thinner, stronger, and more water-absorbing.

Various starches are also mainly composed of different polysaccharides chains. They can be linear macromolecular chains with different lengths, and branched macromolecular chains with different branching numbers and branching lengths, as shown in the figures (a), (b) below.

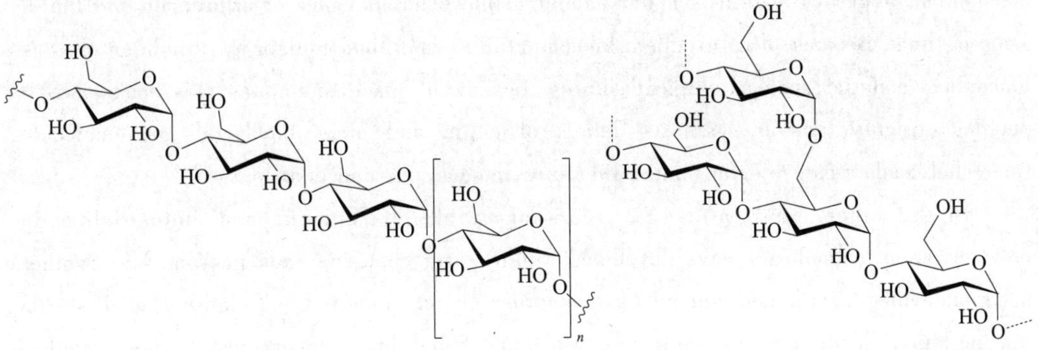

(a)   (b)

Under the synergistic action of a series of enzymes, it can be synthesized in chloroplasts in photosynthetic organs or in non-photosynthetic organs (such as endosperm). Between these two, there are both commonalities and differences. For example, in the final stage of starch synthesis, starch synthase (SS), starch branching enzyme (SBE) and debranching enzyme (SDBE) are three main and key regulatory proteases. Under their synergistic catalysis, the ratio of straight and branched chains in the polysaccharides chain not only depends on plant seeds and species (such as wheat, corn, potatoes, etc.), but also varies with the weather and environment (such as temperature, precipitation, sunlight, soil, fertilizer, etc.). The details are omitted, and interested readers can get the detailed information from some textbooks of plant molecular biology.

The ratio of linear and branched chains in a polysaccharides material also directly affects the texture of starch foods (mechanical properties, namely viscoelasticity, not taste). Among them, the content of straight chain and branched chain are positively correlated with the elasticity and viscosity of starch foods. For a given linear-to-branched chain ratio, the length of the branched chain is also positively related to the elasticity. Of course, the viscoelasticity of food is also related to the orientation, aggregation and cross-linking of macromolecules during processing, and the protein content of the raw materials. What needs to be pointed out is that Chinese people are obsessed with the food texture. This is why Chinese people like ramen and special noodles from all over the regions. On the other hand, the branched chain content in potato starch (~80%) is similar to rice (75%-82%) and wheat (74%-81%) but higher than corn (~70%). However, due to its shorter branched chains, cooked potatoes are soft and inelastic, so that they are still not able to become the staple food on our table. Therefore, making it a staple food through modification will directly affect the food security of our country. The profound understanding of macromolecular chemistry and physics is precondition of the modification.

Chitin has a structure similar to celluloses, but its branched chains contain nitrogen atoms, which can form stronger interchain hydrogen bonds, making its materials with some better mechanical properties. Chitin is mainly found in the external bones of arthropods and the cell walls of fungi. Because of its excellent biocompatibility and biodegradability, it is often used as a biomedical material, such as surgical sutures. In view of space limitations, this book omits the detailed structure and biosynthesis details of chitin and other biological macromolecules. Interested readers can refer to related literature, monographs and textbooks.

On the other hand, after ~ 80 years of rapid development and industrialization, polymerization technologies have developed rapidly, ever changing each passing day. Synthetic macromolecules have penetrated into every corner of our modern life (clothing, food, shelter, transportation, medicine, etc.), within reaching, benefiting humankind. Various synthetic macromolecular materials are inexpensive and good quality. It is no exaggeration to say that everyone who comes to this world today will never be able to deviate her/his sight from synthetic macromolecular materials from the first instance when they open their eyes at birth to the last moment when they pass away. Although many people are beginning to care and worry about the environmental pollution of synthetic plastics and various other synthetic macromolecular materials, it is unimaginable of a world without synthetic macromolecules. For example, if there were no polyester-based synthetic fibers, how could more than seven billion people on the planet is sheltered from the cold?

Common polymerization methods forthe preparation of synthetic macromolecules include radical polymerization, ionic polymerization, coordination polymerization, polycondensation, ring-opening polymerization, addition polymerization, etc. They can be divided into different categories according to different methods. Any one good classification method should include at least the following two points. First, there is no overlap between different categories; second, a sum of all the categories includes all types of polymerization, no missing. Various current classification methods in polymer chemistry textbooks are not satisfactory, and do not satisfy the above two conditions.

This book suggests reclassifying different polymerization methods into two new categories that are completely different from the traditional classifications, not only covering all polymerization, but also no overlapping. The first category is based on whether a polymerization reaction involves other small molecules except monomers: a) radical, ionic, coordination, and certain ring-opening polymerizations, etc.; b) condensation, some ring-opening polymerization, etc. The second category depends on whether each growing macromolecular chain has only one initiation point: a) free radicals, ionic polymerization, coordination, some ring-opening polymerization, etc.; b) condensation, some ring-opening polymerization, etc. In the actual use, one of the classification methods can be selected. Logically, these two new classification methods avoid the overlap or incomplete coverage of existing methods in traditional polymer textbooks.

In the first category, when the initiator causes the monomer to polymerize, the monomers are attached to one chain end one by one. This type of reaction can be further subdivided into living and non-living polymerization. In living polymerization, the chain growth rate is lower than the initiation rate, so that the polymer chain length produced is more uniform and the chain length distribution is narrow; on the contrary, In non-living polymerization, the chain growth rate is much greater than the initiation rate, and the chain length distribution of the resulting polymer is often very broad. Macromolecules obtained in the same polymerization reaction have various lengths. The difference depends on the conditions and type of reaction, the purity of the monomer, etc. However, the size of each kind of small molecules is constant, regardless of its preparation method and source.

Using these polymerization methods, synthetic chemists have prepared thousands of different polymers. It should be noted that there are only about twenty types of large-scale produced polymers, including polyolefin, polyurethane, polystyrene, polyvinyl chloride, polycarbonate, and polyethylene oxide. Regarding the reaction conditions, reaction mechanisms and industrial processes of various polymerization methods, this book will not go into details due to a limited space. Interested readers can refer to relevant textbooks and process monographs.

Since this chapter deals with natural and synthetic macromolecules, it will be very interesting to discuss "natural" and "unnatural" further. The two are opposed and unified, rely on each other, and are indispensable. Almost all modern human civilizations are based on unnatural fabricated materials. In today's world, people enjoy all the excitement in the city, but yearn for the desolation and silence in the wilderness; enjoy the artificial fine materials, and sigh the primitive and rough beauty; enjoy the cheap food with good quality provided by modern agriculture, but complain about different soil and environmental problems caused by chemical fertilizers and pesticides; enjoy the convenience and comfort of various means of transportation, but worry about the depletion of energy, such as petroleum and coal, in the future;.... In reality, few people are willing to live in the wilderness for a long time and return to the years of slash-and-burn cultivation, just because of scarcity and yearning!

Protecting the environment while developing is aconstant issue, but it is not opposing "natural" and "man-made"! Recycling is an important part of protecting the environment. The emphasis is on saving resources and avoiding waste, rather than opposing "natural" and "man-made"! Taking synthetic macromolecules as an example, the production of fully biodegradable polymers and replacement of current non-degradable plastics have no obvious technical hurdles nowadays, only the cost is slightly higher. Therefore, this is only a question of market and politics. The market and politics depend on consumers (people)! What kind of people are there, there will be what kind of market and politics, especially in a democratic society. Therefore, perhaps people should not just only complain about the production enterprises and governments at all levels. Unfortunately, attributing all mistakes and failures to others has become the reality nowadays, and the world may be just so absurd! Perhaps, people should think about whether it is necessary to revisit the "nothing comes easy" to avoid waste? Do we need to stuff a hundred clothes in the closet? Do we need to replace a new mobile phone every year? These are not absurd questions.

Summary

Macromolecules are made up of hundreds or even thousands of identical or different small molecules (monomers) that are connected by covalent chemical bonds, and are divided into natural and artificially synthesis. Natural macromolecules are usually produced by cells in various and ever-changing ways. Many natural macromolecules have a certain chemical composition and special three-dimensional spatial structure in cells of animals and plants, while synthetic macromolecules are prepared by artificial polymerization. Even under identical conditions of polymerization, the resulting synthetic macromolecules still vary in length. On the other hand, for a given synthetic macromolecular chain without special intra-chain interactions, it still has an ever-changing three-dimensional shape. There are dozens of polymerization types, classified in different ways.

Materials composed of macromolecules and small molecules have different physical properties. The most obvious is that as the environment (temperature, pressure, etc.) changes, materials composed of small molecules are either elastic solid or viscous liquid or gas, only one of the two. In macromolecular materials, when the chains are longer than the entanglement length, as the concentration increases, the long chains will entangle with each other, resulting in their unique physical properties, which can occur at a certain temperature, leading to their special physical property. Namely, they can appear to be viscous and elastic simultaneously within a range of temperature or time, referred to as "viscoelasticity" for short. In fact, only when the chain length of a polymer is five to ten times the entanglement length, it can exhibit obvious viscoelasticity.

Strictly speaking, synthetic macromolecules made of dozens or even hundreds of small monomers, with the chain length shorter than the entanglement length, should be called oligomers or low polymers, not high polymers. The materials composed of them have physical property similar to small molecular materials (e.g., glasses and water), either viscous or elastic, lacking special visco-elasticity of macromolecular materials. A short summary is listed below.

Summary 1

小分子
small molecules

　结构单一、组成确定、易结晶、非弹即黏……
　unique and defined composition and structure, easily
　crystallization, elastic or viscous, ...

　低聚物
　low polymer

大分子
macromolecules

　来自合成
　synthetic

　高聚物
　high polymer

　　聚合反应分两类polymerization types

　　a. 涉及或不涉及其他小分子
　　involve other small molecules or not

　　b. 链上有一或多个反应引发起点
　　one or more reaction starting points

　源自天然
　natural

　　脱氧核糖核酸、核糖核酸、蛋白质、
　　多糖、几丁质、树脂……
　　DNA, RNA, proteins, polysaccharose,
　　chitins, resins, ...

有构象和高级结构、多分散、链长依赖性、黏弹性……
with conformation and different structures, polydispersity,
viscoelasticity, chain length dependent properties...

第二章 大分子的结构

小分子和大分子均由不同的原子排列组合而成。为了进一步区分这两种不同的分子,以及深入地讨论它们之间的异同性,先引入以下几个不同的重点概念。

组成

一个分子中原子的种类和每种原子的数目。不同小分子有不同的组成。同种小分子的组成完全相同,与来源无关。例如,水分子含有一个氧原子和两个氢原子(H_2O);食盐分子含有一个钠离子和一个氯离子($NaCl$);二氧化碳分子含有一个碳原子和两个氧原子(CO_2);以及二十种氨基酸($H_2NCHRCOOH$)含有二十种不同的侧基(R)。

与小分子一样,改变聚合单体的组成可得不同的大分子。不同的是,如果聚合反应仅涉及一种给定的单体,大分子链上的原子种类和不同原子间的比例不随聚合反应变化,但链上每种原子的数目仍然随着链的增长成比例增加。对小分子而言,己烷、庚烷、辛烷等已是不同的分子。但在大分子科学中,长度不同的大分子链仍为同一种分子。譬如,由小分子乙烯单体($CH_2{=}CH_2$)聚合而成的大分子聚乙烯[—(CH_2—CH_2)$_n$—]链的长短不一,其中氢原子和碳原子的比例保持不变(2:1,除了末端可忽略的细小差别)。但是,两种原子的数目均随着链长(聚合度)增加。聚合反应一定导致链的长短不一,聚合度不定。严格地说,合成聚合物链一定没有确定的组成,仅有一个统计平均组成。

组构

对一个给定的组成,各单元(单体)按一定的方式以化学键组织构造成一个三维结构,简称组构,即初级或一级结构。小分子中的单元是各种原子,连接这些原子的方式称为"结构形式"("构型"),不同的连接方式对应着不同种的分子。然而,一种大分子可有不同的连接方式,如线型或支化连接的聚苯乙烯。即使给定了一种连接方式以后,还可存在不同的结构。将会详细讨论。

本书将英文"configuration"翻译成"组构,configurational structure"而不是常用的"构型"主要有以下两点考虑:首先,对应的英文"configuration"源于拉丁文的 con(together,一起)和 figurare(to form,构成或形成)两个词根。采用的"组"和"构"可以分别很好地对应着二者。其次,"configuration"的中文翻译有多种,除了常用的"构型"以外,还包括组态、构态等。在小分子合成化学中,采用"构型"习以成俗,毫无疑义。这是因为小分子中的原子连接和空间分布方式之间存在一一对应的关系。然而,对一个给定的连接方式和结构,大分子链上各个单元还有不同的空间分布形式(称为构象)。

尽管"组构/构型"和"构象/构形"二者的定义明确,分别是"如何连接单体"和"如何在空间中分布单体"，但"构型"较易和"构形"混淆。主要原因是中文的"型"和"形"读音一样、写法相似，但二者的内涵完全不同。此"型"非彼"形"也！"型"指一个物体所属的"分类或类型"，"形"则指一个物体的"形状、外形、形象"，空间分布形状。

采用"构型"较易和"构象/构形"混淆。如果教师在讲课时没有每次都反复强调二者之别，或学生们将"构型"一旦误听或误解为"构形"，就会混淆这两个不同的概念，陷入困境。在英文中，没有这一问题。也许，当初将小分子化学里的"构型"习以成俗地当作"configuration"后，已不能再将"conformation"翻译成"构形"，因为二者极易混淆。于是，不得不采用了日常生活中少用的"构象"一词。这只是一个猜想而已。在详细解释了大分子链的"构象"后，再进一步举例说明为何本书采用"组构"来替代"构型"。

对给定组成的小分子而言，不同的连接方式导致不同的分子，如正丁烷和异丁烷。在小分子合成化学中，也把有不同连接方式的两个小分子称为同分异构体，如两个手性对称的镜像分子，以及顺式和反式同分异构体，如右图所示。它们也许具有相同的化学性质。严格地说，它们之间因一些性质差别，已不是"同一个分子"，而是物理上不同的分子，故有不同的命名，如"左旋"和"右旋"。

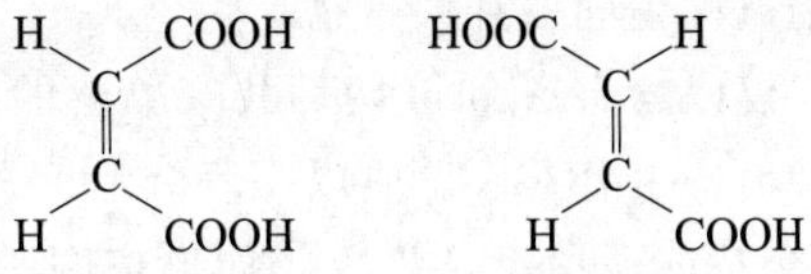

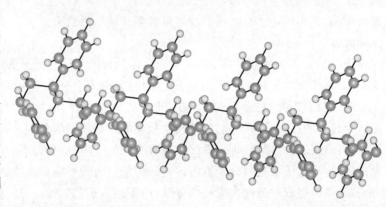

顺丁烯二酸　　　　　反丁烯二酸
maleic acid　　　　　fumaric acid

即使是只由一种单体聚合而成的均聚物，大分子也具有远比小分子丰富和复杂的组构。对一给定的组成（单体种类和数目），各个单体之间还有不同的、具体的连接方式，如线型或支化型，形成不同的组构。对大分子而言，在忽略了链内的连接细节（如总聚合度和支化链的长短）后，大分子可被归类于不同的组构，如线型、支化型和星型。在大分子化学中，它们被当作同一种聚合物，只是在前面加上组构的名称，如线型聚乙烯、支化型聚乙烯、星型聚乙烯等。

为了阐明即使给定了一个组成和连接方式（组构），仍有不同的方式排列链上不同的单体（结构），仅以由苯乙烯组成的聚苯乙烯为例，如右图所示。如果将一个不对称单体的两端分别当作"头"和"尾"，通常情况下，单体将以"头—尾"的方式连接在一起，组成一条线型的大分子链。但是，一个苯乙烯单体仍有一个较小的概率以"尾—头"的方式连接在一条增长的链末端，使得随后的链增长变成"尾—头"相接，聚合中可能还有多次反转。一个活性基团在增长的链上还可出现一定概率的转移，从而使得原本线型增长的主链上出现支化。每条支化链（也称子链)还可继续支化,形成千变万化的各种结构。

另外,如果沿着碳-碳主链观察,所有苯乙烯单体上的苯环有三种可能的排列:有序地朝着一个方向旋转、交替地向着相反的方向旋转和无规地向着两个相反的方向旋转,分别称为等规、间规和无规聚合物。注意:它们具有相同的线型组构,不同的结构。一个实际的聚合反应不可能产生和小分子一样的均一结构,总是得到各种结构的混合物。通过控制反应条件,可以让其中某一个结构占多数。

涉及两种或两种以上单体的共聚反应可产生组构更加丰富和复杂的共聚物。此时,除了链长分布和上述的各种不同组构以外,即使对给定的组成、组构和链长,不同共聚单体在一条线型共聚物大分子链上的不同分布也会导致不同的结构。在大分子化学中,它们仍被称为同一种共聚物。

仅以含有 A 和 B 两种共聚单体的共聚物为例,A 和 B 可无规地(AABAAA…BAA)、交替地(ABAB…AB)或以一段 A 和一段 B 地镶嵌在链上(A…AB…BA…A),分别称为无规、交替或嵌段型共聚物。如果依嵌段数目分类,组构还可被细分为两嵌段、三嵌段和多嵌段型共聚物。如果 A 和 B 以支化的方式连接,即给定了支化组构,二元共聚支化链仍有极大量的不同结构。大分子链的组构和结构种类繁多,不胜枚举,故不详述。

因此,与蛋白质等生物大分子不同,合成大分子没有确定的组构,仅有统计平均组构。如果忽略单体之间的连接细节,大分子链的组构可被粗略地分为线型、支化型、环型等类型。

天然大分子的结构更极端丰富。DNA 和 RNA 链具有相同的线型组构,各含有四个不同的碱基。对于一给定的碱基组成,它们的相对排列和组合可导致具有不同的功能和作用的 DNA 和 RNA。蛋白质分子由二十种不同的氨基酸组成,全部具有线型组构。即使对一给定的氨基酸组成,理论上,一条对应的蛋白质链也有天文数字的不同结构。不同的结构直接导致它们是否具有某些生物活性。天然植物大分子也具有十分复杂的结构。鉴于篇幅受限,恕不详述。有兴趣的读者可参考相应的教科书。

构象

在按一给定的方式组构后,大分子链内各单元的不同空间分布形式构成了其特有的二级结构,称作大分子构象。在小分子化学里,如果一个分子里的原子有两种相对稳定的空间分布,它们被称为构象异构,如船式和椅式。值得注意:这两个构象之间还同时存在着无数个过渡的空间分布,它们只是相对稳定的概率较大的两个分布。除此以外,每个原子都在不停地转动和振动,每一瞬间也会有不同的空间分布,此小分子构象非彼大分子构象也!因此,小分子没有大分子里所讨论的构象。大分子构象仅考虑聚合单体,甚至更大的、由若干单体组成的链段在空间的分布,而不是原子们如何在空间分布的细节。

例如,一条由成百上千个没有特殊相互作用的聚合单体构成的柔软线型大分子链的一个极端构象是不改变键角的完全伸直。另一个极端构象是一个蜷缩的单链均匀小球。在溶液中或熔融状态里,即使没有外力或能量输入,在给定的绝对温度($T > 0$)下,体系也有热能($k_{\mathrm{B}}T$)。在热能的激发下,链上的每一个链段都在不停地无规运动。如果向着两个极端构象中的任何一个变化,其构象熵都会减少,而其链内排除体积效应增大,引致其自由能增加。这违背了在一个恒温和恒压下自发过程的热力学判据故不会自动发生。

因此，在热能驱动下的无规运动导致线型大分子形成一个无规的线团构象。其尺寸和形态不停地变化（涨落），趋于最大构象熵。需要注意：每一条链无论是在溶液还是在熔融或本体（固体）中都具有一个扩展的无规线团构象。在稀溶液中，链段之间充满了溶剂分子；随着浓度的增加，除了溶剂以外，它们之间还有许多与之缠结、来自其他链的链段；而在熔体或本体中，链段之间充满了来自其他链的链段，互相贯穿和彼此缠结。因此，由聚合反应得到的没有特殊链内相互作用的大分子没有确定的构象。下面，将会用一章详细地讨论如何对大分子链千变万化的构象做统计平均以获得其平均构象和平均尺寸。

为了更清楚地说明"组构/构型"和"构象/构形"二者间的本质差别，可想象用刚性细杆和柔软细绳串接八个珠子分别代表小分子和大分子，得到线型和支化型两类组构，如下图所示。对小分子而言，在每个连接点处不允许旋转，一种连接方式只对应着一种结构，一种结构对应着一种分子，一种分子只有一个构象。而大分子的每种连接方式对应着同种大分子的不同结构，每个结构又有无数个构象（空间分布）。换而言之，对小分子而言，其连接方式和其空间分布一一对应；而具有一种给定连接方式的大分子的一个结构可对应着不同的构象。切记这里"组构"和"构象"分别描绘的是单元（原子、聚合单体或链段）"如何连接"和"如何分布"的方式，千万不要混淆这两个完全不同的概念。

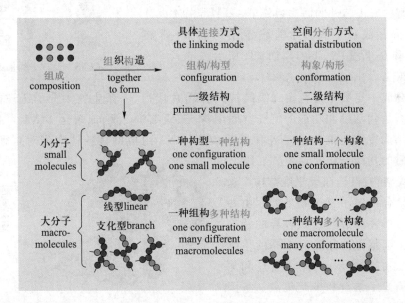

用"构型"描绘连接方式时，因中文里的"构型"与"构形"读音相同，极易将"型"与"形"混淆。讲课的过程中，教师很难每次都强调"有'土'的型，而不是形式的'形'"，外加"构象"和"构形"同义，极易在中文里将"构型"与"构形/构象"混淆。英文"configuration"的其他翻译"组态和构态"的重心均不在"结构"上，也无结构的含义。简单地使用"结构"也有问题，除了每次均要加上累赘的"一级"或"二级"二字，以区别其他大分子独有的高级结构以外，一个组构包含了多种结构。例如，不同的支化结构都属于支化组构。正是由于中文中的此"型"非彼"形"也，本书采用中文翻译"组构"而非习以成俗的

"构型"来表达"configuration"，希望可减少一些混淆。

另外，同样的英文"linear"在中文翻译里对应着三个内涵和外延完全不同的词："线型""线形"和"线性"。英文中"linear polymers or chains"本来源于一个副词"线性连接地"。其英文原意和全称应是"linearly formed（connected）polymers or chains"。这里看似形容词，但描绘了连接方式，并非空间分布形式。在中文里应为"线型"，不可与"线形"混淆。在大分子物理中，对一给定的连接方式，具体的连接细节（如单体内化学基团在链上排列）并不那样重要。故在本书中，仅用"线性"表示两个变量之间的依赖关系。当意指一类组构时，采用"型"字，如线型链、支化型链、星型链等。

对给定的组成和组构，大分子链内或链间的单元之间存在的一些特殊相互作用（如离子键、氢键、疏水作用等）可导致大分子链中的一段链节折叠成特定的三维空间分布，形成一些非无规线团的特殊构象，即一些具有确定空间分布的二级结构，特别是众多的天然生物大分子。与合成高聚物链相比，像蛋白质和 RNA 这样的生物大分子链一般比较短，故不应关注它们因链缠结而导致的特殊物理性质，而应关注其具有生物功能的特殊构象。在细胞内或细胞外的适当条件下，许多蛋白质大分子链都可折叠成一个或多个有确定结构的 α-螺旋（α-helix）和 β-折叠（β-sheet），如右图所示。

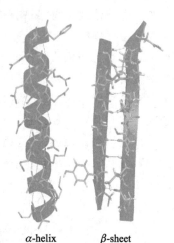

α-helix   β-sheet

高级结构

在折叠成一定数量的特定二级结构（构象）片段后，一条生物大分子链还能进一步折叠成具有生物活性和功能的特殊空间结构：三级结构。例如，输运 tRNA 可折叠成特定的三级结构在细胞液中运送各种氨基酸到 rRNA 组成的核糖体内配对信使 mRNA。

一个蛋白质的初级结构决定了其二级结构和三级结构，甚至四级三维结构。这些结构的集合决定了其生物功能。由各种蛋白质链折叠而成的形状各异的特殊三级结构在细胞里往往具有不同的生物催化功能，如右图所示，其中浅蓝色和深蓝色分别标记了结合和催化位点。基因

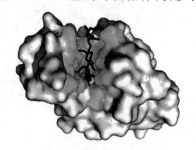

图片来源

突变（影响一级结构）或某些内部干扰可影响蛋白质的二级结构（构象），常常导致其丧失特定的生物功能。

因此，蛋白质可以有效地在常温和常压下，在水溶液中催化许多在化学实验室里根本无法完成的化学反应。例如，在化学实验室里或化工生产线中，一个水溶液非碱性即酸性。因此，仅可以在酸性或者碱性的条件下进行一个水解反应，而不可能获得一个既是酸性又是碱性的反应条件。然而，在细胞里，蛋白水解酶则可以将其链上的氨基和羧基折叠在其三维空间结构中的适当位置上，使它们可从两边作用在一条淀粉链上，同时

实现酸和碱的催化反应。因此,淀粉可在常温和常压的条件下被有效地分解,以提供所需的能量。

利用特殊合成技术,也可在链上引入相互作用的基团,故单条合成长链也能折叠成一个较有序的构象。例如,在窄分布的热敏性聚 N-异丙基丙烯酰胺(PNIPAM)长链上统计地接枝上多条亲水性聚环氧乙烷(PEO)短链或均匀地嵌入疏水的聚苯乙烯(PS)短链节。水溶性的 PNIPAM 在 ~32 ℃ 有一个下临界溶液温度,即当高于这一临界温度时,PNIPAM 开始疏水不溶。因此,通过在 25~40 ℃ 范围内调节溶液温度,可分别引致上述共聚长链折叠。其构象从无规线团分别变成以疏水 PNIPAM 或聚集的疏水 PS 短链节为核和以亲水 PEO 短链或小的相对亲水的 PNIPAM 圈链为壳的"准三级结构",如下图(a)和(b)所示。

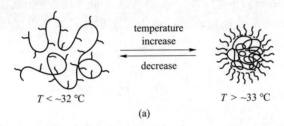

Wu C,Qiu XP. Physical Review Letters,1998,80(3): 620-622.(Figure 3,permission was granted by publisher)

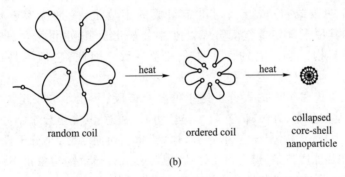

Zhang GZ,Winnik F,Wu C. Physical Review Letters,2003,90(3), 35506-1-4.(Figure 3,permission was granted by publisher)

应该强调:对一给定的组成,聚合反应的本质决定了人工合成大分子的链长和组构的非均一性。它们由单链构成的三维空间结构也不可避免地具有不同的大小、形状和内部结构。因此,与天然大分子确定和均一的三级结构相比,这样比较有序的空间分布并非严格意义上的三级结构。

多条相同或不同的具有特殊二级和三级结构的天然大分子还可以进一步组装成确定的多分子四级结构。例如,两条 DNA 链组成的双螺旋结构、染色体内的碟状组蛋白八聚

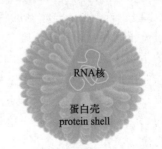

体,以及由 RNA 或 DNA 为核、核衣蛋白为壳、外加其他蛋白组装而成的各种病毒。大众往往是谈疫色变,从大分子角度看去,病毒只是一种介于生命体和非生命体之间的由不同大分子链组装而成的简单四级结构。上图描绘了一个 RNA 病毒。在细胞里,由多个生物大分子组装而成的四级结构,种类繁多,不胜枚举。有兴趣的读者可以参考生物学教科书。

值得指出的是,在过去的几十年里,有关聚合物链(尤其是各种嵌段聚合物)"自组装"的研究方兴未艾。文献中有各种各样漂亮的卡通结构。但是,"聚合物链的自组装"非"天然大分子链的自组装"也!所得结构的尺寸和链数绝非均匀,它们只是一个较有序的集合体。严格地说,其形成过程只能称为聚集,而不是真正的"自组装"。由聚合物链聚集而成的较有序的结构与由生物大分子自组装形成的四级结构有着天壤之别。不可有意或无意地混淆二者之间的本质区别。正因为这一根本的差别,聚合物链没有三级和四级结构。当然,许多研究聚合物"自组装"的人可能不认可这样有关大分子链高级结构的严格定义。

小　　结

每个大、小分子都有组成(小分子:原子种类和数目;大分子:单体种类和数目)。小分子内的原子们以化学键按一定方式连接,形成不同的构型。在一个大分子内,其单体们以化学键按一定方式(线型地、支化地……)组织构造成一个空间结构,简称"组构",也称作初级或一级结构;对给定组成的小分子,一种构型就对应着一种分子、一个结构和其原子确定的空间分布。在小分子化学中,也将两个具有不同组构的分子称为"同分异构",但二者的性质并非完全一致。因此,它们不是严格物理意义上的同样分子。

而一条给定组成和组构的大分子链中的各个链段(不是小分子里原子的转动和振动)还有不同的空间分布,称为"构象"(二级结构)。注意:其完全异于小分子光谱学中论及的"构象",此大分子"构象"非彼小分子"构象"也。换而言之,小分子没有大分子中定义的"构象"。在热能激发下,链上每个链段无规地运动,导致链构象永不停息地变化,链段间无特殊相互作用的线型大分子链的构象为无规线团。

一些天然大分子(如蛋白质)则可因为链内不同单元间的特殊相互作用(如氢键等)形成局部有序的链内构象(二级结构单元,如蛋白质链上的 α-螺旋和 β-折叠)。这些特定的结构单元还可在三维空间中精确和有序地排列成一种稳定和确定的空间分布(三级结构)。若干个确定数目的相同或不同的具有特定三级结构的生物大分子还可进一步通过分子间相互作用在三维空间中组装成一个稳定和确定的四级结构,在细胞里实现特定的生物功能。

由此可见,小分子没有大分子中定义的一级以上的高级结构;部分生物大分子可形成三级和四级结构。严格地说,合成大分子是"三无"分子,无确定的组成、无确定的组构、无确定的构象。对一给定的聚合物样品,仅有统计的平均组成、平均组构和平均构象。

一些特殊合成的大分子可折叠成相对有序的单链结构和形成相对有序的聚集体。由于合成大分子链的组成和组构并非均匀,所以它们没有严格意义上的三级、四级结构。下图总结了大、小分子有关"组成""组构"和"构象"的区别。

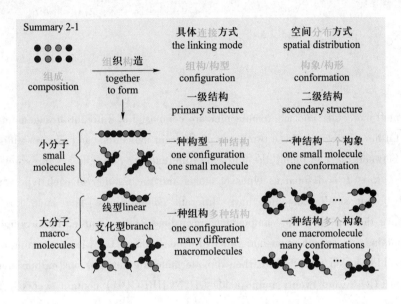

源自维基百科的下图(Holger87)很好地总结了一个蛋白质分子从一级至四级的结构。

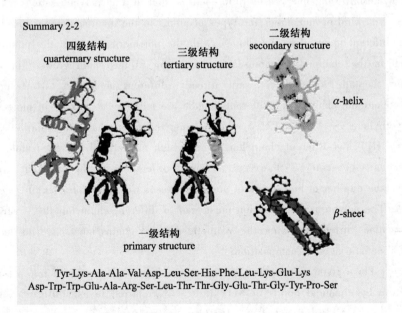

图片来源

Chapter 2. Macromolecular Structures

Both small molecules and macromolecules are composed of different arrangements of atoms. In order to further distinguish these two kinds of different molecules, and discuss similarities and differences between them in depth, the following different key concepts are introduced first.

Composition: It is defined as types of atoms and the number of each type of atom in a molecule. Different small molecules have different compositions. The same kind of small molecules have the exactly same composition, regardless of the source. For example, a water molecule contains one oxygen atom and two hydrogen atoms (H_2O); salt contains one sodium ion and one chloride ion (NaCl); a carbon dioxide molecule contains one carbon atom and two oxygen atoms (CO_2); and twenty amino acids ($H_2NCHRCOOH$) contain twenty different side groups (R).

Like small molecules, different macromolecules can be obtained by changing the composition of polymerized monomers. The difference is that if the polymerization reaction involves only one given kind of monomers, the types of atoms in the macromolecular chain and the ratio between different atoms does not change with the polymerization, but the number of each type of atom in the chain still increases in proportion as the chain grows. For small molecules, hexane, heptane, octane, etc. are already different molecules. However, in macromolecular science, the chains with different lengths are still the same kind of molecules. For example, polyethylene ($-(CH_2-CH_2)_n-$), polymerized from small molecular ethylene monomers ($CH_2=CH_2$), has different chain lengths, in which the ratio of hydrogen atoms and carbon atoms remains a constant (2:1, except for small differences at the end that can be ignored). However, the number of both types of atoms increases with the chain length (degree of polymerization). The polymerization reaction must lead to different chain lengths, a variable degree of polymerization. Strictly speaking, the synthetic polymer chains must have no definite composition, only a statistic average composition.

Configuration: For a given composition, each unit (monomer) is organized into a three-dimensional structure by chemical bonds in a certain way, referred to as configuration in short, i.e., primary or the 1st order structure. The units in small molecules are various atoms. The way of connecting these atoms is called "structural form" ("configuration"), different ways of connection correspond to different molecules. However, one kind of macromolecules can have different ways of connection, e.g., linear or branched polystyrenes. Even for a given way of connection, there still exist different structures. It will be discussed in details.

This book translates the English "configuration" into "organized (configurational) structure" instead of the commonly used "structural form" mainly for the following two considerations. First, the word "configuration" is derived from two Latin root words, "con" (together) and "figurare" (to form). The adoptions of "organized" and "formation" respectively corresponds to them well. Next, there are different translations of "configuration" in Chinese, including "organization state, structural state, etc." besides the commonly used "structure form". In small molecular synthetic chemistry, it has become customary to use the "structure form", no ambiguity. This is because there is a one-to-one correspondence between how atoms are connected and spatially distributed in a small molecule. However, for a given connection mode and structure, each unit on the macromolecular chain still has different spatial distributions (called conformation).

Although the "organized structure/structural form" and "conformation/structural shape" are clearly defined for "how monomers are specifically connected" and "how monomers are spatially distributed", respectively, the "structural form" is easily mixed up with the "structural shape". The main reason for the confusion is that the Chinese words "form" and "shape" has the same pronunciation and a similar writing, but their connotations are completely different. This "form" is not that "shape"! The "form" implies to which "category and kind" an object belongs, and the "shape" refers to the "shape, appearance, and image" of an object, its spatial distribution shape.

The adoption of "structural form" is easily mixed up with "structural imagine or structural shape". If a teacher does not emphasize their differences each time repeatedly when lecturing in classroom, or if students mishear and misunderstand the "structural form" as the "structural shape", they will mix up the two different concepts, leading to a difficult situation. In English, there is no such a problem. Perhaps, after the "structural form" was customarily used as the "configuration" in small molecule chemistry, it is not possible to translate the "conformation" into the "structural shape" because it is extremely easy to mix them up. Therefore, the term "structural imagine" that is rarely used in daily life has to be adopted. This is just a guess. More examples will be used to further explain why the "organized structure" is used instead of "structural form" in this book after the macromolecular "conformation" is examined in details.

For small molecules with a given composition, different ways of connection lead to different molecules, such as maleic acid and fumaric acid. In synthetic chemistry of small molecules, two molecules with different ways of connection are called isomers, e. g., two chiral symmetric mirror molecules; and the *cis*- and *trans*-isomers, as shown on the right. They might have the same chemical properties. Strictly speaking, due to some property differences between them, they are no

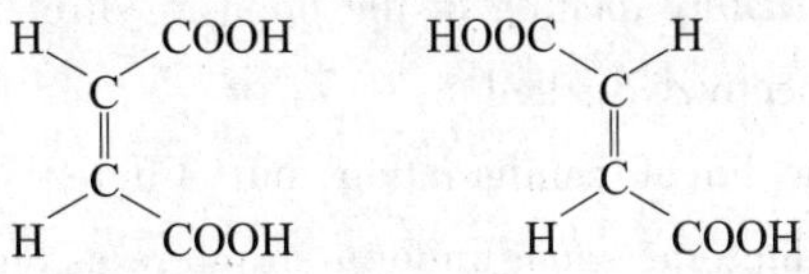

顺丁烯二酸
maleic acid

反丁烯二酸
fumaric acid

longer "the same molecule", but different molecules physically, so they have different names, such as "left-handed" and "right-handed".

The macromolecules already have far richer and more complicated configurations than small molecules even if they are homopolymers formed by the polymerization of only one kind of monomers. For a given composition (monomer type and number), there exist different, specific connection modes between the monomers, such as linear or branched, forming different configurations. After ignoring the connection details in the chain (such as the total degree of polymerization and the subchain length in the branched chain), macromolecules are catalogued into different configurations, such as linear, branched and star chains. In macromolecular chemistry, they are treated as the same kind of polymers, but with an added configuration name, such as linear polyethylene, branched polyethylene, star polyethylene, etc.

In order to clarify that even for a given composition and way of connection (configuration), there are still different ways to arrange different monomers on a chain (structure). Polystyrene made of styrene is only used as an exam-

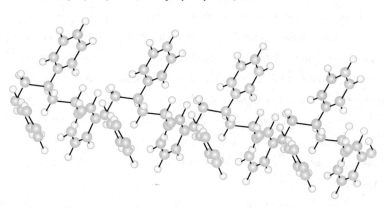

ple, as shown on the right. If two ends of an asymmetric monomer are taken as "head" and "tail", under normal circumstances, monomers will be linked together in a "head-to-tail" fashion, forming a linear macromolecular chain. However, a styrene monomer still has a small chance of being linked to the end of a growing chain in a "tail-to-head" manner, so that the subsequent chain growth becomes a "tail-to-head" connection, there may also be multiple reversals during polymerization. An active group on a growing chain can also transfer with a certain probability, so that a branching appears on the originally linearly growing main chain. Each branched chain (also called a subchain) can continue to branch, forming ever changing various structures.

In addition, if viewed along the carbon-carbon backbone, there are three possible arrangements of benzene rings on all styrene monomers: orderly rotating in one direction, alternately rotating in the opposite direction, and randomly rotating in two opposite directions, respectively, called as isotactic, syndiotactic, and random polymers. Note: they have the same linear configuration, but different structures. An actual polymerization reaction cannot produce the same uniform structure as small molecules, always resulting in a mixture of various structures. By controlling the reaction conditions, one of the structures can be dominant.

The copolymerization involving two or more kinds of monomers can produce copolymers with much richer and more complex structures. Now, in addition to the chain length distribution and various above mentioned different structures, even for a given composition, structure and chain length, different distributions of different comonomers on a linear copolymer macromolecular chain will also lead to different structures. In macromolecular chemistry, they are still called the same copolymer.

Taking copolymers with only two kinds of comonomers A and B as an example, A and B can be embedded in the chain randomly (AABAAA...BAA), alternately (ABAB...AB), or with a segment of A and a segment of B (A...AB...BA...A), respectively, called random, alternating or block copolymers. If catalogued by the block number, the configuration can also be subdivided into di-block, tri-block and multi-block copolymers. If A and B are connected in a branching manner, i.e., for a given configuration, the binary copolymerized branch chains still have a huge number of different structures. The configuration and structure of macromolecular chains are various in kinds, innumerous in number, so that no details will be given here.

Therefore, unlike biological macromolecules such as proteins, synthetic macromolecules have no definite configuration and structure in a strict sense, only a statistically averaged configuration and structure. If ignoring the connection details between monomers, the macromolecular configuration can be roughly catalogued into linear, branched, cyclic and other types.

The structure of natural macromolecules is extremely rich. DNA and RNA strands have the same linear configuration, and each contains four different bases. For a given base composition, their relative arrangement and combination can lead to DNA and RNA with different functions and roles. The protein molecules are composed of twenty different amino acids, and all of them have a linear configuration. Even for a given composition of amino acids, in theory, a corresponding protein chain has an astronomical number of different structures. Different structures directly lead to whether they have some biological activities. Natural plant macromolecules also have very complex structures. Due to a limited space, they will not be elaborated here. Interested readers can refer to the corresponding textbook.

Conformation: After being configured in a given way, different spatial distribution forms of each unit in the macromolecular chain constitute its different unique secondary structures, called macromolecular conformation. In small molecule chemistry, if the atoms in a molecule have two relatively stable spatial distributions, they are called as conformational isomers, such as the boat and chair forms. It is worth noting that there are countless transitional spatial distributions between these two conformations, and they are only relatively stable distributions with a higher probability. In addition, each atom is constantly rotating and vibrating, leading to a different spatial distribution every moment. This kind of conformations of small molecule is not that conformation of macromolecules! Therefore, small molecules do not have the conformation discussed in macromolecules. The macromolecular conformation considers only the spatial distribution of polymerized monomers, even larger segments composed of several monomers, not the details of how atoms are distributed.

For example, one extreme conformation of a flexible linear macromolecular chain composed of hundreds or thousands of polymerized monomers with no special interaction is completely straightened with no change of the bond angle. Another extreme conformation is a collapsed single-chain uniform sphere. In the solution or in the molten state, even if there is no external force or energy input, there is a thermal energy ($k_B T$) in a system at a given absolute temperature ($T > 0$). Agitated by the thermal energy, every segment on the chain is ever moving randomly. If it changes to either of the two extreme conformations, its conformational entropy will decrease, while its intrachain excluded volume effect will increase, resulting in an increase in its free energy. This violates the thermodynamic criterion of a spontaneous process under a constant temperature and pressure, so it will not happen spontaneously.

Therefore, the random movement driven by the thermal energy causes a linear macromolecule to form a random-coil conformation. Its size and shape constantly change (fluctuate), tending to the maximum conformational entropy. It is necessary to note that every chain either in solution or in melt or bulk (solid) has an extended random-coil conformation. In a dilute solution, solvent molecules fill between the segments; as the concentration increases, in addition to solvent, there are also many entangled segments from other chains between; and in the melt or bulk, there are full of the segments from other chains, interpenetrating and entangled with each other. Therefore, macromolecules with no special intrachain interactions obtained by polymerization have no defined conformation. In the following, a chapter will be used to discuss in detail how to average the ever-changing conformations of macromolecular chains statistically to obtain their average conformation and average size.

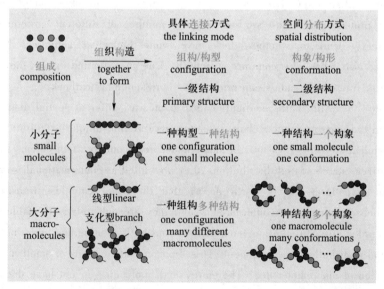

In order to more clearly illustrate the essential differences between the "organized structure/structural form" and "conformation/structural shape", it can be imagined that rigid rods and flexible strings are used to connect eight beads to, respectively, represent small molecules and macromolecules, leading to two configurations, linear and branched, as shown above. For small molecules, no rotation is allowed at each connection point, one way of connecting atoms inside a small molecule corresponds to one structure, one structure corresponds to one kind of molecule, and one kind of molecules have only one conformation. While each way of connecting monomers inside one macromolecule corresponds to one type of macromolecules with different structures, each structure has countless conformations (spatial distribution). In other words, for a small molecule, there is a one-to-one correspondence between its way of connection and its spatial distribution; while for a given configuration, one macromolecular structure (see figure) can correspond to different conformations. It should be remembered that here the "configuration" and "conformation" respectively describe the "how to connect" and "how to distribute" of units (atoms, polymerized monomers or chain segments), never confusing these two completely different concepts.

If using "structural form" to describe the connection mode, "form" easily confuse with "shape" because the pronunciations of "structural form" and "structural shape" are the same in Chinese. During lectures, it is difficult for a teacher to emphasize the "form" with an "earth" root, not "shape" every time. In addition, "conformation" and "structural shape" are synonymous; it is extremely easy to confuse "structural form" with "structural shape/conformation". The other translations of "configuration" in English, "organized state and structural state", are not focused on "structure", nor do they have structural meaning. If using "structure", it is also problematic. Besides the cumbersome words "first" or "second" should be added each time to distinguish it from other unique macromolecular high-level structures, one configuration includes different structures. For example, different branched structures all belong to the branched configuration. It is precisely because this "form" is not that "shape" in Chinese, this book uses the Chinese translation "organized structure" instead of the conventional Chinese translation "structural form" to express "configuration", hoping to reduce some confusion.

In addition, the same English "linear" corresponds to three words with completely different connotations and extensions in Chinese translation: "line type", "line shape" and "linear". "Linear polymers or chains" in English originates from an adverb "linearly connected". Its original meaning and full name in English should be "linearly formed (connected) polymers or chains". It looks like an adjective here, but it describes the connection mode, not the spatially distributed shape. It should be "linear type" in Chinese and should not be confused with "linear shape". In macromolecular physics, for a given configuration, the specific connection detail (for example, the arrangement of chemical groups in monomers on the chain) is not that important. Therefore, in this book, "linear" is only used to express the dependency between two variables. When referring to a type of configuration, the word "type" is used, such as linear chain, branched chain, star chain, etc.

For given the composition and organization, some special interactions (such as ionic bonds, hydrogen bonds, hydrophobic interaction, etc.) between the intrachain and interchain units can make a chain link to fold into a specific three-dimensional spatial distribution, forming some special non-random-coil conformations, i. e., some secondary structures with a certain spatial distribution, especially numerous natural biological macromolecules. In comparison with synthetic polymer chains, biological macromolecular chains, such as proteins and RNA, are generally shorter, so that one should not focus on their special physical properties due to the chain entanglement,

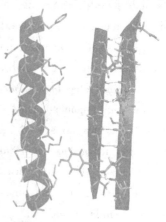

α-helix   β-sheet

but their special conformations with biological functions. Under appropriate conditions inside or outside the cell, many protein macromolecular chains can fold into one or more α−helices and β−sheets with a certain structure, as shown in the above figure.

High-order structures: After folded into a certain number of specific secondary structural (conformational) segments, a biological macromolecular chain can further fold into a special spatial structure with biological activity and function: the tertiary structure. For example, the transport tRNA can fold into a specific tertiary conformation and transport various amino acids in the cytosol to

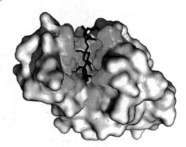

the paired messenger mRNA inside the ribosome made of rRNA. The primary structure of a protein determines its secondary, tertiary, and even quaternary three−dimensional structure. A combination of these structures determines its biological functions. The special tertiary structures of various shapes made of folded protein chains often have different biocatalytic functions in cells, as shown on the right, where light blue and dark blue color mark the binding and catalytic sites, respectively. The gene mutation (affecting the primary structure) or some internal interferences can influence the secondary structure (conformation) of a protein, often causing it to lose specific biological functions.

Therefore, proteins can effectively catalyze many chemical reactions in aqueous solutions that cannot be completed in a chemistry laboratory under the normal temperature and pressure. For example, in a chemical laboratory or production line, an aqueous solution is either alkaline or acidic. Therefore, one can only carry out a hydrolysis reaction under an acidic or basic condition, and it is impossible to obtain a reaction condition that is both acidic and basic. However, in the cell, proteolytic enzymes can fold the amino and carboxylic acid groups on the chain at appropriate positions in their three-dimensional structure, making them to act on a starch chain from the both sides, thereby simultaneously achieving acid and alkali catalysis. Therefore, starch can be effectively decomposed into sugars at normal temperature and pressure to provide energy.

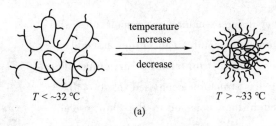

(a)

Wu C, Qiu XP. Physical Review Letters, 1998, 80(3):
620-622. (Figure 3, permission was granted by publisher)

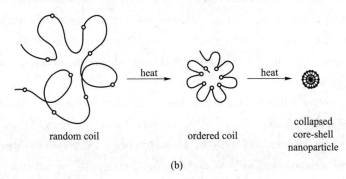

(b)

Zhang GZ, Winnik F, Wu C. Physical Review Letters, 2003, 90(3),
35506-1-4. (Figure 3, permission was granted by publisher)

Using special synthesis technology, one can also introduce interacting groups on the chain, so that a single synthetic long chain is also able to fold into a more ordered conformation. For example, a number of short hydrophilic polyethylene oxide (PEO) chains or hydrophobic polystyrene (PS) chain links are statistically grafted onto or uniformly embedded in a narrowly distributed thermally sensitive poly-N-isopropyl acrylamide (PNIPAM) long chain. Water-soluble PNIPAM has a low critical solution temperature at ~ 32 ℃. When higher than this critical temperature, PNIPAM becomes hydrophobic and insoluble. Therefore, by adjusting the solution temperature in the range 25-40 ℃, one can make the above-mentioned long copolymer chains to fold. Its conformation changes from a random coil to a "quasi-tertiary structure", respectively, with the collapsed hydrophobic PNIPAM or the aggregated short PS chain links as the core and the short hydrophilic PEO or the small relatively hydrophilic PNIPAM looping chains as the shell, as shown in the figures (a), (b) above.

It should be emphasized that for a given composition, the polymerization nature determines the chain length and structure heterogeneities of synthetic macromolecules. Their three-dimensional spatial structures made of single chains inevitably have different size, shape and internal structures. Therefore, in comparison with the definite and uniform tertiary structure of natural macromolecules, this relatively ordered spatial distribution is not strictly a tertiary structure.

Multiple identical or different natural macromolecules with special secondary and tertiary structures can further assemble to form a multi-molecular defined **quaternary structure**. For example, the double helix structure composed of two DNA strands, the disc-shaped histone octamer in the chromosome, and various viruses assembled from RNA or DNA as the core, nucleo-coat protein as the shell, and other proteins. The public  often talks about different epidemics. From the perspective of macromolecules, viruses are just a simple quaternary structure composed of different macromolecular chains between living and non-living bodies. The picture on the right depicts an RNA virus. In cells, there are many types of quaternary structures assembled from multiple biological macromolecules. Interested readers can refer to biology textbooks.

It is worth pointing out that in the past few decades, the research related to the "self-assembly" of polymer chains (especially various block polymers) has been in the ascendant. There are various beautiful cartoon structures in the literature. However, the "self-assembly of polymer chains" is not the "self-assembly of natural macromolecular chains"! The size and chain number of the resultant structure are by no means homogeneous, just a more ordered association. Strictly speaking, its formation process can only be called aggregation, rather than true "self-assembly". The more orderly structure formed by the aggregation of polymer chains is very different from the quaternary structure formed by the self-assembly of biological macromolecules. The essential difference between these two should not be confused intentionally or unintentionally. It is due to this fundamental difference that polymer chains have no tertiary and quaternary structures. Of course, many people who study the polymer "self-assembly" may not accept such a strict definition of the high-level structure of macromolecular chains.

Summary

Each macro-and small molecule has a **composition** (small molecules: type and number of atoms; macromolecule: type and number of monomers). **Atoms** in a small molecule are connected by chemical bonds in **a certain way**, forming different **configurations**, called "structural forms". In a macromolecule, its **monomers** are **organized to form** a three-dimensional spatial structure by chemical bonds in **a certain way** (linearly, branched, etc.), referred to as the "organized structure", also called primary or **the first order structure**. For small molecules with a given composition, each configuration corresponds to a molecule, a structure and a certain spatial distribution of its atoms. In small molecular chemistry, two molecules with different configurations are also called "isomerism", but their properties are not identical. Physically, they are not the same molecules.

While for a given composition and configuration, every chain segment (not those rotation and vibration of atoms in small molecules) in a macromolecular chain has different spatial distributions, called "**conformations**" (**secondary structure**). Note: it is completely different from the "conformation" discussed in the spectroscopy of small molecules. This macromolecular "conformation" is not that small molecular "conformation". In other words, small molecules do not have the "conformation" defined in macromolecules. Under the agitation of thermal energy, each chain segment on the chain randomly moves, which causes the chain conformation changes constantly. The conformation of linear macromolecular chains with no special interaction among different chain segments is a random coil.

Some natural macromolecules (such as proteins) can form a partially ordered intrachain conformation (**secondary structural units**, such as α-helices and β-sheets on protein chains) due to special interactions among different units in the chain (such as hydrogen bonds, etc.). These specific structural units can also be arranged accurately and orderly in a three-dimensional space into a stable and definite spatial distribution (**tertiary structure**). **A certain number** of the same or different biological macromolecules with specific tertiary structures can be further assembled into a stable and definite **quaternary structure** in three-dimensional space through intermolecular interactions, achieving specific biological functions in cells.

It can be seen that small molecules do not have high-order structures above the first level defined in macromolecules; some biological macromolecules can form tertiary and quaternary structures. Strictly speaking, synthetic macromolecules are molecules with no definite composition, no definite configuration, and no definite conformation. For each given polymer sample, there are only the statistical average composition, average configuration and average composition.

Some specially synthesized macromolecules can be folded into relatively ordered single-chain structures and form relatively ordered aggregates. Because synthetic macromolecular chains are not uniform in their composition and configuration, so that they do not have the tertiary or quaternary structures in the strict sense. The picture below summarizes differences among "composition", "configuration" and "conformation" of macro-and small molecules.

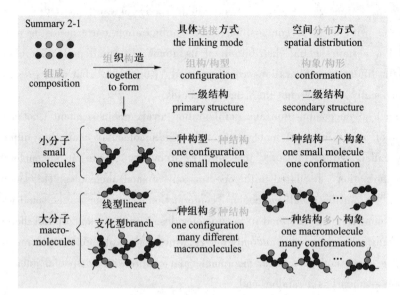

The picture on the below (Holger87) from Wikipedia provides a good summary of protein molecules from first to fourth structures.

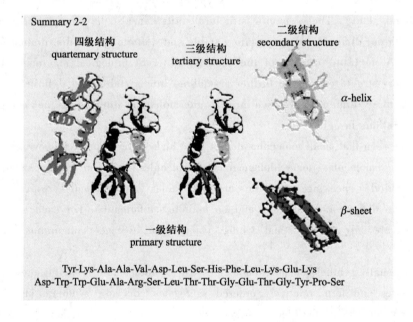

第三章　大分子溶液制备中的热力学

为了获得单个大分子链的性质(如链的尺寸、构象、摩尔质量等),首先需要将在本体中互相缠结的大分子链分开成一根一根的单链。将一种或多种物质以单个分子的形式分散到另一种物质中形成一个均匀混合物的过程谓之溶解,形成的混合物称作溶液。通常,较少的组分为溶质,而较多的则是溶剂。例如,在日常生活中,酒精、醋、盐或糖溶于水中形成一个均匀的水溶液;在实验中,聚苯乙烯常常溶于苯、甲苯或四氢呋喃中形成一个有机溶液。物理学上,将这样由两种、三种或多种分子组成的溶液统称为二元、三元或多元共混物或混合物。本书将仅讨论含有一种溶质的二元共混物。

大、小分子的溶液制备并无天壤之别。对一给定的绝对温度(T),一种物质能否溶解在一个溶剂中,仅取决于溶解前后体系的自由能变化($\Delta G_{mix} = \Delta H_{mix} - T\Delta S_{mix}$),其中 ΔH_{mix} 和 ΔS_{mix} 分别是混合前后的焓变和熵变。由热力学可知,在一个等温和等压条件下的自发过程中,自由能降低。因此,仅当 $\Delta G_{mix} < 0$ 时,一个溶质方可自发地溶解在一个溶剂中形成一个均匀的二元共混物;反之,该共混物将自发地朝着相反的方向,形成具有一个物理界面的两层(相)溶液,分别是对方在己方中的饱和溶液。一般而言,混合两种不同分子时,熵增加,即 $\Delta S_{mix} > 0$。因此,当 $\Delta H_{mix} \leqslant 0$ 或 $\Delta H_{mix} < |T\Delta S_{mix}|$ 时,溶解才会自动发生。

然而,因质量和体积的差别,大、小分子的溶解也有差异。焓变和熵变的相对贡献明显不同。首先,大分子链上被化学键连在一起的聚合单体无法像小分子溶解时那样完全地分开。每条链只可作为一个整体在溶液中无规运动。与小分子相比,大分子在溶解中获得的平动熵增加相对较少,故在其溶解中,焓变的作用更大。

对给定的混合焓变,即单体和聚合物分别和溶剂的作用相近时,常常出现单体可溶,而达到一定长度的聚合物则不溶的情况。沉淀聚合反应正是基于这一热力学原理来控制合成大分子的聚合度。另外,小分子混合时,单个溶质和溶剂的分子具有相近的体积,混合中熵的变化对组成的依赖性接近对称,最大混合熵对应的组成位于50%附近。一个大分子所占据的体积远大于一个溶剂分子,故混合熵的变化对大分子组成的依赖性完全不对称,向低浓度方向的左边倾斜。

除此以外,一个小分子溶质溶解时,分子从其表面一个又一个地逐渐扩散到溶剂之中,搅拌可降低其表面浓度,促进扩散,加速溶解。由于大分子链之间的互相缠结,它们的溶解包括两个步骤:首先是溶剂小分子向链中扩散,导致其溶胀,链段间距增加,相互作用减弱;然后,大分子链方可慢慢地互相解开缠结,就像小分子一样逐一地从其表面向溶剂中扩散。大、小分子的尺寸相差百倍,所以大分子扩散也要慢百倍,缓慢搅拌可促进大分子溶解,但溶胀和解缠结过程远慢于扩散,决定了溶解速率。因此,制备大分子溶液时,除了心理安慰以外,快速搅拌基本上不会加速溶解。对于超长链聚合物样品,高速和加力搅拌不仅无助,还可能使大分子链断裂。即使学习高聚物的师生们也常常会犯类似的错误。

另一方面,与小分子不同,一条大分子链与其周围的链互相缠绕,故不同的大分子链之间存在多点相互作用。相对较小的混合熵变有时无法驱动溶剂分子渗入互相作用的大分子链,溶胀和解开作用的链段,故高聚物的溶解温度有时必须高于其熔化温度。例如,著名的塑料王"聚四氟乙烯"(熔点为 227 ℃)仅在 330 ℃ 时才可溶于其低聚物中[Chu B,Wu C,Zuo J. Macromolecules,1987,20:700]。

为了方便讨论大分子与溶剂分子在混合中的热力学变化,让我们先回顾一下在热力学中学习过的在等温和等压条件下混合 N_A 个 A 粒子和 N_B 个 B 粒子的过程。假定 A 和 B 具有同样的大小,占据相等的空间(称为一个格子),以及粒子之间没有任何相互作用。

在混合前的初始状态 (i) 中,它们均为纯物质。N_A 个相同的 A 粒子占据了 N_A 个格点;N_B 个相同的 B 粒子占据了 N_B 个格点。A 粒子和 B 粒子的空间排列方式 (Ω) 分别仅有一种,即 $\Omega_{i,A}=1$ 和 $\Omega_{i,B}=1$。按照熵的定义,$S=k_B\ln\Omega$,混合前的总熵为

$$S_i = k_B\ln\Omega_{i,A} + k_B\ln\Omega_{i,B} = 0 \tag{3.1}$$

在混合后的最终状态 (f) 中,宏观性质是"N_A 个相同的 A 粒子和 N_B 个相同的 B 粒子共占据了 (N_A+N_B) 个格点"。对这样一个宏观性质,这些 A 粒子和 B 粒子在 (N_A+N_B) 个格点上有着不同的微观排列方式,其总数为 $\Omega_f=(N_A+N_B)!/N_A!\,N_B!$。因此,混合物的熵为

$$\begin{aligned}
S_f &= k_B\ln\Omega_f = k_B\ln\frac{(N_A+N_B)!}{N_A!\,N_B!} \\
&\cong k_B\big[(N_A+N_B)\ln(N_A+N_B)-(N_A+N_B)-(N_A\ln N_A-N_A+N_B\ln N_B-N_B)\big] \\
&= -k_B\Big(N_A\ln\frac{N_A}{N_A+N_B}+N_B\ln\frac{N_B}{N_A+N_B}\Big) \\
&= -N\,k_B(X_A\ln X_A+X_B\ln X_B) \\
&= -(n_A+n_B)N_{AV}k_B\Big(\frac{n_A}{n_A+n_B}\ln\frac{n_A}{n_A+n_B}+\frac{n_B}{n_A+n_B}\ln\frac{n_B}{n_A+n_B}\Big) \\
&= -nR(x_A\ln x_A+x_B\ln x_B)
\end{aligned} \tag{3.2}$$

其中,当 $N\gg1$ 时,$\ln N!\approx N\ln N-N$;$X_A=N_A/N$,$X_B=N_B/N$,$N=N_A+N_B$;$n_A=N_A/N_{AV}$,$n_B=N_B/N_{AV}$,$n=n_A+n_B$ 和 $R=N_{AV}k_B$;N_{AV} 是 Avogadro 常数。注意 A 和 B 粒子具有的相等体积的假定,所以,$x_A=X_A$ 和 $x_B=X_B$。综上所述,混合前后,熵的变化为,

$$\Delta S_{mix}=k_B(\ln\Omega_f-\ln\Omega_i)=-nR(x_A\ln x_A+x_B\ln x_B) \tag{3.3}$$

由热力学已知,Gibbs 自由能 (G) 的变化依赖于体积 (V)、熵 (S) 以及压强 (p) 和温度 (T) 的变化,即 $dG=Vdp-SdT$。因此,对一等压过程,$dG=-SdT$ 或 $(\partial G/\partial T)_p=-S$,进而写成 $[\partial(G_f-G_i)/\partial T]_{p,n}=-(S_f-S_i)$,或简写为 $(\partial\Delta G/\partial T)_p=-\Delta S$。选择绝对零度作为纯物质和混合物的参考点,可得

$$\Delta G_{mix}=-\int_0^T\Delta S_{mix}\,dT=nRT(x_A\ln x_A+x_B\ln x_B) \tag{3.4}$$

在式(3.3)和式(3.4)中,$x_A<1$,$x_B<1$,$x_A+x_B=1$。因此,$\Delta S_{mix}>0$ 和 $\Delta G_{mix}<0$,即混合

两种没有相互作用的粒子总是导致熵的增加和自由能的减少。一个等温和等压过程可自动发生的热力学判据为 $\Delta G < 0$，故两种没有相互作用的粒子总是可以自发地混合成一个完全均匀的一相体系（溶液），由熵的增加驱动。用 nRT 和 nR 约化的 ΔG_{mix} 和 ΔS_{mix} 仅依赖于体积分数（组分）。ΔG_{mix} 和 ΔS_{mix} 随组分的变化完全对称；它们相应的最低值和最高值都出现在 $x_A = 1/2$ 和 $x_B = 1/2$ 处，源于等体积的假定。

由热力学已知，焓包含了与功和热有关的自由能和熵，即 $H = G + TS$。因此，在一个等温过程中，$\Delta G = \Delta H - T\Delta S$。一般而言，真实分子之间存在"相互作用"，$\Delta H_{mix}$ 非零。混合前，粒子 A 和 B 均为纯物质。依据前述格子模型，每个粒子占据一个格点，体系共有 $(N_A + N_B)$ 个格点。在热力学中：$H = U + pV$ 和 $G = A + pV$，这里 U 和 A 分别是内能和 Helmholtz 自由能。如果在混合中压强不变且体积不可压缩，内能和 Helmholtz 自由能的变化（ΔU_{mix} 和 ΔA_{mix}）就分别与焓和 Gibbs 自由能的变化（ΔH_{mix} 和 ΔG_{mix}）无异。

在一个真实的二元混合物（溶液）中，两个真实分子不可同时占据同一个物理空间，每个分子具有一定的硬核体积。微观上，一个粒子（分子）自身的平均排除体积（v_e）包含两部分：其硬核体积（v_0）和其因粒子间相互作用引致的虚拟体积（Δv），即 $v_e = v_0 + \Delta v$。如果将排斥体积看作一个小球，其质心之间的平均距离对应着两个粒子之间相互作用势能的最小点，而与 v_0 有关的直径则可近似地对应着两个粒子之间相互作用势能从吸引（负）变为排斥（正）的距离。

链与链之间和一条链上的不同链段之间还存在着各种"相互作用"。例如，两个中性分子互相靠近时，带正电荷的原子核会吸引对方带负电荷的电子向己方偏移，分别产生两个方向相反、互相吸引的诱导偶极矩，称为色散作用。相关能量称为色散能。在气体和液体相变中，也被称为凝聚能（热）或蒸发能（热）。

一方面，硬核体积的存在减少了每个分子可达的体积，即增加了每个分子到达体系内任何一点的概率（减少了熵）。另一方面，色散相互作用引致的吸引减慢了分子的运动，减少了前述概率（增加了熵），产生了一个增加可达及体积的效应，而不是一个真的增加了的体积。换而言之，色散吸引等效于一个负的排除体积，但不是一个真的负排除体积。其类似"离心力是非惯性坐标系中一个虚拟力（惯性）"，产生了一个力的效应，但不是一个真的力。

因此，在一个真实的二元混合物中，有真实的硬核体积产生的效应，还有另一个源于色散吸引的、虚拟的负排除体积的反效应。将源于一个真的硬核体积和一个虚拟的排除体积的效应综合起来就是"排除体积效应"，如右图所示。对于吸引作用，排除体积效应可以为正、为零或为负。如果相互作用为排斥（如聚电解质链），其就会产生一个虚拟的正排除体积效应。这样，排除体积效应就总是为

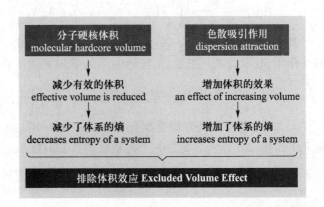

正。排除体积效应比没有相互作用时更强,从而导致溶液中的链构象更为扩展。

混合前,在$(N_A + N_B)$个格点上,仅有N_A个A粒子(分子)或N_B个B粒子(分子)。在$(N_A + N_B)$个格点上的任何一个格点上,发现第一个粒子为A或者B的概率为

$$\frac{N_A}{N_A + N_B} \quad \text{或者} \quad \frac{N_B}{N_A + N_B}$$

选择第二个粒子为A或者B的概率则为

$$\frac{N_A - 1}{N_A + N_B - 1} \quad \text{或者} \quad \frac{N_B - 1}{N_A + N_B - 1}$$

在混合前各自的纯物质中,在体系中产生一对A-A或者B-B二体相互作用的概率为

$$\frac{N_A(N_A - 1)}{(N_A + N_B)(N_A + N_B - 1)} \quad \text{或者} \quad \frac{N_B(N_B - 1)}{(N_A + N_B)(N_A + N_B - 1)}$$

上述概率乘以格点数$(N_A + N_B)$可得A和A或B和B之间二体相互作用的总对数,即

$$\frac{N_A^2}{2(N_A + N_B)} \quad \text{或者} \quad \frac{N_B^2}{2(N_A + N_B)}$$

此处,利用了$N_A \gg 1$和$N_B \gg 1$。上述推导中,两个粒子产生了两对相互作用,但实际上它们只能产生一对相互作用,故除以2可避免重复计算。

混合后,体系中多了A和B以及B和A之间的二体相互作用。注意:二者相同,一个A粒子和一个B粒子之间只能产生一对,而不是两对相互作用。总交叉作用对的数目为

$$\frac{N_A N_B}{2(N_A + N_B)} + \frac{N_B N_A}{2(N_A + N_B)} = \frac{N_A N_B}{N_A + N_B}$$

同时,原先A-A和B-B二体相互作用对的数目减少,减少的数目分别对应着上述增加的A-B和B-A二体相互作用对的数目。换而言之,分别打开一对A-A和一对B-B二体相互作用可产生两对A-B二体相互作用。假定每对不同的相互作用的能量分别为ε_{AA}、ε_{BB}和ε_{AB},混合前后的焓分别为

$$H_{\text{before}} = \frac{Z}{N_A + N_B}\left(\frac{N_A^2}{2}\varepsilon_{AA} + \frac{N_B^2}{2}\varepsilon_{BB}\right)$$

$$H_{\text{after}} = \frac{Z}{N_A + N_B}\left[\left(\frac{N_A^2}{2}\varepsilon_{AA} - \frac{N_A N_B}{2}\varepsilon_{AA}\right) + N_A N_B \varepsilon_{AB} + \left(\frac{N_B^2}{2}\varepsilon_{BB} - \frac{N_B N_A}{2}\varepsilon_{BB}\right)\right] \quad (3.5)$$

其中,假定格子模型中每个格点有Z个邻近的相互作用点,即一个粒子可产生Z对相互作用,故乘以Z。注意,在实验中,Z应该随着浓度变化。由以上两式可得

$$\Delta H_{\text{mix}} = H_{\text{after}} - H_{\text{before}} = \frac{Z N_A N_B}{N_A + N_B}\left(\varepsilon_{AB} - \frac{\varepsilon_{AA} + \varepsilon_{BB}}{2}\right)$$

上式中,括号内的常数是混合后形成一对A-B二体相互作用的平均焓变。如果用热能$k_B T$约化,$\Delta H_{\text{mix}}/k_B T$成为一个无量纲的状态函数。定义

$$\chi = \frac{Z}{k_B T}\left(\varepsilon_{AB} - \frac{\varepsilon_{AA} + \varepsilon_{BB}}{2}\right) \quad (3.6)$$

上式可重新写成

$$\frac{\Delta H_{\mathrm{mix}}}{k_{\mathrm{B}}T} = \frac{N_{\mathrm{A}}N_{\mathrm{B}}}{N_{\mathrm{A}} + N_{\mathrm{B}}}\chi \qquad (3.7)$$

注意：取决于 $\varepsilon_{\mathrm{AA}}$、$\varepsilon_{\mathrm{BB}}$ 和 $\varepsilon_{\mathrm{AB}}$ 的相对数值，χ 可为正、为负或为零。不同数值的 $\varepsilon_{\mathrm{AA}}$、$\varepsilon_{\mathrm{BB}}$ 和 $\varepsilon_{\mathrm{AB}}$ 可以组合得到一个完全相同值的 χ。因此，χ 与粒子间的相互作用并不一一对应。一组给定的 $\varepsilon_{\mathrm{AA}}$、$\varepsilon_{\mathrm{BB}}$ 和 $\varepsilon_{\mathrm{AB}}$ 仅是 χ 的充分非必要条件，只反映了因粒子相互作用引致的平均焓变。

$\chi < 0$ 时，混合中，排除体积效应减少，利于混合；反之，不利于混合。$\chi = 0$ 时，体系为无热混合气体或液体，即温度对该体系毫无影响。注意，在式（3.6）中，T 位于分母上，因此，χ 通常随着温度升高而变小。对一个真实的溶液（二元混合物），χ 对温度的依赖性与 $\varepsilon_{\mathrm{AA}}$、$\varepsilon_{\mathrm{BB}}$ 和 $\varepsilon_{\mathrm{AB}}$ 三者与温度的依赖性有关。将 $k_{\mathrm{B}}T$ 移到右边，再在右边同时乘以和除以 N_{AV}^2 以及利用 $n = (N_{\mathrm{A}} + N_{\mathrm{B}})/N_{\mathrm{AV}}$，式（3.7）可被重新写成

$$\Delta H_{\mathrm{mix}} = nRT\, X_{\mathrm{A}}(1 - X_{\mathrm{A}})\chi = nRT\, x_{\mathrm{A}}(1 - x_{\mathrm{A}})\chi \qquad (3.8)$$

其中，$X_{\mathrm{A}} = x_{\mathrm{A}}$ 和 $X_{\mathrm{B}} = x_{\mathrm{B}}$；$n_{\mathrm{A}} = N_{\mathrm{A}}/N_{\mathrm{AV}}$；$n_{\mathrm{B}} = N_{\mathrm{B}}/N_{\mathrm{AV}}$；$n = n_{\mathrm{A}} + n_{\mathrm{B}}$；$x_{\mathrm{A}} = n_{\mathrm{A}}/n$；$x_{\mathrm{B}} = n_{\mathrm{B}}/n$ 和 $x_{\mathrm{A}} + x_{\mathrm{B}} = 1$。对一个等温过程，将 ΔH_{mix} 和 ΔS_{mix} 代入 $\Delta G = \Delta H - T\Delta S$ 可得

$$\Delta G_{\mathrm{mix}} = nRT\left[x_{\mathrm{A}}(1 - x_{\mathrm{A}})\chi + x_{\mathrm{A}}\ln x_{\mathrm{A}} + (1 - x_{\mathrm{A}})\ln(1 - x_{\mathrm{A}}) \right] \qquad (3.9)$$

注意 $\chi x_{\mathrm{A}}(1 - x_{\mathrm{A}})$ 和 $x_{\mathrm{A}}\ln x_{\mathrm{A}} + (1 - x_{\mathrm{A}})\ln(1 - x_{\mathrm{A}})$ 分别与焓变和熵变有关。上式可被进一步写成

$$\Delta G_{\mathrm{mix}} = RT\left[x_{\mathrm{A}}n_{\mathrm{B}}\chi + n_{\mathrm{A}}\ln x_{\mathrm{A}} + n_{\mathrm{B}}\ln(1 - x_{\mathrm{A}}) \right] \qquad (3.10)$$

或

$$\Delta G_{\mathrm{mix}} = RT\left[x_{\mathrm{B}}n_{\mathrm{A}}\chi + n_{\mathrm{B}}\ln x_{\mathrm{B}} + n_{\mathrm{A}}\ln(1 - x_{\mathrm{B}}) \right] \qquad (3.11)$$

即用 x_{B} 代替 x_{A} 作为变量，同样也可得到式（3.9）。ΔG_{mix} 对 x_{A} 或 x_{B} 作图完全对称，源于 A 粒子和 B 粒子具有相同体积的假定。

显然，大分子溶液中已无上述对称关系。但是，上述讨论对大分子溶液仍然成立。唯一的修正是，将一个大分子的体积等价于 n_v 个溶剂分子的体积。换而言之，如果将溶液体积按照一个溶剂分子的体积分割成 N_{box} 个小盒子，每个溶剂分子占据一个小盒子，而每个大分子则占据 n_v 个小盒子，即一条大分子链"等同于" n_v 个溶剂分子。

令 A 和 B 分别代表溶剂（小分子）和溶质（大分子），记为"s"和"p"；其分子数分别为 N_s 和 N_p；$N_{\mathrm{box}} = N_s + n_v N_p$。溶质和溶剂占据的格子分数，即体积分数，分别为 $X_p = n_v N_p/(N_s + n_v N_p) = n_v n_p/(n_s + n_v n_p) = n_v n_p/n_{\mathrm{box}}$ 和 $X_s = N_s/(N_s + n_v N_p) = n_s/(n_s + n_v n_p) = n_s/n_{\mathrm{box}}$；而溶质和溶剂的摩尔分数分别为 $x_p = N_p/(N_s + N_p) = n_p/(n_s + n_p)$ 和 $x_s = N_s/(N_s + N_p) = n_s/(n_s + n_p)$。很清楚，溶质和溶剂的体积分数已不等于它们对应的分子或摩尔分数。

再强调一次：在小分子溶液的讨论中，$n_v = 1$；体积分数等于摩尔分数。在大分子溶液中，$n_v \gg 1$；体积分数与摩尔分数完全不等！由式（3.2）可得下式。

$$S_f = k_{\mathrm{B}}\ln\Omega_f = k_{\mathrm{B}}\ln\frac{(N_s + n_v N_p)!}{N_s!\,(n_v N_p)!} = -k_{\mathrm{B}}\left(N_s\ln\frac{N_s}{N_s + n_v N_p} + N_p\ln\frac{n_v N_p}{N_s + n_v N_p} \right)$$

$$\Delta S_{\mathrm{mix}} = -N_{\mathrm{box}}k_{\mathrm{B}}\left(X_s\ln X_s + \frac{X_p}{n_v}\ln X_p \right)$$

$$\frac{T\Delta S_{\text{mix}}}{n_{\text{box}}} = -RT\left[(1-X_p)\ln(1-X_p) + \frac{X_p}{n_v}\ln X_p\right] \tag{3.12}$$

其中，n_{box} 为小盒子的摩尔数（$N_{\text{box}}/N_{\text{AV}}$）和 $R = k_B N_{\text{AV}}$。两边同乘以绝对温度（T）使得混合熵（ΔS_{mix}）成为一部分混合自由能（ΔG_{mix}）。其物理意义是，平均到每摩尔小盒子上因混合熵导致的混合自由能。注意：n_v 正比于大分子的聚合度，所以混合熵也随着聚合度的增加而变小，具有链长依赖性，这是大、小分子溶液之间的一个显著差别。大分子的链越长（n_v 越大），混合熵对体积分数或摩尔分数的依赖性就越不对称。在极稀溶液中，上式中的第一项可以略去，故混合熵与聚合度成反比。

在真实大分子溶液中，必须考虑由硬核体积和相互作用合起来的"排除体积效应"。与讨论小分子溶液时稍有不同，在大分子溶液中，ε_{BB} 代表了一个与溶剂分子等效的"链段"和另一个"链段"相互作用时涉及的能量；ε_{AB} 代表了一个"链段"和一个溶剂分子相互作用时涉及的能量。注意：这里的"链段"不是 Kuhn 链段，而是与一个溶剂分子具有相同等效体积的一小段链。换而言之，如果忽略端基效应，一个这样的"链段"除了和链上另外的两个"链段"连接外，还可以和（$Z-2$）个溶剂分子作用。假定一条大分子链可同时与 n_v 个溶剂分子产生排除体积效应，拆开一对"链-链"排除体积效应和 n_v 对"溶剂-溶剂"排除体积效应可形成 $2n_v$ 对"链段-溶剂"排除体积效应。因此，沿用前述式（3.6）的定义，将 χ_{AB} 写成 χ_{sp}，以及有关小分子混合焓能的讨论，可得与式（3.8）一样的结论。只是因为大分子和溶剂小分子的体积之差，无法再用摩尔分数（x_{B}）代替体积分数（X_{B}）。

$$\Delta H_{\text{mix}} = n_{\text{box}} RT X_p (1-X_p) \chi_{sp} \tag{3.13}$$

对一个等温过程，将 ΔH_{mix} 和 ΔS_{mix} 代入 $\Delta G = \Delta H - T\Delta S$ 可得与式（3.9）类似的下式：

$$\frac{\Delta G_{\text{mix}}}{n_{\text{box}} RT} = X_p(1-X_p)\chi_{sp} + \frac{X_p}{n_v}\ln X_p + (1-X_p)\ln(1-X_p)$$

$$\frac{\Delta G_{\text{mix}}}{RT} = X_p n_s \chi_{sp} + n_p \ln X_p + n_s \ln X_s \tag{3.14}$$

其中，利用了 $1-X_p = X_s$，$X_s = n_s/n_{\text{box}}$ 和 $X_p = n_v n_p/n_{\text{box}}$。$\Delta G_{\text{mix}}/RT$ 已无量纲。

注意，由于 $n_v \gg 1$，与小分子溶液相比，ΔG_{mix} 对 X_p 的依赖极度地不对称。混合前，溶剂或溶质为纯物质（$X_p = 0$ 或 1），$\Delta G_{\text{mix}} = 0$。当 X_p 充分小或接近 1 时，混合导致自由能下降，$\Delta G_{\text{mix}} < 0$，溶解发生。换而言之，理论上，没有绝对不溶的二元混合物。任何物质在任何溶剂中都可溶，仅是溶解度有别而已。通常所讲的"不溶解"，仅是指溶解度很低而已。根据国际纯粹与应用化学联合会的定义，当溶解度低于万分之一时，即可认为是一个物质在一个溶剂中"实际上的不溶"或简称"不溶"。从统计物理学的观点出发，世界上任何事都可能发生，只是其发生的概率不同！

依据式（3.14），对给定的温度，当 $\chi_{sp} \leq 0$ 时，$\Delta G_{\text{mix}} < 0$；溶解自发地进行，大分子可以任何比例溶于溶剂，形成溶液。如果 $\chi_{sp} > 0$，但小于一个临界值（χ_c），ΔG_{mix} 仍为负，混合自发地进行。当 $\chi_{sp} > \chi_c$ 时，ΔG_{mix} 则在 X_p 的一定范围内为正，体系朝混合的相反方向，自发地分开成具有一个界面的两层（相）：稀相为大分子在溶剂中的饱和溶液；浓相则是溶剂在大分子中的饱和溶液（被溶剂分子渗透和溶胀的互相缠结的链）。

右图显示了当 $n_v = 2$、3 和 100 以及 $\chi_{sp} = 1 < \chi_c$ 和 $\chi_{sp} = 2 > \chi_c$ 时，$\Delta G_{\mathrm{mix}} / (n_{\mathrm{box}} RT)$ 如何随着大分子的体积分数变化。据式（3.14），当 $\chi_{sp} > \chi_c$ 时，无论 χ_{sp} 为何值，在 X_p 足够小时，ΔG_{mix} 总会先从零下降，为负值，到一最低点后上升，越过零点，继续上升，直至达到一个峰值。在右图中，也有一个类似的变化，即随着溶剂含量的增加（从右向左），ΔG_{mix} 先下降再上升。从左或右两边出发（将大分子不断地加入溶剂或将溶剂不断地加入大分子），ΔG_{mix} 都汇

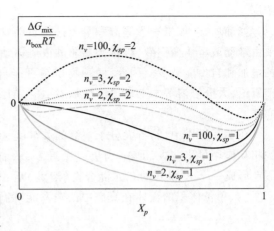

合在同一个峰值。由于大分子和溶剂分子的体积悬殊，变化完全不对称。当 $n_v \gg 1$ 时，该变化在图中靠近左边纵轴无法显示（如图中 $n_v = 100$ 时），仿佛 $\Delta G_{\mathrm{mix}} / (n_{\mathrm{box}} RT)$ 没有初始的下降，一开始就从零上升。实际上，这只是一个错觉。不幸的是，传统的高聚物教科书并没有特别地对学生和读者们强调该点，会造成混淆和误解。

实验上，混合后溶质和溶剂的体积分数无法直接测量，故常用摩尔分数，其可从溶质和溶剂的质量获得。因此，利用 $n_{\mathrm{box}} = N_{\mathrm{box}} / N_{\mathrm{AV}}$，$N_{\mathrm{box}} = N_s + n_v N_p$ 和 $n = n_s + n_p$，可将上式里的体积分数换算成摩尔分数，得到上式里的混合焓（ΔH_{mix}）对混合自由能的贡献，

$$\frac{\Delta H_{\mathrm{mix}}}{nRT} = x_p (1 - x_p) \frac{n_v}{1 + (n_v - 1) x_p} \chi_{sp} \tag{3.15}$$

其值可正可负，取决于 χ_{sp}，其随温度升高而变小。混合熵变（ΔS_{mix}）对自由能的贡献为

$$\frac{\Delta S_{\mathrm{mix}}}{nR} = -\left\{ x_p \ln \left[x_p \frac{n_v}{1 + (n_v - 1) x_p} \right] + (1 - x_p) \ln (1 - x_p) \right\} \tag{3.16}$$

其中 $x_p < 1$，$(1 - x_p) < 1$ 和 $n_v x_p < 1 + (n_v - 1) x_p$。上式中的两项总为负值，故在一个混合过程中，体系的熵总是增加。与式（3.3）和式（3.8）（小分子）相比，上两式中各多了一个因子：

$$\frac{n_v}{1 + (n_v - 1) x_p}$$

显然，当 $n_v = 1$ 时，它们还原成式（3.3）和式（3.8）。对高聚物而言，$n_v \approx 10^2 \sim 10^4 \gg 1$。与式（3.3）中小分子混合熵相比，源自大分子链的混合熵（上式中的第一项）贡献相对较小。与不通过化学键连接的相应 n_v 个小分子相比，一个大分子的溶解导致平移熵的增加较少。因此，大分子溶解时混合熵主要来自第二项中溶剂小分子的混合熵。将上两式合并可得大分子溶解后的混合自由能（$\Delta G_{\mathrm{mix}} = \Delta H_{\mathrm{mix}} - T \Delta S_{\mathrm{mix}}$），即著名的 Flory-Huggins 关系式。

大分子实验中，通常采用质（重）量浓度（克/毫升），即单位体积里的大分子的质量，而不是摩尔分数。在第五章中，将会讨论如何将体积分数和摩尔分数换算成质量

浓度。

如前所述,如果一个等温和等压过程自动发生,体系的自由能必定降低。该降低可主要由焓变或者熵变引致,也称为焓驱动或熵驱动的过程。通常,混合两种物质总是引起熵的增加和自由能减少。因此,两种物质是否能自发地形成一个溶液主要取决于焓变。如前所论,大分子溶解中平动熵的增加对 ΔG_{mix} 的影响相对较小,而且 ΔS_{mix} 随着链长的增加递减。

在实际操作中,制备大分子溶液的关键就是寻获一个适当的溶剂,使其与大分子之间的相互作用尽量减少,即 χ_{sp} 为负或接近零,从而导致 $|T\Delta S_{mix}| > |\Delta H_{mix}|$,即 $\Delta G_{mix} < 0$。假定形成一个 A 和 B 的二元混合物(溶液)后,总体积保持不变,χ_{sp} 则与它们的色散能量密度(ε,单位体积的液体在真空中的蒸发能)的开方和溶剂的摩尔体积(V_m)相关,即

$$\chi_{sp} \approx V_m \frac{(\varepsilon_A^{1/2} - \varepsilon_B^{1/2})^2}{RT} \tag{3.17}$$

其中,$\varepsilon^{1/2}$ 也被称为 Hildebrand 溶解度参数,常记为 δ。对非液体材料(如大分子),其也被称为色散参数,更反映了其物理本质。其与分子蒸发焓(ΔH_{vap})的关系如下:

$$\varepsilon = \frac{\Delta H_{vap} - RT}{V_m} \tag{3.18}$$

注意:在上式的推导中假定了混合时总体积不变。常见分子的蒸发焓可从物理化学手册获得。然而,大分子在高温下分解,无法蒸发,因此,可以通过选择一个和聚合单体有相似化学组成和组构的小分子,或者将聚合单体想象地拆分成几个化学官能团的组合,寻获它们各自的蒸发焓后再加和。此法可估计大分子样品的色散参数,有助选用一个溶剂使得它们的溶解度参数相近,减少溶剂与大分子之间的相互作用(χ_{sp})。实验中,常用的一个溶解判断标准为 $|\varepsilon_p^{1/2} - \varepsilon_s^{1/2}| \leq 3.6\ MPa^{1/2}$。这就是化学中有名的"相似相溶"原理。注意,前提是溶质和溶剂分子之间没有特殊相互作用。但是,它不是一个大分子特有的物理性质,与溶质的尺寸无关,不属于大分子物理!

利用该原理的一个著名例子是制备聚四氟乙烯溶液。自从 1938 年杜邦公司的 Roy Plunkett 博士发现四氟乙烯可以聚合后,聚四氟乙烯因其耐腐蚀特性获得了极其广泛的应用,即不溶于任何溶剂。其应用不胜枚举,覆盖了从不粘锅到防油防水材料,从普通水管接口的密封带到制造核武器所需的特殊密封垫圈,被誉为"塑料之王"。然而,正是其"不溶"性质使得人们在它问世多年后,仍无法窥视其分子特性。20 世纪 80 年代中期,作者和合作者在其导师朱鹏年教授的指导下,利用"相似相溶"的原理,终于将聚四氟乙烯在 325 ℃ 的高温下溶于含有 22 个四氟乙烯单体的低聚物中。进而,利用特别设计、自制的高温激光光散射仪成功地测得其平均摩尔质量和平动扩散系数分布等大分子参数〔Chu B,Wu C,Zuo J. Macromolecules,1987,20:700〕。

小　结

尽管大、小分子的溶解均受制于热力学原理,二者的尺寸之差仍然导致了溶液形成中的差别。首先,因为链的缠绕,大分子在溶解时先要经过一个溶胀过程,而小分子的溶解

不涉及该过程。在熵的驱动下,溶剂小分子慢慢地渗入大分子溶质之中,增加链间距离,减弱不同链段之间的相互作用,从而使得链可解开缠绕,逐一地扩散到溶剂之中形成溶液。该"溶胀-解缠"过程可长至数天甚至数周,取决于链长和链间相互作用。一般而言,升温可有助和加速溶解,但受限于溶剂的沸点。其次,与小分子溶解不同,溶胀后的大分子从样品表面慢慢地向溶剂中扩散,其并不主导溶解。轻微晃动或搅拌有助溶解,但剧烈搅拌对大分子溶解无助。认识和牢记这些差别可避免可能的溶解不充分或加速加力搅拌的错误。对于超长链的大分子,加速加力搅拌可能使链断裂。

对一给定的质量浓度,大分子的数目仅为小分子的百分之一至万分之一,故其在溶解中获得的平动熵远远小于小分子。因此,选择一个大分子的溶剂时,"相似相溶"原理就显得格外重要。应该先选择摩尔蒸发能与单体相近的溶剂,再进行溶解实验,可事半功倍! 需要强调的是,真实分子的硬核体积永远存在,不会消失,更不会为负。分子之间的"相互作用"可产生一个效果,等效于真实分子的排除体积,一个虚拟的排除体积,但不是真的。读者千万不要将二者混淆。硬核体积和相互作用对溶液的综合影响称作"排除体积效应",其可为正、为负或为零。

另外,大分子溶解时的混合熵对摩尔浓度的依赖性不再对称。并随着链长或聚合度的增加,不对称性加剧,反映到大分子溶液的相图上,就是其临界相变温度和浓度均有链长依赖性,后面将会详叙。而小分子溶液的相图则没有这一链长依赖性。下图总结了本章讨论的问题和大分子和小分子在溶解中焓和熵的变化。

Summary 3

$$H = G + TS \xrightarrow{\Delta T = 0} \Delta G_{\text{mix}} = \Delta H_{\text{mix}} - T\Delta S_{\text{mix}}$$

小分子
small molecules
$$\begin{cases} \dfrac{\Delta H_{\text{mix}}}{nRT} = x_B(1 - x_B)\chi \\[2mm] \dfrac{\Delta S_{\text{mix}}}{nR} = -(x_A \ln x_A + x_B \ln x_B) \end{cases}$$

溶解度参数 $\left(\varepsilon_A^{1/2}\right)$
solubility parameter
(Joel H.Hildebrand)

大分子
macro-molecules
$$\begin{cases} \dfrac{\Delta H_{\text{mix}}}{nRT} = x_B(1 - x_B)\chi_{sp}\dfrac{n_\nu}{1 + (n_\nu - 1)x_B} \\[2mm] \dfrac{\Delta S_{\text{mix}}}{nR} = -\left[x_A \ln x_A + x_B \ln x_B \dfrac{n_\nu}{1 + (n_\nu - 1)x_B}\right] \end{cases}$$

$$\chi_{sp} = Z\frac{\varepsilon_{AB} - \dfrac{\varepsilon_{AA} + \varepsilon_{BB}}{2}}{k_B T} = \frac{V_m\left(\varepsilon_A^{1/2} - \varepsilon_B^{1/2}\right)^2}{RT}$$

Chapter 3. Thermodynamics in Preparation of Macromolecular Solutions

In order to obtain the properties of individual macromolecular chains (such as the chain size, conformation, molar mass, etc.), it is first necessary to separate the entangled macromolecular chains into single chains in bulk. The process of dispersing one or more substances in the form of individual molecules into another substance to form a homogeneous mixture is called dissolution, and the resulting mixture is called a solution. Usually, the fewer components are solutes and the most are a solvent. For example, in daily life, alcohol, vinegar, salt or sugar are dissolved in water to form a uniform aqueous solution; in experiments, polystyrene is often dissolved in benzene, toluene or tetrahydrofuran to form an organic solution. Physically, such solutions consisting of two, three or more molecules are collectively referred to as binary, ternary or multi-component blends or mixtures. This book will only discuss binary mixtures with only one solute.

There is no fundamental difference in preparing solutions of macro-and small molecules. For a given absolute temperature (T), whether a substance is soluble in a solvent depends only on the change in the free energy of the system after and before dissolution ($\Delta G_{mix} = \Delta H_{mix} - T\Delta S_{mix}$), where ΔH_{mix} and ΔS_{mix} are the changes of enthalpy and entropy after and before the mixing, respectively. It has been known from thermodynamics that in a spontaneous process under isothermal and isobaric conditions, free energy decreases. Therefore, only when $\Delta G_{mix} < 0$, a solute can spontaneously dissolve in a solvent to form a uniform binary mixture; on the contrary, the mixture will spontaneously move in the opposite direction, forming two solution layers (phase) with a physical interface, which are the saturated solutions of one in another, respectively. Generally speaking, when two different molecules are mixed, entropy increases, i. e., $\Delta S_{mix} > 0$. Therefore, when $\Delta H_{mix} < 0$ or $\Delta H_{mix} < \mid T\Delta S_{mix} \mid$, the dissolution will occur automatically.

However, due to differences in mass and volume, the dissolutions of macro- and small molecules are also different. The relative contributions of the changes enthalpy and entropy are obviously different. First, the polymerized monomers that are chemically bonded together on the macromolecular chain cannot be completely separated as small molecules in the dissolution. Each chain can only move randomly in the solution as a whole. Compared with small molecules, the increase in translational entropy of macromolecules during the dissolution is relatively small, so the enthalpy change plays a bigger role in its dissolution.

For a given mixing enthalpy change, i.e., when monomers and polymers interact similarly to solvent molecules, respectively, monomers are often soluble but not polymers after they reach a certain length. The precipitation polymerization is based on this thermodynamic principle to control the degree of polymerization of synthetic macromolecules. In addition, when small molecules are mixed, individual solute and solvent molecules have a similar volume, so that the dependence of the change of entropy in the mixing on the composition is nearly symmetric, i.e., the composition that corresponds to the maximum mixing entropy is around 50%. A macromolecule occupies a much larger volume than a solvent molecule, so the dependence of the mixing entropy change on the composition of macromolecules is completely asymmetric, skewing towards the left of the low concentration direction.

In addition, when a small molecule solute is dissolved, the molecules gradually diffuse from its surface into the solvent one by one. Stirring can reduce its surface concentration, promoting the diffusion, and accelerating the dissolution. Due to the entanglements among macromolecular chains, their dissolution involves two steps: first, small solvent molecules diffuse into the chains, causing them to swell, increasing the distance between the chain segments, and weakening the interaction. The macromolecular chains can slowly disentangle from each other, individually diffusing from its surface into solvent just like small molecules. The sizes of macro-and small molecules are a hundred times different, so that macromolecules diffuse a hundred times slower. Slow stirring can promote the dissolution of macromolecules, but the swelling and untangling processes are much slower than the diffusion, determining the dissolution rate. Therefore, when preparing a macromolecular solution, in addition to the psychological comfort, the rapid stirring essentially does not accelerate the dissolution. For polymer samples with the ultra-long chains, the high-speed and forced stirring is not only helpless, but may also break the macromolecular chain. Even teachers and students who study polymers often make similar mistakes, too.

On the other hand, unlike small molecules, each macromolecular chain entangles with its surrounding chains, so that there are multiple interactions among different macromolecular chains. The relatively small change of the mixing entropy is sometimes unable to drive solvent molecules to penetrate into a polymer sample, swelling and disentangling the strongly interacted segments, so that the dissolution temperature of a polymer sometimes must be higher than its melting temperature. For example, the famous plastic king "polytetrafluoroethylene" (its melting temperature is 327 ℃) dissolves in its oligomers only at 330 ℃ [Chu B, Wu C, Zuo J, Macromolecules, 1987, 20: 700].

In order to facilitate the discussion of thermodynamic changes in the mixing of macromolecules and solvent molecules, let us first review the mixing of N_A particles of A and N_B particles of B under isothermal and isobaric conditions, which was learned in thermodynamics. Let us also assume that A and B have the same size, occupy an equal space (called a lattice), and there is no any interaction among the particles.

In the initial state (i) before mixing, they are all pure substances. N_A identical particles A occupy N_A lattice points; N_B identical particles B occupy N_B lattice points. There are respectively only one spatial arrangement (Ω) for particles A and B particles; namely, $\Omega_{i,A} = 1$ and $\Omega_{i,B} = 1$. According to the definition of entropy, $S = k_B \ln\Omega$, the total entropy before mixing is

$$S_i = k_B \ln\Omega_{i,A} + k_B \ln\Omega_{i,B} = 0 \qquad (3.1)$$

In the final state (f) after mixing, the macroscopic property is "N_A identical particles A and N_B identical particles B occupy ($N_A + N_B$) lattice points". For such a macroscopic property, these particles A and B have different microscopic arrangements on those ($N_A + N_B$) lattices. The total number of different arrangements is $\Omega_f = (N_A + N_B)!/N_A!N_B!$. Therefore, the mixing entropy is

$$S_f = k_B \ln\Omega_f = k_B \ln \frac{(N_A + N_B)!}{N_A! N_B!}$$

$$\cong k_B [(N_A + N_B)\ln(N_A + N_B) - (N_A + N_B) - (N_A \ln N_A - N_A + N_B \ln N_B - N_B)]$$

$$= -k_B \left(N_A \ln \frac{N_A}{N_A + N_B} + N_B \ln \frac{N_B}{N_A + N_B} \right)$$

$$= -N k_B (X_A \ln X_A + X_B \ln X_B) \quad \text{or written as}$$

$$= -(n_A + n_B) N_{AV} k_B \left(\frac{n_A}{n_A + n_B} \ln \frac{n_A}{n_A + n_B} + \frac{n_B}{n_A + n_B} \ln \frac{n_B}{n_A + n_B} \right)$$

$$= -nR (x_A \ln x_A + x_B \ln x_B) \qquad (3.2)$$

where $\ln N! \approx N \ln N - N$ when $N \gg 1$; $X_A = N_A/N$, $X_B = N_B/N$, $N = N_A + N_B$; $n_A = N_A/N_{AV}$, $n_B = N_B/N_{AV}$, $n = n_A + n_B$ and $R = N_{AV} k_B$; N_{AV} is the Avogadro constant. Note the assumption that particles A and B have the same volume, so that $x_A = X_A$ and $x_B = X_B$. In summary, before and after mixing, the change in entropy is,

$$\Delta S_{mix} = k_B (\ln\Omega_f - \ln\Omega_i) = -nR (x_A \ln x_A + x_B \ln x_B) \qquad (3.3)$$

It has been known that the change of Gibbs free energy (G) depends on volume (V), entropy (S) as well as and the changes of pressure (p) and temperature (T), i.e., $dG = Vdp - SdT$. Therefore, for an isobaric process, $dG = -SdT$ or $(\partial G/\partial T)_p = -S$, further written as $[\partial(G_f - G_i)/\partial T]_{p,n} = -(S_f - S_i)$, or briefly written as $(\partial \Delta G/\partial T)_p = -\Delta S$. The choice of the absolute zero as a reference point for pure substances and the mixture leads to

$$\Delta G_{mix} = -\int_0^T \Delta S_{mix} dT = nRT (x_A \ln x_A + x_B \ln x_B) \qquad (3.4)$$

In eqs. (3.3) and (3.4), $x_A < 1$, $x_B < 1$, $x_A + x_B = 1$. Therefore, $\Delta S_{mix} > 0$ and $\Delta G_{mix} < 0$; namely, mixing two kinds of non-interacting particles always leads to an increase of entropy and a decrease of free energy. The thermodynamic criterion whether an isothermal and isostatic process can occur spontaneously is $\Delta G < 0$, so that two kinds of particles without any interaction can always spontaneously mix into a completely uniform one-phase system (solution), driven by the increase of entropy. ΔG_{mix} and ΔS_{mix} normalized by nRT and nR, respectively, only depend on the volume fraction (composition). ΔG_{mix} and ΔS_{mix} are completely symmetrical; their corresponding minimum and maximum values appear at $x_A = 1/2$ and $x_B = 1/2$, originating from the equal volume assumption.

It has been known from thermodynamics that enthalpy includes free energy and entropy related to work and heat, respectively, i.e., $H = G + TS$. Therefore, in an isothermal process, $\Delta G = \Delta H - T\Delta S$. In generally speaking, there are "interactions" among real molecules and ΔH_{mix} is non-zero. Before mixing, particles A and B are pure substances. According to the aforementioned lattice model, each particle occupies a lattice point, and the system has a total of $(N_A + N_B)$ lattice points. In thermodynamics: $H = U + pV$ and $G = A + pV$, where U and A are internal energy and Helmholtz free energy, respectively. If the pressure is constant and the volume is incompressible during mixing, the changes of internal energy and Helmholtz free energy (ΔU_{mix} and ΔA_{mix}) are no different from the changes of enthalpy and Gibbs free energy (ΔH_{mix} and ΔG_{mix}), respectively.

In a real binary mixture (solution), two real molecules cannot simultaneously occupy the same physical space. Each molecule has a certain hardcore volume. Microscopically, the average excluded volume (v_e) of a particle (molecule) itself consists of two parts: its hardcore volume (v_0) and a virtual volume (Δv) caused by the interaction between particles, i.e., $v_e = v_0 + \Delta v$. If the excluded volume is regarded as a small ball, the distance between the centroids corresponds to the minimum point of the interaction potential energy between two particles, while the diameter related to v_0 corresponds to the distance between the centroids at which the interaction potential energy changes from attraction (negative) to repulsion (positive).

There are also various "interactions" among different chains and different segments on one chain. For example, when two neutral molecules are approaching each other, the positively charged nucleus will attract the negatively charged electrons on the opposing party towards themselves, generating two induced dipole moments in opposite directions, attracting each other, called dispersion effect. The related energy is called the dispersion energy. In the phase transitions of gas and liquid, it is also called condensation energy (heat) or evaporation energy (heat).

On the one hand, the existence of the hardcore volume reduces the volume available for every molecule, i.e., increasing the probability of every molecule reaching any point in the system (decreasing entropy). On the other hand, the dispersion interaction induced attraction slows down the motion of molecules, and decreases the above-mentioned probability (increases entropy), producing an "effect" of increasing the available volume, but not a really increased volume. In other words, the dispersion induced attraction is equivalent to a negative excluded volume, but not a real negative excluded volume. It is similar to the "centrifugal force is a virtual force (inertia) in a non-inertial coordinate system", which produces an effect of force, but not a real force.

Therefore, in a real binary mixture, there is an effect generated by the real hardcore volume, and another opposite effect of virtual negative excluded volume originated from dispersion attraction. A combination of effects from one real hardcore volume and one virtual excluded volume is "the excluded volume effect", as shown below. For the attractive interaction, the excluded volume effect can be positive, zero or negative. If the interaction is repulsive (e.g., polyelectrolytes), it generates an effect of virtual positive excluded volume. Therefore, the excluded volume effect is always positive. The excluded volume effect is stronger than without the interaction, so that the

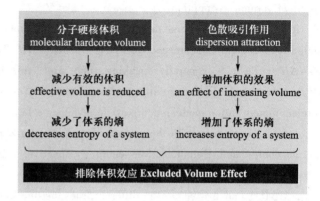

chain conformation in the solution expands more.

Before mixing, there are only N_A particles (molecules) A or N_B particles (molecules) B at $(N_A + N_B)$ lattice points. The probability of finding the first particle as A or B on any one of $(N_A + N_B)$ lattice points is

$$\frac{N_A}{N_A + N_B} \quad or \quad \frac{N_B}{N_A + N_B}$$

The probability offinding the second one as A or B on any one of $(N_A + N_B - 1)$ lattice points is

$$\frac{N_A - 1}{N_A + N_B - 1} \quad or \quad \frac{N_B - 1}{N_A + N_B - 1}$$

In the respective pure substances before mixing, the probability ofgenerating a pair of A—A or B—B interaction in the system is

$$\frac{N_A(N_A - 1)}{(N_A + N_B)(N_A + N_B - 1)} \quad or \quad \frac{N_B(N_B - 1)}{(N_A + N_B)(N_A + N_B - 1)}$$

The total pair number of the two body interaction between A and A or between B and B can be obtained by multiplying the above probabilities by the number of lattice points $(N_A + N_B)$, i.e.,

$$\frac{N_A^2}{2(N_A + N_B)} \quad or \quad \frac{N_B^2}{2(N_A + N_B)}$$

where, $N_A \gg 1$ and $N_B \gg 1$ are used. In the above derivation, two particles produce two pairs of interactions, but in fact they can only produce one pair of interaction, so that a division of 2 can avoid the repeated counting.

After mixing, there areadditional two−body interactions between A and B as well as between B and A in the system. Note: both of them are the same, only one pair of interaction, not two pairs, can be produced between one particle A and one particle B. The total number of crossing pairs is

$$\frac{N_A N_B}{2(N_A + N_B)} + \frac{N_B N_A}{2(N_A + N_B)} = \frac{N_A N_B}{N_A + N_B}$$

At the same time, the number of original A–A and B–B two–body interaction pairs are reduced, and the reduced number corresponds to the above–mentioned increased number of A–B and B–A two–body interaction pairs. In other words, respectively opening one pair of A–A and one pair of B–B interactions can produce two pairs of A–B interactions. With the assumption that the energies of each pair of different interactions are ε_{AA}, ε_{BB} and ε_{AB}, the enthalpy energy before and after mixing are respectively

$$H_{before} = \frac{Z}{N_A + N_B}\left(\frac{N_A^2}{2}\varepsilon_{AA} + \frac{N_B^2}{2}\varepsilon_{BB}\right) \quad \text{and}$$

$$H_{after} = \frac{Z}{N_A + N_B}\left[\left(\frac{N_A^2}{2}\varepsilon_{AA} - \frac{N_A N_B}{2}\varepsilon_{AA}\right) + N_A N_B \varepsilon_{AB} + \left(\frac{N_B^2}{2}\varepsilon_{BB} - \frac{N_B N_A}{2}\varepsilon_{BB}\right)\right] \quad (3.5)$$

Among them, it is assumed that each lattice point in the lattice model has Z adjacent interaction points, i.e., one particle can produce Z pairs of interactions, so that Z is multiplied. It should be noted that Z **should change with concentration in experiments**. The above two equations lead to

$$\Delta H_{mix} = H_{after} - H_{before} = \frac{Z N_A N_B}{N_A + N_B}\left(\varepsilon_{AB} - \frac{\varepsilon_{AA} + \varepsilon_{BB}}{2}\right)$$

In the above equation, the constant in parentheses is the average enthalpy change of forming one pair of A–B interaction after mixing. If reduced by thermal energy $k_B T$, $\Delta H_{mix}/k_B T$ becomes a dimensionless state function. Using the following definition

$$\chi = \frac{Z}{k_B T}\left(\varepsilon_{AB} - \frac{\varepsilon_{AA} + \varepsilon_{BB}}{2}\right) \quad (3.6)$$

The above equation can be rewritten as

$$\frac{\Delta H_{mix}}{k_B T} = \frac{N_A N_B}{N_A + N_B}\chi \quad (3.7)$$

Note: depending on the relative values of ε_{AA}, ε_{BB} and ε_{AB}, χ can be positive, negative or zero. Different values of ε_{AA}, ε_{BB} and ε_{AB} can be combined to get the same value of χ. Therefore, there is no one–to–one correspondence between χ and the interaction energies between different particles. A given set of ε_{AA}, ε_{BB} and ε_{AB} are only a sufficient and non–essential condition for χ, only reflecting the average enthalpy change due to particle interaction.

When $\chi < 0$, the excluded volume effect is reduced during mixing, conducive to mixing; on the contrary, it is not conducive to mixing. When $\chi = 0$, the system is **an athermal mixture of gases or liquids**; namely, the temperature has no effect on the system. Note that in eq. (3.6), T is on the denominator, so that χ usually decreases as the temperature increases. For a real solution (binary mixture), the dependence of χ on temperature is related to the specific dependence of ε_{AA}, ε_{BB} and ε_{AB} on temperature. Moving $k_B T$ to the right, then multiplying and dividing the right side by N_{AV}^2 and using $n = (N_A + N_B)/N_{AV}$, eq. (3.7) can be rewritten as

$$\Delta H_{mix} = nRT\, X_A(1 - X_A)\chi = nRT\, x_A(1 - x_A)\chi \quad (3.8)$$

where $X_A = x_A$ and $X_B = x_B$; $n_A = N_A/N_{AV}$, $n_B = N_B/N_{AV}$, $n = n_A + n_B$, $x_A = n_A/n$, $x_B = n_B/n$ and $x_A + x_B = 1$. For an isothermal process, substituting ΔH_{mix} and ΔS_{mix} into $\Delta G = \Delta H - T\Delta S$ results in

$$\Delta G_{mix} = nRT[x_A(1 - x_A)\chi + x_A \ln x_A + (1 - x_A)\ln(1 - x_A)] \qquad (3.9)$$

Note that $\chi x_A(1 - x_A)$ and $x_A \ln x_A + (1 - x_A)\ln(1 - x_A)$ are related to the changes of enthalpy and entropy, respectively. The above equation can be rewritten as

$$\Delta G_{mix} = RT[x_A n_B \chi + n_A \ln x_A + n_B \ln(1 - x_A)] \qquad (3.10)$$

or

$$\Delta G_{mix} = RT[x_B n_A \chi + n_B \ln x_B + n_A \ln(1 - x_B)] \qquad (3.11)$$

Namely, using x_B instead of x_A as a variable, eq. (3.9) can also be obtained. The plot of ΔG_{mix} versus x_A or x_B is completely symmetrical, originating from the assumption that particles A and B have the same volume.

Obviously, there is no such symmetry relationship in macromolecular solutions. However, the above discussion still holds true for macromolecular solutions. The only correction is that the volume of one macromolecule is equivalent to the volume of n_v solvent molecules. In other words, if the solution volume is divided into N_{box} small boxes according to the volume of one solvent molecule, each solvent molecule occupies one small box, while each macromolecule occupies n_v small boxes, i.e., a macromolecular chain is "equivalent" to n_v solvent molecules.

Let A and B denote solvent (small molecules) and solute (macromolecules), respectively, written as "s" and "p"; their molecular numbers are N_s and N_p, respectively; $N_{box} = N_s + n_v N_p$. The lattice fractions respectively occupied by the solute and the solvent, i. e., the volume fraction, are $X_p = n_v N_p/(N_s + n_v N_p) = n_v n_p/(n_s + n_v n_p) = n_v n_p/n_{box}$ and $X_s = N_s/(N_s + n_v N_p) = n_s/(n_s + n_v n_p) = n_s/n_{box}$; while the mole fractions of solute and solvent $x_p = N_p/(N_s + N_p) = n_p/(n_s + n_p)$ and $x_s = N_s/(N_s + N_p) = n_s/(n_s + n_p)$. It is clear that the **volume fraction of solute and solvent is no longer equal to their corresponding molecular or molar fraction.**

Emphasizing once more: in the discussion of small molecular solutions, $n_v = 1$; the volume fraction equals the molar fraction. In a macromolecular solution, $n_v \gg 1$; the volume fraction and molar fractions are completely different! The following equation is obtainable from eq. (3.2).

$$S_f = k_B \ln \Omega_f = k_B \ln \frac{(N_s + n_v N_p)!}{N_s!(n_v N_p)!} = -k_B \left(N_s \ln \frac{N_s}{N_s + n_v N_p} + N_p \ln \frac{n_v N_p}{N_s + n_v N_p} \right)$$

$$\Delta S_{mix} = -N_{box} k_B \left(X_s \ln X_s + \frac{X_p}{n_v} \ln X_p \right)$$

$$\frac{T\Delta S_{mix}}{n_{box}} = -RT \left[(1 - X_p)\ln(1 - X_p) + \frac{X_p}{n_v} \ln X_p \right] \qquad (3.12)$$

where n_{box} is the molar number of small boxes (N_{box}/N_{AV}) and $R = k_B N_{AV}$. The multiplication of both sides by the absolute temperature (T) enables the mixing entropy (ΔS_{mix}) to become one part of the mixing free energy (ΔG_{mix}). Its physical meaning is the average mixing free energy caused by the mixing entropy per molar small boxes. Note: n_v is directly proportional to the degree of polymerization of the macromolecule, so that the mixing entropy also decreases as the degree of polymerization increases, **depending on the chain length**. This is a significant difference between macro-and small molecular solutions. The longer the macromolecular chain (the larger the n_v), the more asymmetrical the dependence of the mixing entropy on the volume or molar fraction. In extremely dilute solutions, the first term in the above equation can be omitted, so that the mixing entropy is inversely proportional to the degree of polymerization.

In real macromolecular solutions, the "exclusion volume effect" from a combination of the hardcore volume and interaction must be considered. Slightly different from the discussion of small molecular solutions, in a macromolecular solution, ε_{BB} represents the interaction energy of **one "segment"** equivalent to a solvent molecule with **another "segment"**; ε_{AB} represents the interaction energy of **one "segment" with one solvent molecule**. Note: the "segment" here is not a Kuhn segment, but a short chain segment with the same equivalent volume as a solvent molecule. In other words, if the end group effect is ignored, one such a "segment" can interact with $(Z-2)$ solvent molecules in addition to its connection to the other two "segments" in the chain. If one macromolecular chain can simultaneously produce the excluded volume effect with n_v solvent molecules, disassembling a pair of "chain-chain" excluded volume effect and n_v pairs of "solvent-solvent" excluded volume effect can form $2n_v$ pairs of "segment-solvent" excluded volume effect. Therefore, following the definition of the aforementioned eq. (3.6), where χ_{AB} is written as χ_{sp}, and the discussion related to the mixing enthalpy energy of small molecules, the same conclusion can be obtained as eq. (3.8). Just because of the difference in macro-and small solvent molecular volumes, the mole fraction (x_B) can no longer be used to replace the volume fraction (X_B).

$$\Delta H_{mix} = n_{box} RT X_p (1 - X_p)\chi_{sp} \tag{3.13}$$

For an isothermal process, substituting ΔH_{mix} and ΔS_{mix} into $\Delta G = \Delta H - T\Delta S$, an equation similar to eq. (3.9) is obtained as follows.

$$\frac{\Delta G_{mix}}{n_{box} RT} = X_p (1 - X_p)\chi_{sp} + \frac{X_p}{n_v}\ln X_p + (1 - X_p)\ln(1 - X_p)$$

$$\frac{\Delta G_{mix}}{RT} = X_p n_s \chi_{sp} + n_p \ln X_p + n_s \ln X_s \tag{3.14}$$

where $1 - X_p = X_s$, $X_s = n_s/n_{box}$ and $X_p = n_v n_p/n_{box}$ were used. $\Delta G_{mix}/RT$ is dimensionless.

Note that due to $n_v \gg 1$, the dependence of ΔG_{mix} on the volume fraction X_p is extremely asymmetric in comparison with small molecular solutions. Before mixing, the solvent or solute is a pure substance ($X_p = 0$ or 1), $\Delta G_{mix} = 0$. When X_p is sufficiently small or close to 1, mixing always leads to a decrease in free energy, $\Delta G_{mix} < 0$, the dissolution occurs. In other words, theoretically, no binary mixture is absolutely insoluble. Any substance is soluble in any solvent, only different in solubility. Generally speaking, "insoluble" only means that the solubility is very low. According to the definition of the International Union of Pure and Applied Chemistry, when the solubility is less than one part in 10,000, it can be regarded as "actually insoluble" or simply "insoluble" in a solvent. From the statistical physics point of view, everything in the world can happen, just different probabilities of its occurrence!

According to eq. (3.14), for a given temperature, when $\chi_{sp} < 0$, $\Delta G_{mix} < 0$; the dissolution occurs spontaneously, and macromolecules can be dissolved in the solvent in any proportion to form a solution; when $\chi_{sp} > 0$, but less than a critical value (χ_c), ΔG_{mix} is still negative, and the mixing is spontaneous. When $\chi_{sp} > \chi_c$, ΔG_{mix} becomes positive within a certain range of X_p, and the system spontaneously moves in the opposite direction of mixing, separating into two layers (phases) with an interface. The dilute phase is a saturated solution of macromolecules in solvent; the concentrated phase is a saturated solution of solvent in macromolecules (the entangled chains were penetrated and swollen by solvent molecules).

The figure on the right shows how $\Delta G_{mix}/(n_{box}RT)$ changes with the volume fraction of macromolecules when $n_v = 2, 3$ and 100 and $\chi_{sp} = 1 < \chi_c$ and $\chi_{sp} = 2 > \chi_c$. According to eq. (3.13), when $\chi_{sp} > \chi_c$, no matter what the value of χ_{sp} is, when X_p is sufficiently small, ΔG_{mix} always falls first from zero and becomes negative, then rises after reaching a lowest point, crossing the zero point, until reaching a peak value. On the right side, there is a similar change;

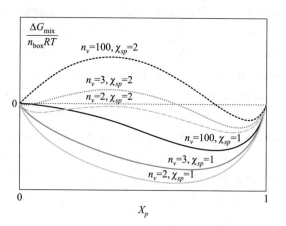

namely, ΔG_{mix} first drops and then rises as the solvent content increases (from right to left). Starting from the left or right sides (the macromolecule is continuously added to the solvent or the solvent is continuously added to the macromolecule), ΔG_{mix} converges at the same peak value. Due to the large disparity in volume between macromolecules and solvent molecules, the change of ΔG_{mix} is completely asymmetric. When $n_v \gg 1$, the change near the left vertical axis cannot be displayed in the figure (when $n_v = 100$ in the figure), as if there was no initial decrease in $\Delta G_{mix}/(n_{box}RT)$ and started to rise from zero. In fact, this is just a misleading. Unfortunately, traditional polymer textbooks do not specifically emphasize this point to students and readers, which may cause confusion and misunderstanding.

In experiments, the volume fraction of solute and solventafter mixing cannot be measured directly, so that the molar fraction is often used, which is obtainable from the masses of solute and solvent. Therefore, using $n_{box} = N_{box}/N_{AV}$, $N_{box} = N_s + n_v N_p$ and $n = n_s + n_p$, the volume fraction in the above equation can be converted into the molar fraction, resulting the contribution of the mixing enthalpy (ΔH_{mix}) to the mixing free energy.

$$\frac{\Delta H_{mix}}{nRT} = x_p(1 - x_p)\frac{n_v}{1 + (n_v - 1)x_p}\chi_{sp} \tag{3.15}$$

Its value can be positive or negative, depending on χ_{sp}, which decreases as the temperature increases. The contribution of the mixed entropy (ΔS_{mix}) to the mixing free energy is

$$\frac{\Delta S_{mix}}{nR} = -\left\{ x_p \ln\left[x_p \frac{n_v}{1 + (n_v - 1)x_p} \right] + (1 - x_p)\ln(1 - x_p) \right\} \qquad (3.16)$$

where $x_p < 1$, $(1 - x_p) < 1$ and $n_v x_p < 1 + (n_v - 1)x_p$. The two terms in the above equation are always negative, so that the entropy of the system always increases in a mixing process. In comparison with eqs. (3.3) and (3.8) (small molecules), There is one more factor in each of the above two equations

$$\frac{n_v}{1 + (n_v - 1)x_p}$$

Obviously, when $n_v = 1$, they are reduced to eqs. (3.3) and (3.8). For polymers, $n_v \sim 10^2 - 10^4 \gg 1$. Compared with the mixing entropy of small molecules in eq. (3.3), the contribution of the mixing entropy derived from the macromolecular chains (the first term in the above equation) is relatively small. The dissolution of one macromolecule leads to a less gain in the translational entropy than that of corresponding n_v small molecules that are not connected by chemical bonds. Therefore, the mixing entropy of macromolecules is mainly from the mixing entropy of small solvent molecules in the second term. A combination of the above two equations results in the mixing free energy after the dissolution of macromolecules ($\Delta G_{mix} = \Delta H_{mix} - T\Delta S_{mix}$); namely, the famous Flory–Huggins relationship.

In macromolecular experiments, the mass (weight) concentration (g/ml) is usually used, i.e., the mass of macromolecules per unit volume, not the mole fraction. In Chapter 5, how to convert a volume fraction or a mole fraction into a mass concentration will be discussed in details.

As mentioned before, if an isothermal and isobaric process occurs spontaneously, the free energy of the system must decrease. Such a decrease can be mainly caused by an enthalpy or entropy change, called as an enthalpy–or entropy–driven process. Generally, mixing two substances always leads to an increase in entropy and a decrease in free energy. Therefore, whether two substances are able to form a solution spontaneously depends mainly on the enthalpy change. As discussed earlier, the increase in the translational entropy in the dissolution of macromolecules has a smaller effect on ΔG_{mix}, and ΔS_{mix} decreases with an increase of the chain length.

In an actual operation, the key to preparing a macromolecule solution is to find an appropriate solvent to minimize the interaction with the macromolecule, i.e., χ_{sp} is negative or close to zero, resulting in $|T\Delta S_{mix}| > |\Delta H_{mix}|$, i.e., $\Delta G_{mix} < 0$. Assuming that after a binary mixture (solution) of A and B is formed, the total volume remains a constant, χ_{sp} is then related to the square root of their dispersion energy densities (ε, the evaporation energy per unit volume of liquid in vacuum) and the molar volume (V_m) of solvent; namely

$$\chi_{sp} \approx V_m \frac{(\varepsilon_A^{1/2} - \varepsilon_B^{1/2})^2}{RT} \tag{3.17}$$

where $\varepsilon^{1/2}$ is also referred to as the Hildebrand solubility parameter, often denoted as δ. For non-liquid materials (such as macromolecules), it is also called the dispersion parameter, reflecting its physical nature more. It is related to the molecular enthalpy of evaporation (ΔH_{vap}) as follows.

$$\varepsilon = \frac{\Delta H_{vap} - RT}{V_m} \tag{3.18}$$

Note: In the derivation of the aboveequation, a constant total volume during mixing was assumed. The evaporation enthalpy of common molecules can be found from the Handbooks of Physical Chemistry. However, macromolecules are decomposed at high temperature and not able to evaporate. Therefore, one can choose a small molecule with a chemical composition and structure similar to the polymerized monomer, or imaginatively split the polymerized monomer into a combination of several chemical functional groups. Find their respective evaporation enthalpies and then sum then together. This method can estimate the dispersion parameter of a macromolecular sample, helping to select a solvent to make their solubility parameters similar, reducing the interaction between the solvent and the macromolecule (χ_{sp}). In the experiment, a commonly used soluble criterion is $| \varepsilon_p^{1/2} - \varepsilon_s^{1/2} | \leqslant 3.6$ MPa$^{1/2}$. This is the well-known principle of "similar dissolves similar" in chemistry. Note: there is no special interaction between solute and solvent molecules. However, it is not a specific physical property of macromolecules, independent of the size of solute, and not belonging to macromolecular physics!

A famous example of using this principle is the preparation of polytetrafluoroethylene solution. Since Dr. Roy Plunkett of DuPont discovered that tetrafluoroethylene could be polymerized in 1938, polytetrafluoroethylene has been widely used due to its corrosion resistance properties, insoluble in any solvent. Its applications are innumerous, covering from non-stick pans to oil and water resistant materials and from the sealing tape of ordinary water pipe joints to the special sealing gaskets required to manufacture nuclear weapons, is known as the "king of plastics." However, it is its "insoluble" property that makes it impossible for people to peek into its molecular properties many years after its advent. In the mid-1980s, the author and collaborators, under the guidance of his supervisor, Professor Benjamin Chu, used the "similar dissolves similar" principle and finally dissolved polytetrafluoroethylene in an oligomer containing 22 tetrafluoroethylene monomers at a high temperature of 325 ℃. Further, using a specially designed, self-made high-temperature laser light scattering instrument, they successfully measured its weight average molar mass and dynamic diffusion coefficient distribution and other macromolecular parameters [Chu B, Wu C, Zuo J, Macromolecules, 1987, 20:700].

Summary

Although the dissolution of macro-and small molecules is subject to thermodynamic principles, their size difference still leads to differences in formation of solutions. First, due to the chain entanglement, macromolecules must undergo a swelling process when they are dissolved, while the dissolution of small molecules involves no such a process. Driven by entropy, small solvent molecules slowly penetrate into a macromolecular solute, increasing the distance between chains and weakening the interaction between different chain segments, so that the chains can unwind and diffuse into solvent one by one to form a solution. This "swelling-unwinding" process can take as long as several days to several weeks, depending on the chain length and the interaction between the chains. Generally speaking, increasing the temperature can help and accelerate the dissolution, but is limited by the boiling point of solvent. Secondly, unlike the dissolution of small molecules, the swollen macromolecules diffuse slowly from the sample surface into the solvent, which is not a dominant process in the dissolution. Slight shaking or stirring will help the dissolution, but vigorous stirring is helpless. Recognizing and remembering these differences can avoid possible mistakes of insufficient dissolution or accelerated stirring. For macromolecule with ultra-long chains, accelerated stirring may break the chains.

For a given mass concentration, the number of macromolecules is only one hundredth to one ten thousandth of that of small molecules, so that macromolecules gain less translational entropy than small molecules in the dissolution. Therefore, when choosing a solvent for macromolecules, the "similar dissolves similar" principle is particularly important. One should first choose a solvent with a molar evaporation energy similar to that of monomer, and then conduct the dissolution experiment, which is much more effective! It is necessary to emphasize that the hardcore volume of real molecules always exist, never disappear, let alone be negative. The "interaction" between molecules can produce an effect equivalent to the excluded volume, a virtual excluded volume, but not real. Readers should never confuse them. The combined influence of the hardcore volume and the interaction on a solution is called the "excluded volume effect", which can be positive, negative or zero.

In addition, the dependence of the mixing entropy on the molar concentration is no longer symmetrical, reflecting in the phase diagram of macromolecular solution. Its critical phase transition temperature and concentration are dependent on the chain length, while the phase diagram of small molecule solutions has no such a chain length dependence. The following figure summarizes the issues discussed in this chapter and the mixing enthalpy and entropy of macro- and small molecules.

Summary 3

$$H = G + TS \xrightarrow{\Delta T = 0} \Delta G_{\text{mix}} = \Delta H_{\text{mix}} - T\Delta S_{\text{mix}}$$

小分子
small
molecules
$$\begin{cases} \dfrac{\Delta H_{\text{mix}}}{nRT} = x_{\text{B}}(1 - x_{\text{B}})\chi \\[3mm] \dfrac{\Delta S_{\text{mix}}}{nR} = -(x_{\text{A}}\ln x_{\text{A}} + x_{\text{B}}\ln x_{\text{B}}) \end{cases}$$

溶解度参数 $\left(\varepsilon_{\text{A}}^{1/2}\right)$
solubility parameter
(Joel H.Hildebrand)

大分子
macro-
molecules
$$\begin{cases} \dfrac{\Delta H_{\text{mix}}}{nRT} = x_{\text{B}}(1 - x_{\text{B}})\chi_{sp}\dfrac{n_v}{1 + (n_v - 1)x_{\text{B}}} \\[4mm] \dfrac{\Delta S_{\text{mix}}}{nR} = -\left[x_{\text{A}}\ln x_{\text{A}} + x_{\text{B}}\ln x_{\text{B}}\dfrac{n_v}{1 + (n_v - 1)x_{\text{B}}}\right] \end{cases}$$

$$\chi_{sp} = Z\frac{\varepsilon_{\text{AB}} - \dfrac{\varepsilon_{\text{AA}} + \varepsilon_{\text{BB}}}{2}}{k_{\text{B}}T} = \frac{V_{\text{m}}\left(\varepsilon_{\text{A}}^{1/2} - \varepsilon_{\text{B}}^{1/2}\right)^2}{RT}$$

第四章 大分子溶液中链的构象

对于给定的组成,最简单的线型组构通过单体的线性聚合获得。因此,首先讨论在溶液中链段间无特殊相互作用的、线型长链的构象和其尺寸对链长的依赖性。每条链含有 N_0 个聚合单体。k 个相邻的聚合单体可集合形成一个长度为 b 的 **Kuhn** 链段(以下简称"链段")。每一个链段可围绕着其与前一链段的连接点在 $360°$ 的球面空间里自由旋转,包括 $180°$ 折回原处,其指向任一方向的概率均等。

换而言之,任一链段的空间指向与其他链段的空间指向毫无关系(也称"不相干");也与其在前一个时刻的指向不相干,即其指向没有任何记忆。由 N 个这样的链段构成的大分子链被称作自由连接理想链或简称理想链。若干个相邻的链段还可组成一个短的链节。显然,$N = N_0/k$。不同的高聚物具有不同的 k 值,k 值取决于化学结构和链段之间的相互作用。k 值反映了聚合物链的柔软性,其越小,链则越柔软。对于一条完全刚性的链,$k \to N_0$ 和 $N \to 1$。

在一个稀溶液中,扩展的大分子链彼此之间相互独立,故链间的相互作用较弱。在热能的扰动下,每一条链上的每个链段都在不停地、无规地运动(扩散和旋转),完全独立于其他链段的运动,形成一个无规线团构象。因此,在三维空间中的链构象也不断地随着时间变化(涨落);链的质心也在溶液中无规地扩散。为了量度一根给定长度链的尺寸并且寻获链的尺寸对链长的依赖性,有必要对一条链的构象作时间平均或者同时对无数条链的构象作系综平均。基于统计物理学,对一个完全随机的体系,二者的结果应该完全一致。

线型高聚物链的平均末端距

如右图所示,一条线型大分子链的末端距是一个从其始端指向其终端的空间矢量($\vec{R}$),由 N 个长度一致(b)的短的空间矢量 $\vec{b}$ 构成,表达为 $\vec{R} = \sum_{i=1}^{N} \vec{b}_i$,其中每个 $\vec{b}_i$ 的指向都无规,没有一个方向特殊。对同一个 $\vec{b}_i$,其指向也随时间无规地变化;不同时间的指向毫不相干,即没有一个链段有关于其方向的"记忆"。当时间(t)足够长时,在一个时刻指向某个方向的链段定会在另一个时刻指向相反的方向。换而言之,在一条无限长的链上,如果有一个指

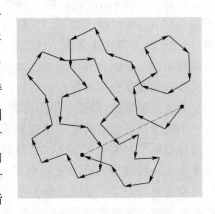

向某个方向的链段,就一定有一个与其对应的指向相反方向的链段。因此,对一个链段的指向作无穷长的时间平均或对无限多个链段作指向平均(即无限多步无规行走)的结果就是回到原点。

数学表达是,当$t \to \infty$或$N \to \infty$时,对任何一个链段$\vec{\mathbf{b}}_i$的时间平均或对无限多个链段的系综平均为零,即$\langle \vec{\mathbf{b}}_i \rangle = 0$。因此,$\langle \vec{\mathbf{R}} \rangle = \sum_{i=1}^{N} \langle \vec{\mathbf{b}}_i \rangle = N \langle \vec{\mathbf{b}}_i \rangle = 0$。换而言之,无法用这样的平均方法来区分两根长度(聚合度n)完全不同的大分子链。然而,直觉告诉我们长度不同的链应该占据着大小不同的空间,具有不同的构象尺寸。在统计学上有一个解决这类问题的有效办法,那就是先平方$\vec{\mathbf{R}} \cdot \vec{\mathbf{R}}$、后做统计平均$\langle \vec{\mathbf{R}} \cdot \vec{\mathbf{R}} \rangle$,简称为"方均",最后再做一个开方$\langle \vec{\mathbf{R}} \cdot \vec{\mathbf{R}} \rangle^{1/2}$,记为$RMS$,得到一个完全随机变量的平均涨落幅度,即

$$R_{RMS,0} = \langle \vec{\mathbf{R}} \cdot \vec{\mathbf{R}} \rangle^{1/2} = \langle \sum_{i=1}^{N} \vec{\mathbf{b}}_i \cdot \sum_{j=1}^{N} \vec{\mathbf{b}}_j \rangle^{1/2} = \left(\sum_{i=1}^{N} \sum_{j=1}^{N} \langle \vec{\mathbf{b}}_i \cdot \vec{\mathbf{b}}_j \rangle \right)^{1/2} \quad (4.1)$$

这里$\sum_{i=1}^{N} \vec{\mathbf{b}}_i \cdot \sum_{j=1}^{N} \vec{\mathbf{b}}_j$代表着两个"$N$项之和"的乘积。其展开后可共得$N \cdot N = N^2$项,包括$N$个完全相关的、平均值为$\langle \vec{\mathbf{b}}_i \cdot \vec{\mathbf{b}}_i \rangle = \langle b^2 \rangle = b^2$的自身项($i=j$);以及($N^2-N$)个平均值为$\langle \vec{\mathbf{b}}_i \cdot \vec{\mathbf{b}}_j \rangle$的交叉项($i \neq j$)。对于一条自由连接理想链而言,一个链段的运动与其他链段毫不相关,故它们乘积的平均应等于各自平均后的乘积,即$\langle \vec{\mathbf{b}}_i \cdot \vec{\mathbf{b}}_j \rangle = \langle \vec{\mathbf{b}}_i \rangle \langle \vec{\mathbf{b}}_j \rangle$。由于每个链段的运动和指向完全无规,对一个无限长的时间,等效于一条无限长的链,$\langle \vec{\mathbf{b}}_i \rangle = \langle \vec{\mathbf{b}}_j \rangle = 0$。因此,如果$R_{RMS,0}$记为$R_0$,上式成为

$$R_0 = \langle \vec{\mathbf{R}} \cdot \vec{\mathbf{R}} \rangle^{1/2} = (b^2 N)^{1/2} = b N^{1/2} \quad (4.2)$$

一条无限长($N \to \infty$)的自由连接(理想)线型大分子的平均末端距正比于每个链段的长度(b)以及链段数的开方($N^{1/2}$)。这一标度关系的物理含义是,N步无规行走后离开出发原点的RMS平均距离等效于朝着一个方向走了步长为b的$N^{1/2}$步。这等效于一条含有$N^{1/2}$个链段的刚性链,因为其末端距正比于链段数。换而言之,就平均末端距而言,一条含有$N^{1/2}$个链段的刚性链等效于一条含有N个链段的自由连接链。

一条真实大分子链的长度一定有限。链上的每个链段各自占有一定的三维物理体积(称为排除体积,$\sim b^3$),故整条链也占据着一定的三维物理空间,其平均排除体积与平均末端距有关,$\sim R_0^3$。如前所述,一条链或一个链段的硬核体积总是存在,不会消失,更不可能为负!其存在减少了粒子们实际可达体积,增加了一个粒子抵达体系内任意一点的概率。另外,化学键连接的各个链段之间和链与链之间存在着近、远程相互作用。这些相互作用可影响一条链和其链段抵达体系内任意一点的概率,产生一个改变排除体积的效应。硬核体积和相互作用对体系的综合作用就是前述的"排除体积效应"。

为了方便讨论,先假定排除体积效应可被忽略。在热能激发扰动下,一条大分子链上每一个链段在三维空间里不停地和无规地运动,不受其他链段的影响。含有x个链段的一段线型链节的两个末端之间的直接距离(末端距)也一定是一个无规变量。该链节可被看作从一端(i点)出发,无规行走了步长为b的x步以后,抵达了另一端(j点)。显然,在每次无规行走了x步后,i和j两点之间的直接距离无规地变化。假定每个这样的直线距离出现的概率遵循由自由连接理想链推而广之所得的高斯分布$G(\vec{\mathbf{r}}_i, \vec{\mathbf{r}}_j, x)$,对两个任

意链段间的直线距离的平方作统计平均就可得到 x 步无规行走后的方均行走距离（方均末端距）。等于每步（链段）长度（b）的平方和行走的步数（二者间的链段数,x）之积,即

$$\langle (\vec{r}_i - \vec{r}_j)^2 \rangle = b^2 x \qquad (4.3)$$

略去详细计算,依据简单的物理图像可以理解上述结论的正确性。如果有两个自由连接的链段,固定一个链段,并让另外一个链段自由地折叠,向前与第一个链段成一条直线为 0°,完全折回与第一个链段重合为 180°,经过无数次（步）随机折叠,第二个链段的平均位置一定是与第一个链段成直角。它们之间的方均末端距就是一个边长为 b 的等边直角三角形斜边长的平方,即 $b^2 + b^2 = 2b^2$。如果加上第三个链段,前两个可被等效地处理成一个长度为 $2^{1/2}b$ 的新"链段",故这三个链段的方均末端距为 $(b^2+2b^2) = 3b^2$;以此类推,可得上式。

特别需要强调:在无环状结构的大分子链上,对给定的任意两个链段（i 和 j）,一定存在且只存在一条在二者中间含有 x 个链段的线型连接路径。然而,对于非线型组构,往往不能简单地从两个链段的位置获得 x。在含有环状结构的链上,可能有多条连接两个给定链段的线型路径,更加复杂。例如,对于由一条有 N 个链段的线型链首尾相接而成的一个环状链,任意两个链段间的链段数可是 $|i - j|$ 或者 $N-|i - j|$,取决于如何计数。所以,对具体的链结构,要具体分析任意两个链段间的 x。有时,并非易事。

对于一条线型大分子链,$x = |i - j|$,较为简单;一条线型链上任意一段链节的方均末端距为 $\langle (\vec{r}_i - \vec{r}_j)^2 \rangle = b^2 |i - j|$。对整条线型链,$\vec{r}_1 - \vec{r}_N = \vec{R}$,$\langle (\vec{r}_1 - \vec{r}_N)^2 \rangle = \langle \vec{R} \cdot \vec{R} \rangle = b^2 (N - 1)$。由此可见,当 $N \gg 1$ 时,高斯链和自由连接链具有一致的平均末端距和链长（N）之间的标度律,$\langle \vec{R} \cdot \vec{R} \rangle^{1/2} = b N^{1/2}$。然而,二者基于完全不同的假定。对高斯链而言,$|i - j|$ 可大可小,甚至可为零,与链长无关;N 仅是一个有限的变数。而在自由连接（理想）链上,N 一定要趋于无穷大。也要注意,二者均无链长的概念。

真实的大分子链的聚合度总是有限的,故高斯链模型可以更好地描绘真实的链。大分子物理与小分子物理的根本差别就是大分子的某些物理性质的链长依赖性。注意,在后面的讨论中,常常混用链的伸直长度（L_c,简称链长）,链上的链段数（N）,链的聚合度（n）和链的摩尔质量（M）。它们之间互相都成正比,即 $L_c = bN = bn/k = b(M/M_o)/k$,其中 k 和 M_o 分别是 Kuhn 链段的聚合度和聚合单体的摩尔质量。在实验中,n 和 M 是两个直接可测的物理参数。下面,让我们进一步考察链段之间的排除体积效应（溶剂性质）如何影响一条线型链的平均末端距,以及其如何随着链段数（链长）变化。

在良溶剂中溶解时,被热能搅动的小分子在熵的驱动下都尽量互相远离,最后以单个分子的形式充满整个溶液。同样,大分子链的链段也都试图尽可能地互相远离,以便减少相互作用,从而引致链的扩展,即平均末端距（R）增大。与小分子不同,大分子链上的聚合单体由化学键连接在一起,因此,热能不可将它们断开成一个个的单体,分散到整个溶液中。链扩展的极限就是在不改变键角的前提下完全地伸直,其末端距为伸直长度（L_c）。

如果将一条链看成一根弹簧,其能量随着弹簧伸展而升高。能量变化正比于弹簧长度的平方。伸展的弹簧总是倾向于收缩,以降低其能量。可能的构象数（构象熵）随着链伸展减少。如果链完全伸展,其就只剩下了一个伸直构象。依据热力学,在一个隔绝体系

里或一个绝热过程中,体系总是自发地朝着熵增加的方向移动。因此,上述排除体积效应和熵的弹性效应对链扩展(R增加)有着相反的依赖关系。它们的竞争使得每条大分子链仅仅可以扩展到一定的尺寸,远小于L_c。大分子链上的各个链段都只能在其平均体积($\sim R^3$)内无规运动。

可想象地将一条真实大分子链的平均体积($\sim R^3$)按照一个链段的排除体积($\sim b^3$)分割成R^3/b^3个小盒子。一个链段在任意一个小盒子中的概率反比于小盒子数,即$1/(R^3/b^3) = b^3/R^3$。N个链段落入任意一个小盒子里的概率则为其N倍,即Nb^3/R^3。$(N-1)$个剩余的链段落入第二个小盒子里的概率应是$(N-1)b^3/R^3$。因此,链上两个链段之间产生二体排除体积效应的概率为$N(N-1)b^6/R^6$,乘以小盒子数(R^3/b^3),可得一条链上共可产生多少对二体排除体积效应,即$\sim b^3N^2/R^3$,其中,利用了$N \gg 1$和$N-1 \cong N$。

以一条理想链的平均末端距($R_0 = bN^{1/2}$)为参照零点,与链伸展相关的自由能变化正比于$R^2/(b^2N)$。上述两个相反的、对自由能变化均贡献均与熵有关,其数学表达如下

$$\frac{\Delta G}{k_B T} \sim \frac{b^3 N^2}{R^3} + \frac{R^2}{R_0^2} \tag{4.4}$$

其中第一项和第二项前面分别与二体排除体积和熵"弹性"效应有关的常数系数(ε_2和γ)已被略去。上式两边分别对R微分得量纲为力的($\partial \Delta G/\partial R$)$_{T,p}$,等效于牛顿力学里的一个"力的效应",但不是真正的力。从($\partial \Delta G/\partial R$)$_{T,p}=0$,可得自由能变化的极小值和线型链的平均末端距($R$)对链段长度($b$)和链段数($N$)的依赖性,即

$$\left(\frac{\partial \Delta G}{\partial R}\right)_{T,p} \sim -\frac{3b^3 N^2}{R^4} + \frac{2R}{b^2 N} = 0 \quad \rightarrow \quad R \sim bN^{\frac{3}{5}} \quad \text{(良溶剂)} \tag{4.5}$$

其显示大分子链扩展至一个稳定平衡状态。如果链因热能扰动扩展,熵"弹性"就会驱动其收缩;相反时,排除体积效应就会使链扩展,增大其尺寸。由统计热力学可知,在此平衡时,最大的链尺寸涨落对应着自由能变化半份热能($k_B T/2$)。

当二体排除体积效应完全消失时,溶液进入"准理想"状态,即大分子研究中常称的Θ状态。在此状态,就有必要考虑初时忽略的链上三个链段之间相对较弱的三体排除体积效应。依据上述讨论,在一条链上产生三体排除体积效应的概率应为$N(N-1)(N-2)/(R^9/b^9)$,乘以小盒子数(R^3/b^3)可得链上产生三体作用对的总数(b^6N^3/R^6)。因此,

$$\frac{\Delta G}{k_B T} \sim \frac{b^6 N^3}{R^6} + \frac{R^2}{R_0^2} \tag{4.6}$$

这里,利用了$N \gg 1$和$N-1 \cong N-2 \cong N$。类似地,右边第一项和第二项前面分别与三体排除体积效应(ε_3)和熵弹性(γ)有关的常数系数已被略去。可从上式求得自由能变化在Θ状态下或Θ溶剂中的极小值,具体如下:

$$\left(\frac{\partial \Delta G}{\partial R}\right)_{T,p} \sim -\frac{6 b^6 N^3}{R^7} + \frac{2R}{b^2 N} = 0 \quad \rightarrow \quad R \sim b N^{\frac{1}{2}} \quad \text{(}\Theta\text{ 溶剂)} \tag{4.7}$$

由此可见,随着溶剂性质变差,具有一个扩展的无规线团构象的链会收缩,其极限是一个内部含有溶剂的均匀小球,即$R \sim bN^{1/3}$。标度指数从3/5降至1/3,该"线团至小球"的链构象转变是自20世纪60年代以来最著名的一个大分子物理命题,留待第八章中讨论。

读者可能会注意到,在式(4.2)、式(4.7)以及高斯链的讨论中,线型自由连接理想链、在 Θ 状态下(二体排除体积效应消失)的线型真实链和线型高斯链具有同样的 R 与 N 的标度依赖关系($R \propto N^{1/2}$)。三者完全相同吗?答案是否定的!在推导这些标度关系中,它们各自关于链段数(N)的假定不同。前者假定了链长无限,$N \to \infty$,否则就不会有 $\langle \vec{b}_i \rangle = 0$;高斯链上的 N 很大,但有限,即 $1 \ll N \ll \infty$。真实链的长度有限,合成大分子的聚合度(n)通常小于 10^5。如果真实链太长时($N \gg 1$),就有必要考虑在 Θ 状态下可能残留的痕量二体排除体积效应,即

$$\frac{\Delta G}{k_B T} \sim \delta \frac{b^3 N^2}{R^3} + \frac{b^6 N^3}{R^6} + \frac{R^2}{R_0^2} \qquad (4.8)$$

其中 $\delta \ll 1$,在 Θ 状态附近,引入微小涨落;右边第一项和第二项及第三项前面与二体和三体排除体积效应(ε_2 和 ε_3)及熵弹性(γ)有关的常数系数也被略去。在上式中,当大分子的链长有限时,二体排除体积效应可被忽略,故式(4.8)还原成式(4.7)。反之,则可从上式求出自由能变化的极小值如下,

$$\left(\frac{\partial \Delta G}{\partial R}\right)_{T,p} \sim -\delta \frac{3\, b^3 N^2}{R^4} - \frac{6\, b^6 N^3}{R^7} + \frac{2R}{b^2 N} = 0 \quad \to \quad R \sim b\, N^{\frac{1}{2}} (1 + \delta\, N^{\frac{1}{2}})^{\frac{1}{8}} \qquad (4.9)$$

推导中,利用了 $R \cong R_0 = b\, N^{1/2}$。当 $N \gg 1$ 时,式(4.9)中的 $\delta\, N^{1/2} \gg 1$,标度指数从 1/2 略微地增至 9/16,即 $R \sim b\, N^{9/16}$;显示,残留的痕量二体排除体积效应使得链稍稍地扩展,这在物理上很容易理解和解释。

回转半径(R_g)

在不同假定下,线型链的平均末端距和链段数之间的不同标度关系可从理论上导出。然而,目前仍然无法直接测量得一个溶液中每根大分子链的末端距或测出一根大分子链的末端距随时间的变化,然后对它们统计平均。因此,链的平均末端距必须和一个实验上可测的物理量关联,方可验证由理论推导得到的不同标度关系。

大分子链的平均回转半径(R_g)就是这样一个可测的物理量,关乎其质量(m)的空间分布。正如其名所示,mR_g^2 是一个物体围绕着其质(重)心的转动惯量,定义如下:

$$R_g = \sqrt{R_g^2} = \sqrt{\frac{\sum_{i=1}^N m_i (\vec{r}_i - \vec{r}_G)^2}{\sum_{i=1}^N m_i}} \qquad (4.10)$$

其中,m_i、$\vec{r}_i$ 和 $\vec{r}_G$ 分别是第 i 个单元的质量、位置和质心位置的空间矢量,$\vec{r}_G = \sum_{i=1}^N \vec{r}_i / N$。如果每一个单元具有相同的质量,即 m_i 为常数,式(4.10)可进一步简化为

$$R_g^2 = \left[\sum_{i=1}^N (\vec{r}_i - \vec{r}_G)^2 \right] / N \qquad (4.11)$$

对形状简单的物体,不难利用上式直接相加或积分求出其回转半径。对一个具有薄球壳的空心球而言,$|\vec{r}_i| \equiv R$ 和 $\vec{r}_G = 0$,其回转半径就是其半径,$R_g = R$;一个均匀球体的回转半径小于其半径,$R_g = (3/5)^{1/2} R \cong 0.774R$;以及一根刚性细杆的回转半径与其长度($L$)的依赖关系为 $R_g = (1/12)^{1/2} L$。对于一个形状复杂的物体,就要进行一些具体的计算。

对有复杂组构的一条大分子链,即使给定了链段的空间分布,其回转半径的计算也十

分不易。况且,在热能的搅动下,链上的每个链段都不停地无规折叠和旋转,导致一个不停变化的链构象和一个无规移动的质心。换而言之,长度相同的不同的链在每一瞬间都具有不同的构象。因此,类似计算链的平均末端距,先求得一条大分子链的各种可能构象的回转半径的平方,再作统计平均,可得到**方均回转半径**($\langle R_g^2 \rangle$)。其开平方可得方均根回转半径($\langle R_g^2 \rangle^{1/2}$),记为 R_g,称作平均回转半径,量度了链段在空间的平均分布。依照定义,

$$R_g = \sqrt{\langle R_g^2 \rangle} \ \text{以及} \ \langle R_g^2 \rangle = \langle [\sum_{i=1}^{N}(\vec{r}_i - \vec{r}_G)^2]\rangle/N = [\sum_{i=1}^{N}\langle(\vec{r}_i - \vec{r}_G)^2\rangle]/N \quad (4.12)$$

其中,$\vec{r}_i$ 是第 i 个链段的空间矢量,假定了每个链段具有相同的质量(m_i)和利用了"加和后的统计平均"等于"统计平均后的加和"这一引理。然而,对于一个质心不断移动的物体,就很难利用上式直接地算出其平均回转半径。利用另一条引理,可将式(4.11)中的 $\sum_{i=1}^{N}(\vec{r}_i - \vec{r}_G)^2$ 转换成任意两点之间的方距加和,即 $\sum_{i=1}^{N}\sum_{j=1}^{N}(\vec{r}_i - \vec{r}_j)^2$,具体如下。

对于一条具有任意组构的大分子链(或一个运动的复杂物体),可以想象链上的任意两个链段(i 和 j)分别为一条高斯链节的始端和终端,故($\vec{r}_i - \vec{r}_j$)就是两个链段间的直接距离。将链上所有可能的两点间的直接距离先平方,即($\vec{r}_i - \vec{r}_j$)^2,再作时间(构象)统计平均得出所有可能的方均距,即 $\langle(\vec{r}_i - \vec{r}_j)^2\rangle$,最后,将它们全部相加,即 $\sum_{i=1}^{N}\sum_{j=1}^{N}\langle(\vec{r}_i - \vec{r}_j)^2\rangle$。在 $\vec{r}_i$ 和 $\vec{r}_j$ 之间同时加上和减去一个质心矢量 $\vec{r}_G$ 对二者之差没有影响,即

$$\sum_{i=1}^{N}\sum_{j=1}^{N}(\vec{r}_i - \vec{r}_j)^2 = \sum_{i=1}^{N}\sum_{j=1}^{N}[(\vec{r}_i - \vec{r}_G)-(\vec{r}_j - \vec{r}_G)]^2$$

$$= \sum_{i=1}^{N}\sum_{j=1}^{N}(\vec{r}_i - \vec{r}_G)^2 - 2\sum_{i=1}^{N}\sum_{j=1}^{N}(\vec{r}_i - \vec{r}_G)(\vec{r}_j - \vec{r}_G) + \sum_{i=1}^{N}\sum_{j=1}^{N}(\vec{r}_j - \vec{r}_G)^2$$

其中,右边第一和第三个加和中分别含有 N 个与 j 和 i 无关的项,两个加和相等。在中间一项中,加和里的($\vec{r}_i - \vec{r}_G$)和($\vec{r}_j - \vec{r}_G$)分别与 j 和 i 无关。因此,上式可被进一步写成

$$\sum_{i=1}^{N}\sum_{j=1}^{N}(\vec{r}_i - \vec{r}_j)^2 = 2N\sum_{i=1}^{N}(\vec{r}_i - \vec{r}_G)^2 - 2\Big(\sum_{i=1}^{N}\vec{r}_i - N\vec{r}_G\Big)\Big(\sum_{j=1}^{N}\vec{r}_j - N\vec{r}_G\Big)$$

依据其定义,$\vec{r}_G = \sum_{i=1}^{N}\vec{r}_i/N$。第二项中的两个加和($\sum_{i=1}^{N}\vec{r}_i$ 和 $\sum_{j=1}^{N}\vec{r}_j$)均为 $N\vec{r}_G$,故第二项为零,即

$$\Big[\sum_{i=1}^{N}(\vec{r}_i - \vec{r}_G)^2\Big] = \Big[\sum_{i=1}^{N}\sum_{j=1}^{N}(\vec{r}_i - \vec{r}_j)^2\Big]/2N \quad (4.13)$$

将其代入式(4.11),可得下列一个相当有用的引理(关系式):

$$R_g^2 = \Big[\sum_{i=1}^{N}\sum_{j=1}^{N}\langle(\vec{r}_i - \vec{r}_j)^2\rangle\Big]/2N^2 \quad (4.14)$$

对一条**线型链**而言,式(4.3)中的 $x = |i-j|$,故上式中的方均距加和可具体计算如下:

$$\sum_{i=1}^{N}\sum_{j=1}^{N}\langle(\vec{r}_i - \vec{r}_j)^2\rangle = \sum_{i=1}^{N}\sum_{j=1}^{N}b^2|i-j| = 2b^2\sum_{i=1}^{N}\sum_{j=1}^{i}(i-j)$$

$$= 2b^2\sum_{i=1}^{N}\frac{i(i+1)}{2} = b^2\Big(\sum_{i=1}^{N}i^2 + \sum_{i=1}^{N}i\Big) = b^2\Big[\frac{N(N+1)(2N+1)}{6} + \frac{N(N+1)}{2}\Big]$$

$$\sum_{i=1}^{N} \sum_{j=1}^{N} \langle (\vec{\mathbf{r}}_i - \vec{\mathbf{r}}_j)^2 \rangle = b^2 \left[\frac{N(N+1)(N+2)}{3} \right] \cong b^2 \frac{N^3}{3} \qquad (4.15)$$

其中,第一行中的系数 2 源于 $|i-j|$ 中的 i 可大于或小于 j,且二者对称。有了 $j \le i$ 的限制,绝对值符号移去。仅考虑了原本加和的一半,故乘以 2。第二行中,$\sum_{j=1}^{i}(i-j) = \sum_{j=1}^{i} i - \sum_{j=1}^{i} j$。前者为 i^2;后者是对 j 从 1 到 i 的求和,即从 0 到 $i-1$ 之和,得 $i(i-1)/2$。二者合一,结果为 $i(i+1)/2$。最后的近似利用了 $N \gg 1$。将其代入式(4.14),可得

$$\langle R_{g,0}^2 \rangle = \frac{b^2 N}{6} \quad \text{或} \quad \langle R_{g,0}^2 \rangle = \frac{\langle R_0^2 \rangle}{6} \qquad (4.16)$$

可见,对于一根线型自由连接的理想链,其方均回转半径为其方均末端距的六分之一。后面,将会详细地讨论如何利用激光光散射测量大分子链的平均回转半径。将上式代入式(4.5)和式(4.7)可分别得到可测的柔软线型链在溶液中的平均回转半径(R_g)和摩尔质量(M)之间的普适标度律,即

$$R_g = k_{R_g} M^{\alpha_{R_g}} \left(\alpha_{R_g} = \frac{3}{5}, \text{良溶剂和} \ \alpha_{R_g} = \frac{1}{2}, \Theta \ \text{溶剂} \right) \qquad (4.17)$$

需要强调的是,k_{R_g} 包含了链段长度(b),并随大分子和溶剂的化学组成而变;而标度指数(α_{R_g})则是普适的,与溶液的化学组成无关。显然,当 N 个链段连成一条刚性直链时,其平均回转半径和链段数成正比,即 $R_g \sim N$ 或写成 $R_g = k_{R_g} M$,此处 $\alpha_{R_g} = 1$。

实验上,先制备一套(五个以上)窄分布、具有不同的平均摩尔质量(两个数量级以上)的大分子样品,然后利用激光光散射测得它们各自的 R_g 和 M。从"R_g 对 M"双对数作图的截距和斜率分别可得 k_{R_g} 和 α_{R_g}。在过去五十多年里,对柔性线型大分子而言,这两个标度关系已被大量的实验数据所验证。

在良溶剂中,α_{R_g} 的实验值约为 0.59,略小于基于高斯链模型得到的值。后来,通过重整化群理论进一步修正,α_{R_g} 从 3/5 降至 0.59,与实验值更吻合。这一修正仅仅略微降低了式(4.16)中 $R_{g,0}$ 与 $R_{RMS,0}$ 之比。有关理论的推导细节,恕不赘述。有兴趣的读者可进一步参考一些相关的原始文献。

对一给定的组成,除了最简单的线型组构以外,将一条线型链的两端首尾相接可形成的环状组构,需要特殊合成。在聚合反应中,由于一些反应基团的自发转移或添加的支化共聚单体,可形成不同的支化组构。如果每条支化链都不再进一步支化,可得到"梳状"构型,如右图所示,其中主链和支化链通常由同种单体构成,但也可由两种或甚至多种单体共聚而成。

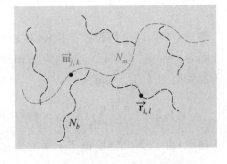

图中的不同颜色并不代表不同的单体,只是为了清楚地标出主链和支化链。

对长接枝支化链,有时很难区分主链和支链。对于环状、支化和其他复杂组构而言,大分子链已经没有平均末端距,但所有的链都有一个平均回转半径。下面,将逐一详细地

讨论具有不同支化组构的大分子链的平均回转半径以及其对整条支化母链以及紧邻支化点之间的支化子链的聚合度的标度依赖关系。

即使对于一给定的化学组成,聚合反应也常常会产生主链和接枝支化链长短不一的梳型链。因此,对给定组成和梳状组构的链,其链结构也可千变万化。为了方便讨论,先假定每条支化链含有相同数目的链段(N_b),每个链段的长度为b;以及主链上均匀地接枝着n_b条支化链;两个紧邻接枝点之间的链节含有N_m个链段。因此,每条梳型链上的链段总数为$N=(N_b+N_m)n_b$。将第i条支化子链上的第l个链段和第j条主链链节上的第k个链段的空间矢量分别记为$\vec{r}_{i,l}$和$\vec{m}_{j,k}$,如上图所示。

如前所叙,即使对一给定组成和支化结构的梳型链,热能也会激发链上的每个链段作无规运动,形成数之不尽的构象。对于一条无环状结构的链,有且只有一条线型路径连接链上的任何两个链段。由式(4.12)和式(4.14)已知,平均回转半径与"方均距求和"有关,其中方均距求和为先将任意两个链段之间的(直接)距离平方得到方均距、再对方均距作构象平均、最后对链上所有可能两点的方均距求和。

对一条给定组成和结构的梳型链,给定了一套l,k,i和j的链上两点之间直接距离的构象平均(简称方均距,写成$\langle(\vec{r}_{i,l}-\vec{m}_{j,k})^2\rangle$)为$b^2x$。链上任何可能的两点之间的方均距加和的计算包括两步。首先,寻获x是如何具体地与l,k,i和j相关;其次,分别通过将l和k从1变到N_b以及将i和j从1变到n_b对所有可能的$\langle(\vec{r}_{i,l}-\vec{m}_{j,k})^2\rangle$求和,其数学表达为$\sum_{i,j=1}^{n_b}\sum_{l,k=1}^{N_b}\langle(\vec{r}_{i,l}-\vec{m}_{j,k})^2\rangle$,自此以后,称作方均距加和。

为了求出梳型链复杂的"方均距加和",不得不寻获四种不同的情况下的两个给定链段之间线型路径上的链段数(x):(Ⅰ)二者均位于同一条接枝支化链上;(Ⅱ)二者位于不同的接枝支化链上;(Ⅲ)一个位于主链,另一个位于一条接枝支化链上;(Ⅳ)二者均位于主链上。

可先利用式(4.15)分别求得每一类中的方均距加和,然后再相加。整个计算较为繁琐;读者可以跳过下面用两条虚线隔开的部分而直接采用推导结论。然而,学会下列数学推导和计算原理,读者将能够算出其他组构链的平均回转半径。因此,仍建议读者潜心消化下列内容,掌握其中的推导逻辑和理解背后的物理图像,而不是仅仅知道每一步的数学推导。

情况(Ⅰ),两个链段位于同一条支化链上($i=j;x=|l-k|$):先求出同一条支化链上任意隔着$|l-k|$个链段的方均距加和,只需用N_b取代式(4.15)中的N;n_b条接枝链的方均距加和则为n_b倍。因此,

$$(\text{Ⅰ})=\sum_{i=1}^{n_b}\sum_{l,k=1}^{N_b}\langle(\vec{r}_{i,l}-\vec{r}_{i,k})^2\rangle=\sum_{i=1}^{n_b}\frac{b^2N_b(N_b+1)(N_b+2)}{3}\cong b^2\frac{N_b^3}{3}n_b$$

情况(Ⅱ),两个链段位于不同的支化链上($i\neq j$):两个链段之间隔着$|i-j|$个主链链节。每个主链链节含有N_m个链段,所以二者之间隔着$N_m|i-j|$个主链上的链段,外加分别在两个不同支化链上的l个和k个链段,即$x=|i-j|N_m+l+k$。注意:主链链节上的链段与l和k无关;以及两条支化链上的链段则与i和j无关。因此,主链段上间隔着$|i-j|N_m$个链段的两个链段之间的方均距为$b^2|i-j|N_m$;每条支化链的方均距为

$b^2 N_b$。它们的加和为

$$\sum_{i \neq j}^{n_b} \sum_{l,k=1}^{N_b} \langle (\vec{r}_{i,l} - \vec{r}_{j,k})^2 \rangle = \sum_{i \neq j}^{n_b} \sum_{l,k=1}^{N_b} b^2 (|i-j|N_m + N_b) = b^2 \sum_{i \neq j}^{n_b} \left[N_m N_b^2 |i-j| + N_b^3 \right]$$

其中，先分别变化 l 和 k 求和。右边第一项中的 $|i-j|N_m$ 与 l 和 k 无关，故求和可得 N_b^2 倍的 $|i-j|N_m$，即 $N_m N_b^2 |i-j|$；而第二项 N_b 为一个常数，故求和也得 N_b^2 倍的 N_b，结果为 N_b^3；也可对 $l+k$ 求和，即分别对 l 和 k 求和，每个求和可得 $N_b(N_b+1)/2$，共有 N_b 项，且 $N_b \gg 1$，故结果仍为 N_b^3。然后，分别变化 i 和 j 求和。由第一项中的 $\sum_{i \neq j}^{n_b} |i-j|$ 可得 $n_b(n_b+1)(n_b+2)/3 \cong n_b^3/3$；因为 $i \neq j$，第二项求和仅得 $(n_b^2 - n_b)$ 个相同的交叉项，每项均为 N_b^3。最终结果为

$$(\text{II}) = \sum_{i \neq j}^{n_b} \sum_{l,k=1}^{N_b} \langle (\vec{r}_{i,l} - \vec{r}_{j,k})^2 \rangle \cong b^2 \left[\frac{N_m N_b^2 n_b^3}{3} + (n_b^2 - n_b) N_b^3 \right]$$

情况（III），一个链段位于主链，另一个在支化链上：两个链段间隔着 $|i-j|$ 个主链链节，外加分别位于主链和支化链上的 l 和 k 个链段。i 和 j 之间的链段数 $x = N_m |i-j| + l + k$。与情况（II）中的加和相比，一条主链链节和一条接枝链上的链段数（N_m 和 N_b）不同，即

$$\sum_{i,j=1}^{n_b} \sum_{l=1}^{N_m} \sum_{k=1}^{N_b} \langle (\vec{r}_{i,l} - \vec{m}_{j,k})^2 \rangle = \sum_{i,j=1}^{n_b} \sum_{l=1}^{N_m} \sum_{k=1}^{N_b} b^2 (N_m |i-j| + l + k)$$

包含了主链加和以及侧链加和两个部分。先分别变化 l 和 k 求和。右边的第一项 $|i-j|N_m$ 与 l 和 k 无关。求和分别得到 N_m 项和 N_b 项，每项均为 $|i-j|N_m$，结果为 $|i-j|N_m$ 的 $N_m N_b$ 倍，即 $|i-j|N_m^2 N_b$。第二项和第三项 $\sum_{k=1}^{N_b} \sum_{l=1}^{N_m} l$ 和 $\sum_{l=1}^{N_m} \sum_{k=1}^{N_b} k$ 分别与 N_b 和 N_m 无关，故两个求和分别为 $N_b \sum_{l=1}^{N_m} l$ 和 $N_m \sum_{k=1}^{N_b} k$。$\sum_{l=1}^{N_m} l = N_m(N_m+1)/2 \cong N_m^2/2$ 和 $\sum_{k=1}^{N_b} k = N_b(N_b+1)/2 \cong N_b^2$。因此，

$$\sum_{i,j=1}^{n_b} \sum_{l=1}^{N_m} \sum_{l,k=1}^{N_b} \langle (\vec{r}_{i,l} - \vec{m}_{j,k})^2 \rangle = b^2 \sum_{i,j=1}^{n_b} \left[N_b N_m^2 |i-j| + \left(\frac{N_b N_m^2 + N_m N_b^2}{2} \right) \right]$$

第一项 $\sum_{i,j=1}^{n_b} |i-j|$ 的求和结果为 $n_b^3/3$。第二项与 i 和 j 无关，故求和共得 n_b^2 个项，每项均为 $N_m N_b(N_m + N_b)/2$，即 $N_m N_b(N_m + N_b) n_b^2/2$。因此，

$$(\text{III}) = \sum_{i,j=1}^{n_b} \sum_{l=1}^{N_m} \sum_{l,k=1}^{N_b} \langle (\vec{r}_{i,l} - \vec{m}_{j,k})^2 \rangle \cong b^2 \left[\frac{N_b N_m^2 n_b^3}{3} + \frac{N_b N_m (N_b + N_m) n_b^2}{2} \right]$$

情况（IV），两个链段均位于主链上：较为简单，先求得隔着 $|i-j|$ 个链节的两个点间的方均距，结果为 $b^2 |i-j|$，然后求和。注意：主链上的链段数为 $N_m n_b$。由式（4.15）可导出

$$(\text{IV}) = \sum_{i,j=1}^{n_b} \sum_{l,k=1}^{N_m} \langle (\vec{m}_{i,l} - \vec{m}_{j,k})^2 \rangle = \sum_{i,j=1}^{N_m n_b} b^2 |i-j| \cong b^2 \frac{(N_m n_b)^3}{3}$$

上述四种情况的加和涵盖了一条梳型链上任意两个链段之间的方均距加和，结果为

$$\sum_{i,j=1}^{n_b} \sum_{l,k=1}^{N_b} \langle (\vec{r}_{i,l} - \vec{m}_{j,k})^2 \rangle = (\text{I}) + (\text{II}) + (\text{III}) + (\text{IV})$$

$$= \frac{b^2 n_b^2}{3} \left[\frac{N_b^3}{n_b} + n_b N_m N_b^2 + 3\left(1 - \frac{1}{n_b}\right) N_b^3 + n_b N_b N_m^2 + \frac{3N_b N_m (N_b + N_m)}{2} + n_b N_m^3 \right]$$

$$= \frac{b^2 n_b^2}{3} \left[n_b N_m (N_b + N_m)^2 - n_b N_b N_m^2 + \left(3 - \frac{2}{n_b}\right) N_b^3 + \frac{3N_b N_m (N_b + N_m)}{2} \right]$$

其中,同时加上和减去了 $n_b N_b N_m^2$,以得到第一项中的 $(N_b + N_m)^2$。依据式(4.12)中链的方均回转半径的定义,利用式(4.14),并在上式中代入链段的总数 $N = (N_b + N_m)n_b$,可得下列梳型链在理想状态下的方均回转半径的表达式

$$\langle R_{g,0}^2 \rangle_{\text{comb}} = \frac{b^2}{6} \left[n_b N_m - \frac{n_b N_b N_m^2}{(N_b + N_m)^2} + \frac{3n_b - 2}{n_b (N_b + N_m)^2} N_b^3 + \frac{3N_b N_m}{2(N_b + N_m)} \right] \qquad (4.18)$$

其可被用来考察一些具有特殊结构的"梳型链"。一个极端的结构是接枝支化链很短的梳型链,即主链上两个紧邻支化点之间的链节远长于每条接枝支化链,$N_m \gg N_b$。式(4.18)中的方括号里的第一项远大于其余各项,以及 $N = (N_b + N_m)n_b \cong N_m n_b$,故

$$\langle R_{g,0}^2 \rangle \cong \frac{b^2 n_b N_m}{6} \cong \frac{b^2 N}{6} \propto b^2 N \qquad (4.19)$$

其中,$N_m n_b$ 是主链上的链段数,显示具有短支化链节的梳型链和主链具有相似的方均回转半径。换而言之,可以忽略短接枝支化链节对整条梳型链的方均回转半径的贡献。依据回转半径有关一个物体的质量在空间分布的定义,其在物理上合理。如果每条接枝支化链和每个主链链节具有相近的长度($N_m \cong N_b$),且 $n_b \gg 1$,式(4.18)则成为

$$\langle R_{g,0}^2 \rangle \cong \frac{b^2 N_b n_b}{8} \left(1 + \frac{2}{n_b} - \frac{2}{3n_b^2}\right) \approx \frac{b^2 N_b n_b}{8} = \frac{b^2 N}{16} \propto b^2 N \qquad (4.20)$$

其中,利用了 $N = (N_b + N_m)n_b \approx 2N_b n_b$。另一个极端的结构是支化链很长的梳型链,即 $N_b \gg N_m$。式(4.18)中的方括号里的第二项和第四项可以略去;式(4.18)可近似成

$$\langle R_{g,0}^2 \rangle \cong \frac{b^2}{6} \left[n_b N_m + N_b \left(3 - \frac{2}{n_b}\right) \right] \qquad (4.21)$$

可见主链和接枝支化链对具有长接枝支化链的梳型链的方均回转半径均有贡献。贡献大小取决于接枝支化链的数目(n_b)。当主链和一条接枝支化链具有相似的长度时,$n_b N_m \sim N_b$,上式可被进一步写成,

$$\langle R_{g,0}^2 \rangle \cong \frac{b^2}{6} \left[n_b N_m + N_b \left(3 - \frac{2}{n_b}\right) \right] \cong \frac{1}{3}\left(2 - \frac{1}{n_b}\right) b^2 N_b \propto b^2 N \qquad (4.22)$$

长支化梳型链最极端的例子是所有的接枝支化长链都连接在一个中心点上,形成一个星型组构(星型链),$N_m \cong 0$,即

$$\langle R_{g,0}^2 \rangle_{\text{star}} \cong \frac{b^2}{6} N_b \left(3 - \frac{2}{n_b}\right) \propto b^2 N \qquad (4.23)$$

其中,每条接枝支化长链也称作一条"臂"。显然,当一条星型链上臂的数目(n_b)足够大时,$\langle R_{g,0}^2 \rangle_{\text{star}} \cong b^2 N_b / 2$,三倍于一条臂的方均回转半径。

另一方面,当 $n_b = 2$ 或 $N_b = 0$ 时,一条星型或支化链就成了一条线型链,其链段总数为一条臂上链段数的两倍,即 $N = 2N_b$,或等于主链上的链段数,即 $N = N_m n_b$,故上式还原成了线型链的式(4.16),即 $\langle R_{g,0}^2 \rangle \cong b^2 N/6$。

将一条线型链的两端相接可形成一条具有一个环型组构的大分子链。对该链上的任何两个链段(i 和 j),式(4.3)中的 x 等于 $|i - j|$ 或 $N - |i - j|$,取决于从何边开始计数。但是,方均距求和对一条给定的链唯一,所以它们的结果必须一致,等效于 $x = |i - j|/2$,与如何计数无关。代入式(4.15),所得的方均距求和为 $b^2 N^3/6$。进一步利用式(4.14),可得

$$\langle R_{g,0}^2 \rangle_{\text{ring}} \cong \frac{b^2 N}{12} \propto b^2 N \qquad (4.24)$$

环状链的方均回转半径仅为线型链的一半。该结果符合不用数学推导的下列物理图像:一条线型链的平均末端距($bN^{1/2}$)等效于与一条含有 $N^{1/2}$ 个链段的刚性链的末端距,即所有的 $N^{1/2}$ 个链段朝着一个方向行走。如果 $N^{1/2}$ 个链段无规地行走,朝所有方向行走的概率均等。统计上,意味着 $N^{1/2}$ 链段中的一半向着与原本垂直的平均方向行走,一条直链弯成了一个等边直角,故其末端距就是

$$\left(bN^{\frac{1}{2}}/2\right)^2 + \left(bN^{\frac{1}{2}}/2\right)^2 = \frac{b^2 N}{2}$$

其显示了物理直觉与数学推导同样的重要。从式(4.19)到式(4.24),对梳型链的不同极端结构,$\langle R_{g,0}^2 \rangle$ 总是正比于 $b^2 N$,导致一个一般性的结论:对具有任何结构的梳型链,$\langle R_{g,0}^2 \rangle \propto b^2 N$。这一结论还可被进一步推广到具有任何组构的链。不同的链结构仅导致不同的正比常数。该推广基于下列统计物理学原理:一个无规变量(x)经过 N 步无规行走后的方均变化($\langle x^2 \rangle$)正比于步长(b)的平方和步数(N),即 $\langle x^2 \rangle \propto b^2 N$ 或 $\sqrt{\langle x^2 \rangle} = bN^{1/2}$。这里,$x$ 是回转半径;步长和步数分别为链段长度和链段数。对一个给定的链结构,每个链构象都是步长为 b 的 N 步(N 个链段)无规行走的结果,故有上述的一般推论。

综上所述,"支化"组构可被看作一个基本的"根"组构。取决于链中有无多环结构,支化链可被分为两类。(Ⅰ)无多环:在这一类支化链中,如果支化链的每一条支化子链不再进一步支化,它就成了梳型组构。如果梳型链上的所有支化子链都连接在一个中心上,它就成了星型组构。如果星型链上只有两条支化子链(两条臂),它就成了线型组构。如果线型链的两个末端连在一起,它就成了仅有一个大环的环型组构。(Ⅱ)有多环:如果一条支化链含有多环,它成为一个三维空间网络,被溶剂充分溶胀后形成一个凝胶。凝胶可看作一类介于液体和固体之间的特殊物质。一方面,凝胶在无施加外力(重力除外)时,可保持其形状,类似固体;另一方面,施加很弱的外力通常会使凝胶破碎并流动,又好像液体。所以,凝胶也被称为"湿"物质。

如前所述,在二官能团单体的线性聚合中,引入少量含有 f 个反应基团的支化共聚单体可获得在一条生长的链上不断支化而成的支化组构,其中,$f \geq 3$。支化频率随着共聚单体的浓度增加,形成含有大量短链节的支化大分子。它们的结构远比梳型链复杂,常被称作超支化链。在一个实际的聚合反应中,一个支化共聚单体上的每一个反应基团并非都会参与共聚反应。即使参与,它们也并非一定同时反应。因此,所有 $(2 + f)$ 的支化聚合

反应一定会导致两个紧邻支化点之间的链节长短不一。除了链节长度的宽分布以外，每条链上的支化点数和位置也各异，故对于一给定的共聚组成，所得支化结构也会完全不同，千变万化！即使两条支化链上含有完全相同数目的二聚单体和支化点，它们的支化结构也可各有秋色。

为简化讨论，仅以含有三个反应基团的支化共聚单体为例。假定支化共聚单体上的每个反应基团都同时反应，各形成一条长度相同的线型链节。如果一条链节的末端和另一个支化单体反应，该链节就继续支化，形成一个新的支化点以及 $(f-1)$ 条新的链节；反之，该链节就连在一个支化点上，成为一个悬挂链节（末端），如右图所示。同一支化聚合反应可形成不同的支化结构。下面，整条支化链和两个紧邻支化点之间的链节（包括那些悬挂末端链节）分别被称为母链和子链。每一条子链含有 N_b 个链段。

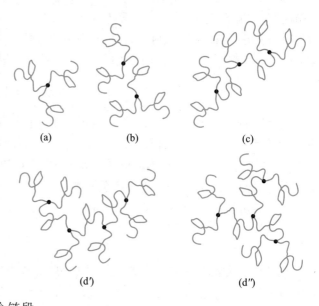

（a）只有一个支化点：三臂星型链：基于式（4.23），$\langle R_g^2 \rangle \cong 7b^2 N_b/18$。（b）和（c）分别有两个和三个支化点：分别含有两条和三条悬挂子链的梳型链，其中 $N_b = N_m$，$n_b = 2$ 或 3；基于式（4.20），$\langle R_g^2 \rangle \cong 11b^2 N_b/24$ 或 $\langle R_g^2 \rangle \cong 43b^2 N_b/72$。（d'）和（d''）有四个支化点：两条具有不同结构的支化链，变得复杂。一条为有四条接枝悬挂子链的梳型链，$n_b = 4$，基于式（4.18），可得 $\langle R_g^2 \rangle \cong 35b^2 N_b/48$，而另一条已不是梳型链，故式（4.20）不再适用。以此类推，当支化点多于四个时，具备相同组成和组构的链可有许多不同的支化结构。

一个极端的结构就是详细讨论过的梳型链。树突型的支化链另一个极端的结构，其具有一个支化中心，从其出发，每条支化子链以相同的概率进一步支化。需要强调，具有多个悬挂子链的支化链已无末端距，但仍有方均距加和与方均回转半径。与线型链相比，计算支化链的方均回转半径时，先要对一给定结构的支化链的所有可能的构象进行统计平均得到其均方距加和，然后再对所有可能的链的结构进行统计平均，相当繁杂。

早在 1949 年，通过对给定组成的支化链的各种可能的结构和构象进行统计平均，Bruno H. Zimm 和 Walter H. Stockmayer［Journal of Chemical Physics，1949，17：1301］就已得到双重标度律，$R_{g,0} = \sqrt{\langle R_g^2 \rangle} \propto b N_b^{1/4} N^{1/4}$，其中 $R_{g,0}$ 是在无扰状态下支化链的平均回转半径，N 和 N_b 分别为母链和子链上的链段数。推导的核心是使用了 Kramers 定理，想象地将一条含有 N 个链段且无环的支化链在第 i 个链段处剪成两部分。它们分别含有 N_i 和 $(N - N_i)$ 个链段。统计平均后得到其方均距：$\langle R^2 \rangle \cong b^2 \langle N_i(N - N_i) \rangle / N$，其中 $N_i(N - N_i)/N$ 反映了第 i 个链段对 $\langle R^2 \rangle$ 的统计贡献。而平均 $\langle N_i(N - N_i) \rangle$ 则覆盖了所有可能的剪切

方式。

约二十年后,P.-G. de Gennes[Biopolymers,1968,6:715]利用了一个相对简明方法,即在高斯分布的倒格子空间里对链段长度进行拉普拉斯变换,也得到$R_{g,0}$与N之间的相同标度关系,$R_{g,0} \propto N^{1/4}$。即使在七十多年后的今天,除了少数理论学家以外,绝大部分的大分子研究者仍然很难理解 Zimm 和 Stockmayer 原文中所用的艰深物理原理和繁杂数学推导,故本书略去他们的详细推导过程。读者仅需记住上述标度关系。以下另辟捷径的推演仅利用了一个简化的图像,外加统计原理和有关的标度理论。

为方便讨论,可将每条含有N_b个链段的子链被处理成一个小球或新"链段",其尺度或长度为$\xi = b N_b^{1/2}$。这样,每个小球只有两个选择:进一步连接($f-1$)个小球形成一个新的支化点,或者成为一个悬挂的末端。因此,一条含有N个链段的支化链被简化成一个由N/N_b个小球或新"链段"连接而成的结构,如右图所示,这里$f=3$。注意,这只是多种可能的支化结构中的一种。可以想象这样的 65 个小球能够形成各种各样的支化结构。

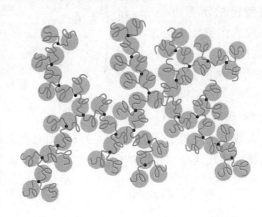

在这一描绘中,一条给定组成和支化组构的链的不同结构就被处理成了N/N_b步无规行走后的不同轨迹。除了那些悬挂的小球以外,中间的每个小球都进一步连接着($f-1$)个球,形成一个支化点。如果支化点数目为n_b,则有$[(f-2)n_b+2]$个悬挂小球和(n_b-1)个处于两个邻近支化点之间的小球。二者之和就是小球的总数:

$$\frac{N}{N_b} = [(f-2)n_b+2] + (n_b-1) = (f-1)n_b+1 \quad \rightarrow \quad n_b = [(N/N_b)-1]/(f-1)$$

因此,上图中共有 32 个支化点。这样描绘的优点是不用考虑每条子链的细节;换而言之,因为每条子链本身就是一个"链段",在计算其方均距加和时,无须再考虑一个链段是子链上的第l个还是第m个链段。

对给定的小球数目(组成),小球们可互相连接形成不同的组构,同一个组构还可能有不同的结构。一个极端的结构就是所有的小球们线型地连接在一起成为一条线型链。其无扰方均回转半径为$\langle R_{g,0}^2 \rangle = \xi^2(N/N_b)/6 = b^2 N_b(N/N_b)/6 = b^2N/6$。如果$N/N_b$个小球支化地连接成一条梳型链,其中,式(4.18)里的$N_b = N_m$,即$\langle R_{g,0}^2 \rangle = \xi^2(N/N_b)/16 = b^2 N_b(N/N_b)/16 = (1/16)b^2N$。如果$N/N_b$个小球连接成有$k$条臂的星型链,每条臂上的小球数则为$(N/N_b)/k$。依据式(4.23),$\langle R_{g,0}^2 \rangle = [(3k-2)/6k^2]b^2N$,或当$k \gg 1$时,$\langle R_{g,0}^2 \rangle = (1/2k)b^2N$。注意:不同的结构均有同样的$R_{g,0} \propto bN^{1/2}$,但是有不同的正比系数。

对具有同样数目(N,相同组成)、不同结构的支化链,对任意一个给定的支化结构的所有可能的构象的统计平均可得其方均距加和,其正比于小球尺寸的平方(ξ^2)和小球数目(N/N_b),对应着一个构象平均的回转半径$\langle R_g \rangle_c$,如下图所示。不同的支化结构具有不同的正比系数。每个支化结构类似一个无规行走了$(N/N_b)^{1/2}$步的轨迹,故每个$\langle R_g \rangle_c$对应着一条想象的、含有$(N/N_b)^{1/2}$个尺寸为ξ的小球的线型链构象。

进一步,对含有N/N_b个小球的支化链的所有可能结构的统计平均等效于对一条上述想象的线型链的所有可能的构象作统计平均,如右图所示。注意:这里小球的数目为$(N/N_b)^{1/2}$,而不是原来的N/N_b。已知,

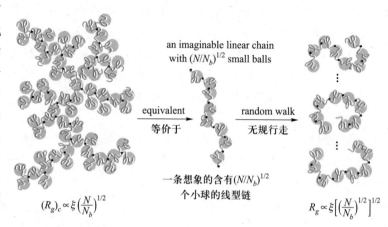

an imaginable linear chain with $(N/N_b)^{1/2}$ small balls

equivalent
等价于

random walk
无规行走

一条想象的含有$(N/N_b)^{1/2}$个小球的线型链

$(R_g)_c \propto \xi \left(\dfrac{N}{N_b}\right)^{1/2}$

$R_g \propto \xi \left[\left(\dfrac{N}{N_b}\right)^{1/2}\right]^{1/2}$

一条线型链的构象平均回转半径与链段长度和链段数目的开方成正比。因此,如果可忽略排除体积效应,对具有给定组成的支化链,构象和结构双重统计的结果为

$$R_{g,0} = \sqrt{\langle R_{g,0}^2 \rangle} \propto \xi \left[\left(\frac{N}{N_b}\right)^{1/2}\right]^{1/2} \propto b\,N_b^{1/2}\left[\left(\frac{N}{N_b}\right)^{1/2}\right]^{1/2} \propto bN_b^{1/4}N^{1/4} \quad (4.25)$$

其与 Zimm 和 Stockmayer 以及 de Gennes 经过一些复杂数学推导得到的双重标度关系完全一致,殊途同归。

通过运用上述一般统计原理,也可以获得相同的双重标度关系。即任意一个随机物理变量的波动幅度的方均根与每次波动的平均幅度以及波动次数(样本数)的平方根成比例。

因此,对一个给定的组成(小球数),存在着变化各异的支化结构。对一条给定结构的支化链,其构象仍然不停地和随机地变化。为了求出给定组成的支化链的平均回转半径,就不得不做两次统计平均。一次是对一个给定支化结构的所有可能的构象作平均。另一次是对所有可能的支化结构作平均,等效于一条想象的、仅由$(N/N_b)^{1/2}$小球组成的线型链的构象平均。每次平均都导致小球数的一次开方。故如果排除体积效应可被忽略,给定组成的支化链的平均回转半径$(R_{g,0})$应该正比于小球的尺寸(ξ)和小球数(N/N_b)的两次开方,即$R_{g,0} \propto b\,N_b^{\frac{1}{4}}N^{\frac{1}{4}}$,完全一致。

显然,对给定的组成,支化链平均比线型链小了$(N/N_b)^{1/2}$倍。其所占据的平均体积缩小了$(N/N_b)^{3/2}$倍,故不同链段之间的平均距离变得更短。因此,在真实的大分子溶液中,一条支化链的内部应存在着更强的排除体积效应。约四十年前,J. Isaacson 和 T. C. Lubensky [Le Journal de Physique-Letters,1980,41:469] 发现,在考虑了良溶剂中的排除体积效应之后,支化链的平均回转半径和链段总数(N)的标度指数从 1/4 增至 1/2。随后不久,M. Daoud 和 J. F. Joanny [Le Journal de Physique-Letter,1980,42:1359] 利用 de Gennes 发展的方法,进一步考察了在良溶剂中,支化链的平均回转半径与子链和母链上的链段数之间的双重标度关系。他们确认了在考虑排除体积效应后,对N的标度指数确

实为 1/2;同时还发现,对 N_b 的标度指数从 1/4 降至 1/10。计算十分烦琐,故在此处略去。

如下所示,一个简单推导也可得到同样的双重标度关系。推导过程与那些得到式 (4.4) 至式 (4.7) 的相似。即当一条大分子链在良溶剂中扩展时,其自由能变化包括两个相反的贡献。两者均和链的扩展有关,分别为二体排除体积效应(其本质也是熵)和熵弹性。与线型链相比,仅需记住对于支化链,$R_0^2 \cong b^2 N_b^{1/2} N^{1/2}$,即

$$\Delta G \sim \frac{b^3 N^2}{R^3} + \frac{R^2}{b^2 N_b^{\frac{1}{2}} N^{\frac{1}{2}}} \quad \rightarrow \quad \left(\frac{\partial \Delta G}{\partial R}\right)_{T,p} \sim -\frac{3b^3 N^2}{R^4} + \frac{2R}{b^2 N_b^{\frac{1}{2}} N^{\frac{1}{2}}}$$

$$\left(\frac{\partial \Delta G}{\partial R}\right)_{T,p} = 0 \quad \rightarrow \quad R \sim b N_b^{\frac{1}{10}} N^{\frac{1}{2}} (\text{良溶剂}) \tag{4.26}$$

随着溶剂性质变差,二体排除体积效应慢慢地消失。因此,有必要考虑原本较弱的三体排除体积效应。沿用有关得到式 (4.6) 的讨论和代入 $R_0^2 \cong b^2 N_b^{1/2} N^{1/2}$,可得

$$\Delta G \sim \frac{b^6 N^3}{R^6} + \frac{R^2}{b^2 N_b^{\frac{1}{2}} N^{\frac{1}{2}}} \quad \rightarrow \quad \left(\frac{\partial \Delta G}{\partial R}\right)_{T,p} \sim -\frac{6b^6 N^3}{R^7} + \frac{2R}{b^2 N_b^{\frac{1}{2}} N^{\frac{1}{2}}}$$

$$\left(\frac{\partial \Delta G}{\partial R}\right)_{T,p} = 0 \quad \rightarrow \quad R \sim b N_b^{\frac{1}{16}} N^{\frac{7}{16}} (\Theta \text{溶剂}) \tag{4.27}$$

这里需要注意,如果将一条线型链视为一个支化度为零、只有一条"支链"的"支化链","支链"就是"主链",$N_b = N$。将其代入以上两式,分别得到 $R \sim b N^{3/5}$ 和 $R \sim b N^{1/2}$,与线型链在两种溶剂中的前述结果完全一致,进一步说明线型链是一种特殊的支化链。

依据上述双重标度关系,$\lg R_g$ 对 $\lg N_b$ 以及 $\lg R_g$ 对 $\lg N$ 双对数作图的斜率分别为两个标度指数。一般而言,支化聚合物的 N_b 和 N 的范围分别为 $10^1 \sim 10^2$ 和 $10^3 \sim 10^4$。因此,$N_b^{1/10}, N_b^{1/16}, N^{1/2}$ 和 $N^{7/16}$ 分别在 1.3 ~ 1.6,1.2 ~ 1.3,30 ~ 100 和 20 ~ 56 的范围内。在 R_g, N_b 和 N 的实际测量中,总是存在着不可避免的实验误差。这样所得的标度指数的相对误差如下。

求每个斜率的过程涉及两个对数之差除以另外两个对数之差,可写作 $\mathrm{d}(\ln y)/\mathrm{d}(\ln x) = (x/y)(\mathrm{d}y/\mathrm{d}x)$;假定 x 和 y 具有同样的相对误差 $\pm a\%$,$\mathrm{d}x$ 和 $\mathrm{d}y$ 的相对误差增至 $\pm 2a\%$。两个变量相乘或相除,误差增加一倍。因此,总的相对误差应为 $\pm 8a\%$。实验上所测的平均回转半径和平均摩尔质量的相对误差($\pm a\%$)不小于 $\pm 3\%$,故所求斜率的总相对误差不会低于 $\pm 25\%$。将每组样品的数目增至五个以上,可降低由作图所得斜率的相对误差,但仍在 $\pm 10\%$ 左右。因此,几乎不可能利用实验数据和以上两个标度关系来判断溶剂性质。

除了实验误差以外,如果一个支化共聚单体上的 f 个反应基团仅有两个参与了聚合,两个紧邻支化点之间的子链就长了一倍;如果连续发生,那么两个紧邻支化点之间的这条子链的长度就为设计长度的三倍或多倍。另一方面,每个支化反应基团并非同时参与聚合,不仅导致子链长短不一,而且也使得每条支化链具有不同的组成和结构,双重的宽分布。

验证多年前理论上已预测的双重标度关系需要两套支化链:(a) 对给定的子链上的链段数(N_b),母链上有不同的链段总数(N);(b) 对给定的母链上的链段总数(N),子链

上有不同的链段数（N_b）。在过去的几十年之间，尽管有许多不懈的尝试和努力，始终无法得到母链和子链聚合度分别均一的一系列完美的支化链。在支化链的以往实验中，不得不用子链和母链的平均聚合度或平均摩尔质量分别代替上式中的 N_b 和 N，导致无法确定地比较实验结果和理论预测的标度关系。

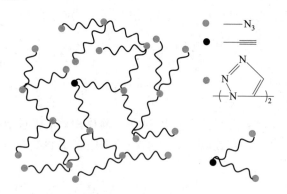

直至最近，利用两端为叠氮（A）、中间为炔基（B）的大单体（A…B…A），其中…代表了一条线型短链节，如右图所示。叠氮和炔基之间的高效点击化学反应可将若干个大单体连接成具有均匀子链长度的接近完美的支化链。每条支化链上仅剩一个未反应的炔基（黑点），其可以进一步点击一个端基叠氮成环（未画），但其概率很低。为了制备这样一个窄分布大单体，先用带有双引发基团的 B 同时引发活性聚合，形成两条连接在 B 处的线型链节，利用两个 A 基团封住链的两个活性端。

使用上述巧妙的设计和不同的引发剂和反应单体的比例，首先利用活性聚合反应制得具有不同长度的、窄分布的大单体，其链段数为 $2N_b$。然后，再用点击化学反应将每种长度的大单体分别连接起来，形成一系列含有不同支链长度（N_b）的"完美"支化链。通过逐步沉淀的方法将每种具有同样 N_b 但不同 N 的支化链进一步分开。最终，可得两个系列的支化链。在一个系列中，N_b 相同，但 N 不同；在另一个系列中，N 相近，但 N_b 不同。

利用激光光散射分别测得每个样品中支化链的平均回转半径（R_g）和平均流体力学半径（R_h）以及初始大单体和支化链的平均摩尔聚合度（n_b 和 n）。依照上述双重标度律，每种链尺寸（R_g 或 R_h）分别对母链和子链的摩尔聚合度作双对数图可得两条直线，其斜率对应着两个标度指数。理论上，一共可得四个标度指数。

链尺寸与子链摩尔聚合度之间的标度指数接近预测的 1/10，不太敏感。链尺寸与母链摩尔聚合度的标度指数分别为 0.46 和 0.48，略小于预测的 1/2，如右图所示。注意，其纵坐标已作了对子链摩尔聚合度的约化（$n_b^{1/10}$），故两张双对数图可合二为一，只剩下两条直线。需要指出：在良溶剂中，所测得的线型链的标度指数为 ~0.59，也小于没有经过重整化群理论修正的理论值（3/5）。经过近七十年不懈地努力，终于在实验上验证了无规支化链的平均回转半径对其子链和母链摩尔聚合度的双重标度依赖性。稍后，将会详细地介绍流体力学半径（R_h）的意义以及相关的测量方法。

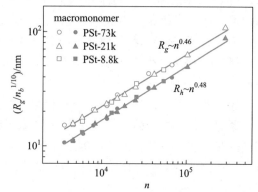

Li L W, Lu Y Y, An L J, et al. Journal of Chemical Physics, 2013, 138: 114908.

（Fiugre 2, permission was granted by publisher）

小　结

对一给定的组成,大、小分子均可具有不同的组构,但大分子具有远比小分子丰富和复杂的组构。即便是只含有一种单体的聚合物也有许多不同的组构和结构。支化连接实际上可被看作最基本的"根"组构,可进一步按链中是否有多个环状结构分为两类。

(Ⅰ) 无多环:如果每条接枝支化子链都不再继续支化,其就成了梳状组构;当一条梳型链上所有的支化子链都连接在一个中心上时,其就成了星状组构;当一条星状链上只有两条支化子链(两条臂)时,就成了线型组构;当一条线型链的两个末端相连时,就成了一个环状组构;

(Ⅱ) 有多环:支化链具有一个三维的空间网络组构。其被溶剂充分溶胀后,形成一个凝胶,可看作一类介于液体和固体之间的特殊物质,也称作湿物质。一方面,凝胶在无外部施力(重力除外)时可保持其形状,类似固体;另一方面,稍加施力通常又会使凝胶破碎并使其变形,好像液体。

含有 N_0 个聚合单体、具有任何组构的大分子链可被理想地看成由 N 个自由连接的Kuhn 链段。每个链段均可在三维空间中任意折叠和旋转,其长度为 b,含有 k 个聚合了的单体,即 $N = N_0/k$。不同大分子链有不同的 k 值,反映了链的柔软性,k 值越小越柔软。一条刚性链可看成只有一个链段。N 与实验上可测的聚合度或者摩尔质量成正比。高斯链是扩展的理想链,其假定从任一链段 i 处出发经过 x 个链段后在 $\vec{r}_i - \vec{r}_j$ 处发现另一个链段 j 的概率遵循高斯分布 $G(\vec{r}_i - \vec{r}_j, x)$。

与小分子完全不同,即使给定了一个组成和组构,在热能的搅动下,一条大分子链上的每个链段无规地运动,引致链的质量在空间中的分布(构象)也不停地无规变化。因此,只有对各种可能构象进行统计平均方可得到一条大分子链的平均尺寸。

一般而言,对非线型无环链上任意的两个链段(i 和 j),存在且只存在一条含有 x 个链段的线型链节(路径)将二者连接。二者间的直接距离随着链的构象不停地变化,统计平均可得二者之间的方均距,即对任意两个链段点间在所有可能的构象中的直线距离的平方作平均。结果为 b^2x,x 随组构变化。

一条线型链或链节的 $x = |i - j|$;而一条非线型链上两个链段之间的 x 往往不可简单地与 i 和 j 关联。对具体组构,需具体分析。将链上所有可能的"均方距"相加求和,即将两个链段的位置分别从 1 变到 N,称作"方均距加和"。其理论上和可测的"平均回转半径"相连。"方均距加和"是一个核心但不易理解的物理概念和统计操作。

对最简单的线型组构而言,每条链有一伸直长度:$L_c = bN$。如果将 i 和 j 分别取作其首尾两端,方均距即为方均末端距 $\langle R^2 \rangle = b^2N$。可由方均距加和得到方均回转半径,即 $\langle R_{g,0}^2 \rangle = b^2N/6$。注意:方均回转半径仅为方均末端距的 $1/6$。线型理想链的平均末端距($R_0 = \langle R^2 \rangle^{1/2}$)和平均回转半径($R_{g,0} = \langle R_{g,0}^2 \rangle^{1/2}$)均正比于 $bN^{1/2}$。非线型组构的链可具有多个末端,故无方均末端距,但其仍有一个方均距。理论上,在无扰状态下,支化链的平均回转半径与支化母链和子链上的链段数(N 和 N_b)存在一个双重标度关系(也称标度律):

$R_{g,0} \propto bN_b^{1/4}N^{1/4}$。

　　真实链中的每个链段占有一定的三维空间($\sim b^3$)，称作排斥体积。链段之间的色散吸引产生一个等效于减小排斥体积的作用。两者结合在一起称作排除体积效应。在良溶剂或没有二体相互排除效应的 Θ 溶剂中，上述双重标度律成为 $R_g \propto bN_b^{1/10}N^{1/2}$ 或 $R_g \propto bN_b^{1/16}N^{7/16}$。如果将一条线型链看成只有一条子链的特殊"支化链"，即 $N_b = N$，上述双重标度律就成为 $R_g \propto bN^{3/5}$ 和 $R_g \propto bN^{1/2}$，与直接从线型链推导出的结果一致。由此可见，支化组构的确是一个基本的"根"组构。

　　在一个聚合反应中，每条高聚物链由不同数目的单体组成，具有不同的尺寸。对非线型组构，即使给定了组成和组构，每条链仍可具有不同的结构，很复杂！例如，在同一条支化链上，子链长度不均；在不同支化链上，支化点数和位置不同。因此，不得不用可测的数均或重均摩尔质量(M_n 和 M_w)代替链段数(N)。因二者成正比，不影响标度律。

　　R_g 与 M_n 或 M_w 之间的标度律被一般性地写成 $R_g = k_{R_g}M_n^{\alpha_{R_g}}$ 或 $R_g = k_{R_g}M_w^{\alpha_{R_g}}$，其中 k_{R_g} 是一个依赖体系的变量，随大分子和溶剂的化学组成等因素而变；而 α_{R_g} 则是一个与化学组成无关、只依赖于组构和溶剂性质的普适标度指数。从良溶剂到 Θ 溶剂，再到不良溶剂的过程中，柔性线型链的 α_{R_g} 逐渐地从 3/5 降至 1/2，再速降至 1/3，即一个扩展的无规线团逐渐地蜷缩成一个均匀的、内部含有大量溶剂分子的小球构象。

　　下图总结了各种大分子组构之间的相互关系，从支化型的"根"组构出发，逐渐地向上"生长"，可得各种常见的大分子组构；还总结了在不同大分子的组构中，平均回转半径分别与母链和子链上的链段数之间的标度律。

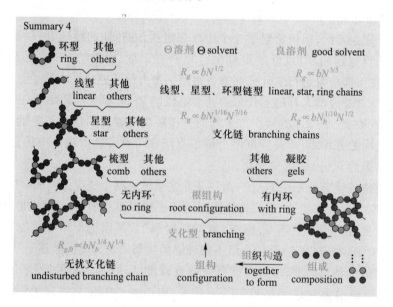

Chapter 4. Conformations of Chains in Macromolecular Solutions

For a given composition, the simplest linear configuration is obtainable by linear polymerization of monomers. Therefore, the chain length dependence of conformation and size of linear chains in solution without special interaction among the chain segments will be discussed first. Each chain has N_0 polymerized monomers. The k adjacent polymerized monomers can be combined to form a **Kuhn segment** (simply called "segment" hereafter) that has a length of b. Each segment can freely rotate in a 360° spherical space around its joint point with the previous segment, including the 180° folding back, the probability of its pointing to any direction is equal.

In other words, the pointing direction of any segment has nothing to do with the pointing directions of other segments (also addressed as "uncorrelated"); also uncorrelated to its pointing direction at the previous moment, i. e., its pointing direction has no memory. A macromolecular chain made of N such segments is called as **free-joint ideal chain** or **simply called ideal chain**. A number of neighboring segments can also form a short **chain link**. Obviously, $N = N_0/k$. Different polymers have different values of k, depending on chemical structures and interaction among the segments. The value of k reflects the flexibility of a polymer chain, the smaller the value, the softer the chain. For a completely rigid chain, $k \to N_0$ and $N \to 1$.

In a dilute solution, the expanded macromolecular chains are independent from each other, so that the interchain interaction is weak. Under the agitation of thermal energy, every segment on each chain is constantly and randomly moving (diffusing and rotating), completely independent on the movements of other segments, forming a random coil conformation. Therefore, the chain conformation in three-dimensional space is also constantly varying (fluctuating) with time; the center of mass of the chain is also randomly diffusing in the solution. In order to measure the size of a chain with a given length and find the dependence of the chain size on the chain length, it is necessary to time-average the conformation of one chain or to assembly-average the conformation of a countless number of chains simultaneously. Based on statistical physics, for a completely random system, their results should be identical.

Average end-to-end distance of linear polymer chains: As shown on the right, the end-to-end distance of a linear chain is a spatical vector pointing from its initial end to its final end ($\vec{R}$), consisting of N short spatial vectors ($\vec{b}$) with an identical length (b), expressed as $\vec{R} = \sum_{i=1}^{N} \vec{b}_i$, where the pointing direction of every $\vec{b}_i$ is random, no special direction. For the same $\vec{\mathbf{b}}_i$, its direction is also randomly changing with time; the directions at different times are uncorrelated, i. e., no segment has a "memory" about its direction. When time (t) is sufficiently long, a segment that points to a certain direction at one time must point to the opposite direction at another time. In other words, in an infinite long chain, if there is one segment that points to a certain direction, there must be another corresponding segment pointing to the opposite direction. Therefore, the result of averaging the directions of one segment

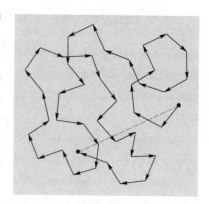

over an infinitely long time or an infinite number of segments at one time (i.e., infinite steps of random walk) is just returning to the original point.

The mathematic expression is that when $t \to \infty$ or $N \to \infty$, the time average of any one of the chain segment $\vec{\mathbf{b}}_i$ or the assembly average of an infinite number of the chain segments is zero, i.e., $\langle \vec{\mathbf{b}}_i \rangle = 0$. Therefore, $\langle \vec{\mathbf{R}} \rangle = \sum_{i=1}^{N} \langle \vec{\mathbf{b}}_i \rangle = N \langle \vec{\mathbf{b}}_i \rangle = 0$. In other words, there is no way to distinguish two macromolecular chains with completely different lengths (degree of polymerization, n) by such an average method. However, the intuition tells us that chains with different lengths should occupy different sizes of space and have different conformational sizes. In statistics, there is an effective way to solve this kind of problems; namely, first to square $\vec{\mathbf{R}} \cdot \vec{\mathbf{R}}$, then do a statistical average $\langle \vec{\mathbf{R}} \cdot \vec{\mathbf{R}} \rangle$, referred as "mean square", and finally take a square root $\langle \vec{\mathbf{R}} \cdot \vec{\mathbf{R}} \rangle^{1/2}$, written as RMS, resulting in an average fluctuation amplitude of a completely random variable, i.e.,

$$R_{RMS,0} = \langle \vec{\mathbf{R}} \cdot \vec{\mathbf{R}} \rangle^{1/2} = \langle \sum_{i=1}^{N} \vec{\mathbf{b}}_i \cdot \sum_{j=1}^{N} \vec{\mathbf{b}}_j \rangle^{1/2} = (\sum_{i=1}^{N} \sum_{j=1}^{N} \langle \vec{\mathbf{b}}_i \cdot \vec{\mathbf{b}}_j \rangle)^{1/2} \qquad (4.1)$$

where $\sum_{i=1}^{N} \vec{\mathbf{b}}_i \cdot \sum_{j=1}^{N} \vec{\mathbf{b}}_j$ represents a multiplication of two "sum of N terms". Its expansion leads to a total of $N \cdot N = N^2$ terms, including N completely correlated self-terms ($i = j$) with an average value of $\langle \vec{\mathbf{b}}_i \cdot \vec{\mathbf{b}}_i \rangle = \langle b^2 \rangle = b^2$; and ($N^2 - N$) cross terms ($i \neq j$) with an average value of $\langle \vec{\mathbf{b}}_i \cdot \vec{\mathbf{b}}_j \rangle$. For a free-joint chain, the movement of one segment is completely uncorrelated with other segments, so that an average of their product should equal a product of their respective averages. i.e., $\langle \vec{\mathbf{b}}_i \cdot \vec{\mathbf{b}}_j \rangle = \langle \vec{\mathbf{b}}_i \rangle \langle \vec{\mathbf{b}}_j \rangle$. For an infinitely long time, equivalent to an infinitely long chain, $\langle \vec{\mathbf{b}}_i \rangle = \langle \vec{\mathbf{b}}_j \rangle = 0$ because the movement and direction of each segment are completely random. Therefore, after written $R_{RMS,0}$ as R_0, the above equation becomes

$$R_0 = \langle \vec{\mathbf{R}} \cdot \vec{\mathbf{R}} \rangle^{1/2} = (b^2 N)^{1/2} = bN^{1/2} \qquad (4.2)$$

The average end-to-end distance of an infinitely long, free-joint (ideal) linear chain is proportional to the length of each segment (b) and the square root of the segment number

($N^{1/2}$). The physical implication of this scaling relationship is that the RMS average distance from the original starting point after N steps random walk is equivalent to walk $N^{1/2}$ steps with a step length of b towards one direction. This is equivalent to a rigid chain with $N^{1/2}$ segments because its end−to−end distance is proportional to the number of its segments. In other word, a rigid chain with $N^{1/2}$ segments is equivalent to a free−joint chain with N segments in terms of the average end−to−end distance.

The length of a real macromolecular chain is finite. Each segment itself occupies a certain three dimensional physical volume (strickly speaking, an excluded volume, $\sim b^3$), so that the entire chain also occupies a three dimensional physical space, its average hardcore volume is related to the average end−to−end distance, $\sim R_0^3$. As mentioned earlier, the hardcore volume of a chain or a segment always exists, will not disappear, and let alone be negative! Its existence reduces the actual reachable volume of particles, and increases the probability of a particle reaching any point in the system. In addition, there exist short−and long−range interactions among different chains and different segments connected by chemical bonds. These interactions affect the actual available volume and the probability of one chain or its segment of reaching any one point in the system, generating an effect of changing the excluded volume. A combination of the hardcore volume and the interactions is the previously discussed "the excluded volume effect".

For the convenient of discussion, the first assumption is to ignore the excluded volume effect. Under the agitation of thermal energy, every segment on a macromolecular chain is constantly and randomly moving in a three−dimensional space, not affected by other segments. The direct distance between two ends (the end−to−end distance) of a linear chain link with x segments is also a random variable. Such a chain link can be viewed as starting from one end (point i), after randomly walking x steps with a step length of b, and reaching another end (point j). Obviously, each time after the x−step random walking, the direct distance between points i and j is randomly varied. With an assumption that the occurring probability of each of such direct distances follows the Gaussian distribution $G(\vec{r}_i, \vec{r}_j, x)$, derived from an extension of free−joint ideal chains, the square of the direct distance between any two segments can be statistically averaged to obtain the mean square walking distance (the mean square end−to−end distance) after an x−step random walk. It equals the square of the step (segment) length multiplied by the walking steps (the number of segments between them, x), i.e.,

$$\langle (\vec{r}_i - \vec{r}_j)^2 \rangle = b^2 x \qquad (4.3)$$

Omitting the calculation details, the correctness of the above conclusion is understandable based on a simple physical picture. If there are two freely jointed segments, fixing one segment, and letting another one to freely fold, forwarding to form a straight line with the first segment is $0°$, completely backwarding on top of the first segment is $180°$, the average position of the second segment after innumerous steps of randomly folding must be in perpendicular to the first segment. The mean square end–to–end distance between them is the square of the hypotenuse of an equilateral right–angled triangle with a side length of b, i.e., $b^2 + b^2 = 2b^2$. If adding the third segment, the first two can be equivalently treated as a new "segment" with a length of $2^{1/2}b$, so that the mean square end–to–end distance of these three segments is $(b^2 + 2b^2) = 3b^2$; by analogy, the above equation is obtainable.

It is necessary to specially emphasize that on a macromolecular chain without ring structures, for given any two segments (i and j), there must exist and only exist one linear connecting path with x segments between them. However, for non–linear configurations, it is often impossible to have x simply from the positions of the two segments. On a chain with ring structures, it is possible to have multiple paths to connect two given segments, more complicated. For example, for a ring chain formed by connecting the head and tail of a linear chain with N segments, the number of segments between any two segments can be $|i - j|$ or $N - |i - j|$, depending on how to count. Therefore, for a specific chain structure, x between two segments has to be specifically analyzed. Sometimes, it is not easy.

For a linear macromolecular chain, $x = |i - j|$, relatively simple. The mean square end–to–end distance of any chain link on a linear chain is $\langle (\vec{\mathbf{r}}_i - \vec{\mathbf{r}}_j)^2 \rangle = b^2 |i - j|$. For an entire linear chain, $\vec{\mathbf{r}}_0 - \vec{\mathbf{r}}_N = \vec{\mathbf{R}}$, $\langle (\vec{\mathbf{r}}_0 - \vec{\mathbf{r}}_N)^2 \rangle = \langle \vec{\mathbf{R}} \cdot \vec{\mathbf{R}} \rangle = b^2(N - 1)$. A Gaussian chain and a free–joint chain have identical scaling relationships between the average end–to–end distances and the chain length (N), $\langle \vec{\mathbf{R}} \cdot \vec{\mathbf{R}} \rangle^{1/2} = bN^{1/2}$. However, two of them are based on completely different assumptions. For Gaussian chains, $|i - j|$ can be large or small, even zero, not related to the chain length, N is only a finite variable, while on a free–joint (ideal) chain, N must approach infinite. Also note: both of them have no concept of the chain length.

The degree of polymerization of real macromolecular chains is always limited, so that the Gaussian chain model can describe real chains better. The fundamental difference between macromolecular and small molecular physics is the chain length dependence of certain physical properties of macromolecules. Note that in the following discussion, we often mix the Conture length (L_C, simply called the chain length), the number of segments (N), the degree of polymerization (n), and the molar mass (M) of the chain. They are proportional to each other, i.e., $L_C = bN = bN_0/k = b(M/M_o)/k$, where k and M_o are the degree of polymerization of a Kuhn segment and the molar mass of polymerization monomers, respectively. Experimentally, N_0 and M are two directly measurable physical parameters. In the following, let us further examine how the excluded volume effect (the solvent quality) influences the average end–to–end distance of a linear chain as well as how it varies with the number of segments (the chain length).

When dissolving in a good solvent, small molecules agitated by the thermal energy are driven away from each other as much as possible by entropy, finally filling the entire solution as individual molecules. Similarly, the segments on a macromolecular chain also attempts to stay away from each other as much as possible to reduce the interaction, leading to an chain expansion, i. e. , an increase of the average end-to-end distance (R). Different from small molecules, the polymerized monomers in a macromolecular chain are connected by chemical bonds, so that thermal energy is not able to break them up into individual monomers and make them to disperse into the entire solution. The limit of chain expansion is completely straight without affecting the bond angle, and its end-to-end distance is the Contour length (L_C).

If a chain is treated as a spring, its energy increases as the chain is stretched. The change of energy is proportional to the square of the spring's length. A stretched spring always tends to contract to reduce its energy. The number of possible chain conformations (conformational entropy) decreases as the chain stretches. If the chain is completely stretched, it has only one straight conformation. According to thermodynamics, in an isolated system or an adiabatic process, the system always spontaneously moves towards the direction of increasing entropy. Therefore, the excluded volume and entropy elasticity effects have opposite dependence on the chain expansion (an increase in R). Their competition allows each macromolecular chain to expand to a certain size only, far smaller than L_C. Every segment in a macromolecular chain can only randomly move inside its average volume ($\sim R^3$).

The average volume ($\sim R^3$) of a real macromolecular chain could be imaginably divided into R^3/b^3 small boxes according to the excluded volume ($\sim b^3$) of one segment. The chance of one segment at any one of small boxes is reversibly proportional to the number of small boxes, i.e., $1/(R^3/b^3) = b^3/R^3$. The chance of N segments into any one of small boxes is just N times, i.e., Nb^3/R^3. The chance of the rest ($N-1$) segments at any one of small boxes is $(N-1)b^3/R^3$. Therefore, the probability of having the two-body interaction between two segments in the chain is $N(N-1)b^6/R^6$, multiplying the number of small boxes (R^3/b^3), resulting in the total number of two-body interaction pairs on the chain, i.e., $\sim b^3N^2/R^3$, where $N \gg 1$ and $N-1 \cong N$ are used.

Using the average end-to-end distance of an ideal chain ($R_0 = bN^{1/2}$) as the reference zero point, The free energy change related to the chain stretching is proportional to $R^2/(b^2N)$. The above two opposite contributions to ΔG are all related to entropy, its mathematic expression is as follows.

$$\frac{\Delta G}{k_B T} \sim \frac{b^3 N^2}{R^3} + \frac{R^2}{R_0^2} \qquad (4.4)$$

where the constant coefficients (ε_2 and γ) in front of the first and second terms, respectively, related to the two-body excluded volume and the entropy "elasticity" effects have been omitted. The differentiation of the both sides of the above equation to R leads to $(\partial \Delta G/\partial R)_{T,p}$ with a dimension of force, equivalent to the effect of a force in Newtonian mechanics, but not a real force. From $(\partial \Delta G/\partial R)_{T,p} = 0$, the minimum of the change of free energy and the dependences of the average end-to-end distance (R) on the length (b) and number (N) of the segments can be obtained, i.e.,

$$\left(\frac{\partial \Delta G}{\partial R}\right)_{T,p} \sim -\frac{3b^3 N^2}{R^4} + \frac{2R}{b^2 N} = 0 \quad \rightarrow \quad R \sim bN^{\frac{3}{5}} \quad (\text{good solvent}) \qquad (4.5)$$

It shows that a macromolecular chain expands to a **stable equilibrant state**. If the chain expands due to the disturbance of thermal energy, the entropy "elasticity" will drive it to shrink; on the contrary, the excluded volume effect will expand the chain, increasing its size. It can be found in statistic thermodynamics that at this equilibrium the maximum fluctuation of the chain size corresponds to a change of half thermal energy ($k_B T/2$) in the free energy.

When the two-body excluded volume effect disappears completely. The solution enters a "pseudo-ideal state", i.e., the Θ state often called in macromolecular research. At this state, it is necessary to consider the initially ignored relatively weaker three-body excluded volume effect among three segments on the chain. According to the above discussion, the probability of generating the three-body excluded volume effect on a chain should be $N(N-1)(N-2)/(R^9/b^9)$, multiplying the number of small boxes (R^3/b^3) leads to the total number of the three-body interaction pairs ($b^6 N^3/R^6$) on the chain. Therefore,

$$\frac{\Delta G}{k_B T} \sim \frac{b^6 N^3}{R^6} + \frac{R^2}{R_0^2} \qquad (4.6)$$

where $N \gg 1$ and $N - 1 \cong N - 2 \cong N$ are used. Similarly, the constant coefficients in front of the first and second terms on the right side, respectively, related to the three-body excluded volume effect (ε_3) and the entropy elasticity (γ), have been omitted. The minimum point of free energy change at the Θ state or in the Θ solvent can be calculated from the above equation as follows

$$\left(\frac{\partial \Delta G}{\partial R}\right)_{T,p} \sim -\frac{6b^6 N^3}{R^7} + \frac{2R}{b^2 N} = 0 \quad \rightarrow \quad R \sim bN^{\frac{1}{2}} \quad (\Theta \text{ solvent}) \qquad (4.7)$$

It shows that as the solvent quality becomes poorer, the chain with an expanded random-coil conformation will shrink, its limit is a uniform globule with solvent molecules trapped inside, i.e., $R \sim bN^{1/3}$. The scaling exponent decreases from 3/5 to 1/3. Such a "coil-to-globule" transition of the chain conformation is one of the most famous propositions of macromolecular physics since 1960s, which will be discussed in Chapter 8.

Readers might note that in eqs (4.2), (4.7) and the Gaussian chain discussion, the linear free-joint ideal chain, the linear real chain at the Θ state and the linear Gaussian chain have the same scaling dependence of R on N ($R \propto N^{1/2}$). Are these three identical? The answer is negative. In the derivations of these scaling relationships, their respective assumptions about N are different. The former assumed an infinitely long chain, $N \rightarrow \infty$, otherwise, there is no $\langle \vec{\mathbf{b}}_i \rangle = 0$; the Gaussian chain has a large, but limited N, i.e., $1 \ll N \ll \infty$. The length of a real chain is limited, the degree of polymerization of synthetic macromolecules (n) is normally less than 10^5. When the real chain is too long ($N \gg 1$), it is necessary to consider the possible trace residual two-body excluded volume effect in the Θ state, i.e.

$$\frac{\Delta G}{k_B T} \sim \delta \frac{b^3 N^2}{R^3} + \frac{b^6 N^3}{R^6} + \frac{R^2}{R_0^2} \qquad (4.8)$$

where $\delta \ll 1$, introducing a small fluctuation nearby the Θ state; the constant coefficients of the first, second and third terms on the right side, respectively, related to the two- and three-body excluded volume effects (ε_2 and ε_3) and the entropy elasticity (γ) are also omitted. In the above equation, the two-body excluded volume effect can be ignored when the macromolecular length is limited, so that eq. (4.8) is reduced to eq. (4.7). Conversely, the minimum of free energy change can be obtained from the above equation as follows,

$$\left(\frac{\partial \Delta G}{\partial R}\right)_{T,p} \sim -\delta \frac{3b^3 N^2}{R^4} - \frac{6b^6 N^3}{R^7} + \frac{2R}{b^2 N} = 0 \quad \rightarrow \quad R \sim bN^{\frac{1}{2}}(1 + \delta N^{\frac{1}{2}})^{\frac{1}{8}} \qquad (4.9)$$

In the derivation, $R \cong R_0 = bN^{1/2}$ was used. When $N \gg 1$, $\delta N^{1/2} \gg 1$ in eq. (4.9), the scaling exponent increases slightly from 1/2 to 9/16, i.e., $R \sim bN^{9/16}$, showing that the trace residual two-body excluded volume effect makes the chain expends slightly, which is easily understandable and explainable in physics.

Radius of Gyration (R_g) : Under different assumptions, different scaling relationships between the average end-to-end distance of linear chains and the segment number can be theoretically deduced. However, at present, it is still impossible to directly measure the end-to-end distance of every macromolecular chain in a solution or the variation of the end-to-end distance of one macromolecular chain with time, and then statistically average them. Therefore, in order to verify theoretically derived different scaling relationships, the average end-to-end distance of the chain must be related to an experimentally measurable physical quantity.

The average **radius of gyration** (R_g) of a macromolecular chain is just such a measurable physical quantity, related to the spatial distribution of its mass (m). Just as its name suggested, mR_g^2 is the rotational inertia of an object around its center of mass (gravity), defined as follows.

$$R_g = \sqrt{R_g^2} = \sqrt{\frac{\sum_{i=1}^N m_i (\vec{\mathbf{r}}_i - \vec{\mathbf{r}}_G)^2}{\sum_{i=1}^N m_i}} \qquad (4.10)$$

Where m_i, $\vec{\mathbf{r}}_i$ and $\vec{\mathbf{r}}_G$ are the mass, the spatial vectors of the ith unit and the center of mass, respectively, $\vec{\mathbf{r}}_G = \sum_{i=1}^N \vec{\mathbf{r}}_i / N$. If each unit has the same mass, i.e., m_i is a constant, eq. (4.10) is further simplified as

$$R_g^2 = [\sum_{i=1}^N (\vec{\mathbf{r}}_i - \vec{\mathbf{r}}_G)^2]/N \qquad (4.11)$$

For an object with a simple shape, it is not difficult to sum or integrate directly to obtain its radius of gyration by using the above equation. For a hollow sphere with a thin shell, $|\vec{\mathbf{r}}_i| \equiv R$ and $\vec{\mathbf{r}}_G = 0$, its radius of gyration is its radius, $R_g = R$; the radius of gyration of a uniform sphere is smaller than its radius, $R_g = (3/5)^{1/2} R \cong 0.774R$; and the dependence of the radius of gyration of a thin rigid rod on its length (L) is $R_g = (1/12)^{1/2} L$. For an object with a complicated shape, it is necessary to do some specific calculation.

For a macromolecular chain with a complicated configuration, it is not easy to calculate its radius of gyration even its segment spatial distribution is given. Moreover, under the agitation of thermal energy, each segment in the chain is constantly folding and rotating, leading to a constantly varying chain conformation and a randomly moving center of mass. In other words, different chains with the same length have different conformations at each moment. Therefore, similar to the calculation of the average end−to−end distance of a chain, the radii of gyration of all possible conformations of a macromolecular chain are calculated first, and then statistically averaged, resulting in the **mean square radius of gyration** ($\langle R_g^2 \rangle$). Its square root is the root mean square radius of gyration ($\langle R_g^2 \rangle^{1/2}$), denoted as R_g and written as the average radius of gyration, which measures the average spatial distribution of every segments. According to the definition,

$$R_g = \sqrt{\langle R_g^2 \rangle} \text{ and } \langle R_g^2 \rangle = \langle [\sum_{i=1}^{N} (\vec{\mathbf{r}}_i - \vec{\mathbf{r}}_G)^2] \rangle / N = [\sum_{i=1}^{N} \langle (\vec{\mathbf{r}}_i - \vec{\mathbf{r}}_G)^2 \rangle] / N \quad (4.12)$$

where $\vec{\mathbf{r}}_i$ is the spatial vector of the ith segment, every segment with the same mass (m_i) was used, and a lemma of the equivalent of "statistical average after the summation" and "summation after statistic average" was applied. However, for an object with a constantly moving center of mass, it is rather difficult to use the above equation to calculate its average radius of gyration directly. Using another lemma, the calculation of $\sum_{i=1}^{N} (\vec{\mathbf{r}}_i - \vec{\mathbf{r}}_G)^2$ in eq. (4.11) can be converted to a sum of the square distance between any two points, i. e., $\sum_{i=1}^{N} \sum_{j=1}^{N} (\vec{\mathbf{r}}_i - \vec{\mathbf{r}}_j)^2$. The detail is as follows.

For a macromolecular chain with any configuration (or a moving complex object), it can be imagined that any two segments (i and j) in the chain are the starting and ending points of a Gaussian chain link, respectively, so that ($\vec{\mathbf{r}}_i - \vec{\mathbf{r}}_j$) is the direct distance between them. The direct distance between all possible two points on the chain is squared first, i. e., $(\vec{\mathbf{r}}_i - \vec{\mathbf{r}}_j)^2$, then do a time (conformation) statistical average, i. e., $\langle (\vec{\mathbf{r}}_i - \vec{\mathbf{r}}_j)^2 \rangle$, and finally sum all of them together; namely, $\sum_{i=1}^{N} \sum_{j=1}^{N} \langle (\vec{\mathbf{r}}_i - \vec{\mathbf{r}}_j)^2 \rangle$. The simultaneous addition and subtraction of the spatial vector of the center of mass between $\vec{\mathbf{r}}_i$ and $\vec{\mathbf{r}}_j$ has no effect on their difference, i. e.,

$$\sum_{i=1}^{N} \sum_{j=1}^{N} (\vec{\mathbf{r}}_i - \vec{\mathbf{r}}_j)^2 = \sum_{i=1}^{N} \sum_{j=1}^{N} [(\vec{\mathbf{r}}_i - \vec{\mathbf{r}}_G) - (\vec{\mathbf{r}}_j - \vec{\mathbf{r}}_G)]^2$$

$$= \sum_{i=1}^{N} \sum_{j=1}^{N} (\vec{\mathbf{r}}_i - \vec{\mathbf{r}}_G)^2 - 2 \sum_{i=1}^{N} \sum_{j=1}^{N} (\vec{\mathbf{r}}_i - \vec{\mathbf{r}}_G)(\vec{\mathbf{r}}_j - \vec{\mathbf{r}}_G) + \sum_{i=1}^{N} \sum_{j=1}^{N} (\vec{\mathbf{r}}_j - \vec{\mathbf{r}}_G)^2$$

where the first and third sums on the right has N terms, respectively, unrelated to j and I; and the two sums are equal. In the middle term, $(\vec{\mathbf{r}}_i - \vec{\mathbf{r}}_G)$ and $(\vec{\mathbf{r}}_j - \vec{\mathbf{r}}_G)$ in the summation are not related to i and j, respectively. Therefore, the above equation can be further written as

$$\sum_{i=1}^{N} \sum_{j=1}^{N} (\vec{\mathbf{r}}_i - \vec{\mathbf{r}}_j)^2 = 2N \sum_{i=1}^{N} (\vec{\mathbf{r}}_i - \vec{\mathbf{r}}_G)^2 - 2 \left(\sum_{i=1}^{N} \vec{\mathbf{r}}_i - N \vec{\mathbf{r}}_G \right) \left(\sum_{j=1}^{N} \vec{\mathbf{r}}_j - N \vec{\mathbf{r}}_G \right)$$

According to its definition, $\vec{\mathbf{r}}_G = \sum_{i=1}^{N} \vec{\mathbf{r}}_i / N$. The two sums ($\sum_{i=1}^{N} \vec{\mathbf{r}}_i$ and $\sum_{j=1}^{N} \vec{\mathbf{r}}_j$) in the second term are $N \vec{\mathbf{r}}_G$, so that the second term is zero, i.e.,

$$\left[\sum_{i=1}^{N}(\vec{\mathbf{r}}_i - \vec{\mathbf{r}}_G)^2\right] = \left[\sum_{i=1}^{N}\sum_{j=1}^{N}(\vec{\mathbf{r}}_i - \vec{\mathbf{r}}_j)^2\right]/2N \tag{4.13}$$

Substituting it into eq, (4.11), a rather useful lemma (relationship) is obtained as follows.

$$R_g^2 = \left[\sum_{i=1}^{N}\sum_{j=1}^{N}\langle(\vec{\mathbf{r}}_i - \vec{\mathbf{r}}_j)^2\rangle\right]/2N^2 \tag{4.14}$$

For a **linear** chain, $x = |i - j|$ in eq. (4.3), so that the sum of the mean square distance in the above equation can be specifically calculated as follows.

$$\sum_{i=1}^{N}\sum_{j=1}^{N}\langle(\vec{\mathbf{r}}_i - \vec{\mathbf{r}}_j)^2\rangle = \sum_{i=1}^{N}\sum_{j=1}^{N}b^2|i-j| = 2b^2\sum_{i=1}^{N}\sum_{j=1}^{i}(i-j)$$

$$= 2b^2\sum_{i=1}^{N}\frac{i(i+1)}{2} = b^2\left(\sum_{i=1}^{N}i^2 + \sum_{i=1}^{N}i\right)$$

$$= b^2\left[\frac{N(N+1)(2N+1)}{6} + \frac{N(N+1)}{2}\right]$$

$$\sum_{i=1}^{N}\sum_{j=1}^{N}\langle(\vec{\mathbf{r}}_i - \vec{\mathbf{r}}_j)^2\rangle = b^2\left[\frac{N(N+1)(N+2)}{3}\right] \cong b^2\frac{N^3}{3} \tag{4.15}$$

where the coefficient 2 in the first line was derived from $|i-j|$ in which i can be larger or smaller than j, and the two are symmetrical. With a limitation of $j \leq i$, the absolute symbol was removed. Only half the original sum was considered, so that 2 is multiplied. In the second line, $\sum_{j=1}^{i}(i-j) = \sum_{j=1}^{i}i - \sum_{j=1}^{i}j$. The former is i^2; the latter is the sum with j varying from 1 to i, i. e., the sum with j varying from 0 to $i-1$, resulting in $i(i-1)/2$. A combination of them leads to $i(i+1)/2$. $N \gg 1$ was used in the last approximation. Substituting it into eq. (4.14) leads to

$$\langle R_{g,0}^2\rangle = \frac{b^2 N}{6} \quad \text{or} \quad \langle R_{g,0}^2\rangle = \frac{\langle R_0^2\rangle}{6} \tag{4.16}$$

It can be seen that for a linear free-joint ideal chain, its mean square radius of gyration is one sixth of its mean square end-to-end distance. Later, how to use laser light scattering to measure the average radius of gyration of macromolecular chains will be discussed in details. Substituting the above equation, respectively, into eqs. (4.5) and (4.7) can result in two measurable universal scaling laws between the average radius of gyration (R_g) and molar mass (M) for linear flexible chains in solutions, i.e.,

$$R_g = k_{R_g} M^{\alpha_{R_g}}\left(\alpha_{R_g} = \frac{3}{5}, \text{good solvent}\right) \text{ and } \left(\alpha_{R_g} = \frac{1}{2}, \Theta \text{ solvent}\right) \tag{4.17}$$

It is necessary to emphasize that k_{R_g} contains the chain length (b), and varies with the chemical compositions of macromolecules and solvent, while the scaling exponent (α_{R_g}) is universal, independent of the chemical composition of solution. Obviously, when N segments are linked into a rigid straight chain, its average radius of gyration is proportional to the number of segments, i.e., $R_g \sim N$ or written as $R_g = k_{R_g} M$, where $\alpha_{R_g} = 1$.

Experimentally, one set (five or more) of narrowly distributed macromolecular samples with different average molar masses (more than two orders of magnitude) are first prepared, then their respective R_g and M are measured by laser light scattering. k_{R_g} and α_{R_g} can be obtained, respectively, from the intercept and slope of a double logarithmic plot of "R_g versus M". In the past fifty years, for linear flexible macromolecules, these two scaling relationships have been confirmed by a large amount of experimental data.

In good solvents, the experimental value of α_{R_g} is ~ 0. 59, slightly smaller than the value obtained based on the Gaussian chain model. Later, using a further modification of the renormalization group theory, α_{R_g} decreases from 3/5 to 0. 59, more consistent with the experimental results. Such a modification only slightly decreases the ratio of $R_{g,0}$ to R_0 in eq. (4. 16). The details of the related theoretical derivation will not be discussed. Interested Readers can further refer to some related original literature.

For a given composition, besides the simplest linear configuration, a ring configuration can be formed by linking two ends of a linear chain, requiring a special synthesis. In a polymerization reaction, various configurations can be produced due to some spontaneous transferring of reactive groups or some added branching comonomers. If each branching subchain is no long to branch 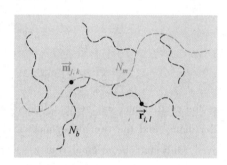 further, a "comb" configuration is formed, as shown on the right, where the main chain and the branching subchains are often composed of the same kind of monomer, but they can also be made of two or more kinds of monomers by copolymerization. Different colors in the figure do not represent different monomers, only for clearly marking out the main and the branching subchains.

For long grafted branching chains, it is sometimes difficult to distinguish the main chain and the branching subchains. For the ring, branched and other complicated configurations, macromolecular chains have no average end-to-end distance, but all chains have an average radius of gyration. In the following, the scaling relationships between the average radius of gyration of macromolecular chains with different branched configurations and the degrees of polymerization of the entire parent branched chain and the subchains between the adjacent branching points will be discussed.

Even for a given chemical composition, the polymerization can also produce comb chains in which the main and the grafted branching subchains have different lengths. Therefore, for the chains with a given composition and comb configuration, their structures can also vary dramatically. For the convenience of discussion, it is assumed that each grafted branching subchain has the same number of segments (N_b), each segment has a length of b; and the main chain is evenly grafted with n_b branching subchains; the chain link between two adjacent grafting points has N_m segments. Therefore, the total number of segments on every comb chain is $N = (N_b + N_m) n_b$. The spatial vectors of the lth segment on the ith branching subchain and the kth segment on the jth **chain link on the main chain** are written as $\vec{\mathbf{r}}_{i,l}$ and $\vec{\mathbf{m}}_{j,k}$, respectively, as shown above.

As discussed before, even for a comb chain with a given composition and branching structure, the thermal energy can also agitate every segment to move randomly, forming countless conformations. For a chain with no ring structure, there exists and only exists one linear path to connect any two segments on the chain. It has been known from eqs. (4.12) and (4.14) that the average radius of gyration is related to "the sum of mean square (direct) distance of any two segments on a chain". The mean square distance of any two segments on a Gaussian chain has been given in eq. (4.3), i.e., $\langle(\vec{\mathbf{r}}_i - \vec{\mathbf{r}}_j)^2\rangle = b^2 x$.

For a comb chain with a given composition and structure, the conformational average of the square direct distance between any two points on the chain with a given set of i, j, l, and k (simply called the mean square distance, written as $\langle(\vec{\mathbf{r}}_{i,l} - \vec{\mathbf{m}}_{j,k})^2\rangle$) is $b^2 x$. The calculation of the sum of mean square distance of all possible two points on a chain includes two steps. First, find how x is specifically related to i, j, l, and k; and second, sum all possible $\langle(\vec{\mathbf{r}}_{i,l} - \vec{\mathbf{m}}_{j,k})^2\rangle$ by varying l and k from 1 to N_b as well as i and j from 1 to n_b, respectively, which is mathematically expressed as $\sum_{i,j=1}^{n_b} \sum_{l,k=1}^{N_b} \langle(\vec{\mathbf{r}}_{i,l} - \vec{\mathbf{m}}_{j,k})^2\rangle$, called as "the sum of mean square distance" hereafter.

In order to calculate the complicated "sum of mean square distance" of a comb chain, the number of segments (x) on the linear path between two given segments has to be found in each of four different cases as follows. I) both of them are on the same grafted branching subchain; II) both of them are on different grafted branching subchains; III) one is on the main chain, and another is on a grafted branching subchain; IV) both of them are on the main chain.

The sum of mean square distance in each case can be calculated using eq. (4.15) first, and then, added together. The whole calculation is cumbersome; readers may directly use the deducted results by jumping the part separated by two dash lines. However, after learning the following mathematic deduction and calculating principle, readers will be able to calculate the average radius of gyration of chains with other configurations. Therefore, it is still recommended that readers digest the following content with concentration, master the derivation logic and understand physical picture behind it, not only know the mathematic derivation in each step.

Case I : **two segments are on the same branching subchain** ($i = j$; $x = |l - k|$). First, the sum of mean square distance separated by $|l - k|$ segments on the same branching chain is calculated, just replacing N in eq. (4.15) with N_b; the sum of mean square distance of n_b branching subchains is just n_b times. Therefore,

$$(\text{I}) = \sum_{i=1}^{n_b} \sum_{l,k=1}^{N_b} \langle(\vec{\mathbf{r}}_{i,l} - \vec{\mathbf{r}}_{i,k})^2\rangle = \sum_{i=1}^{n_b} \frac{b^2 N_b(N_b + 1)(N_b + 2)}{3} \cong b^2 \frac{N_b^3}{3} n_b$$

Case II : **two segments are on different branching subchain** ($i \neq j$). The two segments are separated by $|i-j|$ main chain links. Each main chain link has N_m segments, so that they are separated by $N_m|i-j|$ segments, plus l and k segments on two different branching chains, respectively. Namely, $x = |i-j|N_m + l + k$. Note: the segments on the main chain are unrelated to l and k; and the segments on two branching subchains are unrelated to j and j. Therefore, the mean square distance of two segments separated by $|i-j|N_m$ segments on the main chain is $b^2|i-j|N_m$; and the mean square distance of each branching subchain, $b^2 N_b$. Their sum is

$$\sum_{i\neq j}^{n_b}\sum_{l,k=1}^{N_b}\langle(\vec{\mathbf{r}}_{i,l}-\vec{\mathbf{r}}_{j,k})^2\rangle = \sum_{i\neq j}^{n_b}\sum_{l,k=1}^{N_b}b^2(|i-j|N_m + N_b) = b^2\sum_{i\neq j}^{n_b}[N_m N_b^2|i-j| + N_b^3]$$

where the sums of varying l and k are first calculated. $|i-j|N_m$ in the first term on the right is unrelated to l and k, so that the sum leads to N_b^2 times $|i-j|N_m$, i.e., $N_m N_b^2|i-j|$; while N_b in the second term is a constant, so that the sum also leads to N_b^2 times N_b, resulting in N_b^3. It is also possible to sum $l + k$, that is, to sum l and k separately. Each sum can get $N_b(N_b + 1)/2$. There are N_b terms in total, and $N_b \gg 1$, so that the result is still N_b^3. Then, the sum of varying i and j are calculated. The calculation of $\sum_{i\neq j}^{n_b}|i-j|$ in the first term results in $n_b(n_b + 1)$ $(n_b + 2)/3 \cong n_b^3/3$; since $i \neq j$, the sum of the second term only leads $(n_b^2 - n_b)$ identical cross-terms, and each term is N_b^3. The final result is

$$(\text{II}) = \sum_{i\neq j}^{n_b}\sum_{l,k=1}^{N_b}\langle(\vec{\mathbf{r}}_{i,l}-\vec{\mathbf{r}}_{j,k})^2\rangle \cong b^2\left[\frac{N_m N_b^2 n_b^3}{3} + (n_b^2 - n_b)N_b^3\right]$$

Case III: one segment is on the main chain and another is on a branching subchain: Two segments are separated by $|i-j|$ main chain links, plus l and k segments, respectively, on the main chain and the branching subchains. The number of segments between i and j is $x = N_m|i-j| + l + k$. In comparison with the sum in the second case, the numbers of segments (N_m and N_b) in the main chain link and in the branching subchain are different, i.e.,

$$\sum_{i,j=1}^{n_b}\sum_{l=1}^{N_m}\sum_{l,k=1}^{N_b}\langle(\vec{\mathbf{r}}_{i,l}-\vec{\mathbf{m}}_{j,k})^2\rangle = \sum_{i,j=1}^{n_b}\sum_{l=1}^{N_m}\sum_{k=1}^{N_b}b^2(N_m|i-j| + l + k)$$

Including the two parts, the sums of the main chain and the subchain. The sum of varying l and k are first calculated. The first term $|i-j|N_m$ on the right is unrelated to l and k. The sums leads to N_m and N_b terms, respectively, each term is $|i-j|N_m$. The result is $N_m N_b$ times $|i-j|N_m$, i.e., $|i-j|N_m^2 N_b$. The second and third terms $\sum_{k=1}^{N_b}\sum_{l=1}^{N_m}l$ and $\sum_{l=1}^{N_m}\sum_{k=1}^{N_b}k$ are not related to N_b and N_m, respectively, so that the two sums are $N_b\sum_{l=1}^{N_m}l$ and $N_m\sum_{k=1}^{N_b}k$, respectively. $\sum_{l=1}^{N_m}l = N_m(N_m + 1)/2 \cong N_m^2/2$ and $\sum_{k=1}^{N_b}k = N_b(N_b + 1)/2 \cong N_b^2$. Therefore,

$$\sum_{i,j=1}^{n_b}\sum_{l=1}^{N_m}\sum_{l,k=1}^{N_b}\langle(\vec{\mathbf{r}}_{i,l}-\vec{\mathbf{m}}_{j,k})^2\rangle = b^2\sum_{i,j=1}^{n_b}\left[N_b N_m^2|i-j| + \left(\frac{N_b N_m^2 + N_m N_b^2}{2}\right)\right]$$

The sum of the first term $\sum_{i,j=1}^{n_b}|i-j|$ results in $n_b^3/3$. The second term is unrelated to i and j, so that the sum gets n_b^2 items, each term is $N_m N_b(N_m + N_b)/2$, i.e., $N_m N_b(N_m + N_b) n_b^2/2$. Therefore,

$$(\text{III}) = \sum_{i,j=1}^{n_b} \sum_{l=1}^{N_m} \sum_{l,k=1}^{N_b} \langle (\vec{\mathbf{r}}_{i,l} - \vec{\mathbf{m}}_{j,k})^2 \rangle \cong b^2 \left[\frac{N_b N_m^2 n_b^3}{3} + \frac{N_b N_m (N_b + N_m) n_b^2}{2} \right]$$

Case IV: two segments are on the main chain: relatively simple, the mean square distance between two points separated by $|i - j|$ main chain links is calculated first, and then the sum. Note: the number of segments on the main chain is $N_m n_b$. Eq. (4.15) leads to

$$(\text{IV}) = \sum_{i,j=1}^{n_b} \sum_{l,k=1}^{N_m} \langle (\vec{\mathbf{m}}_{i,l} - \vec{\mathbf{m}}_{j,k})^2 \rangle = \sum_{i,j=1}^{N_m n_b} b^2 |i - j| \cong b^2 \frac{(N_m n_b)^3}{3}$$

The sum of the above four cases covers the mean square distance of any two segments on a comb chain, resulting in

$$\sum_{i,j=1}^{n_b} \sum_{l,k=1}^{N_b} \langle (\vec{\mathbf{r}}_{i,l} - \vec{\mathbf{m}}_{j,k})^2 \rangle = (\text{I}) + (\text{II}) + (\text{III}) + (\text{IV})$$

$$= \frac{b^2 n_b^2}{3} \left[\frac{N_b^3}{n_b} + n_b N_m N_b^2 + 3\left(1 - \frac{1}{n_b}\right) N_b^3 + n_b N_b N_m^2 + \frac{3 N_b N_m (N_b + N_m)}{2} + n_b N_m^3 \right]$$

$$= \frac{b^2 n_b^2}{3} \left[n_b N_m (N_b + N_m)^2 - n_b N_b N_m^2 + \left(3 - \frac{2}{n_b}\right) N_b^3 + \frac{3 N_b N_m (N_b + N_m)}{2} \right]$$

where $n_b N_b N_m^2$ was simultaneously added and subtracted, in order to get $(N_b + N_m)^2$ in the first term. According to the definition of mean square radius of gyration in eq. (4.12), using eq. (4.14), substituting the total segment number $N = (N_b + N_m) n_b$ in the above equation, an expression of the mean square radius of gyration of the comb chain in the ideal state is obtained as follows.

$$\langle R_{g,0}^2 \rangle_{\text{comb}} = \frac{b^2}{6} \left[n_b N_m - \frac{n_b N_b N_m^2}{(N_b + N_m)^2} + \frac{3 n_b - 2}{n_b (N_b + N_m)^2} N_b^3 + \frac{3 N_b N_m}{2 (N_b + N_m)} \right] \tag{4.18}$$

It can be used to examine "comb chains" with special structures. A comb chain with very short grafted branching subchains is one extreme structure, i.e., the main chain links between two adjacent branching points are much longer than each grafted branching subchain, $N_m \gg N_b$. The first term in the square parentheses in eq. (4.18) is far larger than other terms, and $N = (N_b + N_m) n_b \cong N_m n_b$,

$$\langle R_{g,0}^2 \rangle \cong \frac{b^2 n_b N_m}{6} \cong \frac{b^2 N}{6} \propto b^2 N \tag{4.19}$$

where $N_m n_b$ is the number of segments on the main chain, revealing that a comb chain with short branching chains and the main chain have a similar average radius of gyrations. In other words, the contribution of short grafted branching subchains to the radius of gyration of the entire comb chain can be ignored. It is physically reasonable according to the definition of the radius of gyration related to the spatial distribution of mass of an object. If each branching subchain and each main chain link have a similar length ($N_m \cong N_b$), and $n_b \gg 1$, eq. (4.18) becomes

$$\langle R_{g,0}^2 \rangle \cong \frac{b^2 N_b n_b}{8}\left(1 + \frac{2}{n_b} - \frac{2}{3n_b^2}\right) \approx \frac{b^2 N_b n_b}{8} = \frac{b^2 N}{16} \propto b^2 N \qquad (4.20)$$

where $N = (N_b + N_m) n_b \approx 2N_b n_b$ was used. A comb chain with long branching subchains is another extreme structure, i.e., $N_b \gg N_m$. The second and fourth terms in the square parentheses in eq. (4.18) can be ignored, so that eq. (4.18) is approximated as

$$\langle R_{g,0}^2 \rangle \cong \frac{b^2}{6}\left[n_b N_m + N_b\left(3 - \frac{2}{n_b}\right)\right] \qquad (4.21)$$

It can be seen that both the main chain and grafted branching subchains contribute to the mean square radius of gyration of a comb chain with long grafted branching subchains. The contribution depends on the number of grafted branching subchains (n_b). When the main chain and one grafted branching subchain have a similar length, $n_b N_m \sim N_b$, the above equation can be further written as

$$\langle R_{g,0}^2 \rangle \cong \frac{b^2}{6}\left[n_b N_m + N_b\left(3 - \frac{2}{n_b}\right)\right] \cong \frac{1}{3}\left(2 - \frac{1}{n_b}\right) b^2 N_b \propto b^2 N \qquad (4.22)$$

The most extreme example of a comb chain is that all of long grafted branching subchains are connected to a center point, forming a star configuration (star chains), $N_m \cong 0$, i.e.,

$$\langle R_{g,0}^2 \rangle_{\text{star}} \cong \frac{b^2}{6}N_b\left(3 - \frac{2}{n_b}\right) \propto b^2 N \qquad (4.23)$$

where each grafted branching subchain is also called as one "arm". Obviously, when the number of arms (n_b) on a star chain is sufficiently larger, the above equation becomes $\langle R_{g,0}^2 \rangle_{\text{star}} \cong b^2 N_b/2$, three times of the mean square radius of gyration of one arm.

On the other hand, when $n_b = 2$ or $N_b = 0$, a star or comb chain becomes a linear chain, whose total number of segments is twice that in one arm, i.e., $N = 2N_b$, or equals that in the main chain, i.e., $N = N_m n_b$, so that eq. (4.23) is reduced to eq. (4.16) for linear chains, i.e., $\langle R_{g,0}^2 \rangle \cong b^2 N/6$.

A connection of the two ends of a linear chain can form a macromolecular chain with a ring configuration. For any two segments (i and j) on such a chain, x in eq. (4.3) equals $|i - j|$ or $N - |i - j|$, depending on from which side one starts to count. However, the sum of mean square distance is unique for a given chain, so that their results must be the same, equivalent to $x = |i - j|/2$, independent of how to count. Substituting it into eq. (4.15), the obtained mean square distance is $b^2 N^3/6$. Further using eq. (4.14), the following can be obtained,

$$\langle R_{g,0}^2 \rangle_{\text{ring}} \cong \frac{b^2 N}{12} \propto b^2 N \qquad (4.24)$$

The mean square radius of gyration of a ring chain is only half that of a linear chain. This result agrees with the following physical picture with no mathematic derivation; namely, the average end-to-end distance of a linear chain is equivalent to the end-to-end distance of a rigid chain with $N^{1/2}$ segments, all $N^{1/2}$ segments walk towards one direction. If $N^{1/2}$ segments randomly walk, the probability of moving towards any direction is equal. Statistically, it means that half of

$N^{1/2}$ segments move towards an average direction perpendicular to the original one, i.e., a straight chain is bent into an equilateral right angle, so that its square end-to-end distance is

$$(bN^{\frac{1}{2}}/2)^2 + (bN^{\frac{1}{2}}/2)^2 = \frac{b^2 N}{2}$$

It shows that the he physical intuition is as important as mathematic derivation. From eq. (4.19) to (4.24), $\langle R_{g,0}^2 \rangle$ is always proportional to $b^2 N$ for different extreme structures of comb chains, leading to a general conclusion: $\langle R_{g,0}^2 \rangle \propto b^2 N$ for comb chains with any structures. Such a conclusion can also be further generalized to chains with any configuration. Different chain structures only lead to different proportional constants. This generalization is based on the following principle of statistic physics. The mean square ($\langle x^2 \rangle$) of a random variable (x) after N steps of random walks is proportional to a square of the step length (b) and the step number (N), i.e., $\langle x^2 \rangle = b^2 N$ or $\sqrt{\langle x^2 \rangle} = bN^{1/2}$. Here, x is the radius of gyration; the step length and number are the segment length and number, respectively. For a given chain structure, each chain conformation is the result of N steps (N segments) of random walks, so that there is the above general corollary.

In summary, the "branched configuration" can be considered as a basic "root" configuration. The branched chains can be divided into two categories, depending on whether there are many rings in the chain. I) without many ring: In this kind of branched chains, if each branching subchain of a branching subchain no long branch further, it becomes a comb configuration. If all branching subchains on a comb chain are linked to a center point, it becomes a star configuration. If a star chain has only two branching subchains (arms), it becomes a linear configuration If two ends of a linear chain are connected together, it becomes a ring configuration, only one big ring. II) with many rings: If a branching subchain contains many rings, it becomes a three-dimensional spatial network, forming a gel after sufficiently swollen by solvent molecules. Gels are viewed as a kind of special matters between liquids and solids. On the one hand, a gel can keep its shape when there is no external exerting force (besides the gravity), similar to solids. On the other hand, exerting a weak force can usually make a gel broken and flowing, also similar to liquids. Therefore, gels are called as "wet" matters.

As discussed before, in the linear polymerization of bifunctional monomers, an addition of a small amount of branching comonomers with f-reactive groups can lead to continuous branching on a growing chain to form a branching configuration, where $f \geqslant 3$. The branching frequency increases with the comonomer concentration, forming branching macromolecules with lots of short chain links. Their structures are much more complicated than comb chains, often called as hyperbranched chains. In an actual polymerization, not every reactive group on a branching comonomer participates in the copolymerization. Even if participating, they do not necessarily react at the same time. Therefore, all $(2 + f)$ branching polymerization must lead to chain links between two adjacent branching points with different lengths. Besides a broad distribution in the chain links, the number and position of branching points on every chain are also different, so that for a given composition, the resultant branched structures are also completely different, ever changing! Even for two branched chains with identical numbers of bifunctional monomers and branching points, their branched structures can also be different.

To simplify discussion, a branching comonomer with tri-reactive groups is used as an example. The assumption is that every reactive group on the branching comonomer simultaneously reacts and each forms a linear chain link with the same length. If the end of a chain link reacts with another branching comonomer, this chain link continues to branch, forming a new branching point and $(f-1)$ new chain links; on the contrary, this chain link is attached to a branching point, becoming one dangling chain link (end), as

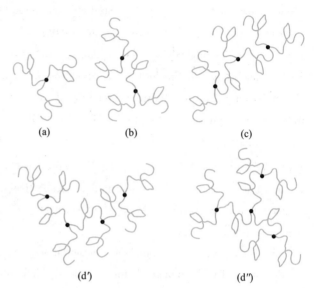

(a)   (b)   (c)

(d')   (d'')

shown on the right. The same branching polymerization can form different structures. The entire branched chain and short chain link between two neighboring branching points (including the dangling chain ends) are called as the parent and sub-chains hereafter, respectively. Each branching subchain contains N_b segments.

(a) with only one branching point: a 3-arm star chain, $\langle R_g^2 \rangle \cong 7b^2 N_b/18$ based on eq. (4.23). (b) or (c), respectively, with two and three branching points: a comb chain, respectively, with two and three dangling subchains, in which $N_b = N_m$, $n_b = 2$ or 3; $\langle R_g^2 \rangle \cong 11b^2 N_b/24$ or $\langle R_g^2 \rangle \cong 43b^2 N_b/72$ based on eq. (4.20). (d') and (d'') with four branching points: two branched chains with different structures, becoming complicated. One is a comb chain with four dangling subchains, $n_b = 4$, $\langle R_g^2 \rangle \cong 35b^2 N_b/48$ based on eq. (4.20); another is not a comb chain, so that eq. (4.18) is no long applicable. In a similar derivation, with four or more branching points, the chains with the same composition and configuration have many different branched structures.

One extreme structure is the comb chains that have been discussed in details. The dendritic branched chain is another extreme structure, which has one branching centrum from which every subchain further branches with the same probability. It is necessary to emphasize that a branched chain with many dangling subchains has no end-to-end distance, but still has a sum of the mean square distance and the mean square radius of gyration. In comparison with linear chains, when calculating the mean square radius of gyration of branched chains, one has to first statistically average all possible conformations of a branched chain with a given structure to obtain its sum of mean square distance, and then, statistically average all possible chain structures, rather complicated.

As earlier as in 1949, by statistically averaging all possible chain conformations and branched structures of branched chains with a given composition, Bruno H. Zimm and Walter H. Stockmayer [Journal of Chemical Physics, 1949, 17: 1301] already derived a double scaling laws, $R_{g,0} = \langle R_{g,0}^2 \rangle^{1/2} \propto b N_b^{1/4} N^{1/4}$, where $R_{g,0}$ is the radius of gyration of branched chains in the unperturbed state, and N and N_b are the numbers of segments of the parent and sub-chains, respectively. The core of the derivation is to apply the Kramers principle, imaginably cutting a branched chain with N segments and no ring at the ith segment into two parts. They contain N_i and $(N - N_i)$ segments, respectively. The statistical average leads to its mean square distance: $\langle R^2 \rangle \cong b^2 \langle N_i(N - N_i) \rangle / N$, where $N_i(N - N_i)/N$ reflects the statistical contribution of the ith segment to $\langle R^2 \rangle$; while the average $\langle N_i(N - N_i) \rangle$ covers all possible ways of cutting the chain.

About twenty years later, P.-G. de Gennes [Biopolymers, 1968, 6: 715] used a relatively concise method, i.e., a Laplace inversion of the segment length in the reciprocal lattice space of Gaussian distribution, to obtain the same scaling relationship between $R_{g,0}$ and N, too, i.e., $R_{g,0} \propto N^{1/4}$. Even today, more than seventy years later, besides few theoreticians, it is still difficult for most of polymer researchers to comprehend difficult physical principles and complicated mathematical derivation used in the original publication of Zimm and Stockmayer, so that their detailed derivation is omitted in this textbook. Readers only need to remember the above scaling relationships. The following shortcut deduction without involving difficult mathematics only uses a simplified picture together with the statistical principle and the relevant scaling theory.

For the convenience of discussion, each subchain with N_b segments is treated as a small ball or a new "segment", its size or length is $\xi = b N_b^{1/2}$. In this way, each ball has only two choices: further linking $(f - 1)$ small balls to form one new branching point, or becoming one dangling end. Therefore, a branched chain with N segments is simplified as a structure with N/N_b interconnected small balls or new "segments", as shown on the right, where $f = 3$. Note that this is only one of many possible branched structures. It is imaginable that those 65 small balls are able to form various branched structures.

In such a description, different structures of a chain with a given composition and branched configuration has been treated as different traces of N/N_b steps random walks. Besides those dangling ends, each small ball in the middle is further connected to $(f - 1)$ balls, forming one branching point. If the number of branching points is n_b, there are $[(f - 2) n_b + 2]$ dangling small balls and

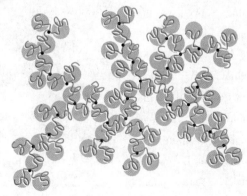

$(n_b - 1)$ small balls between two neighboring branching points. The sum of them is the total number of small balls, i.e.,

$$\frac{N}{N_b} = [(f - 2) n_b + 2] + (n_b - 1) = (f - 1) n_b + 1 \quad \rightarrow \quad n_b = [(N/N_b) - 1]/(f - 1)$$

Therefore, there are 32 branching points in the figure above. The advantage of such a description is no need to consider the detail of every subchain. In other words, it is not necessary to consider whether a segment is the ith or the jth one on a subchain in calculating its sum of mean square distance because each subchain itself is a "segment".

For a given number of small balls (composition), small balls can connect to each other to form different configurations. The same configuration can also have different structures. One extreme structure is that all small balls are linearly connected together to form a linear chain. Its unpertubated mean square radius of gyration is $\langle R_{g,0}^2 \rangle = \xi^2 (N/N_b)/6 = b^2 N_b (N/N_b)/6 = (1/6) b^2 N$. If N/N_b small balls are branchingly connected to form a com chain, where $N_b = N_m$ in eq. (4.18), i.e., $\langle R_{g,0}^2 \rangle = \xi^2 (N/N_b)/16 = b^2 N_b (N/N_b)/16 = (1/16) b^2 N$. If N/N_b small balls are connected to form a star chain with k arms, the number of small balls on each arm is $(N/N_b)/k$. According to eq (4.23), $\langle R_{g,0}^2 \rangle = [(3k-2)/6k^2] b^2 N$ or $\langle R_{g,0}^2 \rangle = (1/2k) b^2 N$ for $k \gg 1$. Note: different structures all have the same $R_{g,0} \propto bN^{1/2}$, but different proportional constants.

For branched chains with an identical number of small balls (N, the same composition) but different structures, the statistical average of all possible conformations of any one of given branched structure leads to its sum of mean square distance. It is proportional to the square size of small balls (ξ^2) and the number of balls (N/N_b), which corresponds to a conformation average radius of gyration $\langle R_g \rangle_c$, as shown on the right. Different branch structures have different proportional constants. Each branched structure resembles a trace of $(N/N_b)^{1/2}$ steps of random walk, so that each $\langle R_g \rangle_c$ corresponds to one conformation of an imaginable linear chain composed of $(N/N_b)^{1/2}$ small balls with a size of ξ.

Further, the statistical average of **all possible structures** of branched chains made of N/N_b small balls becomes equivalent to another statistical average of **all possible conformations** of one

above imaginable linear chain. Note: here the number of small balls is $(N/N_b)^{1/2}$, not the original N/N_b. It has been known that the conformation average radius of gyration of a linear chain is proportional to its segment

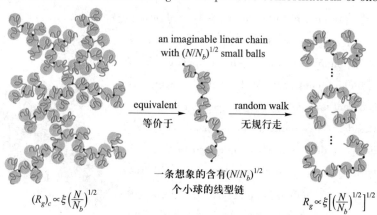

an imaginable linear chain with $(N/N_b)^{1/2}$ small balls

equivalent
等价于

random walk
无规行走

一条想象的含有$(N/N_b)^{1/2}$个小球的线型链

$(R_g)_c \propto \xi \left(\dfrac{N}{N_b}\right)^{1/2}$

$R_g \propto \xi \left[\left(\dfrac{N}{N_b}\right)^{1/2}\right]^{1/2}$

length and the square root of its segment number, as shown above. Therefore, if the excluded volume effect is ignored, for branched chains with a given composition, the double conformational and structural statistical averages result in

$$R_{g,0} = \sqrt{\langle R_{g,0}^2 \rangle} \propto \xi \left[\left(\frac{N}{N_b} \right)^{1/2} \right]^{1/2} \propto b N_b^{1/2} \left[\left(\frac{N}{N_b} \right)^{1/2} \right]^{1/2} \propto b N_b^{1/4} N^{1/4} \qquad (4.25)$$

It is identical to the scaling relationships obtained by Zimm and Stockmayer as well as by de Gennes via some complicated mathematical derivations, different approaches but the same result.

The same double scaling relationship can also obtained by applying the above-mentioned general statistical principle. Namely, the root mean square of the fluctuation amplitude of any one of the random physical variables is proportional to the average amplitude of each fluctuation and the square root of the total number of fluctuations (times of sampling).

Therefore, for a given composition (the number of small balls), there exist various branched structures. For a chain with a given branched structure, its conformation still constantly and randomly changes. In order to obtain the average radius of gyration of branched chains with a given composition, one has to do the statistical average twice. One time is to average all-possible conformations of a given branched structure, equivalent to the average of conformations of an imaginable linear chain composed of only $(N/N_b)^{1/2}$ small balls. Another time is to average all-possible branched structures. Each average leads to a square root of the number of small balls, so that if the excluded volume effect is ignored, the average radius of gyration $(R_{g,0})$ of branched chains with a given composition should be proportional to the size of small balls (ξ) and twice square root of the number of small balls (N/N_b), i.e., $R_{g,0} \propto b \, N_b^{1/2} N^{1/2}$, completely identical.

Obviously, for a given composition, branched chains on average are $(N/N_b)^{1/2}$ times smaller than linear chains and their occupied average volume decreases $(N/N_b)^{3/2}$ times, so that the average distance between different chain segments becomes shorter. Therefore, in real macromolecular solutions, there exists a stronger excluded volume effect inside a branched chain. About forty years ago, J. Isaacson and T. C. Lubensky [Le Journal de Physique-Letters, 1980, 41: 469] found that the scaling exponent between the average radius of gyration and the total number of segments (N) of the branched chains increases from 1/4 to 1/2 after considering the exclude volume effect in a good solvent. Soon after, M. Daoud and J. F. Joanny [Le Journal de Physique-Letters, 1981, 42: 1359] further examined the double scaling relationship between the average radius of gyration of the branched chains and the numbers of segments of sub- and parent chains by using the method developed by de Gennes. They confirmed that the scaling exponent on N is indeed 1/2 after considering the excluded volume effect, and at the same time, uncovered that the scaling exponent on N_b decreases from 1/4 to 1/10. The calculation is cumbersome, so it is omitted here.

As shown below, a simple derivation can also lead to the same double scaling relationships. The derivation is similar to those used to obtain eqs. (4.4) – (4.7). Namely, when a macromolecular chain expands inside a good solvent, its free energy change includes two opposite contributions. Both of them depend on the chain expansion, which are the two–body excluded volume effect (its nature is also entropy) and the entropy elasticity, respectively. In comparison with linear chains, one only needs to remember $R_0^2 \cong b^2 N_b^{1/2} N^{1/2}$ for branched chains, i.e.

$$\Delta G \sim \frac{N^2}{R^3} + \frac{R^2}{b^2 N_b^{\frac{1}{2}} N^{\frac{1}{2}}} \quad \rightarrow \quad \left(\frac{\partial \Delta G}{\partial R}\right)_{T,p} \sim -\frac{3b^3 N^2}{R^4} + \frac{2R}{b^2 N_b^{\frac{1}{2}} N^{\frac{1}{2}}}$$

$$\left(\frac{\partial \Delta G}{\partial R}\right)_{T,p} = 0 \quad \rightarrow \quad R \sim b N_b^{\frac{1}{10}} N^{\frac{1}{2}} \, (\text{good solvent}) \tag{4.26}$$

As the solvent quality becomes poorer, the two – body excluded volume effect slowly disappears. Therefore, it is necessary to consider the originally weaker three – body excluded volume effect. Following the relevant discussion in obtaining eq. (4.6), one has

$$\Delta G \sim \frac{b^6 N^3}{R^6} + \frac{R^2}{b^2 N_b^{\frac{1}{2}} N^{\frac{1}{2}}} \quad \rightarrow \quad \left(\frac{\partial \Delta G}{\partial R}\right)_{T,p} \sim -\frac{6b^6 N^3}{R^7} + \frac{2R}{b^2 N_b^{\frac{1}{2}} N^{\frac{1}{2}}}$$

$$\left(\frac{\partial \Delta G}{\partial R}\right)_{T,p} = 0 \quad \rightarrow \quad R \sim b N_b^{\frac{1}{16}} N^{\frac{7}{16}} \, (\Theta \text{ solvent}) \tag{4.27}$$

Here it is necessary to note that if considering a linear chain as a "branched macromolecule" with a zero degree of branching and only one subchain, the "subchain" is just the "main" chain and $N_b = N$. Substituting it into the above two equations, one obtains $R \sim bN^{3/5}$ and $R \sim bN^{1/2}$, respectively, completely identical to the previous results of linear chains in the two kinds of solvents, further illustrating that linear chains are a special kind of branched chains.

According to the above double scaling relationship, the slopes of double logarithmic plots of $\lg R_g$ versus $\lg N_b$ and $\lg R_g$ versus $\lg N$ are two scaling exponents, respectively. In general, N_b and N of branched polymers are in the ranges of $10^1 - 10^2$ and $10^3 - 10^4$, respectively. Therefore, $N_b^{1/10}, N_b^{1/16}, N^{1/2}$ and $N^{7/16}$ are in the ranges of $1.3 - 1.6, 1.2 - 1.3, 30 - 100$ and $20 - 56$, respectively. There are always unavoidable experimental errors in real measurements of R_g, N_b and N. The relative errors of such obtained scaling exponents are as follows.

The process of finding each slope involves one difference between two logarithms divided by another difference between other two logarithms, can be written as $\mathrm{d}(\ln y)/\mathrm{d}(\ln x) = (x/y)$ $(\mathrm{d}y/\mathrm{d}x)$. With an assumption that x and y have the same relative error of $\pm a\%$, the relative errors of $\mathrm{d}x$ and $\mathrm{d}y$ increase to $\pm 2a\%$. The multiplication or division of two variables doubles the relative errors; Therefore, the total relative error should be $\pm 8a\%$. The relative errors of the average radius of gyration and average molar mass measured by experiments ($\pm a\%$) are not less than $\pm 3\%$, so that the total relative errors of the obtainable slopes should not be less than $\pm 25\%$. With an increase of the number of each set of samples to five or more, the relative errors of the obtainable slopes can be reduced, but still around $\pm 10\%$. Therefore, it is nearly impossible to use the experimental data and the above two scaling relationships to judge the solvent property.

Besides the experimental errors, if only two of f reactive groups on a branching comonomer participate in the polymerization, the length of the subchain between two adjacent branching points doubles. If this consecutively happens, the length of this subchain will be three or more times longer than the designed length. On the other hand, every branching reactive group does not simultaneously participate into the polymerization, not only resulting in different subchain lengths, but also making every branched chain with different compositions and structures, double broad distributions.

To validate the double scaling relationship predicted long time ago requires two sets of branched chains: a) for a given number of segments on the subchain, the parent chains have different numbers of the total segments (N); and b) for a given number of the total segments on the parent chains, the subchains have different numbers of segments (N_b). In the past few decades, despite many unremitting attempts and efforts, a series of perfectly branched chains, respectively, with uniform parent chains and subchains have never been prepared. In previous experiments of branched chains, the average degrees of polymerization or the average molar masses of sub- and parent chains have to be used to replace N_b and N, so that the scaling relationships from the experimental result and the theoretical prediction could not be compared with certainty.

Until recently, using macromonomers ($A \cdots B \cdots A$) with two azide ends (A), one alkynyl middle group (B), where $\cdots$ is a short linear chain link, as shown on the right. The effective click chemical reaction between azide and alkynyl groups can connect a number of macro-monomers to form nearly perfect branched chains with a uniform subchain length. Only one unreacted 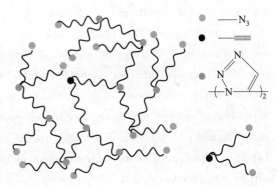 alkynyl group (black dot) is left on each branched chain, which can further click with one azide end to form a ring (not drawing), but its probability is very low. To make such a macromonomer, B with double initiators was used to simultaneously trigger living polymerization, forming two short linear chain links connected at B, and two A groups were used to seal two active chain ends.

Using the above clever design and different initiator-and-reactive monomer ratio, the narrowly distributed macromonomers with different lengths ($2N_b$) were prepared by living polymerization first; and then, the click chemistry was used to link the macromonomers of each length separately to form a series of "perfect" branched chains with different subchain lengths (N_b). Each kind of branched chains with the same N_b but different N can be further separated by the gradual precipitation method. Finally, two sets of the branched chains can be obtained. In one set, N_b is the same, but N is different, while in another set, N is similar, but N_b is different.

The average radius of gyration (R_g) and the average hydrodynamic radius (R_h) of each sample as well as the average degrees of polymerization (n_b and n) of macromonomers and the branched chains were measured by LLS, respectively. According to the above double scaling laws, the double logarithmic plots of each kind of the chain size (R_g or R_h) versus the molar degrees of polymerization of the parent and sub-chain, respectively, leads to two straight lines, whose slopes correspond to two scaling exponents. In theory, four scaling exponents can be obtained.

The scaling exponents between the chain size and the molar degree of polymerization of the subchain are close to 1/10, not too sensitive. The scaling exponents between the chain sizes and the molar degree of polymerization of the parent chain are 0.46 and 0.48, respectively, smaller than the predicted 1/2, as shown on the right. Note: its y-axis has been normalized by the molar degree of polymerization of the subchain, so that two double logarithmic plots can be merged into one, only two straight lines are

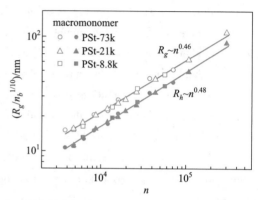

Li L W, Lu Y Y, An L J, et al. Journal of Chemical Physics, 2013, 138:114908.

(Fiugre 2, permission was granted by publisher)

left. It is necessary to state that in good solvents, the measured scaling exponent of linear chains is ~ 0.59, smaller than the theoretical value (3/5) without a modification of using the renormalization group theory. After nearly seventy years of unremitting efforts, the double scaling dependences between the average radius of gyration of the random branched chains and the molar degrees of polymerization of the parent and sub-chains have been finally confirmed. Later, the significance of the hydrodynamic radius (R_h) and its related measurement methods will be introduced in detail.

Summary

For a given composition, both small and macro-molecules can have different configurations, but macromolecules have much richer and more complicated configurations and structures than small molecules. Even polymers made of only one kinds of monomers already have many different configurations. The branched configuration can be regarded as the mostly fundamental "root" configuration, which can be further divided into two kinds, depending on whether there are many ring structures in the chain.

I) Without many rings: if every branching subchain no long continues to branch, it becomes a comb configuration. If all branching subchains on a comb chain are connected to a center point, it becomes a star configuration. When a star chain has only two branching subchains (two arms), it becomes a linear configuration. When two ends of a linear chain are connected, it becomes a **ring configuration**.

Ⅱ) With many rings, the branched chain has a three-dimensional **spatial network** configuration. After sufficiently swollen by solvent, it forms a gel, which can be viewed as a special kind of matters between solids and liquids, also called wet matters. On the one hand, a gel can keep its shape without an exserting external force (besides the gravity), similar to a solid; on the other hand, a weak exserting force usually breaks a gel and makes it deforming, similar to a liquid.

Macromolecular chains with N_0 polymerized monomers and any configurations can be ideally viewed as freely linked N Kuhn segments. Every segment can freely fold and rotate in the three-dimensional space, its length is b, with k polymerized monomers, i.e., $N = N_0/k$. Different macromolecules have different values of k, reflecting the flexibility of a chain, the smaller the more flexible. A rigid chain can be viewed to have only one segment. N is proportional to the experimentally measurable degree of polymerization or molar mass. The Gaussian chain is an extension of the ideal chain. It assumes that starting from any one segment i, the probability of finding another segment j at the position $\vec{\mathbf{r}}_i - \vec{\mathbf{r}}_j$ after passing x segments follows the Gaussian distribution $G(\vec{\mathbf{r}}_i - \vec{\mathbf{r}}_j, x)$.

Completely different from small molecules, even for a given composition and configuration, under the agitation of thermal energy, each segment in a macromolecular chain randomly moves, leading to a constantly random variation of the mass spatial distribution (conformation) of the chain. Therefore, only a statistical average of all possible conformations can get the average size of a macromolecular chain.

Generally speaking, for any two segments (i and j) in a non-linear chain with no ring, there exist and only exists one linear chain link (path) with x segments to connect them. The direct distance between them changes constantly with the chain conformation, the statistical average can get the "mean square distance" between them, i.e., averaging the square direct distance between any two segments in all possible chain conformations. The result is $b^2 x$. x varies with the configuration.

For a linear chain or chain link, $x = |i - j|$; while x between two segments on a non-linear chain is often not simply related to i and j. For a specific configuration, a specific analysis is required. A sum of all possible mean square distances, i.e., respectively varying the positions of the two segments from 1 to N, is called the "sum of mean square distance". It is theoretically related to the measurable "average radius of gyration". The "sum of mean square distance" is a core but difficult to understand physical concept and statistical operation.

For the simplest linear configuration, each chain has a stretched length: $L_C = bN$. If taking i and j as its first and the last segments, the mean square distance is the mean square end-to-end distance $\langle R^2 \rangle = b^2 N$. The average mean square radius of gyration can be obtained from the sum of the mean square distances, i.e., $\langle R_{g,0}^2 \rangle = b^2 N/6$. Note: the mean square radius of gyration is only $1/6$ of the mean square end-to-end distance. Both the average end-to-end distance ($R_0 = \langle R^2 \rangle^{1/2}$) and the average radius of gyration ($R_{g,0} = \langle R_{g,0}^2 \rangle^{1/2}$) of a linear ideal chain are proportional to $bN^{1/2}$. The chains with nonlinear configurations might have a number of ends, so that there is no mean square end-to-end distance, but it still has a mean square distance. Theoretically, in the unperturbed state, there exists a double scaling relationship (also called scaling law) between the average radius of gyration of branched chains and the numbers of segments on the parent and sub-chain (N and N_b): $R_{g,0} \propto bN_b^{1/4}N^{1/4}$.

Every segment on a real chain occupies a certain three-dimensional space ($\sim b^3$), called as the excluded volume. The dispersion attraction between the segments generates an equivalent effect of a virtual negative excluded volume. A combination of them together is called the excluded volume effect. In a good solvent or a Θ solvent with no two-body interaction, the above double scaling law becomes $R_g \propto bN_b^{1/10}N^{1/2}$ or $R_g \propto bN_b^{1/16}N^{7/16}$. If a linear chain is viewed as a special branched chain with only one subchain, i. e., $N_b = N$, the above double scaling law becomes $R_g \propto bN^{3/5}$ or $R_g \propto bN^{1/2}$, identical to the results directly derived from linear chains. It shows that the branched configuration is indeed a basic "root" configuration.

In a polymerization, each polymer chain is made of a different number of monomers, with different sizes. For nonlinear configurations, even with a given composition and configuration, every chain still have different structures, very complicated! For example, in one branching chain, the subchain lengths are not uniform; in different branching chains, the number and the position of branching points are different. Therefore, it is necessary to replace the number of segments (N) with the measurable number or weight average molar mass (M_n or M_w). Since N and M_n or M_w are proportional to each other, not affecting the scaling laws.

The scaling laws between R_g and M_n or M_w are generally written as $R_g = k_{R_g}M_n^{\alpha_{R_g}}$ or $R_g = k_{R_g}M_w^{\alpha_{R_g}}$, where k_{R_g} is a system-dependent variable, varying with the chemical compositions of macromolecules and solvent and other factors; while α_{R_g} is a universal scaling exponent that is unrelated to chemical compositions, but only dependent of the configuration and solvent quality. In a process from a good solvent to a Θ solvent, and then to a poor solvent α_{R_g} of a linear flexible chain first gradually decreases from $3/5$ to $1/2$, then quickly to $1/3$; namely, an expanded random coil collapses into a uniform globular conformation with lots of solvent molecules inside.

The following figure summarizes the relations between different macromolecular configurations. Staring from the branched "root" configuration, gradually "growing" upwards, various common macromolecular configurations are obtainable. The figure also summarizes, the scaling laws between the average radius of gyration and the number of the chain segments in different macromolecular configurations.

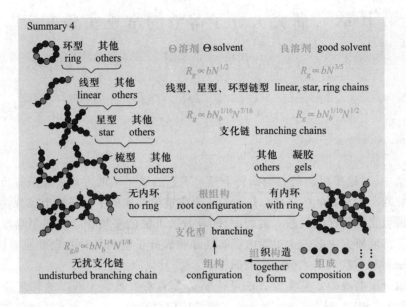

Summary 4

环型　其他
ring　others

线型　其他
linear　others

Θ 溶剂 Θ solvent　　　　良溶剂 good solvent

$R_g \propto b N^{1/2}$　　　　$R_g \propto b N^{3/5}$

线型、星型、环型链型 linear, star, ring chains

星型　其他
star　others

$R_g \propto b N_b^{1/16} N^{7/16}$　　　$R_g \propto b N_b^{1/10} N^{1/2}$

支化链 branching chains

梳型　其他
comb　others

其他　凝胶
others　gels

无内环　　根组构　　有内环
no ring　root configuration　with ring

支化型 branching

$R_{g,0} \propto b N_b^{1/4} N^{1/4}$

无扰支化链
undisturbed branching chain

组构　　←组织构造→　　组成
configuration　together to form　composition

第五章　研究大分子溶液的常见实验方法

第四章中,讨论了可测的方均回转半径与理论的线型链的方均末端距或具有其他组构的链的方均距加和之间的关联。还讨论了支化链的可测的平均回转半径分别与母链和子链的链段数之间的标度关系。实践中,链段数被可测的平均聚合度或平均摩尔质量取代。大分子溶液研究中的一个重要内容就是寻获大分子链的各种物理性质对链长(摩尔质量)的依赖性,其构成了大分子物理的核心。

研究大分子溶液的常见实验方法可按照是否需要依赖事先获得的标定曲线分为两组:"绝对"方法和"相对"方法。每条标定曲线的建立需要一套具有不同摩尔质量的窄分布标准样品。端基测量、渗透压差、激光光散射、微分折射法和超速离心是绝对方法;而尺寸排除(凝胶渗透)色谱、特性黏度等方法则依赖于每个仪器的标定曲线。

随着各种快速和灵敏的光电检测器,以及计算机的存储和运算速度的飞速发展,在过去的三十年内,绝大部分的仪器都实现了电子化和自动化操作。各种仪器的生产公司和销售人员都试图告诉使用者不必担心很多实际使用中可能存在的问题。很多使用者也将各种仪器当作一个黑盒子,放进样品、按下按钮、取得结果、写进报告,而不知每种仪器的基本原理、注意事项、潜在问题、仪器的精度和确度,等等。因此,本章十分重要。

对于任何一个实验方法,始终需要掌握以下三个要点:其基本的化学或物理原理;直接测量的物理量和测量的精度和确度;以及如何通过适当地处理所测数据获得最终所需的物理参数。因此,将按照这三个要点逐一地介绍上述各种常用实验方法。

端　基　测　量

每条大分子链的两个末端基团(简称末端),其化学结构往往与链上那些已聚合的单体不同。例如,每条蛋白质链或聚氨酯链,其两端分别为酸性的羧基和碱性的氨基。一个末端对应着一条大分子链。利用链末端和链上聚合单元之间化学或物理性质的差异,溶液中大分子链两个末端基团中任意一个的摩尔浓度(n),即链的摩尔浓度,是可测的。从溶液配制中已知的单位体积内的质量(C,质量浓度),可利用 $n = C/M$,求出大分子链的平均摩尔质量(M)。在大分子实验中,C 通常采用的单位是"克/毫升"。注意,大分子教科书中通常沿用小分子中的使用习惯,称摩尔质量为分子量。考虑到后续有关大分子介观球相和胶体的讨论,本书将采用物理上更准确的摩尔质量,而不是分子量。尽管二者具有相等的数值,但内涵有异。

如前所述,在一个聚合反应中,所得的线型链有长有短、支化型链内部结构不均、有大有小,所以,它们只有平均摩尔质量。显然,端基法只可用来测量带有一个或两个端基的线型链的摩尔质量,而对端基数目不确定的大分子链不适用。另外,如果聚合度大于100,末端基团的浓度将很低,测量精度和确度均达不到要求。因此,端基法一般适用于测量寡聚物和低度聚合物的平均摩尔质量。测量端基数目的常见方法为经典的化学滴定法,其精度和确度均可达到千分之一;也可采用核磁共振法,其精度和确度仅为百分之一或更差。

对一个多分散的大分子样品,不同长度的线型链均会对所测的端基摩尔浓度和质量浓度有贡献。因此,需要将具有不同摩尔质量的所有链的贡献相加求和,即 $n = (\sum_{i=1}^{N} n_i)/V$ 和 $C = (\sum_{i=1}^{N} W_i)/V$,其中 n_i 和 W_i 分别是摩尔质量为 M_i 的链末端的摩尔数和质量。显然,$W_i = M_i n_i$,因此

$$M_n = \frac{C}{n} = \frac{\sum_{i=1}^{N} M_i n_i}{\sum_{i=1}^{N} n_i} \tag{5.1}$$

端基分析得到的是数量权重的数均摩尔质量,用下标"n"标明。以此类推,还有其他权重的平均摩尔质量,如重均摩尔质量(M_w)和 z 均摩尔质量(M_z),分别定义如下:

$$M_w = \frac{\sum_{i=1}^{N} M_i W_i}{\sum_{i=1}^{N} W_i} = \frac{\sum_{i=1}^{N} M_i^2 n_i}{\sum_{i=1}^{N} M_i n_i} \quad \text{和} \quad M_z = \frac{\sum_{i=1}^{N} M_i^3 n_i}{\sum_{i=1}^{N} M_i^2 n_i} \tag{5.2}$$

从数学的角度看,从给定的摩尔质量分布计算出的各种平均值,M_n,M_w 和 M_z 分别只是摩尔质量的一次距与零次距,二次距与一次距和三次距与二次距之比而已。需要强调,这不是一个人为的数学游戏或运算,而是源于实验;我们将会逐步地了解,不同的实验方法只能别无选择地测得大分子链的不同的平均摩尔质量。

小　　结

基本原理:线型链具有两个不同或相同的端基,故可由端基数得到大分子链的数目。

可测物理量:利用一个化学或物理方法,如化学滴定法或核磁共振法,测得溶液中的大分子链末端的摩尔数(n)。

数据处理:利用配制溶液时已知的质量(重量)浓度(C,通常用"克/毫升")与所测得摩尔数(n)之比可得到大分子链的数均摩尔质量(M_n)。

渗透压差(π)

传统教科书中的"渗透压"实际上是溶液和溶剂之间的压强之差。它是溶解过程中,从溶剂到溶液时的压强变化,而不是一个压强。严格地说,与渗透无关。"渗透"仅是测量该压差的一个方法而已。因历史原因写成了易生误解、不符原本内涵的"渗透压"。其将本质上的压强差和一个测量该压强差的方法混为一谈。本书沿用"渗透压"这一习惯

用词,但加一个"差"字,即"渗透压差",以正其名。讨论渗透压差有不同的途径。可从讨论溶剂和溶液的压强出发讨论二者之差。也可运用热力学中的自由能和化学势变化直接导出。

提及压强(p),可先回顾一下其物理本质。在物理化学或普通物理教科书中,常常将压强描绘成因气体分子对容器壁的撞击,施加在单位表面积上的平均力。这是又一个将物理性质和其测量方法混为一谈的例子。一瓶气体的中间也存在压强,与器壁无关,但测量一个压强则需要一个表面。这就如同每个人都有一定的质量(重量),其称量可利用一个磅秤或其他工具,但该质量与磅秤和称量与否毫无关系。

为了更好地讨论压强的本质,让我们首先考察一下其量纲:p 为单位作用面积上的作用力。常用的单位是帕斯卡(Pa,牛顿/平方米,$N \cdot m^{-2}$)。压强还有其他单位,如标准大气压(atm)和巴(bar)。它们之间以及和帕斯卡的关系是 $1.00 \ bar = 1.00 \times 10^5 \ Pa = 0.990 \ atm$。将作用力和作用面积分别乘以一个长度可得能量($E$)和体积($V$)。因此,压强也是单位体积里的能量,$p = E/V$,即能量密度。

如果没有源于环境的外来能量,体系中每个分子仅具有热能($k_B T$),这里 k_B 和 T 分别为 Boltzmann 常数和体系的绝对温度。因此,体系中每个气体分子贡献的能量密度(产生的压强)为 $k_B T/V$。对 N 个分子而言,$p = N k_B T/V = n R T/V$,其中,$R = k_B N_{AV}$(摩尔气体常数)和 $n = N/N_{AV}$。这就是理想气体的状态方程,最简单的状态方程。

微观上,可用一个小的体积(v)作为一个单位将一瓶气体的体积(V)分割成 n_{box} 个小盒子,即 $n_{box} = V/v$。假定气体分子既无尺寸、也无相互作用,在热能搅动下,每一个气体分子作无规运动,到达每个小盒子的概率(ψ)均等。显然,$\psi = 1/n_{box} = v/V$,反比于 V。可想象在任意一个小盒子处放置一个微小的压强检测器,所测压强将正比于一个气体分子到达该小盒子的概率。正比常数 p_v 的物理意义是当 $\psi = 100\%$ 时,一个气体分子所产生的压强。当 V 小至 v 时,即体系里仅有一个小盒子,$\psi = 100\%$。注意 p_v 也是体积小至 v 时的能量密度 $k_B T/v$。因此,$p = k_B T/V$,其中 p_v 和 ψ 中的 v 互相约去,可见如何选择 v 的大小并不重要。对 N 个气体分子而言,可得同样的理想气体状态方程。

历史上,Robert Boyle 在 1662 年发表了一篇论文,描绘了一罐封闭空气的气压和体积的乘积接近一常数。一百多年后的 1780 年,Jacques Charles 发现体积和压强二者均正比于绝对温度 T。现在有了微观上的解释:它们均和一个粒子到达体系内任意一点的概率有关。该概率反比于体积,但正比于压强。另一方面,对于一给定的体积,随着温度的升高,粒子获得更多的热能量,故运动加快,在给定的时间内,也增加了该概率。因此,对于一给定的体积,压强必定随绝对温度线性地增加。否则,体积就会改变。更重要的是,该概率随着气体分子数线性地增加。因此,压强有着依赖分子数的性质,简称为"依数性"。

在真实气体中,每一个分子具有一个硬核体积(v_0),分子间存在着与色散能(ε)有关的吸引作用。二者的综合产生了一个排除体积效应。真实气体的状态方程变得复杂。在一瓶真实气体中,N 个气体分子占据的总排除体积为 $v_e N$,每个气体分子实际可达及的体积从 V 减至 $V - v_e N$。因此,每个气体分子抵达任意一个小盒子的概率增至

$$\psi = 1 \bigg/ \frac{V - v_e N}{v_e} \cong \frac{v_e}{V} + \frac{v_e^2 N}{V^2} \tag{5.3}$$

其中,利用了 $v_e N \ll V$ 和 $1/(1 - v_e N/V) \cong 1 + v_e N/V$。上式中的首项是一个在热能搅动下作无规运动的粒子到达瓶内任意一点的概率。次项则是源于排除体积效应的额外概率,导致一个额外的压强,二者间的正比常数为体积小至 $2v_e$、概率为 100% 时的色散能密度($\varepsilon/2v_e$)。因此,N 个真实气体分子产生的压强为

$$p = \frac{nRT}{V} + \frac{\varepsilon v_e N^2}{2 V^2} = \frac{nRT}{V} + \frac{\varepsilon v_e N_{AV}^2}{2} \left(\frac{n}{V}\right)^2 \tag{5.4}$$

用 $V - v_e N$ 代替理想状态中的 V,上式可重写成

$$\left[p - a \left(\frac{n}{V}\right)^2\right] (V - bn) = nRT \tag{5.5}$$

这就是著名的真实气体的 van der Waals 状态方程,其中 $a = (\varepsilon N_{AV} v_e N_{AV})/2 = (\varepsilon_m v_{e,m})/2$ 和 $b = v_e N_{AV} = v_{e,m}$。"ε_m"和"$v_{e,m}$"一起代表了摩尔排除体积效应。分子之间的色散吸引作用产生了一个减小排除体积的效应,其增大了分子可达及的体积,降低了一个气体分子到达体系内任意一点的概率。而分子的硬核体积本身起到一个相反的效果。需要注意:源于互相吸引"a"为负值。然而,在物理化学教科书中,其通常被列为正数。这就是为何与上式相比,在物理化学教科书中,a 的前面为一"+"号。

上式还可从两个气体分子间产生相互作用的概率直接导出。沿用上述讨论,对一个含有 N 个分子的体系,在总共 V/v_e 个小盒子中任意一个小盒子上出现第一个分子的概率为 $N/(V/v_e)$;在第二个小盒子上出现第二个分子的概率为 $(N-1)/(V/v_e - 1)$。因此,发生两体排除体积效应的概率为 $[N/(V/v_e)][(N-1)/(V/v_e - 1)] = v_e^2 N^2/V^2$,其中利用了 $N \gg 1$ 和 $V/v_e \gg 1$。除了由热能引致的压强(nRT/V)外,还有源于色散能(ε)的一个额外负压效应。由于每一对气体分子占据的体积为 $2v_e$,故能量密度为 $\varepsilon/2v_e$。所以,由排除体积效应造成的总额外负压强效应为能量密度和概率的乘积,即 $[\varepsilon/(2v_e)][N/(V/v_e)^2] = \varepsilon v_e N^2/(2V^2)$,与式(5.4)中的第二项一致,殊途同归。

上述有关气体压强的讨论可直接运用到液体中。气体压强和溶液压强有着完全相同的物理本质,都正比于分子在热能驱动下通过无规平动扩散到达体系内任意一点的概率,或更深入地讲,与一个体系的熵有关,反比于体积,但正比于分子数。沿用前面讨论大分子溶解时使用的术语,并假定溶液中的分子们之间没有相互作用,以及一条大分子链的排除体积($v_{e,p}$)是一个溶剂分子排除体积($v_{e,s}$)的 n_v 倍。因此,可将一个含有 n_p 摩尔大分子链的溶液视为大分子链取代了 $n_v n_p$ 摩尔溶剂分子,而不需考虑其余的溶剂分子。注意:对高聚物而言,$n_v \gg 1$。

由于 π 是溶液压强和溶剂压强之差,而压强又正比于分子数,故渗透压差正比于溶液和溶剂中的分子数之差,即

$$\pi = p_{solution} - p_{solvent} = \frac{(n_p - n_v n_p) RT}{V} = \frac{n_p RT}{V} - \frac{n_v n_p RT}{V}$$

π 既然是压强之差,可将纯溶剂当作参照状态,并定义其压强 $n_v n_p RT/V = 0$,物理上等

于将溶液中的溶剂视为真空。所以，n_p 摩尔大分子链在体积 V 内产生的压强在数值上等于渗透压差：

$$\pi = \frac{n_p RT}{V} - 0 = \frac{CRT}{M}$$

这里 M 是大分子的摩尔质量，并利用了 $n_p = W/M$ 和 $C = W/V$。对于多分散体系，

$$\pi = \sum_i \pi_i = \sum_i \frac{n_i RT}{V} = \frac{CRT}{W/\sum_i n_i} = \frac{CRT}{M_n} \tag{5.6}$$

其中，利用了 $W = \sum_i M_i n_i$ 和 $M_n = \sum_i M_i n_i / \sum_i n_i$。由此可见，利用渗透压差所测得的也是数均摩尔质量。

在真实溶液中，不得不考虑排除体积效应。沿用有关真实气体的讨论，只需注意在一给定的二元液体混合物中，共有三对分子间的排除体积效应：溶剂–溶剂（ε_{AA}）、溶质–溶质（ε_{BB}）和溶质–溶剂（ε_{AB}）。重复第三章中的论述如下，ε_{AB} 代表了一个链段和一个溶剂分子的作用，即一条链可同时与 n_v 个溶剂分子产生排除体积效应。

一对链溶解时，同时断开 n_v 对"溶剂–溶剂"相互作用和一对"链–链"相互作用，造成 $2n_v$ 对"链段–溶剂分子"排除体积效应。因此，形成每一对"链段–溶剂分子"排除体积效应的平均能量变化为 $\varepsilon_{AB} - (\varepsilon_{AA} + \varepsilon_{BB}/n_v)/2$，用热能 $k_B T$ 归一，并乘以作用点数 Z 后，可如前记为 χ_{sp}。

在理想溶剂中引入微弱的二体排除体积效应（ε_{AA}）和借用上面讨论真实气体的结果，可知，即将在溶解中被大分子链取代的 $n_v n_p$ 摩尔溶剂分子所产生的压强为

$$p_{\text{solvent}} = \frac{n_v n_p RT}{V} + \frac{\varepsilon_{AA} v_{e,s} N_{AV}^2}{2} \left(\frac{n_v n_p}{V} \right)^2$$

在溶解前，固体大分子样品中的链在常温下无法在热能的搅动下发生平动扩散。链段只能围绕链的质心扰动（密度涨落），到达体系中任意一点的概率为零，故无平动扩散引致的压强。但是，仍存在二体排除体积效应，其对应的压强为

$$p_{\text{polymer}} = \frac{\varepsilon_{BB} v_{e,p} N_{AV}^2}{2} \left(\frac{n_p}{V} \right)^2$$

其中，$v_{e,p} = v_{e,s} n_v$。溶解后，n_p 摩尔的链取代了 $n_v n_p$ 摩尔的溶剂分子。每一条链产生了 n_v 对"链段–溶剂分子"排除体积效应（ε_{AB}）；所以，溶液的压强为

$$p_{\text{solution}} = \frac{n_p RT}{V} + n_v \varepsilon_{AB} v_{e,p} N_{AV}^2 \left(\frac{n_p}{V} \right)^2$$

因此，溶液形成前后的压强（渗透压）差为

$$
\begin{aligned}
\pi &= p_{\text{solution}} - (p_{\text{solvent}} + p_{\text{polymer}}) \\
&= RT(1 - n_v)\frac{n_p}{V} + n_v \varepsilon_{AB} v_{e,p} N_{AV}^2 \left(\frac{n_p}{V} \right)^2 - \left(\frac{\varepsilon_{AA} v_{e,s} n_v^2}{2} + \frac{\varepsilon_{BB} v_{e,p}}{2} \right) N_{AV}^2 \left(\frac{n_p}{V} \right)^2 \\
&= RT(1 - n_v)\frac{n_p}{V} + \left[\varepsilon_{AB} - \left(\frac{\varepsilon_{AA} + \dfrac{\varepsilon_{BB}}{n_v}}{2} \right) \right] n_v v_{e,p} N_{AV}^2 \left(\frac{n_p}{V} \right)^2
\end{aligned}
$$

$$= RT(1 - n_v)\frac{n_p}{V} + RTX_{sp}n_vv_{e,p}N_{\mathrm{AV}}\left(\frac{n_p}{V}\right)^2 \qquad (5.7)$$

其中,对第二项的系数作了热能(k_BT)归一,即同时乘以和除以热能。继续定义 n_vn_p 摩尔溶剂分子产生的压强(RTn_vn_p/V)为零,上式成为

$$\pi = RT\frac{n_p}{V} + RTX_{sp}n_vv_{e,p}N_{\mathrm{AV}}\left(\frac{n_p}{V}\right)^2$$

为了方便讨论,下面引入偏比容(v)的概念。v 被定义为 $(\partial V/\partial m)_{T,p}$,其物理意义是体系的体积随其中某一个组分质量增加的变化,是体积变化率,不是体积,也不是单位质量的体积(比容),千万不要混淆!对于纯物质,"偏比容"和"比容"二者相同。

对二组分的大分子溶液,大分子和溶剂有各自的偏比容 v_p 和 v_s。体积分数、质量浓度和偏比容互相依赖,由它们的定义,不难推出 $X_s = C_sv_s$ 和 $X_p = Cv_p$。理解和记住它们的定义和互相之间的依赖关系极为重要。将一个体积分数转换成一个实验上常用的质(重)量浓度时,常常用到它们。上式可被进一步写成如下形式:

$$\pi = RT\frac{C}{M} + RTX_{sp}n_vv_p\left(\frac{C}{M}\right)^2 \quad \text{或} \quad \frac{\pi}{RTC} = \frac{1}{M}\left(1 + \frac{X_{sp}n_vv_p}{M}C\right) = \frac{1}{M}(1 + A_2C) \quad (5.8)$$

其中, $A_2 = X_{sp}n_vv_p/M$,为二阶维里系数,可由实验确定。在高浓度时,可能不得不加上三体排除体积效应,即附加一项 A_3C^2,其中, A_3 是三阶维里系数。对一个多分散的大分子样品,上式中的 M 应被换成 M_n。

式(5.8)也可由热力学原理直接推出:等温和等压时,自由能随着溶剂和大分子含量变化的变化率(偏摩尔自由能, $[\partial G/\partial n_s]_{T,p}$ 和 $[\partial G/\partial n_p]_{T,p}$)分别称为溶剂和大分子的化学势。依据定义,在压强($p + \pi$)下,体系混合前的自由能为纯溶剂的自由能 $G^*(p + \pi)$;混合后,溶剂中多了大分子链,自由能降为 $G(p + \pi, x_s)$,自由能的变化为

$$G(p + \pi, x_s) - G^*(p + \pi) = \Delta G_{\mathrm{mix}}(p + \pi)$$

$$\left[\frac{\partial G(p + \pi, x_s)}{\partial n_s}\right]_{T,p,n_p} - \left[\frac{\partial G^*(p + \pi)}{\partial n_s}\right]_{T,p,n_p} = \left[\frac{\partial \Delta G_{\mathrm{mix}}(p + \pi)}{\partial n_s}\right]_{T,p,n_p}$$

$$\mu_s(p + \pi, x_s) - \mu_s^*(p + \pi) = \left[\frac{\partial \Delta G_{\mathrm{mix}}(p + \pi)}{\partial n_s}\right]_{T,p,n_p}$$

当溶液的压强从 p 升至 $p + \pi$ 达到平衡时,溶液中溶剂的化学势等于纯溶剂的化学势,即 $\mu_s(p + \pi, x_s) = \mu_s^*(p)$。其物理意义是当溶剂和溶液交换一个或一摩尔溶剂分子时,溶剂和溶液二者的自由能保持不变。将 $\mu_s(p + \pi, x_s) = \mu_s^*(p)$ 代入上式,

$$\mu_s^*(p) - \mu_s^*(p + \pi) = \left[\frac{\partial \Delta G_{\mathrm{mix}}(p + \pi)}{\partial n_s}\right]_{T,p,n_p} \qquad (5.9)$$

只要求出上式左边的纯溶剂在两个不同压强下的化学势之差,以及右边混合自由能对溶剂摩尔数的微分,即可得到 π 与浓度之间的关系式。先求右边的偏微分。在第三章中,已详细讨论混合自由能对大分子体积分数 X_p 的依赖关系。为了方便讨论,将式(3.15)重写如下:

$$\Delta G_{\mathrm{mix}} = RT\big[\, n_s \ln(1 - X_p) + n_p \ln X_p + n_s X_p \chi_{sp} \,\big]$$

为了求出 ΔG_{mix} 对 n_s 的偏微分,可先将 ΔG_{mix} 对 X_p 偏微分,再乘以 X_p 对 n_s 的偏微分。因温度和压强保持不变,为了书写简洁,移去所有偏微分中的下标。对 n_s 求偏微分时,只需记住 n_p 为常数,即

$$\frac{\partial \Delta G_{\mathrm{mix}}(p + \pi)}{\partial n_s} = RT \frac{\partial\big[\, n_s \ln(1 - X_p) + n_p \ln X_p + n_s X_p \chi_{sp} \,\big]}{\partial X_p} \frac{\partial X_p}{\partial n_s}$$

$$= RT\left[\frac{\partial n_s}{\partial X_p} \ln(1 - X_p) - \frac{n_s}{1 - X_p} + \frac{n_p}{X_p} + \frac{\partial n_s}{\partial X_p} X_p \chi_{sp} + n_s \chi_{sp} \right] \frac{\partial X_p}{\partial n_s}$$

$$= RT\left[\ln(1 - X_p) - \frac{n_s}{1 - X_p} \frac{\partial X_p}{\partial n_s} + \frac{n_p}{X_p} \frac{\partial X_p}{\partial n_s} + X_p \chi_{sp} + n_s \chi_{sp} \frac{\partial X_p}{\partial n_s} \right]$$

$$\left(\frac{\partial \Delta G_{\mathrm{mix}}(p + \pi)}{\partial n_s} \right)_{T,p,n_p} = RT\left[\ln(1 - X_p) + \left(1 - \frac{1}{n_v}\right) X_p + X_p^2 \chi_{sp} \right] \qquad (5.10)$$

在进行最后一步推导时,利用了下面一个十分有用的公式

$$\left(\frac{\partial X_p}{n_s} \right)_{T,p,n_p} = \frac{\partial \left(\dfrac{n_v n_p}{n_s + n_v n_p} \right)}{\partial n_s} = -\frac{n_v n_p}{(n_s + n_v n_p)^2} = -\frac{X_p(1 - X_p)}{n_s} = -\frac{X_p^2}{n_v n_p}$$

下一步,寻获纯溶剂自由能对压强的依赖性。由热力学第一定律(能量守恒)已知,内能的变化($\mathrm{d}U$)等于做功(w)和传热(q)之和,内能的变化与具体路径无关。因此,可通过一个假设的可逆路径($w_{\mathrm{rev}} = -p\mathrm{d}V$ 和 $q_{\mathrm{rev}} = T\mathrm{d}S$)求得,即 $\mathrm{d}U = T\mathrm{d}S - p\mathrm{d}V$。内能和焓能($H = U + pV$)包括分子定向运动可自由做功的能量(自由能,$A$ 或 G)和分子无规运动产生热量(TS)的能量。即理论学家偏爱 $U = A + TS$,因体积不变;而实验学家则喜欢 $H = G + TS$,因压强不变。

将上述关系式合起可得,$\mathbf{d}G = \mathbf{V}\mathbf{d}p - \mathbf{S}\mathbf{d}T$。在等温条件下,$\mathrm{d}G = V\mathrm{d}p$ 或 $(\partial G/\partial p)_{T,n} = V$,其中体积总为正值,故自由能随压强增加。将其两边在等温和等压的条件下对 n 微分可得

$$\left[\frac{\partial \left(\frac{\partial G}{\partial n} \right)_{T,p}}{\partial p} \right]_{T,n} = \left(\frac{\partial V}{\partial n} \right)_{T,p} \qquad (5.11)$$

依据定义,$(\partial G/\partial n)_{T,p}$ 为化学势(μ)。对纯物质,其就是摩尔自由能。$(\partial V/\partial n)_{T,p}$ 为纯物质的摩尔体积($V_{s,m}$)。因此,$(\partial \mu/\partial p)_{T,n} = V_{s,m}$。假定溶剂的摩尔体积不依赖于压强,当压强从 p 升至 $p + \pi$ 时,化学势的变化为,

$$\mu_s^*(p + \pi) - \mu_s^*(p) = \int_p^{p+\pi} V_m \mathrm{d}p = V_{s,m} \pi \qquad (5.12)$$

将式(5.10)和式(5.12)代入式(5.9)的结果为

$$-\pi V_{s,m} = RT\left[\ln(1 - X_p) + \left(1 - \frac{1}{n_v}\right) X_p + \chi_{sp} X_p^2 \right]$$

当 $X_p \ll 1$ 时,$\ln(1 - X_p) \approx -X_p - X_p^2/2$;利用 X_p 与大分子偏比容(v_p)和质量浓度(C)之间的关系,$X_p = C v_p$,以及一条大分子链与一个溶剂分子的体积之比为 $n_v, v_p = n_v V_{s,m}/M =$

$V_{p,m}/M$ 。因此,上式可重写成

$$\pi \cong \frac{RT}{V_{s,m}}\left[\frac{X_p}{n_v} + \left(\frac{1}{2} - \chi_{sp}\right)X_p^2\right] = \frac{RTC}{M}\left[1 + \left(\frac{1}{2} - \chi_{sp}\right)\frac{v_p n_v}{M}C\right]$$

$$\frac{\pi}{RTC} = \frac{1}{M}(1 + A_2 C) \tag{5.13}$$

这里,$A_2 = (1/2 - \chi_{sp})v_p n_v/M$。注意:1/2 源于 $\ln(1 - X_p)$ 的展开,来自混合自由能里的熵变,硬核体积的存在减少了分子可达及的体积,增加了分子到达体系里任意一点的概率;另一方面,χ_{sp} 则反映了二体色散吸引作用。该吸引作用减少了分子在给定的时间内到达体系里任意一点的概率,一个负排除体积效应,好像有一个虚拟力在压迫分子,增加了分子可达及的体积。χ_{sp} 随温度升高而变小。因此,A_2 可为正、为负或为零,取决于 1/2 和 χ_{sp} 的相对大小。通常,随着温度下降,A_2 逐渐地由正变负,溶剂从良溶剂,经过 Θ 状态,变成不良溶剂。

与式(5.8)比较,除了讨论中没有考虑熵变,二阶维里系数的表达略有不同以外,渗透压差对大分子摩尔质量和浓度的依赖性一致,殊途同归。实验上,二者没有区别,A_2 只是一个待定的物理量。细心的读者会发现,大、小分子溶液压强对分子数的依赖性也一致。如前所述,对一个多分散的大分子样品,用 M_n 代替上式中的 M。

溶液与溶剂之间的压差可用一个只让溶剂分子渗透通过的半透膜测得,如下图所示。半透膜的孔径小于单个大分子的尺寸。实验中,先在左、右的两个容器分别注入溶液和溶剂。如前所述,溶剂的压强大于溶液。该压强差造成一个溶剂分子从右向左的净流动。渗透膜使得右边始终保持了纯溶剂状态。由于从右向左溶剂分子的净流动,伴随着右边液面的降低,左边液面逐渐升高,达到一定高度时,左边溶液的压强等于右边纯溶剂的压强。溶剂分子以相同的迁移速率向两边移动,达到动态平衡。

当设计传统的半透膜仪器测量渗透压差时,必须考虑到溶剂分子的扩散时间。通常采用面积尽可能大的膜、尽可能薄的空腔和尽可能细的测液面高度的玻璃管,以缩短溶液和纯溶剂达到平衡的时间(通常为 1~2 天)。对于一个多分散的大分子样品,膜的选择也很重要。一是孔径要均匀;二是孔径要尽量大,以缩短平衡时间,但孔径又要小于最短链的尺寸,以免漏算短链在测量时的贡献,得到一个偏大的平均摩尔质量。

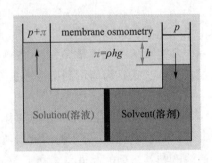

在测量渗透压差时,可先在稀溶液范围内配制一个较高的浓度,测量溶液和溶剂液面之间的高度差 h_0;然后定量地稀释溶液,再次测得高度差 h_1;继续稀释,重复实验,测得一套随浓度变化的液面高度,直至无法准确地测量 h;再将高度差换算成 π;最后,将 $\pi/(CRT)$ 对 C 作图得一直线。其截距的倒数和斜率分别对应着大分子的数均摩尔质量和二阶维里系数。

考虑到实际测量温度($\pm 1\%$)、浓度($\pm 1\%$)和液面高度差($\pm 1\%$)等实验参量时的误差,所得的数均摩尔质量和二阶维里系数的误差通常大于 $\pm 3\%$。为了缩短达至平衡(测

量)时间,人们已设计出不同的仪器。例如,可先测量溶液液面上升的速度,再换算成渗透压差。随着灵敏的压强传感器问世,现在可直接测量在一个密封的、不易变形的容器里的因溶剂分子渗透所引起的溶液压强变化,其测量精确度取决于压强传感器的绝对标定和灵敏度。

小　　结

　　渗透压是溶液与其溶剂之间的压强差。该压差总是客观存在的,与测量与否和半透膜或其测量仪器毫无关系。半渗透膜和与其相关的仪器仅是测量一个溶液和其溶剂之间压强差的一种方法而已。希望读者永远不要将该压强差和它的测量方法混淆。就像气体压强与表面无关一样,但测量气体压强时,一般需要一个表面;也如同每个人都有质量(重量),与是否称量和任何称量装置无关。

　　测量大分子的数均摩尔质量仅是渗透压差的一个应用。在实验室里,透析也是利用了这一压差;在人体内,因血液里含有大分子蛋白质,压强较低,故组织里多余的水可渗透进血管,经过肾排出体外。这也是为何医生需要给无法摄取蛋白质营养的患者补充血液里的蛋白质含量,以减少身体浮肿;在厨房里,通过加盐,可降低蔬菜或肉类细胞外的压强,达到脱水的目的,等等,不胜枚举。

　　基本原理:和气体压强一样,液体压强也正比于分子在热能的搅动下平动扩散到达体系内任意一点的概率。正比常数为概率达到100%时的能量密度。该概率也正比于液体里的分子数。读者应该认识到压强的依数性;即对给定的体积,粒子数多的液体一般具有较高的压强。与一个溶剂小分子相比,一条大分子链占据的空间较大,故对给定的体积,纯溶剂里的分子数大于大分子溶液中链与溶剂分子的数目之和,故纯溶剂具有较高的压强。因历史的原因,溶液与其溶剂之间的压强差被定义为渗透压,但其与渗透毫无关系。本书沿用习惯的定义,但加了一个字,即"渗透压差",以指出其本质。读者也要注意,该压强差和溶质的摩尔质量无关,只依赖于溶质的分子数。对给定的浓度,分子数和摩尔质量成反比。这正是为何可利用渗透压差测定一个溶质的摩尔质量。读者千万不要以为渗透压差与摩尔质量直接相关!

　　可测物理量:利用一个只让溶剂分子渗透而过的半透膜或其他方法,测得一系列(至少五个)不同浓度的大分子稀溶液与纯溶剂的压强差。传统上,通过量度被一个半透膜隔开的溶液和其溶剂的高度差来测定渗透压差,现代的膜渗透压仪器通常利用敏感和精确的压强传感器测量溶剂和溶液之间的压强差。

　　数据处理:如果直接测量的物理量不是压强差,就要先将所测得的物理量(如高度差)转换成压强差。将压强差除以其对应的大分子浓度,再对浓度作图。当大分子溶液足够稀时,该作图应得一条直线,其截距和直线斜率分别对应着大分子的数均摩尔质量和二阶维里系数,实验误差通常大于±3%。

超 速 离 心

如右图所示,当一个质量为 m 的物体,围绕着一个固定点,在半径为 r 的圆周上以 $\vec{\omega}\,(=\mathrm{d}\vec{\theta}/\mathrm{d}t$,与纸面垂直,向下)的角速度旋转时,沿着圆周切线方向的线速度为 $\vec{\omega}\times\vec{r}$,其数值 ωr 不变,但方向却沿着切线方向一直变化,因该方向变化导致的向心加速度为 $(\vec{\omega})\times\vec{\omega}\times\vec{r}=\omega^2r$,指向旋转中心的方向,而具有同样大小的离心加速度则指向相反的方向。由经典力学可知,离心加速度和质量的乘积等于力。因此 $-m\omega^2r$ 被称为

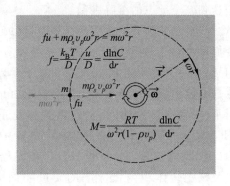

背离圆心方向的"离心力"。然而,离心力不是一个"力"!其实际上根本不存在,只是一种惯性的表现。该命名基于其产生了一个力的效应。类似前面反复讨论和提及的"排除体积效应",色散吸引不是体积,造成了一个效应等效于一个负排除体积。

可想象用手旋转一端拴着物体的绳子。对一个站在边上的观察者而言(在惯性参考系中),旋转者通过绳子对物体施力,不断地改变其运动方向,这不违背经典力学。如果观察者变成旋转者,站在中心并以同样的速度旋转,其就会发现物体明显受力(手的感觉),但其却相对地静止。"物体受一外力而不动",这就明显地违背了经典力学。正是为了在一个非惯性的旋转参考系中使用经典力学,就不得不引进了一个虚拟的"离心"力。

旋转前,大分子作为单链均匀地分布在溶液中。旋转后,每条大分子链在离心力的驱动下朝着背离旋转中心的方向加速移动。对给定的旋转速度,离旋转中心越远,移动速度($u=\mathrm{d}r/\mathrm{d}t$)越快。速度越快,链所受的摩擦阻力($fu$)也越大,与移动方向相反,其中 f 是链与溶剂的摩擦系数,为热能(k_BT)与链的平动扩散系数(D)之比,即 $f=k_BT/D$。链的体积为链的偏比容与质量之积,即 v_pm。每条链在溶液中排除了同等体积但质量为 ρ_sv_pm 的若干个溶剂分子,故会感受到一个与离心加速度成正比,但方向相反的"浮力"。如上图所示,每条链承受了三个"力";达到平衡时,三"力"之和为零,即

$$fu+\rho_sv_pm\,\omega^2r-m\,\omega^2r=0 \quad\rightarrow\quad m=\frac{fu}{\omega^2r(1-\rho_sv_p)}$$

其中,ρ_s 和 v_p 可由与密度相关的仪器分别测出。达到这一动态平衡后,大分子链并非处于静止状态,而是匀速地向溶液底部移动。另一方面,由于每个大分子都从溶液顶部向底部移动,在后方留下质量较低、移动缓慢的溶剂分子,形成由顶至底递增的大分子浓度梯度。这一梯度驱使着链朝顶部反向移动。实际上,这是处于压强较高处的溶剂分子向处于压强较低处的大分子层移动,驱动着大分子向上移动。物理图像如下。

旋转前,有一层处于溶液表面的大分子(界面清晰,外部和内部浓度分别为 0 和 C)。旋转后,它们和溶液里其他大分子一样在离心场中朝底部移动,先加速,后匀速。由于浓度梯度造成它们的反向移动(平动扩散),原本清晰的顶部大分子层逐渐地变得模糊,浓度分布逐渐地变宽,但反向移动并不改变其中心的位置。该中心位置继续向底部移动,直至达到一个动态的沉降平衡。

问题是如何将上式中的微观变量(m、f 和 u)转变成可测得宏观物理量?先考察一下初步的流体力学,在一个流动场中,流量(J)定义为单位时间内流过单位面积的物体质量,即 $J = dQ/(Adt)$,如右图所示;在本书讨论的问题中,物体是大分子链。显然,在两个面积为 A、且无限靠近($dr \to 0$)的面板之间(为了清楚起见,放大了 dr),体积变化 $dV = Adr$,以及该变化体积内大分子的质量变化 $dQ = Q_{out} - Q_{in}$;换算成大分子浓度(C),可得 $dQ = CdV = CAdr$,即 $J = Cdr/dt$。

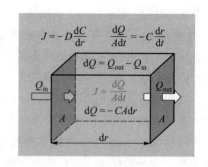

另外,由物理直觉可知,大分子自动地从高浓度向低浓度平动扩散时,流量应随着浓度差变大而增加,正比于浓度梯度(dC/dr)。Adolf Fick 早在 1855 年已指出正比常数为平动扩散系数(D),即 Fick 第一定律:$J = -DdC/dr$,引入负号是由于当 dC/dr 为正时,流出量大于流入量,J 必定为一个负值。将 $J = Cdr/dt$ 代入 Fick 第一定律可得 $fu = k_B T dlnC/dr$。将其进一步代入上式,并在两边分别乘以 N_{AV},可得

$$M = \frac{RT}{\omega^2(1 - \rho_s v_p)} \frac{dlnC}{rdr} = \frac{2RT}{\omega^2(1 - \rho_s v_p)} \frac{dlnC}{dr^2} \to \frac{\omega^2(1 - \rho_s v_p)}{2RT} \frac{dr^2}{dlnC} = \frac{1}{M} \quad (5.14)$$

对于给定的大分子溶液、仪器和转速,上式左边的前置因子为一常数,如果可测得一系列在不同 r 处的对应浓度 C,可从"lnC 对 r^2"的斜率得到大分子的摩尔质量。然而,与小分子不同的是,大分子链的长短不一、大小不同。

对于一个多分散的样品,在任意一个给定的位置(r),具有摩尔质量 M_i 的第 i 个组分的浓度($C_{i,r}$)并不等于旋转前其在起始溶液中的浓度(C_i);上式表达了每一组分的浓度 $C_{i,r}$ 随 r 的变化,注意:M_i 则是一个不依赖于 r 的常数。因此,在计算平均摩尔质量时,需要先算出每个组分在溶液中的总值(C_i),即对 r 从顶部至底部积分(整个溶液),再将所有组分的摩尔质量经过不同的浓度权重后加和,得到不同的平均摩尔质量。

先将上式中的 $dlnC_{i,r}$ 写成 $dC_{i,r}/C_{i,r}$,再将 $C_{i,r}$ 移至左边。将左边从顶部(r_T)至底部(r_B)和对应地将右边从顶部浓度(C_{i,r_T})至底部浓度(C_{i,r_T})积分,可得

$$\int_{r_T}^{r_B} M_i C_{i,r} dr^2 = \frac{2RT}{\omega^2(1 - \rho_s v_p)} \int_{C_{i,r_T}}^{C_{i,r_B}} dC_{i,r}$$

其中,$C_{i,r}$ 随 r 变化,但是其平均值 C_i 则是一个常数,故左边的积分等于 C_i 乘以相应的宽度:

$$M_i C_i(r_B^2 - r_T^2) = \frac{2RT}{\omega^2(1 - \rho_s v_p)}(C_{i,r_B} - C_{i,r_T})$$

两边求和,再分别除以 $C = \sum_i C_i$。右边,$\sum_i C_{i,r_B} - \sum_i C_{i,r_T} = C_{r_B} - C_{r_T}$。左边的常数 $(r_B^2 - r_T^2)$ 移至右边,可得

$$M_w = \frac{\sum_i M_i C_i}{\sum_i C_i} = \frac{2RT(C_{r_B} - C_{r_T})}{\omega^2(1 - \rho_s v_p)(r_B^2 - r_T^2)C} \tag{5.15}$$

在离心平衡的实验中,只要测出达到平衡后溶液表面和底部的浓度,即可结合其他实验参数求出大分子的**重均摩尔质量**,这对于直接量度浓度的光学检测方法而言,十分简便。

达到沉降平衡时,大分子浓度(C_r)在整个池子里沿着半径(r)从溶液顶部(r_T)至溶液底部(r_B)逐渐增大,形成一个动态稳定的分布。将式(5.14)重写,并对左边从 r_T 至 r 和右边从 $C_{r,T}$ 至 C 积分,可得该浓度分布如下,

$$\int_{r_T}^{r} \frac{M\omega^2(1 - \rho_s v_p)}{2RT} dr^2 = \int_{C_{r_T}}^{C_r} d\ln C_r$$

$$\frac{M\omega^2(1 - \rho_s v_p)}{2RT}(r^2 - r_T^2) = \ln\frac{C_r}{C_{r_T}}$$

$$C_r = C_{r_T} \exp\left[\frac{M\omega^2(1 - \rho_s v_p)}{2RT}(r^2 - r_T^2)\right] \tag{5.16}$$

对于一个多分散的样品,上式只是描绘了其中每一个组分在溶液中浓度对位置的依赖,即在每一个 r 处,有一个对应于 M_i 的浓度 $C_{i,r}$。对上式的两边分别从 r_T 至 r_B 积分,囊括了一个组分在整个溶液中所有的链,等于该组分在溶液中原本的浓度(C_i)。另一方面,对一个组分,即将 M 和 C_r 分别换成 M_i 和 $C_{i,r}$,式(5.14)可被重写成

$$\frac{\omega^2(1 - \rho_s v_p)}{2RT} M_i = \frac{dC_{i,r}}{C_{i,r} dr^2}$$

对给定的实验条件和质量为 M_i 的大分子链,其左边是一个独立于 r 的量。将 M_i 移到右边后,$(dC_{i,r}/dr^2)/(M_i C_{i,r})$ 是一个独立于 r 和 M_i 二者的量;如果将 $dC_{i,r}/dr$ 定义为 $z_{i,r}$,$(z_{i,r}/r)$ 正比于 $M_i C_{i,r}$,以及 M_i 不随 r 变化,故 $d\ln C_{i,r} = d\ln(z_i/r)_r$。注意,$C_r$ 包括了位于 r 处的所有组分,故 $C_{i,r}$ 不等于 C_r。上式可被重写成

$$M_i = \frac{2RT}{\omega^2(1 - \rho_s v_p)}\frac{d\ln C_{i,r}}{dr^2} = \frac{2RT}{\omega^2(1 - \rho_s v_p)}\frac{d\ln\left(\frac{z_{i,r}}{r}\right)}{dr^2} = \frac{2RT}{\omega^2(1 - \rho_s v_p)}\frac{d\left(\frac{z_{i,r}}{r}\right)}{\frac{z_{i,r}}{r} dr^2}$$

$$M_i \frac{z_{i,r}}{r} dr^2 = \frac{2RT}{\omega^2(1 - \rho_s v_p)} d\left(\frac{z_{i,r}}{r}\right)$$

两边分别从顶部到底部积分,左边从 r_T 到 r_B,右边从 $(z_i/r)_{r_T}$ 到 $(z_i/r)_{r_B}$,即

$$M_i \int_{r_T}^{r_B} 2z_{i,r} dr = \frac{2RT}{\omega^2(1 - \rho_s v_p)} \int_{\left(\frac{z_i}{r}\right)_{r_T}}^{\left(\frac{z_i}{r}\right)_{r_B}} d\left(\frac{z_{i,r}}{r}\right)$$

其中,$z_{i,r}$ 在溶液中是一个变量,但它的平均值 z_i 是常数。如前所述,左边的积分等于一个维度的平均值乘上另一个对应维度的宽度,即

$$M_i z_i \int_{r_{\mathrm{T}}}^{r_{\mathrm{B}}} \mathrm{d}r = \frac{RT}{\omega^2(1 - \rho_s v_p)}\left[\left(\frac{z_i}{r}\right)_{r_{\mathrm{B}}} - \left(\frac{z_i}{r}\right)_{r_{\mathrm{T}}}\right]$$

将两边求和,左边的积分与对 i 的求和无关,为一常数。因此,可得

$$\sum_i M_i z_i \left(\int_{r_{\mathrm{T}}}^{r_{\mathrm{B}}} \mathrm{d}r\right) = \frac{RT}{\omega^2(1 - \rho_s v_p)} \sum_i \left[\left(\frac{z_i}{r}\right)_{r_{\mathrm{B}}} - \left(\frac{z_i}{r}\right)_{r_{\mathrm{T}}}\right]$$

在左边分别乘以和除以一个常数, $z = \sum_i z_i$。注意其与位置无关,以及 $\mathrm{d}C/\mathrm{d}r = z$,即 $z\mathrm{d}r = \mathrm{d}C$,可得

$$\frac{\sum_i M_i z_i}{\sum_i z_i}\left(\int_{C_{r_{\mathrm{T}}}}^{C_{r_{\mathrm{B}}}} \mathrm{d}C\right) = \frac{RT}{\omega^2(1 - \rho_s v_p)}\left[\left(\frac{z}{r}\right)_{r_{\mathrm{B}}} - \left(\frac{z}{r}\right)_{r_{\mathrm{T}}}\right]$$

其中,左边的积分为 $(C_{r_{\mathrm{B}}} - C_{r_{\mathrm{T}}})$,其可被移至右边。因此,

$$M_z = \frac{\sum_i M_i z_i}{\sum_i z_i} = \frac{\sum_i M_i^2 C_i}{\sum_i M_i C_i} = \frac{2RT}{\omega^2(1 - \rho_s v_p)}\frac{\left[\left(\dfrac{\mathrm{d}C}{\mathrm{d}r^2}\right)_{r_{\mathrm{B}}} - \left(\dfrac{\mathrm{d}C}{\mathrm{d}r^2}\right)_{r_{\mathrm{T}}}\right]}{(C_{r_{\mathrm{B}}} - C_{r_{\mathrm{T}}})} \tag{5.17}$$

其中,利用了 $z_i \propto M_i C_i$,以及前述关于平均摩尔质量的一般定义和 z 均摩尔质量的具体定义。这也许正是当初为何将 $\sum_i M_i^2 C_i / \sum_i M_i C_i$ 定义为 z 均的原因。

其显示,如果在超速离心实验中,测得 C_r 对 r 的依赖性,用 C_r 对 r^2 作图,求得每点的斜率 $\mathrm{d}C_r/\mathrm{d}r^2$ 后,再将 $\mathrm{d}C_r/\mathrm{d}r^2$ 对 C_r 作图,获得在溶液顶部和底部的两个斜率。将这两个斜率和对应的浓度以及其他实验参数结合即可获得 z 均摩尔质量。

超速离心实验中常用的光学纹影(Schlieren)检测器可直接测得每处的浓度梯度。与其对应的浓度可由紫外吸收或光学干涉检测器获得。这样,利用上式计算 z 均摩尔质量就十分方便。如果只是采用一个紫外吸收或光学干涉检测器,所测的是浓度对位置的依赖性,可较易地算出浓度的对数,基于式(5.15),可得重均摩尔质量。这充分地显示,不同的平均摩尔质量不是源于数学定义和操作,而取决于不同的实验和检测方法。

也可从热力学原理直接推出式(5.14)。具体如下:对大分子而言,混合前,大分子是一个纯物质,其自由能为 $G^*(0)$;混合后,自由能降低为 $G(X_p)$, X_p 为大分子的体积分数。自由能的变化(混合自由能 ΔG_{mix})为

$$G(X_p) - G^*(0) = \Delta G_{\mathrm{mix}}(X_p)$$

如前所述,在离心力 $m\omega^2 r$ 的驱动下,质量为 m 的大分子沿着径向朝着溶液底部运动。越远受力越大,速度更快,形成一个密度梯度(压强梯度),由旋转中心向外递增。其直接导致一个量纲为力的自由能梯度($\mathrm{d}G/\mathrm{d}r$)。因此,处于该压强梯度中的大分子链必定向内(溶液顶部)移动,以减小该密度梯度。

讨论渗透压差时,已用到 $\mathrm{d}G = V\mathrm{d}p - S\mathrm{d}T$;在等温条件下, $\mathrm{d}T = 0$。当离心力和熵增引致的驱动力达至平衡时, $\mathrm{d}G/\mathrm{d}r + m\omega^2 r = 0$ 或 $V\mathrm{d}p/\mathrm{d}r = -m\omega^2 r$,也可写成 $\mathrm{d}p/\mathrm{d}r = -\rho\omega^2 r$, ρ 为体系的密度。在一个等温和等压过程中,将上式中的每一项对 r 作偏微分,

$$\left[\frac{\partial G(X_p)}{\partial r}\right]_{T,p} - \left[\frac{\partial G^*(0)}{\partial r}\right]_{T,p} = \left[\frac{\partial \Delta G_{\mathrm{mix}}(X_p)}{\partial r}\right]_{T,p}$$

将 $dG/dr = Vdp/dr$ 代入,并用上标" $*$ "区别混合物(溶液)和纯物质(大分子),得

$$V\frac{dp}{dr} - V^*\frac{dp}{dr} = \left[\frac{\partial \Delta G_{mix}(X_p)}{\partial r}\right]_{T,p} \tag{5.18}$$

将上式中的每一项对 n_p 作偏微分。依据定义 $n_p = m/M$ 和 $(\partial V/\partial m)_{T,p,n_s} = v_p$,不难得到 $(\partial V/\partial n_p)_{T,p,n_s} = v_p M$,其中 $(\partial V^*/\partial n_p)_{T,p}$ 是大分子的摩尔体积 (V_m^*)。代入上式可得

$$v_p M\frac{dp}{dr} - V_m^*\frac{dp}{dr} = \left[\frac{\partial \Delta G_{mix}(X_p)/\partial n_p}{\partial r}\right]_{T,p} = \left[\frac{\partial \Delta \mu(X_p)}{\partial r}\right]_{T,p} \tag{5.19}$$

采用纯大分子为参照状态(初始状态), $\Delta\mu(X_p) = \mu(X_p) - \mu^*(0)$。对 n_p 作偏微分的计算可分成两步:先对 X_p 微分,再乘以 X_p 对 n_p 的微分。结果为

$$\left(\frac{\partial \Delta G_{mix}}{\partial n_p}\right)_{T,p,n_s} = \mu_p - \mu_p^* = RT\frac{\partial[n_s\ln(1-X_p) + n_p\ln X_p + n_s X_p \chi_{sp}]}{\partial X_p}\frac{\partial X_p}{\partial n_p}$$

$$= RT\left[-\frac{n_s}{1-X_p} + \frac{\partial n_p}{\partial X_p}\ln X_p + \frac{n_p}{X_p} + n_s\chi_{sp}\right]\frac{\partial X_p}{\partial n_p}$$

$$= RT\left[-\frac{n_s}{1-X_p}\frac{\partial X_p}{\partial n_p} + \ln X_p + \frac{n_p}{X_p}\frac{\partial X_p}{\partial n_p} + n_s\frac{\partial X_p}{\partial n_p}\chi_{sp}\right]$$

$$\mu(X_p) - \mu^*(0) = RT[\ln X_p + (1-n_v)(1-X_p) + \chi_{sp}n_v(1-X_p)^2] \tag{5.20}$$

在以上的最后一步推导中,利用了 $(\partial \Delta G/\partial n_p)_{T,p,n_s} = \mu(X_p) - \mu^*(0)$ 和下式

$$\left(\frac{\partial X_p}{n_p}\right)_{T,p,n_s} = \frac{\partial\left(1-\frac{n_s}{n_s+n_v n_p}\right)}{\partial n_p} = \frac{n_v n_s}{(n_s+n_v n_p)^2} = \frac{X_s(1-X_s)}{n_p} = \frac{n_v(1-X_p)^2}{n_s}$$

注意:式(5.10)和式(5.20)是两个重要的公式,由它们可推导出许多可测的物理量。仔细考察这两个方程可知,自由能和化学势对组分含量的依赖性对称;物理上,在一个二元混合物中,溶剂和溶质的定义原本就是相对的,故合理。将 $dp/dr = -\rho\omega^2 r$(大分子溶液)和 $dp/dr = -\rho^*\omega^2 r = -(M/V_m^*)\omega^2 r$(纯大分子)代入式(5.18),可得

$$-v_p\rho M\omega^2 r - (-M\omega^2 r) = \left[\frac{\partial \Delta\mu(X_p)}{\partial X_p}\right]_{T,p}\left[\frac{\partial X_p}{\partial C}\right]_{T,p}\left[\frac{\partial C}{\partial r}\right]_{T,p} \tag{5.21}$$

因此,在离心场中,利用式(5.20)可将上式重写成,

$$(1-v_p\rho)M\omega^2 r = RT\frac{\partial[\ln X_p + (1-n_v)(1-X_p) + \chi_{sp}n_v(1-X_p)^2]}{\partial X_p}\frac{\partial X_p}{\partial C}\frac{\partial C}{\partial r}$$

其中,已略去等温和等压下标; ρ 是溶液密度; v_p 为上节定义和讨论过的大分子偏比容,其与大分子体积分数的关系为 $X_p = Cv_p$,即 $(\partial X_p/\partial C)_{T,p} = v_p$。注意 $X_p = n_v n_p/(n_s + n_v n_p)$ 和 $X_s + X_p = 1$,上式可被重写成

$$\omega^2 r(1-\rho v_p)M = RT\left[\frac{1}{X_p}-(1-n_v)-2\chi_{sp}n_v(1-X_p)\right]v_p\frac{\partial C}{\partial r}$$

$$=RT\left[\frac{1-X_p}{X_p}+n_v-2\chi_{sp}n_v(1-X_p)\right]v_p\frac{\partial C}{\partial r}$$

$$\omega^2 r(1-\rho v_p)=\frac{RT\,X_s}{M}\left[\frac{1}{X_p}+\left(\frac{1}{X_s}-2\chi_{sp}\right)n_v\right]v_p\frac{\partial C}{\partial r}$$

在稀溶液中,$1/X_s=1/(1-X_p)\approx 1+X_p$。如果在溶解时体积不变,$(1-\rho v_p)/X_s=1-\rho_s v_p$,即将溶液密度换算成溶剂密度,一个已知量。在温度、压强和组分不变时,偏微分 $\partial C/\partial r$ 可被写成 dC/dr。在稀溶液中,$1/X_s=1/(1-X_p)\approx 1+X_p$。上式成为

$$\omega^2 r(1-\rho_s v_p)=\frac{RT}{M}\left[\frac{1}{C}+(1-2\chi_{sp})n_v v_p+n_v v_p^2 C\right]\frac{dC}{dr}$$

将已知和可测得物理量移至左边,未知的放在右边,其可被进一步写成实验上常用的形式,

$$\frac{\omega^2 r(1-\rho_s v_p)C}{RT\frac{dC}{dr}}=\frac{1}{M}+2\left(\frac{1}{2}-\chi_{sp}\right)\frac{n_v v_p}{M}C+n_v v_p^2 C^2$$

$$\frac{\omega^2(1-\rho_s v_p)}{2RT}\frac{dr^2}{d\ln C}=\frac{1}{M}+2A_2 C+3A_3 C^2 \tag{5.22}$$

其中,$A_2=(1/2-\chi_{sp})n_v v_p/M$,与先前定义的一致,以及 $A_3=v_p^3/3$。在物理上,n_v 等于一个大分子与一个溶剂分子的体积之比。对于给定的溶液,n_v 和 v_p 为两个常数。在稀溶液中,可略去二次项,甚至一次项,回到式(5.14)。

如前所述,对于一个多分散的大分子样品,达到离心平衡时,实验上可测得一系列在不同径向位置(r)处的大分子浓度(C_r)。如果,直接测量的是浓度本身(C_r),由溶液顶部和底部的浓度,再结合其他实验参数,可得重均摩尔质量(M_w)。如果直接测量的是溶液里每处的浓度梯度(dC_r/dr)和 C,则可由在溶液顶部和底部的两个浓度梯度,结合它们对应的浓度和其他实验参数,可获 z 均摩尔质量(M_z)。

基于上述讨论,离心平衡实验主要取决于溶液中大分子在重力场下的平动扩散。平动扩散所需的时间和距离的平方成正比。大分子的平动扩散系数一般位于 $10^{-6}\sim10^{-7}$ cm^2/s 之间,扩散一个典型样品池长(~3 mm)、达到平衡的时间需以天计。设计仪器时,希望转速越快越好和样品离旋转轴心越远越好,但其始终受限于转子材料的力学性能和质量。

为了在高速旋转时测量沿着径向的大分子浓度,需要在样品池上设计两个可以透光的窗口,并配上可承受极大压强的石英玻璃,有时甚至采用更硬更强的蓝宝石。这一系列的实验要求都限制了转速、延长了实验所需的时间和降低了测量的精确度。紫外吸收和光学折射(Schlieren 检测装置或利用通过溶液和溶剂的两束光的干涉),外加底片曝光等方法在过去常被用来测量大分子的浓度梯度。本书略去测量方法和原理的详细讨论,有兴趣的读者可参考有关超速离心仪的专著。

为了缩短测量时间,实验上往往采用沉降速度,而不是沉降平衡的方法。如前所述,

旋转前,样品池里各处的浓度均匀,记为 C,也永远是整个池子的平均浓度。在顶部,从溶液外向溶液内,浓度从 0 跃至 C,界限分明。旋转后,所有的分子均向底部移动,但大分子链移动较快,溶液顶部只剩下溶剂分子,造成一个大分子的浓度梯度,其驱动大分子链向回(顶部)扩散,熵增驱动。

原本清晰的界面变得模糊,但界面的中心位置(r_p)不受反向扩散的影响,仍然加速向底部移去。随着速度($u = \mathrm{d}r_p/\mathrm{d}t$)增加,阻力增强,速度很快趋于一个常数。但是,加速度随着 r 增加,速度也逐渐增加。之后,中心位置继续移向底部,直至在底部附近达到前述的平衡状态。在这一过程中,界面因为扩散和沉降变得越来越模糊。在 Schlieren 检测中,界面是一个前行的、不断变宽的小峰($\mathrm{d}C/\mathrm{d}r$)。峰值的位置对应着界面的中心位置,可较精确地测得其随时间的变化。

理论上,方程(5.18)显示,达到动态平衡时,左右两边相等。大分子在离心场中的受到的净"力"为零。在达到动态平衡之前,有一个净的离心力施加在每摩尔大分子上,使得大分子向溶液底部移动,即

$$F = (1 - v_p\rho)M\omega^2 r - \left[\frac{\partial \Delta\mu}{\partial C_{r_p}}\right]_{T,p}\left[\frac{\partial C_{r_p}}{\partial r_p}\right]_{T,p} \tag{5.23}$$

在旋转的初始阶段,即界面移动不超过溶液高度的 1/4,界面基本保持清晰,无明显模糊。因此,界面浓度(C_{r_p})与起始的溶液浓度非常接近,可近似地当作不变,即 $\partial C_{r_p}/\partial r_p \cong 0$,故上式的第二项可略去。每摩尔大分子所受之"力"为 $F = fu = RTu/D$;即

$$\frac{\mathrm{d}r_p}{\mathrm{d}t} = \frac{(1 - v_p\rho)MD}{RT}\omega^2 r_p = S\,\omega^2 r_p$$

其中,S 是定义的摩尔沉降系数:

$$S = \frac{(1 - v_p\rho)MD}{RT} = \frac{\mathrm{d}\ln r_p}{\omega^2 \mathrm{d}t} \quad \rightarrow \quad \frac{S}{D} = \frac{(1 - v_p\rho)}{RT}M \tag{5.24}$$

右边的表达形式是一个极其有用和实用的公式。在一个实验中,可先测得离心初期峰值位置 r_p 随时间的变化,再将 $\ln r_p$ 对 t 作图。对给定的胶体或大分子溶液和其他实验条件,该图应是一条直线,由其斜率可得摩尔沉降系数 S。如果 S 不随时间变化,对上式两边分别从 r_0 至 r_p 和从 t_0 至 t 积分可得

$$\ln r_p = S\,\omega^2 t + \ln r_{p,0} - S\,\omega^2 t_0$$

其中,$r_{p,0}$ 和 r_p 分别是峰值位置在 t_0 和 t 时离开旋转轴心的距离。对给定的大分子溶液和转速,右边的两项均为常数。

由于 S 可由超速离心实验获得,对于一个已知摩尔质量 M 的大分子,利用超速离心实验可得到大分子在给定溶剂中的平动扩散系数 D,算出其摩擦系数 $f = RT/D$。反之,如果利用其他方法(如动态激光散射)先测得 D,则可利用超速离心实验测定大分子的摩尔质量 M。前提是实验温度和溶剂相同,以及所用的浓度相同或相近。对于真实溶液,存在着排除体积效应,所以有以下两个一般性的方程。

$$\frac{1}{S} = \frac{1}{S_0}(1 + A_{2,s}C + A_{3,s}C^2 + \cdots) \text{ 和 } D = D_0(1 + A_{2,D}C + A_{3,D}C^2 + \cdots)$$

其中，A 分别是沉降系数和平动扩散系数的不同维里系数。对给定的大分子和溶剂，它们均为常数。实验中，先配制一系列不同浓度的溶液，分别测得每个溶液中的 S 和 D，再将 $1/S$ 和 D 分别对浓度作图。外推至浓度为零可获 S_0 和 D_0。溶液中 S 和 D 的二阶维里系数具有相同的正负。在求二者比值时，浓度效应部分地互相抵消。因此，对稀溶液而言，可用 S/D 代替 S_0/D_0 而不会造成太大的误差。对于一个多分散的样品，

$$\frac{RT}{(1 - v_p\rho)} \sum_i S_i = \sum_i D_i M_i = \sum_i k_D M_i^{1-\alpha_D}$$

$$M_s = (\sum_i M_i^{1-\alpha_D})^{\frac{1}{1-\alpha_D}} = \left[\frac{RTS}{(1 - v_p\rho) k_D} \right]^{\frac{1}{1-\alpha_D}} \tag{5.25}$$

其中，利用了大分子尺寸和摩尔质量之间的标度关系，即 $D = k_D M^{-\alpha_D}$，$1/3 \leqslant \alpha_D \leqslant 3/5$。因此，$S$ 均摩尔质量(M_s)源自一种特殊的平均，其处于数均和重均摩尔质量之间。

Theodor Svedberg 早在 1924-1925 年就发明了世界上第一台超速离心仪。当时的转速已达到 12,000 rpm，可产生 7,000 倍的重力加速度(g)。不到两年后，他又成功地将转速提升了近四倍，产生 100,000 g。利用这些超速离心仪，他研究了一系列胶体和蛋白质，并因这些研究和成果于 1926 年荣获 Nobel 化学奖。十年后，Edward Greydon Pickels 发明了真空超速离心仪，既减少了空气对转速的影响，又避免了转子与空气摩擦而发热。离心仪内的温度梯度可造成空气对流、不精确控温，直接影响浓度梯度的建立、测量和分析。

早期，铝合金是转子的主要材料。现在，较昂贵的钛合金也是一个选择。目前，新颖、质轻高强的碳纤维也进入了仪器制造者的视线。在相当长的时间里，超速离心仪是研究大分子溶液的一个非常重要和珍贵的、且不需要标定曲线的仪器。正因为实验周期很长，数据处理烦琐，和一些其他缺点，其逐步地被一些后来发展的快速分析仪器所取代，尤其是便宜、快速和简单的凝胶渗透（尺寸排除）色谱在 20 世纪 60 年代出现之后。

然而，作为一个无须标定曲线和按照粒子质量分级的方法，在大分子研究中，超速离心仪仍然有着不可替代的作用，尤其是在生物大分子的研究中。近年来，材料的发展使得转子可做得更轻、半径更大、转速更高，可达一百万倍重力加速度。目前，即使在一个转子上铣出十几个样品池空洞，也不影响转子在高速旋转时的力学性能。这样，就可在一次实验中测量两种、多个浓度的溶液，以满足浓度依赖性的研究。在过去三十几年内，由于光敏元件的检测精度和速率的极大提高，以及计算机计算和数据处理速度的飞速发展，现代的超速离心仪早已经摆脱了制作底片的烦恼。分析型超速离心仪自 20 世纪 90 年代以来，逐步地回到仪器市场，开始重新进入一些大分子研究实验室。这是为何本书要用一节特地介绍这一重要方法。

最后，需要强调超速离心实验中必须注意的安全事项。由于高达几万转每分钟的转速，转子具有极大的动能。在其制作的过程中，厂家已经竭尽全力地使得转子质量分布均匀。然而在实际使用中，尤其是在化学实验室里，不小心溅出的液体可能腐蚀转子表面，还有无意地磕碰也会造成微小的质量分布不均。对其他实验，这可能并不重要。但在一个超速离心实验中，这是需要高度重视的。否则，轻则机毁（例如，因受力不均，转轴弯曲），重则人伤（裂开的转子犹如炮弹）。每次实验前，均需认真检查转子的每个部分，仔细查看是否有损坏和裂纹；每一对应位置上的两个样品池要精准称量，确保二者一致。还

要严格遵守仪器的其他操作规程。

小　结

在一个离心场中,液体里的大分子或粒子在一个虚拟离心力的作用下,远离旋转中心、加速而去(沉降)。运动导致的摩擦阻力随速度增加而变大,方向相反。另外,由于大分子(粒子)和分散介质(溶剂)的密度之差,其还受到一个反方向的"浮力"。随着阻力逐渐变大,三个力最终达至平衡。之后,大分子或粒子继续匀速向前,直至抵达液体底部。

基本原理:

沉降平衡:大分子(粒子)和溶剂分子均移向液体底部。但移动速度不同,造成一个沿径向的浓度梯度,驱使大分子(粒子)反向移动。达至平衡时,形成一个宏观稳定的浓度梯度(dC_r/dr),即浓度(C_r)随着离旋转中心的距离(r)增加,其对 r 的积分等于原始浓度(C)。该过程也可被想象成在离心重力场下,一层大分子(粒子)从底部向上作缓慢的扩散,时间以"天"为单位。应该注意:此平衡非前述的三力平衡也!

沉降速度:旋转前,在大分子溶液或胶体表面的内外,大分子(粒子)浓度从零突跃成 C,为一阶梯函数,清晰可见。旋转后,大分子或粒子从界面向液体底部移动,在后方(顶部)留下一个溶剂层。浓度差驱使大分子(粒子)反向扩散,模糊了原本清晰的移动界面,但其中心位置(r_p)仍为原本界面的位置,并不受扩散影响。在 Schlieren 光学测量中,该浓度梯度对应着一个移动的小峰(dC/dr)。峰值位置对应着界面位置。界面的移动(沉降)速度(dr_p/dt)依赖于转速、大分子(粒子)与分散介质的密度差、大分子(粒子)的摩尔质量,以及摩擦系数或平动扩散系数。对于给定的大分子溶液(胶体)和实验条件,前两项为常数。因此,该方法可被用来测量大分子(粒子)的摩尔质量或平动扩散系数,前提是先用其他方法测得两个变量中的一个。对单分散的蛋白质分子,可从峰的增宽得到平动扩散系数,故仅用超速离心即可得到其摩尔质量。

可测物理量:利用紫外吸收、Schlieren 光学检测装置或两束分别通过溶液和溶剂的光干涉,在沉降平衡实验中,测量在距离转动中心 r 处的大分子(粒子)浓度(C_r)或者直接测量浓度梯度(dC_r/dr);或在沉降速度实验中,测量表面一层大分子(粒子)在虚拟离心力的作用下向液体底部的移动速度,即其位置(r_p)随时间(t)不断地向液体底部运动轨迹。测量范围是实验的初期阶段,即移动距离短于溶液高度的 1/4,尽管现在的拟合软件可以测量整个沉降过程。

数据处理:

沉降平衡:对于一个给定原始浓度(C)的大分子溶液或胶体,检测器可直接测量达到沉降平衡时不同 r 处的 C_r。结合液体顶层和底层的浓度以及其他实验参数,依据式(5.15),可得在该浓度时的表观重均摩尔质量。也可从顶部和底部的浓度梯度和位置,依据式(5.17)算出表观 z 均摩尔质量。获得在不同浓度(最好五个以上)中的表观摩尔质量后,将其倒数对浓度作图,依据式(5.20),可从其截距的倒数和斜率分别获得大分子(粒子)的重均摩尔质量或 z 均摩尔质量以及二阶维里系数。

沉降速度:利用 Schlieren 光学检测装置测得大分子(粒子)的顶层界面位置(r_p)对时

间(t)的依赖性后,将 $\ln r_p$ 对 r 作图,可得一条直线。依据 $S = [\,\mathrm{d}\ln r_p/\mathrm{d}t\,]/\omega^2$,从其斜率可得摩尔沉降系数 S。基于式(5.23),如果已知大分子(胶体粒子)的摩尔质量,可算出其平动扩散系数和在液体中的摩擦系数。反之,可得到一个介于数均摩尔质量和重均摩尔质量之间的平均摩尔质量。

激光光散射

静态激光光散射:也称为经典光散射。光散射是一个每时每刻都发生在人们身边的物理现象,普罗大众只是"不识庐山真面目,只缘身在此山中"而已。如果没有空气中的气体、灰尘、水等分子和颗粒散射入射太阳光,世界就真的是"黑白分明"了,所照之处,炙热难熬,光亮耀眼;遮蔽之地,暗无天日,不见五指。正是由于它的存在,人们可通过窗户采集太阳光的散射光,而不用白日点灯;也正是由于它的存在,大家方可欣赏美丽灿烂的日出和晚霞,引发了诸多广泛流传和脍炙人口的诗篇。

利用这一物理现象,通过光和分子或粒子的相互作用,可窥探许多微观的分子参数和性质。在过去的三十多年里,由于激光、快电子、单光子检测器和计算机存储和运算能力的飞速发展,原本昂贵的研究型激光光散射仪逐渐地从专门的激光光散射实验室走进越来越多的大分子研究室,成为常规的实验仪器。今天,商品化的现代激光光散射仪一般都包括静态和动态光散射。

在静态激光光散射中,对一个给定浓度的大分子溶液或粒子分散液(以下将大分子也称为"粒子"),测量其在每个散射角度的时间平均散射光强。从该时间平均散射光强对角度的依赖性可得到粒子的重均摩尔质量、平均回转半径和溶液的二阶维里系数。在动态激光光散射里,散射光强在一个散射角度随时间的变化(涨落)可被一个单光子检测器捕获。信号经过放大和整形后,再利用一个数字时间相干仪得到"光强-光强"时间相关函数。

对一个多分散体系,通过分析所测的时间相关函数,可得体系(其中的大分子或粒子)的一个特征弛豫时间分布,在一定的条件下,该特征弛豫时间分布可转换成一个平动扩散系数分布,或进一步转换成一个流体力学半径分布。这也是为何市场上小型的单角度(通常为 90°)动态光散射仪常被称为粒度分析仪,或简称为粒度仪。

经典光散射源远流长,早在 1802 年,德国化学家 Jeremias Benjamin Richter 就注意到一束光线射入金的溶胶时所形成的较亮的光束。然而,第一个在 1860 年代对该现象做系统性研究的则是英国剑桥的 John Tyndall,他观察到散射光呈现很淡的蓝色,并发现,如果采用偏振的入射光,散射光也具有同样的偏振;他还提出了 19 世纪气象学中的有趣的两大谜题:为何天空是蓝色的? 为何来自天空的散射光是偏振的?

他的同事,剑桥 Cavendish 实验室第二任主任 John William Strutt, 3rd Baron Rayleigh(常被称为 The Lord Rayleigh)于 1871 年成功地解开了这两个谜题。进而,他在 1881 年又利用其前任 James Clerk Maxwell 的电磁场理论阐明,在无吸收和无相互作用时,光学各向同性的小粒子的散射光强(I)与入射光波长的四次方成反比(λ_0),即 $I \propto \lambda_0^{-4}$。

1944 年，Peter Debye 利用散射光强测得稀溶液中高聚物的重均摩尔质量。后来，Bruno Hasbrouck Zimm 于 1948 年提出了一个将散射光强随散射角度和浓度变化的三维曲面投影到平面上，从而可在一张图中同时将散射角度和浓度外推至零，即著名的 Zimm 作图。从此，静态光散射成为测量大分子重均摩尔质量的一个经典方法。早期，光源为汞灯；自 1960 年代激光问世后，光源逐步地被各种相干的激光取代。

自量子力学问世以来，光的波粒二象性已逐渐地被广泛接受。在阐述相关理论前，先定性地从"粒"和"波"两个角度考察光散射。如果将入射光当作一串众多平行地、连续不断地射入的"光（粒）子"，当它们与液体中数目庞大的粒子弹性撞击（无能量吸收）时，因每个"光子"与每个粒子撞击的角度不同，光子就会散射到整个球面空间，形成散射光。

另一方面，如果将入射光当作一个不断地向前传播的电磁波，由于可见光的振动频率高达 10^{15} Hz（每秒振动的次数），质量较重的原子核和分子几乎无法以此频率振动，处于相对静止状态。而较轻的电子则在该入射电磁场的作用下，沿着电磁波的振动方向不停地加速和减速，产生一个额外的诱导偶极矩。由物理已知，一个带电的粒子在加速或减速时会朝着整个球面空间发出（韧致辐射）电磁波，这就是散射光的来源。

无论是基于上述入射光的"粒性"还是"波性"的简单讨论，不用任何理论推导和数学演算，就可得出散射光强（I）的某些性质：其反比于给定检测面积的检测器到散射中心之距（r）的平方。这是因为散射光的能量均匀地分布在一个 $4\pi r^2$ 的球面上，越远接收的能量越少，就如同离一个热源越远，我们面积固定的身体感受到的热量就越小。另外，其正比于入射光强（I_0）、液体里的粒子数（"粒子"碰撞或电磁波作用的概率）和粒子的体积（"粒子"碰撞概率或一个粒子内的电子数）。显然，$I r^2/I_0$（记为 R，即 Rayleigh 比值）就是一个与散射装置无关，而只与散射体系里面粒子的数目和体积有关的变量。上述基于物理描绘和直觉得到的关系可通过下面的电动力学理论获得。

如右图所示，一束圆频率为 ω 的入射光的交变电场为方向朝右的矢量（$\vec{E}$）。$\omega = 2\pi f$，这里，f 为频率。其与在真空和传播介质中的波长（λ_0 和 λ）、光速（c）和传播介质的折射指数（n）之间的关系为 $c = \lambda f$ 和 $\lambda = \lambda_0/n$。ϕ 为初始位相。作为一个相干光源，一束激光中的每条光线均具有相同的初始位相。真空中，一个各向同性的粒子里的所有电子都将随着电场的交替变化而出现周期性的振荡，不停地加速和减速，形成一个

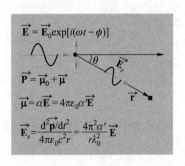

诱导偶极子（$\vec{\mu}$），其方向不断改变。如果入射光（电场）不是超强，仅需考虑诱导偶极矩对电场的线性依赖，可略去非线性项，其中的正比常数 α，α' 和 ε_0 分别是极化率、分子的体积极化率（量纲与体积相同）和真空电容率。

诱导偶极子的振荡使得每一个粒子成为一个二次点光源，向各个方向辐射同样频率的电磁波，形成散射光。依据 Maxwell 方程组可解出，在距离散射粒子 $\vec{r}$ 处的检测平面上，散射电场（$\vec{E}_s$）正比于入射电场，反比于 $|\vec{r}|$ 和入射光在真空中的波长（λ_0）的平方。粒子的偶极矩包括其永久偶极矩（$\vec{\mu}_0$）和诱导偶极矩。在常温下，$|\vec{\mu}_0| > |\vec{\mu}|$；在 ~ 10^{15} Hz（可

见光)的高频下,因分子的质量太重,$\vec{\mu}_0$的取向无法随着交变电场变化,故其在对时间的微分中消失。

除了交变电场以外,入射光的另一半能量储存在一个与垂直的、指向纸面的磁场($\vec{B}$)中。光的传播方向由右手定律决定,即伸开手掌,先用右手四指指向磁场方向,然后向电场方向旋转,拇指的指向就是光的传播方向;而入射光的时间平均光强(I_0)正比于$\vec{E}$和$\vec{B}$的矢量之积,即$I_0 \propto \vec{E} \times \vec{B}$。依据电磁理论,可算出一个粒子的散射光的时间平均光强(i_s)为

$$i_s = \varepsilon_0 c \langle \vec{E}_s \times \vec{E}_s^* \rangle = \frac{16\,\pi^4 \alpha'^2}{\lambda_0^4 r^2}[\,\varepsilon_0 c \langle \vec{E} \times \vec{E}^* \rangle\,] = \frac{16\,\pi^4 \alpha'^2}{\lambda_0^4 r^2}(\varepsilon_0 c\,E_0^2) = \frac{16\,\pi^4 \alpha'^2}{\lambda_0^4 r^2}I_0$$

散射光强对粒子体积、入射光强和到散射中心距离的依赖性之前已从物理直观获得。唯一新获的是与入射波长的四次方成反比的依赖性。这也是当初获得此式后,Rayleigh 可解释为何天空呈现蓝色。在可见光中,蓝光的波长较短,故散射较强,天空的颜色主要由散射光构成。也可据此解释为何日出和晚霞会呈现红色。由于地球为球形,清晨和黄昏时,太阳光要经过更厚的大气层方可到达眼睛,当蓝光被散射后,进入视线的就只剩下偏红的光线了。大气中水蒸气的含量越高,晚霞就会越红;故有农业谚语:"早烧不出门、晚烧行千里",因地球自转,朝霞意味着出行者迎着水汽而行,而晚霞表示水汽已离去。

注意:散射体积(V)仅有 $200\ \mu m^3$,即约 10^{-5} mL,远小于溶液体积(约 10^0 mL)。假定散射体积内有 N_p 个独立的粒子。每个粒子的尺寸小于波长的 5%,即小于 25 nm,具有相同 α'。所有粒子的散射电场之和为 $\vec{E}_s = \sum_{i=1}^{N_p} \vec{E}_{s,i}$,故单位体积的散射光强为

$$I_s = \frac{16\,\pi^4 \alpha'^2}{\lambda_0^4 r^2 V}[\,\varepsilon_0 c \langle \sum_{i=1}^{N_p} \vec{E}_{s,i} \times \sum_{j=1}^{N_p} \vec{E}_{s,j}^* \rangle\,] = \frac{16\,\pi^4 \alpha'^2}{\lambda_0^4 r^2 V}\varepsilon_0 c \sum_{i=1}^{N_p} \sum_{j=1}^{N_p} \langle \vec{E}_{s,i} \times \vec{E}_{s,j}^* \rangle$$

求和展开后共有 N_p^2 项;其中,有 N_p 个相同项 $\langle \vec{E}_{s,i} \times \vec{E}_{s,j}^* \rangle = E_{s,i}^2 (i=j)$ 和 $(N_p^2 - N_p)$ 个交叉项 $\langle \vec{E}_{s,i} \times \vec{E}_{s,j}^* \rangle = \langle \vec{E}_{s,i} \rangle \langle \vec{E}_{s,j}^* \rangle = 0(i \neq j)$。所以,上式可被重写为

$$I_s = \frac{16\,\pi^4 \alpha'^2 N_p}{\lambda_0^4 r^2 V}I_0$$

依据前述 Rayleigh 值(R)的定义,上式可再被重写成一个只与散射体系相关的下式。

$$R = \frac{I_s r^2}{I_0} = \frac{16\,\pi^4 \alpha'^2 N_p}{\lambda_0^4 V} \tag{5.26}$$

α' 和 N_p/V 分别对应着粒子的物理本性、体积(质量)和散射体积内的粒子密度(浓度)。因此,测得一个已知密度的纯液体或已知浓度的溶液的绝对时间平均散射光强可求出粒子体积(质量),这正是利用光散射测量大分子或粒子摩尔质量的物理基础。

对体积为 V' 的大粒子,可将其分割成 N 个具有相同 α'_0 的散射单元,即 $\alpha'_0 = \alpha'/N$。入射光里的两条光线分别与第 i 个和第 j 个单元发生散射,如下图所示。作为一个相干的激光光源,所有光线的初始位相一致,$\phi_{0,i} = \phi_{0,j}$。然而,被一个透镜聚焦在检测平面上的这

两条光线却经过了不同的光程,二者之间的光程差,即两个蓝色矢量之差,为 $\vec{r}_{l,m} = \vec{l} - \vec{m} = \vec{r}_l - \vec{r}_m$,注意:位于一个粒子内的两个矢量差的最大值也不会大于粒子尺寸,远远小于散射中心到检测平面的距离($\vec{r}$)。

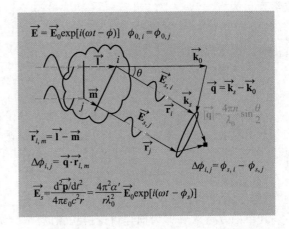

在弹性散射中,无能量损失,故入射光和散射光的波矢量($\vec{k}_0$ 和 $\vec{k}_s$)方向不同,但大小一样($2\pi n/\lambda_0$),其中 n 为介质的折射指数。二者之差为散射矢量($\vec{q}$),大小为一个等边三角形的底部,即两倍的"斜边乘以半个顶角的正弦值"(见上图)。注意,$\vec{q}$ 的量纲为长度的倒数。物理上,其等效于量度粒子尺寸的一把微观光学尺子。再由几何关系可知,$\vec{q} \cdot \vec{r}_{l,m}$ 等于两条散射光线的位相差 $\Delta \phi_{i,j} = \phi_{s,i} - \phi_{s,j}$,是一个无量纲的物理量。一个大粒子的散射电场应为其所有单元的散射电场的矢量之和,即 $\vec{E} = \sum_{i=1}^{N} \vec{E}_{s,i}$。注意,这里加和的总数为一个粒子内的散射单元数($N$)。所以,一个粒子的时间平均散射光强为

$$i_s = \frac{16\,\pi^4 \alpha_0'^2}{\lambda_0^4 r^2} I_0 \sum_{i=1}^{N} \sum_{j=1}^{N} \exp\left(i\Delta\phi_{i,j}\right) = \frac{16\,\pi^4 \alpha_0'^2}{\lambda_0^4 r^2} I_0 \sum_{l=1}^{N} \sum_{m=1}^{N} \exp\left(i\,\vec{q}\cdot\vec{r}_{l,m}\right)$$

即使粒子刚性,$\vec{r}_{l,m}$ 的大小不变,但由于每个粒子都在无序地平动和转动,其方向也在不停地变化,故还要对所有可能的取向作统计平均,结果如下

$$i_s(q) = \frac{16\,\pi^4 \alpha_0'^2}{\lambda_0^4 r^2} I_0 \sum_{l=1}^{N} \sum_{m=1}^{N} \left\langle \exp\left(i\,\vec{q}\cdot\vec{r}_{l,m}\right) \right\rangle = \frac{16\,\pi^4 \alpha_0'^2}{\lambda_0^4 r^2} I_0 \sum_{l=1}^{N} \sum_{m=1}^{N} \frac{\sin\left(q\,r_{l,m}\right)}{q\,r_{l,m}}$$

此处的 $i_s(q)$ 是源自一个粒子的散射光强,N_p 个互相独立、毫不相干的粒子的散射光强即为 N_p 倍。上式中的双重求和展开后共有 N^2 项。数学上已知,当 $x \to 0$ 时,$\sin(x)/x \to 1$。所以,当粒子或者散射角度很小时,$\sin(qr_{l,m})/qr_{l,m} \to 1$,故双重求和的结果为 N^2。注意:$\alpha_0'^2 N^2 = \alpha'^2$,上式成为先前讨论的结果。依据 Rayleigh 值的定义,可得一个粒子的 $R(q)$,如果将其用散射角度为零时的 $R(0)$ 归一,可得到一个变量 $P(q)$,其只和粒子或大分子的结构有关,而与粒子的摩尔质量无关,故其称作粒子或大分子的结构因子,也曰散射函数,如右图所示。顾名思义,归一后的结构因子,只与粒子或大分子的结构有关。在稀溶液里,其也不依赖于粒子或大分子浓度。其

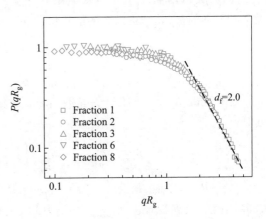

Li L W, He C, He W D, et al. Macromolecules, 2011,44:8195-8206.

(Figure 12, permission was granted by publisher)

数学表达为

$$P(q) = \frac{R(q)}{R(0)} = \frac{\dfrac{i_s(q)r^2}{I_0}}{\dfrac{i_s(0)r^2}{I_0}} = \frac{1}{N^2} \sum_{l=1}^{N} \sum_{m=1}^{N} \langle \exp(i\vec{\mathbf{q}} \cdot \vec{\mathbf{r}}_{l,m}) \rangle = \frac{1}{N^2} \sum_{l=1}^{N} \sum_{m=1}^{N} \frac{\sin(qr_{l,m})}{qr_{l,m}}$$

这里定义的结构因子已经归一，和其他大分子教科书上定义的 $S(q)$ 略有不同，多除了一个单元数 (N)。将 $\sin(qr_{l,m})$ 用 Maclaurin 级数展开，即 $\sin x = x - x^3/3! + x^5/5! - \cdots$ 得到

$$P(q) = 1 - \frac{q^2}{6N^2} \sum_{l=1}^{N} \sum_{m=1}^{N} r_{l,m}^2 + \cdots$$

其中的双重求和即为前述讨论链构象时一个物体内任意两点的方距加和。依据刚性物体回转半径的定义，$R_g^2 = \sum_{l=1}^{N} \sum_{m=1}^{N} r_{l,m}^2 / 2N^2$，故 $P(q) = 1 - q^2 R_g^2/3 + \cdots$ 对于柔性大分子链，$\vec{\mathbf{r}}_{l,m}$ 随着链的构象变化，方向和大小都在不停地、无规地变化。只有经过构象统计平均后，$\vec{\mathbf{r}}_{l,m}^2$ 才有意义，即粒子的方距被构象统计平均后的方均距取代，将上式中的 $\vec{\mathbf{r}}_{l,m}^2$ 简单地换成 $\langle \vec{\mathbf{r}}_{l,m}^2 \rangle$ 即可。因此，对于柔性大分子链，

$$P(q) = 1 - \frac{q^2 \langle R_g^2 \rangle}{3} + \cdots \tag{5.27}$$

当粒子或散射角度较小时，如果 $q^2 \langle R_g^2 \rangle \ll 1$，式 (5.27) 中的高次项可以略去。该结果对所有形状的散射粒子均适用。当粒子或散射角度较大时，$q^2 \langle R_g^2 \rangle \geqslant 1$。在此情形下，就需要用具体的模型分析粒子结构。对半径为 R、密度均匀的球形粒子，$P(q)$ 有一个解析解，

$$P_{\text{sphere}}(q) = \frac{9[\sin(qR) - qR\cos(qR)]^2}{(qR)^6}$$

随着 qR 的增加，$P_{\text{sphere}}(q)$ 从 1 降至一个最低值，然后逐步地振荡衰减。对长度为 $2L$、无限或很细的棒，其结构因子的解析结果包含一个已知的积分，

$$P_{\text{rod}}(q) = \frac{\displaystyle\int_0^{2qL} \frac{\sin x}{x} \mathrm{d}x}{qL} - \left[\frac{\sin(qL)}{qL}\right]^2$$

高斯链的 $P(q)$ 也有一个解析结果，称为 Debye 函数，对任何 $q^2 \langle R_g^2 \rangle$ 均适用，无须近似，

$$P_{\text{Gaussian}}(q) = \frac{2}{q^2 \langle R_g^2 \rangle}\left[1 - \frac{1 - \exp(-q^2 \langle R_g^2 \rangle)}{q^2 \langle R_g^2 \rangle}\right]$$

当 $q^2 \langle R_g^2 \rangle \gg 1$ 时，$P_{\text{Gaussian}}(q)$ 趋于 $2/q^2 \langle R_g^2 \rangle$。可将"$q^2 \langle R_g^2 \rangle P(q)$ 对 $q^2 \langle R_g^2 \rangle$"作图，即 Kratky 作图，其在物理上等效于半径为 qR_g 的球面上的面密度。由这样的作图可知，当 $q^2 \langle R_g^2 \rangle \gg 1$ 时，$q^2 \langle R_g^2 \rangle P(q) \rightarrow 2$。

在 20 世纪 90 年代中期，Walther Burchard [Macromolecules, 1997, 10:919] 指出，由含有三个反应基团的单体 A_2B 通过 A—B(2 + 1) 反应得到的支化型大分子具有下列结构

因子：

$$P_{\text{branched}}(q) = \frac{1 + \dfrac{q^2 \langle R_g^2 \rangle}{3\,n_B}}{\left[1 + \left(\dfrac{1 + n_B}{6 n_B}\right) q^2 \langle R_g^2 \rangle\right]^2}$$

采用 Kratky 作图法，两边同时乘以 $q^2 \langle R_g^2 \rangle$，当 $q^2 \langle R_g^2 \rangle \gg 1$ 时，$P(q \to \infty) \to 12\, n_B /(1 + n_B)^2$，由此可得支化型大分子链上平均的支化点数目（$n_B$），如右图所示。由一个支化大分子样品得到的五个不同级分（不同摩尔质量）的散射因子互相重叠，证实，$P(q)$ 确实不依赖于链的摩尔质量。另外，正如式（5.27）所示，当 $qR_g < 1$ 时，具有不同支化结构的支化链的结构因子互相重叠，已无法区分大分子结构细节。因此，只有在角度或粒子足够大时，方可采用由散射实验测得的结构因子导出结构信息。还要注意：粒子或大分子应比较均一。如非，拟合质量就会较差，结论也不十分确定。

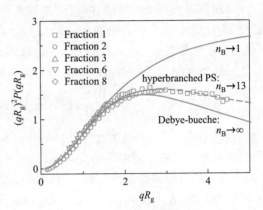

Li L W, He C, He W D, et al. Macromolecules, 2011, 44:8195-8206.

(Figure 13, permission was granted by publisher)

当 $q^2 \langle R_g^2 \rangle \to 0$ 时，依照定义，上述各种散射结构因子 $P(q) \to 1$；当 $q^2 \langle R_g^2 \rangle \ll 1$ 或者 $qL \ll 1$ 时，上述各种散射结构因子都还原成式（5.27）。对于普通高聚物链而言，在实验上，总可找到满足 $q^2 \langle R_g^2 \rangle < 1$ 的角度范围；在这一范围内，在足够多的角度上测得大分子溶液的时间平均散射光强，就可得到较精确的平均回转半径。如果散射体积（V）内有 N_p 个大粒子，由时间平均散射光强得到的 Rayleigh 比值为

$$R(q) = \frac{I_s r^2}{I_0} = \frac{16\,\pi^4 \alpha'^2 N_p}{\lambda_0^4 V} P(q) = \frac{16\,\pi^4 \alpha'^2 N_p}{\lambda_0^4 V}\left[1 - \frac{q^2 \langle R_g^2 \rangle}{3} + \cdots\right] \tag{5.28}$$

基于上述讨论，对一个均匀纯物质而言，其中任意一个与散射中心相距（$\vec{r}$）的散射单元都有一个对应的位于 $-\vec{r}$ 处的散射单元。它们的散射电场在检测平面的位相差正好是 π，故会干涉相消。因此，均匀纯物质的散射光强理应为零。

如果一个均匀的、毫无杂质的光学玻璃或其他透明的固体材料含有一些微小的缺陷，如小气泡（即使内部为真空），该材料也会产生散射，甚至很强，常被误解成气泡散射。气泡里的分子数很少，如果真空，更没有分子，何来散射？其实，这是每一个与气泡相对应的等体积的固体的散射电场无法被抵消的缘故，即散射源于一个对应的小的固体单元。物理上，散射源于体系里相对的密度涨落。如果将固体材料本身当作参照点，定义其密度为零，那么每个气泡就等效于一个具有负密度的、等体积的固体粒子，但气泡本身并不散射。

另一方面，当一束激光通过纯净气体或液体时，从侧面仍可见一条微弱的散射光

柱。这是由于在热能的搅动下,气体和液体中的分子不停地无规运动,引致内部的物理性质(包括密度)围绕着其平衡值随时间和空间的涨落。即使在一个纯气体或液体中,其与密度有关的体积极化率(α')总是围绕着其平均值$\langle \alpha' \rangle$涨落,即 $\alpha' = \langle \alpha' \rangle + \delta\alpha'$。将其平均值当作参考零点,每一次小的涨落就等效于一个小的"粒子",造成散射,尽管涨落本身并没有产生真正的"粒子"。

沿用前述大粒子的讨论,可较严格地推导纯液体散射,不过此处是将散射体积,而不是粒子,分割成 N 个小体积单元,$\alpha'_0 = \alpha'/N = \langle \alpha'_0 \rangle + \delta\alpha'_0$。另外,将体积极化率的平均值 $\langle \alpha'_0 \rangle$ 当作参考零点,则 $\alpha'_0 = \delta\alpha'_0$,最后,利用 Rayleigh 比值的定义和"加和后的平均等于平均后的加和",对无规涨落的 $\delta\alpha'_0$ 进行时间统计平均,可得

$$R(q) = \frac{16\,\pi^4}{\lambda_0^4 V} \sum_{l=1}^{N} \sum_{m=1}^{N} \langle \delta\alpha'_{0,l}\, \delta\alpha'_{0,m} \rangle \exp(i\vec{\mathbf{q}} \cdot \vec{\mathbf{r}}_{l,m})$$

上式中时间平均可分为自身项($l = m$):$\langle \delta\alpha'_{0,l}\, \delta\alpha'_{0,m} \rangle = \langle (\delta\alpha'_0)^2 \rangle$ 和 $\exp(i\vec{\mathbf{q}} \cdot \vec{\mathbf{r}}_{l,m}) = 1$;和两个互相独立的散射单元的交叉项($l \neq m$):$\langle \delta\alpha'_l \delta\alpha'_m \rangle = \langle \delta\alpha'_l \rangle \langle \delta\alpha'_m \rangle = 0$,其中利用了"乘积和平均可以彼此交换"和"一个无规变量的涨落的平均值一定为零"。上式双重加和的结果是 N^2 倍的 $\langle (\delta\alpha'_0)^2 \rangle$,而 $N\delta\alpha'_0 = \delta\alpha'$,所以

$$R(q) = \frac{16\,\pi^4}{\lambda_0^4 V} \langle (\delta\alpha')^2 \rangle \tag{5.29}$$

上式的左边可由测量的散射光强、入射光强等仪器参数获得。利用上式可得纯液体的方均密度涨落。以上有关粒子在液体中光散射的方程也可利用热力学原理导出。具体如下。

在二元混合物中,如在一个稀的大分子溶液或胶体分散液中,α' 除了依赖于分散介质的密度(ρ)以外,还取决于组成(浓度,C)。因此,需要寻获 $\langle (\delta\alpha')^2 \rangle$ 与摩尔质量和其他物理量之间的关系。可先考察体积极化率对介质密度和粒子浓度的依赖性,即

$$\delta\alpha' = \left(\frac{\partial\alpha'}{\partial\rho} \right)_{T,p,C} \delta\rho + \left(\frac{\partial\alpha'}{\partial C} \right)_{T,p,\rho} \delta C$$

对一个给定密度或浓度,α' 随密度或随浓度的变化率,上式中的每一个偏微分,均为常数。密度和浓度随时间的涨落互相独立。将两边分别平方后,右边的交叉项在平均后消失,故

$$\langle (\delta\alpha')^2 \rangle = \left(\frac{\partial\alpha'}{\partial\rho} \right)_{T,p,C}^2 \langle (\delta\rho)^2 \rangle + \left(\frac{\partial\alpha'}{\partial C} \right)_{T,p,\rho}^2 \langle (\delta C)^2 \rangle \tag{5.30}$$

从实验的角度出发,左边对应着溶液的散射;右边第一项的密度涨落对应着溶剂的散射;二者之差等于溶液和溶剂的可测的时间平均散射光强之差,一个实验上可测的物理量。右边第二项对应着大分子或胶体粒子的净散射光强。其与待测物理量的关系才是重要的。下面,将先求出 α' 随浓度的变化率($\partial\alpha'/\partial C$)$_{T,p,\rho}$,再计算方均浓度涨落$\langle (\delta C)^2 \rangle$。

早在 1850 年,意大利物理学家 Ottaviano-Fabrizio Mossotti 就发现了物质的电容率(ε)和真空电容率(ε_0)的比值(ε_r,相对电容率,即介电常数)与体积极化率(α')之间的关系。后来,德国物理学家和数学家 Rudolf Julius Emanuel Clausius 在 1879 年也独立地得到这一看似简单的关系。因此,其被命名为 Clausius-Mossotti 关系式,

$$\frac{\varepsilon_r - 1}{\varepsilon_r + 2} = \frac{4\pi\alpha'}{3}$$

其与 Ludvig Lorenz 于 1869 年和后来 Hendrik Lorentz 于 1878 年独立地发现的折射指数(n)与体积极化率(α')之间的 Lorentz-Lorenz 关系一致,异功同曲,缘于 $\varepsilon_r = n^2$,故

$$\left(\frac{\partial\alpha'}{\partial C}\right)_{T,p,\rho} = \left(\frac{\partial\alpha'}{\partial\varepsilon_r}\right)_{T,p,\rho}\left(\frac{\partial\varepsilon_r}{\partial n}\right)_{T,p,\rho}\left(\frac{\partial n}{\partial C}\right)_{T,p,\rho} = \frac{9nV}{2\pi(\varepsilon_r + 2)^2}\left(\frac{\partial n}{\partial C}\right)_{T,p,\rho}$$

$$\left(\frac{\partial\alpha'}{\partial C}\right)_{T,p,\rho} \approx \frac{nV}{2\pi}\left(\frac{\partial n}{\partial C}\right)_{T,p,\rho} \qquad (5.31)$$

最后一步的近似基于,对于一个给定的散射体系,$9/(\varepsilon_r + 2)^2$ 是一个略小于 1 的常数。也可在一开始就将 $\varepsilon_r + 2 \approx 3$,并用 n^2 代替 ε_r,即直接采用近似的 Lorentz-Lorenz 关系,这样也可直接得到相同的结果。下面,再利用热力学原理求出 $\langle(\delta C)^2\rangle$。

对给定的实验条件(温度、压强、浓度⋯⋯),随着体系抵达一个宏观稳定的动态平衡状态,自由能降至最低(G_0)。然而,在热能的搅动下,微观上,粒子仍在不停地运动,使得体系略微地偏离 G_0,引起自由能升高,造成一个永远为正(增加)的涨落,选择浓度(C)为一个涨落变量,将 G 在平衡浓度处作 Taylor 展开,可得

$$\delta G = G - G_0 = \left(\frac{\partial G}{\partial C}\right)_{T,p,\rho}\delta C + \frac{1}{2}\left(\frac{\partial^2 G}{\partial C^2}\right)_{T,p,\rho}(\delta C)^2 + \cdots$$

其中,在给定浓度 C 处的自由能的偏微分为常数;浓度的无规涨落的时间平均为零。因此,经时间平均后,右边的第一项消失,如果不考虑浓度涨落的三次方贡献,上式成为

$$\langle\delta G\rangle = \frac{1}{2}\left(\frac{\partial^2 G}{\partial C^2}\right)_{T,p,\rho}\langle(\delta C)^2\rangle$$

另外,从相当一般的统计考虑出发(略去推导),可知在热能的搅动下,对任何一个物理变量(Y)而言,其时间平均涨落($\langle\delta Y\rangle$)对应着半份热能($k_B T/2$)的变化。因此,每个分子的自由能的时间平均涨落 $\langle\delta G\rangle = k_B T/2$。将 k_B 换成摩尔气体常数 R,每摩尔粒子的自由能的时间平均涨落为

$$\langle(\delta C)^2\rangle = RT\Big/\left(\frac{\partial^2 G}{\partial C^2}\right)_{T,p,\rho} \qquad (5.32)$$

依据第三章和前述有关渗透压差的讨论,一个溶液的自由能包括纯溶剂的自由能,两者之差为混合自由能,$\Delta G_{\mathrm{mix}} = G - G^*$。依据其定义,$G^*$ 独立于粒子浓度,为一常数。所以,

$$\left(\frac{\partial^2\Delta G_{\mathrm{mix}}}{\partial C^2}\right)_{T,p,\rho} = \left[\frac{\partial^2(G - G^*)}{\partial C^2}\right]_{T,p,\rho} = \left(\frac{\partial^2 G}{\partial C^2}\right)_{T,p,\rho}$$

可将混合自由能对 C 的偏微分变成混合自由能对溶剂分子摩尔数(n_s)的偏微分乘上 n_s 对 C 的偏微分,即

$$\left(\frac{\partial^2 G}{\partial C^2}\right)_{T,p,\rho} = \left\{\frac{\partial\left[\left(\dfrac{\partial\Delta G_{\mathrm{mix}}}{\partial n_s}\right)_{T,p,\rho}\left(\dfrac{\partial n_s}{\partial C}\right)_{T,p,\rho}\right]}{\partial C}\right\}_{T,p,\rho}$$

依据定义,混合自由能对 n_s 的偏微分为溶液和溶剂的化学势之差($\Delta\mu$),以及

$$\left(\frac{\partial n_s}{\partial C}\right)_{T,p,\rho} = \left(\frac{\partial n_s}{\partial X_p}\right)_{T,p,\rho}\left(\frac{\partial X_p}{\partial C}\right)_{T,p,\rho} = -\frac{n_s}{X_s X_p}v_p = -\frac{V}{V_{s,m}C}$$

在上述推导中,利用了在前述讨论渗透压差中得到的下列四个关系式:$(\partial X_p / \partial n_s)_{T,p,C} = -X_s X_p / n_s$,$X_p = C\,v_p$,$X_s = V_s / V$ 和 $V_{s,m} = V_s / n_s$。注意,C 是粒子浓度。对一个给定的散射实验,其为一个定值。将上式代入前式可得

$$\left(\frac{\partial^2 G}{\partial C^2}\right)_{T,p,\rho} = -\frac{V}{V_{s,m}C}\left(\frac{\partial \Delta\mu}{\partial C}\right)_{T,p,\rho} = -\frac{V}{V_{s,m}C}\left(\frac{\partial \Delta\mu}{\partial \pi}\right)_{T,p,\rho}\left(\frac{\partial \pi}{\partial C}\right)_{T,p,\rho} = \frac{V}{C}\left(\frac{\partial \pi}{\partial C}\right)_{T,p,\rho}$$

最后一步的推导利用了在讨论渗透压差中已获得的式(5.12),即在纯溶剂中,化学势随着压强增加,$\Delta\mu = \mu^*(p+\pi) - \mu^*(p) = V_m p$ 和 $p = -\pi$。另外,在渗透压差的讨论中,已知其如何随浓度变化,即

$$\pi \cong \frac{RTC}{M}(1 + A_2 C)$$

求出 $(\partial\pi / \partial C)_{T,p,C}$,即可得到 $(\partial^2 G / \partial C^2)_{T,p,C}$。代入式(5.32),可得

$$\langle(\delta C)^2\rangle = \frac{CM}{V(1 + 2A_2 C)} \tag{5.33}$$

将式(5.31)和式(5.33)代入式(5.30),再代入式(5.29),扣除溶剂分子的散射光,得下列结果:

$$R(q) = R_{\text{solution}}(q) - R_{\text{solvent}}(q) = \frac{16\,\pi^4}{\lambda_0^4 V}\left(\frac{\partial \alpha'}{\partial C}\right)_{T,p,\rho}^2 \langle(\delta C)^2\rangle$$

$$= \frac{16\,\pi^4}{\lambda_0^4 V}\left(\frac{nV}{2\pi}\right)^2\left(\frac{\partial n}{\partial C}\right)_{T,p,\rho}^2 \frac{CM}{V(1+2A_2 C)} = \frac{4\,\pi^2 n^2}{\lambda_0^4}\left(\frac{\partial n}{\partial C}\right)_{T,p,\rho}^2 \frac{CM}{(1+2A_2 C)}$$

其中,$(\partial n / \partial C)_{T,p,\rho}$ 是微分折射指数,可由微分折射仪单独测获(将会详细介绍)。定义

$$K = \frac{4\,\pi^2 n^2}{\lambda_0^4}\left(\frac{\partial n}{\partial C}\right)_{T,p,\rho}^2$$

对大粒子而言,必须考虑结构因子 $P(q)$。当 $q^2\langle R_g^2\rangle \leqslant 1$ 时,$R(q)$ 可被重写成

$$\frac{KC}{R(q)} = \frac{(1+2A_2 C)}{MP(q)} \approx \frac{1}{M}\left(1 + \frac{q^2\langle R_g^2\rangle}{3} + 2A_2 C\cdots\right)$$

对于一个多分散的粒子或大分子样品,上式可再被重写成

$$\frac{K}{(1+2A_2 C)}\sum_i C_i M_i\left(1 - \frac{q^2\langle R_g^2\rangle_i}{3} + \cdots\right) = \frac{K}{(1+2A_2 C)}\left(\sum_i C_i M_i - \frac{q^2}{3}\sum_i C_i M_i\langle R_g^2\rangle_i + \cdots\right)$$

$$= \frac{KC}{(1+2A_2 C)}\frac{\sum_i C_i M_i}{\sum_i C_i}\left(1 - \frac{q^2}{3}\frac{\sum_i C_i M_i\langle R_g^2\rangle_i}{\sum_i C_i M_i} + \cdots\right)$$

$$= \sum_i R_i(q) = R(q)$$

在最后一步的运算中,左边的分子和分母同时乘了一个浓度 C,依据 $C = \sum_i C_i$,以及重均摩尔质量的定义,上式可重写成一个实验上极其有用的如下形式:

$$\frac{KC}{R(q)} \approx \frac{1}{M_w}\left(1 + \frac{q^2\langle R_g^2\rangle_z}{3} + 2A_2 C\cdots\right) \tag{5.34}$$

在前述讨论中,均假设不同粒子之间互不相关、彼此独立。它们散射光电场乘积的时间平均等于各自时间平均后的乘积。各自时间平均的结果为零,故不影响散射光强。然而,当粒子浓度很大或粒子之间存在着很强的相互作用时,就要进一步考虑源于不同粒子的两个单元的散射电场之间的干涉。换而言之,对这样的大粒子体系就需要考虑粒子内外双重的散射电场之间的干涉。所得到的将是溶液的结构因子或散射函数,包括了一个粒子内的两个不同单元和不同粒子间的两个单元的散射电场之间的相互干涉。一个粒子内有 N 个单元(链段),散射体积内有 N_p 个粒子或大分子链。因此,前述的粒子结构因子就成为溶液的结构因子。其一般的数学表达如下,

$$P(q) = \frac{1}{(NN_p)^2} \sum_{j=1}^{N_p} \sum_{k=1}^{N_p} \sum_{l=1}^{N} \sum_{m=1}^{N} \exp\left[i\,\vec{q}\cdot(\vec{r}_{j,l} - \vec{r}_{k,m})\right]$$

$$= \frac{1}{(NN_p)^2} \left\{ \begin{array}{l} N_p \sum_{l=1}^{N} \sum_{m=1}^{N} \exp\left[i\,\vec{q}\cdot(\vec{r}_{j,l} - \vec{r}_{k,m})\right] + \\ \sum_{j \neq k=1}^{N_p} \sum_{l=1}^{N} \sum_{m=1}^{N} \exp\left[i\,\vec{q}\cdot(\vec{r}_{j,l} - \vec{r}_{k,m})\right] \end{array} \right\}$$

读者不要被该公式吓倒。物理上并不太难。大括号里的第一项反映了 N_p 个粒子自身的静态结构因子;第二项代表了所有不同粒子间的静态结构因子。当 $q \to 0$ 时,$P(q) \to (N^2 N_p + N^2 N_p (N_p - 1))/(NN_p)^2 = 1$。

结构因子与液体中局部的散射单元(链段)的密度有关,定义为 $\rho(\vec{r}) = \delta(\vec{r} - \vec{r}_{j,l})$,其中矢量 $\vec{r}_{j,l}$ 代表了第 j 个粒子(大分子链)的第 l 个单元(链段)的位置。将其对整个散射体积(V)加和,得到液体的宏观平均密度(ρ_0)。对给定的液体,为一常数,即

$$\rho_0 = \langle \rho(\vec{r}) \rangle = \frac{1}{V} \sum_{j=1}^{N_p} \sum_{l=1}^{N} \rho(\vec{r}) = \frac{1}{V} \sum_{j=1}^{N_p} \sum_{l=1}^{N} \delta(\vec{r} - \vec{r}_{j,l}) = \frac{NN_p}{V}$$

密度偏离其平均值的无规涨落,$\Delta\rho(\vec{r}) = \rho(\vec{r}) - \rho_0$ 的平均一定为零。可是,这样就无法区分两个涨落振幅(波及的范围)完全不同的体系。如前所述,可采用先平方,随后再平均(方均)的方法。具体到一个体系中的密度涨落振幅,统计上的解决办法是求其距离密度-密度自相关函数,即

$$\langle \Delta\rho(\vec{r}) \Delta\rho(0) \rangle = \langle \rho(\vec{r})\rho(0) \rangle - \rho_0^2 \quad \to \quad \frac{\langle \Delta\rho(\vec{r}) \Delta\rho(0) \rangle}{\rho_0^2} = \frac{\langle \rho(\vec{r})\rho(0) \rangle}{\rho_0^2} - 1$$

这里,$\vec{r} = \vec{r}_j - \vec{r}_k$,任意两点间的距离矢量,平均代表了对体系内所有任意两点的密度乘积求加和。对于一个各向同性的宏观均匀体系,矢量可被标量代替,即 $\langle \rho(\vec{r})\rho(0) \rangle = \langle \rho(r)\rho(0) \rangle$。计算 $\langle \rho(r)\rho(0) \rangle$ 包括下列步骤:先选取体系中的任意一点作为第一点,其密度为 $\rho(0)$,再选取距离 r 处的另一密度为 $\rho(r)$ 的第二点。将两个密度相乘,再对所有距离为 r 的第二点求加和。再任意选取另一个点为第一点,重复操作,直至第一点选遍了体系内所有的点。由于密度涨落无规,与宏观平均密度(ρ_0)相比,$\rho(0)$ 和 $\rho(r)$ 可或高或低;它们与 ρ_0 的差值也可有正有负。

当 $r = 0$ 时,$\langle \rho(r)\rho(0) \rangle = \langle \rho(0)^2 \rangle$,为方均密度。可证明,$\langle \rho(0)^2 \rangle = 2\rho_0^2$。此时,$\Delta\rho(r)$ 和 $\Delta\rho(0)$ 是 100% 的相关。当 $r \to \infty$ 时,$\Delta\rho(r)$ 和 $\Delta\rho(0)$ 应该变得毫不相关。换而言之,二者乘积的平均等于二者分别平均后的乘积,$\langle \Delta\rho(r \to \infty) \Delta\rho(0) \rangle = \langle \Delta\rho(r \to \infty) \rangle \langle \Delta\rho(0) \rangle = 0$,零相关度。随着 r 的增加,$\langle \rho(r)\rho(0) \rangle$ 逐渐地从 100% 相关递减成 100% 不相关。存在一个特

征弛豫长度(ξ),反映了密度涨落平均可波及的距离,对应着粒子的平均回转半径。严格地说,$\xi^2 = \langle R_g^2 \rangle / 3$,大约 1.7 个平均回转半径。

可以证明,单元(链段)的距离密度自相关函数的 Fourier 变换就是体系的静态结构因子:

$$P(q) = \frac{1}{V\rho_0^2} \int \langle \rho(r)\rho(0) \rangle \exp(i\,\vec{q} \cdot \vec{r}) \mathrm{d}\vec{r} \quad 和 \quad \langle \rho(r)\rho(0) \rangle = \frac{V\rho_0^2}{(2\pi)^3} \int P(q) \exp(-i\,\vec{q} \cdot \vec{r}) \mathrm{d}\vec{q}$$

换而言之,正格子空间里的 $\rho(r)$ 和倒格子空间里的 $P(q)$ 互为共轭。对一个密度均匀、没有涨落的体系,$\langle \rho(r)\rho(0) \rangle = \langle [\rho(0)]^2 \rangle = \rho_0^2$。因此,$P(q) = \delta(\vec{q})$,即体系除了在前行方向以外,均不散射。关于这点,先前已从干涉相消给出了定性的解释。这里只是进一步的理论阐明。在下节讨论动态激光光散射时,还会遇到另一个相关函数。

现代的 LLS 光谱仪均使用偏振、单色和相干激光光源,并且偏振方向通常设置为垂直于光学平台。另外,在检测器的前面还安装了垂直偏振器,即仅检测与入射光具有相同偏振方向的散射光,因此 $R(q)$ 通常写为 $R_{vv}(q)$,其中下标"vv"表示垂直偏振的入射光和垂直偏振的散射光。如果不遵循基于电动力学的完整推导,化学专业或其他专业的大多数学生应该能够理解上述讨论。掌握经典 LLS 光谱仪的基本原理相对容易。经典 LLS 光谱仪的实验看似简单,但却相当困难。

下图显示了一台现代激光光散射仪的平面图,包括了两台不同波长的激光(HeNe:633 nm 和 Nd:YAG:532 nm)和一台基于新型设计的微分折射仪(将在下节中详细讨论)。该仪器利用了两个反射镜将激光折回 $180°$,一是为了缩小光学平台的长度;二是在不大幅度移动激光的前提下,方便地调节激光高度和方向;图中也列出了现代激光光散射仪的可测量大分子和胶体粒子的范围。在使用现代激光光散射仪研究一个大分子溶液或胶体分散液时,读者们应该注意以下一些具体的要点。

(1)仪器调整:其一,调整入射激光与光学台面平行;穿过焦距为 $30 \sim 50$ cm 的聚焦镜的中心,并聚焦在旋转中心,束径约为 200 μm,仍然平行于台面,与台面的夹角小于 4 个微弧度(即每公里偏差小于 4 mm),通过位于零散射角度上的检测圆孔。其二,调整转动平台,使转动平面与台面平行,即转臂(长约 50 cm)在 $10° \sim 155°$ 范围内旋转时,末端上下波动小于 2 μm(头发丝的直径约为 60 μm)。其三,调节折射指数匹配池(直径约为 10 cm)和样品池(直径为 $1.0 \sim 2.0$ cm)与旋转中心吻合,且垂直于光学平台,故样品池的插入不影响光束通过在零散射角度上的检测圆孔。其四,在样品池注入光学纯度的无尘甲苯,将转臂移至 $90°$,调整光学检测器的位置,使其获得最大散射光强。

仪器是否调节到位的最终检验标准是,在 $15° \sim 155°$ 的散射角度范围内,源自甲苯的时间平均散射光强的上下波动小于 $\pm 1\%$。如果大于该误差,校准就需重新开始。即使对于一个有经验的实验者,该校准过程也是以天计算,而非小时。

(2)样品制备:由上式已知,散射光强正比于 $C_i M_i$,而 C_i 又正比于 M_i;因此,散射光强正比于散射粒子摩尔质量的平方。一粒尺寸为 1 μm 的微小灰尘或气泡的等效摩尔质量可高达 10^{12} g/mol,而普通高聚物分子的摩尔质量仅有 10^5 g/mol;换而言之,一粒小灰尘或气泡的散射等效于 10^{14} 条高聚物链。因此,为避免干扰实验结果,除去溶液中灰尘和气泡就是利用激光光散射研究大分子溶液的一个必要和重要条件。

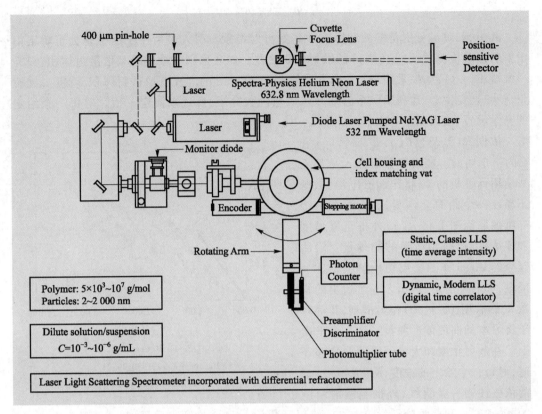

具体而言,首先,需要在一个密封的环境中用冷凝的丙酮反复冲洗每一个散射池内部,以达到完全除尘;然后,通过一个过滤膜(一般情况下,孔径为 0.2 μm)注入除去灰尘颗粒和气泡的溶液。判断除尘效果的标准如下:将溶液置于 15°,观察散射光强在 10~15 min 内的涨落幅度是否小于 ±5%;如果散射光强在此期间出现跳跃,说明除尘不净,需要重新过滤溶液。水溶液的除尘尤为不易,耐心是一个必要条件,还需要特别的技巧和经验。

尽管在静态光散射实验中存在着许多困难,但对大分子的定量研究,往往需要确定一系列样品的绝对摩尔质量。所以,古老的、经典的静态光散射仍是大分子实验室里不可或缺和有价值的进行定量研究的仪器。静态光散射、膜渗透压差和超速离心仪构成了一组无须标定曲线的、测量大分子绝对分子量的表征方法。

对一给定的散射角度,溶液和溶剂的 Rayleigh 比之差,即 $R(q)$,并非通过测量入射激光的光强(I_0)、散射光强(I)和检测平面到散射中心的距离(r)获得。而是,先选择一个已知 Rayleigh 比(R_0)的标准样品,如 R_0(甲苯,25 ℃) = 2.698×10^{-5} cm^{-1}。利用该值,还可从相对的散射光强变化得到甲苯在其他温度或条件下的 R_0。标准样品和纯溶剂的分子很小,它们的散射光强不依赖于散射角度。实验上,仍然测量它们在每个角度的散射光强,以弥补仪器在不同散射角度的微小差异。通过分别测量标准样品、纯溶剂和溶液在不同散射角度的散射光强[$I_0(q)$、$I^*(q)$ 和 $I(q)$],就可依据上式中过量散射(溶液和溶剂的 Rayleigh 比之差)的定义算出 $R(q)$,即

$$R(q) = R_0(q) \frac{I(q) - I^*(q)}{I_0(q)} \left(\frac{n}{n_0} \right)^{\gamma} \tag{5.35}$$

其中, n 和 n_0 分别是溶液和标准样品的折射指数。引入最后一个因子是为了矫正由于折射造成检测器所"见(测)"的散射体积之差。如果采用一维狭缝截取散射体积,只需一维矫正, $\gamma = 1$;如果采用小于聚焦光束直径(200 μm)的圆孔获取散射体积,就需二维矫正, $\gamma = 2$。实验上,常利用一个已知摩尔质量的大分子,反过来测定 γ 值。对每一个给定的光学设置,仅需测量一次。

依据式(5.33), $R(q)$ 是依赖于 q 和 C 的一个曲面。在大多数研究人员都无法使用计算机的 20 世纪四五十年代,既不易在一个曲面上作图,也无法在一个三维的曲面图上将 $R(q)$ 外推至零角度和零浓度,以获得所需的物理量。沿用至今的 Zimm 作图法成功地解决了这一问题,如右图所示;即引入一个适当的常数 k,然后用 $(q^2 + kC)$ 作为横轴,其等效于将原本的曲面伸展并投影到一个平面上。在散射角度和大分子浓度均足够小时,对每一个给定的浓度, $KC/R(q)$ 对 q^2 的依赖性为一条直线,外推至散射角度

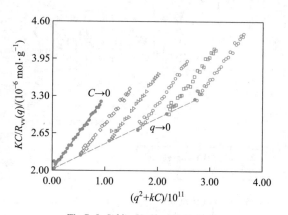

Wu P Q, Siddiq M, Chen H Y, et al.
Macromolecules, 1996, 29:277.
(Figure 1, permission was granted by publisher)

为零,其斜率和截距分别对应着二阶维里系数(A_2)和重均摩尔质量(M_w)。另一方面,对每一个给定的角度,外推至浓度为零,也得一条直线,其斜率和截距分别对应着 z 均回转半径($\langle R_g^2 \rangle_z$)和重均摩尔质量。理论上,两条直线的外推应相交于一点,这样得到的两个 M_w 相同,这也被用作检验实验质量和数据处理的一个判据。

采用 Zimm 作图法时,选择适当的 k 值很关键,太小分不开不同浓度的数据;太大又将同一浓度、不同角度的数据压缩成一条几乎垂直的线,看似失去角度依赖性。早期,这是一件非常费时的工作,取决于经验。今天,在计算机作图的辅助下,数据处理再已不是问题。但是,在报告数据时,图形的适当与否将直接影响读者的观感。上图可作为一个好的典范。

利用激光光散射测得一系列窄分布、具有不同摩尔质量的大分子的 z 均回转半径、二阶维里系数,以及重均摩尔质量后,利用下列标度关系,可将 z 均回转半径和二阶维里系数分别对重均摩尔质量进行双对数作图可得不同的前置因子和标度指数,即

$$R_g = (\langle R_g^2 \rangle_z)^{1/2} = k_{R_g} M_w^{\alpha_{R_g}} \quad \text{和} \quad A_2 = k_{A_2} M_w^{\alpha_{A_2}}$$

其中,对线型柔性链, $1/3 \leqslant \alpha_{R_g} \leqslant 3/5$ 和 $\alpha_{A_2} = 1/4$。然而, A_2 的测量误差可高达 $\pm 20\%$。

小　结

亘古以来,光散射现象就伴随着人类。约在两百年前,就有正式的疑问"天空为何是

蓝色?"。更早的农耕时代,农民们就知道,早见朝霞会下雨,日落晚霞预天晴。也正是由于太阳光的散射,我们的世界才不是那么"黑白分明"。太阳光的散射光透过窗户,照亮室内。然而,真正地理解光散射的物理原理仅始于 1881 年。在电磁理论问世以后,利用光散射原理测量粒子和分子的微观性质则是更晚。现今,激光光散射已成为研究胶体和大分子一个不可或缺的、重要的实验方法。

基本原理:可见光是一个交变的高频($\sim 10^{15}$ Hz)电磁波,其能量不足以使得胶体粒子或大分子中较重的原子核发生这样一个高频的移动,仅有较轻的电子们可在该高频电场下振荡,产生诱导偶极矩,其大小和方向不断地变化,向周围(球面)发出韧致辐射。对给定的检测面积,接收到的散射光强自然地与散射源距离(r)的平方成反比;也不难理解散射光强与入射光强和粒子体积内的电子数成正比。对于给定的粒子,电子数正比于其体积或质量,这正是可从散射光强获得粒子摩尔质量的物理基础。

当散射角度为零时,体系内所有单元的散射光都具有相同的位相,干涉相增,散射光强达到最大。对于多分散体系,零角度的时间平均散射光强正比于重均摩尔质量。在一个给定的非零散射角度,即使作为相干光源的入射激光束里的每条光线都具有相同的初始位相,但因大粒子内部各个单元的空间位置不同,散射电场到达检测器时,也会有不同的位相,产生干涉相消,使得时间平均散射光强减弱,减弱的程度分别与粒子的平均回转半径和散射矢量有关。另外,良溶剂中的排除体积效应也使得散射光强变弱。

可测物理量:首先测量标准样品(甲苯)在不同散射角度(θ)的时间平均散射光强,乘以 $\sin\theta$ 作出散射体积矫正。其时间平均散射光强随散射角度变化的波动应小于±1%。再对于给定的粒子或大分子浓度,测量胶体分散液或大分子溶液、分散介质或溶剂在不同散射角度的时间平均散射光强,并作出散射体积矫正。对不同的浓度,重复以上实验。

数据处理:依据式(5.35),用所测得的时间平均散射光强,先算出在不同浓度和不同角度下的 Rayleigh 比,再利用 Zimm 作图法,分别外推出散射角度和浓度为零的两条直线。由它们的斜率分别可得粒子或大分子链的 z 均回转半径和二阶维里系数;从它们的截距可获粒子或大分子链的**重均摩尔质量**。对于分布较窄的样品,通过不同的作图,还可从时间平均散射光强对散射角度的依赖性获得散射粒子的形状和构象。

动态激光光散射:采用激光作为光源后,才发展出了近代的动态激光光散射。顾名思义,该方法是动态地测量散射光强,而非静态地测量时间平均散射光强。其原理基于激光是一个单色的相干光源,其中的每一条光线具有相同的起始位相。每个散射的粒子(单元)处于不同的空间位置,类似前述大粒子的不同部分。它们各自的散射光电场,在经历过不同的光程抵达检测平面后的位相已经完全不同。取决于位相差,互相之间的干涉可增强或减弱散射光强,在检测器的感光平面上形成无数个明暗交替的小光斑,如右图所示。由于散射体积里互相独立的粒子都在热能的搅动下不停和无规地运动,这些小光斑出现的位置也在不停地变化。如果盯住一点观察,就会发现该点忽暗忽明、不停地和无规

地闪烁,频率可达百万赫兹。

如果检测所用的感光元件覆盖了很多小光斑,尽管每个点都在无规地闪烁,但检测本身等于对众多小光斑取了平均,检测的只是一个系综(样本)平均散射光强。另一方面,即使感光元件小到仅覆盖几个光斑,但不够灵敏和快速,在测量时间内,每个光斑已经闪烁了成千上万次,那么测量的也只是一个时间平均散射光强。依据统计学定理,一个无规变量的系综平均等于时间平均。除了要有强的单色、相干激光光源以外,在散射光强变化的动态测量中,检测元件还必须具备以下特征:第一,感光面积要小到只覆盖几个光斑,理想的是仅覆盖一个小光斑,方可有效地跟踪光强的涨落。第二,检测器噪声低、灵敏度高,可测量极小感光面积上极弱的信号。第三,采样速度足够快,短于 20 ns,足以分辨和追踪每秒一百万次的光强涨落。

普通汞灯或氙灯中的每条光线,即使经过光栅单色后,仍有微小的波长差别,不同的起始位相和不停变化的强度,引致源自不同光束的散射光之间的干涉。仅当激光在 20 世纪 60 年代问世后,动态地测量散射光强方才成为现实。在过去的几十年间,随着快速单光子检测器(光电倍增管,现为雪崩型光电二极管)、电子放大器、整形器和计算机的飞速发展,动态激光光散射才得以逐步地在 60 年代完善理论,在 70 年代发展仪器,在 80 年代日趋成熟,最终,在 20 世纪 90 年代进入普通大分子研究室,成为一个普通仪器。对应于一百多年前发展的"经典"光散射,其也被称作"近代"光散射。

从另一个观点出发,也可用多普勒效应来描绘动态激光光散射。当一束单一频率(ω)的激光被一个粒子弹性散射时,如果粒子不动,检测到的散射光频率就仍为 ω。然而,每一个粒子在热能的搅动下不停地、无规地运动。依据多普勒效应,当该粒子朝向或背向检测器运动时,所测得的散射光频率就会稍微增加或减少一点($\Delta\omega$),其幅度随运动速度增加。粒子朝任意一个方向的运动都可被分解成平行和垂直检测方向的两个分量,与检测方向垂直的运动对散射光的频率毫无影响。由于粒子的运动方向无规、速度不定,即使入射光具有单一圆频率,散射光的圆频率呈现出一个在入射圆频率($\Delta\omega = 0$)处对称的 Lorentz 分布,如下图所示。具体的数学表达如下:

$$S(\omega) = \frac{2\Gamma}{\Gamma^2 + (\Delta\omega)^2}$$

其中,峰高为 $2/\Gamma$,当 $\Delta\omega = \pm\Gamma$ 时,峰高降为一半,如右图所示。$S(\omega)$ 是一个功率分布函数(与散射光强有关)。Γ 常被称作线宽。线宽随着粒子的运动速度增加。对给定的实验条件(如液体温度和黏度),越小的粒子运动越快。如果只考虑平动扩散运动,测得线宽,就可间接地知道粒子的平动扩散系数(D,在超速离心一节中已讨论过)。对多分散体系,一种粒子尺寸对应着一个 Lorentz 分布和一个线宽(Γ)。峰下

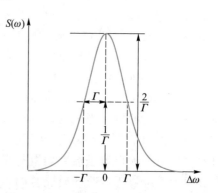

的面积对应着该尺寸粒子的数目,即在体系中的权重。实际测量的是一个分布(峰),其包含了体系中不同尺寸的粒子分别对应的经光强权重后的 Lorentz 分布。测量过程相当

于一个积分,其线宽对应的为光强平均线宽($\langle \Gamma \rangle$),权重为线宽的光强分布 $G(\Gamma)$,即

$$\langle \Gamma \rangle = \int_0^\infty \Gamma G(\Gamma)\,\mathrm{d}\Gamma$$

因散射光的频率出现微小变化,动态激光光散射已不是严格意义上的弹性散射,故从事弹性散射的物理学家更倾向于称其为"准弹性散射"。可见范围内的激光频率约为 10^{15} Hz,因粒子运动引致的频率变化(线宽)约为 10^6 Hz,相对变化仅约为十亿分之一。理论上,可利用 Fabry-Perot 干涉器在频率空间里直接测量每一个频率处的功率(散射光强)。平均线宽可由所得的功率分布算出。对小粒子,运动较快,Γ 较大,勉强可行。因实验难度,这方面的历史文献十分有限。对运动很慢的大分子,在频率空间直接测量其线宽则无法实行。同其他光谱一样,早期无选择时,只能在频率空间中测量。

在数学和物理上,频率和时间是共轭的。在一个空间中的函数可通过 Fourier 变换至另一个空间。因此,在频率空间较难实现的测量可转换到时间空间中去做。随着激光、快电子、计算机的发展,大部分曾经在频率空间里的测量现今都已通过 Fourier 变换到时间空间中操作。先测量一个(延迟)时间相关函数,再通过 Fourier 反演算出频率空间的功率(光强)分布,如红外、拉曼等 Fourier 光谱仪。

早在 1930 年,Norbert Wiener 就对可微分的确定性函数证明了"功率与时间相关函数是一对 Fourier 共轭函数"的定理。Aleksandr Khintchine 随后在 1934 年针对平稳随机过程得到一个概率类似的结果。Albert Einstein 于更早的 1914 年就在一个两页的备忘录中解释了这一想法。数学上可以严格证明动态激光光散射里希望获得的 Lorentz 功率分布函数 $[S(\omega)]$ 和散射电场的时间自相关函数 $[g^{(1)}(\tau) = \langle \vec{E}_s^*(t)\vec{E}_s(t + \tau) \rangle]$ 是一对 Fourier 共轭函数。在动态激光光散射中,$S(\omega)$ 是所需的物理量。测量 $g^{(1)}(\tau)$ 是一个对 $S(\omega)$ 积分过程,而反演则是从测得的 $g^{(1)}(\tau)$ 计算出 $S(\omega)$。具体的数学表达如下,

实验测量   反演计算

$$\langle \vec{E}_s^*(t)\vec{E}_s(t + \tau) \rangle = \frac{1}{2\pi}\int_{-\infty}^{\infty} S(\omega)\mathrm{e}^{i\omega\tau}\,\mathrm{d}\omega \qquad S(\omega) = \int_{-\infty}^{\infty} \langle \vec{E}_s^*(t)\vec{E}_s(t + \tau) \rangle\,\mathrm{e}^{-i\omega\tau}\,\mathrm{d}\tau$$

其中,$\vec{E}_s^*(t) = \vec{E}_0 \exp[-i(\omega t - \phi(t))]$ 和 $\vec{E}_s(t + \tau) = \vec{E}_0 \exp[i(\omega(t + \tau) - \phi(t + \tau))]$,$\tau$ 是延迟时间。假定散射体积里有 N_p 个粒子,散射电场就是 N_p 个电场的矢量之和。每个粒子散射的电场到达检测平面时有自己的位相(ϕ)。沿用前述有关 N_p 个独立小粒子散射的讨论,或者将散射体积看作一个"大粒子",那么 N_p 个"粒子"可被看作"大粒子"中 N_p 个独立的散射单元。散射电场的时间自相关函数为

$$\langle \vec{E}_s^*(t)\vec{E}_s(t + \tau) \rangle = |\vec{E}_0|^2 \langle [\sum_{i,j=1}^{N_p} \exp[i(\phi_{s,i}(t) - \phi_{s,j}(t + \tau))]] \rangle$$

用正比于平均散射光强的 $\langle \vec{E}_s^*(t)\vec{E}_s(t) \rangle$ 归一上式,可得

$$g^{(1)}(\tau) = \frac{\langle \vec{E}_s^*(0)\vec{E}_s(\tau) \rangle}{\langle \vec{E}_s^*(0)\vec{E}_s(0) \rangle} = \frac{\langle \sum_{i,j=1}^{N_p} \exp[i(\phi_{s,i}(0) - \phi_{s,j}(\tau))] \rangle}{\langle \sum_{i,j=1}^{N_p} \exp[i(\phi_{s,i}(0) - \phi_{s,j}(0))] \rangle}$$

在一个稳态体系中,从一个时间(t)到另一个时间($t + \tau$)只是间隔着一个延迟时间 τ 而已,与起点 t 无关,故可定义 $t = 0$。每个求和共有 N_p^2 项,包括 N_p 个自身项($i = j$)和

$(N_p{}^2 - N_p)$ 个交叉项($i \neq j$)。独立的粒子在作无规运动时、彼此之间毫不相干,互不影响,故所有交叉项的时间平均值均为零,只剩下 N_p 个自身项,故上式成为

$$g^{(1)}(\tau) = \frac{\langle \vec{\mathbf{E}}_s^*(0)\vec{\mathbf{E}}_s(\tau) \rangle}{\langle \vec{\mathbf{E}}_s^*(0)\vec{\mathbf{E}}_s(0) \rangle} = \langle \exp[i(\phi_{s,j}(0) - \phi_{s,j}(\tau))] \rangle$$

上式的分子有 N_p 个相同的项,分母中求和的每一项均为 1,故求和得 N_p。上下的 N_p 相互抵消。利用了统计学定理:求和后的时间平均等于时间平均后的求和。$\langle \vec{\mathbf{E}}_s^*(0)\vec{\mathbf{E}}_s(\tau) \rangle/N_p$ 也被称为动态结构因子,记为 $P(q,\tau)$;$g^{(1)}(\tau) = P(q,\tau)/P(q,0)$,其中,$P(q,0) = 1$。

满足下列任何一个条件,上式就和静态激光光散射中 N_p 个小粒子的散射一致。其一,延迟时间为零($\tau = 0$),求和中的每一项均为 1,时间平均后仍为 1,故求和结果为一个小粒子散射的 N_p 倍。其二,粒子都静止不动,$\phi_{s,j}(0) = \phi_{s,j}(\tau)$,求和中的每一项的时间平均也均为 1,同样回到前述的 N_p 个小粒子散射的结果。在一个激光光散射实验中,在热能的搅动下,每个粒子都在散射体积内一直不停地、无规地运动。即使经过一段极短的延迟时间 τ 后,粒子们也已离开原位,无规地移动到了另一个的地方。$\phi_{s,j}(\tau)$ 与粒子位置有关,因此,一般而言,$\phi_{s,j}(\tau) \neq \phi_{s,j}(0)$。

换个观点思考散射电场的自身时间相关函数,当 $\tau = 0$ 时,无论每个粒子的运动有多快,上式求和中的每一项都为 1,求和结果为 N_p,$g^{(1)}(\tau) = 1$,100% 地相关;当 $\tau \to \infty$ 时,无规运动的粒子可达散射体积内的任何一处,且到达每处的概率完全相同。其位置与其起始点已毫无关系。因此,两个指数项乘积的时间平均等于每项时间平均的乘积:$\langle \exp[i(\phi_{s,j}(0) - \phi_{s,j}(\tau))] \rangle = \langle \exp[i\phi_{s,j}(0)] \rangle \langle \exp[-i\phi_{s,j}(\tau)] \rangle$。由于每个粒子的位相随粒子位置无规地在 0 至 2π 之间变化,$\langle \exp[i\phi_{s,j}(0)] \rangle = \langle \exp[-i\phi_{s,j}(\tau)] \rangle = 0$,即在延迟时间足够长时,散射电场自身是 100% 的不相关。

这两个极端情况的讨论清楚地显示,当延迟时间有限时,散射电场自身的相干性一定处于 0% 和 100% 之间,随着延迟时间的增加递减。衰减的速度随着粒子的平均无规运动(平动扩散)速度递增。换而言之,对于给定的散射体系(粒子大小、温度、压强和分散介质等),其必定具有一个特征弛豫时间(τ_c)。依据定义,第 j 个粒子的散射电场具有一个位相 $\phi_{s,j} = \vec{\mathbf{q}} \cdot \vec{\mathbf{r}}_j$。因此,

$$g^{(1)}(\tau) = \langle \exp[i\vec{\mathbf{q}} \cdot (\vec{\mathbf{r}}_j(0) - \vec{\mathbf{r}}_j(\tau))] \rangle \tag{5.36}$$

其也称作单个粒子的动态结构因子。如果 $\vec{\mathbf{q}} \cdot (\vec{\mathbf{r}}_j(0) - \vec{\mathbf{r}}_j(\tau)) < 1$,即散射角度较小或延迟时间较短,其 Taylor 展开为

$$g^{(1)}(\tau) = \langle 1 + i\vec{\mathbf{q}} \cdot [\vec{\mathbf{r}}_j(0) - \vec{\mathbf{r}}_j(\tau)] - \frac{1}{2}\vec{\mathbf{q}}^2 \cdot [\vec{\mathbf{r}}_j(0) - \vec{\mathbf{r}}_j(\tau)]^2 + \cdots \rangle$$

其中,利用了 $i^2 = -1$。在热能搅动下,粒子无规行走,位置不断变化,时间平均位置保持不变,故 $\langle i\vec{\mathbf{q}} \cdot [\vec{\mathbf{r}}_j(0) - \vec{\mathbf{r}}_j(\tau)] \rangle = i\vec{\mathbf{q}} \cdot \langle \vec{\mathbf{r}}_j(0) - \vec{\mathbf{r}}_j(\tau) \rangle = 0$。上式可近似成

$$g^{(1)}(\tau) \approx \exp\left[-\frac{1}{2}\vec{\mathbf{q}}^2 \cdot \langle [\vec{\mathbf{r}}_j(0) - \vec{\mathbf{r}}_j(\tau)]^2 \rangle\right] \tag{5.37}$$

其中，$\langle [\vec{r}_j(0) - \vec{r}_j(\tau)]^2 \rangle$就是一个粒子无规行走了一段时间$\tau$后的方均距。依据扩散方程，一个粒子平动扩散一段距离(x)所需的时间(t)与平动扩散系数(D)之间的依赖关系为$2Dt = x^2$。这里的$t = \tau$和$x^2 = \langle [\vec{r}_j(0) - \vec{r}_j(\tau)]^2 \rangle$。所以，上式可被重写成

$$g^{(1)}(\tau) = \exp(-q^2 D\tau)$$

其中，利用了$\vec{q}^2 = q^2$。将上式代入前述 Fourier 反演计算的公式，可得$\Gamma = q^2 D$。对于一个多分散样品，上式可被进一步写成

$$g^{(1)}(\tau) = \int_0^\infty g^{(1)}(\Gamma, \tau) \mathrm{d}\Gamma = \int_0^\infty G(\Gamma) \exp(-\Gamma\tau) \mathrm{d}\Gamma \tag{5.38}$$

上式为一个 Laplace 积分。其显示，测量的过程等于右边的积分。依据上式，通过对所测的$g^{(1)}(\tau)$进行 Laplace 反演即可得到所需的线宽分布$G(\Gamma)$。

还可采用下列方法演绎式(5.37)。如前所述，散射矢量的倒数$(1/q)$在静态激光光散射中是量度散射粒子尺寸的一把"光学尺子"。在动态激光光散射里，它同样是一把"光学尺子"，测量的是无规运动的粒子在给定时间内迁移(扩散)的平均距离。换而言之，粒子平均扩散一个$1/q$距离时，散射光电场的位相就改变2π，完成一次振荡，其对应的时间就是前述的特征弛豫时间(τ_c)，即$2D\tau_c = (1/q)^2$。因此，式(5.37)可被重写成

$$g^{(1)}(\tau) = \exp\left[-\frac{\tau}{\tau_c} \right]$$

与前式比较可知，$\Gamma = 1/\tau_c = q^2 D$；其在物理上显示，$\Gamma$和$\tau_c$分别代表了频谱和时间空间里散射体系同一个的特征参数(性质)，而后一个等式表明，如果只考虑体系内部粒子们的平动扩散时，该特征参数(性质)与平动扩散系数和度量该参数的"光学尺子"相关。物理上，τ_c也可被理解成一个无规变化体系的"记忆时间"；即一个体系可以"记住"自己任一状态的时间。

光学实验使用的均是"平方"检测器，测得的只是光强，而不是光的电场。因此，实验上无法直接测量散射电场时间自相关函数。A. J. F. Siegert 在 1943 年指出，如果体系内粒子的无规运动互相独立、密度涨落遵循 Gaussian 分布，归一的散射电场时间自相关函数$g^{(1)}(\tau)$与归一的散射光强时间自相关函数$g^{(2)}(\tau) = \langle I(0)I(\tau) \rangle / \langle I(0)I(0) \rangle$有关，其后来常被称为 Siegert 关系，即

$$g^{(2)}(\tau) = \frac{\langle \{ \sum_{i=1}^{N_p} \sum_{j=1}^{N_p} \exp[i(\phi_{s,i}(0) - \phi_{s,j}(0))] \} \{ \sum_{l=1}^{N_p} \sum_{m=1}^{N_p} \exp[i(\phi_{s,l}(0) - \phi_{s,m}(\tau))] \} \rangle}{\langle \{ \sum_{i=1}^{N_p} \sum_{j=1}^{N_p} \exp[i(\phi_{s,i}(0) - \phi_{s,j}(0))] \} \{ \sum_{l=1}^{N_p} \sum_{m=1}^{N_p} \exp[i(\phi_{s,l}(0) - \phi_{s,m}(0))] \} \rangle}$$

读者不要被看似复杂的数学表达式吓到，只要记住背后的物理图像。在$t = 0$时，N_p个粒子散射了N_p条光线，源自第i个粒子的散射电场具有自己的位相$\phi_{s,i}(0)$。经过一小段延迟时间τ后，N_p个粒子移动了以及散射了另外N_p条光线。第i个粒子的散射电场的位相从$\phi_{s,i}(0)$变成$\phi_{s,i}(\tau)$。上式中共有四组求和，每组求和有N_p^2项，故在分子和分母上各有N_p^4项。

上述公式中的每个指数项都可写成两个指数项的乘积。由于粒子的无规运动互不相关，统计上有"先乘积后平均等于先平均后乘积"。因其位置完全无规，每个粒子的时间

平均 $\langle\exp[i\,\phi_{s,j}(t)]\rangle=\langle\exp[i\,\vec{q}\cdot\vec{r}_j(t)]\rangle=\delta(q)$。当 $q\neq0$ 时（实验上总是成立），所有不同粒子之间的交叉项（$i\neq j;l\neq m;i\neq l;j\neq m$）均为零。剩下的仅有自身项，即源自每个粒子散射电场与散射电场的乘积，其可分为以下两类。

其一，对一个粒子在同一个时间和同一个位置而言，$i=j$ 和 $l=m$，每项都为 1。分子和分母上的两个求和分别各有 N_p 项，故结果各为 N_p^2。二者之比为 1。

其二，对一个粒子在间隔为 τ 的两个不同时间（不同的位置）上，$i=l$ 和 $j=m$，但 $i\neq j$ 或 $l\neq m$。依据定义，每一对 $i=l$ 对应着一个散射电场时间自相关函数 $g^{(1)}(\tau)$，共有 N_p 对。每一对 $j=m$ 也对应着一个 $g^{(1)}(\tau)$，但由于 $i\neq j$ 或 $l\neq m$ 限制，仅有 (N_p-1) 个 $g^{(1)}(\tau)$。二者相乘得 $N_p(N_p-1)\,|g^{(1)}(\tau)|^2$，故上式成为

$$g^{(2)}(q,\tau)=\frac{N_p^2+N_p^2\,|g^{(1)}(q,\tau)|^2+N_p\,|g^{(1)}(q,\tau)|^2}{N_p^2}=1+|g^{(1)}(q,\tau)|^2+\frac{|g^{(1)}(q,\tau)|^2}{N_p}$$

当 $N_p>10^3$ 时，右边最后一项小于千分之一，与实验中的噪声相似，故可忽略不计。然而，如果 $N_p<10^2$ 时，就不得不考虑这最后一项。散射体积的尺寸约为 200 μm。粗略估计显示，对尺寸为 1 μm 的粒子，质量浓度的下限是 10^{-5} g/mL；以此类推。

上述讨论显然也适用于一个大粒子（大分子链）内的各个散射单元或者较浓的分散液（溶液）中的不同粒子（大分子），唯一的区别就是不同的散射单元或粒子（大分子）之间存在一定的相关性，故前式中所有交叉项的时间平均不会完全消失为零。情况相当复杂，将在大分子链的动力学一章中详细讨论。

上式中最后一项对应的松弛过程的物理本质是粒子扩散进和出散射体积而引致的粒子数涨落。散射体积的尺寸是 $1/q$ 的一千倍，扩散所需的时间与距离的平方成正比。所以，与在 $1/q$ 的尺度上因位相涨落而引致的松弛相比，这一由于粒子数涨落而引致的松弛要慢一百万倍，弱 N_p 倍。在通常动态光散射的测量时间窗口中，其可被认作一个很小的、基本上不松弛的常数，仅微不足道地、微弱地贡献了一点测量的基线，可忽略。在通常的动态光散射中，实验上直接测量的是没有用 $I(q)^2$ 归一的散射光强时间自相关函数，

$$G^{(2)}(q,\tau)\cong I(q)^2[1+\beta\,|g^{(1)}(q,\tau)|^2] \tag{5.39}$$

其中，忽略了前式中的最后一项，并引入了一个仪器参数（$0<\beta<1$），在一个激光光散射仪中，检测系统永远达不到理想的完善状态（如检测器的感光面积不可能无限小，光路上所有的表面、限制光束的圆孔等都会散射而出现杂散光）。现代激光光散射仪的检测相关性（β）已可高达 95%，永远也不会达到 100%。β 是一个待测的物理量，通常在数据处理时当作一个拟合参数。设计仪器时，提高 β 可增加信噪比和缩短测量时间。但是，同时也会减少所测的散射光强，延长测量时间。因此，二者之间需要取

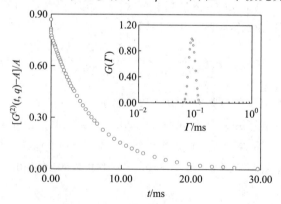

Lau A C W, Wu C. Macromolecules, 1999, 32: 581-584.

(Figure 4, permission was granted by publisher)

得适当的平衡。

依据上式,通过在数字时间相关器上设置若干个延迟时间(τ)超长的延迟记录通道,所得的$g^{(1)}(q,\tau)$应该已经完全不相关,衰减至零,这样获得的散射光强的时间自相关函数就是时间平均散射光强的平方($I(q)^2$),也称实验为"基线",记为A。上图显示了一个典型的、近乎完美的、已用测得的基线归一的散射光强时间自相关函数。$G^{(2)}(q,\tau)$和$I(q)^2$之差的开方为$\beta^{1/2}|g^{(1)}(q,\tau)|$。依据式(5.37),可以利用 Laplace 反演算出线宽分布$G(\Gamma)$。图中的插图是利用通常使用的 CONTIN 计算软件所算出的线宽分布,显示这是一个窄分布的大分子样品。

由于式(5.37)中散射电场时间自相关函数正比于散射光强,因此,线宽分布$G(\Gamma)$是散射光强的分布($G(\Gamma_i) \propto M_i C_i$)。因此,一个多分散样品的平均线宽($\langle \Gamma \rangle$)为

$$\langle \Gamma \rangle = \frac{\int_0^\infty \Gamma G(\Gamma) \mathrm{d}\Gamma}{\int_0^\infty G(\Gamma) \mathrm{d}\Gamma} = \frac{\sum_{i=0}^\infty \Gamma_i M_i C_i}{\sum_{i=0}^\infty M_i C_i}$$

依据定义,这是一个z均线宽。还可从线宽分布$G(\Gamma)$计算出一个多分散样品的线宽方差($\mu_2 = \langle (\Gamma - \langle \Gamma \rangle)^2 \rangle$)和相对方差$\langle (\Gamma - \langle \Gamma \rangle)^2 \rangle / \langle \Gamma \rangle^2$),即

$$\langle (\Gamma - \langle \Gamma \rangle)^2 \rangle = \int_0^\infty (\Gamma - \langle \Gamma \rangle)^2 G(\Gamma) \mathrm{d}\Gamma = \langle \Gamma^2 \rangle - \langle \Gamma \rangle^2 \text{ 和 } \frac{\mu_2}{\langle \Gamma \rangle^2} = \frac{\langle \Gamma^2 \rangle - \langle \Gamma \rangle^2}{\langle \Gamma \rangle^2} = \frac{\langle \Gamma^2 \rangle}{\langle \Gamma \rangle^2} - 1$$

当高聚物摩尔质量分布的多分散指数(M_w/M_n)小于 2.5 时,$M_w/M_n \cong 1 + 4\mu_2/\langle \Gamma \rangle^2$。上述的理论看上去都顺理成章。注意,该 Laplace 反演是求解一个数学上的病态方程。实验上存在无可避免的误差。在误差范围内,同一个测得的$\beta^{1/2}|g^{(1)}(q,\tau)|$的 Laplace 反演对应着无数多种可能的结果。因此,在反演计算中,必须先选定一个误差判据标准。当反演的计算结果与所测数据之间的相对方差低于这一误差判据标准时,反演计算终止。然而,在这一误差判断值判据标准内,仍有不同的计算结果。如果没有由从其他方法获得的补充信息,计算机和操作者均无法知道哪一个反演结果更好。只能在误差判断值的范围内选取相对方差较小的结果。

因此,选定一个恰如其分的误差判据标准就变得十分重要。定宽了,可能将一些测量误差和噪声当作信息计入所得的线宽分布;选窄了,又可能将一些体系信息当作误差和噪声从所得的线宽分布删去。在一个真实的实验中,这一误差判据标准本身也是未知的。在商品化的激光光散射仪的分析软件中已根据一般情况设定了一个误差标准。有经验的实验操作者,可根据具体体系和其他额外的信息,调节和重新设定该误差判据标准,以获得更好的分析结果。注意,这是一个实验问题,而不是计算问题!因此,采用干净样品、长的测量时间、获得高质量数据更为重要。

除了采用 Laplace 反演方法外,还可利用 cumulant 多项式展开来分析所测得的散射电场时间自相关函数,即

$$\log \left[\frac{G^{(2)}(\tau) - A}{A} \right]^{1/2} = \log \beta^{1/2} - \langle \Gamma \rangle \tau + \frac{\mu_2}{2!} \tau^2 + \frac{\mu_3}{3!} \tau^3 + \cdots$$

其中,第一项为常数,作为一个分析时的拟合参数。数学上,$\mu_i (i = 1, 2, 3, \cdots)$称为一个

分布的 i 次距。对随机变量,一次距为零;二次距为分布宽度;三次距反映了分布的不对称性。实验上,一般最多展开到三次距。可先将所测得的散射光强时间自相关函数取对数,再作半对数图。当延迟时间较短时,得一递减直线,斜率为 z 均线宽;当延迟时间较长时,斜率对应着线宽分布的方差,进而可算出相对方差。当样品分布很宽时,还可算出三次距。

在早期没有计算机或计算速度不够快时,cumulant 分析是唯一的选择。看上去,它好像不如 Laplace 反演那样得到一个可直观表达的线宽分布。实际上,二者得到的信息相差无几:"平均线宽"和"线宽分布的相对方差"。读者需要注意,Laplace 反演对所测的时间自相关函数的质量要求甚高;如果数据质量达不到 Laplace 反演的要求,cumulant 分析得到的结果可能更加可靠,故其并未过时。

实验上,由于记忆单元的价格下降和存储速度提升,现代的研究型动态激光散射光谱仪已实现了全数字化。采用高灵敏检测器直接测量散射光的光子数,并利用硬件数字时间相关器实时地获得散射光子数的时间自相关函数 $G^{(2)}(q,\tau)$。小型的基于动态激光散射原理的粒度仪则可能使用便宜的软件数字时间相关器,即先记录和存储一连串变化的光子数,再利用软件计算时间相关函数,而不是一边采集散射光子数,一边用硬件直接计算时间自相关函数。由于在每次存储光子数时均具有一个很短的死时间,软件相关器可能会失去小部分信息。由于现代存储器的极快速度,使用者一般已很难区分二者的差别。

在一个动态激光光散射的测量中,一个又一个测得的 $G^{(2)}(q,\tau)$ 不断地叠加在一起,即一个对时间的统计平均过程。理论上,无规的噪声会在叠加的过程中互相抵消,不会随着测量时间增长,而真正的信息会随着测量时间增加。因此,为了提高信噪比,应充分地延长测量时间,而非数据分析的一些操作。对于极稀的大分子溶液,测量时间可达数小时甚至数天,这又要求溶液要绝对的无尘。判断一个动态激光光散射的测量是否合格的一个必要非充分标准是,测量基线(平均散射光强)与反演后计算基线之间的相对误差小于千分之一。很少有其他实验和光谱方法可达到这一测量精度和确度。

如前所述,一粒细小灰尘或气泡的散射光强相当于数以十亿计的大分子链的散射光强,稍有不慎,所测得的结果就包含了杂质的贡献。测量大粒子的胶体时,由于大粒子的散射较强,痕量杂质的影响相对较小,但是适当地清洗散射池和除尘仍是必不可少的步骤。然而,粒度分析仪的销售员通常不会对使用者强调该点。不幸的是,欠缺基本光散射常识和操作经验的使用者往往也将基于动态激光光散射原理的粒度仪当作一个黑盒子,即放入没有适当除尘的样品,按下按钮,得到扭曲的结果,故目前的文献中充斥了许多似是而非的结论。

除了要精心地制备溶液,正确地进行了动态激光光散射测量,以及适当地设置分析参数以外,使用者还要清楚地认识到动态激光光散射测量的是散射体系在时间空间里的特征弛豫时间(τ_c)或等效的在频率空间中的线宽(Γ)。对一个多分散的大分子或胶体样品,所测得就是一个 τ_c 或 Γ 的分布。

一个散射体系的弛豫可源于各种不同的原因,如棒状粒子的转动、柔性链中各个链段之间相对位置的涨落、带电粒子在水溶液中的长程相互作用、温度或浓度梯度引致的扩散,等等。平动扩散只是众多弛豫原因中的一种。因此,仅当体系的弛豫可完全归于粒子或大分子的平动扩散时,方可将 τ_c 和 Γ 或它们的分布与平动扩散系数(D)或平动扩散系数分布($G(D)$)关联。右图显示了几个由线宽分布算出的具有不同摩尔质量的大分子的平动扩散系数分布。进而可将 D 或 $G(D)$ 进一步转变成粒子或大分子的流体力学半径(R_h)或其分布($f(R_h)$),相关理论可追溯到很久以前。

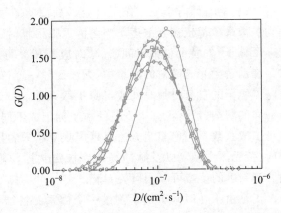

Wu P Q,Siddiq M,Chen H Y,et al. Macromolecules,1996,29:277.

(Figure 3,permission was granted by publisher)

早在 1827 年,植物学家 Robert Brown 就指出花粉在水中的无规抖动与花粉无关,灰尘也有同样的抖动性质,后被称为 Brownian 运动。尽管当时有人也关注了这一奇怪的现象,但其真正的解释却在近八十年后的 1904 年,William Sutherland 在一个学术会议上公开了他的解释,并于 1905 年三月正式发表了其结果。同年,Albert Einstein 独立地指出平动扩散系数等于热能除以摩擦系数(f)。另一方面,George Gabriel Stokes 在 1851 年就已发现,对一个在黏度为 η 的液体中半径为 R 的圆球而言,非湍(层)流中的 $f = 6\eta\pi R$。两者结合即得著名的 Stokes-Einstein 关系式, $D = (k_B T)/(6\eta\pi R)$。Marian Smoluchowski 在大约一年后,也独立地发现了此公式。也许是因为当时信息交流的缓慢,Einstein 在其文章中并没提及 Sutherland 的工作。即使不考虑后者的作用,作为一个后来者,他也不应抹杀前者的贡献,上式的正确表达应为 Stokes-Sutherland-Einstein 关系式,可见历史有时并不太公平。利用上式,不难从平动扩散系数分布 $G(D)$ 得到流体力学半径的分布 $f(R_h)$。由于在作 Laplace 反演时,为了覆盖较宽的分布,常采用对数坐标($\log \tau_c$ 或 $\log \Gamma$),使得上述从 D 到 R_h 的转换更易,成了线性空间里的简单平行转换。

由于动态激光光散射测量粒子的运动,一个粒子自身的运动速度和两个粒子相对的运动速度显然不同。假定在散射体积中有 N_p 个粒子,散射光强的时间自相关函数应写为

$$g^{(1)}(\tau) = \langle \exp[i\,\vec{q} \cdot (\vec{r}_i(0) - \vec{r}_j(\tau))]\rangle \tag{5.40}$$

这就是动态结构因子,包括两部分:因粒子自身的运动和由粒子相对运动引起的松弛。其右边的平均是两次求和,分别从 $i = 1$ 变到 N_p 和从 $j = 1$ 变到 N_p;共有 N_p^2 项,包括 N_p 个自身项($i = j$)和 $N_p^2 - N_p$ 个交叉项($i \neq j$),即

$$g^{(1)}(\tau) = \frac{\sum_{i=1}^{N_p}\exp[i\,\vec{q}\cdot(\vec{r}_i(0) - \vec{r}_i(\tau))]}{N_p} + \frac{\sum_{i=1}^{N_p}\sum_{j\neq i=1}^{N_p}\exp[i\,\vec{q}\cdot(\vec{r}_i(0) - \vec{r}_j(\tau))]}{N_p^2}$$

因此,动态激光光散射测量的是粒子间的特征相互松弛时间(τ_m),对应着粒子的相互平动扩散系数(D_m)。由上式的第一项可得特征自身松弛时间(τ_s),对应着平动自身扩散系数(D_s)。

在上述的讨论中,假定了溶液极稀,粒子之间没有相互作用。但在任何一个真实的分散液或溶液中,粒子间总是存在着一定的相互作用。即使在中性粒子之间没有特殊相互

作用,长程的流体力学相互作用也会影响彼此的平动扩散。上面的两个特征松弛时间或平动扩散系数都依赖于浓度。当溶液无限稀释时,二者均趋向于一个孤立粒子(体系里仅有一个粒子)的特征松弛时间(τ_0),对应着其平动扩散系数(D_0)。

在动态激光光散射的测量中,无法区分一个粒子和另一个粒子的散射光强,故不可直接测得粒子的自身松弛时间或自身扩散系数。变通的办法是,先寻找一个折射指数和粒子一样的溶剂,然后将其中少量的粒子标记成不同的折射指数,即可测得自身松弛时间或自身扩散系数。一些实验方法可直接测量粒子的自身扩散系数,包括但不限于,强迫瑞利散射、荧光漂白恢复以及脉冲场梯度核磁共振。限于篇幅,本书就不详细地介绍这些方法。有兴趣的读者可参阅有关专著。

正是由于在时间空间里测量一个体系的特征松弛时间,动态激光光散射也常被称为时间相关光谱,特别是那些习惯于其他谱学的人。基于同一个物理原理的同一个实验方法有三个完全不同的名称:动态光散射、准弹性光散射和时间相关光谱。奇怪吗? 才不呢,就如不同的人用不同的名字称呼我们每个人一样,有大名,也有昵称,取决于其与我们的关系。这里也一样,曾经从事静态光散射的人将其称为动态光散射;而在物理里从事弹性散射的人自然地从原理出发将其称为准弹性散射。这些不同的称呼也说明其是一个相对"年轻"的方法,至今仍然没有一个统一的名称。

研究胶体和大分子溶液的散射方法还包括小角中子和 X 光散射。从依赖角度的散射光强也可得到大分子的摩尔质量和尺寸,以及结构因子,其中散射矢量和散射光角度依赖性的数学表达完全相同。不同的是,光散射源于溶质和溶剂折射指数之差,即微分折射指数(dn/dC);中子散射取决于溶质和溶剂分子散射长度(b)之别,即与散射截面($4\pi b^2$)之差;X 光散射依赖于溶质和溶剂分子的电子密度之差。

在同步辐射和强中子光源出现之前,大分子溶液的中子和 X 光散射太弱,小角中子和X 光散射主要用于研究固体。自 20 世纪 90 年代起,随着同步辐射和强中子源的兴建,它们在研究大分子溶液中的应用逐渐增多。

在小角 X 光散射中,散射矢量的倒数($1/q$,"光学尺子")对应着 0.2~5.0 nm;在激光散射中,$1/q \approx 35~300$ nm;而小角中子散射的 $1/q$ 处于小角 X 光散射和激光散射之间。三者结合,就可覆盖和获得从聚合单体到链段,再到整条大分子的信息。本书略去了小角X 光散射和中子散射的细节。

小　　结

基本原理:粒子(大分子)在一个分散液或溶液中不停地运动,包括平动扩散、转动和构象的涨落。这些运动都可改变散射光的电场的位相。具有不同位相的光在检测平面上彼此干涉,形成无数个或明或暗的、闪烁频率高达百万赫兹的光斑。如果采用具有很小检测面积的、灵敏的、快速的单光子检测器跟踪几个光斑,就可测因粒子无规运动而引致的散射光强的涨落。速度越快,涨落的频率越高。

换个角度从频率空间考察同样的现象,因多普勒效应,粒子运动会轻微地改变散射光的频率,从入射时的单一频率变成一个频率的分布。运动越快,频率分布越宽。这里的光强涨落和频谱增宽都和粒子的运动速度有关,反映了一个体系的特征弛豫时间,也曰"记忆"时间。通过在时间空间里测量散射光强的时间自相关函数可获得该特征松弛时间。只有平动扩散运动时,由特征松弛时间可得平动扩散系数,进而可再通过 Stokes-Sutherland-Einstein 关系式获得流体力学半径。市场上测量小粒子的粒范围小于 2000 nm 的粒度仪均基于这一原理。

可测物理量:现代的动态激光光散射仪一般已带有一个数字时间相关器,利用其可直接测量具有不同粒子(大分子)浓度的分散液(溶液)在不同散射角度的散射光强时间自相关函数 $G^{(2)}(q,\tau)$。需要注意:首先,要彻底除去测量液体里的灰尘和气泡。判断标准为散射光强在小散射角度(15°)的涨落小于百分之五。其次,采样(测量)时间要足够长。判断标准为测量和计算基线之间的相对误差小于千分之一。

数据处理:依据式(5.37)和式(5.38),对每个测得的散射光强时间自相关函数 $G^{(2)}(q,\tau)$ 进行 Laplace 反演可得在一个给定测量角度(q)和浓度(C)中的线宽 $\Gamma(q,C)$ 或特征松弛时间 $\tau_c(q,C)$ 的分布。分别外推 $q\to0$ 和 $C\to0$ 可得 $G(\Gamma)$ 真正的散射线宽或特征松弛时间。如果只有平动扩散运动,可利用 $\Gamma=Dq^2$,将线宽分布转换成平动扩散系数分布 $G(D)$,进而可利用 Stokes-Sutherland-Einstein 关系式算出流体力学半径分布 $f(R_h)$。现代的动态激光光散射仪已包括了所有的计算软件。好处:方便;坏处:使用者往往不求甚解,错误百出。

微分折射法

从静态激光光散射的讨论中已知,散射光强正比于微分折射指数,定义为 $(\partial n/\partial C)_{T,p}$ 的平方。在散射光强中,微分折射指数的误差会被放大一倍。因此,精确地测定微分折射指数对于后续利用光散射测定大分子重均摩尔质量、z 均回转半径和二阶维里系数至关重要。"微分"意味着通过一个方法在实验上扣除溶剂对折射指数的贡献,得到溶液中大分子的折射指数对浓度的依赖性。通常的微分折射仪主要基于两种不同的原理:干涉和折射。

先讨论基于干涉原理的微分折射仪,如右图所示:从左向右,先利用一个透镜收集和准直源自汞灯的光成一道平行光束;一个滤光片使得入射光单色(未画);一个狭缝或小圆孔从平行光中截取一束细光和一个偏振片起偏。再利用一个由 William Hyde Wollaston 于 1802 年发明的棱镜(称作 Wollaston 棱镜),将一束入射的偏振光分解成偏振方向互相垂直的两束光,分别通过溶剂和溶液。两束光经过一

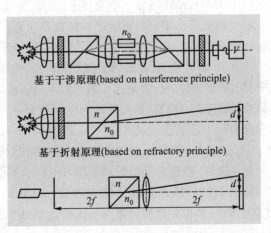

基于干涉原理(based on interference principle)

基于折射原理(based on refractory principle)

个透镜会聚在另一个 Wollaston 棱镜上之后，又变成一束偏振光。溶剂和溶液折射指数的细微差别(Δn)导致不同的光程，故两束会聚光的偏振方向略微偏离先前的偏振方向。利用随后的一个偏振片和光电二极管，可测出偏离角度。其与 Δn 成正比，Δn 又进一步与浓度和样品池的长度成正比。实验上，先在两个池中注满同样的溶剂，调节第二个偏振片，使得二者正交，即最低透光光强。然后，将一个池子中的溶剂换成一个浓度为 C 溶液。从所测的偏离角度获得 Δn。逐步地从稀到浓地置换溶液，可得到一系列与 C 对应的 Δn。Δn 对 C 作图为一条直线。其斜率对应着微分折射指数，$(\partial n/\partial C)_{T,p}$。

基于折射原理的微分折射仪相对简单明了，如上图中间所示：同样从左到右，用滤光片单色、用透镜准直、用狭缝截取光和用偏振片起偏。然后，将这束准直、偏光的光射入一个分割成两部分的样品池；如果两个池子都充满同样的溶剂，光束在理论上不会偏离原来的方向，仅会因为池子中 45° 的分割玻璃出现一点平移。然而，样品池永不完美。例如，其内外的四个玻璃面并不平行。实际上，光束会既平移又稍微偏离原来的方向。为了在实验上消除这些缺陷，通常在显微标尺上先读一次光束的位置，将池子旋转 180° 后，再读一次光束位置，两次读数的平均值即为光束在不偏不移时的原点位置。将一个半池里的溶剂换成溶液，先恒温，后读取光束的偏离距离。同样需要将池子旋转 180°，再次读取偏离距离，两次读数的平均即为该溶液的偏离距离，其与 Δn 成正比。逐步地从稀到浓地置换大分子溶液，得到一系列与 C 对应的 Δn。将二者作图得一条直线，斜率为 $(\partial n/\partial C)_{T,p}$。

利用微分折射仪测量光束偏离位置的要求估计如下。通常，大分子溶液的浓度和微分折射指数分别约为 10^{-3} g/mL 和 10^{-1} RI，故 $\Delta n \sim 10^{-4}$。为了达到 ±1% 的精度，一台微分折射仪必须可测 10^{-6} RI 的变化。即使显微标尺与样品池相距 1 m，测量精度也要达到 1 μm，不易。另外，折射指数随温度的变化，dn/dT，约为 10^{-4} RI/℃，故样品池的温度波动应小于 ±0.01 ℃，需要一台好的控温装置。

读者应该知道，以上的折射测量看似原理简单、实行不难。但是，实际的测量过程却是极其漫长和"痛苦"。首先，由于偏离的距离很小，以至于每次都必须使用显微标尺读数三次以上，求其平均。测量不同浓度时，必须要彻底地取出先前的溶液，然后用下一个稍浓的溶液反复地清洗三次，以避免先前残留的溶液影响测量结果。为了达到恒温，每次均需等待半个小时以上。即使采取了所有必要的措施和小心地操作，所测的偏差距离对浓度的作图仍然往往不在一条直线上，不通过理论预测的原点。这样，耗时数天的整个实验必须推倒重来。实验的精确度始终存在问题。这就是为何高聚物手册中所列的同一个高聚物溶液的微分折射指数会因不同实验室的测量而出现较大的偏差。

20 世纪 80 年代，位置敏感检测器问世后，传统的显微标尺就被其取而代之，极大地简化了痛苦的数据采集。位置敏感检测器的原理非常简单，它是一个特殊设计的、光敏面积为一维长条形的光电二极管，在其两端和中间各有一个引线，可测两端的电压差($V_+ - V_-$)以及两端分别与中间的电压差($V_+ - V_0$)和($V_- - V_0$)。如果一束光正好射到正中的位置，($V_+ - V_0$) = ($V_- - V_0$)。换而言之，将位置敏感检测器调到一个二者相等的位置，即为原点位置。随后，测量光束在某点的电压(V)，即可从 $L(V_+ - V)/(V_+ - V_-)$ 获得偏离的距离，其中的 L 为感光长度。如果 $L = 10$ mm，两端电压读数为 ±10 V，测量精度为 ±0.001 V，所测位置的精度就是 1 μm，远超过普通显微标尺的精度。

另外,折射指数依赖于光的波长,并与波长的平方成反比。采用汞灯时,无法得到和光散射中所用激光具有相同波长的滤光片。因此,必须采用两个波长分别长于和短于激光波长的滤光片,测量微分折射指数,再用内插法获得在激光波长时的微分折射指数。读者也许会问,为何不直接采用光散射中的激光作为微分折射仪的光源。其实许多人都问过同样的问题,也尝试过。然而,均没有成功。为回答这一问题,必须先考察以下激光产生的原理。

　　激光是在一个由一个全反射镜和一个半反射、半透射镜组成的空腔里通过诱导其中介质的电子同时跃迁产生的。在制造激光时,设计者已采用热稳定性极好的材料固定两面镜子和其他部件,以达到激光光束极好的准直性和稳定性。但与上述经典设计中的固定狭缝或小圆孔相比,激光光束仍会因镜子的热涨落而出现极小的漂移。通常,对许多的应用,这样的漂移可以忽略不计;但是在微分折射仪里,偏离距离的测量精度就是微米量级,因此,自激光在 20 世纪 60 年代问世以后,光束漂移就阻止了将激光作为微分折射仪的光源。

　　笔者在测量微分折射指数的痛苦中反复思考是否有一个更好的办法。以下是一个真实的小故事。直至 20 世纪 90 年代初的一个周六,我突然意识到可以利用一个简单的光学方法来克服激光光束漂移的问题,立即前往实验室,然而公司的实验室因安全考量,周末全部关闭。吃了闭门羹,不得不无奈地回家,苦苦地等待周一的到来。当时我的心情是恨不得一把将太阳从地平线下抓上来。等到周一开门,我立即冲进实验室,按照设想开始实验,得到预期的结果。无法形容笔者当时的心情。回想这一过程,笔者由衷地感谢在中国科学技术大学受到的本科物理训练,可见基础课是何等的重要。这一简单设想的详情如下。

　　用激光照亮一个小圆孔,然后选择一个会聚透镜,将小圆孔和位置检测器分别置于透镜的两边,相距透镜中心的光程差均为两倍的焦距。在光学上,这等价于将小圆孔贴在位置检测器的光敏平面上。这样一来,无论激光如何漂移,均不会影响小圆孔的位置,如右图所示。因溶剂和溶液折射指数的差别而引致的光束偏移,光学上等于将小圆孔在检测平面上移到一

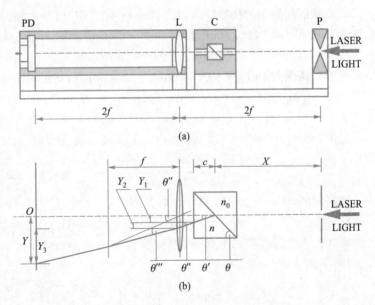

(a)

(b)

Wu C,Xia K Q. Rev. Sci. Instrum.,1994,65:587.

(Figure 1,permission was granted by publisher)

个新的位置。激光光束的漂移也不会影响小圆孔新的位置。上述的描绘可通过几何光学原理证明,如上图所示。注意,此时激光光源画在右边,从右向左,

$$Y = Y_1 + Y_2 + Y_3 = c\tan\theta' + (2f - X - c)\tan\theta'' + 2f\tan\theta'''$$

以及

$$f\tan\theta'' = c\tan\theta' + (2f - X - c)\tan\theta'' + f\tan\theta'''$$

其中,c,X 和 θ 均为常数。依据上图或者将上面两式合并,不难得到下式,

$$Y = f\tan\theta'' + f\tan\theta''' = f(\tan\theta'' + \tan\theta''')$$

另一方面,在 1621 年,Willebrord Snellius 发现,一束光从一个介质穿过一个界面进入另一个介质时,折射指数和其与法线夹角的正弦值之积为一个不变量。数学上,

$$n_1\sin\theta_1 = n_2\sin\theta_2$$

现被称为 Snell 定律。实际上,该定律的发现可追溯到 984 年,波斯科学家 Ibn Sahl 当时已精确地描绘了这一规律。在 1602 年,Thomas Horriot 又重新发现了这一规律。接着在 1637 年,René Descartes 独立地利用启发式动量守恒也得到了这一规律,并用其解决了一系列问题。由此可见,科学研究的延续性和承前启后的发展。早期,信息传递和来往通信并非今天这样发达,所以,一个科学定律往往被多人独立地发现。有些仅存在于私人交流和来往信件之中,并未正式发表,当然,也有完全不知其他文章的重复发表。幸运的是,早期的科学研究纯属学术和个人兴趣,并不太像当今这样功利。依据该定律,

$$n_0\sin(90° - \theta) = (n_0 + \Delta n)\sin(90° - \theta - \theta')$$

在样品池和透镜之间是空气,其折射指数为 1。依据上图,还有以下关系:

$$(n_0 + \Delta n)\sin\theta' = \sin\theta''$$

溶液和溶剂的折射指数之差很小($\Delta n \sim 10^{-4}$ RI),故上述几个式子中的角度 θ',θ'' 和 θ''' 实际上很小。因此,$\sin\theta' \cong \theta'$,$\sin\theta'' \cong \theta''$,$\tan\theta'' \cong \theta''$,$\tan\theta''' \cong \theta'''$。由上式可得

$$(n_0 + \Delta n)\theta' \cong \theta''$$

利用三角函数 $\sin(x - y) = \sin x \cos y - \cos x \sin y$,前式可被重写成

$$n_0\sin(90° - \theta) = (n_0 + \Delta n)\left[\sin(90° - \theta)\cos\theta' - \cos(90° - \theta)\sin\theta' \right]$$

$$(n_0 + \Delta n)\theta'\cos(90° - \theta) \cong \Delta n\sin(90° - \theta)$$

其中,$\theta' \ll 1$,$\sin\theta' \cong \theta'$ 和 $\cos\theta' \cong 1$。上式还可被重写成

$$\frac{(n_0 + \Delta n)\theta'}{\Delta n} \cong \frac{\sin(90° - \theta)}{\cos(90° - \theta)} = \tan(90° - \theta)$$

或利用 $\theta'' \cong (n_0 + \Delta n)\theta'$,进一步将其写成

$$\theta'' = \Delta n\tan(90° - \theta)$$

将其代入上述的第二个式子,并利用 $\tan\theta'' \cong \theta''$,$\tan\theta''' \cong \theta'''$,以及将 θ''' 表达式与 θ'' 关联,可得下式:

$$f(\tan\theta'' + \tan\theta''') = \left[X + c\left(1 - \frac{1}{n_0}\right) \right]\tan(90° - \theta)\Delta n$$

将 Δn 前面的正比常数因子定义为 K。并代入前面 Y 的表达式,可得最终结果:

$$Y = K\Delta n$$

依据上图,如果将透镜紧贴着样品池的外壁,$(X+c) \cong 2f$。在笔者以前的一个设计中,$c = 5.0 \pm 0.1$ mm;$X = 155.0 \pm 0.5$ mm;$\theta = 45°$,故 K 为 $156 \sim 157$ mm。光束偏离的距离(Y)与溶液和溶剂折射指数之差($\Delta n \sim 10^{-4}$)成正比,即 Y 仅约为几百微米。这就是为何测量精度必须达到微米级,方可保证所测的微分折射指数的相对误差低于 $\pm 1\%$。

显然,选择长焦距的透镜可以提高测量精度,但是受限于仪器所准许的长度,以及长基座的稳定性,除非是在一个光学平台上。如果作为在线的检测器,其长度不可能太长。基于这一新颖设计的微分折射仪具有极高的稳定性、精度和确度,以及可重复性,如下图所示;源自四个浓度的聚苯乙烯甲苯溶液的数据点均落在一条过原点的直线上。其斜率对应着聚苯乙烯在甲苯中的微分折射指数。精度之高,原则上仅需测量一个浓度,即可得到微分折射指数。当然,仍然选择测量四个以上的浓度,以进一步提高测量的精度和确度。

目前,基于这一原理的微分折射仪已被德国 ALV 公司商品化,作为其激光光散射仪器的配套设备;约十五年前,国内的健科仪器公司在笔者的协助下,也将其商品化,除了可以用作激光光散射的配套仪器以外,为了不同的应用,其还可被设计成手提式的、超灵敏的差分浓度检测器。这样,仪器不能太长。一个解决的办法就是在透镜后面加一个全反射镜,可将仪器长度缩短一半,精度提高一倍。据此原理,笔者开发出了可装入一个普通手提箱的便携式微分折射仪,如下图所示。这一新颖的微分折射仪可用于研究所有涉及浓度或折射指数变化的化学和物理过程。

与其他浓度检测器相比,如 UV,这一基于新型设计的微分折射仪不仅可以稳定地、精确地测定每毫升溶液中每毫克组分的变化,而且还可利用另一个没有变化的溶液作为参照体系。因此,复杂体系里所有不变的组分在实验上被完全扣除,该仪器只观察变化的组分。例如,在研究和开发中,可用来在线地测定聚合反应过程中浓度的变化、在酶

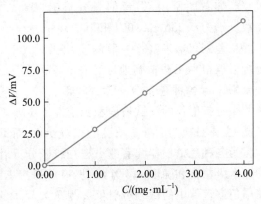

Wu C,Xia K Q. Rev. Sci. Instrum.,1994,65:587.

(Figure 5,permission was granted by publisher)

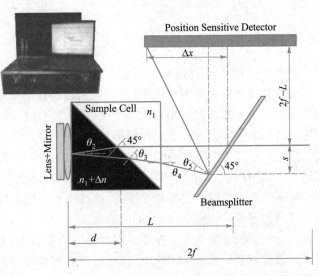

Gong X J,Ngai T,Wu C. Rev. Sci. Instrum.,2013,84:114103.

(Figure 3,permission was granted by publisher)

的反应过程中聚合物的降解速率等。在生产和质量控制中,可以在两个池子中分别注入已知质量和待测的溶液,利用微分的方法获得一个未知样品的纯度和质量,包括但不限于,利用该微分折射仪快速地辨别高级香水和酒的真伪。

据笔者所知,这是目前唯一利用激光光源和位置检测器的高精度、超稳定微分折射仪,可以测量细微的微分折射指数差别。例如,利用这一微分折射仪可以研究通过不同方法制备的聚苯乙烯纳米粒子,了解这些粒子里是否含有不同数量的聚苯乙烯链,即它们的微观密度,因为已知它们的尺寸。如右图所示,二者的微分折射指数分别为 0.236 ± 0.001 和 0.256 ± 0.001,差别仅有 ~8%;一般的商用微分折射仪根本无法精确地区分这么小的差别。为各种色谱仪配套的、商品

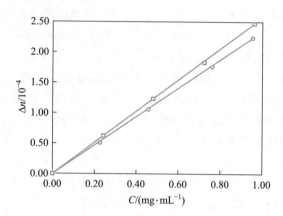

Wu C,Chan K K,Woo K F,et al.Macromolecules,1995,28:1592.

(Figure 1,permission was granted by publisher)

化的微分折射仪的测试精度和稳定性则更差。这一差别并不奇怪,作为色谱仪器配套的仪器,主要是检测大分子浓度的相对变化,而不是大分子微分折射指数的绝对值。

为各种色谱仪器配套的小型在线微分折射仪根本达不到测量位相偏差或光束偏移时的微米级精度和确度。不幸的是,有些研究者并不全然了解微分折射的基本原理。他们使用色谱仪器中的微分折射仪去测量大分子的微分折射指数,然后再代入静态激光光散射实验中的计算。他们所得大分子的重均摩尔质量的误差已经翻倍。现在,这些糟糕的数据充斥目前的文献之中。

小　结

基本原理:在干涉型的微分折射仪中,利用光的传播速度反比于折射指数。将一束偏振的光分解成互相正交的两束光,并分别通过溶液和溶剂。二者微小的折射指数之差造成传播过程中位相之差。当它们再合并成一束光时,其偏振会略微地偏离原先的方向。利用一个与原本偏振方向正交的偏振片,可测得微小的折射指数之差。该折射指数之差正比于透过第二个偏振片的光强。在折射型的微分折射仪中,让一束准直平行的光束连续通过溶剂和溶液,二者间用一个约45°角度的玻璃隔板分开。利用光束因溶液和溶剂之间的折射指数之差偏离的折射原理,可通过偏离距离获得与之成正比的折射指数之差。

可测物理量:制备四个以上不同浓度的大分子溶液。由稀到浓地测量每一个溶液与溶剂之间的折射指数之差。在干涉型的微分折射仪中,测量通过第二个正交偏振片的光强。而在折射型的微分折射仪中,测量一束入射准直平行光束连续经过样品池中分割的溶液和溶剂后的偏离距离。实验中,需要首先利用已知折射指数的溶剂和溶液,通常采用水和氯化钠或氯化钾的水溶液,标定仪器;即得到透射光强或偏离距离与折射指数之差之

间的正比常数。

数据处理：利用标定的正比常数，先将由每个给定浓度(C)的溶液测得的透射光强或偏离距离换算成折射指数之差(Δn)；再将Δn对C作图，得一条过原点的直线，其斜率对应着微分折射指数。如果数据不在一条直线上，或直线不过原点，实验中一定存在问题，包括但不限于：浓度配制不准、恒温时间不够长、溶液池没完全清洗，或仪器标定有误。整个实验需要从头开始。

黏　度

黏度是日常生活中常常使用的一个名词，和液体流动有关。依据经验，给定作用力后，材料不同，形变速度有异。例如，利用重力倾倒一个液体时，取决于其流动的快慢，可判断该液体是否黏稠。又如，分别倾斜一瓶水和一瓶蜂蜜时，水的倾斜速度远快于蜂蜜，在较冷的天气，二者的差别更为明显；与水相比，蜂蜜的黏度随温度降得较快。

在严格和详细地讨论黏度的定义和来源之前，先熟悉一下几个与材料性质有关的宏观变量。材料单位面积上承受的作用力，称作应力($\vec{\sigma}$)。当力的作用方向与表面垂直或者平行时，分别称作法向应力($\vec{\sigma}_{\perp}$)和剪切应力($\vec{\sigma}_{=}$)。任何方向的作用力($\vec{F}$)总是可以分解成垂直和平行的两个分力($\vec{F}_{\perp}$和$\vec{F}_{=}$)。一个材料的相对形变称作应变(γ)，如右图所示。在小形变下，应力正比于应变。正比常数分别被称作杨氏模量($\vec{E}$)和剪切模量($\vec{G}$)，即$\vec{\sigma}_{\perp}=\vec{E}\gamma$和$\vec{\sigma}_{=}=\vec{G}\gamma$。由于

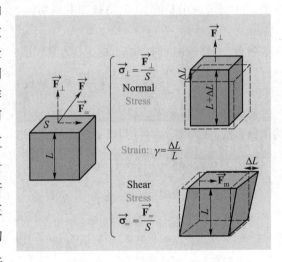

方向已经给定，所以矢量符号可以移去，写成标量形式：$\sigma_{\perp}=E\gamma$和$\sigma_{=}=G\gamma$，反映了一个物体的"弹性"。

实验早已证明，在较小的应力下，应力也成正比形变速度(或曰速率，$\mathrm{d}\gamma/\mathrm{d}t$)。正比常数被称为黏度($\eta$)，即$\sigma=\eta\mathrm{d}\gamma/\mathrm{d}t$，反映了物体的"黏性"。在给定应力后，黏度和形变速度成反比。黏度越大，形变速度越小。反之，在给定形变速度后，黏度也反映了物体抗拒形变的能力。黏度越大，就需要更大的应力才可达到同样的形变速度。

严格地说，任何物体都既具有弹性，又具有黏性。通常，在熔点之下时，小分子固体的黏度极大，在给定的较小应力下，形变速率极慢，故其被认为仅有弹性。反之，在熔点之上时，小分子液体的弹性模量极小，形变极大，故其又被认为仅有黏性。而大分子材料，无论是处于固体还是液体状态，则可既黏又弹，称作"黏弹性"。如果将"弹性(elastic)"和"黏性(viscous)"看作大分子材料的两个"元件"，二者并联时，总应力(σ_T)为它们的应力之

和。将其代入应力定义,可得

$$\sigma_T = \sigma_e + \sigma_v = E\gamma + \eta \frac{d\gamma}{dt} \quad \text{或} \quad \sigma_T = \sigma_e + \sigma_v = G\gamma + \eta \frac{d\gamma}{dt}$$

在应变松弛(蠕变)实验中,总应力为一常数(σ_T),不依赖时间。从上述两个简单的微分方程,不难分别得到以下两个称为应变松弛的解:

$$\gamma(t) = \frac{\sigma_T}{E} \left[1 - \exp\left(-\frac{t}{\tau_c}\right) \right] \quad \text{或} \quad \gamma(t) = \frac{\sigma_T}{G} \left[1 - \exp\left(-\frac{t}{\tau_c}\right) \right]$$

其中,特征松弛时间(τ_c)等于η/E或η/G。对一交变应力,$\sigma_T = \sigma_{T,0}\cos(\omega t)$,可先假定$\gamma(t) = \gamma_{c,0}\cos(\omega t) + \gamma_{s,0}\sin(\omega t)$,代入前式,比较两边,可得$\sigma_{T,0} = E\gamma_{c,0} + \eta\omega\gamma_{s,0}$和$0 = E\gamma_{s,0} - \eta\omega\gamma_{c,0}$或$\sigma_{T,0} = G\gamma_{c,0} + \eta\omega\gamma_{s,0}$和$0 = G\gamma_{s,0} - \eta\omega\gamma_{c,0}$,分别导出

$$\gamma_{c,0} = \frac{\gamma_0}{1 + (\omega\tau_c)^2} = \gamma_0 G'(\omega) \quad \text{和} \quad \gamma_{s,0} = \frac{\gamma_0 \omega\tau_c}{1 + (\omega\tau_c)^2} = \gamma_0 G''(\omega)$$

其中,$G'(\omega)$和$G''(\omega)$分别代表了材料储存和耗散能量的两个部分,分别称作储存模量和耗散模量。$G''(\omega)/G'(\omega) = \omega\tau_c = \text{tg}\delta$,这里$\delta$是耗散角。将上述两个关系代入$\gamma(t)$的假定,并注意$\tau_c = \eta/E$或$\eta/G$,可推出其随时间的周期性变化如下:

$$\gamma(t) = \gamma_0 [G'(\omega)\cos(\omega t) + G''(\omega)\sin(\omega t)]$$

两个"元件"串联时,总应变(γ_T)为它们应变之和,代入应变定义,可得

$$\frac{d\gamma_T}{dt} = \frac{d(\gamma_e + \gamma_v)}{dt} = \frac{d\sigma}{Edt} + \frac{\sigma}{\eta} \quad \text{或} \quad \frac{d\gamma_T}{dt} = \frac{d(\gamma_e + \gamma_v)}{dt} = \frac{d\sigma}{Gdt} + \frac{\sigma}{\eta}$$

在应力松弛实验中,总应变为一常数(γ_T),不依赖时间,$d\gamma_T/dt = 0$。从上述两个简单的微分方程,也不难分别得到以下两个称为应变(蠕变)松弛的解:

$$\sigma(t) = \sigma_0 \exp\left(-\frac{Et}{\eta}\right) = \sigma_0 \exp\left(-\frac{t}{\tau_c}\right) \quad \text{或} \quad \sigma(t) = \sigma_0 \exp\left(-\frac{Gt}{\eta}\right) = \sigma_0 \exp\left(-\frac{t}{\tau_c}\right)$$

其中,特征松弛时间(τ_c)也等于η/E或η/G。对一个交变应变,$\gamma_T = \gamma_{T,0}\sin(\omega t)$,假定$\sigma(t) = \sigma_{c,0}\cos(\omega t) + \sigma_{s,0}\sin(\omega t)$,代入前式,比较两边,可得$\sigma_{s,0}/\eta - \sigma_{c,0}\omega/E = 0$和$\gamma_{T,0}\omega = \sigma_{c,0}/\eta + \sigma_{s,0}\omega/E$或$\sigma_{s,0}/\eta - \sigma_{c,0}\omega/G = 0$和$\gamma_{T,0}\omega = \sigma_{c,0}/\eta + \sigma_{s,0}\omega/G$,可得

$$\sigma_{c,0} = \frac{\sigma_0(\omega\tau_c)^2}{1 + (\omega\tau_c)^2} = \sigma_0 G'(\omega) \quad \text{和} \quad \sigma_{s,0} = \frac{\sigma_0 \omega\tau_c}{1 + (\omega\tau_c)^2} = \sigma_0 G''(\omega)$$

其中,$G'(\omega)$和$G''(\omega)$的物理意义仍然是分别代表了材料储存和耗散能量的两个部分,故仍然分别称为储存模量和耗散模量;注意,与应力松弛时相比,$G'(\omega)$的分子上多了$(\omega\tau_c)^2$;仍有$G''(\omega)/G'(\omega) = 1/\omega\tau_c = \text{tg}\delta$,这里$\delta$仍是耗散角。将上述两个关系代入$\sigma(t)$的假定,并记住$\tau_c = \eta/E$或$\eta/G$,可推出其随时间周期性地的变化如下:

$$\sigma(t) = \sigma_0 [G'(\omega)\cos(\omega t) + G''(\omega)\sin(\omega t)]$$

在一个真实的应用中,一个材料所经历的既不是纯粹的应力松弛,也不是纯粹的蠕变松弛,而是它们复杂的排列和组合,涉及多个应力和蠕变松弛,但万变不离其宗。只要理解了上述两个松弛的物理本质,任何一个复杂的应力-应变过程都可被分解成二者的不同组合。

黏性和弹性是相对的概念,取决于温度和时间。在极短的时间内或极低的温度下,分

子没有时间或没有足够的热能松弛,故物体显出弹性。反之,物体则呈现黏性。譬如,在常温和通常的时间尺度上,水是黏性液体。但是,如果用力将一片石头以一个很小的角度、快速地掠过水面,石片可能在水面上弹起数次,形成一串水花,仿佛水是一个弹性材料。

两千五百多年前,孙武就在《孙子兵法》中论及这一现象:"激水之疾,至于漂石者,势也"。此处之"势":极快之速、极短之时也。另一方面,如果用"千年"作为时间单位,固体的玻璃也是可以缓慢流动的液体。温度的效果则更明显。从分子松弛运动的物理角度出发,时间和温度是等效的。一个较长的松弛时间对应着一个较低的温度;反之,一个较低的温度对应着一个较长的松弛时间。正因有着这一等效关系,工程师们方可通过在高温下进行一个材料的"老化"实验来模仿其经过长期使用后的性质。

一般而言,小分子液体(如水、食用油、酒精、汽油等)的模量趋近于零,仅有黏度,稍加施力,液体就会有很大的变形。日常生活中,利用重力来倾倒一个液体。也正是利用了这一极易流动的特性,可以通过施压将自来水送进千家万户。黏度的物理本质是分子在运动时,互相之间产生的摩擦。与小分子溶剂相比,大分子长链可互相缠结,其随着链长增加。有时,加入少量大分子长链即可消除湍流、减少流阻和增加流动稳定性,这在许多实际的应用中都至关重要。例如,穿越阿拉斯

加的原油管道长达 1300 km,如右图所示。添加长链烃类聚合物减少了流动阻力,使得流量从 1.44 百万桶增至 2.14 百万桶,增加了近 50%,相当于多了半条油管的线路。

熟悉电工电路的读者不难发现,如果将应力和应变分别当作"电容"和"电阻",那么上述的所有数学公式都同电学里的一致。实际上,上述数学公式可描绘任何包括储存和消耗能量的物理过程。物理通过数学揭示了万物的变化本质,其精巧和美妙之处,溢于言表。

如前所述,George Gabriel Stokes 早在 1851 年就发现,一个半径为 r 的小球在一个黏度为 η^* 的纯液体里扩散运动时所受的阻力系数,$f = 6\pi\eta^* r$;后来,Albert Einstein 于 1851 年在其博士论文中指出,含有小球的二元混合体系的液体黏度(η)与液体里所有小球的体积分数(X_p)遵循一个简单的关系式,如下

$$\eta = \eta^* (1 + 2.5X_p)$$

依据物理直觉可知,假定一个质量为 m,半径为 r 的大粒子被分成 n 个同样的小粒子。r 和 m 之间的标度关系为 $r = k_R m^{\alpha_R}$。在同样的质量浓度时,分割粒子将如何影响粒子总截面积和总体积? 分割前,总截面积和总体积分别就是一个大粒子的截面积和体积,即 $S_{before} = 4\pi r^2 = 4\pi k_R^2 m^{2\alpha_R}$ 和 $V_{before} = 4\pi r^3/3 = 4\pi k_R^3 m^{3\alpha_R}/3$。分割成 n 个小粒子后,每个小粒子的半径成为 $k_R(m/n)^{\alpha_R}$,故 $S_{after} = 4\pi r^2 n = 4\pi k_R^2 m^{2\alpha_R} n^{1-2\alpha_R}$ 和 $V_{after} = 4\pi k_R^3 m^{3\alpha_R} n^{1-3\alpha_R}/3$。因此,$S_{after}/S_{before} \sim n^{1-2\alpha_R}$ 和 $V_{after}/V_{before} \sim n^{1-3\alpha_R}$。粒子密度均匀时,$\alpha = 1/3$,$S_{after}/S_{before} \sim n^{1/3}$。分

割增加总截面积,但如预期的,对总体积没有影响。在良溶剂中的线型柔性大分子,$1/2<\alpha_R<3/5$,所以,分割使得总截面积和总体积均减少。反之,将短链连接成长链可增加大分子链的体积分数。

如第四章所述,线型柔性大分子链在溶液中的构象为无规线团。先假定每条线型大分子链等效于一个小球,以及溶液中共有N_p条链,再将上式中的体积分数($X_p=V_p/V$)换算成质量浓度($C=W/V$),即

$$X_p=\frac{v_p N_p}{V}=\frac{4\pi}{3}\frac{k_R^3 M^{3\alpha_R}N_{AV}}{M}\frac{W}{V}=\frac{4\pi}{3}k_R^3 N_{AV}M^{3\alpha_R-1}C$$

其中,利用了$N_p=N_{AV}(W/M)$,一条链的体积($v_p\propto R^3$)和$R\propto M^{\alpha_R}$。将其代入前式可得

$$\eta=\eta^*\left(1+\frac{4\pi}{3}k_R^3 N_{AV}M^{3\alpha_R-1}C\right)\tag{5.41}$$

其中,k_R是一个依赖于大分子和溶剂性质的常数。显然,在良溶剂中,线型柔性大分子稀溶液的黏度(η)随着大分子的摩尔质量和质量浓度增加。物理解释如下,对给定的质量浓度,长的大分子链的总体积较大,故分子间的摩擦增加,导致溶液黏度变大。对给定的大分子链,链的总体积随浓度增加,也导致分子间的摩擦增加和溶液黏度变高。

因此,如果为了增加一个给定质量浓度的溶液黏度,应该添加线型长链。另一方面,对给定的大分子和质量浓度,也可通过改变链构象来变化溶液中的大分子体积分数和溶液黏度。一个有趣的例子是在汽车发动机的润滑机油里添加一种大分子。大分子链在低温下处于蜷缩状态,但无聚集。蜷缩的大分子链仅略微增加机油黏度,不影响发动机在低温时的启动。在运行的高温下,机油本身的黏度急剧下降,不利于润滑。但添加的大分子可在热机

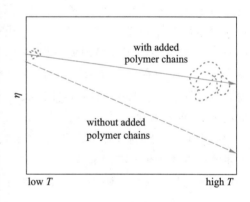

油里充分溶解,形成扩展的链构象,增大体积分数,增加机油黏度,所以其仍可润滑发动机,如右图所示。

可分别测量溶剂和大分子溶液的黏度。二者之比(η/η^*)是相对黏度(η_r)。二者之差除以溶剂黏度为比黏度[$\eta_{sp}=(\eta-\eta^*)/\eta^*=\eta/\eta^*-1$]。进而,比黏度与大分子浓度之比得约化黏度[$\eta_{red}=(\eta-\eta^*)/\eta^*/C$],也称作黏度数。在无限稀释的大分子溶液中,约化黏度成为与浓度无关的特性黏度([η]$=\lim_{C\to0}\eta_{red}$)。各种黏度之间的关系如下:

$$\eta_r=1+[\eta]C+B_2C^2+\cdots;\quad \eta_{sp}=[\eta]C+B_2C^2+\cdots;\quad \frac{\eta_{sp}}{C}=[\eta]+B_2C+\cdots$$

其中,特性黏度的单位是浓度的倒数;B_2表征着黏度对浓度的非线性依赖,取决于溶剂和大分子的特性,反映了排除体积效应,故其可为正、为负或为零,类似前述的二阶维里系数。依据η_{red}和[η]定义,可将式(5.41)重写如下:

$$[\eta] = \lim_{C \to 0} \frac{\eta - \eta^*}{\eta^* C} = \frac{10\pi}{3} \frac{k_R^3 N_{AV}}{M^{3\alpha_R - 1}} = K_\eta M^{\alpha_\eta} \qquad (5.42)$$

其中,$\alpha_\eta = 3\alpha_R - 1$,称作 Mark-Houwink-Sakurada 指数;K_η包括了所有常数,定义如下:

$$K_\eta = \frac{10\pi}{3} \frac{k_R^3 N_{AV}}{3}$$

其只与大分子和溶剂的本质有关,与浓度无关。由此可见,特性黏度与大分子在溶液中的体积分数有关,其又进一步和摩尔质量相关。已知,对良溶剂中的线型柔性大分子而言 $1/2 < \alpha_\eta < 4/5$。具有相同摩尔质量的半刚性链的体积比线型柔软链大,$4/5 < \alpha_\eta < 2$。对完全刚性的链,$\alpha_\eta = 2$。至于具有其他组构的大分子链,α_η 也可从类似的标度律获得。依据上式,如果知道链的组构,就可估计出 α_η。理论上,仅要测量一个已知摩尔质量的大分子样品,即可获得K_η。对于一个同种,但未知摩尔质量的大分子,可先测量其约化黏度对浓度的依赖性,再算出其特征黏度,最后利用上式决定其摩尔质量。

即使在诸多电子仪器极端发达的今天,毛细管黏度计仍然是实验室中一个常规的装置。其原理基于,不可压缩的牛顿液体在一个管道里层流,即流速相对较低,无湍流,液面为中心流速大、管壁处流速为零的抛物线形。管道两端的压差(Δp)与管道半径和长度(r 和 l)、液体黏度(η),以及流动的体积和时间(V 和 t)有关。Jean Léonard Matie Poiseuille 和 Gotthilf Heinrich Ludwig Hangen 分别于 1838 年和 1839 年独立地在实验中发现了这一依赖关系。Poiseuille 于 1840 年正式将其发表,George Gabriel Stokes 在 1845 年完成了其论证,即如下的 Hangen-Poiseuille 方程:

$$\eta = \frac{\pi r^4 \Delta p}{8Vl} t$$

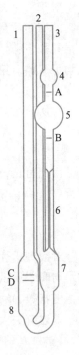

测量液体黏度时,通常采用长度为 l 的毛细管和重力(g)产生的压强差(Δp),即 $\Delta p = \rho g l$,其中的 ρ 为液体密度。将其代入上式可得 $\eta = K\rho t$,这里 K 是一个仪器常数。将 η/ρ 定义为动力学黏度(η_k),$\eta_{k,1}/\eta_{k,2} = t_1/t_2$。因此,用一个已知密度和黏度的液体即可标定一个毛细管黏度仪的仪器常数。之后,仅要知道待测液体的密度和测得其流动时间便可算出其黏度。

为了测量大分子溶液相对较大的黏度,Leo Ubbelohde 于 1936 年发明了悬浮液位黏度计,现被常称为 Ubbelohde 黏度计,如右图所示。黏度计的长度约为 28 cm,毛细管(6)的长度约为 9 cm,球(5)的直径为 3~4 cm;管(2)的作用是保证上下的气压差始终如一,剩下的净压差(Δp)仅依赖于液体密度和毛细管长度。使用时,将整个黏度计置于一个温控的、充满液体(水)的玻璃槽中,只有三个管口在外。毛细管(6)必须垂直。刻度线 A 和 B 要朝外放置,使得易于从外部观察。

从高浓度开始测量。经管(1)注入溶液。达到一个恒定温度后,用洗耳球从管(3)将溶液吸至球(4)的上端。松开吸球,让液体沿毛细管(6)在重力作用下自然流入下部容器(7 和 8)。当弯曲液面的最低点分别经过刻度线 A 和 B 时,记录两个时间,二者之差即是流动时间(t),至少重复测量 3 次,得平均流动时间,以减少测量误差。完成一个浓度的黏度测量后,经管(3)加入适量的溶剂稀释溶液。用洗耳球将溶液在管(3)中吸上和放下数次,清洗两个球(4 和 5)和毛细管,以及充分混合溶液。随后,一个浓度接另一个浓度地重复测量。

测完所有不同浓度的溶液后,用溶剂彻底清洗黏度仪,尤其是两个球(4 和 5)和毛细管。最后,测量溶剂的流动时间(t_0)。在稀溶液中,溶液和溶剂密度近似地相等,故黏度正比于流动时间,$\eta_{red} = (t - t_0)/t_0$。在稀溶液中,非线性项很小。可直接将相对黏度对浓度作图,如下图所示,从斜率可得特性黏度($[\eta]$)。如果浓度较大,相对黏度对浓度作图不呈直线,故不得不用约化黏度对浓度作图得一直线,其截距和斜率分别对应着特性黏度和系数(B_2)。目前,这些都已成为许多聚合物研究室和工业界的常规操作,包括控制聚合反应和聚合物表征。

由上述 Hangen - Poiseuille 方程已知,流动时间与 r^4 成反比。采用一套不同管径的 Ubbelohde 黏度计,就可测量具有不同黏度的液体。选用适当的管径,使得流动时间在 100～200 s 之间,即保证了测量精度(可达 0.1%),又避免了长的测量时间。由于黏度的测量便宜、简单和易行,其常被用在聚合物的生产线上。通过只测量一个给定浓度的溶液黏度,即可控制反应进程和摩尔质量。

对于一个多分散的样品,由特性黏度得到的摩尔质量是一个特殊的平均值。具体推导如下:对一个足够稀的溶液,非线性项均可忽略。第 i 个组分和整个体系的比黏度分别为

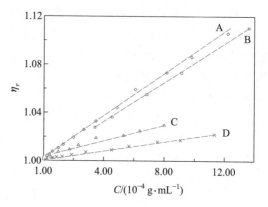

Liu M Z,Cheng R S,Wu C,et al. J. Polym. Sci. B:Polym. Phys,1997,345:2421.

(Figure 3,permission was granted by publisher)

$$\eta_{sp,i} = [\eta]_i C_i + \cdots \quad \text{和} \quad \eta_{sp} = [\eta] C + \cdots$$

对所有组分求和。左边求和为 $\eta_{sp} = \sum_i \eta_{sp,i}$;右边的求和得 $\sum_i [\eta]_i C_i$。因此,

$$[\eta] C = \sum_i [\eta]_i C_i \quad \rightarrow \quad [\eta] = \frac{\sum_i [\eta]_i C_i}{C}$$

由此可见,对一个多分散样品,所测的是一个重均特性黏度。将式(5.42)代入上式,可得

$$k_\eta M_\eta^{\alpha_\eta} = \frac{\sum_i k_\eta M_i^{\alpha_\eta} C_i}{C} \quad \rightarrow \quad M_\eta = \left(\frac{\sum_i M_i^{\alpha_\eta} C_i}{C} \right)^{1/\alpha_\eta}$$

对线型柔性大分子链,$1/2 \leqslant \alpha_\eta \leqslant 4/5$。因此,对一个多分散的样品,由特性黏度获得的平均摩尔质量(M_η)在理论上应该介于数均和重均摩尔质量之间。对于半刚性链,取决于标度指数,M_η 一般大于 M_w。如果链是完全刚性,$\alpha_\eta = 2$,则 $M_\eta = (M_z M_w)^{1/2}$。

上述有关大分子链摩尔质量与尺寸的标度律的讨论假定了大分子链是一个密度不均匀、尺寸为 R 的"实心硬球",故其标度指数小于 3,以及内部的所有溶剂均随链一起扩散。

然而,其内部的一小部分溶剂分子并非和链一起运动,即存在一个"排水"效应。因此,链段会感受到一些摩擦阻力。整条大分子链的摩擦系数大于尺寸为 R 的实心小球。换而言之,溶液黏度强烈地依赖于大分子链的摩尔质量。

文献中有许多有关线型大分子链的特征黏度和它们的摩尔质量依赖性的数据,但是关于支化型大分子链的则较少。正如在第四章中所述,制备具有均一支化子链长度的支化链较难,制备两套分别是只有支化子链长度变化或只有母链摩尔质量变化的支化链则更难。理论上,在良溶剂中,链尺寸(R)与支化子链和母链摩尔质量(M_s 和 M)之间标度关系为 $R = k_R M_s^{\alpha_{R,s}} M^{\alpha_R}$,其中,$\alpha_{R,s} = 1/10$ 和 $\alpha_R = 1/2$;实验上,已证实 $\alpha_{R,s} = 1/10$ 和 $\alpha_R = 0.46$。对给定的摩尔质量,支化链比线型链具有较小的尺寸,所以支化链的特性黏度对浓度的依赖性也较弱。对给定质量浓度($C = W/V$),支化大分子链在溶液中的体积分数为

$$X_p = \frac{10\pi}{3} k_R^3 N_{AV} M_s^{3\alpha_{R,s}} M^{3\alpha_R - 1} C$$

将其代入上述 Einstein 有关小球在液体中黏度的方程,以及沿用前述 K_η 的定义,可得下式

$$\eta = \eta^* (1 + K_\eta M_s^{3\alpha_{R,s}} M^{3\alpha_R - 1} C) \quad 或 \quad [\eta] = K_\eta M_s^{3\alpha_{R,s}} M^{3\alpha_R - 1}$$

由此可见,支化子链和母链的摩尔质量均可影响支化链的特性黏度。在真实的实验中,由于不可避免的链长多分散性,摩尔质量通常被数均或重均摩尔质量代替。注意:由于统计平均中的权重不同,由黏度和动态激光光散射分别得到的平均流体力学半径不等。

右图(a)显示,在良溶剂中,对给定的支化子链摩尔质量($M_{w,s}$)、支化链的特性黏度随着母链的重均摩尔质量(M_w)增加。如果采用测得的 $\alpha_R = 0.4$,标度指数为 0.39,非常接近($3\alpha_R - 1$)。相反,给定母链摩尔质量时,支化链的特性黏度在良溶剂中也随着支化子链的摩尔质量略有增加(图中沿纵向上移),拟合的结果为 $3\alpha_{R,s} = 0.31$,也非常接近理论预期值(3/10)。而右图(b)则显示,如果用支化子链的摩尔质量修正所测的特性黏度($[\eta]/M_s^{3\alpha_{R,s}}$),图(a)中所有的实验结果都塌缩成一条直线,进一步证实了在第四章中所讨论的支化链的尺寸与支化子链和母链摩尔质量之间的标度关系。

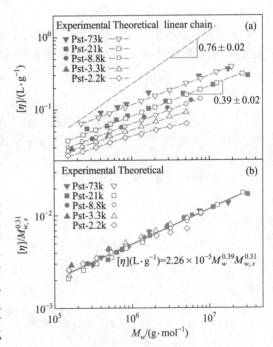

Li L W, Lu Y Y, An L J, et al. Journal of Chemical Physics, 2013, 138: 114908.

(Figure 3, permission was granted by publisher)

小　结

__基本原理__：材料的黏度源于分子运动时的相互摩擦，反映了对一给定的外力，材料形变的速率。在熔融温度以下或以上，一个由小分子构成的固体的黏度很大或很小，仅呈现弹性或黏性。在熔融状态，施加一个微弱的外力，该材料便会流动。而大分子材料，无论是在固态还是液态，都可同时显示足够的黏性和弹性，即其具有独特的黏弹性。这就是为何高聚物可被吹塑或拉伸成一片很薄的膜、纺成很细的丝，如食品保鲜膜和由超细合成纤维制成的高档仿丝服装。

大分子稀溶液仅有黏性，就如同由小分子构成的液体一样。但是，添加少量的长链大分子可极大地提高溶液的黏度、减少高速流动时的湍流、改善流动性质。其物理本质是，液体黏度与其中粒子或大分子所占的体积分数成正比。如将十条短链连成一条长链，溶液的比黏度理论上就会增加六倍。在浓度极稀时，比黏度与浓度的比值就成为一个仅与大分子和溶剂性质有关，而与浓度无关的特性黏度。其直接与大分子的体积（正确地说，流体力学体积）成正比。特性黏度与大分子摩尔质量也有一个标度关系。

__可测物理量__：采用 Instron 拉伸测试仪可直接测得固体材料的模量（E 或 G）和在不同剪切频率下的存储模量（ω'）和耗散模量（ω''）；对于纯液体，则可利用黏度计，常用一个标定好的 Ubbelohde 黏度计，分别测得一定体积的已知密度（ρ_0 和 ρ）的标准和待测液体流过一段毛细管所需的时间（t_0 和 t）。对于大分子稀溶液，分别测得一定体积的溶剂和不同浓度的溶液流过一段毛细管所需的时间（t^* 和 t）。

__数据处理__：对于固体材料，可利用应力或蠕变松弛实验中测得的特征松弛时间（τ_c）和模量之间的关系（η/E 或 η/G）算出黏度。也可将动态剪切中测得的 ω'' 和 ω' 相除得到耗散角，乘以或除以剪切频率后，再外推到零频率，即可得特征松弛时间，进而从其算出黏度。

对于纯液体，利用 $\eta = \eta^* (\rho/\rho^*)(t/t^*)$，就可算出其黏度。对于大分子稀溶液，先算出相对黏度（t/t^*），再将相对黏度对浓度作图，从斜率可算出特性黏度。如果浓度较大，则要先算出比黏度 $[(t - t^*)/t^*]$，再除以浓度得到约化黏度。将约化黏度对浓度作图可得一条直线，其截距和斜率分别对应着特性黏度和非线性系数，类似二阶维里系数，包含了排除体积效应。

尺寸排除色谱

顾名思义，尺寸排除色谱就是在空心的不锈钢色谱柱里加压填满了多孔颗粒填料，每个填料颗粒含有许多互相贯穿、平均口径为 r 的小孔。在一个稳速的流动场中，大小不同的大分子链被同时注入，如下图所示。假定大分子链和填料表面，包括那些小孔内部的表面，之间无特殊的相互吸引作用，较短的大分子链可以进入小孔，停留更长的时间，而较长的大分子链进入小孔的机会较低。在流动场中长链的停留时间会较短，故它们会被较早地冲出,也如图中所示。换而言之,小孔对尺寸不同的大分子链具有不同排除效应,所以色谱柱可按照尺寸将不同的大分子链分开,这就是为何称为尺寸排除色谱。正确地说,这是体积排除,将在下一章中讨论。

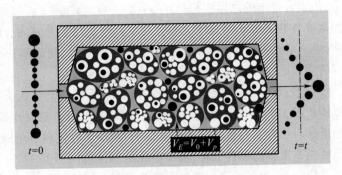

显然,对于一个体积不可压缩的硬粒子,它只能进入孔径较大的孔。对一条线型柔性大分子链,其尺寸和构象不停地变化,链可以毫无阻碍地进出较大的孔。如果进入一个尺寸小于其平均尺寸的孔,链就会受压,偏离其自由能最低的稳定状态,其自由能将升高。这一自由能变化(ΔA)决定了其进入小孔的概率。详情如下。

在良溶剂中,大分子链的尺寸(R)和其摩尔质量(M)标度关系为 $R = k_R M^{\alpha_R}$。如果一节大分子链的尺寸和一个柱状小孔的孔径(r)相等,该链节的摩尔质量(M_b)和孔径标度关系则为 $r = k_R M_b^{\alpha_R}$。如果将一条长链挤压进一个小孔,它定会沿着小孔的纵向伸展,变成一连串直径为孔径小圆柱,其总长度为 l,等效于 l/r 个小球连在一起,$l/r = M/M_b$。因此,$R/r = (M/M_b)^{\alpha_R}$,即 $l/r = (R/r)^{1/\alpha_R}$。在溶液中,每条链的自由能为一个 $k_B T$;每个小球的自由能也是一个 $k_B T$;进入柱状小孔后,链的自由能增加了 l/r 倍,自由能的变化(ΔA)为 $(l/r - 1)k_B T$。因此,进入小孔的概率(ψ)为

$$\psi = \exp\left(-\frac{\Delta A}{k_B T}\right) = \exp\left[1 - \left(\frac{R}{r}\right)^{\frac{1}{\alpha_R}}\right] \quad (R \geqslant r,\text{柱状小孔})$$

显然,当 $R = r$ 时,$\psi = 100\%$;在 Θ 溶剂和良溶剂中,$1/\alpha_R$ 分别为 2 和 5/3。因此,在 Θ 溶剂和良溶剂中,当 $R = 2r$ 时,ψ 就从 100% 降为 4.98% 和 8.80%。很清楚,随着链尺寸 R 的增加,大分子进入小孔的概率急剧地下降,趋于零。如果小孔是尺寸为 r 的空心球,$r \leqslant R$;那么大分子链在其中的压强就从外面的 $k_B T/R^3$ 增加为 $k_B T/r^3$;为了将一条链挤进空心小球,就需要额外的能量,$\Delta A = k_B T[(R/r)^3 - 1]$,代入上式,可得

$$\psi = \exp\left(-\frac{\Delta A}{k_B T}\right) = \exp\left[1 - \left(\frac{R}{r}\right)^3\right] \quad (R \geqslant r,\text{球状小孔})$$

当 $R = 2r$ 时,ψ 就从 100% 降为 0.1%。对给定的孔径,一条大分子链进入球状小孔的概率更低。在尺寸排除色谱柱中,小孔互相贯穿。在每个靠近表面的小孔近似为一个短的柱状小孔,大分子链可以在小孔的贯穿网络中停留更长的时间。

上图显示,除了填料本身占据的体积以外,色谱柱里的空洞体积(V_E,排除体积)包括两部分。一部分为所有填料颗粒之间的间隙(V_0,远大于链尺寸,也称为死体积);另一部分是填料粒子中所有小孔的体积之和(V_p)。因此,$V_E = V_0 + V_p$。所有的链,无论其尺寸大小,均可在 V_0 中运动,但只可进出孔径大于或接近其尺寸的小孔。对一条给定尺寸的大分子链,其可及的小孔总体积等于链进入小孔的概率(ψ)乘以小孔的总体积,即 ψV_p。那

么,该链可进出的总体积为$V_E = V_0 + \psi V_p$,或写成$\psi = (V_E - V_0)/V_p$。在稀溶液中,进入小孔的概率ψ也表征了大分子在小孔和填料间隙二者之间的分配系数。其仅取决于R/r。显然,进入小孔的概率越大,链在色谱柱内的停留时间就越长,反之亦然。链的尺寸越大,停留时间就越短,故长的大分子链较先流出。

显然,对于给定的色谱柱和多孔填料,不同尺寸的大分子链具有不同的ψ。理论上,死体积(V_0)越小,ψV_p所占的体积分数就越大,分离效果也就越好。这解释了为何在制备色谱柱时通常需要施加高压。一方面,高压可将填料尽量压实,减少死体积;另一方面,高压可填入更多的多孔填料,增加小孔的总体积(V_p)。

理论上,所有尺寸大于最大的孔径(r_{max})的大分子链均同时流出,它们具有最短的停留时间(t_0);所有尺寸小于最小的孔径(r_{min})的大分子链也都同时流出,它们的停留时间最长(t_m)。实验上,在这两种极端的条件下,前者和后者并非真地同时流出,在每种情况下,停留时间仍略有差别。

小孔具有一个平均孔径r。当大分子链的尺寸(R)大于r时,其在小孔和填料间隙中的分配比例(k)等于分配系数乘以二者的体积之比,即$k = \psi V_p/V_0$;因此,大分子链在小孔和填料间隙中的概率分别就是它们各自对大分子链可提供的体积(V_0或ψV_p)与总的排除体积($V_0 + \psi V_p$)之比。

$$\frac{\psi V_p}{V_0 + \psi V_p} = \frac{k}{1 + k} \quad \text{和} \quad \frac{V_0}{V_0 + \psi V_p} = \frac{1}{1 + k}$$

另一方面,一条大分子链在色谱柱中的总停留时间(t)等于其在小孔(可能进出多个小孔)中的停留时间(t_p)加上其在填料间隙中的停留时间(t_0),即$t = t_p + t_0$。与之对应的是,分配比例等于二者之比,$k = t_p/t_0$。将它们和前述的ψ表达式代入上式,可得

$$\exp\left[1 - \left(\frac{R}{r}\right)^{\frac{1}{\alpha_R}}\right] = \frac{V_0}{V_p t_0}(t - t_0) \quad \text{或写成} \quad t = t_0 + \frac{V_p t_0}{V_0}\exp\left[1 - \left(\frac{R}{r}\right)^{\frac{1}{\alpha_R}}\right] \quad (5.43)$$

如果孔径(r)分布很宽,对尺寸为R的大分子,就必须考虑一条链进入所有孔径小于链尺寸(R)的小孔的概率。即将上式左边从孔径为零积分到孔径为R;右边从时间t_0积分到时间t。使用具有宽分布小孔尺寸填料的好处是,仅用一根色谱柱就可分析具有不同平均摩尔质量的大分子样品,不用更换色谱柱,缺点是分辨率较差。利用$R = k_R M^{\alpha_R}$将上式中的R换成实验上使用的大分子摩尔质量(M),可得

$$t = t_0 + \frac{V_p t_0}{V_0}\exp\left[1 - \left(\frac{k_R}{r}\right)^{\frac{1}{\alpha_R}}M\right] \quad (5.44)$$

停留时间非线性地依赖于摩尔质量。为了使用方便,通常会将右边的摩尔质量放进对数空间,写成$A + B \log M$,这里,A和B为两个需要通过实验标定的常数。标定则需要一系列(最少两个)已知摩尔质量、窄分布的标准样品。因此,上式成为

$$t = A + B \log M \quad (5.45)$$

从热力学原理出发,也可得到上式,具体如下。在色谱柱里,填料之间的空间(间隙)和填料中小孔的空间代表了两个相,称为流动相和固定相。将注入色谱柱前浓度为C_0的大分子溶液选作一个参考状态,其化学势为μ^0。注入后,一部分大分子链留在间隙,另一

部分进入小孔。在流动相和固定相的大分子浓度分别为 C_m 和 C_s。对尺寸大于孔径的大分子链,如果挤进小孔(即在固定相中),与在流动相中时相比,它们将感受到一个额外的压强差(如同渗透压差,也记为 π)。平衡时,每摩尔大分子在两相的化学势必须相等,即

$$\mu^0 + RT\ln\frac{C_m}{C_0} = \mu^0 + RT\ln\frac{C_s}{C_0} + \pi V_m \quad \rightarrow \quad -\pi V_m = RT\ln\frac{C_s}{C_m}$$

其中,利用了 $dG = Vdp$。在一个等温过程中,先将两边分别对大分子摩尔数(n_p)取偏微分,得 $d\mu = V_m dp$。如果大分子的摩尔体积(V_m)不依赖压强,将两边分别左边从 μ^0 到 μ 和右边从 p 到 $(p + \pi)$ 积分,结果为 $\Delta\mu = \mu_s - \mu_m = V_m \pi$,反映了大分子的摩尔自由能随压强差的增量变化。在前述讨论中已知,每摩尔大分子链挤进孔径为 r 小孔的自由能增量为 $(l/r - 1)RT$,以及 $l/r = (R/r)^{1/\alpha_R}$。所以

$$\psi = \frac{C_s}{C_m} = \exp\left[1 - \left(\frac{R}{r}\right)^{\frac{1}{\alpha_R}}\right]$$

再引用前述 ψ 和分配比例的关系,进行同样的推导,即可得到式(5.44),恕不赘述。物理上,其意味着停留时间依赖于大分子尺寸。而对于给定的摩尔质量,大分子的尺寸随着溶剂性质而变。因此,对一种给定的大分子和溶剂,就要有一系列(最少两个)具有不同摩尔质量的窄分布标准样品,测定它们的停留时间,标定色谱柱、得到 A 和 B。随后,方可利用该标定曲线将一个未知摩尔质量、同种的大分子样品在同样溶剂中测得的停留时间换算成一个摩尔质量。

如前所论,极稀大分子溶液的黏度和其摩尔质量之间有着标度关系。因此,在线测量溶液的黏度,可确定在每一个停留时间的摩尔质量,这导致了黏度检测器。但是,其应用也需要一条事先标定的曲线。自 20 世纪 90 年代以来,各种在线激光光散射检测器也逐渐普及。作为一个绝对方法,其不需要标定,是其一主要优势。然而,必须注意,其中的散射池是一个细长的圆柱,入射的激光光束沿着柱子的中心向前。玻璃的折射率(~1.50)与有机溶剂通常有异。因此,在每个散射角度均需对散射体积进行折射矫正。水的折射率为~1.33,是最糟糕的情形。仪器的设计者假定了经过体积修正后的小分子溶剂的散射没有角度依赖性。因此,计算机程序人为地在每个角度添加了一个正比常数(标定常数)以强迫在每个角度的散射光强相等。这样,无法知道出现的误差是由于折射矫正不妥,仪器调节有误、还是仪器本身的缺陷。在静态激光光散射一节中,已经详细地讨论过仪器调节的困难。不幸的是,在推广这类在线激光光散射检测器时,常将其描绘成一个简单的黑盒子。

在实际的操作中,一般采用恒定流速。为了保证恒定流速,并减少脉冲,一般由两个或两个以上的推进泵交替进行推动。对一个调好的尺寸排除色谱仪,当流速稳定后,测量基线应是一条水平线。此时,可注入大分子溶液。浓度不可太高,使柱子饱和,降低分辨能力。但也不可太低,使得每个流出组分的信号太弱,不易检测。需要根据体系,调节浓度。原则上,在满足检测灵敏度的前提下,溶液应尽量稀释。注意,对给定的流速,停留时间和流出(停留)体积成正比,故实验上常用更直观的、更直接的停留体积。

通过标定色谱柱,将流出(停留)体积(停留时间)转换成摩尔质量,或者利用在线激

光光散射检测器直接测量每个流出体积(每个级分)对应的摩尔质量。这只解决了摩尔质量分布中的横坐标问题。至于每个摩尔质量对应的权重(浓度或质量)则需要由另外一个可检测该组分含量的检测器,包括一台在线的紫外吸收光谱或微分折射仪。利用紫外吸收测量溶液浓度的前提是大分子具有紫外吸收,这并不普适。因此,在实际使用中,微分折射仪则更为普适。当然,如果溶剂分子和大分子的折射指数相近时,微分折射测量的误差将会较大。下图显示了几个典型的流出体积的浓度分布图。

理论上,每条曲线下的面积正比于注射前大分子溶液的浓度,用 $C(V)$ 除以面积可得归一化的分布,$C(V)/\sum_i C_i$。将停留体积或停留时间换算成摩尔质量时,其与摩尔质量的对数($\log M$)成正比,$\mathrm{d}V \propto \mathrm{d}(\log M)$。换算只是在线性空间里的平移。在作摩尔质量分布图时,横坐标必须是摩尔质量的对数($\log M$),纵坐标只需用新的面积归一即可。如果作图时,横坐标采用摩尔质量(M),由于 $\mathrm{d}V \propto \mathrm{d}M/M$,每个纵坐标就必须除以对应的摩尔质量,然后再归一。否则,摩尔质量较高的组分就会权重太大。该错误常在文献中出现。

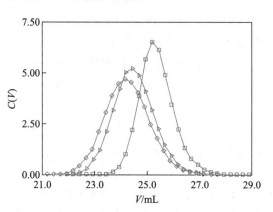

Wu C,Siddiq M,Jiang S H,et al.J.Appl.
Polym.Sci.,1995,58:1779.
(Figure 2,permission was granted by publisher)

显然,如果所测的权重是浓度(质量),直接算出的平均摩尔质量将是重均摩尔质量:$M_w = \sum_i M_i C_i / \sum_i C_i$。也可依据定义,$\sum_i C_i / \sum_i (C_i/M_i)$,间接地算出数均摩尔质量。两者之比可得摩尔质量分布的宽度,也称作分散指数。读者应该清楚地记住,源自尺寸排除色谱的重均摩尔质量更可靠,数均摩尔质量仅是一个计算结果。

将两个或多个不同的在线检测器联用就涉及连接问题。一般有两种连接方式:串联和并联。通常采用串联方式,好处是安装和操作简单。但是,不同的检测器对摩尔质量的敏感度不同,如紫外吸收和微分折射均正比于每一组分对应的质量(重量),而散射光强则正比于摩尔质量的平方。在每个组分通过不同的检测器后如何校正和对齐保留体积并不是一件容易的事。它直接关系到在真空紫外吸收或微分折射所测的浓度是否对应着其在黏度仪或激光光散射光强中所测的摩尔质量。

例如,对一个宽分布的样品,当一个具有高摩尔质量的组分呈现在激光光散射检测器的检测信号上时,紫外吸收和微分折射可能还检测不到这一组分;反之亦真。销售仪器的推销员同样不会告诉你这些麻烦和问题,而是告诉你不用担心。其实这些恰恰是一个认真的使用者需要担心的。否则,将只会得到扭曲、甚至是错误的结果。现今,仪器电子化带来的是便捷,但也导致了文献上许多似是而非和误导的结论。

具有互穿网络结构的多孔填料一般是交联聚合物。它们通常会在有机溶剂中溶胀,形成凝胶。因此,如果流动液体为有机溶剂,尺寸排除色谱也常被称为凝胶渗透色谱(GPC)。当溶剂为水时,填料虽然没有溶胀,但仍然沿用"凝胶渗透",称作水相 GPC。在上述讨论中,假定了大分子与填料和小孔的表面没有相互作用,即 $\Delta H = 0$。如果 $\Delta H > 0$,大分子与填料和小孔的表面互相排斥,减少了可及的体积,一般不会制备这类色谱柱。

反之,如果通过化学或物理方法修饰柱子和小孔表面使其可强烈地吸附大分子,$\Delta H < 0$ 和 $|\Delta H| > |T\Delta S|$,就可制得另一类型的分离色谱柱。例如,在蛋白质工程中,通常会在合成蛋白质链的一端加上一小段具有特定氨基酸序列多肽标记,使其可以和分离色谱柱里对应的化学基团产生特异的作用。这样,先将所需蛋白质吸附在色谱柱上,再通过淋洗除去其他蛋白质,最后再用可以破坏吸附的淋洗溶剂将所需蛋白质从色谱柱上剥离下来。

小　结

基本原理:在一个空心的不锈钢色谱柱子里填满多孔填料。每个填料颗粒里含有互相贯穿的柱状小孔。当溶剂稳速地流过色谱柱时,将一定体积的大分子溶液注入色谱柱。大分子链可以一定的概率进入小于其尺寸的小孔。而对给定尺寸的小孔,该概率随着链尺寸增大而迅速地降低。显然,随着溶剂的稳速流动,较易进入小孔的那些链将不易被溶剂冲出色谱柱,即较长的停留时间,或曰较大的停留体积。利用这一物理现象,可将尺寸不同的大分子链分开。利用一系列已知摩尔质量的窄分布标准样品可标定一根色谱柱,得到停留时间或停留体积与摩尔质量之间的关系。也可利用在线的激光光散射检测器直接测量每个停留体积对应的散射光强。利用另一个在线浓度检测器(紫外吸收或微分折射仪)测得对应的浓度,即可得到大分子的摩尔质量分布和重均摩尔质量,求出数均摩尔质量。

可测物理量:对一个多分散样品,可用一个在线的黏度仪或激光光散射仪测得与每个组分在色谱柱中的停留时间(停留体积)对应的黏度或散射光强。再用紫外吸收或微分折射仪测出每个停留时间或流出体积对应的每个组分的浓度(质量)。

数据处理:利用事先标定好的标准曲线,将每一停留时间或流出体积换算成摩尔质量;或利用测得的每一个黏度或散射光强直接算出摩尔质量。同时,还要将紫外吸收或微分折射仪测得的转化成浓度(质量)。目前的仪器已可毫不费力地进行这些运算,故数据处理十分简易,但并非无忧。理解其背后的原理和计算仍然十分重要。

将所得摩尔质量分布作图时,摩尔质量坐标应该是其对数($\log M$),而不可画作线性坐标(M)。如果要线性作图,那么每一个得到的浓度(纵坐标)均需除以其对应的摩尔质量,即 $C(V_i)/M_i$。否则,依据该分布就会算出错误的重均摩尔质量。这是文献中常犯的一个错误,归于使用者没有全面理解基本原理。

Chapter 5. Common Experimental Methods for Study of Macromolecular Solutions

In the previous chapter, the relationships between the measurable square radius of gyration and the theoretical mean square end-to-end distances of linear chains or to the sum of the mean square distances of chains with other configurations were discussed. The scaling relationships between the measurable average radius of gyration of branched chains and the number of segments of the parent chain and subchain were also discussed. In practice, the number of segments is replaced by the measurable average degree of polymerization or the measurable average molar mass. An important content in studying macromolecular solutions is to find the chain length (molar mass) dependence of various physical properties of macromolecular chains in solutions, which forms the core of macromolecular physics.

According to whether they need to rely on the calibration curve obtained in advance, common experimental methods for studying macromolecular solutions can be divided into two groups: the "absolute" and "relative" methods. Each calibration curve requests a set of narrowly distributed standard samples with different molar masses. The end-group analysis, solution osmotic pressure, laser light scattering, differential refractometer and ultracentrifugation are the absolute methods; while the methods of the size exclusion (gel permeation) chromatography, intrinsic viscosity, etc., depend on the calibration curve of each instrument.

With rapid developments of various fast and sensitive photodetectors as well as the computer storage and calculation speed, most of the instruments have become electronic, automated, and foolish operative. Companies of producing various instruments and sales staffs try to tell users that there is no need to worry about many possible problems in actual uses. Many users also treat various instruments as a black box, putting in samples, pressing buttons, obtaining results, and writing reports, without knowing the basic principles, precautions, potential problems, precision and accuracy of an instrument, etc. Therefore, this chapter is very important.

For any experimental method, we always need to master the following three main points: its basic chemical or physical principles; the physical quantity directly measured and the precision and accuracy of the measurement; and how to obtain the finally required physical parameters by properly processing the measured data. Therefore, we will introduce the above-mentioned common experimental methods one by one according to these three key points.

Analysis of Chain Ends

The two end groups (referred to as ends) of each macromolecular chain are normally different in chemical structure from those polymerized monomers on the chain. For example, each protein chain or polyurethane chain has an acidic carboxyl and a basic amino group, respectively, at its ends. One end corresponds to a macromolecular chain. Using chemical or physical differences between the chain ends and the polymerized monomers on the chain, the molar concentration (n) of any one of the two ends, i.e., the molar concentration of the chains is measurable. From the mass per unit volume (C, the mass concentration) known in the solution preparation, the average molar mass (M) of macromolecular chain can be obtained by using $n = C/M$. In macromolecule experiments, C usually has a unit of " g/mL ". Note that macromolecular textbooks often follow the habits of small molecules, calling " molar mass " as " molecular weight ". Considering the subsequent discussion about the mesoglobular phase and colloids of macromolecules, this book adopts the physically accurate " molar mass " instead of " molecular weight ". Although both of them have the same value, their connotations are different.

As mentioned above, in a polymerization, the resultant linear chains have different lengths, and branched chains have non-uniform internal structures and different sizes, so that they only have an average molar mass. Obviously, the end group analysis can only be used to measure the molar mass of a linear chain with one or two end groups, and it is not applicable to macromolecular chains with an uncertain number of end groups. In addition, if the degree of polymerization is greater than 100, the end group concentration will be very low, and the accuracy and accuracy of measurement will not meet the requirements. Therefore, the end-group method is generally suitable for measuring the average molar mass of oligomers and low polymers. The common method for measuring the number of end groups is classical chemical titration, whose precision and accuracy can reach one thousandth; nuclear magnetic resonance can also be used, whose precision and accuracy is only one-hundredth or worse.

For a polydisperse macromolecular sample, linear chains with different lengths all contribute to the measured number and mass concentrations of the end groups. Therefore, it is necessary to sum the contributions of all the chains with different molar masses, i.e., $n = (\sum_{i=1}^{N} n_i)/V$ and $C = (\sum_{i=1}^{N} W_i)/V$, where n_i and W_i are the number and mass of the chains with a molar mass of M_i, respectively. Obviously, $W_i = M_i n_i$, so that

$$M_n = \frac{C}{n} = \frac{\sum_{i=1}^{N} M_i n_i}{\sum_{i=1}^{N} n_i} \qquad (5.1)$$

The end analysis leads to a number weighted number-average molar mass, indicated by the subscript " n ". By analogy, there are other weighted average molar masses, such as the weight average molar mass (M_w) and the z-average molar mass (M_z), respectively defined as follows,

$$M_w = \frac{\sum_{i=1}^{N} M_i W_i}{\sum_{i=1}^{N} W_i} = \frac{\sum_{i=1}^{N} M_i^2 n_i}{\sum_{i=1}^{N} M_i n_i} \quad \text{and} \quad M_z = \frac{\sum_{i=1}^{N} M_i^3 n_i}{\sum_{i=1}^{N} M_i^2 n_i} \tag{5.2}$$

From a mathematical point of view, various average values (M_n, M_w and M_z) calculated from a given molar mass distribution are only the ratios of the first to the zeroth moments, the second to the first moments and the third to the second moments of the molar mass, respectively. It is necessary to emphasize that this is not an artificial mathematical game or calculation, but derived from experiments. We will gradually understand that different experimental methods can only measure different average molar masses of macromolecular chains without a choice.

Summary

Basic Principle: Linear chains with two same or different end groups, so that the number of macromolecular chains can be obtained from the number of end groups.

Measurable Physical Quantity: Using a chemical or physical method, such as chemical titration or nuclear magnetic resonance, one can measure the molar number of macromolecular chain ends in solution (n).

Data Processing: The number average molar mass (M_n) of macromolecular chains can be obtained from the ratio of the mass concentration (C, g/mL) known in the solution preparation to the measured molar number (n).

Osmotic Pressure Difference (π)

The "osmotic pressure" in traditional textbooks is actually a pressure difference between a solution and a solvent. It is a pressure change from solvent to solution during the dissolution process, not a pressure. Strictly speaking, it has nothing to do with penetration. The "penetration" is only a method of measuring such a pressure difference. Due to historical reasons, it has been written as "osmotic pressure", easily leading to misunderstanding and not fitting the original connotation. It confuses the essential pressure difference with a method of measuring the pressure difference. This book continues to use the accustomed word "osmotic pressure", but add a word "difference", i.e., "osmotic pressure difference" to correct its name. There are different ways to discuss the osmotic pressure difference. It can start from the pressures of solvent and solution to discuss their difference. It can also be directly derived from the changes of free energy and chemical potential in thermodynamics.

Speaking of pressure (p), we can review its physical nature first. In physical chemistry or general physics textbooks, pressure is often described as the average force exerted on a unit surface area due to the impact of gas molecules on the container wall. This is another example of conflating physical properties with their measurement methods. There also exists a pressure in the middle of a bottle of gas, which has nothing to do with the wall, but a surface is needed to measure a pressure. It is just as if each of us has a certain mass (weight), which can be weighed with a scale or other tools, but this mass has nothing to do with the scale and whether it is weighed.

In order to better discuss the nature of pressure, let us first examine its dimension: p is the force per unit area and its commonly used unit in physics is Pascal (Pa, Newton/square meter, $N \cdot m^{-2}$). Pressure has other units, such as the standard atmospheric pressure (atm) and bar (bar). The relationship between them and Pascal is 1.00 bar $= 1.00 \times 10^5$ Pa $= 0.990$ atm. Multiplying the force and area by a length leads to an energy (E) and a volume (V), respectively. Therefore, a pressure is also energy per unit volume, $p = E/V$, i. e., the energy density.

If there is no external energy from the environment, each molecule in the system only has thermal energy ($k_B T$), where k_B and T are the Boltzmann constant and the absolute temperature of the system, respectively. Therefore, the energy density (generated pressure) contributed by each gas molecule in the system is $k_B T/V$. For N molecules, $p = N k_B T/V = nRT/V$, where $R = k_B N_{AV}$ (gas constant) and $n = N/N_{AV}$. This is the equation of state for an ideal gas, the simplest equation of state.

Microscopically, we can use a small volume (v) as a unit to divide the volume (V) of a bottle of gas into n_{box} small boxes, i.e., $n_{box} = V/v$. Assuming that the gas molecules have neither size nor interaction, under the agitation of thermal energy, each gas molecule moves randomly, and has an equal probability (ψ) of reaching each small box. Obviously, $\psi = 1/n_{box} = v/V$, inversely proportional to V. One can imaginably place a tiny pressure detector at any small box. The pressure measured would be proportional to the probability of reaching the small box for one gas molecule, i.e., $p = p_v \psi$. The physical meaning of the proportional constant p_v is the pressure generated by one gas molecule when $\psi = 100\%$. When V is as small as v, i.e., only one small box in the system, $\psi = 100\%$. Note that p_v is the energy density ($k_B T/v$) when the volume is as small as v. Therefore, $p = k_B T/V$, where v in p_v and v in ψ are cancelled, so that how to choose the size of v is not important. For N gas molecules, the same ideal gas state equation is obtained.

Historically, Robert Boyle published a paper in 1662, describing the product of pressure and volume of a tank of closed air close to a constant. More than a hundred years later, in 1780, Jacques Charles discovered that both volume and pressure are proportional to the absolute temperature T. There is now a microscopic explanation. They are related to the probability of a particle reaching any point in the system. The probability is inversely proportional to volume, but proportional to pressure. On the other hand, for a given volume, as the temperature increases, the particles gain more thermal energy, so that the movement speeds up, increasing the probability for a given time, too. Therefore, for a given volume, the pressure must linearly increase with the temperature. Otherwise, the volume will change. More importantly, the probability increases linearly with the number of gas molecules. Therefore, the pressure has the property of relying on the number of molecules, referred to as "**colligativity**" for short.

In real gas, each molecule has a hardcore volume (v_0), and there is the intermolecular attraction related to the dispersion energy (ε). A combination of these two produces an excluded volume effect. The equation of state of real gas becomes complicated. In a bottle of real gas, the total excluded volume occupied by N gas molecules is $v_e N$, and the actual reachable volume of each gas molecule is reduced from V to $V - v_e N$, so the probability of each gas molecule reaching any small box increases to

$$\psi = 1 \Big/ \frac{V - v_e N}{v_e} \cong \frac{v_e}{V} + \frac{v_e^2 N}{V^2} \tag{5.3}$$

where $v_e N \ll V$ and $1/(1 - v_e N/V) \cong 1 + v_e N/V$ are used. The first term in the above equation is the probability of a randomly moving particle reaching any point in the bottle under the agitation of thermal energy. The second term is the additional probability originated from the excluded volume effect, resulting in an additional pressure. The proportional constant between them is the dispersion energy density ($\varepsilon/2v_e$) when the volume is as small as $2v_e$ and the probability is 100%. Therefore, the pressure generated by N real gas molecules is

$$p = \frac{nRT}{V} + \frac{\varepsilon v_e N^2}{2V^2} = \frac{nRT}{V} + \frac{\varepsilon v_e N_{AV}^2}{2} \left(\frac{n}{V}\right)^2 \tag{5.4}$$

Using $V - v_e N$ to replace V in the ideal state, the above equation can be rewritten as

$$\left[p - a \left(\frac{n}{V}\right)^2 \right] (V - bn) = nRT \tag{5.5}$$

This is the famous Van der Waals equation of state for real gases, where $a = (\varepsilon N_{AV} v_e N_{AV})/2 = (\varepsilon_m v_{e,m})/2$ and $b = v_e N_{AV} = v_{e,m}$. "ε_m" and "$v_{e,m}$" together represent the molar excluded volume effect. The dispersion attraction between molecules produces an **effect** of reducing the excluded volume, it increases the accessible volume of molecules and reduces the probability of a gas molecule reaching any point in the system. The hardcore volume of the molecule itself has the opposite effect. It is necessary to note that "a" from the mutual attraction is negative. However, in physical chemistry textbooks, it is usually listed as a positive number. This is why compared with the above equation, there is a "+" in front of "a" in physical chemistry textbooks.

The above equation can also be directly derived from the probability of the interaction generated between two gas molecules. Following the above discussion, for a system with N molecules, the probability of the first molecule appearing on any small box in the total V/v_e small boxes is $N/(V/v_e)$; and the probability of the second molecule appearing on the second small box is $(N - 1)/(V/v_e - 1)$. Therefore, the probability of generating the two-body excluded volume effect is $[N/(V/v_e)][(N - 1)/(V/v_e - 1)] = v_e^2 N^2/V^2$, where $N \gg 1$ and $V/v_e \gg 1$ are used. In addition to the pressure (nRT/V) caused by thermal energy, there is an additional negative pressure effect derived from the dispersion energy (ε). Since the volume occupied by each pair of gas molecules is $2v_e$, the energy density is $\varepsilon/2v_e$. Therefore, The total additional negative pressure effect caused by the excluded volume effect is the product of the energy density and the probability, i.e., $[\varepsilon/(2v_e)][N/(V/v_e)^2] = \varepsilon v_e N^2/(2V^2)$, identical to the second term in eq. (5.4), by different paths.

The above discussion about gas pressure can be directly applied to liquids. Gas pressure and solution pressure have exactly the same physical nature. Both are directly proportional to the probability of molecules reaching any point in the system through random translational diffusion driven by thermal energy, or more deeply, it is related to the entropy of a system, inversely proportional to the volume but proportional to the number of molecules. Use the terminology used in the previous discussion of the dissolution of macromolecules, and assume that there is no interaction among molecules in the solution, and the excluded volume ($v_{e,p}$) of a macromolecular chain is $v_{e,p}$ times of the excluded volume of a solvent molecule ($v_{e,p}$). Therefore, a solution containing n_p moles of macromolecular chains can be regarded as those macromolecular chains have replaced $n_v n_p$ moles of solvent molecules, not necessary to consider the rest of solvent molecules. Note: for polymers, $n_v \gg 1$.

Since π is the difference between the solution pressure and the solvent pressure, and the pressure is proportional to the number of molecules, the osmotic pressure difference is proportional to the difference between the number of molecules in the solution and in the solvent, i.e.,

$$\pi = p_{\text{solution}} - p_{\text{solvent}} = \frac{(n_p - n_v n_p)k_B T}{V} = \frac{n_p k_B T}{V} - \frac{n_v n_p k_B T}{V}$$

Since it is the difference in pressure, the pure solvent can be used as a reference state, defining its pressure $n_v n_p k_B T/V = 0$, which is physically equivalent to treat the solvent in the solution as a vacuum. Therefore, the pressure generated by n_p moles of macromolecular chains in the volume V is equal to the osmotic pressure difference in value,

$$\pi = \frac{n_p RT}{V} - 0 = \frac{CRT}{M}$$

where M is the molar mass of the macromolecule, and $n_p = W/M$ and $C = W/V$ are used. For polydisperse systems,

$$\pi = \sum_i \pi_i = \sum_i \frac{n_i RT}{V} = \frac{CRT}{W/\sum_i n_i} = \frac{CRT}{M_n} \tag{5.6}$$

where $W = \sum_i M_i n_i$ and $M_n = \sum_i M_i n_i/\sum_i n_i$ are used. It can be seen that the osmotic pressure difference also measures **the number average molar mass**.

In a real solution, the excluded volume effect has to be considered. By following the discussion about real gases, one only need to note that in a given binary liquid mixture, there are three pairs of intermolecular excluded volume effects: solvent – solvent (ε_{AA}), solute – solute (ε_{BB}) and solute – solvent (ε_{AB}). The discussion in Chapter 3 is repeated as follows, ε_{AB}

represents the effect of one chain **segment** and one solvent molecule, i. e., one chain simultaneously produces the excluded volume effect with n_v solvent molecules.

The dissolution of one pair of chains simultaneously breaks n_v pairs of the "solvent – solvent" interaction and one pair of the "chain – chain" interaction to form $2n_v$ pairs of the "segment–solvent molecule" excluded volume effect. Therefore, the average energy change of forming each pair of "segments–solvent molecule" excluded volume effect is $\varepsilon_{AB} - (\varepsilon_{AA} + \varepsilon_{BB}/n_v)/2$, which is denoted as χ_{sp} as before after normalized by thermal energy $k_B T$ and multiplied by the number of action points Z.

After introducing a weak two-body exclusion volume effect (ε_{AA}) in the ideal solvent, and borrowing the results of the real gas discussed above, it can be seen that the pressure generated by the $n_v n_p$ moles of solvent molecules that will be substituted in the dissolution is

$$P_{\text{solvent}} = \frac{n_v n_p RT}{V} + \frac{\varepsilon_{AA} v_{e,s} N_{AV}^2}{2}\left(\frac{n_v n_p}{V}\right)^2$$

Before dissolution, the chains in a solid macromolecular sample cannot undergo the translational diffusion under the agitation of thermal energy at room temperature. The chain segments can only perturb around the center of mass of the chain (density fluctuation), and the probability of reaching any point in the system is zero, so that there is no pressure caused by the translational diffusion. But there is still the two-body excluded volume effect, its corresponding pressure is

$$P_{\text{polymer}} = \frac{\varepsilon_{BB} v_{e,p} N_{AV}^2}{2}\left(\frac{n_p}{V}\right)^2$$

where $v_{e,p} = v_{e,s} n_v$. After the dissolution, n_p moles of the chains replace $n_v n_p$ moles of solvent molecules. Each chain generates n_v pairs of "segment–solvent molecule" excluded volume effect (ε_{AB}), so that the solution pressure is

$$P_{\text{solution}} = \frac{n_p RT}{V} + n_v \varepsilon_{AB} v_{e,p} N_{AV}^2 \left(\frac{n_p}{V}\right)^2$$

Therefore, the pressure (osmotic pressure) difference after and before the dissolution is

$$\pi = P_{\text{solution}} - (P_{\text{solvent}} + P_{\text{polymer}})$$

$$= RT(1 - n_v)\frac{n_p}{V} + n_v \varepsilon_{AB} v_{e,p} N_{AV}^2\left(\frac{n_p}{V}\right)^2 - \left[\frac{\varepsilon_{AA} v_{e,s} n_v^2}{2} + \frac{\varepsilon_{BB} v_{e,p}}{2}\right] N_{AV}^2\left(\frac{n_p}{V}\right)^2$$

$$= RT(1 - n_v)\frac{n_p}{V} + \left[\varepsilon_{AB} - \left(\frac{\varepsilon_{AA} + \dfrac{\varepsilon_{BB}}{n_v}}{2}\right)\right] n_v v_{e,p} N_{AV}^2\left(\frac{n_p}{V}\right)^2$$

$$= RT(1 - n_v)\frac{n_p}{V} + RT\chi_{sp} n_v v_{e,p} N_{AV}\left(\frac{n_p}{V}\right)^2 \tag{5.7}$$

where the coefficient of the second term is normalized the thermal energy ($k_B T$), i. e., multiplying and dividing by thermal energy simultaneously. Continuously defining the pressure

$(n_v n_p RT/V)$ generated by $n_v n_p$ solvent molecules as zero, the above equation becomes

$$\pi = RT\frac{n_p}{V} + RTX_{sp}n_v v_{e,p}N_{AV}\left(\frac{n_p}{V}\right)^2$$

For the convenience of discussion, the concept of **partial volume** (v) is introduced as follows. v is defined as $(\partial V/\partial m)_{T,p}$, and its physical meaning is the volume change of a system as the mass of one of the components changes, which is **the rate of volume change**, not volume, also not **the volume per unit mass (specific volume)**, never be confused! For pure substances, "the partial volume" and "specific volume" are the same.

For a two-component macromolecular solution, the macromolecule and the solvent have their respective specific volumes v_p and v_s. The volume fraction, weight concentration and specific volume depend on each other. From their definitions, it is not difficult to deduce $X_s = C_s v_s$ and $X_p = Cv_p$. It is extremely important to understand and remember their definitions and mutual dependencies. They are often used when converting a volume fraction into a mass (weight) concentration commonly used in experiments. The above equation is further written as follows

$$\pi = RT\frac{C}{M} + RTX_{sp}n_v v_p\left(\frac{C}{M}\right)^2 \quad \text{or} \quad \frac{\pi}{RTC} = \frac{1}{M}\left(1 + \frac{X_{sp}n_v v_p}{M^2}C\right) = \frac{1}{M}(1 + A_2C) \quad (5.8)$$

where $A_2 = X_{sp}n_v v_p/M$, the second virial coefficient, which can be experimentally determined. At high concentrations, the three-body excluded volume effect may have to be added, i.e., an additional term A_3C^2, where A_3 is the third virial coefficient. For a polydisperse macromolecule sample, M_n should replace M in the above equation.

The above equation can be directly derived from the principle of thermodynamics, too. The rate of change of free energy with the change of content of solvent and macromolecule (partial molar free energy, $[\partial G/\partial n_s]_{T,p}$ and $[\partial G/\partial n_p]_{T,p}$) are respectively called the **chemical potentials of solvent and macromolecule**. According to the definition, under pressure ($p + \pi$), free energy of the system before mixing is the free energy $G^*(p + \pi)$ of pure solvent; after mixing, there are additional macromolecular chains in solvent, and the free energy is reduced to $G(p + \pi, x_s)$, the change of free energy is,

$$G(p + \pi, x_s) - G^*(p + \pi) = \Delta G_{mix}(p + \pi)$$

$$\left[\frac{\partial G(p + \pi, x_s)}{\partial n_s}\right]_{T,p,n_p} - \left[\frac{\partial G^*(p + \pi)}{\partial n_s}\right]_{T,p,n_p} = \left[\frac{\partial \Delta G_{mix}(p + \pi)}{\partial n_s}\right]_{T,p,n_p}$$

$$\mu_s(p + \pi, x_s) - \mu_s^*(p + \pi) = \left[\frac{\partial \Delta G_{mix}(p + \pi)}{\partial n_s}\right]_{T,p,n_p}$$

When the solution pressure rises from p to $p + \pi$ to reach an equilibrium, the chemical potential of solvent in the solution is equal to the chemical potential of pure solvent, i.e., $\mu_s(p + \pi, x_s) = \mu_s^*(p)$. Its physical meaning is that when the solvent and the solution exchange one or one mole of solvent molecules, there will be no change of free energy in both the solvent and the solution. Substituting $\mu_s(p + \pi, x_s) = \mu_s^*(p)$ into the above equation,

$$\mu_s^*(p) - \mu_s^*(p + \pi) = \left[\frac{\partial \Delta G_{\text{mix}}(p + \pi)}{\partial n_s}\right]_{T,p,n_p} \tag{5.9}$$

As long as the difference between the chemical potential of the pure solvent at two different pressures on the left side and the partial differential of the mixing free energy with respect to the molar number of solvent on the right side are calculated, the relationship between π and the concentration is obtainable. First, let us calculate the partial differential on the right. In Chapter 3, the dependence of the mixing free energy on the volume fraction of macromolecules X_p has been discussed in detail. For the convenience of discussion, eq. (3.15) is rewritten as follows,

$$\Delta G_{\text{mix}} = RT[n_s \ln(1 - X_p) + n_p \ln X_p + n_s X_p X_{sp}]$$

In order to find the partial differential of ΔG_{mix} with respect to n_s, the partial differential of ΔG_{mix} with respect to X_p can be obtained first, and then it is multiplied by the partial differential of X_p with respect to n_s. Because the temperature and pressure remain, in order to write concisely, all the subscripts in partial differentials are removed. When seeking the partial differential with respect to n_s, just remember that n_p is a constant, i.e.,

$$\frac{\partial \Delta G_{\text{mix}}(p + \pi)}{\partial n_s} = RT \frac{\partial[n_s \ln(1 - X_p) + n_p \ln X_p + n_s X_p X_{sp}]}{\partial X_p} \frac{\partial X_p}{\partial n_s}$$

$$= RT\left[\frac{\partial n_s}{\partial X_p}\ln(1 - X_p) - \frac{n_s}{1 - X_p} + \frac{n_p}{X_p} + \frac{\partial n_s}{\partial X_p}X_p X_{sp} + n_s X_{sp}\right]\frac{\partial X_p}{\partial n_s}$$

$$= RT\left[\ln(1 - X_p) - \frac{n_s}{1 - X_p}\frac{\partial X_p}{\partial n_s} + \frac{n_p}{X_p}\frac{\partial X_p}{\partial n_s} + X_p X_{sp} + n_s X_{sp}\frac{\partial X_p}{\partial n_s}\right]$$

$$\left(\frac{\partial \Delta G_{\text{mix}}(p + \pi)}{\partial n_s}\right)_{T,p,n_p} = RT\left[\ln(1 - X_p) + \left(1 - \frac{1}{n_v}\right)X_p + X_p^2 X_{sp}\right] \tag{5.10}$$

In the last step of derivation, the following very useful formula was used

$$\left(\frac{\partial X_p}{n_s}\right)_{T,p,n_p} = \frac{\partial\left(\frac{n_v n_p}{n_s + n_v n_p}\right)}{\partial n_s} = -\frac{n_v n_p}{(n_s + n_v n_p)^2} = -\frac{X_p(1 - X_p)}{n_s} = -\frac{X_p^2}{n_v n_p}$$

Next, let us find the dependence of the free energy of pure solvent. According to the first law of thermodynamics (energy conservation), the change of internal energy (dU) is equal to the sum of work (w) and heat (q), and the change of internal energy is independent of the specific path. Therefore, it can be obtained through a hypothetical reversible path ($w_{\text{rev}} = -pdV$ and $q_{\text{rev}} = TdS$), i.e., $dU = TdS - pdV$. The internal energy and enthalpy energy ($H = U + pV$) include the energy (free energy, A or G) that can freely do work by the directional movement of molecules and the energy (TS) that generates heat, related to the random movement of molecules. Namely, $U = A + TS$, preferred by theoreticians because V remains a constant; and $H = G + TS$, favored by experimentalists because p is a constant.

A combination of the above relations leads to $dG = Vdp - SdT$. Under the isothermal condition, $dG = Vdp$ or $(\partial G/\partial p)_{T,n} = V$, where the volume is always positive, so that free energy increases with the pressure. The differentiation of the two sides with respect to n under the isothermal and isobaric conditions equal pressure results in

$$\left[\frac{\partial\left(\frac{\partial G}{\partial n}\right)_{T,p}}{\partial p}\right]_{T,n} = \left(\frac{\partial V}{\partial n}\right)_{T,p} \tag{5.11}$$

According to the definition, $(\partial G/\partial n)_{T,p}$ is the chemical potential (μ). For pure substances, it is the molar free energy. $(\partial V/\partial n)_{T,p}$ is the molar volume of pure substance $(V_{s,m})$. Therefore, $(\partial\mu/\partial p)_{T,n} = V_{s,m}$. Assuming that the molar volume of solvent is independent of the pressure, when pressure rises from p to $p + \pi$, the change in chemical potential is,

$$\mu_s^*(p + \pi) - \mu_s^*(p) = \int_p^{p+\pi} V_m dp = V_{s,m}\pi \tag{5.12}$$

The substitutions of eqs. (5.10) and (5.12) into eq. (5.9) result in

$$-\pi V_{s,m} = RT\left[\ln(1 - X_p) + \left(1 - \frac{1}{n_v}\right)X_p + \chi_{sp}X_p^2\right]$$

When $X_p \ll 1$, $\ln(1 - X_p) \approx -X_p - X_p^2/2$. Using the relationship between X_p and the specific volume of macromolecules (v_p) and mass concentration (C), $X_p = Cv_p$, and the volume ratio of a macromolecular chain to a solvent molecule is n_v, $v_p = n_v V_{s,m}/M = V_{p,m}/M$. Therefore, the above equation can be rewritten as

$$\pi \cong \frac{RT}{V_{s,m}}\left[\frac{X_p}{n_v} + \left(\frac{1}{2} - \chi_{sp}\right)X_p^2\right] = \frac{RTC}{M}\left[1 + \left(\frac{1}{2} - \chi_{sp}\right)\frac{v_p n_v}{M}C\right]$$

$$\frac{\pi}{RTC} = \frac{1}{M}(1 + A_2 C) \tag{5.13}$$

where, $A_2 = (1/2 - \chi_{sp})v_p n_v/M$. Note: $1/2$ is from the expansion of $\ln(1 - X_p)$, originating from the entropy change in the mixing free energy. The existence of the hardcore volume reduces the volume available for the molecules, increasing the probability of molecules reaching any point in the system. On the other hand, χ_{sp} reflects the two-body dispersion attraction. This attraction reduces the probability of molecules reaching any point in the system for a given time, creating an effect of decreasing the excluded volume, an effect of negative excluded volume, increasing the volume available to the molecules. χ_{sp} becomes smaller as the temperature rises. Therefore, A_2 can be positive, negative or zero, depending on the relative magnitude of $1/2$ and χ_{sp}. Generally, as the temperature drops, A_2 gradually changes from positive to negative, and the solvent changes from a good solvent to a poor solvent by passing the Θ state.

In comparison with eq. (5.8), except that the entropy change is not considered in the discussion, and the expression of the second-order virial coefficient is slightly different, the dependence of the osmotic pressure difference on the molar mass and concentration of macromolecules is identical. Experimentally, there is no difference between these two, and A_2 is just a physical quantity to be determined. Careful readers will find that the dependence of the solution pressure of macro- and small molecules on the molecular number is also identical. As mentioned earlier, for a polydisperse macromolecular sample, M in the above equation is replaced by M_n.

The pressure difference between the solution and the solvent can be measured with a semi-permeable membrane that only allows solvent molecules to penetrate through, as shown on the right. The pore size in the semi-permeable membrane is smaller than the size of individual macromolecules. In the experiment, the solution and the solvent were respectively injected into the two containers on the

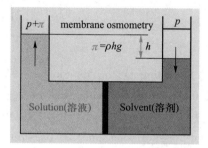

left and right. As mentioned earlier, the solvent has a pressure higher than the solution. This pressure difference creates a net flow of the solvent molecules from right to left. The permeable membrane keeps the right side always in the state of a pure solvent. Due to the net flow of the solvent molecules from right to left, the liquid level on the left gradually rises as the liquid level on the right decreases. When a certain height is reached, the solution pressure on the left is equal to the pure solvent pressure on the right. The solvent molecules move to both the sides at the same migration rate, achieving a dynamic equilibrium.

When designing a traditional semi-permeable membrane instrument to measure the osmotic pressure difference, one has to consider the diffusion time of solvent molecules. The membrane with the largest possible area, the thinnest cavity, and the smallest diameter of the glass tube of measuring the liquid height are usually used to shorten the time for the solution and pure solvent to reach equilibrium (usually 1-2 days). For a polydisperse macromolecular sample, the choice of membrane is also very important. One is to have a uniform pore size; and another is to have the pore size as large as possible to shorten the equilibrium time, but the pore size is also smaller than the size of the shortest chain, so as not to miss the contribution of the short chains in the measurement, resulting in a larger average molar mass.

When measuring the osmotic pressure difference, one can first prepare a higher concentration in the dilute solution region, and measure the height difference h_0 between the liquid surfaces of solution and solvent. Then the solution is quantitatively diluted, and the height difference h_1 is measured again. The process is repeated until it is not able to measure h accurately. A set of the concentration dependent height differences are then converted to the concentration dependent π. The plot of $\pi/(CRT)$ vs C leads to a straight line, whose reciprocal of the intercept and the slope correspond to the number average molar mass and the second virial coefficient, respectively.

Considering the actual measurement errors in the temperature (+1%), concentration (+1%) and height difference (+1%) and other experimental quantities, the error of the resultant number average molar mass and the second virial coefficient is usually larger than +3%. In order to shorten the time of reaching the equilibrium (measurement), people have designed different instruments. For example, one can measure the rising rate of the solution liquid surface first, and then convert it into the osmotic pressure difference. With the advent of sensitive pressure sensors, it is now able to measure the pressure change of the solution caused by the penetration of solvent molecules in a sealed, non-deformable container directly. The measurement accuracy depends on the absolute calibration and sensitivity of the pressure sensor.

Summary

Osmotic pressure is the pressure difference between a solution and its solvent. Such a pressure difference always exists objectively, and has nothing to do with whether it is measured or not and with the semi-permeable membrane or its measurement instruments. The semi-permeable membrane and its related instruments are only a method to measure the pressure difference between a solution and its solvent. I hope readers will never confuse this pressure difference with its measurement method. Just as a gas pressure has nothing to do with a surface, but when measuring a gas pressure, a surface is generally required; also as everyone has a mass (weight), which has nothing to do with whether it is weighed or not and with any weighing device.

Measuring the number average molar mass of macromolecules is only one application of osmotic pressure difference. In the laboratory, dialysis also uses this pressure difference; in our body, because the blood contains macromolecular proteins, the pressure is low, so excess water in the tissue can penetrate into the blood vessels and be eliminated from the body through the kidneys. This is why doctors need to supplement the protein content in the blood of patients who cannot take protein nutrition to reduce the swelling of body. In the kitchen, adding salt can reduce the pressure outside the cells of vegetables or meats to achieve dehydration; etc., countless.

Basic Principles: Like gas pressures, the liquid pressure is also proportional to the probability of molecules diffusing to reach any point in the system under the agitation of thermal energy. The proportional constant is the energy density when the probability reaches 100%. The probability is also proportional to the number of molecules in the liquid. Readers should realize **the colligative nature of pressure**; that is, for a given volume, a liquid with more particles generally has a higher pressure. Compared with one small solvent molecule, one macromolecular chain occupies a larger space. Therefore, for a given volume, the number of molecules in a pure solvent is greater than the sum of the numbers of the chain and solvent molecules in the macromolecular solution, so that pure solvent has a higher pressure. For historical reasons, the pressure difference between a solution and its solvent is defined as osmotic pressure, but it has nothing to do with penetration. This book follows the customary definition, but adds a word, namely the "osmotic pressure difference" to point out its essence. The reader should also note that the pressure difference has nothing to do with the molar mass of solute, but only depends on the molecular number of solute. For a given concentration, the molecular number is inversely proportional to the molar mass. This is why osmotic pressure difference can be used to determine the molar mass of a solute. Readers should never think that osmotic pressure difference is directly related to the molar mass!

<u>Measurable Physical Quantity</u>: Using a semi-permeable membrane of only allowing the solvent molecules to penetrate through or other methods, one measures the pressure differences between a series of (at least five) macromolecular dilute solutions with different concentrations and pure solvent are measured. Traditionally, the pressure difference was measured by the height difference between a solution and its solvent, separated by a semi-permeable membrane. Modern membrane osmotic pressure instruments usually use sensitive and accurate pressure sensors to measure the pressure difference between the solvent and the solution.

<u>Data Processing</u>: If the direct measured physical quantity is not the pressure difference, the measured physical quantity (such as the height difference) is first converted into the pressure difference. The pressure difference is divided by its corresponding macromolecular concentration, and then plot against the concentration. When the macromolecular solution is sufficiently dilute, such a plot should be a straight line, and its intercept and slope correspond to **the number average molar mass** of the macromolecules and the second virial coefficient, respectively. The experimental error is usually larger than ±3%.

Ultracentrifugation

As shown on the right, when an object with a mass of m surrounds a fixed point and rotates at an angular velocity of $\vec{\omega}$ ($= \mathrm{d}\vec{\theta}/\mathrm{d}t$, perpendicular to the paper surface, downward) on a circle with a radius of r. The linear velocity in the tangential direction of the circumference is $\vec{\omega}\times\vec{r}$, its value of ωr remains a constant, but its direction ever changes along the tangent direction. The centripetal acceleration caused by such a direction

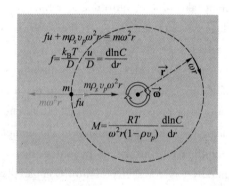

change is ($\vec{\omega}$) $\times\vec{\omega}\times\vec{r} = \omega^2 r$, pointing to the direction of the center of rotation, and the centrifugal acceleration with the same magnitude points to the opposite direction. From classical mechanics, the product of centrifugal acceleration and mass is a force. Therefore, $-m\omega^2 r$ is called a "centrifugal force" away from the center of the circle. However, the centrifugal force is not a "force"! In fact, it does not even exist at all, just a manifestation of inertia. The name is based on that it equivalently creates **an effect of force**. Similar to the "excluded volume effect" repeatedly discussed and mentioned before, the dispersion attraction is not volume, which generates an effect that is an equivalent to the negative excluded volume.

Let us imagine rotating a rope with an object tied at one end by hand. For an observer standing on the side (in the inertial reference frame), the spinner applies a force to the object through the rope, continuously changing its moving direction, not violating classical mechanics. If the observer becomes a spinner, standing in the center and spinning at the same speed, she/he will find that the object is obviously stressed (the hand feeling), but it is relatively static. "A subject is motionless under an imposed force" clearly violates classical mechanics. Just for using classical mechanics in a non-inertial rotating reference frame, a virtual "centrifugal" force has to be introduced.

Before rotating, macromolecules are uniformly distributed in the solution as individual chains. After rotating, every chain is driven by centrifugal force to accelerate in the direction away from the center of rotation. For a given rotation speed, the farther away from the center of rotation, the faster the moving speed ($u = dr/dt$). The faster the speed, the greater the frictional resistance (fu) experienced by the chain, in the opposite direction of the movement, where f is the coefficient of friction between the chain and solvent, a ratio of thermal energy ($k_B T$) to the translational diffusion coefficient (D) of the chain, i.e., $f = k_B T/D$. The volume of the chain is the product of its specific volume and mass, i.e., $v_p m$. Every chain repels a number of solvent molecules with the same volume but a different mass of $\rho_s v_p m$, so that it experiences a "buoyancy" force, whose magnitude is proportional to the centrifugal acceleration, but in an opposite direction. As shown in the figure above, each chain bears three "forces"; when an equilibrium is reached, the sum of three "forces" is zero, i.e.,

$$fu + \rho_s v_p m\omega^2 r - m\omega^2 r = 0 \quad \rightarrow \quad m = \frac{fu}{\omega^2 r(1 - \rho_s v_p)}$$

where ρ_s and v_p can be measured separately by density-related instruments. After reaching this dynamic equilibrium, the macromolecular chain is not in a static state, but is moving toward the bottom of solution at a uniform speed. On the other hand, because every macromolecule moves from the top to the bottom of solution, the slowly moving solvent molecules with a lower mass are left behind to form an increasing concentration gradient of macromolecules from top to bottom. This gradient drives the chain to move backwards towards the top. In fact, this is the movement of solvent molecules at a higher pressure to the macromolecular layer at a lower pressure, driving the macromolecules to move upwards. The physical picture is as follows.

Before the rotating, there is a layer of macromolecules on the solution surface (the interface is clear, the external and internal concentrations are 0 and C, respectively). After the rotating, like other macromolecules in the solution, they move toward the bottom in the centrifugal field, accelerating first, and then at a constant speed. The concentration gradient causes their reverse movement (translational diffusion). The originally clear top macromolecular layer gradually becomes blurred, and the concentration distribution gradually broadens, but its center position does not alternate by the reverse movement. This center position continues to move toward the bottom until dynamic sedimentation equilibrium is reached.

The question is how to transform those microscopic variables (m, f and u) in the above equation into measurable macroscopic physical quantities. Let us look at the preliminary fluid mechanics. In a flow field, the flow rate (J) is defined as the mass of an object flowing through a unit area per unit time, i.e., $J = dQ/$ (Adt), as shown on the right. In the problems discussed in this book, the object is a macromolecular

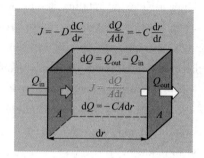

chain. Obviously, between two panels with an area of A and an infinitely close distance ($dr \to 0$) (dr is magnified for clarity), the volume change $dV = Adr$, and the change of macromolecular mass within the volume $dQ = Q_{out} - Q_{in}$, convertible to the macromolecular concentration (C), $dQ = CdV = CAdr$, i.e., $J = Cdr/dt$.

In addition, it can be known from physical intuition that when macromolecules automatically diffuse from a high concentration to low concentration, the flow rate should increase with the concentration difference, proportional to the concentration gradient (dC/dr). As early as 1855, Adolf Fick pointed out that the proportional constant is the translational diffusion coefficient (D), which is Fick's first law: $J = - DdC/dr$. The negative sign is introduced because when dC/dr is positive, the outflow is greater than the inflow, J must be a negative value. The substitution of $J = Cdr/dt$ into Fick's first law results in $fu = k_B T d\ln C/dr$. Further substituting it into the above equation and multiplying N_{AV} on both sides, one can obtain

$$M = \frac{RT}{\omega^2(1 - \rho_s v_p)} \frac{d\ln C}{rdr} = \frac{2RT}{\omega^2(1 - \rho_s v_p)} \frac{d\ln C}{dr^2} \quad \to \quad \frac{\omega^2(1 - \rho_s v_p)}{2RT} \frac{dr^2}{d\ln C} = \frac{1}{M} \quad (5.14)$$

For a given macromolecular solution, instrument and rotating speed, the prefactor on the left side of the above equation is a constant. If a series of corresponding concentrations C at different r can be measured, the molar mass of macromolecule is obtainable from the slope of "$\ln C$ vs r^2". However, unlike small molecules, macromolecular chains have different lengths and sizes.

For a polydisperse sample, at any given position (r), the concentration ($C_{i,r}$) of the ith component with a molar mass M_i is not equal to its concentration (C_i) in the initial solution before rotating. The above equation expresses the concentration change of each component $C_{i,r}$ with r. Note: M_i is a constant, independent of r. Therefore, when calculating the average molar mass, one needs to first calculate the total value of each component in the solution (C_i), i.e., an integration of r from the top to the bottom (the entire solution), and then sum the molar masses of all components after different concentration weights, to obtain different average molar masses.

First, $d\ln C_{i,r}$ in the above equation is written as $dC_{i,r}/C_{i,r}$ first, and then $C_{i,r}$ is moved to the right. The integrations of the left side from the top (r_T) to the bottom (r_B) and correspondingly the right side from the top concentration (C_{i,r_B}) to the bottom concentration (C_{i,r_T}) result in

$$\int_{r_T}^{r_B} M_i C_{i,r} dr^2 = \frac{2RT}{\omega^2(1 - \rho_s v_p)} \int_{C_{i,r_T}}^{C_{i,r_B}} dC_{i,r}$$

where $C_{i,r}$ changes with r, but its average value C_i is a constant, so that the integrations on the left sides is equal to C_i multiplies the corresponding width,

$$M_i C_i(r_B^2 - r_T^2) = \frac{2RT}{\omega^2(1 - \rho_s v_p)}(C_{i,r_B} - C_{i,r_T})$$

The two sides are summed and devided by $C = \sum_i C_i$, respectively. On the right side, $\sum_i C_{i,r_B} - \sum_i C_{i,r_T} = C_{r_B} - C_{r_T}$. After moving the constant $(r_B^2 - r_T^2)$ on the left side to the right,

$$M_w = \frac{\sum_i M_i C_i}{\sum_i C_i} = \frac{2RT(C_{r_B} - C_{r_T})}{\omega^2(1 - \rho_s v_p)(r_B^2 - r_T^2)C} \tag{5.15}$$

In the experiment of centrifugal equilibrium, as long as the concentrations at the surface and bottom of the solution is measured after the equilibrium is reached, the **weight average molar mass** of macromolecules can be obtained by combining other experimental parameters. This is very simple for the optical detection method of directly measuring the concentration.

When the sedimentation equilibrium is reached, the macromolecular concentration (C_r) gradually increases along the radius (r) from the top (r_T) to the bottom (r_B) in the entire pool, forming a dynamically stable distribution. Rewriting eq. (5.14) and integrating the left side from r_T to r and the right side from $C_{r,T}$ to C_r, the concentration distribution is obtained as follows.

$$\int_{r_T}^{r} \frac{M\omega^2(1 - \rho_s v_p)}{2RT} dr^2 = \int_{C_{r_T}}^{C_r} d\ln C_r$$

$$\frac{M\omega^2(1 - \rho_s v_p)}{2RT}(r^2 - r_T^2) = \ln\frac{C_r}{C_{r_T}}$$

$$C_r = C_{r_T}\exp\left[\frac{M\omega^2(1 - \rho_s v_p)}{2RT}(r^2 - r_T^2)\right] \tag{5.16}$$

For a polydisperse sample, the above equation only depicts the position dependence of the concentration of each component (M_i) in the solution, i.e., at each r, there is a concentration $C_{i,r}$ corresponding to M_i. The respective integrations of both sides of the above equation from r_T to r_B include all the chains of one component in the entire solution, equal to the original concentration (C_i) of this component in the solution. On the other hand, for one component, i. e., M and C_r are respectively replaced by M_i and $C_{i,r}$, eq. (5.14) can be rewritten as

$$\frac{\omega^2(1 - \rho_s v_p)}{2RT} M_i = \frac{dC_{i,r}}{C_{i,r} dr^2}$$

For given experimental conditions and a macromolecular chain with a mass of M_i, its left side is a quantity, independent of r. After moving M_i to the right side, $(dC_{i,r}/dr^2)/(M_i C_{i,r})$ becomes a quantity, independent of both r and M_i. If $dC_{i,r}/dr$ is defined as $z_{i,r}$, $(z_{i,r}/r)$ is proportional to $M_i C_{i,r}$, and M_i is independent of r, so that $dlnC_{i,r} = dln(z_i/r)_r$. Note: C_r includes all components at r, so that $C_{i,r}$ is not equal to C_r. The above equation can be rewritten as

$$M_i = \frac{2RT}{\omega^2(1-\rho_s v_p)}\frac{dlnC_{i,r}}{dr^2} = \frac{2RT}{\omega^2(1-\rho_s v_p)}\frac{dln\left(\frac{z_{i,r}}{r}\right)}{dr^2} = \frac{2RT}{\omega^2(1-\rho_s v_p)}\frac{d\left(\frac{z_{i,r}}{r}\right)}{\frac{z_{i,r}}{r}dr^2}$$

$$M_i\frac{z_{i,r}}{r}dr^2 = \frac{2RT}{\omega^2(1-\rho_s v_p)}d\left(\frac{z_{i,r}}{r}\right)$$

Both sides are integrated from top to bottom, respectively; the left side from r_T to r_B, and the right side, from $(z_i/r)_{r_T}$ to $(z_i/r)_{r_B}$; namely,

$$M_i\int_{r_T}^{r_B} 2z_{i,r}dr = \frac{2RT}{\omega^2(1-\rho_s v_p)}\int_{\left(\frac{z_i}{r}\right)_{r_T}}^{\left(\frac{z_i}{r}\right)_{r_B}}d\left(\frac{z_{i,r}}{r}\right)$$

where $z_{i,r}$ is a variable in the solution, but its average values z_i is a constant. As mentioned before, the integrations on the left side is equal to the average value in one dimension multiplies the width in another dimension, i.e.,

$$M_i z_i\int_{r_T}^{r_B} dr = \frac{RT}{\omega^2(1-\rho_s v_p)}\left[\left(\frac{z_i}{r}\right)_{r_B} - \left(\frac{z_i}{r}\right)_{r_T}\right]$$

The both sides are summarized, and the integration on the left side is independent of i, a constant, so that

$$\sum_i M_i z_i\left(\int_{r_T}^{r_B} dr\right) = \frac{RT}{\omega^2(1-\rho_s v_p)}\sum_i\left[\left(\frac{z_i}{r}\right)_{r_B} - \left(\frac{z_i}{r}\right)_{r_T}\right]$$

The left is multiplied and divided by a constant, $z = \sum_i z_i$, respectively, which is independent of the position, and $dC/dr = z$, i.e., $zdr = dC$, resulting in

$$\frac{\sum_i M_i z_i}{\sum_i z_i}\left(\int_{C_{r_T}}^{C_{r_B}} dC\right) = \frac{RT}{\omega^2(1-\rho_s v_p)}\left[\left(\frac{z}{r}\right)_{r_B} - \left(\frac{z}{r}\right)_{r_T}\right]$$

where the left integration is $(C_{r_B} - C_{r_T})$, which can be moved to the right side, so that

$$M_z = \frac{\sum_i M_i z_i}{\sum_i z_i} = \frac{\sum_i M_i^2 C_i}{\sum_i M_i C_i} = \frac{2RT}{\omega^2(1-\rho_s v_p)}\frac{\left[\left(\frac{dC}{dr^2}\right)_{r_B} - \left(\frac{dC}{dr^2}\right)_{r_T}\right]}{(C_{r_B} - C_{r_T})} \tag{5.17}$$

where $z_i \propto M_i C_i$ and the general definition of the average molar mass and the specific definition of z average molar mass are used. This is why $\sum_i M_i^2 C_i/\sum_i M_i C_i$ is originally defined as z average.

It shows that if the r dependence of C_r is measured in an ultracentrifugation experiment, one can plot C_r versus r^2, obtain the slope dC_r/dr^2 at each point, and then plot dC_r/dr^2 against C_r to obtain the two slopes at the top and at the bottom, respectively. A combination of these two slopes and their corresponding concentrations together with other experimental parameters results in the z−average molar mass.

The optical Schlieren detector commonly used in ultracentrifugation experiments directly measures the concentration gradient at each location. Their corresponding concentrations can be obtained by an ultraviolet absorption or optical interference detector. In this way, it is very convenient to use the above equation to calculate the z−average molar mass. If only using a ultraviolet absorption or optical interference detector, the r dependence of the concentration is measured, and the logarithm of the concentration can be easily calculated. Based on eq. (5.15), the weight average molar mass can be obtained. This fully shows that different average molar masses are not derived from mathematical definitions and operations, but depend on different experimental and detecting methods.

Equation (5.14) can also be derived directly from the principle of thermodynamics. The details are as follows: for macromolecules, before mixing, the macromolecule is a pure substance, and its free energy is $G*(0)$; after mixing, the free energy decreases to $G(X_p)$, and X_p is the volume fraction of the macromolecule. The change in free energy (the mixing free energy, ΔG_{mix}) is

$$G(X_p) - G^*(0) = \Delta G_{mix}(X_p)$$

As mentioned earlier, driven by the centrifugal force ($m\omega^2 r$), a macromolecule with a mass of m moves radially toward the bottom of the solution. The farther, the greater the force, the faster the speed, forming a density gradient (pressure gradient), which increases outward from the center of rotation. It directly leads to a free energy gradient (dG/dr) with a dimension of force. Therefore, the macromolecular chain in this pressure gradient must move inward (the top of the solution) to reduce the density gradient.

When discussing the osmotic pressure difference, $dG = Vdp - SdT$ has been used; under an isothermal condition, $dT = 0$. When the centrifugal force and the driving force induced by an increase in entropy reach equilibrium, $dG/dr + m\omega^2 r = 0$ or $Vdp/dr = -m\omega^2 r$, which can also be written as $dp/dr = -\rho\omega^2 r$, ρ is the density of the system. In an isothermal and isobaric process, each term in the above equation is partially differentiated with respect to r,

$$\left[\frac{\partial G(X_p)}{\partial r}\right]_{T,p} - \left[\frac{\partial G^*(0)}{\partial r}\right]_{T,p} = \left[\frac{\partial \Delta G_{mix}(X_p)}{\partial r}\right]_{T,p}$$

The substitution of $dG/dr = Vdp/dr$ and the use of a superscript " * " to distinguish between a mixture (solution) and a pure substance (macromolecule) leads to

$$V\frac{dp}{dr} - V^*\frac{dp}{dr} = \left[\frac{\partial \Delta G_{mix}(X_p)}{\partial r}\right]_{T,p} \tag{5.18}$$

Each item in the above equation is partially differentiated with respect to n_p. According to the definitions of $n_p = m/M$ and $(\partial V/\partial m)_{T,p,n_s} = v_p$, it is not difficult to obtain $(\partial V/\partial n_p)_{T,p,n_s} = v_p M$, where $(\partial V^*/\partial n_p)_{T,p}$ is the molar volume of the macromolecule (V_m^*). Substituting it into the above equation results in

$$v_p M \frac{\mathrm{d}p}{\mathrm{d}r} - V_m^* \frac{\mathrm{d}p}{\mathrm{d}r} = \left[\frac{\partial \Delta G_{\mathrm{mix}}(X_p)/\partial n_p}{\partial r}\right]_{T,p} = \left[\frac{\partial \Delta \mu(X_p)}{\partial r}\right]_{T,p} \qquad (5.19)$$

Using pure macromolecules as a reference state (initial state), $\Delta \mu(X_p) = \mu(X_p) - \mu^*(0)$. The partial differentiation with respect to n_p can be divided into two steps, first to X_p, and then times the differentiation of X_p with respect to n_p. The result is

$$\left(\frac{\partial \Delta G_{\mathrm{mix}}}{\partial n_p}\right)_{T,p,n_s} = \mu_p - \mu_p^* = RT\frac{\partial[n_s \ln(1 - X_p) + n_p \ln X_p + n_s X_p \chi_{sp}]}{\partial X_p}\frac{\partial X_p}{\partial n_p}$$

$$= RT\left[-\frac{n_s}{1 - X_p} + \frac{\partial n_p}{\partial X_p}\ln X_p + \frac{n_p}{X_p} + n_s \chi_{sp}\right]\frac{\partial X_p}{\partial n_p}$$

$$= RT\left[-\frac{n_s}{1 - X_p}\frac{\partial X_p}{\partial n_p} + \ln X_p + \frac{n_p}{X_p}\frac{\partial X_p}{\partial n_p} + \frac{\partial X_p}{\partial n_p}\chi_{sp}\right]$$

$$\mu(X_p) - \mu^*(0) = RT[\ln X_p + (1 - n_v)(1 - X_p) + \chi_{sp}n_v(1 - X_p)^2] \qquad (5.20)$$

In the last step of the above derivation, $(\partial \Delta G/\partial n_p)_{T,p,n_s} = \mu(X_p) - \mu^*(0)$ and the following formula are used,

$$\left(\frac{\partial X_p}{n_p}\right)_{T,p,n_s} = \frac{\partial\left(1 - \dfrac{n_s}{n_s + n_v n_p}\right)}{\partial n_p} = \frac{n_v n_s}{(n_s + n_v n_p)^2} = \frac{X_s(1 - X_s)}{n_p} = \frac{n_v(1 - X_p)^2}{n_s}$$

Note: eqs. (5.10) and (5.20) are two important equations, from which many measurable physical quantities can be derived. A careful examination of these two equations shows that the solution or solvent content dependence of free energy and chemical potential are symmetrical. Physically, in a binary mixture, the definitions of solvent and solute are originally relative, so that it is reasonable. Substituting $\mathrm{d}p/\mathrm{d}r = -\rho \omega^2 r$ (macromolecule solution) and $\mathrm{d}p/\mathrm{d}r = -\rho^* \omega^2 r = -(M/V_m^*)\omega^2 r$ (pure macromolecule) into eq. (5.18) results in

$$-v_p \rho M \omega^2 r - (-M\omega^2 r) = \left[\frac{\partial \Delta \mu(X_p)}{\partial X_p}\right]_{T,p}\left[\frac{\partial X_p}{\partial C}\right]_{T,p}\left[\frac{\partial C}{\partial r}\right]_{T,p} \qquad (5.21)$$

Therefore, in the centrifugal field, using eq. (5.20), the above equation can be rewritten as

$$(1 - v_p\rho)M\omega^2 r = RT\frac{\partial[\ln X_p + (1 - n_v)(1 - X_p) + \chi_{sp}n_v(1 - X_p)^2]}{\partial X_p}\frac{\partial X_p}{\partial C}\frac{\partial C}{\partial r}$$

where the isothermal and isobaric subscripts have been omitted; ρ is the solution density; v_p is the specific volume of macromolecules defined and discussed in the previous section, and its relationship with the volume fraction of macromolecules is $X_p = Cv_p$, i.e., $(\partial X_p/\partial C)_{T,p} = v_p$. Note that $X_p = n_v n_p/(n_s + n_v n_p)$ and $X_s + X_p = 1$. The above equation is rewritten as

$$\omega^2 r(1-\rho v_p)M = RT\left[\frac{1}{X_p}-(1-n_v)-2\chi_{sp}n_v(1-X_p)\right]v_p\frac{\partial C}{\partial r}$$

$$= RT\left[\frac{1-X_p}{X_p}+n_v-2\chi_{sp}n_v(1-X_p)\right]v_p\frac{\partial C}{\partial r}$$

$$\omega^2 r(1-\rho v_p) = \frac{RTX_s}{M}\left[\frac{1}{X_p}+\left(\frac{1}{X_s}-2\chi_{sp}\right)n_v\right]v_p\frac{\partial C}{\partial r}$$

In a dilute solution, $1/X_s = 1/(1-X_p) \approx 1+X_p$. If the volume does not change in the dissolution, $(1-\rho v_p)/X_s = 1-\rho_s v_p$, i.e., the solution density is converted into the solvent density, a known quantity. When the temperature, pressure and composition remain as constants, the partial differentiation $\partial C/\partial r$ can be written as dC/dr. The above equation becomes

$$\omega^2 r(1-\rho_s v_p) = \frac{RT}{M}\left[\frac{1}{C}+(1-2\chi_{sp})n_v v_p + n_v v_p^2 C\right]\frac{dC}{dr}$$

By moving the known and measurable physical quantities to the left, and the unknown ones to the right, it can be further written in the form commonly used in experiments

$$\frac{\omega^2 r(1-\rho_s v_p)C}{RT\dfrac{dC}{dr}} = \frac{1}{M}+2\left(\frac{1}{2}-\chi_{sp}\right)\frac{n_v v_p}{M}C+n_v v_p^2 C^2$$

$$\frac{\omega^2(1-\rho_s v_p)}{2RT}\frac{dr^2}{d\ln C} = \frac{1}{M}+2A_2 C+3A_3 C^2 \tag{5.22}$$

where $A_2 = (1/2-\chi_{sp})n_v v_p/M$, identical to the previously defined one, and $A_3 = v_p^3/3$. n_v is physically equal to the volume ratio of one macromolecules to one solvent molecule. For a given solution, both n_v and v_p are two constants. In a dilute solution, the quadratic term or even the first term can be omitted, back to eq. (5.14).

As mentioned before, for a polydisperse macromolecule sample, when a centrifugal equilibrium is reached, a series of macromolecular concentrations (C_r) at different radial positions (r) can be measured experimentally. If the direct measurement is the concentration itself (C_r), the weight average molar mass(M_w) can be obtained from the concentrations at the top and bottom of the solution and other experimental parameters. If the concentration gradient (dC_r/dr) and C at each point in the solution is measured, the z-average molar mass (M_z) is obtainable from the two concentration gradients at the top and bottom of the solution, their corresponding concentrations and other experimental parameters.

Based on the above discussion, the centrifugal equilibrium experiment mainly depends on the translational diffusion of macromolecules in the solution under the gravity field. The time required for translational diffusion is proportional to the square distance. The translational diffusion coefficient of macromolecules is generally between $10^{-6}-10^{-7}$ cm^2/s, and it takes days to diffuse a typical sample cell length (~ 3 mm) and reach the equilibrium. When designing the instrument, it is hoped hope that the speed is as fast as possible and the sample is as far away from the axis of rotation as possible, but it is always limited by the mechanical properties and weight of the rotor material.

In order to measure the concentration of macromolecules along the radial direction during the high speed rotation, it is necessary to design two light-transmitting windows on the sample cell, and be equipped with quartz glass that can withstand an extreme pressure, and sometimes even harder and stronger sapphire is used. This series of experimental requirements limit the rotating speed, extend the time required for the experiment and reduce the accuracy of the measurement. The ultraviolet absorption and optical refraction (Schlieren detection device or the interference of two beams of light passing through the solution and the solvent) together with the negative film exposure and other methods were often used in the past to measure the macromolecular concentration gradient. The detailed discussion of the measurement method and principle are omitted in this book. Readers can refer to the monograph on ultracentrifuge.

In order to shorten the measurement time, the method of the sedimentation speed instead of the sedimentation equilibrium is often used in experiments. As mentioned earlier, before the rotation, the concentration in the sample cell is uniform everywhere, denoted as C, which is always the average concentration of the entire cell. At the top, from the outside of the solution to the inside, the concentration jumps from 0 to C, and the boundary is clear. After rotation, all the molecules move to the bottom but the macromolecular chains move faster, leaving only solvent molecules on the top of the solution, and creating a macromolecular concentration gradient, which drives the macromolecular chains to diffuse back (the top), driven by an increase in entropy.

The original clear interface becomes blurred, but the central position of the interface (r_p) is not affected by the back-diffusion, and still accelerate to the bottom. As the speed ($u = dr_p/dt$) increases, the resistance increases, and the speed quickly approaches a constant. However, the accelerate speed increases with r, and the speed also gradually increases. After it, the center position moves toward the bottom until reaching the aforementioned equilibrium state nearby the bottom. In this process, the interface becomes blurred more and more due to the diffusion and sedimentation. In the Schlieren detection, it is a small peak (dC/dr), moving forward and continuously widening. The peak position corresponds to the center position of the interface, and its change over time can be measured more accurately.

Theoretically, eq. (5.18) shows that after the dynamic equilibrium is reached, the left and right sides are equal. The net "force" imposed on macromolecules in the centrifugal field becomes zero. Before reaching the dynamic equilibrium, a net centrifugal force (F) imposes on each mole of macromolecules, which makes macromolecules moving toward the bottom of solution, i.e.,

$$F = (1 - v_p\rho)M\omega^2 r - \left[\frac{\partial \Delta\mu}{\partial C_{r_p}}\right]_{T,p}\left[\frac{\partial C_{r_p}}{\partial r_p}\right]_{T,p} \tag{5.23}$$

In the initial stage of rotation, i.e., the move of the interface does not exceed 1/4 of the solution height, the interface basically remains clear with no obvious blur. Therefore, the interface concentration (C_{rp}) is very close to the initial solution concentration, and can be approximately considered as no change, i.e., $\partial C_{rp}/\partial r_p \cong 0$, so that the second term of the above equation can be omitted. The "force" imposed on each mole of macromolecules is $F = fu = RTu/D$; namely,

$$\frac{\mathrm{d}r_p}{\mathrm{d}t} = \frac{(1 - v_p\rho)MD}{RT}\omega^2 r_p = S\omega^2 r_p$$

where S is the defined sedimentation coefficient,

$$S = \frac{(1 - v_p\rho)MD}{RT} = \frac{\mathrm{d}\ln r_p}{\omega^2 \mathrm{d}t} \quad \rightarrow \quad \frac{S}{D} = \frac{(1 - v_p\rho)}{RT}M \qquad (5.24)$$

The expression on the right side is an extremely useful and practical equation. In an experiment, the time dependence of the change of the peak position r_p at the initial stage of centrifugation can be measured first, and then the "$\ln r_p$ versus t" is plotted. For a given colloid or macromolecular solution and other experimental conditions, the plot should be a straight line, and S is obtainable from its slope. If S does not change with time, the integration on both sides of the above equation, respectively, from r_0 to r_p and from t_0 to t leads to

$$\ln r_p = S\omega^2 t + \ln r_{p,0} - S\omega^2 t_0$$

where $r_{p,0}$ and r_p are respectively the distances of the peak position at t_0 and t from the rotating axis. For a given macromolecular solution and speed, the two terms on the right are constants.

Since S is obtainable in an ultracentrifugation experiment, for a macromolecule with a known molar mass M, one can use the ultracentrifugation experiment to get the translational diffusion coefficient D of the macromolecule, and calculate its friction coefficient in a given solvent $f = RT/D$. On the contrary, if D is measured first by other methods (such as dynamic laser scattering), the molar mass M of the macromolecule can be determined by ultracentrifugation experiment. The premise is that the experiment temperature and solvent are the same, and the concentrations used are the same or similar. For real solutions, there exists an excluded volume effect, so that there are two general equations as follows.

$$\frac{1}{S} = \frac{1}{S_0}(1 + A_{2,s}C + A_{3,s}C^2 + \cdots) \quad \text{and} \quad D = D_0(1 + A_{2,D}C + A_{3,D}C^2 + \cdots)$$

where A are different virial coefficients of the sedimentation coefficient and the translational diffusion coefficient. For a given macromolecule and solvent, they are constants. In experiments, a series of solutions with different concentrations are prepared first, S and D in each solution is measured separately, and then $1/S$ and D are plotted against C, respectively. The extrapolation of the concentration to zero in each plot leads to S_0 and D_0, respectively. The second-order virial coefficients of S and D in the solution have the same positive or negative sign. In the calculation of their ratio, the concentration effects partially cancel each other. Therefore, for a dilute solution, S/D can be used instead of S_0/D_0 without causing too much error. For a polydisperse sample,

$$\frac{RT}{(1 - v_p\rho)} \sum_i S_i = \sum_i D_i M_i = \sum_i k_D M_i^{1-\alpha_D}$$

$$M_s = \left(\sum_i M_i^{1-\alpha_D} \right)^{\frac{1}{1-\alpha_D}} = \left[\frac{RTS}{(1 - v_p\rho)k_D} \right]^{\frac{1}{1-\alpha_D}} \qquad (5.25)$$

where the scaling relationship between the size and molar mass of macromolecules was used, i. e., $D = k_D M^{-\alpha_D}$, $1/3 \leqslant \alpha_D \leqslant 3/5$. Therefore, the **S average molar mass** (M_s) is from a special kind of average, **which lies between the number average and weight average molar mass**.

Theodor Svedberg invented the world's first ultracentrifuge as early as 1924 – 1925. The speed at that time has reached 12,000 rpm, which can generate 7,000 times acceleration of gravity (g). Less than two years later, he successfully increased the speed by nearly four times to 100,000g. Using these ultracentrifuges, he studied a series of colloids and proteins, and won the Nobel Prize in Chemistry in 1926 for these studies and results. Ten years later, Edward Greydon Pickels invented a vacuum ultracentrifuge, which not only reduces the influence of air on the speed, but also avoids the friction between the rotor and air, generating heat. The temperature gradient in the centrifuge causes the air convection, an inaccurate temperature control, which directly affects the establishment, measurement and analysis of the concentration gradient.

In the early days, aluminum alloy was the main material of the rotor. Now, the more expensive titanium alloy is also an option. At present, new, lightweight and high–strength carbon fiber has also become one of the choices fore instrument designers. For a long time, the ultracentrifuge is a very important and precious instrument for studying macromolecular solutions, requiring no calibration curve. Due to the long experimental period, cumbersome data processing, and some other shortcomings, it was gradually replaced by some later developed rapid analysis instruments, especially after cheap, fast and simple gel permeation (size exclusion) chromatograph appeared in the middle of 1960s.

However, as a method of separating particles according to masses and without any calibration, it still has an irreplaceable role in macromolecular researches, especially in studying biological macromolecules. In recent years, the development of materials has allowed the rotor to be made lighter, with a larger radius, and a higher speed, up to one million times acceleration of gravity. At present, even if more than a dozen sample cell cavities are milled on a rotor, there is no effect on the mechanical properties of the rotor at high speed. In this way, two types of solutions with different concentrations can be measured in one experiment to satisfy the study of concentration dependence. In the past thirty years, due to the greatly improved detection accuracy and speed of photosensitive elements, and the rapid development of computer calculation and data processing speed, modern ultracentrifugation instruments have long been free from troubles of developing negative film. The analytical ultracentrifuge has gradually backed to the instrument market since the 1990s, and has begun to re-enter some macromolecular research laboratories. This is why this book has dedicated one section to introducing this important method.

Finally, it is necessary to emphasize the safety matters that must be paid attention to in the ultracentrifugation experiment. Due to the speed of tens of thousands of revolutions per minute, the rotor has great kinetic energy. During its producing process, the manufacturer has made every effort to make sure that the rotor mass is uniformly distributed. However, in actual uses, especially in a chemical laboratory, accidentally splashed liquid may corrode the surface of the rotor, and unintentional bumps may also cause a small uneven mass distribution. For other experiments, this may not be important. But in an ultracentrifugation experiment, this needs to be taken seriously. Otherwise, the machine will be destroyed (e. g., due to uneven force, the rotating shaft is bent) or people will be injured (the cracked rotor is acting like a cannonball). Before each experiment, every part of the rotor must be carefully examined to see if there is damage or cracks; the two sample cells at each corresponding positions must be accurately weighed to ensure that they are identical. Other operating procedures of the instrument must also be strictly followed.

Summary

In a centrifugal field, macromolecules or particles in a liquid move away from the rotating center and accelerate (sediment) under the action of a virtual centrifugal force. The motion-induced frictional resistance increases with the moving speed, in an opposite direction. In addition, due to the difference between densities of macromolecules (particles) and the dispersion medium (solvent), they are also subjected to a "buoyancy" in the opposite direction. As the resistance gradually increases, the three forces finally reach the equilibrium. Afterwards, macromolecules or particles continue to move forward at an even speed until they reach the liquid bottom.

Basic Principles: Sedimentation Equilibrium: Macromolecules (particles) and solvent molecules all move towards the liquid bottom. However, the moving speed is different, causing a concentration gradient along the radial direction, driving macromolecules (particles) to move in the opposite direction. When the equilibrium is reached, a macroscopically stable concentration gradient (dC_r/dr) is formed, that is, the concentration (C_r) increases with the distance (r) from the rotating center, and its integration with respect to r is equal to the original concentration (C). This process can also be imagined as a layer of macromolecules (particles) slowly diffuse upwards from the bottom under a centrifugal gravity field, and the time is measured in "days". Note: It should be noted that this balance is not the aforementioned three-force balance!

Sedimentation Speed: Before the rotation, on the inside and outside of a macromolecular solution (colloid) surface, the concentration of macromolecules (particles) jumps from zero to

C, which is a step function, clearly visible. After the rotation, macromolecules (particles) move from the interface to the liquid bottom, leaving a solvent layer behind (on the top). The concentration difference drives the reverse diffusion of macromolecules (particles), blurring the originally clear moving interface, but its center position (r_p) is still the original interface position, not affected by the diffusion. In the Schlieren optical measurement, the concentration gradient corresponds to a small moving peak (dC/dr). The peak position corresponds to the interface position. The interface movement (sedimentation) speed (dr_p/dt) depends on the rotation speed, the density difference between macromolecules (particles) and the dispersion medium, the molar mass of macromolecules (particles), and the coefficient of friction or translational diffusion coefficient. For a given macromolecular solution (colloids) and experimental conditions, the first two items are constants. Therefore, this method can be used to measure the molar mass of macromolecules (particles) or the translational diffusion coefficient if one of them is known by other methods. For monodisperse protein molecules, the translational diffusion coefficient is obtainable from the broadening of the peak, so that its molar mass is obtainable only by ultracentrifugation.

Measurable Physical Quantity: In a sedimentation equilibrium experiment, using the ultraviolet absorption, the optical Schlieren detection device or two interference light beams respectively passing through solution and solvent, one can measure the concentration of macromolecules (particles) at the distance r from the rotating center (C_r) or directly measure the concentration gradient (dC_r/dr). In a sedimentation velocity experiment, one can measure the moving speed of a layer of macromolecules (particles) on the surface to the liquid bottom under the action of virtual centrifugal force, i.e., its position (r_p) continuously moves to the liquid bottom with time (t). The measurement range is in the earlier stage of the experiment, i. e., the moving distance is shorter than one quarter of the liquid height, although the current fitting software can measure the entire settlement process.

Data processing: Sedimentation equilibrium: for a macromolecular solution (colloid) with a given initial concentration (C), the detector can directly measure C_r at different r when the equilibrium is reached. A combination of the concentrations at the top and bottom layer and other experimental parameters can result in an apparent weight average molar mass at this concentration according to eq. (5.15). The apparent z-average molar mass can also be calculated from the concentration gradient and position at the top and bottom according to eq. (5.17). After obtaining the apparent molar masses in different concentrations (preferably more than five), one can plot their reciprocal versus the concentration. According to eq. (5.20), the weight or z-average molar mass of macromolecules (particles) and the second-order virial coefficient are obtainable from the reciprocal of the intercept and slope, respectively.

Sedimentation velocity: After the time (t) dependence of the position of the top interface layer (r_p) of macromolecules (particles) was measured using the Schlieren detection device, one can plot "$\ln r_p$ versus r" to obtain a straight line. According to $S = [d\ln r_p/dt]/\omega^2$, the sedimentation coefficient S can be obtained from its slope. Based on eq. (5.23), if the molar mass of macromolecules (colloid particles) is known, the translational diffusion coefficient and the friction coefficient in the liquid can be calculated. Conversely, an average molar mass between the number and the weight average can be obtained.

Laser Light Scattering (LLS)

Static LLS: It is also known as classical light scattering. Light scattering is a physical phenomenon that occurs around us all the time. The public just "do not know the true face of Mount Lu because they are born in it". If there were no gas, dust, water and other molecules and particles in the air that scatter the incident sunlight, the world would be truly "black and white". The place where it shines is hot and difficult to bear, and dazzling; the sheltered place is so dark that even fingers are invisible. It is precisely because of its existence that people can collect the scattered light of the sun through windows instead of turning on the lights in the daytime; it is also because of its existence that everyone can enjoy the beautiful and splendid sunrise and sunset, leading to many widely spreading and enjoying poems.

Using this physical phenomenon, through the interaction of light and molecules or particles, many microscopic molecular parameters and properties can be peeked. In the past thirty years or so, due to the rapid development of lasers, fast electrons, single-photon detectors, and computer storage and computing capabilities, the originally expensive research-type LLS spectrometers have gradually moved from specialized LLS laboratories into more and more macromolecular research laboratories, becoming a routine experimental instrument. Today, a commercial modern LLS instruments generally include static and dynamic light scattering.

In static LLS, for a given concentration of macromolecule solution or particle dispersion (hereinafter macromolecules also referred to as "particles"), the time-average scattered light intensity at each scattering angle is measured. The angular dependence of the time average scattered light intensity can lead to the weight average molar mass and the average radius of gyration of the particles, and the second-order virial coefficient of the solution. In dynamic LLS, variation (fluctuates) of the intensity of scattered light at a scattering angle over time can be captured by a single photon detector. The signal is amplified and shaped, and then a digital time correlator is used to obtain the "light intensity-light intensity" time autocorrelation function.

For a polydisperse system, by analyzing the measured time autocorrelation function, a characteristic relaxation time distribution of the particles in the system is obtainable. Under a certain condition, the characteristic relaxation time distribution can be converted into a translational diffusion coefficient distribution, or further to a hydrodynamic radius distribution. This is why small single-angle (usually 90°) dynamic light scattering instruments in the market are often called particle size analyzers, or simply particle sizers.

Classical light scattering has a long history. As early as 1802, German chemist Jeremias Benjamin Richter noticed the brighter beam formed when a beam of light hits the gold sol. However, John Tyndall of Cambridge, England was the first one who systematically studied this phenomenon in the 1860s. He observed that the scattered light appeared very pale blue and found that if a polarized incident light is used, the scattered light has the same polarization. He also proposed two interesting puzzles in meteorology in the 19th century: Why is the sky blue? Why is the scattered light from the sky polarized?

His colleague, John William Strutt, 3rd Baron Rayleigh (often called The Lord Rayleigh), the second director of the Cavendish Laboratory in Cambridge successfully solved these two puzzles in 1871. Further, in 1881, by using his predecessor James Clerk Maxwell's electromagnetic field theory, he clarified that in the case of no absorption and no interaction, the scattered light intensity (I) of small optically isotropic particles is inversely proportional to the fourth power of the incident light wavelength (λ_0), i.e., $I \propto \lambda_0^{-4}$.

In 1944, Peter Debye used the scattered light intensity to measure the weight average molar mass of polymers in dilute solutions. Later, in 1948, Bruno Hasbrouck Zimm proposed to project the scattered light intensity with respect to the scattering angle and concentration on a three-dimensional curved surface onto a plane, so that the scattering angle and concentration can be simultaneously extrapolated to zero in one plot, i.e., the famous Zimm plot. Afterwards, static light scattering has become a classic method for measuring the weight average molar mass of macromolecules. In the early days, the light source was a mercury lamp. Since the advent of lasers in the 1960s, various coherent lasers gradually replaced the light source.

Since the advent of quantum mechanics, the wave–particle duality of light has gradually been widely accepted. Before expounding related theories, let us first qualitatively examine the light scattering from two perspectives of "particle" and "wave". If the incident light is regarded as a series of numerous parallel and continuous "photons (particles)", when they elastically collide with a huge number of particles in the liquid elastically with no absorption of energy, they will be scattered over the entire spherical space, forming the scattered light, because the colliding angle between every photon and every particle is different.

On the other hand, if the incident light is regarded as a continuously forward–propagating electromagnetic wave, since the vibration frequency of the visible light is as high as 10^{15} Hz (the number of vibrations per second), the heavier atomic nuclei and molecules can hardly vibrate at such a frequency, in a relatively static state. The lighter electrons are continuously accelerated and decelerated along the vibration direction of the electromagnetic wave under the action of the incident electromagnetic field, generating an additional induced dipole moment. It is known from physics that when a charged particle accelerates or decelerates, it emits (bremsstrahlung) electromagnetic waves toward the entire spherical space, which is the source of scattered light.

Whether it is based on the simple discussion of the "particle" or "wave" of the incident light, without any theoretical derivation and mathematical calculation, some properties of the scattered light intensity (I) can be concluded: it is inversely proportional to the square of the distance (r) between a detector with a given detection surface area and the scattering center. This is because the energy of the scattered light is uniformly distributed on a $4\pi r^2$ spherical surface, the farther away is, the less energy received is, just as our body with a fixed surface area feels less heat if farther away from a heat source. In addition, it is proportional to the incident light intensity (I_0), the number of particles in the liquid (the probability of the " particle" collision or the interaction of electromagnetic wave) and the particle volume (the probability of the "particle" collision or the number of electrons inside a particle). Obviously, Ir^2/I_0 (denoted as R, the Rayleigh ratio) is a variable that has nothing to do with the scattering device, but only with the number and volume of particles in the scattering system. The above – mentioned relationship based on physical description and intuition can be obtained by the following electrodynamic theory.

As shown in the figure on the right, the alternating photoelectric field of an incident light beam with a circular frequency ω is a vector pointing towards the right ($\vec{E}$). $\omega = 2\pi f$, where f is the frequency. Its relationship with the wavelength in the vacuum and propagation medium (λ_0 and λ), the speed of light (c) and the refractive index (n) of the propagation medium are as follows: $c = \lambda f$ and $\lambda = \lambda_0/n$. ϕ is the initial phase; as a

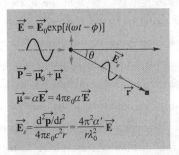

coherent light source, each ray in a laser beam has the same initial phase. In vacuum, all electrons in an isotropic particle will periodically oscillate with the alternation of the photoelectric field, accelerating and decelerating constantly, forming an induced dipole ($\vec{\mu}$). Its direction is constantly changing. If the incident light intensity (photoelectric field) is not super high, one only needs to consider the linear dependence of the induced dipole moment on the photoelectric field, so that the non−linear terms can be ignored, where the proportional constants α, α' and ε_0 are the polarizability, the molecular volume polarizability (the dimension is the same as volume) and the vacuum permittivity.

The oscillation of the induced dipole makes each particle a secondary point light source, which radiates electromagnetic waves of the same frequency in all directions, forming the scattered light. According to the Maxwell equations, it can be solved that on the detection plane at a distance $\vec{r}$ from the scattering particle, the photoelectric field ($\vec{E}_s$) of the scattered light is proportional to the photoelectric field of the incident light and inversely proportional to $|\vec{r}|$ and the square of the wavelength (λ_0) in vacuum. The dipole moment of a particle includes its permanent dipole moment ($\vec{\mu}_0$) and the induced dipole moment. At room temperature, $|\vec{\mu}_0| > |\vec{\mu}|$. At a high frequency of $\sim 10^{15}$ Hz (visible light), the orientation of $\vec{\mu}_0$ is not able to change with the alternating photoelectric field since the molecular mass is too heavy, so that it disappears in a differentiation with respect to time.

In addition to the alternating photoelectric field, the other half of the energy of the incident light is stored in a magnetic field ($\vec{B}$), perpendicular and pointing to the paper surface. The direction of light propagation is determined by the right-hand law, that is, when one stretch her/ his palm, first point the four fingers of right hand to the direction of the magnetic field, and then rotate towards the direction of the photoelectric field. The thumb is pointing towards the light propagation direction. The time average intensity of the incident light (I_0) is proportional the vector product of $\vec{E}$ and $\vec{B}$, that is, $I_0 \propto \vec{E} \times \vec{B}$. According to the electromagnetic theory, the time average light intensity (i_s) of the scattered light of a particle can be calculated as

$$i_s = \varepsilon_0 c \langle \vec{E}_s \times \vec{E}_s^* \rangle = \frac{16\,\pi^4 \alpha'^2}{\lambda_0^4 r^2} [\,\varepsilon_0 c \langle \vec{E} \times \vec{E}^* \rangle\,] = \frac{16\,\pi^4 \alpha'^2}{\lambda_0^4 r^2} (\varepsilon_0 c\, E_0^2) = \frac{16\,\pi^4 \alpha'^2}{\lambda_0^4 r^2} I_0$$

The dependence of scattered light intensity on the particle volume, incident light intensity and distance to the scattering center has been intuitively obtained from physics before. The only new thing here is the dependency that is inversely proportional to the fourth power of the incident wavelength. Rayleigh could explain why the sky is blue after obtaining this equation. In visible light, the blue light has a shorter wavelength so that it scatters more. The color of the sky is mainly composed of the scattered light. It can also explain why the sunrise and sunset glow red. Because the earth is spherical, in the early morning and dusk, the sunlight has to pass through a thicker atmosphere to reach the eyes. When blue light is scattered, only reddish light enters the sight. The higher the content of water vapor in the atmosphere, the redder the sunset will be. Therefore, there is an agricultural proverb: "should not go out if the sky is burning in the morning, but no problem to travel a far distance if the sunset is redish." This is because of the rotation of the earth. The reddish sunrise means that one traveler is moving towards the water vapor, while the reddish sunset means that the water vapor has left.

Note: the scattering volume (V) is only 200 cubic microns, i.e., $\sim 10^{-5}$ mL, much smaller than the solution volume ($\sim 10^0$ mL). Assume that the scattering volume contains N_p independent particles. Every particle has a size less than 5% of the wavelength, i.e., smaller than 25 nm, and the same α'.. The sum of the photoelectric field of the scattered light from all particles is $\vec{E}_s = \sum_{i=1}^{N_p} \vec{E}_{s,i}$, so that the scattered light intensity per unit volume is

$$I_s = \frac{16\,\pi^4 \alpha'^2}{\lambda_0^4 r^2 V} [\,\varepsilon_0 c \langle \sum_{i=1}^{N_p} \vec{E}_{s,i} \times \sum_{j=1}^{N_p} \vec{E}_{s,j}^* \rangle\,] = \frac{16\,\pi^4 \alpha'^2}{\lambda_0^4 r^2 V} \varepsilon_0 c \sum_{i=1}^{N_p} \sum_{j=1}^{N_p} \langle \vec{E}_{s,i} \times \vec{E}_{s,j}^* \rangle$$

After the sum is expanded, there are a total of N_p^2 items; among them, there are N_p identical items $\langle \vec{E}_{s,i} \times \vec{E}_{s,j}^* \rangle = E_{s,i}^2 (i=j)$ and $(N_p^2 - N_p)$ cross terms $\langle \vec{E}_{s,i} \times \vec{E}_{s,j}^* \rangle = \langle \vec{E}_{s,i} \rangle \langle \vec{E}_{s,j}^* \rangle = 0\,(i \neq j)$. Therefore, the above equation can be rewritten as

$$I_s = \frac{16\,\pi^4 \alpha'^2 N_p}{\lambda_0^4 r^2 V} I_0$$

According to the aforementioned definition of Rayleigh ratio (R), the above equation can be rewritten once more as the following equation only related to the scattering system.

$$R = \frac{I_s r^2}{I_0} = \frac{16 \, \pi^4 \alpha'^2 N_p}{\lambda_0^4 V} \tag{5.26}$$

α' and N_p/V respectively correspond to the physical nature of the particle, the volume (mass), and the particle density (concentration) in the scattering volume. Therefore, by measuring the absolute time average scattered light intensity of a pure liquid with a known density or a solution with a known concentration, the particle volume (mass) can be obtained. This is the physical basis for measuring the molar mass of particles or macromolecules by light scattering.

For a larger particle with a volume of V', it can be divided into N small scattering units with the same α'_0, i.e., $\alpha'_0 = \alpha'/N$. The two rays in the incident light beam are scattered by the ith and jth units, respectively, as shown on the right. As a coherent laser light source, the initial phases of all rays of light are the same, $\phi_{0,i} = \phi_{0,j}$. However, the two rays of light focused on the detection plane by a lens have passed different

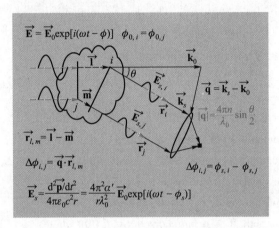

optical paths, and the optical path difference between them, i.e., the difference between the two blue vectors, is $\vec{r}_{l,m} = \vec{l} - \vec{m} = \vec{r}_l - \vec{r}_m$. Note: the maximum difference between these two vectors located in a particle is the particle size, which is much smaller than the distance from the scattering center to the detection plane $(\vec{r})$.

In the elastic scattering, there is no loss in energy, so that the wave vectors $(\vec{k}_0$ and $\vec{k}_s)$ of incident and scattered lights have different directions but the same amplitude size $(2\pi n/\lambda_0)$, where n is the refractive index of the medium. The difference between the two is the scattering vector $(\vec{q})$, whose amplitude is the bottom of an equilateral triangle, i.e., twice the hypotenuse multiplied by the sine of half the vertex angle (see figure). Note that the dimension of $\vec{q}$ is the reciprocal of length. Physically, it is equivalent to a microscopic optical ruler for measuring the particle size. From the geometric relationship, it is known that $\vec{q} \cdot \vec{r}_{l,m}$ is equal to the phase difference of the two scattered rays $\Delta\phi_{i,j} = \phi_{s,i} - \phi_{s,j}$, which is a dimensionless physical quantity. The photoelectric field of the scattered light from a large particle should be a vector sum of the photoelectric field of the scattered light from all its units, i.e., $\vec{E} = \sum_{i=1}^{N} \vec{E}_{s,i}$. Note that here the total number of the sum is the number of the scattering units (N) in a particle. Therefore, the time average scattered light intensity of a particle is

$$i_s = \frac{16\pi^4 \alpha'^2_0}{\lambda_0^4 r^2} I_0 \sum_{i=1}^{N} \sum_{j=1}^{N} \exp(i\Delta\phi_{i,j}) = \frac{16\pi^4 \alpha'^2_0}{\lambda_0^4 r^2} I_0 \sum_{l=1}^{N} \sum_{m=1}^{N} \exp(i\,\vec{q}\cdot\vec{r}_{l,m})$$

Even if the particles are rigid, the amplitude of $\vec{r}_{l,m}$ does not change, but its direction is changing constantly because each particle is randomly moving and rotating, so that one has to statistically average all possible orientations. The result is as follows.

$$i_s(q) = \frac{16\pi^4 \alpha'^2_0}{\lambda_0^4 r^2} I_0 \sum_{l=1}^{N} \sum_{m=1}^{N} \langle \exp(i\,\vec{q}\cdot\vec{r}_{l,m}) \rangle = \frac{16\pi^4 \alpha'^2_0}{\lambda_0^4 r^2} I_0 \sum_{l=1}^{N} \sum_{m=1}^{N} \frac{\sin(qr_{l,m})}{qr_{l,m}}$$

Here $i_s(q)$ is the scattered light intensity from one particle, and the scattered light intensity of N_p independent and uncorrelated particles is N_p times. The double summation in the above equation has a total of N^2 items after expansion. Mathematically, when $x \to 0$, $\sin(x)/x \to 1$. Therefore, when the particle or scattering angle is small, $\sin(qr_{l,m})/qr_{l,m} \to 1$, so the result of the double summation is N^2. Note: $\alpha'^2_0 N^2 = \alpha'^2$, the above equation becomes the result discussed previously. According to the definition of the Rayleigh ratio, $R(q)$ of a particle can be obtained. If it is normalized by $R(0)$ when

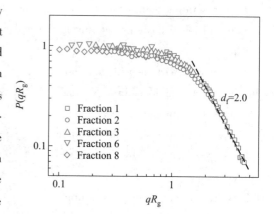

Li L W, He C, He W D, et al. Macromolecules, 2011, 44: 8195-8206.

(Figure 12, permission was granted by publisher)

the scattering angle is zero, a variable $\boldsymbol{P(q)}$ is obtained, which is only related to the particle or macromolecular structure, and independent of the molar mass. It is called **particle or macromolecule structure factor**, or scattering function, as shown on the right. As its name implies, a normalized structure factor is only related to the structure of a particle or a macromolecule. In a dilute solution, it is also not dependent of the particle or macromolecule concentration. Its mathematical expression is

$$P(q) = \frac{R(q)}{R(0)} = \frac{\dfrac{i_s(q)r^2}{I_0}}{\dfrac{i_s(0)r^2}{I_0}} = \frac{1}{N^2} \sum_{l=1}^{N} \sum_{m=1}^{N} \langle \exp(i\,\vec{q}\cdot\vec{r}_{l,m}) \rangle = \frac{1}{N^2} \sum_{l=1}^{N} \sum_{m=1}^{N} \frac{\sin(qr_{l,m})}{qr_{l,m}}$$

The structural factor defined here has been normalized, slightly different from $S(q)$ defined in other macromolecule textbooks, divided by an additional number of units (N). The expansion of $\sin(qr_{l,m})$ with the Maclaurin series, namely $\sin x = x - x^3/3! + x^5/5! - \cdots$, results in

$$P(q) = 1 - \frac{q^2}{6N^2} \sum_{l=1}^{N} \sum_{m=1}^{N} r_{l,m}^2 + \cdots$$

The double summation is the sum of the square distances of any two points in an object in the aforementioned discussion of the chain conformation. According to the definition of radius of gyration of a rigid object, $R_g^2 = \sum_{l=1}^{N} \sum_{m=1}^{N} r_{l,m}^2 / 2\ N^2$, so that $P(q) = 1 - q^2 R_g^2/3 + \cdots$. For flexible macromolecular chains, $\vec{\mathbf{r}}_{l,m}$ changes with the chain conformation, its amplitude and direction are constantly and randomly varying. Only after the conformational statistical average, $\vec{\mathbf{r}}_{l,m}^2$ becomes meaningful, i.e., the square distance of the particles is replaced by the mean square distance after the conformational statistical average, and $\vec{\mathbf{r}}_{l,m}^2$ in the above equation is simply replaced by $\langle \vec{\mathbf{r}}_{l,m}^2 \rangle$. Therefore, for flexible macromolecular chains,

$$P(q) = 1 - \frac{q^2 \langle R_g^2 \rangle}{3} + \cdots \qquad (5.27)$$

When the particle or scattering angle is small, if $q^2 \langle R_g^2 \rangle < 1$, the higher order term in eq. (5.27) can be omitted. This result is applicable to all shapes of scattering particles. When the particle or scattering angle is large, $q^2 \langle R_g^2 \rangle \geqslant 1$. In this case, it is necessary to analyze the particle structure with a specific model. For spherical particles with a radius of R and a uniform density, $P(q)$ has an analytical solution,

$$P_{\text{sphere}}(q) = \frac{9 \left[\sin(qR) - qR\cos(qR) \right]^2}{(qR)^6}$$

As qR increases, $P_{\text{sphere}}(q)$ decreases from one to a minimum value, and then gradually oscillation decay. For a rod with a length of $2L$, infinite or very thin, the analytical result of the structure factor contains a known integral,

$$P_{\text{rod}}(q) = \frac{\int_0^{2qL} \frac{\sin x}{x} dx}{qL} - \left[\frac{\sin(qL)}{qL} \right]^2$$

The $P(q)$ of the Gaussian chain also has an analytical result, called the Debye function, which is applicable to any $q^2 \langle R_g^2 \rangle$, no approximation is required.

$$P_{\text{Gaussian}}(q) = \frac{2}{q^2 \langle R_g^2 \rangle} \left[1 - \frac{1 - \exp(-q^2 \langle R_g^2 \rangle)}{q^2 \langle R_g^2 \rangle} \right]$$

When $q^2 \langle R_g^2 \rangle \gg 1$, $P_{\text{Gaussian}}(q)$ tends to $2/q^2 \langle R_g^2 \rangle$. One can plot "$q^2 \langle R_g^2 \rangle P(q)$ versus $q^2 \langle R_g^2 \rangle$", i.e., the Kratky plot, which is physically equivalent to a surface density on a spherical surface with a radius of qR_g. It can be seen from such a lot that when $q^2 \langle R_g^2 \rangle \gg 1$, $q^2 \langle R_g^2 \rangle P(q) \to 2$.

In the middle of 1990s, Walther Burchard [Macromolecules, 1997, 10:919] pointed out that a branched macromolecule obtained by a A − B (2 + 1) reaction of the monomer A_2B with three reactive groups has the following structure factor

$$P_{\text{branched}}(q) = \frac{1 + \dfrac{q^2 \langle R_g^2 \rangle}{3 n_B}}{\left[1 + \left(\dfrac{1 + n_B}{6 n_B} \right) q^2 \langle R_g^2 \rangle \right]^2}$$

Using the Kratky plot, both sides are multiplied by $q^2 \langle R_g^2 \rangle$ simultaneously, when $q^2 \langle R_g^2 \rangle \gg 1$, $P(q \to \infty) \to 12 \, n_B / (1 + n_B)^2$, which results in the average number of branch points (n_B) in the branched macromolecular chains, as shown on the right. The scattering factors of five different fractions (different molar masses) obtained from a sample of branched macromolecules overlap with each other, confirming that $P(q)$ is independent of the molar mass of the chain. In addition, as shown in eq. (5.27), when $qR_g < 1$, the structural factors of branched chains with

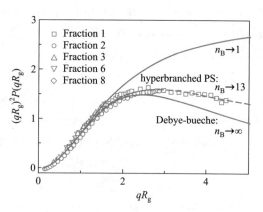

Li L W, He C, He W D, et al. Macromolecules, 2011,44:8195-8206.

(Figure 13, permission was granted by publisher)

different structures overlap with each other, so that it is no longer possible to distinguish the structural details of the macromolecules. Therefore, only when the angle or particle is sufficiently large, can the structural factor measured by a scattering experiment to derive the structural information. Also note that the particles or macromolecules should be relatively uniform. Otherwise, the fitting quality will be poor and the conclusion is not very certain.

When $q^2 \langle R_g^2 \rangle \to 0$, according to the definition, the above various scattering structure factors $P(q) \to 1$; when $q^2 \langle R_g^2 \rangle \ll 1$ or $qL \ll 1$, the above various scattering structure factors are reduced to eq. (5.27). For common polymer chains, experimentally, one can always find an angular range of $q^2 \langle R_g^2 \rangle < 1$; within this range, the time average scattered light intensities at a sufficient number of angles are measured, leading to a more accurate average radius of gyration. If there are N_p large particles inside the scattering volume (V), the Rayleigh ratio obtained from the time average scattered light intensity becomes

$$R(q) = \frac{I_s r^2}{I_0} = \frac{16 \pi^4 \alpha'^2 N_p}{\lambda_0^4 V} P(q) = \frac{16 \pi^4 \alpha'^2 N_p}{\lambda_0^4 V} \left[1 - \frac{q^2 \langle R_g^2 \rangle}{3} + \cdots \right] \qquad (5.28)$$

On the basis of the above discussion, for a uniform and pure substance, any scattering unit with a distance ($\vec{r}$) away from the scattering center has a corresponding scattering unit at $-\vec{r}$. The phase difference between their photoelectric fields at the detection plane is exactly π, so that there is an interference cancellation. Therefore, the scattered light intensity of a uniform and pure substance should be zero.

If a uniform, impurity-free optical glass or other transparent solid material contains some tiny defects, such as small bubbles (even if its inside is vacuum), the material scatters light, even strong, which is often misunderstood as the bubble scattering. The number of molecules in the bubble is very small, and there are no molecules in a vacuum, how can it scatter light? In fact, this is because the photoelectric field of the scattered light from each corresponding solid unit with an equal volume to the bubble cannot be cancelled out, that is, the scattering originates from a corresponding small solid unit. Physically, scattering originates from relative density fluctuations in the system. If the solid material itself is taken as a reference point and its density is defined as zero, then each bubble is **equivalent** to a solid particle with an equal volume and a **negative density**, but the bubble itself scatters no light.

On the other hand, when a laser beam is passing a pure gas or liquid, a faint scattered light column is still visible from the side. This is because under the agitation of thermal energy, the molecules in the gas and liquid keep moving randomly, causing a fluctuation of the internal physical properties (including density) with time and space around its equilibrium value. Even in a pure gas or liquid, its density – related molecular volume polarizability (α') always fluctuates around its average value $\langle \alpha' \rangle$, that is, $\alpha' = \langle \alpha' \rangle + \delta\alpha'$. Taking its average value as the reference zero point, each small fluctuation is **equivalent** to a small "particle", causing scattering, although the fluctuation itself does not generate a real "particle".

Following the previous discussion for large particles, pure liquid scattering can be derived more strictly, but here is to divide the scattering volume, not the particles, into N small volume units, $\alpha_0' = \alpha'/N = \langle \alpha_0' \rangle + \delta\alpha_0'$. In addition, taking the average volume polarizability $\langle \alpha_0' \rangle$ as a reference zero point, then $\alpha_0' = \delta\alpha_0'$. Finally, using the definition of Rayleigh ratio and "the average after the sum is equal to the sum after the average", one can statistically time average the randomly fluctuating $\delta\alpha_0'$ to obtain

$$R(q) = \frac{16\,\pi^4}{\lambda_0^4 V} \sum_{l=1}^{N} \sum_{m=1}^{N} \langle \delta\alpha_{0,l}' \delta\alpha_{0,m}' \rangle \exp(i\vec{\mathbf{q}} \cdot \vec{\mathbf{r}}_{l,m})$$

In the above equation, the time average can be divided into its own term $(l = m)$: $\langle \delta\alpha_{0,l}' \delta\alpha_{0,m}' \rangle = \langle (\delta\alpha_0')^2 \rangle$ and $\exp(i\vec{\mathbf{q}} \cdot \vec{\mathbf{r}}_{l,m}) = 1$; and the cross-term of two independent scattering units $(l \neq m)$: $\langle \delta\alpha_l' \delta\alpha_m' \rangle = \langle \delta\alpha_l' \rangle \langle \delta\alpha_m' \rangle = 0$, where "the product and the average can be exchanged with each other" and "the average of fluctuation of a random variable must be zero" were used. The result of the double addition of the above equation is N^2 times $\langle (\delta\alpha_0')^2 \rangle$, and $N\delta\alpha_0' = \delta\alpha'$, so that

$$R(q) = \frac{16\,\pi^4}{\lambda_0^4 V} \langle (\delta\alpha')^2 \rangle \tag{5.29}$$

The left side can be obtained from the measured scattered light intensity, incident light intensity and other instrument parameters. The above equation can be used to obtain the mean square density fluctuation of a pure liquid. The above equations about the light scattering of particles in a liquid can also be derived using the principle of thermodynamics. The details are as follows.

In a binary mixture, such as a dilute macromolecular solution or colloidal dispersion, α' depends not only on the density (ρ) of the dispersion medium, but also on the composition (concentration, C). Therefore, it is necessary to find the relationships between $\langle (\delta\alpha')^2 \rangle$ and the molar mass and other physical quantities. The dependence of the volume polarizability on the medium density and particle concentration can be examined first, that is,

$$\delta\alpha' = \left(\frac{\partial\alpha'}{\partial\rho}\right)_{T,p,C} \delta\rho + \left(\frac{\partial\alpha'}{\partial C}\right)_{T,p,\rho} \delta C$$

For a given density or concentration, the rate of change of α' with concentration or with density, every partial differential in the above equation, is a constant. Density and concentration fluctuate with time are independent of each other. After respectively squaring the two sides, the cross-terms on the right will disappear after the average, so that

$$\langle(\delta\alpha')^2\rangle = \left(\frac{\partial\alpha'}{\partial\rho}\right)^2_{T,p,C} \langle(\delta\rho)^2\rangle + \left(\frac{\partial\alpha'}{\partial C}\right)^2_{T,p,\rho} \langle(\delta C)^2\rangle. \qquad (5.30)$$

From an experimental point of view, the left side corresponds to the scattering of the **solution**; and the density fluctuation in the first term on the right corresponds to the scattering of the **solvent**; the difference between them is equal to the measurable time average scattered light intensity difference between the solution and the solvent, a measurable physical quantity. The second item on the right corresponds to the net time average scattered light intensity of macromolecules or colloidal particles. Its relation with physical quantities to be measured is important. Next, let us first find the changing rate of α' with concentration, $(\partial\alpha'/\partial C)_{T,p,\rho}$, and then calculate the mean square concentration fluctuation, $\langle(\delta C)^2\rangle$.

As early as 1850, the Italian physicist Ottaviano – Fabrizio Mossotti discovered the relationship between the ratio (ε_r), the relative permittivity, i.e., the dielectric constant of the permittivity (ε) to the permittivity of vacuum (ε_0) of a substance and the volume polarizability (α'). Later, the German physics and mathematician Rudolf Julius Emanuel Clausius also independently obtained this seemingly simple relationship in 1879. Therefore, it is named the Clausius–Mossotti relation,

$$\frac{\varepsilon_r - 1}{\varepsilon_r + 2} = \frac{4\pi\alpha'}{3}$$

It is identical to the Lorentz–Lorenz relationship between the refractive index (n) and the volume polarizability (α') independently discovered by Ludvig Lorenz in 1869 and later by Hendrik Lorentz in 1878, the same song played by different musicians, because $\varepsilon_r = n^2$, so that

$$\left(\frac{\partial\alpha'}{\partial C}\right)_{T,p,\rho} = \left(\frac{\partial\alpha'}{\partial\varepsilon_r}\right)_{T,p,\rho}\left(\frac{\partial\varepsilon_r}{\partial n}\right)_{T,p,\rho}\left(\frac{\partial n}{\partial C}\right)_{T,p,\rho} = \frac{9nV}{2\pi(\varepsilon_r + 2)^2}\left(\frac{\partial n}{\partial C}\right)_{T,p,\rho}$$

$$\left(\frac{\partial\alpha'}{\partial C}\right)_{T,p,\rho} \approx \frac{nV}{2\pi}\left(\frac{\partial n}{\partial C}\right)_{T,p,\rho} \qquad (5.31)$$

The approximation of the last step is based on that for a given scattering system, $99/(\varepsilon_r + 2)^2$ is a constant slightly less than 1. One can also set $\varepsilon_r + 2 \approx 3$ in the initial stage, and use n^2 instead of ε_r, that is, directly use the approximate Lorentz–Lorenz relationship, so that the same result can be directly obtained. Next, let us use the principle of thermodynamics to find $\langle(\delta C)^2\rangle$.

For given experimental conditions (temperature, pressure, concentration, ...), as the system reaches a macroscopically stable dynamic equilibrium state, free energy drops to the lowest G_0.

However, under the agitation of thermal energy, microscopically, the particles are still constantly moving, making the system to deviate from G_0 slightly, causing a rise in free energy, which results in an ever positive (increasing) fluctuation. Choose the particle concentration (C) as a fluctuation variable, one can do the Taylor expansion of G at the equilibrium concentration to get

$$\delta G = G - G_0 = \left(\frac{\partial G}{\partial C}\right)_{T,p,\rho} \delta C + \frac{1}{2}\left(\frac{\partial^2 G}{\partial C^2}\right)_{T,p,\rho} (\delta C)^2 + \cdots$$

where the partial differential of free energy at a given concentration C is constant; and the time average random fluctuation of concentration is zero. Therefore, after time averaging, the first term on the right disappears. If the cubic contribution of the concentration fluctuation is not considered, the above equation becomes

$$\langle \delta G \rangle = \frac{1}{2}\left(\frac{\partial^2 G}{\partial C^2}\right)_{T,p,\rho} \langle (\delta C)^2 \rangle$$

In addition, starting from fairly general statistical considerations (deduction is omitted), it can be known that under the agitation of thermal energy, for any physical variable (Y), its time average fluctuation ($\langle \delta Y \rangle$) corresponds to a change of half of the thermal energy ($k_B T/2$). Therefore, the time average fluctuation of the free energy of each molecule is $\langle \delta G \rangle = k_B T/2$. If k_B is replaced by the gas constant R, the time average fluctuation of free energy per mole of particles is

$$\langle (\delta C)^2 \rangle = RT \Big/ \left(\frac{\partial^2 G}{\partial C^2}\right)_{T,p,\rho} \tag{5.32}$$

According to Chapter 3 and the previous discussion on the osmotic pressure difference, the free energy of a solution includes the free energy of pure solvent, and the difference between the solution and the solvent is the mixing free energy, $\Delta G_{mix} = G - G^*$. By its definition, G^* is independent of the particle concentration, a constant. Therefore,

$$\left(\frac{\partial^2 \Delta G_{mix}}{\partial C^2}\right)_{T,p,\rho} = \left[\frac{\partial^2 (G-G^*)}{\partial C^2}\right]_{T,p,\rho} = \left(\frac{\partial^2 G}{\partial C^2}\right)_{T,p,\rho}$$

The partial differential of the mixing free energy with respect to C can be changed into a partial differential of the mixing free energy with respect to the molar number of solvent molecules (n_s) times the partial differential of n_s with respect to C, namely

$$\left(\frac{\partial^2 G}{\partial C^2}\right)_{T,p,\rho} = \left[\frac{\partial\left[\left(\frac{\partial \Delta G_{mix}}{\partial n_s}\right)_{T,p,\rho}\left(\frac{\partial n_s}{\partial C}\right)_{T,p,\rho}\right]}{\partial C}\right]_{T,p,\rho}$$

According to the definition, the partial differential of the mixing free energy with respect to n_s is the difference between the chemical potentials of the solution and the solvent ($\Delta \mu$), and

$$\left(\frac{\partial n_s}{\partial C}\right)_{T,p,\rho} = \left(\frac{\partial n_s}{\partial X_p}\right)_{T,p,\rho}\left(\frac{\partial X_p}{\partial C}\right)_{T,p,\rho} = -\frac{n_s}{X_s X_p}v_p = -\frac{V}{V_{s,m}C}$$

In the above derivation, the following four relations, obtained in the previous discussion of the osmotic pressure difference, are used: $(\partial X_p/\partial n_s)_{T,p,C} = -X_s X_p/n_s$, $X_p = C v_p$, $X_s = V_s/V$ and

$V_{s,m} = V_s/n_s$ Note that C is the particle concentration. For a given scattering experiment, it is a fixed value. Substituting the above equation into the equation before results in

$$\left(\frac{\partial^2 G}{\partial C^2}\right)_{T,p,\rho} = -\frac{V}{V_{s,m}C}\left(\frac{\partial \Delta\mu}{\partial C}\right)_{T,p,\rho} = -\frac{V}{V_{s,m}C}\left(\frac{\partial \Delta\mu}{\partial \pi}\right)_{T,p,\rho}\left(\frac{\partial \pi}{\partial C}\right)_{T,p,\rho} = \frac{V}{C}\left(\frac{\partial \pi}{\partial C}\right)_{T,p,\rho}$$

The last step of derivation uses eq. (5.12) obtained in the discussion of osmotic pressure difference, i. e., in a pure solvent, the chemical potential increases with the pressure, $\Delta\mu = \mu^*(p + \pi) - \mu^*(p) = V_m p$ and $p = -\pi$. In addition, in the discussion of the osmotic pressure difference, it is known how it changes with the particle concentration, that is, eq. (5.19)

$$\pi \cong \frac{RTC}{M}(1 + A_2 C)$$

Calculating $(\partial \pi/\partial C)_{T,p,C}$, one can get $(\partial^2 G/\partial C^2)_{T,p,C}$. Substituting it into eq. (5.32) result in

$$\langle (\delta C)^2 \rangle = \frac{CM}{V(1 + 2A_2 C)} \tag{5.33}$$

Substituting eq. (5.31) and (5.33) into eq. (5.30), and then substituting eq. (5.30) into eq. (5.29), one can get the following result after subtracting the scattered light intensity from solvent,

$$R(q) = R_{\text{solution}}(q) - R_{\text{solvent}}(q) = \frac{16\pi^4}{\lambda_0^4 V}\left(\frac{\partial \alpha'}{\partial C}\right)_{T,p,\rho}^2 \langle (\delta C)^2 \rangle$$

$$= \frac{16\pi^4}{\lambda_0^4 V}\left(\frac{nV}{2\pi}\right)^2\left(\frac{\partial n}{\partial C}\right)_{T,p,\rho}^2 \frac{CM}{V(1 + 2A_2 C)} = \frac{4\pi^2 n^2}{\lambda_0^4}\left(\frac{\partial n}{\partial C}\right)_{T,p,\rho}^2 \frac{CM}{(1 + 2A_2 C)}$$

where $(\partial n/\partial C)_{T,p,\rho}$ is a differential refractive index that is measurable independently using a differential refractometer (will be introduced later in details). Define

$$K = \frac{4\pi^2 n^2}{\lambda_0^4}\left(\frac{\partial n}{\partial C}\right)_{T,p,\rho}^2$$

For larger particles, the structure factor $P(q)$ has to be considered. When $q^2\langle R_g^2 \rangle < 1$, $R(q)$ can be rewritten as

$$\frac{KC}{R(q)} = \frac{(1 + 2A_2 C)}{MP(q)} \approx \frac{1}{M}\left(1 + \frac{q^2\langle R_g^2 \rangle}{3} + 2A_2 C \cdots\right)$$

For a polydisperse particle or macromolecule sample, the above equation can be rewritten as

$$\frac{K}{(1 + 2A_2 C)}\sum_i C_i M_i\left(1 - \frac{q^2\langle R_g^2 \rangle_i}{3} + \cdots\right) = \frac{K}{(1 + 2A_2 C)}\left(\sum_i C_i M_i - \frac{q^2}{3}\sum_i C_i M_i\langle R_g^2 \rangle_i + \cdots\right)$$

$$= \frac{KC}{(1 + 2A_2 C)}\frac{\sum_i C_i M_i}{\sum_i C_i}\left(1 - \frac{q^2}{3}\frac{\sum_i C_i M_i\langle R_g^2 \rangle_i}{\sum_i C_i M_i} + \cdots\right) = \sum_i R_i(q) = R(q)$$

In the last step of the calculation, the numerator and denominator on the left are multiplied by C at the same time. According to $C = \sum_i C_i$ and the definition of the weight average molar mass, the above equation can be rewritten into an extremely useful form in experiments as follows.

$$\frac{KC}{R(q)} \approx \frac{1}{M_w}\left(1 + \frac{q^2\langle R_g^2\rangle_z}{3} + 2A_2C\cdots\right) \tag{5.34}$$

In the previous discussion, it is assumed that different particles are uncorrelated and independent of each other. The time average of the product of the photoelectric field of the scattered light is equal to the product of their respective time averages. The result of each time average is zero, so they do not affect the scattered light intensity. However, when the particle concentration is large or there is a strong interaction between the particles, the interference between the photoelectric fields of the scattered light of two units originating from different particles must be further considered. In other words, for such a large particle system, it is necessary to consider the double interference between the photoelectric fields of the scattered light from inside and outside the particle. The result will be the structural factor or scattering function of the solution, including the interference between the photoelectric fields of the scattered light of two different units within a particle and two units between different particles. There are N units (chain segments) in a particle, and there are N_p particles or macromolecular chains in the scattering volume. Therefore, the aforementioned particle structure factor becomes the structure factor of the solution. Its general mathematical expression is as follows,

$$P(q) = \frac{1}{(NN_p)^2}\sum_{j=1}^{N_p}\sum_{k=1}^{N_p}\sum_{l=1}^{N}\sum_{m=1}^{N}\exp[i\,\vec{q}\cdot(\vec{r}_{j,l} - \vec{r}_{k,m})]$$

$$= \frac{1}{(NN_p)^2}\left\{\begin{array}{l} N_p\sum_{l=1}^{N}\sum_{m=1}^{N}\exp[i\,\vec{q}\cdot(\vec{r}_{j,l} - \vec{r}_{k,m})] + \\ \sum_{j\neq k=1}^{N_p}\sum_{l=1}^{N}\sum_{m=1}^{N}\exp[i\,\vec{q}\cdot(\vec{r}_{j,l} - \vec{r}_{k,m})] \end{array}\right\}$$

Readers should not be intimidated by this equation. Physically, it is not that difficult. The first item in the big braces reflects the static structure factor of N_p particles; the second item represents the static structure factor between all different particles. When $q \to 0$, $P(q) \to (N^2 N_p + N^2 N_p(N_p - 1))/(NN_p)^2 = 1$.

The structure factor is related to the density of the local scattering unit (segment) in the liquid, defined as $(\vec{r}) = \delta(\vec{r} - \vec{r}_{j,l})$, where the vector $\vec{r}_{j,l}$ represents the position of the lth unit (segment) of the jth particle (macromolecular chain). The sum of the entire scattering volume (V) leads to the macroscopic average density of the liquid (ρ_0). For a given liquid, it is a constant, that is

$$\rho_0 = \langle\rho(\vec{r})\rangle = \frac{1}{V}\sum_{j=1}^{N_p}\sum_{l=1}^{N}\rho(\vec{r}) = \frac{1}{V}\sum_{j=1}^{N_p}\sum_{l=1}^{N}\delta(\vec{r} - \vec{r}_{j,l}) = \frac{NN_p}{V}$$

The random fluctuation of density away from its average value, $\Delta\rho(\vec{r}) = \rho(\vec{r}) - \rho_0$, must be zero. However, in this way, it is impossible to distinguish two systems with completely different fluctuation amplitudes (affecting ranges). As mentioned earlier, the method of the first square and then an average (mean square) can be used. Specific to the density fluctuation amplitude in a system, the statistical solution is to find the distance density-density autocorrelation function, i.e.,

$$\langle \Delta\rho(\vec{r})\Delta\rho(\mathbf{0})\rangle = \langle \rho(\vec{r})\rho(\mathbf{0})\rangle - \rho_0^2 \quad \rightarrow \quad \frac{\langle \Delta\rho(\vec{r})\Delta\rho(\mathbf{0})\rangle}{\rho_0^2} = \frac{\langle \rho(\vec{r})\rho(\mathbf{0})\rangle}{\rho_0^2} - 1$$

where $\vec{r} = \vec{r}_j - \vec{r}_k$, the distance vector between any two points; the average represents the sum of the density products of all any two points in the system. For an isotropic macroscopic uniform system, the vector can be replaced by a scalar, that is, $\langle \rho(\vec{r})\rho(0)\rangle = \langle \rho(r)\rho(0)\rangle$. The calculation of $\langle \rho(r)\rho(0)\rangle$ includes the following steps. Any point in the system is selected as the first point with a density of $\rho(0)$ first, and then another point at a distance r with a density of $\rho(r)$ is chosen as the second point. The two densities are multiplied first, and then all the second points at a distance of r are summed. Afterwards, another first point is chosen arbitrarily and the operation is repeated until all points in the system are selected as the first point. Due to the random density fluctuation, $\rho(0)$ and $\rho(r)$ can be higher or lower than the macroscopic average density (ρ_0), so that the difference between them and ρ_0 can also be positive or negative.

When $r = 0$, $\langle \rho(r)\rho(0)\rangle = \langle \rho(0)^2\rangle$, the mean square density; it can be proved that $\langle \rho(0)^2\rangle = 2\rho_0^2$. At this time, $\Delta\rho(r)$ and $\Delta\rho(0)$ are 100% correlated. When $r\rightarrow\infty$, $\Delta\rho(r)$ and $\Delta\rho(0)$ should become uncorrelated. In other words, the average of their product is equal to the product of their respective averages, $\langle \Delta\rho(r\rightarrow\infty)\Delta\rho(0)\rangle = \langle \Delta\rho(r\rightarrow\infty)\rangle\langle \Delta\rho(0)\rangle = 0$, a zero degree of correlation. As r increases, $\langle \rho(r)\rho(0)\rangle$ gradually decreases from 100% correlation to 100% uncorrelation. There is a characteristic relaxation length (ξ), reflecting the average reachable distance of the density fluctuation, which corresponds to the average radius of gyration of the particles. Strictly speaking, $\xi^2 = \langle R_g^2\rangle/3$, about 1.7 times the average radius of gyration.

It can be proved that the Fourier transform of the distance density autocorrelation function of the unit (chain segment) is the static structure factor of the system,

$$P(q) = \frac{1}{V\rho_0^2}\int\langle \rho(r)\rho(0)\rangle\exp(i\vec{q}\cdot\vec{r})\mathrm{d}\vec{r} \quad \text{and} \quad \langle \rho(r)\rho(0)\rangle = \frac{V\rho_0^2}{(2\pi)^3}\int P(q)\exp(-i\vec{q}\cdot\vec{r})\mathrm{d}\vec{q}$$

In other words, $\rho(r)$ in the lattice space and $P(q)$ in the reciprocal lattice space are conjugate to each other. For a system with a uniform density and no fluctuations, $\langle \rho(r)\rho(0)\rangle = \langle [\rho(0)]^2\rangle = \rho_0^2$. Therefore, $P(q) = \delta(\vec{q})$, i.e., the system does not scatter except in the forward direction. Regarding this point, a qualitative explanation has been given previously from the interference cancellation. This is just a further theoretical clarification. When discussing dynamic laser light scattering in the next section, another correlation function will be encountered.

Modern LLS spectrometer all use polarized, monochromatic and coherent laser light sources, and the polarization direction is generally set to be perpendicular to the optical platform. In addition, a vertical polarizer is also installed in front of the detector, that is, only the scattered light with the same polarization direction as the incident light is detected, so that $R(q)$ is often written as $R_{vv}(q)$, where the subscript "vv" denotes the vertically polarized incident light and the vertically polarized scattered light. If one does not pursue a complete derivation based on electrodynamics, most students in the chemistry department or other departments should be able to understand the above discussion. It is relatively easy to master the basic principle of classical LLS. The experiment of classical LLS seems simple, but it is rather difficult.

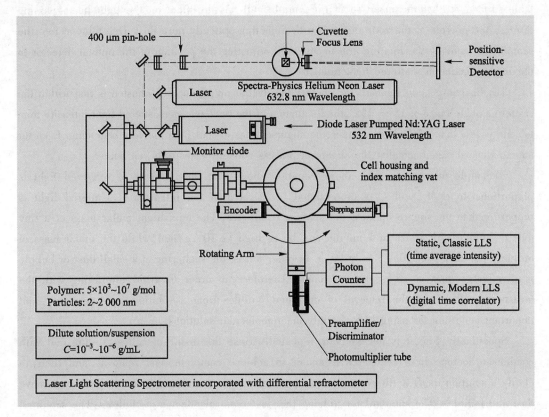

Laser Light Scattering Spectrometer incorporated with differential refractometer

The figure above shows a plan view of a modern LLS instrument, which includes two lasers with two wavelengths (HeNe: 633 nm and Nd: YAG: 532 nm) and a differential refractometer based on a new design (will be discussed in the next section in detail). The figure show that two mirrors are used to fold the laser 180° back, one function is to reduce the length of optical platform; the other is for conveniently adjusting height and direction of the laser without moving the laser much. The measurable range of macromolecules and colloidal particles of modern LLS instruments is also listed. Readers should pay attention to some following specific key points when using LLS to study a macromolecular solution or colloidal dispersion.

1) Instrument adjustment: First, adjusting the incident laser beam parallel to an optical table surface. Inserting a lens with a focal length of 30-50 cm to focus the beam to pass through the center of rotation with a diameter of ~200 μm and make it still parallel to the table surface with an angle between the beam and the table surface less than 4 microradians (i.e., the deviation is less than 4 mm per kilometer). The beam passes through the detection hole at the zero scattering angle. Second, adjusting the rotation plane parallel to the optical table surface so that the end of the rotation arm (~50 cm long) moves up and down less than 2 μm (note: the hair diameter is ~ 60 μm) when in the angular range of 10° - 155°. Third, adjusting the

refractive index match cell (~ 10 cm in diameter) and the sample cell (1. 0 – 2. 0 cm in diameter) to make their rotation centers coincide with the rotation axis and perpendicular to the table surface, so that the insertion of the sample cell has no effect on the light beam passing through the pinhole at the zero angle. Fourth, injecting optically pure dust–free toluene into the sample cell; moving the rotating arm to 90°; and adjusting the position of the optical detector to obtain the maximum scattered light intensity.

The final inspection standard for whether the instrument is properly adjusted is that within the scattering angle range of 15° – 155°, the fluctuation of the time average scattered light intensity from toluene is less than ±1%. If it is larger, the alignment has to start from the very beginning. Even for an experienced experimentalist, the alignment process normally takes days, not hours.

2) Sample preparation: The above equation shows that the intensity of scattered light is proportional to $C_i M_i$, and C_i is proportional to M_i, so that the intensity of scattered light is proportional to the square mass of the scattered particles. The equivalent molar mass of a tiny dust or bubble with a size of 1 micron can be as high as 10^{12} g/mol, while the molar mass of ordinary polymer molecules is only 10^5 g. In other words, the scattering of a small dust or bubble is effectively equal to 10^{14} polymer chains. Therefore, in order to avoid interfering with the measuremental results, the removal of dust and bubbles from a solution is a necessary and important condition for using LLS to study macromolecular solutions.

Specifically, first, it is necessary to repeatedly rinse the inside of each sactering cell with condensed acetone in a sealed environment to achieve complete dust removal, and then, to clarify a solution using a filter membrane (generally, the pore size is 0. 2 microns) to remove dust and bubbles. The standard for judging the dust removal effect is as follows: The scattered light intensity at 15° should be less than ±5% within 10 – 15 minutes; if the scattered light intensity jumps during this period, it means that the dust is not removed properly so that the solution has to be clarified again. The removal of dust from aqueous solutions is particularly difficult. Patience is a necessary condition, and special skills and experience are required.

Although there are many difficulties in static light scattering experiments, the quantitative study of macromolecules often requires the determination of the absolute molar mass of a series of samples. Therefore, the ancient and classic static light scattering is still an indispensable and valuable instrument to conduct quantitative research in macromolecular laboratories. Static light scattering, membrane osmotic pressure difference, and ultracentrifuge constitute a group of characterization methods for the determination of the absolute molar mass of macromolecules with no requirement of a calibration curve.

For a given scattering angle, the difference between the Rayleigh ratio of the solution and the solvent, namely $R(q)$, is not obtained from the intensities of the incident laser light (I_0) and the scattered light (I) and the distance between the detection plane and the scattering center (r). Instead, a standard sample with a known Rayleigh ratio (R_0), e. g. , R_0 (toluene at 25 ℃) = 2. 698×10^{-5} cm^{-1}. Using this value, R_0(toluene) at other temperatures or conditions can also be obtained from the relative change of the scattered light intensity. The molecules of standard samples and pure solvents are very small, so that their scattered light intensity is independent of the scattering angle. Experimentally, the intensity of their scattered light at each angle is still measured to compensate for small differences of the instrument at different scattering angles. By measuring the scattered light intensity ($I_0(q)$, $I^*(q)$ and $I(q)$) of the standard sample, solvent and solution at different scattering angles, one can use the definition of the excessive scattering (the difference between the Rayleigh ratio of solution and solvent) in the above equation to calculate $R(q)$, i.e.,

$$R(q) = R_0(q) \frac{I(q) - I^*(q)}{I_0(q)} \left(\frac{n}{n_0} \right)^{\gamma} \tag{5.35}$$

where n and n_0 are the refractive index of the solution and standard sample, respectively. The last factor is introduced to correct the difference in the scattering volume "seen (measured)" by the detector due to refraction. If a one–dimensional slit is used to intercept the scattering volume, only one–dimensional correction is required, $\gamma = 1$; if a circular pinhole smaller than the focused beam diameter (200 microns) is used to obtain the scattering volume, the two–dimensional correction is required, $\gamma = 2$. Experimentally, it is often to use a macromolecule with a known molar mass to determine the value of γ reversibly. For each given optical set up, only one measurement is required.

According to eq. (5.33), $R(q)$ is a curved surface that depends on q and C. In the 1940s and 1950s when computers were not available for most of researchers, it was not easy to draw on a curved surface, nor was it possible to extrapolate $R(q)$ to the zero angle and zero concentration on a three–dimensional curved surface to obtain the required physical quantity. The Zimm plot, still used nowadays, has successfully solved this problem, as shown on the right; namely, introducing a proper constant k, and then using ($q^2 + kC$) as the horizontal

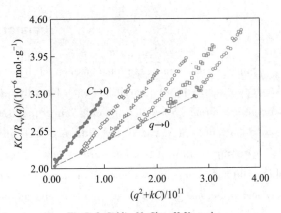

Wu P Q, Siddiq M, Chen H Y, et al.
Macromolecules, 1996, 29:277.

(Figure 1, permission was granted by publisher)

axis, which is equivalent to stretching and projecting the original curved surface onto a plane. When the scattering angle and the macromolecular concentration are sufficiently small, for each given concentration, the q^2 dependence of $KC/R(q)$ is a straight line, after extrapolating to the zero scattering angle, the slope and intercept correspond to the second virial coefficient (A_2) and the weight average molar mass (M_w), respectively. On the other hand, for each given angle, after extrapolating to the zero concentration, a straight line is also obtained, and the slope and intercept respectively correspond to the z–average radius of gyration ($\langle R_g^2 \rangle_z$) and the weight average molar mass. Theoretically, the extrapolations of the two straight lines should intersect at one point, and such obtained two M_w are the same, which is also used as a criterion to check the experimental quality and the data processing.

When using the Zimm plot, it is very important to choose a proper value of k. If too small, the data of different concentrations are not separated; if too large, the data of the same concentration but different angles are compressed into a nearly vertical line, looking like to lose the angular dependence. In the early days, this was a very time-consuming task, depending on experience. Nowadays, with the aid of computer graphics, the data processing is no longer a problem. However, when reporting data, the appropriateness of a graphics will directly affect the viewer's perception. The picture above can serve as a good example.

After using laser light scattering to measure the z-average radius of gyration, second virial coefficient, and weight average molar mass of a series of narrowly distributed macromolecules with different molar masses, one can double-logarithmically plot the average radius of gyration and the second virial coefficient against the weight average molar mass to obtain different prefactors and scaling exponents; namely

$$R_g = (\langle R_g^2 \rangle_z)^{1/2} = k_{R_g} M_w^{\alpha_{R_g}} \quad \text{and} \quad A_2 = k_{A_2} M_w^{\alpha_{A_2}}$$

where $1/3 \leqslant \alpha_{R_g} \leqslant 3/5$ and $\alpha_{A_2} = 1/4$ for linear chains. However, the measurement error of A_2 can be as high as $\pm 20\%$.

Summary

Since ancient times, the phenomenon of light scattering has accompanied human beings. About two hundred years ago, it has been formally questioned "Why is the sky blue?". In earlier farming era, farmers knew that it would rain if they saw the bright red sunrise in the morning, and it would be sunny in the next day if there is a beautiful reddish sunset. It is precisely because of the scattering of sunlight that our world is not so "black and white". The scattered light of the sun shines through the windows and illuminates the room. However, a true understanding of the physical principles of light scattering only began in 1881, after the advent of electromagnetic theory. It is even later to use the principle of light scattering to measure the microscopic properties of particles and macromolecules. Nowadays, laser light scattering has become an indispensable and important experimental method in studying colloids and macromolecules.

Basic Principles: The visible light is an alternating high-frequency ($\sim 10^{15}$ Hz) electromagnetic wave, whose energy is insufficient to cause the heavier nuclei of colloidal particles or macromolecules to move at such a high frequency. Only lighter electrons can oscillate under this high-frequency electric field to produce an induced dipole moment. Its size and direction are constantly changing, and emitting the bremsstrahlung radiation to its surrounding (a spherical surface). The scattered light intensity received by a given detecting area at a distance of r is inversely proportional to r^2. It is also not difficult to understand that the scattered light intensity is proportional to the incident light intensity and to the number of electrons in the particle volume. For a given particle, the number of electrons is proportional to its volume or mass. This is the physical basis that the molar mass of the particles is obtainable from the intensity of scattered light.

When the scattering angle is zero, the scattered light of all units in the system has the same phase; the interference increases the scattered light intensity, reaching the maximum. For a polydisperse system, the time average scattered light intensity is proportional to the weight average molar mass when the scattering angle is zero. For a given non-zero scattering angle, even if each light ray in the incident laser beam as a coherent light source has the same initial phase, due to different spatial positions of the units inside a large particle, the photoelectric fields of the scattered light reach the detector with different phases, resulting in the interference cancellation, so that the time average scattered light intensity is weakened, and the degree of weakening is related to the average radius of gyration of the particles and the scattering vector. In addition, the excluded volume effect in a good solvent also weakens the intensity of scattered light.

Measurable Physical Quantities: First, the time-average scattered light intensities of the standard sample (toluene) at different scattering angles (θ) are measured, which is multiplied by $\sin\theta$ to correct the scattering volume. The variation of the time-average scattered light intensity with the scattering angle should be less than ± 1%. For a given particle or macromolecule concentration, the time average scattered light intensity of the colloidal dispersion or macromolecule solution, dispersion medium or solvent at different scattering angles are measured, making the scattering volume correction. The above experiments are repeated for different concentrations.

Data Processing: According to eq. (5.35), the measured time average scattered light intensity are used to calculate the Rayleigh ratio at different concentrations and different angles first, and then use the Zimm plot to respectively extrapolate the data to the zero scattering angle and the zero concentration to obtain two straight lines. The z-average radius of gyration and the second virial coefficient of the particle or macromolecular chain are obtainable from the slopes of these two lines; and the intercepts lead to the weight average molar mass of the particle or macromolecular chain. For samples with a narrow distribution, using different plots, the shape and conformation of the scattering particles are also obtainable from the angular dependence of the time average scattered light intensity.

Dynamic Laser Light Scattering: After using laser as a light source, modern dynamic laser light scattering was developed. As its name implies, this method measures the scattered light intensity dynamically, rather than statically measuring the time-average scattered light intensity. Its principle based on the fact that laser is a monochromatic coherent light source, each of whose rays has the same initial phase. Each scattered particle (unit) is at a different spatial position, similar to different parts of the aforementioned large particle. The respective photoelectric fields of the scattered light have gone through different optical paths, so that their phases after reaching the detection plane are completely different. Depending on the phase difference, their interference with each other can enhance or weaken the intensity of scattered light, forming countless small spots of alternating light and dark on the detection plane, as shown on the right. Since the independent particles in the scattering volume are constantly randomly moving under the agitation of thermal energy, the positions where these small spots appear are also constantly changing. If staring at a point and observe it, one will find that the point is flickering constantly and randomly, with a frequency of up to 10^6 Hz.

If the photosensitive element used in the detection covers many small spots, although each point is flashing randomly, the detection itself is equivalent to taking an average of many small spots, so that only the an ensemble (sample) average scattered light intensity is measured. On the other hand, even if the photosensitive element is small enough to cover only a few spots, but not sufficiently sensitive and fast, the measurement is only a time average scattered light intensity because each spot has flickered thousands of times during the measurement time. According to the theorem of statistics, the ensemble average of a random variable is equal to the time average. Besides a strong monochromatic, coherent laser light source, in the dynamic measurement of the scattered light intensity, the detection element must have the following characteristics. First, the photosensitive area should be small to cover only a few spots, and ideally only one small spot, and able to effectively track the fluctuation of light intensity. Second, the detector has low noise, high sensitivity, and an ability to measure extremely weak signals on a tiny photosensitive area. Third, the sampling time is sufficiently short, less than 20 nanoseconds, to resolve and track the fluctuation of the scattered light intensity one million times per second.

Every ray in an ordinary mercury or xenon lamp, even monochromatic after the grating, they are slightly different in wavelength, have different initial phases and ever-changing intensity, leading to the interference between the scattered light from different rays. Only after laser came out in the 1960s, the dynamic measurement of the scattered light intensity became a reality. In the last few decades, with the fast developments of fast single-photon detectors (photomultiplier tubes, nowadays avalanche photodiodes), electronic amplifiers, discriminators, and computers, dynamic laser light scattering was gradually perfect in theory in the 1960s, instrumentally developed in the 1970s, matured in the 1980s, and finally entered the common macromolecular laboratories in the 1990s. It has become a common instrument. Corresponding to the "classical" light scattering developed more than a hundred years ago, it is also called as the "modern" light scattering.

From another point of view, the Doppler effect can also be used to describe dynamic laser light scattering. When a beam of laser with a single circular frequency (ω) is elastically scattered by a particle, if the particle does not move, the detected scattered light frequency is still ω. However, each particle moves continuously and randomly under the agitation of thermal energy. According to the Doppler Effect, when the particle moves towards or

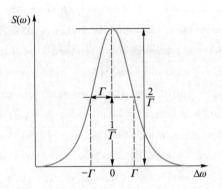

away from the detector, the measured scattered light frequency will increase or decrease slightly ($\Delta\omega$), and its amplitude increases with the speed of movement. The movement of a particle in any direction can be decomposed into two components parallel and perpendicular to the detection direction. The perpendicular movement has no effect on the frequency of scattered light. Due to the random direction and variable speed of the particle movements, the circular frequency of the scattered light even from the incident light with a single circular frequency becomes a symmetrical Lorentz distribution centered at the incident circular frequency ($\Delta\omega = 0$), as shown on the right. Its specific mathematical expression is as follows.

$$S(\omega) = \frac{2\Gamma}{\Gamma^2 + (\Delta\omega)^2}$$

where the peak height is $2/\Gamma$. When $\Delta\omega = \pm\Gamma$, the peak height is reduced to half, as shown in the figure on the right. $S(\omega)$ is a power distribution function (related to the scattered light intensity). Γ is often referred to as the linewidth. The linewidth increases with the movement speed of the particle. For given experimental conditions (such as liquid temperature and viscosity), the smaller the particle, the faster the movement. If only considering the translational diffusive motion, the linewidth is related indirectly to the translational diffusion coefficient (D, discussed in the section of ultracentrifugation). For polydisperse systems, one particle size corresponds to one Lorentz distribution with one linewidth (Γ). The area under the peak corresponds to the number of particles of this size, that is, the weighting in the system. The actually measured is a distribution (peak) composed of the intensity-weighted Lorentz distributions, respectively, corresponding to particles with different sizes in a system. The measurement process is equivalent to an integration, its linewidth corresponds to the intensity average linewidth ($\langle\Gamma\rangle$), and the weighting is an intensity distribution of linewidth, $G(\Gamma)$; namely,

$$\langle\Gamma\rangle = \int_0^\infty \Gamma G(\Gamma)\,\mathrm{d}\Gamma$$

Due to small changes in the frequency of scattered light, dynamic LLS is not strictly the elastic scattering, so physicists engaged in the elastic scattering prefer to call it "quasi-elastic scattering". The laser frequency in the visible range is $\sim 10^{15}$ Hz, and the frequency change (linewidth) caused by particle motion is $\sim 10^6$ Hz. The relative change is only about one part per billion. Theoretically, one can directly use a Fabry−Perot interferometer to measure the power (scattered light intensity) at each frequency in the frequency space. The average linewidth is then calculated from the measured power distribution. For small particles, the movement is faster and Γ is larger, which is barely feasible. Due to the experimental difficulties, historical literature in this area is rather limited. For the slowly moving macromolecules, it is impossible to measure the linewidth in the frequency space directly. Like other spectra, with no choice in the early days, one can only measure spectrum in the frequency space.

In mathematics and physics, frequency and time are conjugate. Functions in one space can be Fourier transformed to another space. Therefore, measurements that are difficult to achieve in the frequency space can be converted to the time space to do. With the developments of lasers, fast electronics, and computers, most of the measurements previously in the frequency space have now been Fourier transferred to the time space to operate. A (delay) time correlation function is measured first, and then the power (light intensity) distribution in the frequency space is calculated by the Fourier inversion, including Fourier infrared, Raman and other spectrometers.

As early as 1930, Norbert Wiener proved the theorem that "power and time related functions are a pair of Fourier conjugated functions" for differentiable deterministic functions. Aleksandr Khintchine then obtained a similar probability result for a stationary random process in 1934. Albert Einstein explained this idea in a two−page memo earlier in 1914. Mathematically, it can be strictly proved that the Lorentz power distribution function $[S(\omega)]$ and "time autocorrelation function of the photoelectric field of the scattered light" $[g^{(1)}(\tau) = \langle \vec{\mathbf{E}}_s^*(t)\vec{\mathbf{E}}_s(t+\tau)\rangle]$ is a pair of Fourier conjugate functions. In dynamic LLS, $S(\omega)$ is desired physical quantity. The measurement of $g^{(1)}(\tau)$ is a process of integrating $S(\omega)$, while the inversion is just to calculate $S(\omega)$ from the measured $g^{(1)}(\tau)$. The specific mathematical expression is as follows,

Experimental measurement   Inversion calculation

$$\langle \vec{\mathbf{E}}_s^*(t)\vec{\mathbf{E}}_s(t+\tau)\rangle = \frac{1}{2\pi}\int_{-\infty}^{\infty} S(\omega)\mathrm{e}^{i\omega\tau}\mathrm{d}\omega \qquad S(\omega) = \int_{-\infty}^{\infty} \langle \vec{\mathbf{E}}_s^*(t)\vec{\mathbf{E}}_s(t+\tau)\rangle \mathrm{e}^{-i\omega\tau}\mathrm{d}\tau$$

where $\vec{\mathbf{E}}_s^*(t) = \vec{\mathbf{E}}_0 \exp[-i(\omega t - \phi(t))]$ and $\vec{\mathbf{E}}_s(t+\tau) = \vec{\mathbf{E}}_0 \exp[i(\omega(t+\tau)-\phi(t+\tau))]$, τ is the delay time. Assuming that there are N_p particles in the scattering volume, the photoelectric field of the scattered light is the vector sum of N_p photoelectric fields. The photoelectric field of the scattered light from each particle has its own phase (ϕ) when it reaches the detection plane. Following the previous discussion on the scattering of N_p independent small particles, or considering the scattering volume as a "large particle", then N_p "particles" an be regarded as N_p independent scattering units in the "large particle". The time autocorrelation function of the photoelectric field of the scattered light is

$$\langle \vec{\mathbf{E}}_s^*(t) \vec{\mathbf{E}}_s(t+\tau) \rangle = |\vec{\mathbf{E}}_0|^2 \langle [\sum_{i,j=1}^{N_p} \exp[i(\phi_{s,i}(t) - \phi_{s,j}(t+\tau))]] \rangle$$

The normalization of the above equation by $\langle \vec{\mathbf{E}}_s^*(t) \vec{\mathbf{E}}_s(t) \rangle$ that is proportional to the average scattered light intensity results in

$$g^{(1)}(\tau) = \frac{\langle \vec{\mathbf{E}}_s^*(0) \vec{\mathbf{E}}_s(\tau) \rangle}{\langle \vec{\mathbf{E}}_s^*(0) \vec{\mathbf{E}}_s(0) \rangle} = \frac{\langle \sum_{i,j=1}^{N_p} \exp[i(\phi_{s,i}(0) - \phi_{s,j}(\tau))] \rangle}{\langle \sum_{i,j=1}^{N_p} \exp[i(\phi_{s,i}(0) - \phi_{s,j}(0))] \rangle}$$

For a steady-state system, there only exists a delay time (τ) from one time (t) to another time $(t+\tau)$, independent of the starting point (t), so that $t = 0$ can be defined. Each summation has N_p^2 terms, including N_p self-terms $(i = j)$ and $(N_p^2 - N_p)$ cross-terms $(i \neq j)$. When the independent particles move randomly, they do not correlate with each other and affect each other, so that the time average of all cross-terms is zero, only N_p self-terms remain. The above equation becomes

$$g^{(1)}(\tau) = \frac{\langle \vec{\mathbf{E}}_s^*(0) \vec{\mathbf{E}}_s(\tau) \rangle}{\langle \vec{\mathbf{E}}_s^*(0) \vec{\mathbf{E}}_s(0) \rangle} = \langle \exp[i(\phi_{s,j}(0) - \phi_{s,j}(\tau))] \rangle$$

The numerator of the above equation has N_p identical items, and each summation in the denominator is 1, so that the sum is N_p,. The upper and lower N_p cancel each other out. The statistical theorem "the time average after the sum is equal to the sum after the time average" is used. $\langle \vec{\mathbf{E}}_s^*(0) \vec{\mathbf{E}}_s(\tau) \rangle / N_p$ is also called dynamic structure factor, denoted as $P(q,\tau)$; $g^{(1)}(\tau) = P(q,\tau)/P(q,0)$, where $P(q,0) = 1$.

If any one of the following conditions is satisfied, the above equation becomes identical to the scattering of N_p small particles in static LLS. First, no delay time $(\tau = 0)$, each item in the summation is 1, and it is still 1 after the time average, so the summing result is N_p times the scattering of a small particle. Second, the particles are all static, $\phi_{s,j}(0) = \phi_{s,j}(\tau)$, the time average of each item in the summation is also 1, going back to the same scattering result of the aforementioned N_p small particles. In a LLs experiment, under the agitation of thermal energy, each particle moves continuously and randomly in the scattering volume. Even after a very short delay time τ, the particles have left their original positions and moved randomly to other places. $\phi_{s,j}(\tau)$ is related to the particle position. Therefore, generally speaking, $\phi_{s,j}(\tau) \neq \phi_{s,j}(0)$.

Let us consider the time autocorrelation function of the photoelectric field of the scattered light from another point of view. When $\tau = 0$, no matter how fast each particle moves, each item in the summation is 1, and the summation result is N_p, $g^{(1)}(\tau) = 1$, 100% correlated. When $\tau \to \infty$, the randomly moving particles can reach any position in the scattering volume, and the probability of reaching each position is exactly the same. Its position has nothing to do is its starting position. Therefore, using the statistic theorem: the time average the product of the two exponential terms is equal to the product of time average each exponential, $\langle \exp[i(\phi_{s,j}(0) - \phi_{s,j}(\tau))] \rangle = \langle \exp[i\phi_{s,j}(0)] \rangle \langle \exp[-i\phi_{s,j}(\tau)] \rangle$. Since the phase changes between 0 and 2π with the particle position randomly, $\langle \exp[i\phi_{s,j}(0)] \rangle = \langle \exp[-i\phi_{s,j}(\tau)] \rangle = 0$, that is, when the delay time is sufficiently long, the photoelectric field of the scattered light itself is 100% uncorrelated.

The discussion of these two extreme cases clearly shows that when the delay time is **finite, the autocorrelation of the photoelectric field of the scattered light must be between 0% and 100%, decreasing as the delay time increases.** The decay rate increases as the average speed of the random motion (translational diffusion) of the particles increases. In other words, for a given scattering system (particle size, temperature, pressure, dispersion medium, etc.), it must have a characteristic relaxation time (τ_c), associated with the system. According to the definition, the photoelectric field of the scattered light from the jth particle has a phase $\phi_{s,j} = \vec{q} \cdot \vec{r}_j$. Therefore,

$$g^{(1)}(\tau) = \langle \exp[i\, \vec{q} \cdot (\vec{r}_j(0) - \vec{r}_j(\tau))] \rangle \qquad (5.36)$$

It is also referred as dynamic structural factor of single particle. If $\vec{q} \cdot (\vec{r}_j(0) - \vec{r}_j(\tau)) < 1$, i.e., the scattering angle is lower or the delay time is shorter, its Taylor expansion is

$$g^{(1)}(\tau) = \langle 1 + i\, \vec{q} \cdot [\vec{r}_j(0) - \vec{r}_j(\tau)] - \frac{1}{2}\vec{q}^2 \cdot [\vec{r}_j(0) - \vec{r}_j(\tau)]^2 + \cdots \rangle$$

where $i^2 = -1$ was used. Under the agitation of thermal energy, the particles randomly walk, their positions constantly change, the time average position remains, so $\langle i\, \vec{q} \cdot [\vec{r}_j(0) - \vec{r}_j(\tau)] \rangle = i\, \vec{q} \cdot \langle \vec{r}_j(0) - \vec{r}_j(\tau) \rangle = 0$. The above equation can be approximated to

$$g^{(1)}(\tau) \approx \exp[-\frac{1}{2}\vec{q}^2 \cdot \langle [\vec{r}_j(0) - \vec{r}_j(\tau)]^2 \rangle] \qquad (5.37)$$

where $\langle [\vec{r}_j(0) - \vec{r}_j(\tau)]^2 \rangle$ is the mean square distance after a particle walks randomly for a period of time τ. According to the diffusion equation, the time (t) required for a particle to diffuse a distance of x is related to the translational diffusion coefficient (D) as $6Dt = x^2$. Here, $t = \tau$ and $x^2 = \langle [\vec{r}_j(0) - \vec{r}_j(\tau)]^2 \rangle$. Therefore, the above equation can be rewritten as

$$g^{(1)}(\tau) = \exp(-q^2 D\tau)$$

where $\vec{q}^2 = q^2$ was used. The substitution of the above equation into the aforementioned Fourier inversion calculation formula results in $\Gamma = q^2 D$. For a polydisperse sample, the above equation can be further written as

$$g^{(1)}(\tau) = \int_0^\infty g^{(1)}(\Gamma, \tau)\, d\Gamma = \int_0^\infty G(\Gamma) \exp(-\Gamma\tau)\, d\Gamma \qquad (5.38)$$

The above equation is the Laplace integration. It shows that the measurement process is equal to the integration on the right. According to the above formula, the required linewidth distribution $G(\Gamma)$ can be obtained by performing the Laplace inversion on the measured $g^{(1)}(\tau)$.

One can also use the following method to deduce eq. (5.37). As discussed before, the reciprocal of the scattering vector ($1/q$) is an "optical ruler" for measuring the size of scattered particles in static LLS. In dynamic LLS, it is also an "optical ruler", which measures the average migration (diffusion) distance of a randomly moving particle in a given time. In other words, when the particle diffuses an average distance of $1/q$, the phase of the photoelectric field of the scattered light changes 2π, completing one oscillation. The corresponding time is the aforementioned characteristic relaxation time (τ_c), i.e., $6D\tau_c = (1/q)^2$. Therefore, eq. (5.37) can be rewritten as,

$$g^{(1)}(\tau) = \exp\left(-\frac{\tau}{\tau_c}\right)$$

In comparison with the previous equation, $\Gamma = 1/\tau_c = q^2 D$. It physically shows that Γ and τ_c respectively represent the same characteristic parameter (property) of the scattering system in the frequency and time spaces. The latter equation shows that if only the translational diffusion of particles in the system is considered, the characteristic parameter is related to the translational diffusion coefficient and the "optical ruler $(1/q)$" that measures the characteristic parameter. Physically, τ_c can also be understood as the "memory time" of a randomly changing system; namely, the time that a system can "remember" any one of its own states.

All optical experiments use "square" detectors, only measuring the light intensity, not the photoelectric field of the scattered light. Therefore, experimentally, it is impossible to measure the time autocorrelation function of the photoelectric field of the scattered light directly. A.J.F. Siegert stated in 1943 that if the random motions of particles in the system are independent of each other and the density fluctuation follows the Gaussian distribution, the normalized time autocorrelation function of the photoelectric field of the scattered light $g^{(1)}(\tau)$ is related to the normalized time autocorrelation function of the scattered light intensity $g^{(2)}(\tau) = \langle I(0)I(\tau)\rangle / \langle I(0)I(0)\rangle$. It is often called as the Siegert relationship later, i.e.,

$$g^{(2)}(\tau) = \frac{\langle \{\sum_{i=1}^{N_p}\sum_{j=1}^{N_p}\exp[i(\phi_{s,i}(0) - \phi_{s,j}(0))]\}\{\sum_{l=1}^{N_p}\sum_{m=1}^{N_p}\exp[i(\phi_{s,l}(0) - \phi_{s,m}(\tau))]\}\rangle}{\langle \{\sum_{i=1}^{N_p}\sum_{j=1}^{N_p}\exp[i(\phi_{s,i}(0) - \phi_{s,j}(0))]\}\{\sum_{l=1}^{N_p}\sum_{m=1}^{N_p}\exp[i(\phi_{s,l}(0) - \phi_{s,m}(0))]\}\rangle}$$

Readers should not be intimidated by the seemingly complicated mathematics, just remembering the physical image behind it. At $t = 0$, N_p particles scatter N_p rays, and the photoelectric field of the scattered light from the ith particle has its own phase $\phi_{s,i}(0)$. After a short delay time τ, N_p particles have moved and scattered another N_p rays. The phase of the photoelectric field of the scattered light from the ith particle has changed from $\phi_{s,i}(0)$ to $\phi_{s,i}(\tau)$. There are four groups of sums in the above equation, and each group has N_p^2 items, so that the numerator and denominator have N_p^4 items, respectively.

Each exponential term in the above equation can be written as the product of two exponential terms. Since the random motions of the particles are not related to each other, statistically, "the average after the multiplication is equal to the multiply after the averages". Because its position is completely random, the time average of each particle is $\langle \exp[i\phi_{s,j}(t)]\rangle = \langle \exp[i\vec{q} \cdot \vec{r}_j(t)]\rangle = \delta(q)$. When $q \neq 0$ (which is always true experimentally), the cross terms $(i \neq j; l \neq m; i \neq l;$ and $j \neq m)$ between all different particles are zero. The only thing left is its own term, that is, the product of two photoelectric fields of the scattered light from every particle itself, which can be divided into the following two categories.

First, for one particle at the same time and at the same position, $i = j$ and $l = m$, and each item becomes 1. Each of the two sums on the numerator and denominator has N_p terms, respectively, so that the results are N_p^2, respectively. Their ratio is 1.

Second, for one particle at two different times (different positions) separated by τ, $i = l$ and $j = m$, but $i \neq j$ or $l \neq m$. According to the definition, each pair of $i = l$ corresponds to a photoelectric field time autocorrelation function $g^{(1)}(\tau)$, there are N_p pairs in total. Each pair of $j = m$ also corresponds to one $g^{(1)}(\tau)$, but due to the limitation of $i \neq j$ or $l \neq m$, there are only $(N_p - 1)$ number of $g^{(1)}(\tau)$. Their multiplication leads to $N_p(N_p - 1)|g^{(1)}(\tau)|^2$, so that the above equation becomes

$$g^{(2)}(q,\tau) = \frac{N_p^2 + N_p^2|g^{(1)}(q,\tau)|^2 + N_p|g^{(1)}(q,\tau)|^2}{N_p^2} = 1 + |g^{(1)}(q,\tau)|^2 + \frac{|g^{(1)}(q,\tau)|^2}{N_p}$$

When $N_p > 10^3$, the last item on the right is less than one thousandth, similar to the noise in the experiment, so it can be ignored. However, if $N_p < 10^2$, this last item has to be considered. The size of the scattering volume is about 200 μm. A rough estimation shows that for particles with a size of 1 μm, the lower limit of the weight concentration is 10^{-5} g/mL; and so on.

The above discussion obviously also applies to each scattering unit in a larger particle (a macromolecular chain) or in different particles (macromolecules) in a concentrated dispersion (solution). The only difference is that there is a certain correlation between different scattering units or particles (macromolecules), so that the time average of all cross terms in the previous equation will not completely disappear to zero. The situation is quite complicated and will be discussed in detail in the Chapter of Dynamics of Macromolecular Chains.

The last term in the above equation corresponds to a relaxation process, whose physical nature is that when the number of particles is small, the number of particles will fluctuate due to the diffusion of particles in and out of the scattering volume. The size of the scattering volume is one thousand times larger than $1/q$, and the time required for diffusion is proportional to the square of the distance. Therefore, in comparison with the relaxation due to the phase fluctuation on the scale of $1/q$, the relaxation due to the particle number fluctuation is one million times slower, N_p times weaker. In the usual measurement time of dynamic LLS, it can be regarded as a small, basically unrelaxed constant, and its contribution to the baseline of the measurement is insignificant, so that it can be ignored. In the usual dynamic LLS, the direct measurement is the time autocorrelation function of scattered light intensity, not normalized by $I(q)^2$.

$$G^{(2)}(q,\tau) \cong I(q)^2[1 + \beta|g^{(1)}(q,\tau)|^2] \tag{5.39}$$

where the last item in the previous equation is ignored, and an instrument parameter $(0<\beta<1)$ is introduced. In a LLS instrument, the detection system never reaches the ideal perfect state (such as, the photosensitive area of a detector is not infinitely small; all surfaces on the optical path, the pinholes that limit the beam, etc. will scatter to generate stray light). The detection correlation (β) of modern LLS instruments is as high as 95%, never reaching 100%. β is a physical quantity to be measured. Usually, it is used as a fitting parameter in the data processing. When designing an instrument, an increase in β can increase the signal−to−noise ratio and shorten the measurement time. However, it will also reduce the measured scattered light intensity and prolong the measurement time. Therefore, a proper balance should be struck between them.

According to the above equation, by setting several delayed recording channels with ultra−long delay time (τ) on the digital time correlator, the resulting $g^{(1)}(q,\tau)$ should have been completely uncorrelated and decay to zero. Such obtained time autocorrelation function of the scattered light intensity is the square of the time average scattered light intensity ($I(q)^2$), also known as the experimental "baseline", written as A. The figure on the right shows a typical, nearly perfect, time autocorrelation function of the scattered

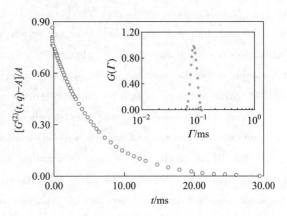

Lau A C W, Wu C. Macromolecules, 1999, 32: 581−584.

(Figure 4, permission was granted by publisher)

light intensity normalized by the measured baseline. The root of the difference between $G^{(2)}(q,\tau)$ and $I(q)^2$ is $\beta^{1/2} \mid g^{(1)}(q,\tau)$. According to eq. (5.37), one can calculate the characteristic linewidth distribution $G(\Gamma)$ from its Laplace inversion. The inset in the figure is the calculated linewidth distribution by using a commonly used CONTIN calculation software, which shows that this is a narrowly distributed macromolecular sample.

Since the time autocorrelation function of the photoelectric field of the scattered light in eq. (5.37) is proportional to the scattered light intensity, the linewidth distribution $G(\Gamma)$ is the distribution of the scattered light intensity ($G(\Gamma_i) \propto M_i C_i$). Therefore, the average line width ($\langle \Gamma \rangle$) of a polydisperse sample is

$$\langle \Gamma \rangle = \frac{\int_0^\infty \Gamma G(\Gamma)\,d\Gamma}{\int_0^\infty G(\Gamma)\,d\Gamma} = \frac{\sum_{i=0}^\infty \Gamma_i M_i C_i}{\sum_{i=0}^\infty M_i C_i}$$

By definition, this is a z−average linewidth. The linewidth variance ($\mu_2 = \langle (\Gamma - \langle \Gamma \rangle)^2 \rangle$) and relative variance ($\langle (\Gamma - \langle \Gamma \rangle)^2 \rangle / \langle \Gamma \rangle^2$) of a polydisperse sample can also be calculated from the linewidth distribution ($G(\Gamma)$), namely

$$\langle (\Gamma - \langle \Gamma \rangle)^2 \rangle = \int_0^\infty (\Gamma - \langle \Gamma \rangle)^2 G(\Gamma)\,d\Gamma = \langle \Gamma^2 \rangle - \langle \Gamma \rangle^2 \text{ and } \frac{\mu_2}{\langle \Gamma \rangle^2} = \frac{\langle \Gamma^2 \rangle - \langle \Gamma \rangle^2}{\langle \Gamma \rangle^2} = \frac{\langle \Gamma^2 \rangle}{\langle \Gamma \rangle^2} - 1$$

When the polydispersity index (M_w/M_n) of the molar mass distribution of polymer is less than 2.5, $M_w/M_n \cong 1 + 4\mu_2/\langle \Gamma \rangle^2$. The theories above seem to be logical. It is necessary to note that the Laplace inversion is to solve an ill-conditioned mathematic equation. There is an inevitable error in the experiment. Within the error range, the Laplace inversion of the same measured $\beta^{1/2} \mid g^{(1)}(q,\tau) \mid$ corresponds to numerous possible results. Therefore, in the inversion calculation, one has to choose an error criterion first. When the relative variance between the inversion result and the measured data is lower than this error criterion, the inversion calculation is terminated. However, within this error criterion, there are still different calculation results. If there is no supplementary information obtained from other methods, neither the computer nor the operator can know which inversion result is better. Only the result with a smaller relative variance within the range of error criterion is selected.

Therefore, it becomes very important to choose an appropriate error criterion. If too large, some measurement errors and noise would be counted as information in the resultant linewidth distribution; if too small, some system information might be omitted as errors and noise in the linewidth distribution obtained. In a real experiment, this error criterion itself is also unknown. An error standard has been set in the analysis software of a commercial LLS instrument according to the general situation. Experienced experimental operators can adjust and reset the error criterion according to specific system and other additional information to obtain better analysis results. Note that this is an experimental problem, not a calculation problem! Therefore, it is more important to use clean samples and long measuring time, and obtain high-quality data.

Besides using the Laplace inversion method, one can also use the cumulants polynomial expansion to analyze the measured time autocorrelation function of the photoelectric field of the scattered light,

$$\log \left[\frac{G^{(2)}(\tau) - A}{A} \right]^{1/2} = \log \beta^{1/2} - \langle \Gamma \rangle \tau + \frac{\mu_2}{2!}\tau^2 + \frac{\mu_3}{3!}\tau^3 + \cdots$$

where the first term is a constant, as a fitting parameter during analysis. Mathematically, μ_i ($i = 1,2,3,\cdots$) is called the ith moment of a distribution. For random variables, the first moment is zero; the second moment is the distribution width; and the third moment reflects the asymmetry of the distribution. In experiments, it is generally expanded to third moments at most. One can take logarithm of the time autocorrelation function of the measured scattered light intensity and make a semi-logarithmic plot. When the delay time is short, a decreasing line is obtainable, and the slope is the z-average linewidth; when the delay time is long, the slope corresponds to the variance of the linewidth distribution, and then the relative variance can be calculated. When the sample is broadly distributed, the third moment can be calculated.

In the early days when there was no computer or the computational speed was not fast enough, the cumulant analysis was the only choice. It seems not getting a visually expressible linewidth distribution like the Laplace inversion. In fact, the information obtained by the two is almost the same: the "average line width" and "relative variance of line width distribution". Readers need to note that the Laplace inversion requires very high quality of the measured time autocorrelation function; if the data quality does not meet the requirements of the Laplace inversion, the results obtained by the cumulant analysis may be more reliable, so that it is not out of date.

Experimentally, due to the decrease in the price of memory units and the increase in storage speed, modern research-type dynamic LLS spectrometers have been fully digitalized. A highly sensitive detector is used to measure the number of scattered photons directly, and a hardware digital time correlator is used to obtain the time autocorrelation function $G^{(2)}(q,\tau)$ of the number of scattered photons in real time. A small particle sizer based on the principle of dynamic LLS may use a cheap software digital time correlator, i.e., first record and store a series of varied photon numbers, and then use the software to calculate the time correlation function instead of directly using hardware to calculate the time autocorrelation function during the collection of the scattered photons. Because there is a short dead time every time the number of photons is stored, the software correlator may lose a small portion of the information. Due to the extremely fast modern memories, it is difficult for users to distinguish the difference between the two generally.

In a dynamic LLS measurement, the measured $G^{(2)}(q,\tau)$ continuously superimpose on top of each other one after another, which is a statistical average process over time. In theory, random noise will cancel each other out during the superposition process, and will not increase with the measurement time, and the real information will increase with the measurement time. Therefore, in order to improve the signal-to-noise ratio, the measurement time should be sufficiently prolonged instead of some manipulation of data analysis. For extremely dilute macromolecular solutions, the measurement time can reach several hours or even days, which in turn requires the solution to be absolute dust-free. A necessary but insufficient criterion to judge whether a dynamic LLS measurement is qualified is that the relative error between the measured baseline (the average scattered light intensity) and the calculated baseline after inversion is less than one thousandth. Few other experiments and spectroscopy methods can achieve such a measuremental accuracy and precision.

As mentioned earlier, the scattered light intensity of a small dust or bubble is equivalent to that of billions of macromolecular chains. If not careful, the measured result includes the contribution of impurities. When measuring the colloid of large particles, the influence of trace impurities is relatively small because the scattering of large particles is relatively strong, but proper cleaning of the scattering cell and the removal of dust from solutions are still essential steps. However, the salesperson of the particle sizers does not normally emphasize this point to users. Unfortunately, users who lack basic light scattering knowledge and operating experience often treat a particle sizer based on the principle of dynamic LLS as a black box: putting in the sample without proper removal of dust, pressing a button, and getting distorted experimental results, so that the current literature is full of specious conclusions.

In addition to carefully preparing the solution, correctly performing dynamic LLS measurements, and properly selecting the analysis parameters, users must also clearly realize that dynamic LLS measured is the characteristic relaxation time (τ_c) in time space or equivalent linewidth (Γ) in frequency space of the scattering system. For a polydisperse macromolecular or colloidal sample, the measured is a distribution of τ_c or Γ.

The relaxation of a scattering system can originate from various different reasons, such as the rotation of rod-like particles, the fluctuation of the relative position between various segments in a flexible chain, the long-range interaction of charged particles in an aqueous solution, the temperature or concentration gradient induced diffusion, etc. The translational diffusion is just one of many sources of relaxation. Therefore, only when the relaxation of a system is completely attributed to the transla-

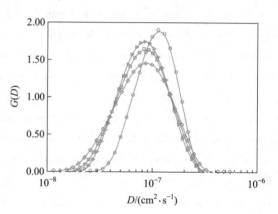

Wu P Q, Siddiq M, Chen H Y, et al. Macromolecules, 1996, 29:277.

(Figure 3, permission was granted by publisher)

tional diffusion of particles or macromolecules, can τ_c and Γ or their distributions are associated with the translational diffusion coefficient (D) or its distribution ($G(D)$). The figure on the right shows translational diffusion coefficient distributions of macromolecules with different molar masses calculated from the linewidth distribution. Further, D and $G(D)$ can be converted to the hydrodynamic radius (R_h) or its distribution ($f(R_h)$) of macromolecules or particles. The relevant theories can be traced back a long time ago.

As early as 1827, the botanist Robert Brown pointed out that the random jigging of pollen in water has nothing to do with pollen. Dust also has the same jigging properties, which was later called the Brownian motion. Although some people were paying attention to this strange phenomenon at the time, the real explanation was in 1904, almost eighty years later, when William Sutherland made his explanation public at an academic conference and officially published his result in March 1905. In the same year, Albert Einstein independently pointed out that the translational diffusion coefficient is equal to the thermal energy divided by the friction coefficient (f). On the other hand, in 1851, George Gabriel Stokes already discovered that for a sphere with a radius of R in a liquid with a viscosity of η, $f = 6\eta\pi R$ in a non-turbulent (laminar) flow. A combination of the two results in the famous Stokes-Einstein relationship, $D = (k_B T)/(6\eta\pi R)$. Marian Smoluchowski independently discovered this formula about a year later. Perhaps because the information exchange was very slow at the time, Einstein did not mention Sutherland's work in his article. Even if not considering the latter's role, as a latecomer, he should not obliterate the former's contribution. The correct expression of the above equation should be the Stokes–Sutherland–Einstein equation, which shows that history is sometimes not fair. Using the above equation, it is not difficult to obtain the hydrodynamic radius distribution $f(R_h)$ from the translational diffusion coefficient distribution $G(D)$. In the Laplace inversion, in order to cover the wider distribution, logarithmic abscissa is usually used ($\log \tau_c$ or $\log \Gamma$), so that the above conversion from D to R_h becomes easier, a simple parallel conversion in a linear space.

Since dynamic LLS measures the particle movement, the velocity of a particle itself is obviously different from the relative velocity of two particles. Assuming that there are N_p particles in the scattering volume, strictly speaking, the time autocorrelation function should be written as

$$g^{(1)}(\tau) = \langle \exp[i\,\vec{\mathbf{q}} \cdot (\vec{\mathbf{r}}_i(0) - \vec{\mathbf{r}}_j(\tau))]\rangle \qquad (5.40)$$

This is the dynamic structural factor, including two parts: the relaxations due to the self-motion of one particle and the relative motion of two particles. Assuming that the scattering volume contains N_p particles, the time autocorrelation function of the scattered light intensity should be written as

$$g^{(1)}(\tau) = \frac{\sum_{i=1}^{N_p} \exp[i\,\vec{\mathbf{q}} \cdot (\vec{\mathbf{r}}_i(0) - \vec{\mathbf{r}}_i(\tau))]}{N_p} + \frac{\sum_{i=1}^{N_p} \sum_{j\neq i=1}^{N_p} \exp[i\,\vec{\mathbf{q}} \cdot (\vec{\mathbf{r}}_i(0) - \vec{\mathbf{r}}_j(\tau))]}{N_p^2}$$

Therefore, dynamic laser light scattering measured is the characteristic mutual relaxation time (τ_m), corresponding to the translational mutual diffusion coefficient (D_m). The first item in the above equation leads to the characteristic self-relaxation time (τ_s), corresponding to the translational self-diffusion coefficient (D_s).

In the above discussion, we assumed that the solution is extremely dilute and there is no interaction between the particles. However, in any real dispersion or solution, there is always a certain interaction between particles. Even there is no special interaction between neutral particles, the long-range hydrodynamic interaction also affects each other's translational diffusion. The above two characteristic relaxation times or translational diffusion coefficients all depend on the concentration. When the solution is infinitely dilute, both of them approach the characteristic relaxation time (τ_0) of an isolated particle (there is only one particle in the system), corresponding to its translational diffusion coefficient (D_0).

Physically, according to their different definitions, when a particle moves relative to another particle, the relative speed increases, so that D_m increases with the particle concentration. On the other hand, the existence of other particles will inevitably hinder the movement of a particle, so that D_s decreases as the concentration increases. Of course, if there is a special interaction between the particles, their dependence on the concentration could be reversed. Experimentally, the mutual diffusion coefficient at a series of concentrations is always measured first, and then extrapolated to the zero concentration to obtain D_0. If only one concentration is measured, users should know that the measured mutual diffusion coefficient is larger, so that the particle size (hydrodynamic radius) obtained is smaller.

Although, in the dynamic LLS measurement, the scattered light intensity of one particle and another cannot be distinguished, so that the self-relaxation time or self-diffusion coefficient of the particle is not directly measurable. The workaround is to find a solvent with the same refractive index as the particles first, and then label a small amount of particles with different refractive indexes to measure their self-relaxation time or self-diffusion coefficient. Some experimental methods can directly measure the particle's self-diffusion coefficient, including but not limited to forced Rayleigh scattering, fluorescence bleaching recovery, and pulsed field gradient nuclear magnetic resonance. Due to the limited space, this book will not introduce these methods in details. Interested readers can refer to the relevant monographs.

It is due to the measurement of the characteristic relaxation time of a system in time space that especially those who are used to other spectroscopies often refer dynamic LLS as photon correlation spectroscopy (PCS). The same experimental method based on the same physical principle has three completely different names; dynamic light scattering, quasi-elastic light scattering, and photon correlation spectroscopy. Is it surprising? Not really, just as different people call each of us by different names, there are formal names and nicknames, depending on their relationship with us. Here too, people who have been engaged in static light scattering call it dynamics light scattering; and those who have been engaged in elastic scattering in physics naturally call it quasi-elastic scattering from the principle. These different names also shows that it is a relatively "young" method, so far there is still no unified name.

The scattering methods of studying colloidal and macromolecular solutions also include small-angle neutron and X-ray scattering. From the angular dependent scattered light intensity, the molar mass and size of the particles and macromolecules, as well as the structural factor can also be obtained, where the mathematical expressions of the scattering vector and the angular dependence of the scattered light intensity are exactly same. The difference is that light scattering originates from the difference between the refractive index of solute and solvent, i.e., the differential refractive index (dn/dC). The neutron scattering depends on the scattering length (b) between solute and solvent molecules, i.e., the difference in the scattering cross-section area ($4\pi b^2$). X-ray scattering relies on the electron density difference between solute and solvent molecules.

Before the advent of synchrotron radiation and strong neutron sources, the neutron and X-ray scattering of macromolecular solutions was too weak, so that they were mainly used to study solids. Since the 1990s, with the construction of synchrotron radiation and strong neutron sources, their applications in the study of macromolecular solutions have gradually increased.

In small-angle X-ray scattering, the reciprocal of the scattering vector ($1/q$, "optical ruler") corresponds to 0.2-5.0 nm; in light scattering, $1/q \sim 35-300$ nm; and while $1/q$ of small-angle neutron scattering is located between that of X-ray and light scattering. The combination of the three can cover and obtain the information from polymerized monomers to the chain segments, and then to the entire macromolecule. This book omits the details of X-ray and neutron scattering.

Summary

Basic Principle: The particles (macromolecules) constantly move inside a solution or dispersion, including translational diffusion, rotation and conformation fluctuations. These motions change the phase of the photoelectric field of the scattered light. The lights with different phases interfere with each other on the detection plane to form countless bright or dark light spots with a jigging frequency up to one million Hz. If using a sensitive and fast single-photon detector with a small detection area of tracking only few light spots, one can thereby measure the fluctuation of the scattered light intensity caused by the particle random movement. The faster the speed, the higher the fluctuation frequency.

Looking at the same phenomenon from another angle in the frequency space, due to the Doppler Effect, the particle movement slightly shifts the frequency of the scattered light, from a single frequency at the time of incidence to a frequency distribution. The faster the movement, the broader the frequency distribution. Here, both the intensity fluctuation and spectral broadening are related to the speed of particle movement, reflecting the characteristic relaxation time of a system, also called the "memory" time. By measuring the time autocorrelation function of the scattered light intensity in time space, one can obtain the characteristic relaxation time. When there is only translational diffusion, the characteristic relaxation time can lead to the translational diffusion coefficient, and then further to the hydrodynamic radius by using the Stokes-Sutherland-Einstein equation. Most of particle sizers with a measuring range less than 2000 nm on market are based on this principle.

Measurable Physical Quantity: Modern dynamic LLS instruments generally have a digital time correlator. One can use it to measure the time autocorrelation function $G^{(2)}(q,\tau)$ of the scattered light intensity of dispersions (solutions) with different particle (macromolecule) concentrations at different scattering angles directly. First, the dust and bubbles in the liquid must be completely removed. The judgment criterion is the fluctuation of the scattered light intensity within 15 minutes less than five percent at at a small scattering angle (15°). Next, the sampling (measurement) time has to be sufficiently long. The criterion is that the relative error between the measured and the calculated baselines is less than one thousandth.

Data Processing: According to eqs. (5.37) and (5.38), the Laplace inversion of each measured time autocorrelation function $G^{(2)}(q,\tau)$ of scattered light intensity leads to an apparent linewidth $\Gamma(q,C)$ or characteristic relaxation time $\tau_c(q,C)$ at a given scattering angle (q) and concentration (C). The respective extrapolation of $q\to0$ and $C\to0$ result in a true linewidth or characteristic relaxation time. If there is only the translational diffusion, using $\Gamma=Dq^2$, one can convert a characteristic linewidth distribution $G(\Gamma)$ to a translational diffusion coefficient distribution $G(D)$, and then further to a hydrodynamic radius distribution $f(R_h)$ the by using the Stokes−Sutherland−Einstein equation. Modern dynamic LLS instruments have included all calculation software. Pros: convenience; Cons: users often do not ask for thorough explanations and make mistakes.

Differential Refraction

From the discussion of static LLS, it has been known that the intensity of scattered light is proportional to the square of differential refractive index, defined as $(\partial n/\partial C)_{T,p}$ in the scattered light intensity, the error of differential refractive index is doubled. Therefore, the accurate determination of differential refractive index is very important for the subsequent use of light scattering to determine the weight average molar mass, z−average radius of gyration and second virial coefficient of macromolecules. The "differentiation" means to use a method to deduct the contribution to the refractive index from solvent to obtain the concentration dependence of refractive index of macromolecules in a given solution. The usual differential refractometer is mainly based on two different principles: interference and refraction.

Let us first discuss the differential refractometer based on the principle of interference, as shown in the figure below. From left to right, first, using a lens to collect and collimate the light from a mercury lamp into a parallel beam; a filter (not shown) to make the incident light monochromic; a slit or a small pinhole to intercept a narrow beam out of the parallel light, and a polarizer for polarization. Then, using one prism (called Wollaston prism) invented by

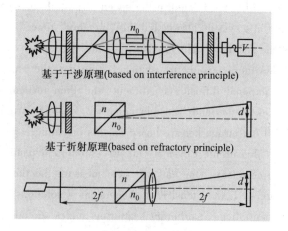

基于干涉原理(based on interference principle)

基于折射原理(based on refractory principle)

William Hyde Wollaston and 1802 to split the incident polarized light into two beams whose directions of polarization are perpendicular to each other, respectively, passing through the solvent and the solution. The two beams become a polarized light beam after converged by a lens and passing through another Wollaston prism. The slight difference between the refractive index of solvent and solution (Δn) leads to different optical paths, so that the polarization direction of the merged light slightly deviates from previous polarization direction. Using a subsequent polarizer and photodiode, one gets the deviation angle. The deviation angle is proportional to Δn, and Δn is further proportional to the concentration and the length of the sample cell. Experimentally, the two cells are filled with the same solvent first, and the second polarizer is adjusted so that the two are orthogonal, i.e., the minimum transmitted light intensity. Then, the solvent in one cell is replaced by a solution with a concentration of C. Δn is obtained from the angular deviation. By replacing the solution with a concentration gradually from low to high, a series of Δn corresponding to C can be obtained. The plot of Δn versus C is a straight line. Its slope corresponds to the differential refractive index, $(\partial n/\partial C)_{T,p}$.

The differential refractometer based on the principle of refraction is relatively simple and clear, as shown in the middle of the figure above. The same from left to right: the monochrome with a filter, the collimation with a lens, the interception with a slit, and the polarization with a polarizer. Then, the polarized and collimated beam enters a divided sample cell. If the divided two sides are filled with the same solvent, the beam will not deviate from the original direction in theory, only shifting a little due to the 45° splitting glass. However, the sample cell is never perfect. For example, the four glass surfaces inside and outside are not parallel. In fact, the beam will shift and slightly deviate from its original direction. In order to experimentally eliminate these defects, one first read the beam position on a microscopic ruler, rotate the cell 180°, and read the beam position once more. The average value of the two readings is the origin position of the beam if it is not deviated. The solvent in one of the half-cells is replaced with a solution. After the thermal equilibrium, the deviation distance of the beam is recorded. One also needs to rotate the cell 180° and read deviation distance once more. The average of the two readings is the deviation distance of the solution, which is proportional to Δn. By replacing the macromolecular solution gradually with a concentration from low to high, a series of Δn corresponding to C is obtainable. A plot of them results in a straight line. whose slope is $(\partial n/\partial C)_{T,p}$.

The requirements of using a differential refractometer to measure the beam deviation position are estimated as follows. Generally, the concentration and the differential refractive index of a macromolecule solution are about 10^{-3} g/mL and 10^{-1} RI, respectively, so that $\Delta n \sim 10^{-4}$. In order to achieve $\pm 1\%$ accuracy, a differential refractometer must be able to measure 10^{-6} RI. Even if the distance between the microscopic ruler and the sample cell is 1 meter, the measurement accuracy must reach 1 micron, not easy. In addition, the refractive index varies with temperature, $dn/dT \sim 10^{-4}$ RI/°C, so that the temperature fluctuation of the sample cell should be less than ± 0.01 °C, which requires a good temperature control device.

Readers should know that the above refraction measurement seems to be simple in principle and not difficult to do. However, the actual measurement process is extremely long and "painful". First, the deviation distance is so small that it is necessary to use a microscopic ruler to read more than three times each time and find an average. When measuring different concentrations, one must remove the previous solution completely, and then use the next slightly concentrated solution to wash three times repeatedly to avoid the residual of the previous solution from affecting the measurement results. To reach the thermal equilibrium, one has to wait for more than half an hour each time. Even if one takes all necessary measures and careful operation, the plot of the measured deviation distance versus the concentration is still often not in a straight line, not passing the origin predicted theoretically. In this way, the entire experiment that takes few days must be overthrown and restarted. The problems always exist with the experimental accuracy and precision. This is why the differential refractive index of the same polymer solution listed in Polymer Handbooks varies greatly due to the measurements in different laboratories.

In the 1980s, after the advent of the position-sensitive detectors, they replaced the traditional microscopic ruler, which greatly simplified the painful data collection. The principle of the position-sensitive detector is very simple. It is a specially designed photodiode with a one-dimensional long strip of photosensitive area. It has a lead at both ends and also at the middle to measure the voltage difference between the two ends ($V_+ - V_-$) and between the two ends and the middle: ($V_+ - V_0$) and ($V_- - V_0$), respectively. If a beam hits the center position exactly, ($V_+ - V_0$) = ($V_- - V_0$). In other words, if the position-sensitive detector is adjusted to a position at which they are equal, i.e., the origin position. Afterwards, by measuring the voltage (V) of the beam at any point, one can get the deviation distance from $L(V_+ - V)/(V_+ - V_-)$, where L is the photosensitive length. For example, if L = 10 mm, the voltage difference between the two ends is +10 volts, and the measurement accuracy is +0.001 volt, the accuracy of the position measured is 1 micron, far exceeding the accuracy of an ordinary microscopic ruler.

In addition, the refractive index depends on the wavelength of light and is inversely proportional to the square of the wavelength. When a mercury lamp was used, the filter with the same wavelength as the laser used in light scattering is not available. Therefore, one has to use two filters with wavelengths longer and shorter than the laser wavelength, respectively, to measure the differential refractive index, and then calculate the differential refractive index at the laser wavelength by the interpolation. Readers may ask why the laser in light scattering is not directly used as the light source in the differential refractometer. In fact, many people had asked the same question and tried. However, none of them succeeded. In order to answer this question, let us first examine how a laser is generated as follows.

The laser is generated in a cavity composed of a total reflection mirror and a semi-reflective and semi-transmitting mirror by inducing the electrons in the medium to transition simultaneously. In manufacturing lasers, designers have used materials with excellent thermal stability to minimalize the thermal drift of the two mirrors and other components, so that the excellent collimation and stability of the laser beam is obtainable. However, in comparison with the fixed slit or small pinhole in the above classic design, the laser beam still exhibits minimal drift due to the thermal fluctuation of the mirrors. For many applications, such small drift can be ignored. However, in a differential refractometer, the measurement accuracy of the deviation distance is in the order of micrometers. Therefore, since the advent of the laser in the 1960s, the beam drift had prevented using a laser as the light source of the differential refractometer.

The author thought repeatedly in the pain of measuring the differential refractive index whether one can find a better way. A true little story is as follows. Until one Saturday in the early 1990s, I suddenly realized that a simple optical method could be used to overcome the problem of laser beam drift, and immediately went to the laboratory. However, the company's all laboratories were closed on weekends due to the safety consideration. Facing the closed doors, I had to go home helplessly and waited bitterly for the arrival of Monday. At that time, I was in the mood to lift the sun up from below the horizon. As soon as the door was open on Monday, I immediately rushed into the laboratory, started the experiment as designed, and got the expected results. The author's mood at that time could not be described. Recalling this process, the author sincerely appreciates the undergraduate physics training received at the University of Science and Technology of China, which shows how the important fundamental courses are. The details of this simple idea are as follows

Use a laser to illuminate a small pinhole, and then select a condensing lens, and place the small pinhole and the position sensitive detector on both sides of the lens. The optical path difference from the center of the lens is twice the focal length. Optically, this is equivalent to attaching the small pinhole on the photosensitive surface of the position sensitive detector. In such a way, no matter how the laser drifts, it will not affect the position of the small

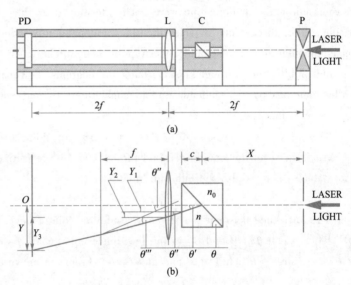

Wu C, Xia K-Q. Rev. Sci. Instrum., 1994, 65:587.

(Figure 1, permission was granted by publisher)

pinhole, as shown on the right. The beam shift caused by the difference in refractive index between solvent and solution is optically equivalent to moving the small pinhole to a new place on the detection plane. The drift of the laser beam will not affect the position of the small pinhole. The above description can be proved by the principle of geometric optics, as shown on the right. At this time, the laser light source is placed on the right, from right to left,

$$Y = Y_1 + Y_2 + Y_3 = c\tan\theta' + (2f - X - c)\tan\theta'' + 2f\tan\theta'''$$

and

$$f\tan\theta'' = c\tan\theta' + (2f - X - c)\tan\theta'' + f\tan\theta'''$$

where c, X and θ are constants. According to the figure above or combining the two above equations, one can easily get the following equation.

$$Y = f\tan\theta'' + f\tan\theta''' = f(\tan\theta'' + \tan\theta''')$$

On the other hand, in 1621, Willebrord Snellius, when a beam from one optical medium passes through an interface and enters another optical medium, the product of the refractive index and the sine of the angle between the beam and the normal is an invariant. Mathematically,

$$n_1\sin\theta_1 = n_2\sin\theta_2$$

It is now called Snell's law. In fact, the discovery of this law is traceable back to 984, when the Persian scientist Ibn Sahl had precisely described this law. In 1602, Thomas Horriot rediscovered this law. Afterwards in 1637, René Descartes independently used the heuristic momentum conservation to get this law, too and use it to solve a series of problems. It shows the continuity of scientific research and its development. In the early days, the information transmission and communication were not as developed as they are today. Therefore, multiple people often discovered a scientific law independently. Some existed only in private communication and correspondence, not published officially. Of course, there were also some repeated publications without knowing other publications. Fortunately, scientific research in the earlier time was purely academic and personal interests, and not so utilitarian as it is today. According to this law

$$n_0\sin(90° - \theta) = (n_0 + \Delta n)\sin(90° - \theta - \theta')$$

Air is between the sample cell and the lens, and its refractive index is 1. According to the figure above, there is also the following relationship,

$$(n_0 + \Delta n)\sin\theta' = \sin\theta''$$

The difference between the refractive index of solution and solvent is very small ($\Delta n \sim 10^{-4}$ RI), so that θ', θ'' and θ''' in the few above equations are actually very small. Therefore, $\sin\theta' \cong \theta'$, $\sin\theta'' \cong \theta''$, $\tan\theta'' \cong \theta''$ and $\tan\theta''' \cong \theta'''$. From the above equation,

$$(n_0 + \Delta n)\theta' \cong \theta''$$

Using the trigonometric function $\sin(x - y) = \sin x\cos y - \cos x\sin y$, the above equation can be rewritten as follows

$$n_0\sin(90° - \theta) = (n_0 + \Delta n)[\sin(90° - \theta)\cos\theta' - \cos(90° - \theta)\sin\theta']$$
$$(n_0 + \Delta n)\theta'\cos(90° - \theta) \cong \Delta n\sin(90° - \theta)$$

where $\theta' \ll 1$, $\sin\theta' \cong \theta'$ and $\cos\theta' \cong 1$. The above equation can be rewritten as

$$\frac{(n_0 + \Delta n)\theta'}{\Delta n} \cong \frac{\sin(90° - \theta)}{\cos(90° - \theta)} = \tan(90° - \theta)$$

Or using $\theta'' \cong (n_0 + \Delta n)\theta'$, one can further write it as

$$\theta'' = \Delta n\tan(90° - \theta)$$

Substituting it in the second equation above, and using $\tan\theta'' \cong \theta''$, $\tan\theta''' \cong \theta'''$, and relating θ''' to θ'', one can get

$$f(\tan\theta'' + \tan\theta''') = \left[X + c\left(1 - \frac{1}{n_0}\right)\right]\tan(90° - \theta)\Delta n$$

The proportional constant factor in front of Δn is defined as K. Its substitution into the previous expression of Y leads to the final result.

$$Y = K\Delta n$$

According to the figure above, if the lens is closely attached to the outer wall of the sample cell, $(X + c) \cong 2f$. In one of the author's previous designs, $c = 5.0 \pm 0.1$ mm; $X = 155.0 \pm 0.5$ mm; and $\theta = 45°$, so that K is about 156~157 mm. The deviated distance (Y) of the beam is proportional to the difference between the refractive index of solution and solvent ($\Delta n \sim 10^{-4}$), i.e., Y is only a few hundred microns. This is why the measurement accuracy must reach the micron level to ensure that the relative error of the differential refractive index measured is less than $\pm 1\%$.

Obviously, choosing a lens with a long focal length can improve the measurement accuracy, but the instrumental length limits it and the stability of a long base is another problem except it is on an optical table. If used as an online detector, its length cannot be too long. The differential refractometer based on this novel design has extremely high stability, accuracy, repeatability and precision, as shown on the right. The data points from four concentrations of the solutions of polystyrene in toluene all fall on

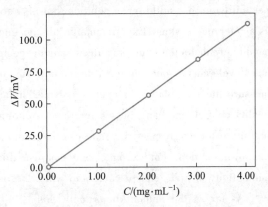

Wu C, Xia K Q. Rev. Sci. Instrum., 1994, 65: 587.

(Figure 5, permission was granted by publisher)

a straight line, passing the origin. Its slope corresponds to differential refractive index of polysty-

rene in toluene. The accuracy is so high that one only need to measure one concentration to get a differential refractive index in principle. Of course, more than four concentrations are still used to further improve accuracy and precision.

At present, a differential refractometer based on this principle has been commercialized by the German ALV company as a supporting equipment for its laser light scattering instrument. About 15 years ago, a domestic Jianke

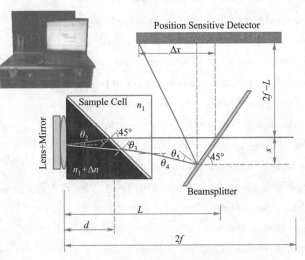

Gong X J, Ngai T, Wu C. Rev. Sci. Instrum., 2013, 84: 114103.

(Figure 3, permission was granted by publisher)

Instrument Company, with the help of the author, also sold its products. Besides as a supporting instrument for laser light scattering, one can also design it as portable and ultra-sensitive differential concentration detectors for various applications. In this situation, the instrument should not be too long. Adding a total reflection mirror behind the lens is one of the solutions, shortening the instrument length by half and doubling its accuracy. Based on this principle, the author developed a portable differential refractometer, which is installed inside an ordinary briefcase, as shown on the right. This novel differential refractometer can be used to study chemical and physical processes that involve changes in the concentration or refractive index.

Compared with other concentration detectors, such as UV, this differential refractometer based on the new design is not only able to stably and accurately measure a component change per milligram per milliliter of solution, but also able to use another solution without any change as a reference system. Experimentally, all the unchanged components in a complex system are completely deducted, and only the changing component is observed. For example, in research and developments, one can use it to monitor the concentration change during polymerization; and measure the degradation rate of a polymer in an enzymatic process, etc. In production and quality control, one can inject a reference solution with a known quality and a testing solution into the two cells, respectively, and use the differential method to get the purity and quality of an unknown solution. For example, one can use this differential refractometer to distinguish the authenticity of high-end perfume and wine.

As far as the author knows, only this high precision, ultra-stable differential refractometer uses a laser as its light source and a position detector to measure subtle differences in refractive index. For example, one could use this differential refractometer to study polystyrene nanoparticles made in different ways and find whether these particles contain different amounts of polystyrene chains, i. e., their microscopic chain density since their sizes were known. As shown on the right, the differential refractive indices of two kinds of polystyrene

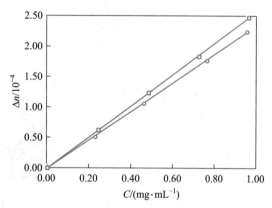

Wu C, Chan K K, Woo K F, et al.
Macromolecules, 1995, 28:1592.
(Figure 1, permission was granted by publisher)

nanoparticles are 0. 236 ± 0. 001 and 0. 256 ± 0. 001, respectively. The difference is only ~8%. Other commercial differential refractometers are not able to distinguish such a small difference accurately. The measurement accuracy and stability of commercial differential refractometers for various chromatographic instruments are even farther worse. Such a difference is not surprising. As a chromatographic instrument, it mainly detects the relative change in the concentration of macromolecules, not the absolute value of the differential refractive index of macromolecules.

The small online differential refractometers for various chromatographic instruments simply cannot achieve the micron-level precision and accuracy when measuring the phase shift or the beam deviation. Unfortunately, some researchers do not fully understand the basic principles of differential refractive index. They used the differential refractometer inside a chromatographic instrument to measure the differential refractive index of macromolecules, and then substitute it into the calculation in static laser light scattering experiment. The error of their obtained weight average molar mass of macromolecules has been doubled. In reality, these terrible data fill the current literature.

Summary

Basic Principle: In the interference type differential refractometer, the propagation speed of light is inversely proportional to the refractive index. A polarized light beam is decomposed into two beams orthogonal to each other to pass through solution and solvent respectively. The tiny difference in refractive index results in a phase difference in the propagation process. When they are combined to form one light beam again, its polarization will slightly deviate from the original direction. Using a polarizer orthogonal to the original polarization direction, the tiny difference in refractive index is measurable. The difference in refractive index is proportional to the intensity of the light passing through the second polarizer. In the refraction type differential refractometer, a collimated light beam is continuously passing through solvent and solution, and the two are separated by a glass partition at an angle of about 45°. Using the refraction principle that a light beam is deviated if there is a difference in refractive index between solution and solvent, the difference in refractive index is obtainable from its proportional deviation distance.

Measurable Physical Quantity: First, four or more macromolecular solutions with different concentrations are prepared. Next, the difference between the refractive index of each solution and solvent is measured from dilute to dense. In the interference type differential refractometer, the intensity of light passing through the second orthogonal polarizer is measured. While in the refraction type differential refractometer, the deviation distance of an incident collimated light beam after continuously passing through solution and solvent in the divided sample cell is measured. Experimentally, it is necessary to use a solvent and solution with a known refractive index, usually water and an aqueous solution of sodium chloride or potassium chloride, to calibrate the instrument; namely, to determine the proportional constant between the transmitted light intensity or the deviation distance and the difference in refractive index.

Data Processing: Using the calibrated proportional constant, the transmitted light intensity or the deviation distance measured in each solution with a given concentration of C is converted into the difference in refractive index (Δn) first, and then, Δn is plotted against C to obtain a straight line passing through the origin, whose slope corresponds to the differential refractive index. If the data is not on a straight line or the straight line does not pass the origin, there must be problems in the experiments, including but not limited, an inaccurate concentration preparation, insufficient thermal equilibrium time, incompletely cleaning of the solution cell or the instrument calibration was incorrect. In this case, one has to start the whole experiment from scratch.

Viscosity

Viscosity is a term often used in daily life, related to the liquid fluidity. According to experience, after a given force, the deformation speed is different for different materials. For example, when using the gravity to pour a liquid, depending on the speed of its flow, one can judge whether the liquid is viscous. For example, when respectively tilting a bottle of water and a bottle of honey, the tilting speed of water is much faster than that of honey. In a colder weather, their difference is more obvious. Compared with water, the viscosity of honey decreases faster with temperature.

Before discussing the definition and source of viscosity in a rigorous and detailed manner, let us first get acquainted with a few macroscopic variables related to material properties. The force sustained per unit area of a material is called **stress** ($\vec{\sigma}$). When the direction of the force is perpendicular or parallel to the surface, it is called the normal stress ($\vec{\sigma}_\perp$) and the shear stress ($\vec{\sigma}_=$). The force in any direction ($\vec{F}$) can always be decomposed into two vertical and parallel components ($\vec{F}_\perp$ and $\vec{F}_=$). The relative deformation of

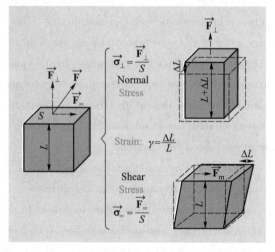

a material is called **strain** (γ), as shown on the right. Under a small deformation, the stress is proportional to the strain. The proportional constants are the Young's modulus ($\vec{E}$) and shear modulus ($\vec{G}$); namely, $\vec{\sigma}_\perp = \vec{E}\gamma$ and $\vec{\sigma}_= = \vec{G}\gamma$. Since the direction has been defined, the vector symbol can be removed, written in scalar form: $\sigma_\perp = E\gamma$ and $\sigma_= = G\gamma$, reflecting the "**elasticity**" of an object.

Experiments have long proven that at smaller stresses, the stress is also proportional to the velocity (rate) of deformation ($d\gamma/dt$). The proportional constant is called **viscosity** (η), i.e. $\sigma = \eta d\gamma/dr$, reflecting the "stickiness" of an object. For a given stress, viscosity is reversibly proportional to the velocity of deformation. The greater the viscosity, the smaller the velocity of deformation. On the contrary, for a given velocity of deformation, viscosity reflects the ability of an object to resist deformation. The greater the viscosity, the greater the stress required to reach the same velocity of the deformation.

Strictly speaking, any object is both elastic and viscous. Generally, below the melting point, the viscosity of small molecular solids is extremely high. Under a given small stress, the velocity of deformation is extremely slow, so that they are considered only elastic. On the contrary, above the melting point, the elastic modulus of small molecular liquids is extremely small, a great deformation, so that they are also considered only sticky. Macromolecular materials, whether in solid or liquid state, can be both viscous and elastic, called "**viscoelasticity**". If regarding "elastic" and "viscous" as two "elements" of macromolecular materials, one can connect them in parallel, so that the total stress (σ_T) is the sum of their stresses. Its substitution into the stress definition leads to

$$\sigma_T = \sigma_e + \sigma_v = E\gamma + \eta\frac{d\gamma}{dt} \quad \text{or} \quad \sigma_T = \sigma_e + \sigma_v = G\gamma + \eta\frac{d\gamma}{dt}$$

In the strain relaxation (creep) experiment, the total stress is a constant (σ_T), independent of time. From the above two simple differential equations, it is not difficult to obtain the following two solutions called strain relaxation respectively.

$$\gamma(t) = \frac{\sigma_T}{E}\left[1 - \exp\left(-\frac{t}{\tau_c}\right)\right] \quad \text{or} \quad \gamma(t) = \frac{\sigma_T}{G}\left[1 - \exp\left(-\frac{t}{\tau_c}\right)\right]$$

where the characteristic relaxation time (τ_c) is equal to η/E or η/G. For an alternating stress, $\sigma_T = \sigma_{T,0}\cos(\omega t)$, first assuming that $\gamma(t) = \gamma_{c,0}\cos(\omega t) + \gamma_{s,0}\sin(\omega t)$, then substituting it into the previous equation and finally comparing the two sides, one has $\sigma_{T,0} = E\gamma_{c,0} + \eta\omega\gamma_{s,0}$ and $0 = E\gamma_{s,0} - \eta\omega\gamma_{c,0}$ or $\sigma_{T,0} = G\gamma_{c,0} + \eta\omega\gamma_{s,0}$ and $0 = G\gamma_{s,0} - \eta\omega\gamma_{c,0}$, respectively; namely,

$$\gamma_{c,0} = \frac{\gamma_0}{1 + (\omega\tau_c)^2} = \gamma_0 G'(\omega) \quad \text{and} \quad \gamma_{s,0} = \frac{\gamma_0\omega\tau_c}{1 + (\omega\tau_c)^2} = \gamma_0 G''(\omega)$$

where $G'(\omega)$ and $G''(\omega)$ represent the storage and dissipation energies of a material, respectively, called the storage modulus and the dissipation modulus, respectively. $G''(\omega)/G'(\omega) = \omega\tau_c = \text{tg}\delta$, where δ is the dissipation angle. Substituting the above two relations into the assumption of γ (t), and noting that $\tau_c = \eta/E$ or η/G, one can derive its periodic variation with time as follows:

$$\gamma(t) = \gamma_0\left[G'(\omega)\cos(\omega t) + G''(\omega)\sin(\omega t)\right]$$

When two "elements" are connected in series, a total strain (γ_T) is the sum of their strains. Substituting it into the strain definition results in

$$\frac{d\gamma_T}{dt} = \frac{d(\gamma_e + \gamma_v)}{dt} = \frac{d\sigma}{Edt} + \frac{\sigma}{\eta} \quad \text{or} \quad \frac{d\gamma_T}{dt} = \frac{d(\gamma_e + \gamma_v)}{dt} = \frac{d\sigma}{Gdt} + \frac{\sigma}{\eta}$$

In the stress relaxation experiment, the total strain is a constant (γ_T), independent of time, $d\gamma_T/dt = 0$. From the above two simple differential equations, it is also not difficult to obtain the following two solutions, called strain (creep) relaxation, respectively,

$$\sigma(t) = \sigma_0 \exp\left(-\frac{Et}{\eta}\right) = \sigma_0 \exp\left(-\frac{t}{\tau_c}\right) \quad \text{or} \quad \sigma(t) = \sigma_0 \exp\left(-\frac{Gt}{\eta}\right) = \sigma_0 \exp\left(-\frac{t}{\tau_c}\right)$$

where the characteristic relaxation time (τ_c) is also equal to η/E or η/G. For an alternating strain, $\gamma_T = \gamma_{T,0}\sin(\omega t)$, first assuming that $\sigma(t) = \sigma_{c,0}\cos(\omega t) + \sigma_{s,0}\sin(\omega t)$, then substituting it into the previous equation and finally comparing the two sides, one has $\sigma_{s,0}/\eta - \sigma_{c,0}\omega/E = 0$ and $\gamma_{T,0}\omega = \sigma_{c,0}/\eta + \sigma_{s,0}\omega/E$ or $\sigma_{s,0}/\eta - \sigma_{c,0}\omega/G = 0$ and $\gamma_{T,0}\omega = \sigma_{c,0}/\eta + \sigma_{s,0}\omega/G$; i.e.,

$$\sigma_{c,0} = \frac{\sigma_0(\omega\tau_c)^2}{1 + (\omega\tau_c)^2} = \sigma_0 G'(\omega) \quad \text{or} \quad \sigma_{s,0} = \frac{\sigma_0\omega\tau_c}{1 + (\omega\tau_c)^2} = \sigma_0 G''(\omega)$$

where $G'(\omega)$ and $G''(\omega)$ still respectively represent the storage and dissipation energies of a material, still called the storage modulus and the dissipation modulus, respectively. Note that in comparison with the stress relaxation, there is an additional $(\omega\tau_c)^2$ in the denominator of $G'(\omega)$. Still $G''(\omega)/G'(\omega) = \omega\tau_c = \text{tg}\delta$, where δ is still the dissipation angle. Substituting the above two relations into the assumption of $\sigma(t)$, and noting that $\tau_c = \eta/E$ or η/G, one can derive its periodic variation with time as follows:

$$\sigma(t) = \sigma_0[G'(\omega)\cos(\omega t) + G''(\omega)\sin(\omega t)]$$

In a real application, what a material experiences is neither pure stress relaxation nor pure creep relaxation, but their complex arrangement and combination, involving multiple stresses and creep relaxations, but the fundamentals never change. As long as the physical natures of the above two relaxation are understood, any complex stress–strain process can be decomposed into their different combinations.

Viscosity and elasticity are two relative concepts, depending on temperature and time. In a very short time or at a very low temperature, molecules do not have time or sufficient thermal energy to relax, so that the object appears elastic. On the contrary, the object looks sticky. For example, at room temperature and the usual time scale, water is a viscous liquid. However, if one throws a stone to hit the water surface with a high speed at a very small angle, the stone could bounce on the surface several times to form a series of splashes, as if water is an elastic material.

More than two thousand five hundred years ago, Sun Wu mentioned such phenomena in "The Art of War by Sun Tzu": "Hitting water in such a speed to make a stone fly is due to a potential", where the "potential" means: an extremely fast speed, an extremely short time. On the other hand, if using "millennium" as the unit of time, one could find that the solid glass is also a slow–flowing liquid. The effect of temperature is even more obvious. From the physical point of view of molecular relaxation motion, time and temperature are equivalent. A longer relaxation time corresponds to a lower temperature; conversely, a higher temperature corresponds to a shorter relaxation time. It is due to this equivalent relationship that engineers can use the "aging" experiment of a material at higher temperatures to mimic its properties after the application for a long period.

The modulus of small molecule liquids (such as water, edible oil, alcohol, gasoline, etc.) is close to zero, with only viscosity, and a weak force will make a liquid to deform greatly. In daily life, one uses gravity to pour a liquid. It is also due to this extremely easy flow feature that tap water can be sent to thousands of households by applying a proper pressure. The physical nature of viscosity is the friction between molecules when they are in motion. In comparison with small molecule solvents, long macromolecular

chains entangle with each other, which increases with the chain length. Sometimes, an addition of a small amount of long macromolecular chains can removes the turbulence, reduces the flow drags and increases the flow stability, which is very important in many applications. For example, the crude oil pipeline that passes through Alaska has a length of 1,300 kilometers, as shown on the right. The addition of long-chain hydrocarbon polymers reduce the flow resistance and increase the flow rate from 1.44 to 2.14 million barrels per day, an increase of nearly 50%, which is equivalent to an additional half of the tubing line.

Readers who are familiar with electrical circuits can easily find that if regarding stress and strain as "capacitance" and "resistance", respectively, all the above mathematical formulas are identical to those in electricity. Actually, the same mathematical formulas can describe any physical process that involves storage and consumption of energy. Physics reveals the nature of changes of all things through mathematics, whose ingenuity and beauty are beyond words.

As mentioned before, as early as in 1851, George Gabriel Stokes discovered that the drag coefficient (f) experienced by a small ball with a radius of r in a pure liquid with a viscosity of η^* is $f = 6\pi\eta^* r$. Later In 1851, Albert Einstein pointed out in his doctoral dissertation that the liquid viscosity (η) of a binary mixture containing small spheres and the volume fraction (X_p) of all small spheres in the liquid follow a simple relationship as follows,

$$\eta = \eta^* (1 + 2.5 X_p)$$

According to physical intuition, it is assumed that a large spherical particle with a mass of m and a radius of r is divided into n identical small particles, and the scale relationship between r and m is $r = k_R m^{\alpha_R}$. For a given mass, let us find how dividing a big particle affects its total cross-sectional area and volume. Before the division, the total cross-sectional area and volume are that of the large particle, respectively: $S_{before} = 4\pi r^2 = 4\pi k_R^2 m^{2\alpha_R}$ and $V_{before} = 4\pi r^3/3 = 4\pi k_R^3 m^{3\alpha_R}/3$. After the division into n small particles, the radius of every small particles is $k_R (m/n)^{\alpha_R}$, so that $S_{after} = 4\pi r^2 n = 4\pi k_R^2 m^{2\alpha_R} n^{1-2\alpha_R}$ and $V_{after} = 4\pi k_R^3 m^{3\alpha_R} n^{1-3\alpha_R}/3$. Therefore, $S_{after}/S_{before} \sim n^{1-2\alpha_R}$ and $V_{after}/V_{before} \sim n^{1-3\alpha_R}$. When the particle density is uniform, $\alpha = 1/3$, $S_{after}/S_{before} \sim n^{1/3}$. The division increases the total cross-sectional area, but as expected, has no effect on the total volume. Linear flexible macromolecules in good solvents, $1/2 < \alpha_R < 3/5$, so that the division reduces the total cross-sectional area and total volume. Conversely, the connection of short chains to long chains increases the volume fraction of macromolecular chains.

As discussed in Chapter 4, the conformation of linear flexible macromolecular chains in solution is random coils. First, let us assume that each linear macromolecular chain is equivalent to a small sphere, and there are a total of N_p chains in the solution, and then the volume fraction in the above equation ($X_p = V_p / V$) is converted to a mass concentration ($C = W/V$), i.e.,

$$X_p = \frac{v_p N_p}{V} = \frac{4\pi \, k_R^3 M^{3\alpha_R} N_{AV}}{3M} \frac{W}{V} = \frac{4\pi \, k_R^3 N_{AV}}{3} M^{3\alpha_R - 1} C$$

where $N_p = N_{AV}(W/M)$, the volume of each chain ($v_p \propto R^3$) and $R \propto M^{\alpha_R}$ were used. Its substitution into the previous equation leads to

$$\eta = \eta^* \left(1 + \frac{4\pi \, k_R^3 N_{AV}}{3} M^{3\alpha_R - 1} C \right) \tag{5.41}$$

where k_R is a constant, depending on properties of macromolecules and solvents. Obviously, in a good solvent, the viscosity of linear flexible macromolecule dilute solution (η) increases with the molar mass and mass concentration of macromolecules. The physical explanation is as follows. For a given mass concentration, the total volume of long macromolecular chains is larger, so that the friction between molecules increases, resulting in a larger solution viscosity. For given macromolecules, the total volume of the chains increases with the concentration, leading to a larger friction between molecules and a higher solution viscosity.

Therefore, in order to increase the viscosity of a solution with a given mass concentration, linear chains with a higher molar mass should be added. On the other hand, for a given macromolecule and weight concentration, the volume fraction of macromolecules in the solution and the solution viscosity can also be changed by the chain conformation. An interesting example is the addition of one kind of macromolecules in the lubricating oil of a car engine. The macromolecular chains collapsed at low temperatures but no aggregation. The collapsed chains only increase in the viscosity of oil slightly, not affecting the engine starting at lower temperatures. At the high operating temperatures, the viscosity of the oil itself drops sharply, no good for the lubrication. But the added macromolecules are fully dissolved in the hot engine oil to form an extended chain conformation, increases the volume fraction and the oil viscosity so that it still lubricates the engine, as shown above.

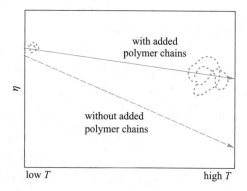

One can measure the viscosities of pure solvent and macromolecular solution, respectively.

Their ratio (η / η^*) is the **relative viscosity** (η_r). Their difference divided by the solvent viscosity is the **specific viscosity** [$\eta_{sp} = (\eta - \eta^*) / \eta^* = \eta / \eta^* - 1$]. Furthermore, the ratio of the specific viscosity to the concentration of macromolecules is the **reduced viscosity** [$\eta_{red} = (\eta - \eta^*) / \eta^* / C$], also known as the **viscosity number**. In an infinitely diluted solution, the reduced viscosity becomes **an intrinsic viscosity** ($[\eta] = \lim\limits_{C \to 0} \eta_{red}$), independent of the concentration. The relationships between various viscosities are as follows.

$$\eta_r = 1 + [\eta] C + B_2 C^2 + \cdots; \quad \eta_{sp} = [\eta] C + B_2 C^2 + \cdots; \quad \frac{\eta_{sp}}{C} = [\eta] + B_2 C + \cdots$$

where the unit of intrinsic viscosity is the reciprocal of concentration; B_2 characterizes the non-linear dependence of viscosity on concentration, dependent of the characteristics of solvent and macromolecules, reflecting the excluded volume effect, so that it can be positive, negative, or zero, similar to the second virial coefficient discussed before. According to the definition of η_{red} and $[\eta]$, eq. (5.41) can be rewritten as follows.

$$[\boldsymbol{\eta}] = \lim_{C \to 0} \frac{\eta - \eta^*}{\eta^* C} = \frac{10 \pi k_R^3 N_{AV}}{3} M^{3\alpha_R - 1} = K_\eta M^{\alpha_\eta} \tag{5.42}$$

where $\alpha_\eta = 3\alpha_R - 1$, called the Mark-Houwink-Sakurada coefficient; K_η includes all the constants and defined as

$$K_\eta = \frac{10 \pi k_R^3 N_{AV}}{3}$$

It is only related to the nature of macromolecule and solvent, not to the concentration. It can be seen that the intrinsic viscosity is related to the volume fraction of the macromolecule (X_p) in the solution, which is further related to the molar mass. It is known that for linear and flexible macromolecular chains in a good solvent, $1/2 < \alpha_\eta < 4/5$. For a given molar mass, a semi-rigid macromolecular chain has a larger volume than a linear flexible chain, $4/5 < \alpha_\eta < 2$. For a completely rigid macromolecular chain, $\alpha_\eta = 2$. As for macromolecular chains with other configurations, α_η is also obtainable from similar scaling laws. According to the above equation, if knowing the chain structure, one can estimate α_η. In theory, only one macromolecule sample with a known molar mass is required to obtain K_η. After knowing K_η and α_η, for the same kind of macromolecule with an unknown molar mass, one can measure its concentration dependence of the reduced viscosity first, then calculate its intrinsic viscosity, and finally use the above equation to determine its molar mass.

Even today with many extremely advanced electronic instruments, the capillary viscometer is still a routine device in laboratory. Its principle is based on the laminar flow of incompressible Newtonian liquid in a pipe, i.e., a relatively low flow rate, no turbulence, a parabolic liquid surface with a higher flow rate in the center and a zero flow rate at the pipe wall. The pressure difference (Δp) between the two pipe ends is related to the pipe radius and length (r and l), the liquid viscosity (η), and the flow volume and time (V and t). Jean Léonard Matie

Poiseuille and Gotthilf Heinrich Ludwig Hangen, respectively, in 1838 and 1839, experimentally discovered this dependence independently. Poiseuille officially published it in 1840, and George Gabriel Stokes completed the argument in 1845, i.e., the following Hangen−Poiseuille equation.

$$\eta = \frac{\pi r^4 \Delta p}{8 V l} t$$

When measuring the liquid viscosity, a capillary tube with a length of l and the pressure difference (Δp) generatded by gravity (g) are usually used, i.e., $\Delta p = \rho g l$, where ρ is the liquid density. Substituting it into the above equation results in $\eta = K \rho t$, where K is an instrument constant. Defining η / ρ as the dynamic viscosity (η_k), $\eta_{k,1} / \eta_{k,2} = t_1 / t_2$. Therefore, using a liquid with a known density and viscosity, one can calibrate the instrument constant of a capillary viscometer. Afterwards, the viscosity of a liquid can be calculated only if knowing its density and measuring its flow time.

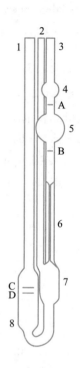

In order to measure the relatively high viscosity of macromolecular solutions, Leo Ubbelohde invented a suspended−level viscometer in 1936, now often called the Ubbelohde viscometer. As shown on the right. The viscometer has a length of about 28 cm. The length of capillary (6) is about 9 cm. The diameter of ball (5) is about 3−4 cm. The function of tube (2) is to ensure that the pressure difference between the upper and lower sides is consistent, so that the remaining net pressure difference (Δp) depends only on the liquid density and the capillary length. In practice, the entire viscometer is immersed in a temperature−controlled glass tank filled with liquid (water), with only three nozzles outside. The capillary (6) must be vertical. The scale lines A and B are positioned towards the outside, so that they can be easily observed from the outside.

The measurement usually starts from a high concentration. The solution is injected through the tube (1). After its temperature reaches a constant, the solution is sucked from the tube (3) to the upper end of the ball (4) by a suction ball. The suction ball is released so that the liquid flows into the lower containers (7 and 8) along the capillary (6) under the gravity. When the lowest point of the curved liquid surface is passing the marks A and B, respectively, two times are recorded, and their difference is the flow time (t). The measurement should be repeated at least 3 times and the flow times are averaged to reduce the measurement uncertainty. After the flow time of one solution is measured, a proper amount of solvent is added to dilute the solution through the tube (3). The solution should be sucked up and down several times in the tube (3) using the suction ball to clean two balls (4 and 5) and the capillary as well as to mix the solution thoroughly. Afterwards, the measurement is repeated one concentration after another.

After all solutions with different concentrations are measured, the viscometer is thoroughly cleaned with solvent, especially two balls (4 and 5) and the capillary, finally measure the flow time of the solvent (t_0). In a dilute solution, the density of the solution and the solvent are approximately equal, so that the viscosity is proportional to the flow time, $\eta_{red} = (t - t_0)/t_0$. In dilute solutions, the nonlinear term is very small. The relative viscosity is plotted against the concentration directly. As shown in the figure below, the intrinsic viscosity ($[\eta]$) is obtainable from the slope. If the concentration is high, the plot of relative viscosity versus concentration is not linear, so that the reduced viscosity versus concentration has to be plotted to obtain a straight line, whose intercept and slope correspond to the intrinsic viscosity and the coefficient (B_2), respectively. At present, these have become routine operations in polymer research laboratories and industries, including the control of polymerization and polymer characterization.

According to the Hangen – Poiseuille equation above, the flow time is inversely proportional to r^4. A set of the Ubbelohde viscometers with different tube diameters can be used to measure liquids with different viscosities. An appropriate tube diameter can be selected to make the flow time between 100–200 seconds, ensuring the measurement accuracy (up to 0.1%) and avoids a long measurement time. Since the viscosity measurement is cheaper, simple and easy to operate, it is often used in the polymer production line. By only measuring the vis-

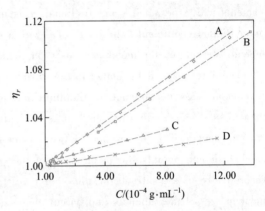

Liu M Z, Cheng R S, Wu C, et al. J. Polym. Sci. B: Polym. Phys, 1997, 345: 2421.
(Figure 3, permission was granted by publisher)

cosity of a solution with a known concentration, one can control the reaction process and molar mass.

For a polydisperse sample, the molar mass obtained from the intrinsic viscosity is a special kind of average. The detailed derivation is as follows: for a sufficiently diluted solution, all non-linear terms can be omitted. The relative viscosities of the ith component and the entire solution are

$$\eta_{sp,i} = [\eta]_i C_i + \cdots \quad \text{and} \quad \eta_{sp} = [\eta] C + \cdots$$

respectively. All components are summed. The sum of the left side is $\eta_{sp} = \sum_i \eta_{sp,i}$; and the sum of the right results in $\sum_i [\eta]_i C_i$. Therefore,

$$[\eta] C = \sum_i [\eta]_i C_i \quad \rightarrow \quad [\eta] = \frac{\sum_i [\eta]_i C_i}{C}$$

It shows that for a polydisperse sample, a weight average intrinsic viscosity is measured. The substitution of eq. (5.42) into the above equation results in

$$k_\eta M_\eta^{\alpha_\eta} = \frac{\sum_i k_\eta M_i^{\alpha_\eta} C_i}{C} \quad \rightarrow \quad M_\eta = \left(\frac{\sum_i M_i^{\alpha_\eta} C_i}{C} \right)^{1/\alpha_\eta}$$

For linear flexible macromolecular chains, $1/2 \leqslant \alpha_\eta \leqslant 4/5$. Therefore, for a polydisperse sample, the average molar mass (M_η) obtained from the intrinsic viscosity should theoretically be between the number and the weight average molar masses. For semi-rigid chains, depending on the scaling exponent, M_η is generally greater than M_w. If the chain is completely rigid, $\alpha_\eta = 2$, then $M_\eta = (M_z M_w)^{1/2}$.

The above discussion about the scaling law between the mass and size of a macromolecular chain has assumed that a macromolecular chain is a "solid hard ball" of size R with a non-uniform density so that its scaling exponent is smaller than 3, as well as all solvent molecules inside diffuse together with the chain. However, a small portion of the solvent molecules inside does not move along with the chain; namely, there is a "drainage" effect. Therefore, the chain segments will experience some frictional resistance. The entire macromolecular chain has a larger friction coefficient than a solid ball with a size of R. In other words, the solution viscosity depends strongly on the molar mass of a macromolecular chain.

There are many data related to the intrinsic viscosity of linear macromolecular chains and their molar mass dependence in literature, but there are few about branched macromolecular chains. As discussed in Chapter 4, it is difficult to prepare branched chains with a uniform subchain length. The preparation of two sets of branched chains with a variation of only the subchain length or only the molar mass of the parent branched chain is even more difficult. Theoretically, in a good solvent, the scaling relationship between the chain size (R) and the molar masses of the subchain and parent chains (M_s and M) is $R = k_R M_s^{\alpha_{R,s}} M^{\alpha_R}$, where $\alpha_{R,s} = 1/10$ and $\alpha_R = 1/2$. Experimentally, it has been confirmed that $\alpha_{R,s} = 1/10$ and $\alpha_R = 0.46$. For a given molar mass, a branched chain has a size smaller than a linear chain, so that the intrinsic viscosity of the branched chains is less dependent on the concentration. For a given weight concentration ($C = W/V$), the volume fraction of branched macromolecular chains in a solution is

$$X_p = \frac{10\pi \, k_R^3 N_{AV}}{3} M_s^{3\alpha_{R,s}} M^{3\alpha_R - 1} C$$

Substituting it into the above Einstein equation about the viscosity of small balls in liquid, and using the above definition of K_η, the following equation can be obtained

$$\eta = \eta^* (1 + K_\eta M_s^{3\alpha_{R,s}} M^{3\alpha_R - 1} C) \quad \text{or} \quad [\eta] = K_\eta M_s^{3\alpha_{R,s}} M^{3\alpha_R - 1}$$

It shows that both the molar masses of the subchain and the parent chain affect the intrinsic viscosity of branched chains. In real experiments, the number or weight average molar mass replaces the molar mass because of the inevitable polydispersity in the chain length. Note: due to the different weightings in the statistical average, the average hydrodynamic radii respectively obtained from the viscosity and dynamic laser light scattering are different.

Figure (a) below shows, in a good solvent, for a given molar mass ($M_{w,s}$) of the subchain, the intrinsic viscosity of the branched chain increases with the weight-average molar mass (M_w) of the parent chain. If using the measured $\alpha_R = 0.46$, the scaling exponent is 0.39, very close to $(3\alpha_R - 1)$. On the contrary, in good solvent, for given molar mass of the parent chain, the intrinsic viscosity also increases slightly with the molar mass of the subchain (upward along the vertical direction in the figure), and the fitting result is $3\alpha_{R,s} = 0.31$, also very close to the value (3/10) predicted by theory. Figure (b) shows that if the measured intrinsic viscosity is modified by the molar mass of the subchain $([\eta]/M_s^{3\alpha_{R,s}})$, all the experimental data in figure (a) collapse into one straight line, further confirming the scaling relationship between the size of branched chains and the molar mass of the subchain and the parent chain discussed in Chapter 4.

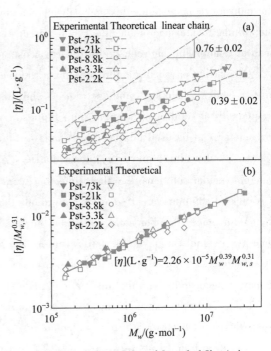

Li L W, Lu Y Y, An L J, et al. Journal of Chemical Physics, 2013, 138:114908.

(Figure 3, permission was granted by publisher)

Summary

Basic principle: The viscosity of a material originates from the friction between molecules in motion, reflecting the deformation rate of a material for a given a given external force. Below or above the melting temperature, the viscosity of a solid made of small molecules is very high or very low, only exhibiting either elastic or viscous properties. In the melting state, the material flows easily under a weak external force. While macromolecular materials, whether in solid or liquid, can show sufficient elastic and viscous properties at the same time, i.e., they have an unique viscoelasticity. This is why high polymers can be blown or stretched into a thin film and spun into a very fine giber, such as the food wrap and the high-end silk-like clothing made of ultrafine synthetic fibers.

Dilute macromolecular solutions are only viscous, just like liquids made of small molecules. However, an addition of a small amount of long-chain macromolecules can greatly increase the solution viscosity, reduce the turbulence during the high-speed flow, and improve the flow properties. This is because the solution viscosity is directly proportional to the volume fraction of particles or macromolecules. If ten short chains were connected into one long chain, the specific viscosity of a solution would increase 6 times theoretically. For an extremely diluted solution, the ratio of the specific viscosity to the concentration becomes an intrinsic viscosity, only related to the natures of macromolecules and solvent, not the concentration. It is directly proportional to the volume of the macromolecule (to be precise, the hydrodynamic volume). There is also a scaling relationship between the intrinsic viscosity and the molar mass of macromolecules.

Measuring physical quantities: Using a rheometer, one can directly measure the modulus (E or G) and the storage modulus (ω') and dissipation modulus (ω'') at different shear frequencies of solid materials. For pure liquids, a viscometer, commonly a calibrated Ubbelohde viscometer, can be used to respectively measure the flow times (t_0 and t^*) of a given volume of standard and testing liquid with known density (ρ_0 and ρ) through a capillary tube. For dilute macromolecular solutions, the flow times (t^* and t) required of a certain given volume of solvent and solutions of different concentrations through a capillary are measured.

Data processing: For solid materials, the relationship between the characteristic relaxation time (τ_c) and the modulus (η/E or η/G) measured in the stress or creep relaxation experiment can be used to calculate the viscosity. One can also divide ω'' and ω' measured in dynamic shear to obtain the dissipation angle, first multiplying or dividing by the shearing frequency, and then extrapolating to the zero frequency to get the characteristic relaxation time, from which the viscosity is further calculated.

For pure liquids, $\eta = \eta^* (\rho/\rho^*)(t/t^*)$ can be used to calculate the viscosity. For dilute macromolecular solutions, the relative viscosity (t/t^*) is calculated first, and then plot the relative viscosity against the concentration to calculate the intrinsic viscosity from the slope. For higher concentrations, one can calculate the specific viscosity $[(t - t^*)/t^*]$ and divide it by the concentration to get the reduced viscosity. Further, a plot of the reduced viscosity versus the concentration can lead to a straight line, whose intercept and slope correspond to the intrinsic viscosity and the nonlinearity coefficient, similar to the second virial coefficient, which is related to the excluded volume effect.

Size Exclusion Chromatography

As the name implies, the size exclusion chromatography is a hollow stainless steel column filled with porous particle fillers under pressure. Each particle contains many interpenetrating pores with an average diameter of r. In a steady flow field, the macromolecular chains with different sizes are injected simultaneously, as shown on the right. If there is no special attraction

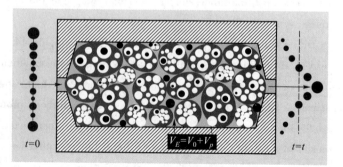

between the chains and the filler surface, including those inner surface inside the pores, shorter chains can enter the pores and stay for a longer time, while longer ones have a less chance to enter small holes. The residence time of long chains in the flow field will be shorter, so that they will be flashed out earlier, also shown in the figure. In other words, small pores have different exclusion effects on macromolecular chains with different sizes, so that a chromatographic column can separate different chains according to their sizes. This is why it is called the size exclusion chromatography (SEC). More precisely, it is the volume exclusion, which will be discussed in the next chapter.

Obviously, for a hard particle with an incompressible volume, it can only enter the pores with a size larger. For a linear flexible macromolecular chain, its size and conformation are constantly varying. The chain can enter and exit larger pores without any hindrance. If entering a pore with a size smaller than its average size, it will be compressed, deviating from the stable state with the lowest free energy, its free energy will increase. This change of free energy (ΔA) determines its probability of entering the pore. The details are as follows.

In a good solvent, the size of the macromolecular chain (R) is scaled to its molar mass (M) as $R = k_R M^{\alpha_R}$. If the size of a chain link is equal to the size of a small cylindrical pore (r), the molar mass (M_b) of the chain link is also scaled to the pore size as $r = k_R M_b^{\alpha_R}$. If a long chain is squeezed into a smaller pore, it will expand along the longitudinal direction of the pore to form a series of small cylinders with a diameter as the pore size, and its total length is l, equivalent to l/r small balls linked together, $l/r = M/M_b$. Therefore, $R/r = (M/M_b)^{\alpha_R}$, i.e., $l/r = (R/r)^{1/\alpha_R}$. In the solution, each chain has a free energy of $k_B T$. Inside the pore, each small cylinder also has a free energy of $k_B T$, so that the free energy of the chain increases to l/r times after it enters the pore. The free energy change (ΔA) is $(l/r - 1)k_B T$. The probability of entering the small pore (ψ) is

$$\psi = \exp\left(-\frac{\Delta A}{k_B T}\right) = \exp\left[1 - \left(\frac{R}{r}\right)^{\frac{1}{\alpha_R}}\right] \quad (R \geqslant r, \text{cylindrical pore})$$

Obviously, when $R = r$, $\psi = 100\%$; in the Θ and good solvents, $1/\alpha_R$ is 2 and 5/3, respectively. Therefore, in the Θ and good solvents, when $R = 2r$, ψ decreases from 100% to 4.98% and 8.80%, respectively. It is clear that the probability of macromolecules of entering the small pores drops sharply and tends to zero as the chain size R increases. If the small pore is a hollow sphere with a size of $r \leqslant R$, the pressure generated by each macromolecular chain increases from $k_B T/R^3$ to $k_B T/r^3$. In order to squeeze a chain into the hollow sphere, an additional energy is required, $\Delta A = k_B T[(R/r)^3 - 1]$. The substitution of it into the above equation results in

$$\psi = \exp\left(-\frac{\Delta A}{k_B T}\right) = \exp\left[1 - \left(\frac{R}{r}\right)^3\right] \quad (R \geqslant r, \text{shperical pore})$$

When $R = 2r$, ψ decreases from 100% to 0.1%. For a given pore size, the probability of a macromolecular chain entering a spherical pore is lower. In a size exclusion chromatography column, small holes penetrate each other. Each small pore near the surface is approximately a short cylindrical column. The macromolecular chains can stay in the penetrating network of small pores for a longer time.

The above figure shows that in addition to the volume occupied by the filler itself, the void volume (V_E, the excluded volume) of the chromatography column consists of two parts. One is the gap between all filler particles (V_0, much larger than the chain size, also called the dead volume); and another is the sum of all the volume of small pores inside the filler particles (V_p). Therefore, $V_E = V_0 + V_p$. All the chains, regardless of its size, can move in V_0, but can only enter and exit small pores with a size larger or close to its size. For a macromolecular chain with a given size, its reachable total volume inside small pores is equal to the probability of the chain entering the small pore (ψ) times the total volume of the small pores, i.e., ψV_p. Then, the total volume that such a chain can enter and exit is $V_E = V_0 + \psi V_p$, or written as $= (V_E - V_0)/V_p$. In a dilute solution, the probability of entering the small pores (ψ) also characterizes the partition coefficient of macromolecules between the small pores and the filler gap. It depends only on R/r. Obviously, the greater the probability of entering the small pores, the longer the residence time of the chain in the column will be, and vice versa. The larger the chain size, the shorter the residence time, so that longer chains flow out first.

Obviously, for a given chromatographic column and porous fillers, macromolecular chains with different sizes have different ψ. Theoretically, the smaller the dead volume (V_0), the larger the volume fraction occupied by ψV_p, and the better the separation effect. This explains why a high pressure is usually applied in manufacturing a chromatographic column. On the one hand, the high pressure compress the fillers, and reduces the dead volume; and on the other hand, the high pressure fills in more porous particles to increase the total volume of small pores (V_p).

Theoretically, all macromolecular chains larger than the largest pore size (r_{max}) flow out at the same time, and they have the shortest residence time (t_0). All macromolecular chains smaller than the smallest pore size (r_{min}) also flow out at the same time, and their retention time is the longest (t_m). Experimentally, under these two extreme conditions, the former and the latter do not flow out really at the same time, and their retention times are still slightly different in each case.

The small pores have an average diameter of r. When the size of macromolecular chains (R) is larger than r, the **distribution ratio** (k) between the gap and the small pores inside the filler is equal to the ratio of the partition coefficient times their volume ratio, i.e., $k = \psi V_p / V_0$. Therefore, the probabilities of macromolecular chains in the gap and inside the pores are the ratios of their respective volumes available to macromolecular chains (V_0 or ψV_p) to the total exclusion volume ($V_0 + \psi V_p$).

$$\frac{\psi V_p}{V_0 + \psi V_p} = \frac{k}{1 + k} \quad \text{and} \quad \frac{V_0}{V_0 + \psi V_p} = \frac{1}{1 + k}$$

On the other hand, the total residence time (t) of a macromolecular chain in the chromatographic column is equal to its residence time (t_p) in the small pores (may enter and exit multiple small holes) plus its residence time in the gap (t_0), i. e., $t = t_p + t_0$. Correspondingly, the distribution ratio is equal to their ratio, $k = t_p/t_0$. Substituting them and the aforementioned ψ expression into the above equation results in

$$\exp\left[1 - \left(\frac{R}{r}\right)^{\frac{1}{\alpha_R}}\right] = \frac{V_0}{V_p t_0}(t - t_0) \quad \text{or written as } t = t_0 + \frac{V_p t_0}{V_0}\exp\left[1 - \left(\frac{R}{r}\right)^{\frac{1}{\alpha_R}}\right] \quad (5.43)$$

If the pore size (r) has a wide distribution, one has to consider the probability of a chain entering all pores smaller than the chain size (R). Namely, the left and right side of the above equation are integrated from 0 to R and from t_0 to t., respectively. The advantage of using the fillers with a broad distribution in the pore size is that one can use only one chromatographic column to analyze macromolecular samples with different average molar masses, not necessary to change the chromatographic column. The disadvantage is a poorer resolution. Using $R = k_R M^{\alpha_R}$ to replace R in the above equation with molar mass (M) used experimentally, one gets

$$t = t_0 + \frac{V_p t_0}{V_0}\exp\left[1 - \left(\frac{k_R}{r}\right)^{\frac{1}{\alpha_R}}M\right] \quad (5.44)$$

The residence (retention) time depends nonlinearly on the molar mass. For ease of use, the molar mass on the right side is usually put into the logarithmic space and written as $A + B\log M$, where A and B are two constants that need to be calibrated experimentally. The calibration requires a series of (at least two) narrowly distributed standard samples with different known molar masses. Therefore, the above equation becomes

$$t = A + B\log M \quad (5.45)$$

From the principles of thermodynamics, one can also get the above equation as follows. In the chromatographic column, the space between the fillers (gaps) and the space inside the small pores represent two different phases, called the mobile and stationary phase. The solution of macromolecule chains with a concentration of C_0 before its injection into the chromatographic column is chosen as a reference state, and its chemical potential is μ^0. After the injection, one part of the macromolecular chain stays in the gap of the packing, and another part enters the small pores. The macromolecule concentrations in the mobile and stationary phase are C_m and C_s, respectively. For the macromolecular chains with a size larger than the pore size, if squeezed into the pore (i. e., in the stationary phase), they will experience an additional pressure difference in comparison with in the mobile phase (like the osmotic pressure difference, also recorded as π). At equilibrium, the chemical potential of each mole of macromolecules in the two phases must be equal, that is

$$\mu^0 + RT\ln\frac{C_m}{C_0} = \mu^0 + RT\ln\frac{C_s}{C_0} + \pi V_m \quad \rightarrow \quad -\pi V_m = RT\ln\frac{C_s}{C_m}$$

where $\mathrm{d}G = V\mathrm{d}p$ was used. In an isothermal process, partially differentiation of both sides of the above equation against the moles of macromolecules (n_p) results in $d\mu = V_m \, dp$. If the molar volume (V_m) of macromolecules is independent of pressure, the integrations of the both side from μ^0 to μ on the left and from p to ($p + \pi$) on the right, respectively, leads to $\Delta\mu = \mu_s - \mu_m = V_m \pi$, reflecting the change of the molar free energy of macromolecules with the pressure difference. In the previous discussion, it is known that the increase of free energy in the process of squeezing one molar macromolecular chains into small pores with a diameter of r is ($l/r - 1$) RT, and $l/r = (R/r)^{1/\alpha_R}$. Therefore,

$$\psi = \frac{C_s}{C_m} = \exp\left[1 - \left(\frac{R}{r}\right)^{\frac{1}{\alpha_R}}\right]$$

Using the relationship between ψ and the distribution ratio and performing the same derivation, one can obtain eq. (5.44), not repeated here. Physically, it means that the residence time relies on the size of the macromolecule. For a given molar mass, the size of a macromolecular chain varies with the solvent quality. Therefore, for a given kind of macromolecules and solvent, one needs a series (at least two) of narrowly distributed standard samples with different molar masses to measure their residence times, calibrate the chromatography column, and get A and B. After it, this calibration curve can be used to convert each measured residence time of the same kind of macromolecular sample with an unknown molar mass in the same solvent into a molar mass.

As discussed before, the viscosity of an extremely dilute macromolecular solution is scaled with its molar mass. Therefore, by measuring the solution viscosity online, the molar mass at each residence time is determinable, leading to a viscosity detector. However, its application also needs a pre-calibrated curve. Since the 1990s, different on-line laser light scattering detectors have gradually become popular. As an absolute method, it requires no calibration, which is its major advantage. However, the scattering cell inside is a slender cylinder, and the incident laser beam moves forward along the center of the column. Therefore, one has to correct the refraction distorted scattering volume each scattering angle. Water with a refractive index of ~1.33 is the worst case. The instrument designer has assumed that the scattered light intensity of small molecular solvent has no angular dependence after the volume correction, so that the computer program artificially adds a proportional (calibration constant) constant to each angle to force the intensities of scattered light at different angles equal to each other. In this way, one has no way to know whether an error is due to an incorrect refraction correction or a misalignment or defects of the instrument itself. In discussing static LLS, all the alignment difficulties were discussed in details. Unfortunately, in promoting this kind of on-line LLS detectors, it is often portrayed as an simple black box.

In its actual operation, a constant flow rate is generally used. In order to ensure a constant flow rate and reduce pulses, two or more propulsion pumps are generally used alternately. For an adjusted size exclusion chromatograph, after the flow rate is steady, the detection baseline should be a horizontal line. At this time, the macromolecular solution can be injected. The concentration should not be too high to saturate the column, reducing the resolution; but it should also not be too low to result in a weak signal of each component flowing out too, making the detection difficult. It is necessary to adjust the concentration according to the system. In principle, under the premise of meeting the detection sensitivity, the solution should be as dilute as possible. Note that for a given flow rate, the residence time is proportional to the outflow (retention) volume, so that a more intuitive and direct flow volume is often used experimentally.

By calibrating the chromatographic column, the effluent (retention) volume (residence time) is converted into molar mass, or using an on-line LLS detector to directly measure the molar mass corresponding to each effluent volume (each fraction). It only solves the problem of the abscissa in the molar mass distribution. The weighting (concentration or mass) corresponding to each molar mass has to be measured by another detector, including an on-line ultraviolet absorption spectrometer or a differential

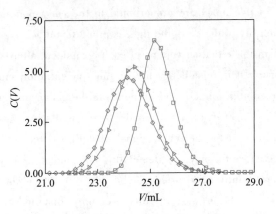

Wu C, Siddiq M, Jiang S H, et al. J. Appl. Polym. Sci., 1995, 58:1779.

(Figure 2, permission was granted by publisher)

refractometer. The prerequisite of using UV absorption to measure the concentration is that macromolecules have UV absorption, not universal. Therefore, in actual use, differential refractometers are universal. Of course, if the refractive indices of macromolecules and solvent are similar, the error of differential refractive index measurement will be larger. The figure on the right shows several typical effluent-volume concentration profiles.

Theoretically, the area under each curve is proportional to the macromolecular concentration before the injection. The division $C(V)$ by the area leave to a normalized distribution, $C(V)/\sum_i C_i$. When converting the retention volume or residence time into molar mass, it is proportional to the logarithm of molar mass ($\log M$), $dV \propto d(\log M)$. The conversion is only a translation in a linear space. When the molar mass distribution is plotted, the abscissa must be the logarithm of molar mass ($\log M$), and the ordinate only needs to be normalized by the new area. If the abscissa is plotted as molar mass (M), each ordinate must be divided by the corresponding molar mass since $dV \propto dM/M$, and then normalized. Otherwise, the component with a higher molar mass will weighted too much. Such a mistake often occurs in literature.

Obviously, it the weighting is measured by the concentration (mass), the directly calculated average molar mass is the **weight average molar mass**: $M_w = \sum_i M_i C_i / \sum_i C_i$. One can also calculate The number average molar mass can also be calculated indirectly according to the definition, $\sum_i C_i / \sum_i (C_i / M_i)$. Their ratio is the molar mass distribution width, called the polydispersity index. Readers should remember clearly that the weight average molar mass from the size exclusion chromatography is more reliable, while the number average molar mass is a calculation result.

A combination of two or more different online detectors involves a connection problem. There are generally two connecting methods: series and parallel. The serial connection is normally used; whose advantages are simple installation and operation. However, different detectors have different sensitivity to molar mass. For example, UV absorption and differential refractive index are proportional to the corresponding mass (weight) of each component. The intensity of the scattered light is proportional to the square of the molar mass. How to correct and align the retention volume of each component after they pass through different detectors is not so trivial. It is related to whether the concentration measured in vacuum ultraviolet absorption or differential refraction corresponds to the molar mass measured in the viscometer or LLS intensity.

For example, for a broadly distributed sample, when a component with a high molar mass is present on the detection signal of a LLS detector, the UV absorption detector and differential refractometer may not detect this component; vice versa. The salesperson will not inform you these troubles and problems, but tell you not to worry. In fact, these are exactly what a serious user needs to worry. Otherwise, only distorted and even wrong results will be obtained. Nowadays, the electronic instrumentation brings convenience, but it also lead to many specious and misleading conclusions in the literature.

The porous fillers with an interpenetrating network structure normally are crosslinked polymers. They usually swell in organic solvents to form gels. Therefore, if the flowing liquid is an organic solvent, size exclusion chromatography is also often called gel permeation chromatography (GPC). When the solvent is water, although the fillers do not swell, the "gel permeation" is still used, called aqueous GPC. In the above discussion, no interaction between macromolecules and the surfaces of the filler and the pores is assumed, i.e., $\Delta H = 0$. If $\Delta H > 0$, macromolecules and the surfaces of the filler and the pore repel each other, reducing the available volume, so that this kind of columns are normally not prepared.

In contrary, if chemically or physically modifying the surfaces of the fillers and pores to make them strongly attract macromolecules, $\Delta H < 0$ and $|\Delta H| > |T\Delta S|$, another type of separation chromatography column can be prepared. For example, in protein engineering, a small peptide tag with a specific amino acid sequence is usually added to one end of synthesized protein chains, so that it can have a specific interaction with the corresponding chemical group in the separation chromatographic column. In this way, the desired proteins are adsorbed on the chromatographic column first, then other proteins are removed by elution, and finally the required protein is stripped from the chromatographic column with an eluting solvent that can break the adsorption.

Summary

Basic principle: A hollow stainless steel chromatographic column is filled with porous fillers. Each filler particle contains interpenetrating cylindrical pores. When a solvent flows through the chromatographic column steadily, a certain volume of macromolecular solution is injected into the chromatographic column. The macromolecular chains can enter those pores smaller than its size with a certain probability. For small pores with a given size, this probability significantly decreases as the chain size increases. Obviously, with a steady flow of solvent, those chains that are more likely to enter the pores will not be easily flushed out chromatography column, i.e., with a longer residence time, or a larger retention volume. Using this physical phenomenon, macromolecular chains with different sizes can be separated. Each chromatographic column can be calibrated by using a series of narrowly distributed standard samples with different molar masses to obtain the relationship between the residence time or retention volume and molar mass. An online LLS detector can also be used to measure the scattered light intensity corresponding to each retention volume directly. Using the corresponding concentration measured by another online concentration detector (ultraviolet absorption or differential refractometer), the molar mass distribution and weight average molar mass of the macromolecule can be obtained, and the number average molar mass can be obtained

Measurable physical quantity: For a polydisperse sample, using an online viscometer or laser light scattering spectrometer, one can measure the viscosity or the scattered light intensity corresponding to the residence time (retention volume) of each component in the column. Using another online UV adsorption or differential refraction, one can also measure the concentration (mass) of each component corresponding to each residence time (retention volume).

Data processing: Using the pre-calibrated standard curve, one can convert each residence time (retention volume) into a molar mass; or directly calculate the molar mass from each measured viscosity or scattered light intensity. At the same time, each measured adsorption or refraction results should be converted to a concentration (mass). The current instruments can do these calculations effortless, so that the data processing is very simple. However, it is not worry free. It is still important to know the principles and calculations behind.

When plotting such obtained molar mass distributions, the molar mass coordinate should be its logarithm ($\log M$), not plotting it as a linear coordinate (M). If a linear plot is required, each of the obtained concentration (ordinate) must be divided by its corresponding molar mass; namely, $C(V_i)/M_i$. Otherwise, a wrong weight average molar mass will be calculated based on such a distribution. This is a mistake often made in the literature because users did not fully understand the basic principles.

第六章　常见实验方法的组合应用

在上述各种需要标定的实验方法中,可利用特性黏度、尺寸排除色谱中的停留体积、动态激光光散射测得的线宽分布与大分子摩尔质量之间的各种标度律或依赖关系,表征大分子摩尔质量分布或平均摩尔质量。在所有的标度律或依赖关系中,均有两个常数:标度指数和前置因子。所以,至少需要两个以上摩尔质量差别较大的同种窄分布大分子标准样品,以及分别测量它们的相关性质,方可得到两个常数和一条标定曲线。

事实上,对大多数聚合物,并不存在一系列具有不同摩尔质量的窄分布样品,获得窄分布标准样品也非易事。除了最常用的商品化窄分布聚苯乙烯标准样品外,仅有为数不多的其他几种标准样品。问题是如何在没有标准样品的时候利用这些需要标定的方法来表征一个多分散的大分子样品。显然,从数学原理上,如果两个标度关系具有内在的联系,可用两种方法分别测量同一个样品,然后将它们的结果结合起来,联解两个方程,得到两个常数,建立标度关系。下面,将逐一地介绍不同方法的联合应用。

特性黏度和尺寸排除色谱:由前述特性黏度的公式(5.39)已知 $[\eta]=(10\pi N_A/3)$ $k_R^3 M^{3\alpha_R-1}$。显然,$[\eta]M\propto R^3\propto V$,这意味着,对不同种类的大分子,特性黏度乘以摩尔质量对应着同样的分子体积,与溶液性质无关。标准样品和未知样品的特性黏度可以用黏度计分别测定。因此,可利用某种一系列已知摩尔质量的窄分布标准样品,分别测得它们的停留体积(V)和特性黏度$[\eta]$,得到一条新的标定曲线,

$$V=A'+B'\log(M[\eta])$$

通常采用一系列商品化的聚苯乙烯标准样品在四氢呋喃中的溶液。注意,这样一条标定曲线对其他种类的大分子"普适"。在实际操作中,对一个未知摩尔质量的大分子,可先利用黏度计测得其溶液的平均特性黏度,再利用尺寸排除色谱测得其流出体积分布。利用上述标定曲线,很容易将流出体积分布换算成 $\log(M[\eta])$ 的分布。在对数空间中,$\log(M[\eta])=\log M+\log[\eta]$,再利用已测得的特性黏度,将 $\log(M[\eta])$ 换算成 $\log M$。注意,这些换算在对数空间中均是线性变换,故纵坐标(浓度)仅需用分布曲线的面积归一即可。如果在作图中将 $\log M$ 换作 M,每一个纵坐标都必须除以其对应的摩尔质量。否则,权重就会出错,从而导致错误的重均摩尔质量和错误的分布宽度。

细心的读者会发现,这里采用了一个近似,即用待测样品的平均特性黏度代替了其中每个具有不同摩尔质量的组分的特性黏度。所幸的是,实验数据显示,这样并没带来太大的误差。对于给定的尺寸排除色谱仪、色谱柱和溶剂,不同组构、不同摩尔质量的高聚物分子的 V 对 $\log(M[\eta])$ 的作图均落在一条主标定曲线上。注意:即使对一个绝对单分散

的样品,所测的停留体积也是一个分布,即色谱柱有增宽效应。虽然,为了解决该增宽问题花费了许多努力,但效果有限。因此,便于操作的尺寸排除色谱作为一个在工业生产中控制反应和质量的仪器毫无问题。但是,大分子领域里的研究者必须清楚地认识到该方法得到的只是一个相对分布;另外,基线的选取也会极大地影响分布宽度的测定。如果进行定量的摩尔质量依赖性研究,尺寸排除色谱绝不是一个可取的方法。

特性黏度和静态激光光散射:在静态激光光散射和特性黏度的讨论中已知以下标度关系,

$$R_g = (\langle R_g^2 \rangle_z)^{1/2} = k_{R_g} M_w^{\alpha_{R_g}} \quad \text{和} \quad [\eta] = K_\eta M_w^{\alpha_\eta}$$

其中,

$$K_\eta = \frac{10\pi N_A}{3} k_{R_g}^3 \quad \text{和} \quad \alpha_\eta = 3\alpha_{R_g} - 1$$

对一个待表征的样品,可由黏度和静态激光光散射分别测得特性黏度、重均摩尔质量和 z 均回转半径,代入上述两个标度关系得包括两个未知数的两个方程,可解,得 k_{R_g} 和 α_{R_g}。

进一步,对于支化型大分子链,从前述特性黏度和静态激光光散射的讨论已知,

$$R_g = k_{R_g} M_s^{\alpha_{R_g,s}} M_w^{\alpha_{R_g}} \quad \text{和} \quad [\eta] = K_\eta M_s^{3\alpha_{R_g,s}} M_w^{3\alpha_{R_g}-1}$$

其中,实验上已证实两个标度常数:$\alpha_{R_g,s}$ 和 α_{R_g}。对一个未知支化细节的支化链样品,可将特性黏度和激光光散射数据结合起来获得支化子链的信息。具体而言,先利用黏度法测得其特性黏度,再通过静态激光光散射测量其 z 均回转半径和重均摩尔质量,在它们各自的标度律中,K_η 与 k_{R_g} 相连,仅有 k_{R_g} 和 M_s 未知;两个方程,两个未知数,可解得 k_{R_g} 和 M_s。可得支化大分子在给定的溶剂中的 R_g 和 $[\eta]$ 对母链和子链平均摩尔质量的标度依赖关系。

尺寸排除色谱和动态激光光散射:两个方法所测得均是与大分子尺寸有关的分布。二者联用应该是一件理所当然的方法,然而在实际的使用中并不普及。这不归于方法本身。尺寸排除色谱的大部分使用者侧重大分子合成,对激光光散射的原理和应用不甚了解。因此,本书特向读者详细介绍这一方法,以便在以后的研究中采用它。

由尺寸排除色谱和动态激光光散射可分别测得同一个样品流出体积分布 $C(V)$ 和线宽分布 $G(\Gamma)$,如果稀溶液体系的松弛仅源于平动扩散,$G(\Gamma)$ 可直接通过 $\Gamma = Dq^2$ 转换成平动扩散系数分布 $G(D)$,其中,散射矢量(q)对给定的散射角度为一常数。流出体积(V)和平动扩散系数(D)以及摩尔质量(M)之间的依赖关系为

$$V = A + B\log M \quad \text{和} \quad D = k_D M^{\alpha_D} \rightarrow \log D = \log k_D + \alpha_D \log M$$

因此,

$$V = A' + B'\log D$$

其中,

$$A' = A - \frac{B\log k_D}{\alpha_D} \quad \text{和} \quad B' = \frac{B}{\alpha_D}$$

将 $V = A' + B'\log D$ 两边平方,得 $V^2 = A'^2 + 2A'B'\log D + B'^2\log^2 D$。再将它们分别求平均,

$$\langle V \rangle = A' + B'\langle \log D \rangle \quad \text{和} \quad \langle V^2 \rangle = A'^2 + 2A'B'\langle \log D \rangle + B'^2\langle \log^2 D \rangle$$

其中,

$$\langle V \rangle = \frac{\int V C(V) \, \mathrm{d}V}{\int C(V) \, \mathrm{d}V} \quad \text{和} \quad \langle V^2 \rangle = \frac{\int V^2 C(V) \, \mathrm{d}V}{\int C(V) \, \mathrm{d}V}$$

$$\langle \log D \rangle = \frac{\int \log D C(V) \, \mathrm{d}V}{\int C(V) \, \mathrm{d}V} \quad \text{和} \quad \langle \log^2 D \rangle = \frac{\int \log^2 D C(V) \, \mathrm{d}V}{\int C(V) \, \mathrm{d}V}$$

为了求出后两个平均,必须将 $C(V)\mathrm{d}V$ 换成对 $G(D)\mathrm{d}D$ 的积分。流出体积分布 $C(V)$ 是一个浓度(重量)分布,且 $\mathrm{d}V \propto \mathrm{dlog}M$。因此,在摩尔质量的对数空间中,$C(V)$ 和重量分布($f_w(M)$)的积分都对应着总浓度。互相之间有着如下正比关系:

$$\int C(V) \, \mathrm{d}V \propto \int f_w(M) \, \mathrm{d}M \propto \int f_w(M) M \mathrm{dlog}M \rightarrow C(V) \mathrm{d}V \propto f_w(M) M \mathrm{dlog}M$$

另一方面,平动扩散系数分布是一个散射光强分布,且 $\mathrm{dlog}D \propto \mathrm{dlog}M$。因此,在摩尔质量的对数空间中,$G(D)$ 和重量分布 $f_w(M)$ 的积分都对应着总散射光强。互相之间存在着下列正比关系:

$$\int G(D) \, \mathrm{d}D \propto \int G(D) D \mathrm{dlog}D \propto \int f_w(M) M \mathrm{d}M \propto \int f_w(M) \, M^2 \mathrm{dlog}M$$

$G(D) D \mathrm{dlog}D \propto f_w(M) M^2 \mathrm{dlog}M$。因此,$C(V)\mathrm{d}V \propto [G(D)D/M] \mathrm{dlog}D$,其中 $D^{1/\alpha_D} \propto M$。将其代入 $\langle \log D \rangle$ 和 $\langle \log^2 D \rangle$ 的积分,可得

$$\langle \log D \rangle = \frac{\int (\log D) G(D) \, D^{1-1/\alpha_D} \mathrm{dlog}D}{\int G(D) \, D^{1-1/\alpha_D} \mathrm{dlog}D} \quad \text{和} \quad \langle \log^2 D \rangle = \frac{\int (\log^2 D) G(D) \, D^{1-1/\alpha_D} \mathrm{dlog}D}{\int G(D) \, D^{1-1/\alpha_D} \mathrm{dlog}D}$$

其中,保留了 D 在对数空间。原因是在 Laplace 反演中,线宽分布中的横坐标线宽在对数空间中间隔相等,故换算而来的平动扩散系数分布在对数空间中也是等间隔的。采用 $\log D$ 后,积分就变成了纵坐标 $G(D)$ 乘上相应的 D 后的简单相加,操作简单。在上两式中,仅有一个未知数 α_D。依据重均摩尔质量的定义,

$$M_w = \frac{\int M f_w(M) \, \mathrm{d}M}{\int f_w(M) \, \mathrm{d}M}$$

可分别从 $C(V)$ 和 $G(D)$ 求得重均摩尔质量,对应同一个大分子样品,二者必须相等,即

$$M_{w,SEC} = \frac{\int M C(V) \, \mathrm{d}V}{\int C(V) \, \mathrm{d}V} \quad \text{和} \quad M_{w,DLS} = \frac{\int G(D) \, \mathrm{d}D}{\int \frac{G(D)}{M} \mathrm{d}D}$$

由 $V = A' + B' \log D$ 和 $D = K_D M^{\alpha_D}$,可得 $M = k_D^{1/\alpha_D} 10^{(V-A')/(B'\alpha_D)}$ 和 $M = k_D^{1/\alpha_D}/D^{1/\alpha_D}$,即

$$M_{w,SEC} = M_{w,DLS} \rightarrow \frac{k_D^{\frac{1}{\alpha_D}} \int 10^{\frac{V-A'}{B'\alpha_D}} C(V) \, \mathrm{d}V}{\int C(V) \, \mathrm{d}V} = \frac{k_D^{\frac{1}{\alpha_D}} \int G(D) \, \mathrm{d}D}{\int G(D) \, D^{1/\alpha_D} \mathrm{d}D}$$

其中,两边的k_D^{1/α_D}互相消去。由第四章大分子构象的讨论已知,对不同组构的大分子,α_D具有不同的值。对线型柔性链,$1/2 \leqslant \alpha_D \leqslant 3/5$;对刚性型链,$\alpha_D = 1$。对良溶剂中的支化型链,母链和子链摩尔质量上的标度指数分别为$\alpha_D = 1/2$ 和$\alpha_{D,s} = 1/10$。所以,根据大分子的化学结构,不难知道α_D的大致范围。

因此,可以先预估一个α_D的值,然后依据定义,从 $G(D)$ 算出$\langle \log D \rangle$和$\langle \log^2 D \rangle$,再结合从 $C(V)$ 算出的$\langle V \rangle$和$\langle V^2 \rangle$,联立求解出 A' 和 B',将它们和预估的α_D一起代入上式。重复以上操作,反复迭代α_D,直至两边的计算值之差达至最小值。获得最佳的α_D后,再利用在静态激光光散射中测得的重均摩尔质量(M_w),就可由上式左边或者右边求出k_D。进而,从 A',B',k_D和α_D算出 A 和 B。如此联用两个方法,就可只用一个多分散的样品,同时标定出由尺寸排除色谱得到的流出体积和由动态激光光

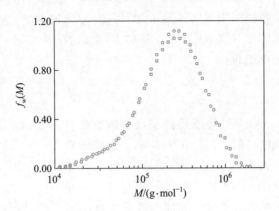

Wu C,Zhang Y B,Yan X H,et al.Acta Polymerica Sinica, 1995,3:343−348.

(Figure 3,permission was granted by publisher)

散射得到的平动扩散系数与摩尔质量之间的标度律。上图显示了从 $G(D)$ 和 $C(V)$ 得到的摩尔质量分布。可见利用这一方法获得的结果与在尺寸排除色谱中利用"普适"标定的结果(小方块符号)相当接近。

静态和动态激光光散射:随着商品化激光光散射仪的普及,许多实验室购置了研究型的包括静态和动态激光光散射的仪器,比小型的、基于动态光散射原理的粒度仪贵十倍。然而,大部分实验室均将静态激光光散射部分闲置,仅将其当作一个普通的粒度仪使用。本书特地介绍将静态和动态激光光散射结合起来的一些应用,使读者可以明白其背后的原理。希望在明白原理之后,读者可在研究中充分发挥仪器的功能,获得更多的信息。

外推到散射角度和浓度为零后,动态激光光散射中测得的平动扩散系数分布($G(D)$)是一个以散射光强为权重的分布,即其横坐标为对数空间中的平动扩散系数,而纵坐标正比于散射光强。$G(D)$ 与时间平均散射光强(I)和摩尔质量的重量分布($f_w(M)$,正比于浓度分布 $C(M)$)有着如下的内在关系:

$$I \propto M_w C = \frac{\int M C(M) \, \mathrm{d}M}{\int C(M) \, \mathrm{d}M} C \propto \int M f_w(M) \, \mathrm{d}M \propto \int G(D) \, \mathrm{d}D$$

鉴于$D = K_D M^{-\alpha_D}$,在对数空间里,$\mathrm{d}\log D \propto \mathrm{d}\log M$,上式可被重写成

$$\int M^2 f_w(M) \, \mathrm{d}\log M \propto \int G(D) D \, \mathrm{d}\log D \propto \int G(D) D \, \mathrm{d}\log M$$

即在对数空间中,$M^2 f_w(M) \propto G(D) D$ 或写成$M f_w(M) \propto G(D) D/M$。将其代入重均摩尔质量的定义,可得

$$M_w = \frac{\int M f_w(M)\,\mathrm{d}M}{\int f_w(M)\,\mathrm{d}M} = \frac{\int M^2 f_w(M)\,\mathrm{dlog}M}{\int M f_w(M)\,\mathrm{dlog}M} = \frac{k^{\frac{1}{\alpha_D}}\int G(D)\,D\mathrm{dlog}D}{\int D^{\frac{1}{\alpha_D}}G(D)\,D\mathrm{dlog}D}$$

注意,由目前仪器自带的 Laplace 反演软件(CONTIN)计算出的平动扩散系数分布在对数空间中等间隔地增加,即 $\Delta\mathrm{log}D = \mathrm{log}D_{i+1} - \mathrm{log}D_i = \mathrm{log}D_i - \mathrm{log}D_{i-1}$ 为一常数。纵坐标 $f(D)$ 已经考虑了权重,为 $G(D)$ 乘以 D,即 $f(D) = G(D)D$,故上式中的积分可化作以下的简单加和,

$$M_w = \frac{k_D^{1/\alpha_D}\sum_i f(D)}{\sum_i D^{1/\alpha_D}f(D)}$$

如果计算所得的平动扩散系数分布已经归一,$\sum_i G(D)D = 1$。上式中存在两个未知数。需要联立两个方程式,方可解出。具体途径有二。一是依据大分子组构和溶剂性质估计 α_D 或通过其他方法获得 α_D。获得 α_D 后,即可利用上式求出 k_D,进而可从扩散系数分布得到摩尔质量的重量分布。上一章讨论动态激光光散射时列举了四个同种线型大分子样品的平动扩散系数分布,右图显示了分别利用这四个平动扩散系数分布获得的摩尔质量的重量分布。在这一具体的换算中,利

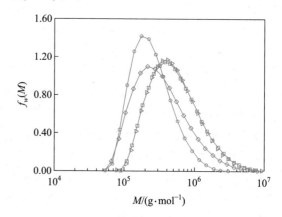

Wu P Q, Siddiq M, Chen H Y, etc. Macromolecules, 1996, 29:277.

(Figure 4, permission was granted by publisher)

用了从黏度测量中得到 α_η 和 $\alpha_\eta = (1 + \alpha_D)/3$。有兴趣的读者可查阅原始文献,从而知道更多的细节。

实验上,可先用静态激光光散射寻获一个大分子溶液在不同温度下的二阶维里系数,再通过内插或者外推获得二阶维里系数为零时的 Θ 温度。在该 Θ 温度下,$\alpha_D = 1/2$。由动态激光光散射获得平动扩散系数分布 $f(D)$,并结合来自静态激光光散射的 M_w,可得 k_D。进而,可将 $f(D)$ 转换成摩尔质量的重量分布。

显然,如果有两个平均摩尔质量不同的同种大分子多分散样品,则可分别由静态和动态激光光散射测得两个重均摩尔质量 M_w 和两个平动扩散系数分布 $f(D)$。依据上述方程,联立求解,可得 k_D 和 α_D。进而,每个 $f(D)$ 可被转换成一个对应的摩尔质量的重量分布 $f_w(M)$。

另一个近似的方法是,假定分别由静态和动态激光光散射测得的重均摩尔质量和 z 均平动扩散系数也依据相同的标度关系,$\langle D \rangle = K_D M_w^{-\alpha_D}$,上式可被重写成

$$\frac{1}{\langle D \rangle^{\frac{1}{\alpha_D}}} = \frac{\sum_i f(D)}{\sum_i D^{\frac{1}{\alpha_D}}f(D)}$$

在估计的范围内,迭代 α_D,使得两边计算结果之差达至最小,获得最佳的 α_D。再利用

前式求出 k_D。随后,可将扩散系数分布换算成摩尔质量的重量分布。细心的读者可以发现,上式实际上就是 $\langle D \rangle = \langle D^{1/\alpha_D} \rangle^{\alpha_D}$。原则上,两边并不完全相等,因此这是一个近似。笔者在 20 世纪 80 年代中期,利用这一方法,从测得的平动扩散系数分布,首次获得了聚四氟乙烯摩尔质量分布等分子参数,揭开了其在 1938 年被在 DuPont 的 Roy J. Plunkett 发现后一直遮盖的面纱。

微分折射仪和激光光散射:最简单的联用就是将前述新型的利用激光作光源和用位置检测器测量折射偏离距离的 $2f$-$2f$ 设计集成到一台激光光散射仪中,可见第五章讨论静态激光光散射时显示的装置图。这里讨论的联用,不是简单地利用常规测量的微分折射指数从散射光强计算出重均摩尔质量。两者结合可以解决一些更困难的问题。一些示例如下。

研究大分子溶液时,常会遇到少量聚集的问题,比如目前在生物化学和分子细胞学的研究中,常用已经被普遍接受的高浓度尿素溶液或 $8 \text{ mol} \cdot L^{-1}$ GdmCl 缓冲液来溶解蛋白质。然而,在这样制备的蛋白质"溶液"中,并非所有的蛋白质均以单个分子的形式完全溶解。严格地说,这些并非真溶液。在许多的蛋白质研究中,这些残存的、少量的蛋白质聚集体并不影响研究和结果。然而,如果研究蛋白质聚集动力学,这些少量的聚集体就成了溶液亚稳态中的小"种子"(核或晶核)。它们的含量多少和大小都会直接地影响实验结果。常用的检查 β-折叠的荧光方法无法检测到这样少量的非结构性聚集体。其他光学方法也束手无策。在许多实验室,研究蛋白质聚集动力学的人都知道很难定量地重复实验得到同样的结果,但往往都选择视而不见,继续发表文章。

由于动态激光光散射测量的线宽分布的权重是散射光强,散射光强又正比于粒子(分子)数目和其摩尔质量的平方,线宽分布对少量的聚集极度敏感。理论上,一个含有十个粒子(分子)的小聚集体散射的光强是单个粒子(分子)的 100 倍。

下图显示了原始朊病毒蛋白(Sup35NM)和经热敏性 PNIPAM 短链在 31 m 位接枝修饰后的朊病毒蛋白(Sup35NM-31m-PNIPAM)在 $8 \text{ mol} \cdot L^{-1}$ GdmCl 缓冲液中的流体力学半径的光强分布(三角形)。存在着比单个朊病毒蛋白大 ~10^2 倍的聚集体。假定每个聚集体的密度均匀,其质量就是单个蛋白链的 10^6 倍。因此,一个这样的小聚集体与 10^{12} 个朊病毒蛋白散射同样的光。将右图换算成重量分布时,聚集体小峰就会变得很小;如果换算成数目分布时,聚集体小峰就看似消失了。用 20 nm 孔径的滤膜可以完全除去这些小聚集体(圆圈)。浓缩 12 倍后,聚集体也不再出现(方形),显示聚集不是一个动态过程。将微分折射仪和激光光散射结合提供了一个绝佳、可能唯一的方法来定量地研究这些数量分数极小的聚集体。

当散射体积中存在少量聚集体时,散

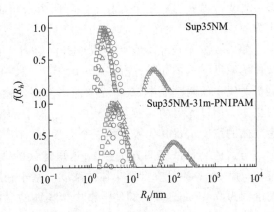

Wang Y J, Wu C.Biochemistry,2017,56:6575.

(Figure 3,permission was granted by publisher)

射光强包含着单链和聚集链二者的贡献。因此,静态和动态激光光散射分别测得的重均摩尔质量和平动扩散系数分布都包括了二者的贡献。假定在二者之间存在一个摩尔质量 M_b,在对数空间可将重均摩尔质量分成两部分,即

$$M_w = \frac{\int_0^\infty M^2 f_w(M)\,\mathrm{d}\log M}{\int_0^\infty M f_w(M)\,\mathrm{d}\log M} = \frac{\int_0^{M_b} M^2 f_w(M)\,\mathrm{d}\log M}{\int_0^\infty M f_w(M)\,\mathrm{d}\log M} + \frac{\int_{M_b}^\infty M^2 f_w(M)\,\mathrm{d}\log M}{\int_0^\infty M f_w(M)\,\mathrm{d}\log M}$$

$$= \frac{\int_0^{M_b} M^2 f_w(M)\,\mathrm{d}\log M}{\int_0^{M_b} M f_w(M)\,\mathrm{d}\log M} \cdot \frac{\int_0^{M_b} M f_w(M)\,\mathrm{d}\log M}{\int_0^\infty M f_w(M)\,\mathrm{d}\log M} + \frac{\int_{M_b}^\infty M^2 f_w(M)\,\mathrm{d}\log M}{\int_{M_b}^\infty M f_w(M)\,\mathrm{d}\log M} \cdot \frac{\int_{M_b}^\infty M f_w(M)\,\mathrm{d}\log M}{\int_0^\infty M f_w(M)\,\mathrm{d}\log M}$$

上式的两项求和中,共有四项。第一和第三项分别为单链和聚集体的重均摩尔质量 $(M_{w,s}$ 和 $M_{w,m})$;第二和第四项分别为单链和聚集体的浓度分数 $(x_s = C_s/C$ 和 $x_m = C_m/C)$,

$$x_s + x_m = 1$$

当浓度和散射角度趋于零时,由时间平均散射光强得到的重均摩尔质量等于二者之和

$$M_w = M_{w,s}x_s + M_{w,m}x_m$$

另一方面,线宽分布正比于散射光强。在对数空间中,$\mathrm{d}\log\Gamma \propto \mathrm{d}\log D \propto \mathrm{d}\log R_h$,三者之间的变换全部线性。当它们互相变换时,仅需对每个分布的积分归一即可。在下列讨论中,选取直觉的流体力学半径分布 $F(R_h)$。单链峰和聚集体的峰对应的面积比 (A_R) 等于单链和聚集体的散射光强之比,即 $A_R = (M_{w,s}x_s)/(M_{w,m}x_m)$,

$$A_R = \frac{\int_0^{R_{h,b}} F(R_h)\,\mathrm{d}R_h}{\int_{R_{h,b}}^\infty F(R_h)\,\mathrm{d}R_h} = \frac{\int_0^{R_{h,b}} R_h F(R_h)\,\mathrm{d}\log R_h}{\int_{R_{h,b}}^\infty R_h F(R_h)\,\mathrm{d}\log R_h} = \frac{M_{w,s}}{M_{w,m}}\frac{x_s}{x_m}$$

如前所述,仪器所带的 Laplace 反演软件(CONTIN)中的 R_h 具有对数间隔,$\Delta\log R_h$ 为一常数。纵坐标为 $R_h F(R_h)$,记作 $f(R_h)$。因此,上述积分就成为对 $f(R_h)$ 的加和,可由动态激光光散射获得。三个方程中共有四个未知数:$M_{w,s}$、$M_{w,m}$、x_s 和 x_m。为了联立求解这三个方程,必须从其他途径知道其中的一个变量。

如果大分子样品本身因合成或其他原因含有聚集体,可先用微分折射仪测得少量聚集体的浓度。具体如下,先配制一个浓度为 10 ~ 100 mg/mL 大分子溶液,将一部分注入微分折射仪中的两个池中,获得光束原点。再用离心或过滤的方法除去另一部分溶液中的聚集体。用过滤后的溶液置换其中一个池中的原始溶液。测量光束从原点的偏离,算出聚集体的浓度。将其代入上面三个方程,可计算出其余的三个变量。

与其他测量浓度的方法相比,这里建议的微分折射方法可以一次扣除溶液中大量单链的贡献,而只测量少量聚集体的浓度。基于新颖 $2f - 2f$ 设计制造的微分折射仪的测量精度约为 0.01 mg/mL。即使溶液中聚集体的含量低至 1%,其浓度也可被精确地测量,误差仅有百分之几。因此,可算出单链的重均摩尔质量、聚集体内的平均链数和其他微观参数。

如果是一个通过改变实验条件(如温度、酸碱度、离子强度等)诱导的聚集,单链的摩尔质量($M_{w,s}$)就是一已知常数。因此,仅需将分别由静态和动态激光光散射测得的重均摩尔质量和流体力学半径分布代入以上三个方程,即可解出$M_{w,m}$和x_m。进而,可算出每个聚集体里单链的平均数目(聚集数,$M_{w,m}/M_{w,s}$)和聚集体的浓度($C_m = x_m C$)。

在蛋白质聚集的研究中,一般已知蛋白单链的摩尔质量。因此,分别由静态和动态激光光散射测得M_w和$f(R_h)$后,即可联立解出方程。过滤除去朊病毒蛋白溶液中极少量的初始小聚集体后,就可通过聚集体大小和分布的变化来研究朊病毒蛋白在溶液中的聚集动力学过程。如下图(a)中散射光强的相对变化所示,将朊病毒蛋白溶液用磷酸盐缓冲盐水溶液(简称 PBS)稀释 200 倍可诱导其中的朊病毒蛋白聚集。即散射光强随时间增加,其中$\langle I \rangle_0$为朊病毒蛋白单链在溶液中无聚集时的初始散射光强。经热敏性 PNIPAM 短链修饰过的朊病毒蛋白聚集较慢,且三次重复实验的结果稳定。将这些散射光强随时间的变化与 Smoluchowski 凝聚方程结合,可得出在不同聚集时间时,含有 i 个朊病毒蛋白的聚集体数目 $n_i(t)$ 随聚集数的变化,如下图(b)所示。即使经过六个小时,溶液中单链($i=1$)数目虽然有所下降,仍占绝大多数。图(b)中的插图则显示了 $n_i(t)$ 随时间的变化。其他方法根本无法定量地跟踪这样的变化。

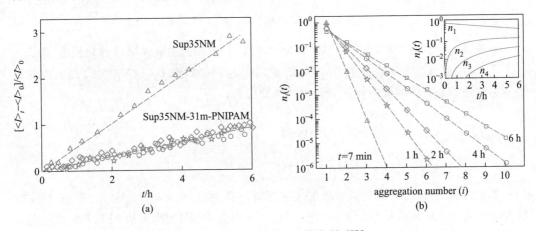

(a) (b)

Wang Y J,Wu C.Biochemistry,2017,56:6575.
(Figures 9 and 11,permission was granted by publisher)

另一个例子,聚集发生在两个不同种的大分子之间。例如,在制备非病毒基因载体时,人们利用各种带正电荷的小分子、大分子或二者络合和"凝聚"一条长的、带负电荷的、含有一个所需基因的脱氧核糖核酸(DNA)质粒。络合使得一条长链质粒蜷缩成一个表面带正电荷的小聚集体。

文献中常把这样一个络合过程归于静电相互作用,大错特错也! 混合前,二者均有各自的小分子反离子和静电作用。况且,一条带电大分子链根本不可能知道一个反离子是小分子还是另一条链上带相反电荷的基团。熵变驱动着这样一个络合。当两根带相反电荷的大分子络合时,虽然它们失去了各自的平动熵,但释放出各自的、大量的小分子反离子,使得它们获得更多的平动熵。体系移向熵增方向:络合。

在这个研究领域,已发表了十万篇以上的文章。但在绝大部分的研究中,小的"DNA+

聚合物"络合体(Polyplex)被当作小的"黑盒子",不知分子细节,如每个络合体里有几条长的质粒,有多少条聚合物短链。只是把这些小"黑盒子"加入含细胞的培养液,观察各种不同细胞(被当作另一个"黑盒子")的内吞和基因转染的宏观效果。直至数年前,微分折射和激光光散射的结合方使得我们有可能研究络合体的组成和打开了这些"黑盒子",定量地揭示了由带负电荷的 DNA 质粒和带正电荷的聚合物链的络合过程。

首先,假定每个络合物含有 i 条长的、带负电荷的、摩尔质量为 M_{DNA} 的 DNA 质粒和 m 条短的、带正电荷的、摩尔质量为 M_{PEI} 的 PEI 链,故络合物的摩尔质量为

$$M_{i,P} = M_{DNA}i + M_{PEI}m$$

其中,下标"i"表示该络合体平均含有 i 条 DNA 质粒。混合前,M_{DNA} 和 M_{PEI} 为已知常数。在络合物中,DNA 和 PEI 的质量分数分别为 $x_{DNA} = M_{DNA}i/(M_{DNA}i + M_{PEI}m)$ 和 $x_{PEI} = M_{PEI}m/(M_{DNA}i + M_{PEI}m)$。络合物的微分折射指数遵循质量加和定律,即

$$\left(\frac{\partial n}{\partial C}\right)_P = x_{DNA}\left(\frac{\partial n}{\partial C}\right)_{DNA} + x_{PEI}\left(\frac{\partial n}{\partial C}\right)_{PEI} = x_{DNA}\left(\frac{\partial n}{\partial C}\right)_{DNA} + (1 - x_{DNA})\left(\frac{\partial n}{\partial C}\right)_{PEI}$$

其中,利用微分折射仪可预先分别测定 DNA 和 PEI 的微分折射指数。另一方面,在静态激光光散射中,络合体和 DNA 的时间平均散射光强(I_P 和 I_{DNA})分别为

$$I_P = \left(\frac{\partial n}{\partial C}\right)_P^2 \sum_{i=1}^{N} C_{i,P}M_{i,P} \quad \text{和} \quad I_{DNA} = \left(\frac{\partial n}{\partial C}\right)_{DNA}^2 C_{DNA}M_{DNA}$$

其中,$C_{i,P}$ 是含有 i 条 DNA 质粒的络合体浓度,其求和为络合体浓度的总浓度,即 $C_P = \sum_{i=1}^{N} C_{i,P}$。注意,$C_{i,P}$ 与络合体中 DNA 的质量($W_{i,DNA}$)和浓度($C_{i,DNA}$)有关。将上式的两边分别相除和并将前面复合体的微分折射指数代入,可得

$$\frac{I_P}{I_{DNA}} = \frac{\left(\dfrac{\partial n}{\partial C}\right)_P^2 \sum_{i=1}^{N} C_{i,P}M_{i,P}}{\left(\dfrac{\partial n}{\partial C}\right)_{DNA}^2 C_{DNA}M_{DNA}}$$

其中,没有络合的 PEI 短链的散射光强相对很弱,故可忽略不计。另外,每条 DNA 质粒可络合的 PEI 短链数目有一上限,记为 m_{max};从混合溶液中 DNA 和 PEI 浓度和它们各自的摩尔质量,也可算出平均可供每条 DNA 质粒络合的 PEI 短链数(m_0)。如果每个络合体内仅有一根 DNA 质粒和 $m_0 > m_{max}$,将有多余的、不可络合的 PEI 短链。这样,每条 DNA 质粒均完全被 PEI 短链覆盖,无法形成一个络合体内有两条 DNA 质粒。因此,上式的求和仅有一项,$i = 1$,$N = 1$ 和 $m = m_{max}$,可被重写成

$$\frac{I_P}{I_{DNA}}\left(\frac{\partial n}{\partial C}\right)_{DNA}^2 C_{DNA}M_{DNA} = \left(\frac{\partial n}{\partial C}\right)_P^2 C_{1,P}M_{1,P}$$

实验上,在上式左边,所有参量均可由实验测得或已知;上式右边中,将微分折射指数、$C_{1,P}$ 和 $M_{1,P}$ 代入,仅有一个未知参量 m_{max},故由数值计算求出。如果 $m_{max,0} < m_{max}$,在极稀的混合物中,则必须考虑且仅需考虑 $N = 2$ 的二聚体存在。假定其质量分数为 x_w,

$$\frac{I_P}{I_{DNA}}\left(\frac{\partial n}{\partial C}\right)_{DNA}^2 C_{DNA}M_{DNA} = \left(\frac{\partial n}{\partial C}\right)_P^2 C_P(1 - x_w)M_{1,P} + C_P x_w M_{2,P}$$

其中,$M_{2,P} = 2M_{1,P}$。近似地采用 $m = m_{max}$,仍可从左边的实验值,通过数值运算,得到

右边唯一的未知量x_w。依据定义，还可求出二聚体的数目分数(x_n)如下：

$$x_n = \frac{N_2}{N_1 + N_2} \quad \text{和} \quad x_w = \frac{2M_{1,P}N_2}{M_{1,P}N_1 + 2M_{1,P}N_2} = \frac{2N_2}{N_1 + 2N_2} \rightarrow x_n = \frac{x_w}{2 - x_w}$$

依据上述方法，获得了带负电荷的 DNA 和不同摩尔质量、不同组构的带正电荷聚合物 PEI 形成的络合体的一系列微观参数，如表 1 所示。当投料中的正、负电荷之比大于 6时，混合溶液中的络合体仅含有一条 DNA 质粒。当摩尔质量相近时，与线型 PEI 链相比，每条 DAN 质粒可络合更多的支化 PEI 链。介绍上述两个与生物大分子有关的例子仅是为了抛砖引玉。希望读者在今后使用激光光散射仪时，可以举一反三地将静态和动态激光光散射以及微分折射仪结合在一起，解决一些其他实验方法无法解决的问题。

表 1　在不同溶液中形成的不同的 DNA/PEI 络合聚集体的实验和计算参数

$(N/P)_{feed}$	sample	$m_{PEI,max}$	$\langle I \rangle_{polyplexes}/\langle I \rangle_{DNA}$	m_{PEI}	x_n	x_w
3	lPEI − 2.5k	557	2.43	759	14	25
	lPEI − 25k	56	2.06	60	2	4
	bPEI − 2.0k	696	2.27	862	9	16
	bPEI − 25k	56	2.30	71	10	18
6	lPEI − 2.5k	1 114	2.72	879	0	0
	lPEI − 25k	112	2.82	92	0	0
	bPEI − 2.0k	1 392	3.33	1 387	0	0
	bPEI − 25k	112	3.62	122	0	0

Dai Z J, Wu C. Macromolecules, 2012, 45: 4346−4353. (Table 1, permission was granted by publisher)

小　结

在表征一个未知大分子样品的摩尔质量分布时，许多的实验方法需要事先标定仪器。标定本身需要一系列（至少两个）不同摩尔质量的、窄分布的大分子标准样品，方可得到一条标定曲线。利用该标定曲线，方可在相同实验条件（溶剂、温度、压强等）下，表征其他同种的大分子。然而，在多数情形下，并无同种的大分子标准样品可用。正因如此，不得不将两种或两种以上、具有内在联系的实验方法结合起来，通过联解两个或多个方程，得到标定曲线。

将黏度和尺寸排除色谱联用得到一条"普适"标定曲线是一个常见的联用。下列是一些不常见的联用。将黏度和静态激光光散射联用得到流体力学尺寸和摩尔质量之间的标度关系，其可被进一步用来将由动态激光光散射测得的线宽分布转换成摩尔质量分布。静态和动态激光光散射的联用可将线宽分布转换成摩尔质量分布。微分折射仪和激光光散射的联用可定量地研究大分子溶液中少量的聚集体结构和聚集动力学。

Chapter 6. Combined Applications of Common Experimental Methods

In the previously discussed calibration-required experimental methods, using various scaling laws and dependences between the intrinsic viscosity, the retention volume (residence time) in the size exclusion chromatography, the linewidth distribution measured by dynamic laser light scattering and the molar mass of macromolecules, one can characterize the macromolecular molar mass distribution or average molar mass. However, in all the scaling laws or dependences, there are two constants: scaling exponent and pre-factor. Therefore, one has to have at least two of the same kinds of narrowly distributed standard samples with a large difference in molar masses first, and then measure their related properties, respectively, to obtain the two constants and a calibration curve.

In reality, for most polymers, there is no series of narrowly distributed samples with different molar masses, and it is not easy to obtain narrowly distributed standard samples. In the market, apart from the most commonly used commercial narrowly distributed polystyrene standard samples, there are only a few other standard samples. The problem is how to use these calibration-required methods to characterize a polydisperse macromolecule sample when there is no standard sample. Obviously, from the mathematical principle, if their scaling relationships are inherently related, one can also use two methods to measure one broadly distributed sample, respectively, and then combine their results to solve two equations to obtain the two constants. The combined applications of different methods will be introduced one by one as follows.

<u>Intrinsic Viscosity and Size Exclusion Chromatography</u>: It has been known from the previous intrinsic viscosity equation (5.39) that $[\eta] = (10\pi N_A/3) k_R^3 M^{3\alpha_R-1}$. Obviously, $[\eta]M \propto R^3 \propto V$, which means that for different kinds of macromolecules, the intrinsic viscosity multiplied by the molar mass corresponds to the same molecular volume, independent on the solution properties. The intrinsic viscosities of the standard samples and the unknown sample can be respectively measured with a viscometer. Therefore, one can use a series of standard samples with known molar masses to measure their retention volume (V) and intrinsic viscosity $[\eta]$ to obtain a new calibration curve.

$$V = A' + B'\log(M[\eta])$$

Usually a series of solutions of commercial polystyrene standards in tetrahydrofuran are used. Note that such a calibration curve is "universal" for other types of macromolecules. In actual operation, for a macromolecule with an unknown molar mass, its average intrinsic viscosity of the solution is measured by a viscometer first, and then the retention volume distribution is measured by size exclusion chromatography. Using the above calibration curve, one can easily convert the retention volume distribution into a distribution of $\log(M[\eta])$. In the logarithmic space, $\log(M[\eta]) = \log M + \log[\eta]$, and then use the measured intrinsic viscosity to convert $\log(M[\eta])$ to $\log M$. Note that these conversions are all linear transformations in logarithmic space, so that the ordinate (concentration) only needs to be normalized by the area under the distribution curve. If exchanging $\log M$ to M in the plot, each ordinate must be divided by its corresponding molar mass. Otherwise, the weights will be wrong, leading to a wrong weight average molar mass and a wrong distribution width.

Careful readers can find that an approximation is used here, that is, the average intrinsic viscosity is used instead of the intrinsic viscosity of each component with different molar mass in the measured sample. Fortunately, the experimental data shows that this brings no big error. For a given size exclusion chromatography instrument, chromatographic column and solvent, the plots of V versus $\log(M[\eta])$ of polymers with different configurations and molar masses fall on one master calibration curve.

Note: even for an absolutely monodisperse sample, the measured retention volume is still a distribution, that is, the column has a broadening effect. Although much effort has been spent to solve this broadening effect, the success is rather limited. Therefore, the easy-to-operate size exclusion chromatography is no problem as an instrument for controlling reaction and quality in industrial production. However, researchers in the field of macromolecules must clearly realize that it only gives a relative distribution. In addition, how to selecting a baseline will greatly affect the measured distribution width (polydispersity index). If a quantitative molar mass dependent study is performed, size exclusion chromatography is by no means a desirable method. However, most people have already pretended to know that there is a problem.

Intrinsic viscosity and static laser light scattering: In the discussion of static laser light scattering and intrinsic viscosity, the following scaling relationship is known

$$R_g = (\langle R_g^2 \rangle_z)^{1/2} = k_{R_g} M_w^{\alpha_{R_g}} \quad \text{and} \quad [\eta] = K_\eta M_w^{\alpha_\eta}$$

where

$$K_\eta = \frac{10\pi N_A}{3} k_{R_g}^3 \quad \text{and} \quad \alpha_\eta = 3\alpha_{R_g} - 1$$

For a sample to be characterized, the intrinsic viscosity, weight-average molar mass, and z-average radius of gyration are measurable by viscosity and static LLS, respectively. Substituting them into the above two scale relationships results in two equations with two unknown variables, so that it can be solved to get k_{R_g} and α_{R_g}.

Furthermore, for branched macromolecular chains, the foregoing discussion of intrinsic viscosity and static laser light scattering show

$$R_g = k_{R_g} M_s^{\alpha_{R_g,s}} M_w^{\alpha_{R_g}} \quad \text{and} \quad [\eta] = K_\eta M_s^{3\alpha_{R_g,s}} M_w^{3\alpha_{R_g}-1}$$

where two scaling exponents have been confirmed experimentally: $\alpha_{R_g,s}$ and α_{R_g}. For a branched chain sample with unknown branching details, one can combine the intrinsic viscosity and laser light scattering data to get the subchain information. Specifically, the intrinsic viscosity is measured by a viscometer method first, and then the z-average radius of gyration and the weight average molar mass are measured by static LLS. In their respective scaling laws, K_η and k_{R_g} are connected. Only k_{R_g} and M_s are unknown. Two equations with two unknowns are solvable to obtain k_{R_g} and M_s. The scaling dependences of R_g and $[\eta]$ on the average molar masses of the parent chain and subchains of branched macromolecules in a given solvent can be obtained.

Size Exclusion Chromatography and Dynamic Laser Light Scattering: Both of them measure the distributions related to the size of macromolecules. The combination of the two should be a natural way, but it is not popular in actual use. This is not attributable to the method itself. Most users of size exclusion chromatography focus on the macromolecular synthesis, and do not know much about the principles and applications of laser light scattering. Therefore, this book introduces this method to readers in detail, so that it can be adopted in research later.

The retention volume distribution $C(V)$ and linewidth distribution $G(\Gamma)$ of the same sample can be measured separately by size exclusion chromatography and dynamic laser light scattering. If the relaxation of a dilute solution only comes from translational diffusion, $G(\Gamma)$ can be directly converted into a translational diffusion coefficient distribution $G(D)$ by $\Gamma = Dq^2$, where the scattering vector (q) is a constant for a given scattering angle. The retention volume (V) is related to the translational diffusion coefficient (D) and the molar mass (M) as

$$V = A + B\log M \quad \text{and} \quad D = k_D M^{\alpha_D} \rightarrow \log D = \log k_D + \alpha_D \log M$$

Therefore,

$$V = A' + B'\log D$$

where

$$A' = A - \frac{B\log k_D}{\alpha_D} \quad \text{and} \quad B' = \frac{B}{\alpha_D}$$

Two sides of $V = A' + B'\log D$ are squared, resulting in $V^2 = A'^2 + 2A'B'\log D + B'^2 \log^2 D$. Their respective averages are

$$\langle V \rangle = A' + B'\langle \log D \rangle \quad \text{and} \quad \langle V^2 \rangle = A'^2 + 2A'B'\langle \log D \rangle + B'^2 \langle \log^2 D \rangle$$

where

$$\langle V \rangle = \frac{\int V C(V)\,\mathrm{d}V}{\int C(V)\,\mathrm{d}V} \quad \text{and} \quad \langle V^2 \rangle = \frac{\int V^2 C(V)\,\mathrm{d}V}{\int C(V)\,\mathrm{d}V}$$

$$\langle \log D \rangle = \frac{\int \log D C(V)\,\mathrm{d}V}{\int C(V)\,\mathrm{d}V} \quad \text{and} \quad \langle \log^2 D \rangle = \frac{\int \log^2 D C(V)\,\mathrm{d}V}{\int C(V)\,\mathrm{d}V}$$

In order to find the last two averages, $C(V)\,\mathrm{d}V$ must be replaced by the integral of $G(D)$

dD. The retention volume distribution $C(V)$ is a concentration (weight) distribution, and $\mathrm{d}V \propto \mathrm{d}\log M$. Therefore, in the logarithmic space of the molar mass, both the integrals of $C(V)$ and weight distribution $(f_w(M))$ correspond to the total concentration. The proportional relationship is

$$\int C(V)\,\mathrm{d}V \propto \int f_w(M)\,\mathrm{d}M \propto \int f_w(M)\,M\mathrm{d}\log M \rightarrow C(V)\,\mathrm{d}V \propto f_w(M)\,M\mathrm{d}\log M$$

On the other hand, the translational diffusion coefficient distribution is a scattered light intensity distribution, and $\log D \propto \mathrm{d}\log M$. Therefore, in the logarithmic space of molar mass, the integrals of $G(D)$ and weight distribution $f_w(M)$ correspond to the total scattered light intensity. There exist the following proportional relationships between them.

$$\int G(D)\,\mathrm{d}D \propto \int G(D)\,D\mathrm{d}\log D \propto \int f_w(M)\,M\mathrm{d}M \propto \int f_w(M)\,M^2\mathrm{d}\log M$$

Namely, $G(D)\,D\mathrm{d}\log D \propto f_w(M)\,M^2\,\mathrm{d}\log M$. Therefore, $C(V)\,\mathrm{d}V \propto [G(D)\,D/M]\,\mathrm{d}\log D$, where $D^{1/\alpha_D} \propto M$. Substituting it into the integrals of $\langle \log D \rangle$ and $\langle \log^2 D \rangle$ results in

$$\langle \log D \rangle = \frac{\int (\log D)\,G(D)\,D^{1-1/\alpha_D}\mathrm{d}\log D}{\int G(D)\,D^{1-1/\alpha_D}\mathrm{d}\log D} \quad \text{and} \quad \langle \log^2 D \rangle = \frac{\int (\log^2 D)\,G(D)\,D^{1-1/\alpha_D}\mathrm{d}\log D}{\int G(D)\,D^{1-1/\alpha_D}\mathrm{d}\log D}$$

where D is retained in the logarithmic space. The reason is that in the Laplace inversion, the abscissa linewidth in the linewidth distribution is evenly divided in the logarithmic space, so that the translated translational diffusion coefficient from the inversion is also evenly divided in the logarithmic space. After using $\log D$, the integral becomes a simple addition of the ordinate $G(D)$ multiplied by the corresponding D, and the operation is simple. In the above two equations, there is only one unknown α_D. According to the definition of weight average molar mass,

$$M_w = \frac{\int M f_w(M)\,\mathrm{d}M}{\int f_w(M)\,\mathrm{d}M}$$

The weight average molar mass can be obtained from $C(V)$ and $G(D)$ respectively, corresponding to the same macromolecule sample, so that they must be equal, namely

$$M_{w,SEC} = \frac{\int M C(V)\,\mathrm{d}V}{\int C(V)\,\mathrm{d}V} \quad \text{and} \quad M_{w,DLS} = \frac{\int G(D)\,\mathrm{d}D}{\int \frac{G(D)}{M}\,\mathrm{d}D}$$

From $V = A' + B'^{\log D}$ and $D = K_D M^{\alpha_D}$, $M = k_D^{1/\alpha_D} 10^{(V-A')/(B'\alpha_D)}$ and $M = k_D^{1/\alpha_D}/D^{1/\alpha_D}$ is obtainable, i.e.,

$$M_{w,SEC} = M_{w,DLS} \rightarrow \frac{k_D^{\frac{1}{\alpha_D}}\int 10^{\frac{V-A'}{B'\alpha_D}}C(V)\,\mathrm{d}V}{\int C(V)\,\mathrm{d}V} = \frac{k_D^{\frac{1}{\alpha_D}}\int G(D)\,\mathrm{d}D}{\int G(D)\,D^{1/\alpha_D}\mathrm{d}D}$$

where k_D^{1/α_D} on the two sides cancel each other. In Chapter 4, the discussion of macromolecular conformation shows that for macromolecules with different configurations, the scaling exponent (α_D) have different values. For linear flexible chains, $1/2 \leqslant \alpha_D \leqslant 3/5$; and for rigid chain, $\alpha_D = 1$. For branched chains in a good solvent, the scaling exponents on the molar masses of the parent and sub chain are $\alpha_D = 1/2$ and $\alpha_{D,s} = 1/10$, respectively. Therefore, according to the chemical structure of a macromolecule, it is not difficult to know the approximate range of α_D.

Therefore, one can estimate a value of α_D and according to the definition, calculate $\langle \log D \rangle$ and $\langle \log^2 D \rangle$ from $G(D)$ first, and then combine $\langle V \rangle$ and $\langle V^2 \rangle$ calculated from $C(V)$, solving equations simultaneously to get A' and B' and substitute them and the estimated α_D into the above equation. One can repeat the above operation, iterate α_D until the difference between the calculated values of the both sides reaches a minimum. After getting the best α_D, one can use the weight average molar mass (M_w) measured in static LLS, one can obtain k_D either from the left or the right side of the above equation. Further, A and B can be calculated from A', B', k_D and α_D. Such a combination of the two methods enables us to use one polydisperse sample to calibrate the scaling laws between the retention volume from size exclusion chromatography and the translational diffusion coefficient from dynamic LLS and the molar mass. The figure on the right shows the molar mass distribution obtained from $G(D)$ and $C(V)$, revealing that the results obtained using this method and using "universal" calibration (small squares) is rather close.

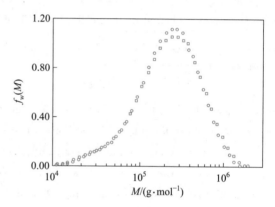

Wu C, Zhang Y B, Yan X H, et al. Acta Polymerica Sinica, 1995, 3: 343–348.

(Figure 3, permission was granted by publisher)

<u>Static and dynamic laser light scattering</u>: As commercial LLS instruments become popular, many research laboratories have purchased research-type instruments, including static and dynamic LLS, ten times more expensive than small particle sizers based on the principle of dynamic LLS. However, most laboratories leave the static LLS part idle and just use it as an ordinary particle size analyzer. This book specially introduces some applications by combining static and dynamic LLS to make readers understand the principles behind them. Hope that after comprehending the principles, readers can fully utilize the instrument functions in research to get more information.

After the scattering angle and concentration are extrapolated to zero, the translational diffusion coefficient distribution ($G(D)$) measured in dynamic LLS is a scattered light intensity weighted distribution, that is, its abscissa is the translational diffusion coefficient in the logarithmic space, and the ordinate is proportional to the scattered light intensity. $G(D)$ and the time average scattered light intensity (I) and the weight distribution of molar mass $(f_w(M))$, proportional to the concentration distribution of molar mass $C(M))$ has the following internal relations:

$$I \propto M_w C = \frac{\int M C(M)\,\mathrm{d}M}{\int C(M)\,\mathrm{d}M} C \propto \int M f_w(M)\,\mathrm{d}M \propto \int G(D)\,\mathrm{d}D$$

Due to $D = K_D M^{-\alpha_D}$, $\mathrm{d}\log D \propto \mathrm{d}\log M$ in the logarithmic space. The above equation can be rewritten as

$$\int M^2 f_w(M)\,\mathrm{d}\log M \propto \int G(D) D\,\mathrm{d}\log D \propto \int G(D) D\,\mathrm{d}\log M$$

Namely, in the logarithmic space, $M^2 f_w(M) \propto G(D) D$ or written as $M f_w(M) \propto G(D)D/M$. Its substitution into to the definition of the weight average molar mass leads to

$$M_w = \frac{\int M f_w(M)\,\mathrm{d}M}{\int f_w(M)\,\mathrm{d}M} = \frac{\int M^2 f_w(M)\,\mathrm{d}\log M}{\int M f_w(M)\,\mathrm{d}\log M} = \frac{k^{1/\alpha_D}\int G(D) D\,\mathrm{d}\log D}{\int D^{1/\alpha_D} G(D) D\,\mathrm{d}\log D}$$

Note: the translational diffusion coefficient distribution calculated by the current Laplace inversion software (CONTIN) that comes with the instrument increases at a regular interval in the logarithmic space, that is, $\Delta \log D = \log D_{i+1} - \log D_i = \log D_i - \log D_{i-1}$ is a constant. The ordinate $f(D)$ has considered the weighting, $G(D)$ multiplied by D; namely, $f(D) = G(D)D$, so that the integral in the above equation can be transformed into a simple summation as follows,

$$M_w = \frac{k_D^{1/\alpha_D} \sum_i f(D)}{\sum_i D^{1/\alpha_D} f(D)}$$

If the calculated translational diffusion coefficient distribution has been normalized, $\sum_i G(D)D = 1$. There exist two unknowns in the above equation. Two equations need to be combined before they can be solved. There are two specific ways. One is to estimate α_D from the macromolecular structure and solvent properties or obtain α_D by other methods. After obtaining α_D, one can use the above equation to get obtain k_D, and further the weight distribution of molar mass from the diffusion coefficient distribution. In the previous discussion of dynamic LLS, the translational

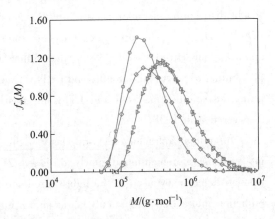

Wu P Q, Siddiq M, Chen H Y, et al. Macromolecules, 1996, 29:277.

(Figure 4, permission was granted by publisher)

diffusion coefficient distributions of four linear macromolecule samples of the same kind were listed. The figure above shows the weight distribution of molar masses obtained from these four translational diffusion coefficient distributions. In this specific transformation, α_η and $\alpha_\eta = (1 + \alpha_D)/3$ were used. Those interested readers can consult the original literature to know more details.

Experimentally, one can use static LLS to find the second virial coefficient of a macromolecular solution at different temperatures first, and then obtain the Θ temperature at which the second virial coefficient is zero by interpolation or extrapolation. At this Θ temperature, $\alpha_D = 1/2$. One can use dynamic LLS to obtain the translational diffusion coefficient distribution $f(D)$, and combine it with M_w from static LLS to obtain k_D. Further, $f(D)$ can be converted to a weight distribution of molar mass.

Obviously, if there are two polydisperse samples of the same kind of macromolecule with different average molar masses, two weight-average molar masses M_w and two translational diffusion coefficient distributions $f(D)$ can be measured by static and dynamic LLS, respectively. According to the above equations, solving them simultaneously, one can get k_D and α_D. Further, each $f(D)$ can be converted to a corresponding molar mass weight distribution $f_w(M)$.

Another approximation method is to assume that the weight average molar mass and z-average translational diffusion coefficient measured by static and dynamic LLS obey the same scaling relationship, $\langle D \rangle = K_D M_w^{-\alpha_D}$, the above equation can be rewritten as

$$\frac{1}{\langle D \rangle^{1/\alpha_D}} = \frac{\sum_i f(D)}{\sum_i D^{1/\alpha_D} f(D)}$$

Within the estimated range, one can iterate α_D to minimize the difference between the calculated results on the two sides to obtain the best α_D. Then, k_D is obtainable using the previous equation. Afterwards, a translational diffusion coefficient distribution converted to a weight distribution of molar mass. Careful readers can find that the above equation is actually $\langle D \rangle = \langle D^{1/\alpha_D} \rangle^{\alpha_D}$. In principle, the two sides are not exactly equal, so that this is an approximation. In the mid-1980s, the author used this method to obtain the molar mass distribution of polytetrafluoroethylene (PTFE) and other molecular parameters for the first time, lifting the veil that had covered PTFE for nearly 50 years since Roy J. Plunkett in DuPont discovered it in 1938.

Differential refractometer and laser light scattering: The simplest combination is to integrate the aforementioned newly designed $2f - 2f$ refractometer in which a laser and a position detector are used as the light source and the device to measure the deviation distance of refraction into a LLS instrument. It was shown in the device diagram in the static LLS section of the previous chapter. The combination discussed here is not to simply use a routinely measured differential refractive index to calculate the weight average molar mass from the scattered light intensity. Some much more difficult problems are solvable by a combination of the two. Some examples are as follows.

When studying macromolecular solutions, one often encounters a small amount of aggregation. For example, in the current research of biochemistry and molecular cytology, the generally accepted high-concentration urea solution or 8M GdmCl buffer is commonly used to dissolve proteins. However, in such prepared protein "solutions", not all protein molecules are completely dissolved as individual molecules. Strictly speaking, these are not true solutions. In many protein studies, these remaining, small amounts of protein aggregates have no effect on the study and results. However, if the protein aggregation kinetics is studied, these small aggregates become small "seeds" (nucleus or crystal nucleus) in a metastable state of solution. Their content and size will directly affect the experimental results. The commonly used fluorescence method of detecting β folding is not able to find a small amount of non-structural aggregates. Other optical methods are also helpless. In many laboratories, people who have studied the protein aggregation kinetics know that it is difficult to repeat their experiments to get the same results quantitatively, but often choosing to ignore it and continuing to publish papers.

Since the linewidth distribution measured in dynamic LLS is the scattered light intensity-weighted, and the scattered light intensity is proportional to the particle (molecule) number and the square of its molar mass, the linewidth distribution is extremely sensitive to a small amount of aggregation. Theoretically, a small aggregate made of 10 particles (molecules) scatter 100 times of light than individual particles (molecules).

The figure below shows the intensity distributions of the hydrodynamic radius (triangles) of the wild prion protein (Sup35NM) and the prion protein modified with a short thermally sensitive PNIPAM chain grafted at position 31 m (Sup35NM-31m-PNIPAM) in 8M GdmCl buffer. There exist aggregates that are $\sim 10^2$ times larger than individual prion protein chains. If each aggregate has a uniform density, its mass is 10^6 times larger than individual protein chains.

Therefore, one such small aggregate scatters the same amount of light as a billion individual prion protein chains. When the figure on the right is converted into a weight distribution, the small aggregates peak is very small; and if converted into a number distribution, the small aggregate peak seem to disappear. These small aggregates (circles) can be completely removed with a membrane filter with 20 nm pores. After being concentrated 12 times, the aggregates no long appear (squares), indicating that the aggregation is not a dynamic process.

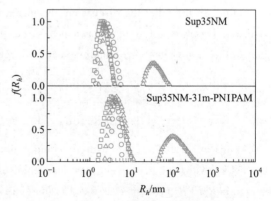

Wang Y J, Wu C.Biochemistry,2017,56:6575.
(Figure 3, permission was granted by publisher)

The combination of differential refractometer and laser light scattering provides an excellent and possibly the only way to study these very small aggregates quantitatively.

When there are a small amount of aggregates in the scattering volume, the intensity of scattered light contains the contributions from both individual chains and the aggregates. Therefore, the weight average molar mass and the translational diffusion coefficient distribution, respectively, measured by static and dynamic LLS include the contributions from both of them. Assuming that there is a molar mass M_b between the two, the weight average molar mass can be separated into two parts in the logarithmic space, namely

$$M_w = \frac{\int_0^\infty M^2 f_w(M)\,\mathrm{d}\log M}{\int_0^\infty M f_w(M)\,\mathrm{d}\log M} = \frac{\int_0^{M_b} M^2 f_w(M)\,\mathrm{d}\log M}{\int_0^\infty M f_w(M)\,\mathrm{d}\log M} + \frac{\int_{M_b}^\infty M^2 f_w(M)\,\mathrm{d}\log M}{\int_0^\infty M f_w(M)\,\mathrm{d}\log M}$$

$$= \frac{\int_0^{M_b} M^2 f_w(M)\,\mathrm{d}\log M}{\int_0^{M_b} M f_w(M)\,\mathrm{d}\log M} \frac{\int_0^{M_b} M f_w(M)\,\mathrm{d}\log M}{\int_0^\infty M f_w(M)\,\mathrm{d}\log M} + \frac{\int_{M_b}^\infty M^2 f_w(M)\,\mathrm{d}\log M}{\int_{M_b}^\infty M f_w(M)\,\mathrm{d}\log M} \frac{\int_{M_b}^\infty M f_w(M)\,\mathrm{d}\log M}{\int_0^\infty M f_w(M)\,\mathrm{d}\log M}$$

There are a total of four items in the two sum of the above equation. The first and third items are the weight average molar masses of individual chains and the aggregate ($M_{w,s}$ and $M_{w,m}$), respectively; the second and the fourth items are the concentration fractions of individual chains and the aggregates ($x_s = C_s/C$ and $x_m = C_m/C$), respectively,

$$x_s + x_m = 1$$

As the concentration and scattering angle approaches zero, the weight average molar mass obtained from the intensity of scattered light is equal to their sum

$$M_w = M_{w,s} x_s + M_{w,m} x_m$$

On the other hand, the line width distribution is proportional to the intensity of scattered light. In the logarithmic space, $\mathrm{d}\log\Gamma \propto \mathrm{d}\log D \propto \mathrm{d}\log R_h$, all transformations between them are linear. When they transform into each other, it is only necessary to normalize the integral of each distribution. In the following discussion, the visualized hydrodynamic radius distribution $F(R_h)$ is chosen. The ratio (A_R) of the two peaks related to individual chains and the aggregates is equal to the ratio of the corresponding scattered light intensities, i.e., $A_R = (M_{w,s} x_s)/(M_{w,m} x_m)$,

$$A_R = \frac{\int_0^{R_{h,b}} F(R_h)\,\mathrm{d}R_h}{\int_{R_{h,b}}^\infty F(R_h)\,\mathrm{d}R_h} = \frac{\int_0^{R_{h,b}} R_h F(R_h)\,\mathrm{d}\log R_h}{\int_{R_{h,b}}^\infty R_h F(R_h)\,\mathrm{d}\log R_h} = \frac{M_{w,s}\ x_s}{M_{w,m}\ x_m}$$

As discussed before, the hydrodynamic radius (R_h) in the Laplace inversion software (CONTIN) with the instrument has logarithmically spaced, and $\Delta\log R_h$ is a constant. The ordinate is $R_h F(R_h)$, denoted as $f(R_h)$. Therefore, the above integrals become a sum of $f(R_h)$, obtainable from dynamic LLS. There are four unknowns in the three equations: $M_{w,s}$, $M_{w,m}$, x_s and x_m. To solve these three equations simultaneously, one of the variables must be known from other ways.

If the macromolecule sample itself contains aggregates due to synthesis or other reasons, we can use a differential refractometer to measure the concentration of a small amount of aggregates. The details are as follows: a macromolecular solution with a concentration of ~10–100 mg/mL is prepared first, and then inject one portion of the solution into the two cells in the differential refractometer to obtain the beam origin. The aggregates in another portion of the solution are removed by centrifugation or filtration. Using the filled solution to replace the original solution in one of the cells, one can measure the beam deviation from the origin, and calculate the concentration of the aggregate. Substituting it into the above three equations, one can calculate the remaining three variables.

In comparison with other concentration measurement methods, the differential refractive index method proposed here can deduct the contribution of a large number of individual chains in the solution at one time, and only measure the concentration of a small amount of the aggregates. The measurement accuracy of the differential refractometer based on the novel $2f$ – $2f$ design is about 0.01 mg/mL. Even if the content of the aggregates in the solution is as low as 1%, their concentration can still be accurately measured with a few percent of error. Therefore, the weight average molar mass of individual chains, the average number of chains per aggregate and other microscopic parameters can be calculated.

If it is an aggregation induced by changing experimental conditions (such as temperature, pH, ionic strength, etc.), the molar mass ($M_{w,s}$) of individual chains is a known constant. Therefore, only substituting the weight average molar mass and hydrodynamic radius distribution measured by static and dynamic LLS, respectively, into the above three equations, one can calculate $M_{w,m}$ and x_m. Moreover, the average number of the chains per aggregate (the aggregation number, $M_{w,m}/M_{w,s}$) and the concentration of the aggregates ($C_m = x_m C$) can be calculated out..

In the study of protein aggregation, the molar mass of individual protein chains is generally known. Therefore, after measuring M_w and $f(R_h)$ by static and dynamic LLS, respectively, one can solve the equations simultaneously. After removing a tiny amount of the initial small aggregates in the prion protein solution, the aggregation kinetics of the prion protein in the solution can be studied in terms of the variation of the size and distribution of the aggregates. The figure (a) below shows the relative change of the scattered light. The dilution of the prion protein solution 200 times with phosphate buffered saline (PBS) induces the aggregation. Namely, the scattered light intensity increases with time, where $\langle I \rangle_0$ is the initial intensity of the light scattered by individual prion chains in the solution with no aggregation. The prion protein modified by the short thermally sensitive PNIPAM chain aggregates slowly, and the results of three repeated experiments are stable. A combination of the time dependent intensity of the scattered light with the Smoluchowski aggregation equation leads to the aggregation number dependence of the number of the aggregates ($n_i(t)$) with i number of prion proteins at different aggregation times, as shown in the figure (b) below. Even after six hours, although the number of individual chains ($i = 1$) in the solution has declined, but still majority. The inset in figure (b) shows the time dependence of $n_i(t)$. Other methods have no way to track such a change quantitatively.

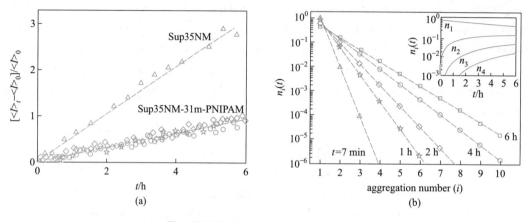

Wang Y J, Wu C.Biochemistry, 2017, 56:6575.

(Figures 9 and 11, permission was granted by publisher)

The second example, the aggregation occurs between two different kinds of macromolecules. For example, in the preparation of non−viral gene vectors, people use various positively charged small molecules, macromolecules, or both to complex and "condense" a long and negatively charged deoxyribonucleic acid (DNA) plasmids with a desired gene. The complexation makes a long plasmid collapse into a small aggregate with a positively charged surface.

The literature often attributed such complexation process to the electrostatic interaction, which is quite wrong! Before mixing, both of them have their own small molecule counter ions and the electrostatic interaction. Moreover, a charged macromolecular chain has no way to know whether the counter ion is a small molecule or a group with an opposite charge on another chain. The entropy change drives such a complexation. When two macromolecules with opposite charges are complexed, although they lose their translational entropy, they release a large number of their own small molecule counter ions, making them gain more translational entropy. The system moves to the direction of increasing entropy: the complexation.

In this research field, hundreds of thousands of papers have been published. In most of these studies, small "DNA + polymers" complexes (polyplex) were treated as small "black boxes" with no molecular detail, such as how many long plasmid and how many short polymer chains are inside each polyplex. These small "black boxes" are add into the cell−containing culture medium to observe some macroscopic effects of the endocytosis and the gene transfection in various cells that were regarded as another "black box". It was only a few years ago, a combination of differential refractory and laser light scattering methods enabled us to study the composition of the polyplexes and open these "black box", quantitatively revealing some molecular details of the complexation process between long anionic DNA plasmid and short cationic PEI chains.

First, assuming that each complex contains i long negatively charged DNA plasmids with a molar mass of M_{DNA}, and m short cationic PEI chains with a molar mass of M_{PEI}, so that the molar mass of the complex ($M_{i,\mathrm{P}}$) is

$$M_{i,\mathrm{P}} = M_{\mathrm{DNA}}i + M_{\mathrm{PEI}}m$$

where the subscript "i" denotes that this polyplex on average has i DNA plasmids. Before mixing, M_{DNA} and M_{PEI} are known constants. In the polyplexes, the mass fractions of DNA and PEI are $x_{\mathrm{DNA}} = M_{\mathrm{DNA}}i / (M_{\mathrm{DNA}}i + M_{\mathrm{PEI}}m)$ and $x_{\mathrm{PEI}} = M_{\mathrm{PEI}}m / (M_{\mathrm{DNA}}i + M_{\mathrm{PEI}}m)$. The differential refractive index of the polyplexes obeys the mass additive law, i.e.,

$$\left(\frac{\partial n}{\partial C}\right)_{\mathrm{P}} = x_{\mathrm{DNA}}\left(\frac{\partial n}{\partial C}\right)_{\mathrm{DNA}} + x_{\mathrm{PEI}}\left(\frac{\partial n}{\partial C}\right)_{\mathrm{PEI}} = x_{\mathrm{DNA}}\left(\frac{\partial n}{\partial C}\right)_{\mathrm{DNA}} + (1 - x_{\mathrm{DNA}})\left(\frac{\partial n}{\partial C}\right)_{\mathrm{PEI}}$$

where one can predetermine the differential refractive indexes of DNA and PEI, respectively. On the other hand, in static LLS, the time average intensity of scattered light of the polyplexes and DNA (I_{P} and I_{DNA}) respectively are

$$I_{\mathrm{P}} = \left(\frac{\partial n}{\partial C}\right)_{\mathrm{P}}^{2} \sum_{i=1}^{N} C_{i,\mathrm{P}} M_{i,\mathrm{P}} \quad \text{and} \quad I_{\mathrm{DNA}} = \left(\frac{\partial n}{\partial C}\right)_{\mathrm{DNA}}^{2} C_{\mathrm{DNA}} M_{\mathrm{DNA}}$$

where $C_{i,\mathrm{P}}$ is the concentration of the polyplex with i DNA plasmids and its sum is the total polyplex concentration, i.e., $C_{\mathrm{P}} = \sum_{i=1}^{N} C_{i,\mathrm{P}}$. Note that $C_{i,\mathrm{P}}$ is related to the weight ($W_{i,\mathrm{DNA}}$) and the concentration ($C_{i,\mathrm{DNA}}$) of DNA. One can respectively divide the two sides of the above equations and substitute the previous differential refractive index of the polyplexes in it, i.e.,

$$\frac{I_{\mathrm{P}}}{I_{\mathrm{DNA}}} = \frac{\left(\frac{\partial n}{\partial C}\right)_{\mathrm{P}}^{2} \sum_{i=1}^{N} C_{i,\mathrm{P}} M_{i,\mathrm{P}}}{\left(\frac{\partial n}{\partial C}\right)_{\mathrm{DNA}}^{2} C_{\mathrm{DNA}} M_{\mathrm{DNA}}}$$

where the intensity of scattered light from short individual PEI chains outside the complexation is so weak that it can be ignored. In addition, the number of short PEI chains that can be complexed with each long DNA plasmid has an upper limit, which is recorded as $m_{\max}$. From the respective concentrations and masses of DNA and PEI in the mixed solution, one can calculate the average number of short PEI chains that available per long DNA plasmid (m_0), too. If there is only one DNA plasmid in each polyplex and $m_0 > m_{\max}$, there will be extra short PEI chains that cannot be complexed. In this case, each long DNA plasmid is completely covered by short PEI chains and cannot form the polyplexes with two long DNA plasmids inside. Therefore, the sum of the above equation has only one term, $i = 1$, $N = 1$, and $m = m_{\max}$ and is rewritten as

$$\frac{I_{\mathrm{P}}}{I_{\mathrm{DNA}}} \left(\frac{\partial n}{\partial C}\right)_{\mathrm{DNA}}^{2} C_{\mathrm{DNA}} M_{\mathrm{DNA}} = \left(\frac{\partial n}{\partial C}\right)_{\mathrm{P}}^{2} C_{1,\mathrm{P}} M_{1,\mathrm{P}}$$

Experimentally, on the left side of the above equation, all parameters can be measured or known. Substituting the differential refractive index, $C_{1,\mathrm{P}}$ and $M_{1,\mathrm{P}}$ on the right side, there is only one unknown parameter $m_{\max}$, calculable by a numerical calculation. If $m_{\max,0} < m_{\max}$, one must consider and only consider the existence of dimers with $N = 2$ and in a very dilute mixture. Assuming that its weight fraction is x_w,

$$\frac{I_{\mathrm{P}}}{I_{\mathrm{DNA}}}\left(\frac{\partial n}{\partial C}\right)_{\mathrm{DNA}}^2 C_{\mathrm{DNA}} M_{\mathrm{DNA}} = \left(\frac{\partial n}{\partial C}\right)_{\mathrm{P}}^2 C_{\mathrm{P}}(1 - x_w) M_{1,\mathrm{P}} + C_{\mathrm{P}} x_w M_{2,\mathrm{P}}$$

$$x_n = \frac{N_2}{N_1 + N_2} \quad \text{and} \quad x_w = \frac{2 M_{1,\mathrm{P}} N_2}{M_{1,\mathrm{P}} N_1 + 2 M_{1,\mathrm{P}} N_2} = \frac{2 N_2}{N_1 + 2 N_2} \rightarrow x_n = \frac{x_w}{2 - x_w}$$

where $M_{2,\mathrm{P}} = 2 M_{1,\mathrm{P}}$. Approximately using $m = m_{\max}$, one can obtain the only unknown quantity x_w on the right by a numerical calculation from the experimental values on the left side. According to the definition, the number fraction of dimers (x_n) is also calculable as follows,

Using the above method, one can find a series of microscopic parameters of the complexation between negatively charged DNA and positively charged PEI chains with different molar masses and different configurations, as shown in Table 1. When N/P in the feed is larger than six, the polyplexes in the solution mixture contains only one DNA plasmid. When the molar mass is similar, in comparison with the linear PEI chains, each DAN plasmid can complex more branched PEI chains. The introduction of the above two specific examples related to biological macro-molecules is for attracting readers to generate more ideas. Hope that readers can use them as an example and skillfully combine static and dynamic LLS and differential refractometer to solve some problems that are not solvable by other experimental methods in the future when using the LLS instrument.

Table 1　在不同溶液中形成的不同的 DNA/PEI 络合聚集体的实验和计算参数

$(\mathrm{N/P})_{\mathrm{feed}}$	sample	$m_{\mathrm{PEI,max}}$	$\langle I \rangle_{\mathrm{polyplexes}} / \langle I \rangle_{\mathrm{DNA}}$	m_{PEI}	x_n	x_w
3	lPEI – 2.5k	557	2.43	759	14	25
	lPEI – 25k	56	2.06	60	2	4
	bPEI – 2.0k	696	2.27	862	9	16
	bPEI – 25k	56	2.30	71	10	18
6	lPEI – 2.5k	1 114	2.72	879	0	0
	lPEI – 25k	112	2.82	92	0	0
	bPEI – 2.0k	1 392	3.33	1 387	0	0
	bPEI – 25k	112	3.62	122	0	0

Dai Z J, Wu C. Macromolecules, 2012, 45:4346–4353(Table 1, permission was granted by publisher).

Summary

When characterizing the molar mass distribution of an unknown macromolecular sample, many experimental methods need to calibrate the instrument in advance. The calibration itself requires a series (at least two) of narrowly distributed macromolecular standard samples with different molar masses, so that a calibration curve can be determined. The calibration curve can be used to characterize other macromolecules of the same kind under the same experimental conditions (solvent, temperature, pressure, etc.). However, in most cases, no standard samples of the same kind of macromolecules are available. It is for this reason that one has to combine two or more experimental methods with internal connections to obtain calibration curves by simultaneously solving two or more equations.

A combination of viscosity and size exclusion chromatography to obtain a "universal" calibration curve is a common combination. The uncommon combinations include the following ones. The viscosity is combined with static LLS to obtain the relationship between the hydrodynamic size and molar mass, which can be further used to convert the linewidth distribution measured by dynamic LLS into a molar mass distribution. A combination of static and dynamic LLS can convert the linewidth distribution into a molar mass distribution. The differential refractometer can be combines with LLS to study the structure and aggregation kinetics of a small amount of aggregates in macromolecular solutions quantitatively.

第七章　大分子稀溶液中链的动态学

如前所述,大分子链上每个链段的无规行走导致其构象不停地变化。换而言之,在每一瞬间,溶液中不同的大分子链有不同的构象。具有一个给定构象的大分子链的数目遵循正态分布。每种链构象都有一个对应的方距,对所有构象的方距进行统计平均可得方均距。对线型柔性大分子链而言,方均距与理论上计算的平均末端距关联。一般而言,其和实验上可测的方均回转半径关联。基于统计学上无规变量的遍历性,对无数条大分子链进行统计平均等于对一条大分子链的构象在无穷长的时间内进行统计平均。因此,大分子链构象的正态分布也可看作每种构象出现的时间分布。

英文"dynamics"和"kinetics"在中文中常常都被翻译成相同的一个词"动力学"。翻译正确! 均指随时间的物理或化学变化。"kinetics"侧重"力和力矩"对物体运动的影响,在力学中也称为"kinematics",在讨论化学反应速率时也常用。"dynamics"则强调能量对变化的影响,范畴更广,包括了 kinetics。自 20 世纪 20 年代中叶,物理学家就逐渐地在教科书中采用"dynamics"一词。与"statics"(静态学,研究体系处于平衡状态时的性质)对应,将"dynamics"翻译成"动态学",尤其是在研究热能对大分子链整体和局部构象运动的影响时,可能更为妥当。故本书将"dynamics"翻译成"动态学"。但在讨论链的构象转变过程时,仍将使用"动力学"一词。在第十至十二章中讨论大分子溶液相变、介观球相和大分子胶体时,也将使用"动力学稳定"描绘体系处于非平衡状态时的亚稳态,以区分处于热力学平衡状态时的"动态稳定"。

大分子动态学所讨论的正是能量对大分子链构象变化的影响。为了简化讨论,先考虑一条链在溶液中的无扰状态。将大分子链中的每个 Kuhn 链段看成一个小珠子和一根可以伸缩的短弹簧。于是,一条大分子线型链就变成了一串 N 个没有大小、没有流体力学相互作用的小珠子[Rouse 模型,J. Chem. Phys.,1953,21:1272]。每两个珠子之间通过一根平均长度为 b 的小弹簧连接在一起。每对弹簧和珠子对应的能量仅有热能(k_BT)。因此,在一个维度方向上,单位长度上的能量密度(k_s,力常数)为 k_BT/b。每个维度均等,故在三个维度上,小弹簧的力常数为一个维度上的三倍,即 $k_s = 3k_BT/b$。在热能的搅动下,每个小弹簧的长度和取向均在无规地变化,每个小珠子的位置也在不停地移动。

为了处理这样一个复杂的由 N 个单元组成大分子链,需要借助三个重要的概念:空间自由度、法向坐标和简正模式。每个空间自由度是一个独立的移动方向。一个珠子在三维空间可朝着三个独立(正交)的方向移动,故其有三个空间自由度,即通常使用的含有 x,y,z 三个正交的方向。其向任意一个方向的移动,总可以分解成朝着这三个方向的分量。换而言之,三个方向上分量的组合可表达一个珠子的移动。两个珠子就有六个空间自由度;以此类推,N 个珠子就有 $3N$ 个空间自由度。从另一个角度,也可以说,对任意一个珠子均需要一个含有三个空间自由度的三维坐标(x_i, y_i, z_i)来表达其位置;表达 N 个珠子的位置就需要有 $3N$ 个空间自由度的三维坐标$(x_i, y_i, z_i, i = 1, 2, \cdots, N)$。

在极坐标系中,三维空间中任意一点与原点的位置可表达为一个从原点指向该点的矢量$(\vec{r}_i)$。表达 N 个珠子的位置就需要有 N 这样的矢量$(\vec{r}_i, i = 1, 2, \cdots, N)$。每个矢量对时间的微分$(\mathrm{d}\vec{r}_i/\mathrm{d}t)$为其对应珠子的运动速度$(\vec{u}_i = \mathrm{d}\vec{r}_i/\mathrm{d}t)$。每个在液体里运动的物体都受到一个阻止其运动的阻力$(f\vec{u}_i)$,其中,$f = 6\pi\eta b$,是摩擦系数,与液体黏度(η)和珠子尺寸(b)相关。运动越快,阻力越大。物理上,每一个具有一定尺寸和质量(m)的珠子在液体里都受到过度阻尼,很快地趋向匀速,即其无法长时间地加速。因此,在小于 10^6 Hz 的低频下或长于 1 μs 的时间尺度,可忽略与加速运动有关的力$(m\mathrm{d}\vec{u}_i/\mathrm{d}t)$。如果仅考虑每个珠子分别与紧邻的两个珠子的相互作用和加上溶剂分子在每个瞬间对每个珠子的无规作用力$(\vec{f}_i(t))$,N 个珠子的运动方程如下:

$$f\vec{u}_i = k_s(\vec{r}_{i-1} - \vec{r}_i) + k_s(\vec{r}_{i+1} - \vec{r}_i) + \vec{f}_i(t), \quad i = 1, 2, \cdots, N$$

细心的读者会注意到,处于两端的两个珠子各只有一个紧邻的珠子。对大分子而言,N 远远大于 1。因此,可以想象在两端各再加上一个珠子,并不影响整条链的运动。上式看上去简洁和完美;然而,这 N 个珠子是连在一起的。一个珠子的运动不仅会影响其紧邻的两个珠子,而且还会影响其附近的其他珠子;反之,其他珠子的运动也会影响其运动。影响的距离取决于珠子之间相互作用的强弱。物理上,称作珠子之间的相关长度。即使只考虑紧邻两个珠子的相互影响,N 个珠子的 $3N$ 个空间自由度的三维坐标也并非完全独立,而是存在着一定程度的互相依赖。正因变量之间的相互依赖,无论是在普通的三维还是在极坐标的直观空间中,基本上无法联立求解上述 N 个方程。

因此,需要一个抽象的多维空间,其维度坐标之间不仅互相独立,而且还能描绘 N 个珠子的耦合运动。这就是所谓的法向坐标,由正常三维空间里的位移坐标线性组合而成,即

$$\vec{p}_n(t) = \frac{1}{N} \sum_{i=1}^{N} \cos\frac{in\pi}{N} \vec{r}_i(t) \quad n = 1, 2, \cdots$$

其描绘的运动就是所谓的简正模式。其相当于线性代数中的矩阵对角化。在通常的三维坐标中,一个含有 $N \times N$ 项矩阵中的每项(前述方程中每项的系数)均不为零。将该矩阵前后分别乘以一个矩阵和其逆矩阵,变换到抽象的法向空间,使得该矩阵中除了对角线上的项(本征值)以外,全部变为零。这样,用 $\vec{p}_n(t)$ 表述的运动方程就可方便地获得求解。

$n = 0$ 时,$\vec{p}_0(t) = [\sum_{i=1}^{N} \vec{r}_i(t)]/N$。依据一个物体质量中心(质心)的定义,$\vec{p}_0(t)$ 代表的就是质心随时间的运动轨迹。当 $n \geqslant 1$ 时,$\vec{p}_n(t)$ 代表着链中的珠子相对于质心的内部运动。$\vec{p}_1(t)$ 等价于一个沿着链的矢量,表达了

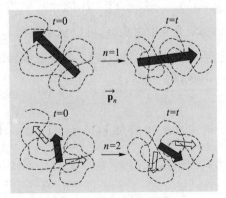

链取向。由于其长度和方向不断变化,可粗略地认作链围绕着其质心的旋转。$\vec{\mathbf{p}}_2(t)$ 等价于两个背向质心、方向基本相反、大小和方向不断变化的矢量随时间的运动轨迹。与 $\vec{\mathbf{p}}_1(t)$ 相比,其对链构象变化的细节更敏感,如上图所示。随着 n 的增大,$\vec{\mathbf{p}}_n(t)$ 涉及越来越小的局部构象变化。$n=N$ 时,$\vec{\mathbf{p}}_N(t)$ 对应着构象的最小变化,两个邻近珠子的相对位移。将以上两式结合可得下列法向坐标中的运动方程。

$$f\frac{\mathrm{d}\vec{\mathbf{p}}_n(t)}{\mathrm{d}t} = \frac{f}{N}\sum_{i=1}^{N}\cos\frac{in\pi}{N}\frac{\mathrm{d}\vec{\mathbf{r}}_i(t)}{\mathrm{d}t} = \frac{f}{N}\sum_{i=1}^{N}\cos\frac{in\pi}{N}\vec{\mathbf{u}}_i \quad (n=1,2,\cdots,N)$$

$$= \frac{k_s}{N}\sum_{i=1}^{N}\cos\frac{in\pi}{N}\left(\vec{\mathbf{r}}_{i-1}(t)+\vec{\mathbf{r}}_{i+1}(t)-2\vec{\mathbf{r}}_i(t)\right) + \frac{1}{N}\sum_{i=1}^{N}\cos\frac{in\pi}{N}\vec{\mathbf{f}}_i(t)$$

$$\cong \frac{k_s}{N}\sum_{i=1}^{N}\left[\cos\frac{n(i-1)\pi}{N}+\cos\frac{n(i+1)\pi}{N}-2\cos\frac{in\pi}{N}\right]\vec{\mathbf{r}}_i(t) + \frac{1}{N}\sum_{i=1}^{N}\cos\frac{in\pi}{N}\vec{\mathbf{f}}_i(t)$$

$$\cong \frac{k_s}{N}\sum_{i=1}^{N}-\left(\frac{n\pi}{N}\right)^2\cos\frac{in\pi}{N}\vec{\mathbf{r}}_i(t) + \frac{1}{N}\sum_{i=1}^{N}\cos\frac{in\pi}{N}\vec{\mathbf{f}}_i(t)$$

$$\frac{\mathrm{d}\vec{\mathbf{p}}_n(t)}{\mathrm{d}t} = -\frac{k_{s,n}}{f}\vec{\mathbf{p}}_n(t) + \frac{1}{fN}\sum_{i=1}^{N}\cos\frac{in\pi}{N}\vec{\mathbf{f}}_i(t)$$

$$\frac{\mathrm{d}\vec{\mathbf{p}}_n(t)}{\mathrm{d}t} = -\frac{1}{\tau_n}\vec{\mathbf{p}}_n(t) + \vec{\Delta}_n(t) \quad n=1,2,\cdots,N \tag{7.1}$$

其中

$$k_{s,n} = \left(\frac{n\pi}{N}\right)^2 k_s; \quad \tau_n = \frac{f}{k_{s,n}} = \frac{2\eta b^3 N^2}{\pi k_B T\, n^2}; \quad \text{和} \quad \vec{\Delta}_n(t) = \frac{1}{fN}\sum_{i=1}^{N}\cos\frac{in\pi}{N}\vec{\mathbf{f}}_i(t)$$

其中,$\langle R^2\rangle = b^2 N$,无扰大分子链的方均末端距:$\langle R_g^2\rangle = \langle R^2\rangle/6$,方均回转半径:$D = k_B T/(fN)$,平动扩散系数;和 $t_{R_g} = \langle R_g^2\rangle/(6D)$,扩散一个 $\sqrt{\langle R_g^2\rangle}$ 距离所需的时间。代入 τ_n,

$$\tau_n = \frac{12\, t_{R_g}}{\pi^2}\frac{1}{n^2} \approx \frac{t_{R_g}}{n^2}$$

τ_n 的量纲是时间,第 n 个内部简正模式运动的特征弛豫时间。当 $n=1$ 时,$\tau_1 \approx t_{R_g}$。物理上,其意义是弛豫一个链的方均回转半径所需的时间。随着 n 的增大,其迅速变短。

由于每一个 $\vec{\mathbf{r}}_i(t)$ 均是无规变量,$\vec{\mathbf{p}}_n(t)$ 也是一个无规变量。$\vec{\mathbf{p}}_n(t)$ 和 $\vec{\Delta}_i(t)$ 二者各自的时间平均都为零,即 $\langle\vec{\mathbf{p}}_n(t)\rangle = 0$ 和 $\langle\vec{\Delta}_i(t)\rangle = 0$;因此,由式(7.1)无法直接解出 τ_n。与前述动态激光光散射类似,可以先求法向坐标的时间相关函数。依据定义,不同的 $\vec{\mathbf{p}}_n(t)$ 之间互不相关;$\vec{\mathbf{p}}_n(t)$ 和 $\vec{\Delta}_i(t)$ 之间也互不相关,故

$$\frac{\mathrm{d}\langle\vec{\mathbf{p}}_n(0)\cdot\vec{\mathbf{p}}_n(t)\rangle}{\mathrm{d}t} = -\frac{1}{\tau_n}\langle\vec{\mathbf{p}}_n(0)\cdot\vec{\mathbf{p}}_n(t)\rangle + \langle\vec{\Delta}_n(0)\cdot\vec{\Delta}_n(t)\rangle \quad n=1,2,\cdots,N \tag{7.2}$$

其中,$\langle\vec{\Delta}_n(0)\cdot\vec{\Delta}_n(t)\rangle = \left[\langle\vec{\mathbf{f}}_n(0)\cdot\vec{\mathbf{f}}_n(t)\rangle\sum_{i=1}^{N}\cos\frac{in\pi}{N}\cos\frac{in\pi}{N}\right]/(fN)^2$。基于无规作用力 $\vec{\mathbf{f}}_n(t)$ 的定义,$\langle\vec{\mathbf{f}}_n(0)\cdot\vec{\mathbf{f}}_n(t)\rangle = A\delta(t)$。可以证明 $A = 6f k_B T$。上式中的求和为

$$\sum_{i=1}^{N}\cos\frac{in\pi}{N}\cos\frac{in\pi}{N} = \frac{1}{2}\sum_{i=1}^{N}\left[\cos\frac{(i+i)n\pi}{N}+\cos\frac{(i-i)n\pi}{N}\right] = \frac{N}{2}$$

推导时,利用了 $2\cos x\cos y = \cos(x+y) + \cos(x-y)$。因 $N \gg 1$,求和中的第一项,仅在 $i = N$ 时,等于 1,其余的项均趋向 0。第二项一直为 1,故在 $t = 0$ 时,

$$\langle \vec{\mathbf{p}}_n(0) \cdot \vec{\mathbf{p}}_n(0) \rangle = \langle p_n^2(0) \rangle = \frac{b^2 N}{\pi^2 n^2} = \frac{6\langle R_g^2 \rangle}{\pi^2 n^2} \approx \frac{2}{3} \frac{\langle R_g^2 \rangle}{n^2}$$

其代表了第 n 个简正模式的方均涨落幅度。当 $n = 1$ 时,$\langle p_1^2(0) \rangle \approx 2\langle R_g^2 \rangle / 3$,即第一个简正模式的方均涨落幅度略小于方均回转半径。随着 n 的增加,简正模式的方均涨落幅度急剧地变小。将 $\langle p_n^2(0) \rangle$ 和 τ_n 代入式 (7.2),

$$\langle \vec{\mathbf{p}}_n(0) \cdot \vec{\mathbf{p}}_n(t) \rangle = \langle p_n^2(0) \rangle \exp\left(-\frac{t}{\tau_n}\right) \approx \frac{2\langle R_g^2 \rangle}{3 n^2} \exp\left(-\frac{t}{\tau_n}\right) \quad n = 1, 2, \cdots, N \quad (7.3)$$

上述时间相关函数的平均涨落幅度 ($\langle q_n^2(0) \rangle$) 和特征弛豫时间 (τ_n) 都反比于 n^2。随着 n 的增大,$\langle \vec{\mathbf{p}}_n(0) \cdot \vec{\mathbf{p}}_n(t) \rangle$ 不仅随时间衰变得更快,而且其对简正模式(链的内部运动)的贡献(平均涨落)也迅速地减少。

法向坐标 $\vec{\mathbf{p}}_n(t)$ 由通常坐标 $\vec{\mathbf{r}}_i(t)$ 线性地组合而成。反过来,$\vec{\mathbf{r}}_i(t)$ 也可由法向坐标的线性组合表达。二者构成两个互相"共轭"的空间。将 $\vec{\mathbf{p}}_n(t)$ 的定义代入下式,可得

$$2\sum_{n=1}^{N} \cos\frac{in\pi}{N} \vec{\mathbf{p}}_n(t) = 2\sum_{n=1}^{N} \cos\frac{in\pi}{N} \left[\frac{1}{N}\sum_{i=1}^{N}\cos\frac{in\pi}{N}\vec{\mathbf{r}}_i(t) \right]$$

$$= \frac{1}{N}\sum_{i=1}^{N}\vec{\mathbf{r}}_i(t) \sum_{n=1}^{N} 2\cos\frac{in\pi}{N}\cos\frac{in\pi}{N}$$

$$= \frac{1}{N}\sum_{i=1}^{N}\vec{\mathbf{r}}_i(t) \sum_{n=1}^{N} \left[\cos\frac{(i+i)n\pi}{N} + \cos\frac{(i-i)n\pi}{N} \right]$$

其中,推导时,也利用了 $2\cos x\cos y = \cos(x+y) + \cos(x-y)$。在方括号里的第一项中,仅当 $(i+i) = 2N$ 时,$\cos[(i+i)n\pi/N] = 1$,其余的项均为零;而第二项总为 1。因此,右边求和的结果为 $\vec{\mathbf{r}}_i(t)(1+N)$,故

$$\vec{\mathbf{r}}_i(t) = 2\sum_{n=1}^{N}\cos\frac{in\pi}{N}\vec{\mathbf{p}}_n(t) \quad i = 1, 2, \cdots, N \quad (7.4)$$

其中,利用了 $(N+1)/N \cong 1$。同样,细心的读者会注意到第 1 个和第 N 个珠子不同于中间的珠子。因为,$N \gg 1$,可在前后各加一粒珠子,然后忽略它们,并不影响结果。利用上述关系可得一些动力学参量。

第一个可求的参量就是第四章中随着时间不断地、无规地变化的链的末端距的松弛,

$$\vec{\mathbf{R}}(t) = \vec{\mathbf{r}}_N(t) - \vec{\mathbf{r}}_1(t) = 2\sum_{n=1}^{N}\left(\cos n\pi - \cos\frac{n\pi}{N}\right)\vec{\mathbf{p}}_n(t)$$

其中,$\cos n\pi = (-1)^n$。随着 n 的增加,$\vec{\mathbf{p}}_n(t)$ 的平均涨落幅度迅速地减少和 $\vec{\mathbf{p}}_n(t)$ 本身急剧地衰减,故通常仅需考虑前几项 ($n < 10$)。故当 $N \gg 1$ 时,$\cos(n\pi/N) \to 1$。因此,仅有 n 为奇数时,该项方不为零。括号里的两个加和之差为 -2。因此,上式成为

$$\vec{\mathbf{R}}(t) = \vec{\mathbf{r}}_N(t) - \vec{\mathbf{r}}_1(t) \cong -4\sum_{n=1:\text{ odd}}^{N}\vec{\mathbf{p}}_n(t)$$

链构象的无规变化导致 $\vec{\mathbf{p}}_n(t)$ 和 $\vec{\mathbf{R}}(t)$ 的无规变化。根据法向坐标的定义,不同的 $\vec{\mathbf{p}}_n(t)$ 之间互不相关。因此,$\vec{\mathbf{R}}(t)$ 和 $\vec{\mathbf{p}}_n(t)$ 各自的时间自相关函数有着以下的关联:

$$\langle \vec{R}(0) \cdot \vec{R}(t) \rangle \cong 16 \sum_{n=1:odd}^{N} \langle \vec{\mathbf{p}}_n(0) \cdot \vec{\mathbf{p}}_n(t) \rangle = \sum_{n=1:odd}^{N} \frac{8Nb^2}{\pi^2 n^2} \exp\left(-\frac{t}{\tau_n}\right)$$

如果仅考虑第一项$(n=1)$,而略去加和中的其余各项,可用式(7.1)中定义的τ_n、无扰链的方均末端距$\langle R_0^2 \rangle = b^2 N \sim M$和伸直链长$(L_c = bN \sim M)$将上式重写成

$$\langle \vec{R}(0) \cdot \vec{R}(t) \rangle = \frac{8\langle R_0^2 \rangle}{\pi^2} \exp\left(-\frac{t}{\tau_1}\right) \approx \langle R_0^2 \rangle \exp\left(-\frac{t}{\tau_1}\right) \tag{7.5}$$

其中,$\tau_1 \propto N^2 \propto M^2$。依据式$(7.1)$中$\tau_1$的定义,可以粗略地估计一下$\tau_1$的长短:在 cgs 体系里,普通聚合物$(M \sim 10^5 \mathrm{g/mol})$的$N \sim 10^3$;溶剂黏度$\eta \sim 10^{-2}$;$k_B T \sim 10^{-14}$;故$\tau_1 \sim 10^{-3}\ \mathrm{s}$。换而言之,与整条链的转动和振动有关的最慢松弛时间约在毫秒量级;其他内部运动模式的弛豫则更快。另一方面,特征松弛时间随链长变长。因此,长的大分子链更易观察和测量。

第二个可求的参量就是大分子链的质心运动。如前所述,法向坐标$\vec{\mathbf{p}}_0(t)$代表了一条链的质心,其运动也可由法向坐标$\vec{\mathbf{p}}_n(t)$的轨迹来描绘。经过了一段给定的时间(t)后,质心位置从初始的$\vec{r}_G(0)$无规地移动到了一个新的位置$\vec{r}_G(t)$,故其方均距为

$$\langle [\vec{r}_G(t) - \vec{r}_G(0)]^2 \rangle = \langle [\vec{\mathbf{p}}_0(t) - \vec{\mathbf{p}}_0(0)]^2 \rangle = \langle \int_0^t \int_0^t \vec{\Delta}_n(t') \vec{\Delta}_n(t'')\, \mathrm{d}t' \mathrm{d}t'' \rangle$$

$$\langle [\vec{r}_G(t) - \vec{r}_G(0)]^2 \rangle = \frac{1}{(fN)^2} \int_0^t \int_0^t \sum_{i=1}^{N} \sum_{j=1}^{N} \langle \vec{\mathbf{f}}_i(t') \vec{\mathbf{f}}_j(t'') \rangle\, \mathrm{d}t' \mathrm{d}t''$$

其中,$\vec{\mathbf{f}}_i(t)$是一个无规作用力,所有的交叉项在时间平均后消失。双重求和只剩下N项相同的自身项:$\langle \vec{\mathbf{f}}_i(t') \vec{\mathbf{f}}_j(t'') \rangle = 6f k_B T \delta(t'' - t')$;且仅在$t'' = t'$时,非零。上式成为

$$\langle [\vec{r}_G(t) - \vec{r}_G(0)]^2 \rangle = \frac{1}{(fN)^2} \int_0^t \sum_{j=1}^{N} 6f k_B T\, \mathrm{d}t = \frac{6k_B T}{fN} t$$

另一方面,在三维空间中,一个无规$\langle [\vec{r}(t) - \vec{r}(0)]^2 \rangle$行走粒子的方均距$(\langle [\vec{r}(t) - \vec{r}(0)]^2 \rangle)$等于其平动扩散系数$(D)$和行走时间$(t)$乘积的六倍,即$\langle [\vec{r}(t) - \vec{r}(0)]^2 \rangle$,故

$$D = \frac{k_B T}{fN} = \frac{k_B T}{6\pi \eta bN} = \frac{k_B T}{6\pi \eta L_c} \sim M^{-1} \tag{7.6}$$

然而,在 Θ 溶剂(准理想无扰状态)中,实验上测得的第一个简正模式的特征松弛时间(τ_1)和质心的平动扩散系数(D)与大分子摩尔质量之间的标度指数分别为 3/2 和 1/2,严重偏离式(7.5)和式(7.6)所预测的 2 和 1。Rouse 模型在稀溶液中的失败,主要归结于其假定了链段之间互相独立,完全忽视了流体力学相互作用,即一个链段的运动可以影响其周围的溶剂分子,而溶剂分子的运动又会影响其他链段的运动。

为了引入流体力学相互作用,克服 Rouse 模型在稀溶液中失效的问题,约三年后,Zimm［J. Chem. Phys.,1956,24:269］利用对 Oseen 张量的预先平均近似,即

$$\langle \mathbf{H}_{ij} \rangle = \frac{\mathbf{I}}{6\pi \eta_s} \left\langle \frac{1}{|\vec{r}_i(t) - \vec{r}_j(t)|} \right\rangle$$

成功地将$\vec{r}_i(t)$的运动方程中的流体力学张量解耦,使得其在法向坐标中有解。物理上,不难看出,$\langle \mathbf{H}_{ij} \rangle$中每一个元素$(H_{ij})$等效于一个细棒状物体的摩擦系数的倒数。细棒

的一端在 $\vec{r}_i(t)$ 处，另一端在 $\vec{r}_j(t)$，$|\vec{r}_n(t)-\vec{r}_m(t)|$ 可以看作链上任意两个链段之间的直接距离，也可看成棒长（$\vec{b}_{nm}(t)$），只是该棒的长度和方向随着链构象不停地、无规地变化，故需要对时间平均。推导过程烦琐，故略去。最终的结果为

$$H_{ij}=\frac{1}{(6\pi^3)^{1/2}\eta_s b\,|i-j|^{\alpha_R}}$$

在法向坐标中，考虑到 $N\gg1$，H_{ij} 可被进一步转换成（推导过程也略去），

$$h_{nm}=\frac{1}{N^2}\sum_{i=1}^{N}\sum_{j=1}^{N}\frac{1}{(6\pi^3)^{\frac12}\eta b\,|i-j|^{\alpha_R}}\cos\frac{in\pi}{N}\cos\frac{jm\pi}{N}$$

$$=\frac{\Gamma(1-\alpha_R)}{2(3\pi^3)^{\frac12}\pi^{1-\alpha_R}\eta b\,N^{\alpha_R}n^{1-\alpha_R}}\delta_{nm}\quad(\text{除了 }i=j=0\text{ 以外})$$

将其代入法向坐标中的运动方程，沿用在 Rouse 模型中的推导，可得式(7.1)中的结果，即

$$\frac{\mathrm{d}\vec{\mathbf{p}}_n(t)}{\mathrm{d}t}=-\frac{1}{\tau_n}\vec{\mathbf{p}}_n(t)+\vec{\Delta}_n(t)\quad n=1,2,\cdots,N \tag{7.7}$$

其中，τ_n 和 $\vec{\Delta}_n(t)$ 的表达和 Rouse 模型里完全相同。区别是，在 Rouse 模型中，摩擦系数 $f=6\pi\eta b$ 和 $k_{s,n}=\left(\dfrac{n\pi}{N}\right)^2 k$，而在 Zimm 模型中，

$$f=\frac{1}{h_{nm}}=\frac{2(3\pi^3)^{\frac12}\pi^{1-\alpha_R}\eta b}{\Gamma(1-\alpha_R)}\left(\frac{n}{N}\right)^{1-\alpha_R}\cong2(3\pi^3)^{\frac12}\eta b\left(\frac{n}{N}\right)^{1-\alpha_R}\quad\text{和}\quad k_{s,n}=\left(\frac{n\pi}{N}\right)^{2\alpha_R+1}k_s$$

在上面的近似中，利用了 $\Gamma(1-\alpha_R)\approx\pi^{1-\alpha_R}$；$1/2\le\alpha_R\le3/5$。将上式和 $k_s=3k_\mathrm{B}T/b^2$ 代入 τ_n，可得在 Zimm 模型和法向坐标中，各个简正模式的特征松弛时间，如下

$$\tau_n=\frac{f}{k_{s,n}}=\frac{2(3\pi)^{\frac12}\eta\,b^3N^{3\alpha_R}}{3\pi^{2\alpha_R}k_\mathrm{B}T\,n^{3\alpha_R}}\approx\frac{44}{\pi^{2\alpha_R+1}}\frac{t_{R_g}}{n^{3\alpha_R}}\approx4\frac{t_{R_g}}{n^{3\alpha_R}}\sim M^{3\alpha_R} \tag{7.8}$$

在第一个简正模式中，$\tau_1\approx4t_{R_g}$。与 Rouse 模型比较，弛豫大约慢了 4 倍，可归于流体力学作用的过度阻尼。在 Θ 溶剂中，$\alpha_R=1/2$ 和 $\tau_1\propto M^{3/2}$，与实验观察值相近。依据在 Rouse 模型中的同样推导，可得第 n 个简正模式的方均涨落幅度，

$$\langle\vec{\mathbf{p}}_n(0)\cdot\vec{\mathbf{p}}_n(0)\rangle=\langle p_n^2(0)\rangle=\frac{b^2N^{2\alpha_R}}{2\pi^2n^{3\alpha_R}}=\frac{6\langle R_g^2\rangle}{2\pi^2n^{3\alpha_R}}\approx\frac{1}{3}\frac{\langle R_g^2\rangle}{n^{3\alpha_R}}$$

与 Rouse 模型相比，第一个简正模式的方均涨落幅度小了约 50%；可归于流体力学作用的过度阻尼。链末端距 $\vec{\mathbf{R}}(t)$ 的时间自相关函数与第 n 个简正模式 $\vec{\mathbf{q}}_n(t)$ 的依赖关系如下：

$$\langle\vec{\mathbf{R}}(0)\cdot\vec{\mathbf{R}}(t)\rangle=\sum_{n=1:odd}^{N}\frac{6N^{2\alpha_R}b^2}{\pi^2n^{3\alpha_R}}\exp\left(-\frac{t}{\tau_n}\right)\approx\frac{2\langle R^2\rangle}{\pi^2}\exp\left(-\frac{t}{\tau_1}\right)\approx\langle R_g^2\rangle\exp\left(-\frac{t}{\tau_1}\right) \tag{7.9}$$

其中，$\langle R^2\rangle=N^{2\alpha_R}b^2$，方均末端距与链段数之间的标度关系；$\langle R^2\rangle=6\langle R_g^2\rangle$；以及

$$\tau_1=\frac{6\eta\,b^3N^{3\alpha_R}}{\pi^{2\alpha_R}k_\mathrm{B}T}\approx4t_{R_g}\sim M^{3\alpha_R}$$

同样，可以粗略地估计一下 τ_1 的长短。在 cgs 体系里，对摩尔质量在 $10^5\sim10^6$ g·mol^{-1} 范围内的大分子和普通溶剂，$\eta\sim10^{-2}$；$b\sim10^{-7}$；$N=10^3\sim10^4$ 和 $k_\mathrm{B}T\sim10^{-14}$；因此，$\tau_1\sim10^{-3}$ s 。

读者应该注意，无论是在 Rouse 模型还是 Zimm 模型中，在没有缠结情形下，通常大分子链的特征松弛时间为毫秒级。换而言之，一条变形的线型大分子链（如一条伸展链）回归其平衡构象（无规线团）的时间只需千分之一秒，极快。依据在 Rouse 模型中的同样推导，可求得的大分子链质心的平动扩散系数，

$$D = \frac{k_B T}{f} \sim \frac{k_B T}{\eta b\, N^{\alpha_R}} \sim M^{-\alpha_R} \tag{7.10}$$

有兴趣了解上述推导详细过程的读者可进一步参考苏联 Lifshits 学派［Grosberg, Khokhlov, "Statistical Physics of Macromolecules", Originally published in Russia in Moscow, 1990; AIP Press (English Ed.), New York, 1994］和英国 Edwards 学派［Doi, Edwards, "The Theory of Polymer Dynamics", Cambridge, New York, 1989］的有关专著。相比之下，法国 de Gennes 学派［de Gennes, "Introduction to Polymer Dynamics", 1990］的处理方法则更加容易被化学、高聚物和其他非物理学系的师生们理解和接受。现简介如下。

如果一条线型柔软大分子链的末端距在时间 $t=0$ 和 $t=t$ 时分别为 $\vec{R}(0)$ 和 $\vec{R}(t)$，如右图所示，并以其无扰的平均构象状态（无规线团）为参照起点，由链构象形变产生的力为 $k_{coil}\vec{R}(t)$ 和链在液体中运动所受的摩擦阻力为 $f_{coil}\,\partial\vec{R}(t)/\partial t$。二者之间必须达至平衡，

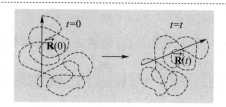

$$\frac{f_{coil}\partial\vec{R}(t)}{\partial t} = k_{coil}\vec{R}(t) \rightarrow \vec{R}(t) = \vec{R}(0)\exp\left(-\frac{t}{\tau_c}\right)$$

其中

$$k_{coil} = \frac{3\,k_B T}{\langle R^2\rangle} = \frac{3\,k_B T}{b^2 N^{2\alpha_R}}; \quad f_{coil} = 6\pi\eta R_h \approx 5\eta b\,N^{\alpha_R} \quad \text{和} \quad \tau_c = \frac{f_{coil}}{k_{coil}} \approx \frac{1.7\eta\,b^3 N^{3\alpha_R}}{k_B T} \approx 5\,t_{R_g}$$

以上推导中，假定了无规线团中的所有溶剂分子均可和大分子链一起运动，即无排水效应。每一条链等效于一个半径为 R_h 的小球；$R_g/R_h \approx 1.5$。并利用了 $R = \sqrt{\langle R^2\rangle} = b\,N^{\alpha_R} = \sqrt{6}R_g$（第四章）。这样得到的结果和上述的 τ_1 非常接近。将 f_{coil} 代入 Stokes 方程，可得与式（7.10）完全相同的链质心的平动扩散系数。

$$D = \frac{k_B T}{6\pi\eta R_h} \approx \frac{k_B T}{4\pi\eta R_g} \approx 0.0786\,\frac{k_B T}{\eta R_g}$$

其中的系数非常接近基于重整化群计算得到的 0.0829。显然，de Gennes 的处理方法相对地简易明了。然而，这样的方法却无法得到链内运动的细节，包括各个简正模式的平均涨落幅度和特征松弛时间。上述的"无排水效应"假定在物理上合理吗？

为了回答这一问题，可借用 Grosberg 和 Khokhlov 专著第 238 页中的讨论，即在一个运动的无规线团链中溶剂分子的速度分布。先假定在一条无规线团链内的链段数密度为 ρ_n，链段具有相同的运动速度（$\vec{u}_0$）和每一个链段与溶剂分子的摩擦系数为 f。如果一个链

段和其邻近溶剂分子的运动存在速度差别($\Delta\vec{u}=\vec{u}_0-\vec{u}$),就会在邻近溶剂分子上施加一个摩擦力($f\Delta\vec{u}$)。忽略惯性的影响、假定其对速度的扰动较小,以及液体体积不可压缩,可利用 Claude-Louis Navier 于 1822 年发表的、后被 George Gabriel Stokes 在 1842 年完善的 Navier-Stokes 方程寻获,在与链运动的垂直方向(y)上,溶剂分子遵循下列方程:

$$\eta\frac{\partial^2\vec{u}}{\partial y^2}+fC\Delta\vec{u}=0$$

其中,$f=6\pi\eta b$;$C=N/(\pi R^3/6)$;$fC=36\eta bN/(b^3 N^{3\alpha_R})=36\eta/(b^2 N^{3\alpha_R-1})$,这里,$R$ 是链的平均末端距;C 为链段密度。物理上不难想象,如果将链中心放在 y 方向的无穷远处,链的边缘接近 $y=0$ 处,上式有两个边界条件:在链的中心($y\to\infty$),溶剂分子和链段具有相同的速度,$\Delta\vec{u}=0$;链外部($y\to0$),溶剂分子不移动,$\vec{u}=0$;在此条件下,上式有解,最后的结果为

$$\vec{u}(y)=\vec{u}_0\left[1-\exp\left(-\frac{y}{\lambda}\right)\right]$$

其中,$\lambda=\left(\frac{\eta}{fC}\right)^{1/2}=\frac{b\,N^{(3\alpha_R-1)/2}}{6}$,为特征衰减(穿透)长度,其与链的平均回转半径($R_g=\sqrt{6}b\,N^{\alpha_R}$)之比如下

$$\frac{\lambda}{R_g}=\frac{N^{(3\alpha_R-1)/2}}{6\sqrt{6}\text{ 或 }7\,N^{\alpha_R}}=\frac{N^{(\alpha_R-1)/2}}{6\sqrt{6}\text{ 或 }7}=0.068N^{(\alpha_R-1)/2}$$

在 Θ 和良溶剂中,$\lambda/R_g=0.068N^{-1/4}$ 和 $\lambda/R_g=0.063N^{-1/5}$。对普通的高聚物而言,$N\sim10^3$;即 $\lambda/R_g<1.7\%$。换算成体积后,一条无规线团链里超过 95% 的溶剂分子都随着链以同样速度移动,特别是在良溶剂中。"无排水效应"的假定无误。

实验上,正如在第五章中已讨论过,在大分子稀溶液中,以及在确定体系的松弛只和平动扩散有关的前提下,可以用动态激光光散射测量不同浓度时和不同散射角度下的相互平动扩散系数,然后,将它们外推到浓度和角度分别为零时,可得到大分子链质心的平动扩散系数。也可利用强迫瑞利散射、荧光漂白恢复,以及脉冲场梯度核磁共振等方法,直接测量标记链的平动扩散系数。

依据上述链的动态学的讨论,还可得到实验可测的特性黏度的一般表达(推导略去),

$$[\eta]=\frac{RT}{2M\eta_s}\sum_{n=1}^{N}\tau_n$$

实验上,也可测量大分子链的特征松弛时间(τ_n)。例如,对沿着链的方向具有永久偶极矩的大分子,可利用介电弛豫谱仪测量链的平均末端距的特征松弛时间,主要为 τ_1。

为了观察一条大分子链的内部运动,观察者必须"站在"链的内部。换而言之,所用尺子的刻度必须远小于链的尺寸和其时间分辨率远快于链段的运动。在第五章中有关激光光散射的讨论中,已定义散射矢量为 $q=4\pi n\sin(\theta/2)/\lambda_0$,其倒数在物理上相当于一把测量物体的光学"尺子"。使用氦氖激光(HeNe,$\lambda_0\approx633$ nm)时,如果假定 $n\approx1.4$,测量范围为 38 nm$\leq1/q\leq$300 nm。

散射角度很小时,$1/q$ 通常大于聚合物链的尺寸($R_g < 100\ nm$)。因此,光散射无法测量一条高聚物链的内部运动。整条大分子链相当于一个小粒子。静态激光光散射仅可测得其整体的尺寸,而动态激光光散射则仅可测得与平动扩散对应的线宽(Γ),其与平动扩散系数(D)的依赖关系为 $\Gamma = Dq^2$。这相当于站在远处观看一列行进中的高速火车。只可看见列车的整体运动,而观察不到车厢内部每个乘客的坐卧行走、千姿百态。如要测量链的内部运动,就需要增大散射矢量 q(即"靠近列车"),使得 $1/q \leqslant R_g$。

当 q 足够大时,激光光散射获得的是尺度为 $1/q$ 的局部信息。物理上,不难想象,此时链本身的尺寸大小已不重要了,如同站在车厢内看乘客们时,车厢外部尺寸已不重要。所测的 Γ 应该和链的尺寸无关。依据 de Gennes 的标度理论,在动态激光光散射中,因为链的内部运动引致的线宽与 qR_g 的标度关系为 $\Gamma \propto (qR_g)^n$。由于整条链一直在无规地扩散,如同列车一直在行进,测得的总线宽应包括两部分:平动和内部运动,为它们的乘积,即 $\Gamma \propto Dq^2(qR_g)^z$。注意,$D \propto 1/R_g$。因此,若要 Γ 与 R_g 无关,z 只能为 1。实验上证实,当 $qR_g \geqslant 4$ 时,所测线宽与散射矢量的标度关系确实为 $\Gamma \propto q^3$。

标度理论没有提供有关体系的细节,而"魔鬼"常常就藏在细节之中!对于一个均匀的物体,无论形状如何,在三维空间中,其体积(V)与线度(l)的标度关系总是三次方,$V \propto l^3$;在二维空间,其面积(S)与 l 的标度关系总是二次方,$S \propto l^2$。在一维空间,其长度(L)与 l 的标度关系总是一次方,$L \propto l$。可是,在计算一个具体物体的体积、面积或长度时,则必须知道形状的细节,即正比常数。

如果希望利用动态激光光散射测量前面几个简正模式的平均涨落幅度和相应的特征松弛时间(τ_n),de Gennes 的标度理论就无法处理了,只能回到较严格的理论计算。先假定链上任意两个链段 i 和 j,其在时间 0 和 t 时的位置分别为 $\vec{r}_i(0)$ 和 $\vec{r}_j(0)$ 以及 $\vec{r}_i(t)$ 和 $\vec{r}_j(t)$。定义 $\vec{r}_{ij}(t) = \vec{r}_i(t) - \vec{r}_j(0)$,其物理意义是在间隔了时间 t 后,第 i 个单元离开第 j 个单元初始位置的距离(矢量)。在法向坐标系中,利用式(7.4)可得

$$\vec{r}_{ij}(t) = [\vec{p}_0(t) - \vec{p}_0(0)] + 2\sum_{n=1}^{N}\left[\cos\frac{ni\pi}{N}\vec{p}_n(t) - \cos\frac{nj\pi}{N}\vec{p}_n(0)\right]$$

不同的简正模式互不相关;因此,对任意给定的两个链段,$\exp[i\vec{q}\cdot\vec{r}_{ij}(t)]$ 的平均值如下

$$\langle\exp[i\vec{q}\cdot\vec{r}_{ij}(t)]\rangle = \langle\exp[i\vec{q}\cdot(\vec{p}_0(t) - \vec{p}_0(0))]\rangle\times$$

$$\prod_{n=1}^{N}\left\langle\exp\left[2i\vec{q}\cdot\left(\cos\frac{ni\pi}{N}\vec{p}_n(t) - \cos\frac{nj\pi}{N}\vec{p}_n(0)\right)\right]\right\rangle$$

在法向坐标中,$\vec{p}_0(t)$ 是质心在经过了一段时间 t 后的位置矢量。注意,这里的 t 实际上是 $\Delta t = t - 0$;因此,右边乘积的第一项(记为 I)意味着,链的质心移动一个 $1/q$ 的距离平均所需的特征松弛时间(τ_0),对应的是特征线宽($\Gamma_0 = 1/\tau_0$);该弛豫过程的表达为

$$\mathrm{I} = \exp\left(-\frac{t}{\tau_0}\right) = \exp(-\Gamma_0 t)$$

经过较复杂的运算(略去具体过程),也可得到右边乘积的第二项(记为 II)的表达,

$$\mathrm{II} = \langle\exp[i\vec{q}\cdot\vec{r}_{ij}(0)]\rangle\left\langle\exp\left[-4q^2\sum_{n=1}^{N}k_n\cos\frac{ni\pi}{N}\cos\frac{nj\pi}{N}\left[1 - \exp\left(\frac{t}{\tau_n}\right)\right]\right]\right\rangle$$

其中，k_n 是一个与模型有关的常数。将上面两式相乘，再将结果代入前式，可得

$$\langle \exp[\,i\,\vec{\mathbf{q}}\cdot\vec{\mathbf{r}}_{ij}(t)\,]\rangle = \mathrm{I}\times\mathrm{II}$$

$$= \exp(-\varGamma_0 t)\Big\langle \exp[\,i\,\vec{\mathbf{q}}\cdot\vec{\mathbf{r}}_{ij}(0)\,]\exp\Big[-4\,q^2\sum_{n=1}^N k_n$$

$$\cos\frac{ni\pi}{N}\cos\frac{nj\pi}{N}\Big[1-\exp\Big(\frac{t}{\tau_n}\Big)\Big]\Big]\Big\rangle$$

与式(5.40)比较，上式是一条大分子链上不同链段之间的动态结构因子；其中的平均是对 i 和 j 的双重求和。归一后，其为散射电场的时间自相关函数，

$$g^{(1)}(q,t)=\frac{\langle \exp[\,i\,\vec{\mathbf{q}}\cdot\vec{\mathbf{r}}_{ij}(t)\,]\rangle}{\langle \exp[\,i\,\vec{\mathbf{q}}\cdot\vec{\mathbf{r}}_{ij}(0)\,]\rangle}=\exp(-\varGamma_0 t)J(q,t)$$

其中，$J(q,t)$ 代表了所有的简正模式(链的内部运动)的弛豫过程。其具体的表达如下

$$J(q,t)=\frac{\sum_{i=1}^N\sum_{j=1}^N\exp[\,i\,\vec{\mathbf{q}}\cdot\vec{\boldsymbol{r}}_{ij}(0)\,]\exp\Big[-4\,q^2\sum_{n=1}^N k_n\cos\dfrac{ni\pi}{N}\cos\dfrac{nj\pi}{N}\Big[1-\exp\Big(-\dfrac{t}{\tau_n}\Big)\Big]\Big]}{\sum_{i=1}^N\sum_{j=1}^N\exp[\,i\,\vec{\mathbf{q}}\cdot\vec{\boldsymbol{r}}_{ij}(0)\,]}$$

当 $t\to0$ 时，$g^{(1)}(q,t)\to1$(大分子链仍未开始移动)。当 $t\to\infty$ 时，II 趋向一个常数。因此，$g^{(1)}(q,t)\propto\exp(-\varGamma_0 t)$；其物理上意味着所有较快的内部运动已完全弛豫，仅剩下较慢的质心弛豫。如果质心弛豫可全部归于平动扩散运动，$\varGamma_0=Dq^2$。物理上，上式中的 $\exp(-\varGamma_0 t)J(q,t)$ 源于同时发生的质心平动扩散和所有链段相对于质心的运动(各种简正模式)。换而言之，所有简正模式的运动都"载在"质心运动之上。各个简正模式(内部运动)的数值解可通过计算获得。

Perico，Piaggio，Cuniberti [J. Chem. Phys.，1975，62:2690；1975，62:4911] 指出，在非排水和 $q^2\langle R_g^2\rangle>1$ 的条件下，$g^{(1)}(q,t)$ 主要取决于前五项弛豫($n=0,1,2,3$ 和 4)，即平动扩散加上四个与链的内部运动有关的主要简正模式。依照平均涨落振幅(A_n，对散射光强的贡献)递减的顺序，上式可被重写成

$$g^{(1)}(q,t)=\frac{\sum_{n=0}^N A_n\exp(-\varGamma_n t)}{\sum_{n=0}^N A_n} \tag{7.11}$$

他们还算出了在 $1\leqslant q^2\langle R_g^2\rangle\leqslant100$ 范围内，所有 A_n 的数值。利用 Zimm 模型可得

$$\varGamma_n=q^2 D+\frac{\pi^{2\alpha_R}k_{\mathrm{B}}T}{6\eta\,b^3 N^{3\alpha_R}}n^{3\alpha_R} \tag{7.12}$$

其中，$\langle R^2\rangle=b^2 N^{2\alpha_R}$ 或写成 $R=\sqrt{\langle R^2\rangle}=b\,N^{\alpha_R}$；$\langle R^2\rangle=6\langle R_g^2\rangle$；$R=\sqrt6\,R_g$。上式中，线宽可用平动扩散对应的线宽($q^2 D$)归一，并将 $D=k_{\mathrm{B}}T/(6\eta\pi R_h)$ 代入，结果如下

$$\frac{\varGamma_n}{q^2 D}=1+\frac{\pi^{2\alpha_R+1}R_h}{6\sqrt6\,R_g q^2\langle R_g^2\rangle}n^{3\alpha_R}\approx1+\frac{0.56}{q^2\langle R_g^2\rangle}n^{3\alpha_R}\qquad n=0,1,2,\cdots,N \tag{7.13}$$

在上述的推导中，利用了线型柔软链在 Θ 溶剂和良溶剂中，分别有 $\alpha_R=1/2$ 和 $R_g/R_h\approx1.3$ 以及 $\alpha_R=3/5$ 和 $R_g/R_h\approx1.5$。为了以后方便书写和讨论，现定义 $x=q^2\langle R_g^2\rangle$。归一后的线宽对 x 作图与链尺寸和链段数无关，只与 n 有关。扣除了大分子链本身质心的平动扩散运动以后，链段相对于质心的内部运动自然和链长无关，故上述结论在物理上合理。

de Gennes 在 20 世纪 70 年代考虑利用动态激光光散射研究高聚物链内部运动时,曾经指出,需要在 $qR_g \sim 2.5$ 的时候才可观察到这些内部的简正模式,并断言对于通常的高聚物($R_g \sim 50$ nm),利用可见光无法测量简正模式,必须要用真空紫外光源。可是当时没有真空紫外激光,即使今天有了这样的光源和高灵敏度的真空紫外单光子检测器,如何在真空紫外下调节整个光路,使其达到激光光散射所要求的精度,仍是一个难题。

细心的读者会想到,既然需要 $qR_g \sim 2.5$,如果不能缩短波长以增加 q 值,为何不用超长的大分子链? 对 HeNe 激光($\lambda_0 \sim 633$ nm)和 150° 的散射角而言,$1/q \sim 40$ nm,即 R_g 需要大于 100 nm 以上,实验上可以实现。但是,一条大分子链的简正模式(内部运动)总是和质心平动扩散耦合在一起。为了将二者分开,必须先在 $qR_g < 1$ 的范围内测得一系列线宽(Γ),然后再将(Γ/q^2)对 q^2 作图。其外推到 $q \to 0$ 可得 D。只有这样,方可在分析散射光强时间自相关函数时,扣除由质心平动扩散引致的弛豫。如果 $R_g \sim 150$ nm,激光光散射实验就必须在小于 15° 的角度下进行。可是绝大部分仪器的角度下限就是 15°。这就是为何实验研究一直远远地落后于理论计算。

直到 20 世纪 90 年代初,Chu 和他的合作者[Macromolecules,1991,24:6832]才通过将普通激光光散射仪和一台特制的带有棱镜散射样品池的小角激光光散射仪结合起来,首次研究了大分子链的内部运动这一复杂问题。但是,他们并未解释为何观察到相对较快和较弱的第二个简正模式,而不是理论上预测的最慢且最强的第一个简正模式。

几年之后,在采用了光束较细的、波长较短的固体激光光源,以及改进了商品仪器的光学检测系统后,笔者实验室的激光光散射仪的最小测量角度可达 5.6°。从而实现了利用激光光散射系统地研究大分子的内部运动。下图(a)显示了动态激光光散射的实验结果,样品为三个具有超高摩尔质量($M_w \sim 10^7$ g/mol)的聚苯乙烯。当 $x < 1$ 时,用 q^2 归一后的线宽分布仅含有一个单峰,且不随 x 变化;而当 $x \geqslant 1$ 时,在 $10^{-7} \sim 10^{-6}$ cm²/s 的范围内,出现了第二个面积很小的峰,远比链质心的平动扩散快。图(a)中的插图放大和清楚地显示了该小峰随着 x 的变化。有兴趣的读者可参考原始文献,进一步了解实验的详情。

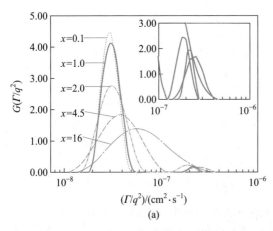
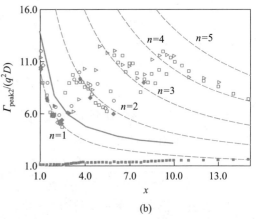

Wu C,Chan K K,Xia K-Q. Macromolecules,1995,28:1032.　　Wu C,Chan K K,Xia K-Q. Macromolecules,1995,28:1032.

(Figure 4,permission was granted by publisher)　　(Figure 5,permission was granted by publisher)

一方面,使用一系列窄分布、超长的聚苯乙烯链($R_g = 130 \sim 310$ nm)使得 qR_g 高达4,从而可观察到更快和更弱的高阶简正模式(在更小尺度上链的内部运动)。另一方面,由于可在极小的散射角度范围内精确地测得这些超长大分子链的平动扩散系数,所以可用 q^2D 将第二个与内部运动有关的小峰的线宽归一,如上图(b)所示;其中,靠近底部横坐标的不同的实心小符号代表了归一后的第一个与平动扩散有关的大峰的线宽($\Gamma_{\text{peak1}}/q^2D$)。由此可见,在 $x < 2$ 时,$\Gamma_{\text{peak1}}/q^2D \cong 1$,没有链长依赖性,归因于链长依赖性已反映在平动扩散系数里。随着 x 的增加,该比值仅增加了一点,表明少量的内部运动慢慢地耦合进平动扩散。

　　正如理论预期的,归一后的三个不同的聚苯乙烯样品的第二个峰的线宽($\Gamma_{\text{peak2}}/q^2D$)基本上与摩尔质量无关。上图(b)中的实线代表的是依据文献中 A_n 的数值解,对式(7.12)中的 Γ_n/q^2D 进行权重求和($n = 1, 2, \cdots, N$)的结果,代表了理论上预测的归一后的第二个峰的线宽($\Gamma_{\text{peak2}}/q^2D$)对 x 的依赖。然而实验结果却大相径庭,在不同的 x 范围内,仅观察到理论预测的单个不同的简正模式,如不同的虚线所示。

　　有趣的是,在 $1 < x < 3$、$3 < x < 6$、$6 < x < 10$ 和 $10 < x < 15$ 的范围内,实验上仅分别观察到第一、第二、第三和第四个简正模式。定性的解释是,只有当简正模式涉及的尺度与光散射中的观察尺度 $1/q$ 相近时,动态光散射方可"看到"这一弛豫模式,类似物理上的"共振",不依赖于一个简正模式涨落的幅度(能量)。后续的不同大分子溶液的实验重复地显示了这一现象。因此,现存的理论必须修改。

　　如前所述,在 $x \gg 1$ 时,无论是严格推导,还是 de Gennes 的标度理论,都预测 Γ/q^3 趋向一个常数。实验结果和理论预测的趋势相符,但是实验上所趋向的常数小于理论预测值,仅有理论预测值的 $70\% \sim 80\%$,如下图(a)所示。这一结果与上述讨论完全一致:理论预测的是所有简正模式的权重加和,而动态激光光散射依其不同的观察尺度($1/q$),仅可依次逐渐地"看到"不同的单个简正模式的弛豫。需要指出的是,当初建立有关大分子链的内部运动的理论时,并未考虑动态激光光散射的测量。到目前为止,还没有一个实验方法可以一次测到权重加和后所有的简正模式或可以逐个测量每个简正模式。简而言之,理论上需要考虑为何动态激光光散射不可测到所有的简正模式。

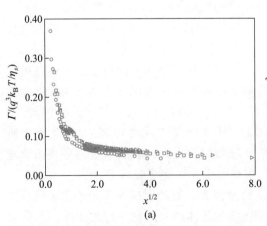

Wu C, Chan K K, Xia K Q. Macromolecules, 1995, 28:1032.

(Figure 6, permission was granted by publisher)

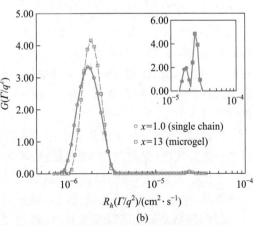

Wu C, Zhou S Q. Macromolecules, 1996, 29:1574.

(Figure 5, permission was granted by publisher)

20 世纪 70 年代,de Gennes 基于当时动态激光光散射仪的状况,预测"$qR_g \sim 2.5$"是一个测量大分子链简正模式运动的必要条件。然而,现代的动态激光光散射仪采用了功率更大的激光、更快的单光子检测器、更大的记忆体和更快的计算机,其具有更好的测量精度和更快的数据处理速度。老的仪器已不可同日而语。近二十年来,测量不同大分子溶液的实验结果已反复地证实,只要达到"$qR_g \geq 1$"这一条件,动态激光光散射就可测出第二个与简正模式运动有关的小峰,不管其面积(正比于其对应的散射光强)多少,如上图(b)所示,其中的插图显示了放大的小峰。

因此,利用动态激光光散射测量不同散射角度下极稀大分子溶液的线宽分布,可从第二个小峰出现的散射角度,反过来估算大分子链的方均回转半径,即$\langle R_g^2 \rangle \sim 1/q^2$,可与由静态激光光散射测量的$\langle R_g^2 \rangle$相互验证。除了大分子链以外,动态激光光散射原则上可以测量任何小尺度、快速的弛豫。上图(b)中还显示了在溶胀状态下,聚 N-异丙基丙烯酰胺球型微凝胶的测量结果;其中为了与同样的线型链比较,横坐标($\Gamma/q^2 = D$)乘以各自不同的流体力学半径(180 nm 和 160 nm,不考虑量纲),等效于扣除了因二者尺寸不同对 D 的影响。

放大后的第二个小峰显示仅当 $x = 13$,即$1/q \sim R_g/3.6 \sim 160/3.6 \approx 44$nm 时,动态激光光散射方可检测到溶胀微凝胶交联网络在热能激发下的涨落。换而言之,该微凝胶网络最大的密度涨落的相关长度约为 44 nm。这是首次在实验上得知,热能不可激发整个微凝胶网络的涨落,仅是局部的网络。已知在充分溶胀的状态下,该微凝胶网络两个邻近交联点之间的平均距离约为 5 nm,这意味着相关长度大概跨越 10 个交联点,即液体的过度阻尼限制了一个邻近交联点之间链节的涨落,其仅可波及每边各 5 个链节。

另外,在 $x \gg 1$ 时,Γ/q^3 趋向的一个常数,仅有理论预测的 ~6%。进一步说明交联限制了链的大部分内部运动,也间接地证实了上述有关动态激光光散射不能测得理论预测的全部简正模式。这个例子清楚地说明,利用动态激光光散射不仅可以测得大分子链或胶体粒子的流体力学半径分布,还可得到其他方法无法获得的微观性质。因此,需要善用此方法!

另一个例子是利用动态激光光散射确定大分子溶液的 Θ 条件。在第四章中讨论大分子链的构象时,曾定义了二阶维里系数(A_2)为零时,溶液处于 Θ 状态;即因链段的硬核体积造成的链扩展与综合色散引力作用引致的、等效的"排斥体积收缩效应"互相抵消,这里的"综合色散作用"包括了链段之间、溶剂分子之间和链段和溶剂分子之间的色散引力作用在溶解中的综合变化。显然,这里论及的是单根大分子链在一条链的尺度上的整体性质。

然而,在所有大分子教科书中论及如何测定一个大分子溶液的 Θ 条件时,均是先测量某一物理性质(散射光强、约化黏度、渗透压差等)对浓度的依赖性,然后再寻获该物理性质不依赖大分子浓度时的条件(温度、压强、溶剂组成等),即溶液的 Θ 条件。

注意,在这一已被广泛接受的方法中测量的是包含了链的硬核体积和色散吸引作用(链之间、溶剂分子之间以及链和溶剂分子之间)的排斥体积效应消失时的条件。所有的教科书都没有交代在测定大分子溶液的 Θ 条件时,为何可以用实验上测得的溶液中链间的排除体积效应代替理论上论及的一条链内链段之间的排除体积效应。

在 20 世纪四五十年代发展大分子物理理论时,根本无法测量单链的物理性质和溶液行为。这也许就是为何用实验上可测的溶液中分子间的排除体积效应代替理论上论及的一条链上链段之间的排除体积效应,仅是一个权宜之计。如上所述,利用现代实验手段已可测量单链的一些物理性质,包括利用动态激光光散射测量单链的内部运动(各种简正模式)。

问题为是否可以利用单链的某一物理性质直接确定大分子溶液的 Θ 条件。定性地已知,在良溶剂中,源于链段硬核体积产生的链扩展主导了色散吸引作用引致的链收缩。链的平均构象比在 Θ 状态下扩展,链段之间受到一个净的张力;而在不良溶剂中,色散吸引作用主导了排除体积效应,故大分子链收缩,即链段之间存在一个净的吸引力。在 Θ 状态下,二者相互抵消,所以链段受到净力为零。理论上,热能应该能够激发链内更多的内部运动(更多的简正模式)。实验上,应该可以观察到源自内部运动的散射光强的增加。换而言之,动态激光光散射中得到的两个峰的面积之比(A_r,散射光强贡献)应该达至一个拐点。基于式(7.11),该面积之比为

$$A_r = \frac{\sum_{n=1}^{N} A_n}{\sum_{n=0}^{N} A_n}(\text{理论上}) \quad \text{和} \quad A_r = \frac{A_{\text{peak2}}}{A_{\text{peak1}} + A_{\text{peak2}}}(\text{实验上})$$

在 $1/q < R_g$ 的条件下,利用动态激光光散射测量不同的大分子溶液和微凝胶分散液的链内简正模式。确实都观察到在 Θ 温度附近,A_r 出现一个拐点,如下图所示。

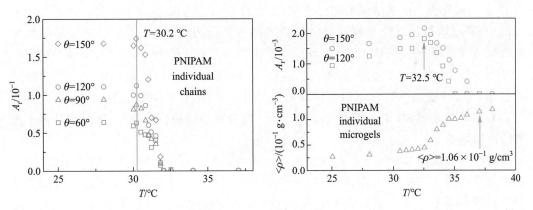

Dai Z J,Wu C. Macromolecules,2010,43:10064.

(Figures 6 and 11,permission was granted by publisher)

在良溶剂中,对给定的温度,随着观察尺度($1/q$)变小,动态激光光散射测得更多的简正模式,反映在 A_r 的显著增加。对给定的散射角度,随着温度的增加,热敏性 PNIPAM 链逐步蜷缩,R_g 变小。一方面,随着温度的增加,排除体积效应变弱,所以热能可激发更多的简正模式;另一方面,链的蜷缩等效于 q 变小,导致动态激光光散射可测的简正模式减少。两个相反的效应互相抵消,定性地解释了为何 A_r 基本上不依赖于温度。在 30℃附近,A_r 变化明显,拐点确定,基本不随观察尺度改变,比 Θ 温度稍低。

后续聚苯乙烯在环己烷中的研究也进一步证明了拐点(~ 36℃)接近,但略微高于用

浊度法测量的 Θ 温度（34.5℃），这是合理的。溶液中链与链之间的链间作用弱于链上的链段与链段之间的链内作用，故需更低的温度方可达至同样的排除体积效应。上述有关大分子链在 Θ 条件下最"软"的讨论，后来被温度依赖的链拉伸实验证实。

当冷却速度不是无限慢时，对具有上或下临界温度的溶液，利用浊度法所测的 Θ 温度总是低于或高于真正的 Θ 温度。不难理解，当链完全塌缩时，链的内部运动消失。在微凝胶中，链之间的交联已经制约了其内部的链段运动，故右上图中微凝胶的 A_r 比左边单链的 A_r 小了两个数量级。虽然 A_r 仅有千分之一二左右，但依然可测，可见现代动态激光光散射之灵敏。右上图中微凝胶的链密度随着温度上升（溶剂性质变差）增加。在溶胀的交联网络中，热能只能激发较少的简正模式。A_r 的拐点明显、同样对应着 Θ 温度。邻近交联点之间的链节远短于大分子单链（几百倍），故其相变温度比单链稍高。

小　　结

稀溶液中的大分子链，除了其质量中心（质心）的平动扩散以外，链上的 N 个链段在热能的搅动下也会相对于质心无规地运动（内部运动）。每个链段都有一个自己的三维坐标 (x, y, z) 或位置矢量 $(\vec{r})$。因此，共有 N 个位置矢量（$\vec{r}_i, i = 1, 2, \cdots, N$），对应着 N 个运动方程。由于链段的相互连接，N 个位置矢量之间也相互依赖。在通常的坐标系中，联立求解这 N 个方程是非常复杂和困难的。为此，可将 N 个位置矢量线性地组合成 N 个互相独立的法向坐标（$\vec{p}_n, n = 0, 1, 2, \cdots, N$），其中，$\vec{p}_0$ 代表了质心的运动。数学上，从一个坐标系 $\vec{r}_i$ 变换到另一个 $\vec{p}_n$ 是线性代数里一个矩阵的对角化。在此新的坐标系中，N 个运动方程在不同的特殊模型中可联解，得到 N 个与一条大分子链的内部运动有关的简正模式（$\vec{p}_n(t), n = 1, 2, \cdots, n$）。随着 n 的增加，简正模式逐渐涉及更小尺度的内部运动。

复杂和烦琐的理论推导有着其完美和优美的一面，但也让大部分实验工作者眩晕。撇开数学方程，仅基于一些物理图像，理解大分子链内的简正模式弛豫的涨落振幅和特征弛豫时间并不复杂。如果可忽略分子（链段）之间的流体力学相互作用，对一条平均构象为无规线团的线型柔软大分子链，虽然链段相对于质心的位置是千变万化的，链的内部运动的最大振幅（即整条链的转动/振动）应该接近但不会超过链尺寸本身（其平均回转半径，$R_g = \sqrt{\langle R_g^2 \rangle}$）；最慢松弛时间应和链扩散一个自身尺寸的距离所需的时间相近。最小的变化尺度（即 $n = N$）就是一个链段的长度（b），两个相邻链段之间的相对运动。如果考虑流体力学相互作用的过度阻尼，最大的涨落振幅度应小于其平均回转半径；最慢的松弛时间应该比链运动一个其自身尺寸的距离所需的时间更长。虽然这样基于物理图像的推理无法给出简正模式对链长依赖性的细节，但提供了一个基本的数量级概念。

在 Rouse 模型中，忽略了流体力学相互作用，适用于流体力学相互作用已被屏蔽掉了的较浓的溶液或熔体。$\vec{p}_n(t)$ 的时间自相关函数的平均涨落幅度：$\langle \vec{p}_n(0) \cdot \vec{p}_n(0) \rangle =$

$\langle p_0^2(0)\rangle \approx \langle R_g^2\rangle/n^2$,正比于$\langle R_g^2\rangle$,反比于$n^2$;特征弛豫时间:$\tau_n \approx t_{R_g}/n^2$,正比于大分子链扩散一个$R_g$所需的平均时间$(t_{R_g})$,反比于$n^2$。可见,当$n=1$时,严格推导的结果和基于物理图像的推理相同,随着$n$的增加,简正模式迅速地变弱、变快!

在Zimm模型中,考虑了流体力学相互作用,更适用于大分子稀溶液或大分子和溶剂分子相互作用较强的体系。$\vec{p}_n(t)$的时间自相关函数的平均涨落幅度:$\langle p_0^2(0)\rangle \approx 0.3\langle R_g^2\rangle/n^{\alpha_{R_g}}$;特征弛豫时间:$\tau_n \approx 4t_{R_g}/n^{\alpha_{R_g}}$,其中,$1/2 \leqslant \alpha_{R_g} \leqslant 3/5$。与Rouse模型比较,由于过度阻尼,第一个简正模式的涨落幅度较小,特征弛豫时间变长。但是,简正模式随着n的变化速度相对较慢。同样,$n=1$时的结果与基于物理图像的推测吻合。

实验上,大分子链内部的简正模式可测,尤其是第一简正模式。方法包括,但不限于,介电弛豫、双折射弛豫和动态激光光散射。前两种方法对待测的大分子属性均有要求,但利用动态激光光散射测量简正模式运动则较为普适,也是本章介绍的重点内容。

基本原理:

动态激光光散射利用大分子或粒子的散射光测量它们的运动。"尺子"的分辨率与波长有关,为散射矢量的倒数$(1/q)$。当$1/q$远远大于平均回转半径(R_g)时,动态激光光散射只可测到一条大分子质心的扩散,而无法窥知其内部的链内运动。正如站在远处观看一列高铁,只可看到列车整体的行进,而不能了解每一个车厢内部乘客的活动。当$1/q \leqslant R_g$时,从散射光强的时间自相关函数,除了仍然可以得到与平动扩散有关的特征弛豫时间以外,还可得到与内部运动(简正模式)有关的特征弛豫时间分布,等同于站在车厢内部观看乘客的不同活动。

可测物理量:

利用数字时间相关器测量大分子溶液在不同浓度时和不同散射角度下散射光强的时间自相关函数。先在$1/q > R_g$的范围内,将表观平动扩散系数外推到无限稀的浓度和零散射角后,就可获得质心的平均平动扩散系数。然后,在$1/q \leqslant R_g$的范围内,测得的散射光强的时间自相关函数以得到与内部运动有关的特征弛豫时间分布。一般而言,只有在$R_g > 50$ nm或$M \sim 10^6$ g/mol时,方可满足$1/q \leqslant R_g$的条件。

数据处理:

对每一个所测的散射光强的时间自相关函数作Laplace反演可得一个对应的线宽(特征弛豫时间)分布$G(\Gamma)$。对于较窄分布的大分子样品,在$1/q > R_g$的范围内,线宽或特征弛豫时间分布为单峰。当$1/q \leqslant R_g$时,会出现一个位于右边较大线宽(低频)区域和大分子链的内部简正模式运动有关的第二个小峰。将其对应的线宽(Γ_{peak2})用第一个大峰对应的线宽(q^2D)归一,可得简正模式运动的平均线宽。另外,第二个小峰和第一个大峰的面积比(A_r)代表了所有简正模式对散射光强的贡献。除了测量大分子链的内部运动(简正模式)以外,利用所测的Γ_{peak2}和A_r,还可得到其他实验方法难以获得的微观信息。本章仅列举了利用动态激光光散射寻获微凝胶内部的涨落尺寸和溶液的Θ条件的两个例子。

因此,仅仅利用昂贵的研究型激光光散射仪测量大分子链或胶体粒子的流体力学半径分布,岂不是杀鸡焉用牛刀! 将静态和动态激光光散射以及其他方法结合起来,可得到许多其他实验方法无法获得的微观性质。因此,建议读者理解激光光散射的原理,尤其是动态激光光散射的理论,然后善用! 下图总结了如何利用动态激光光散射测量链的简正模式;比较了在 Rouse 模型和 Zimm 模型中简正模式的时间自相关函数的平均涨落幅度和特征弛豫时间,其中,D 为平动扩散系数。稀溶液中的实验结果与 Zimm 模型预测的吻合。

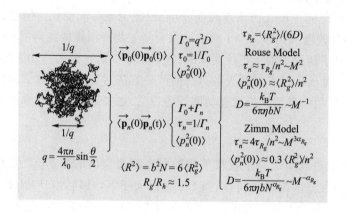

Chapter 7. Dynamics of Chains in Dilute Macromolecular Solutions

As mentioned before, the random walk of each segment on a macromolecular chain leads to an everchanging conformation. In other words, different macromolecular chains in the solution have different conformations at each moment. The number of the chains with a given conformation follows a normal distribution. Each conformation has a corresponding square distance. The statistical average of all conformations results in the mean square distance. For a linear flexible chain, the mean square distance is related to the theoretically calculable mean end distance, and generally speaking, it is related to the experimentally measurable mean gyration radius. Based on the ergodicity of statistical random variables, the statistical average of the conformation of countless macromolecular chains is equal to the statistical average of the conformation of one macromolecular chain over an infinite period of time. Therefore, the normal distribution of the chain conformation can also be regarded as the time distribution of each conformation.

The English "dynamics" and "kinetics" are often translated as the same "kinetics" in Chinese. The translation is correct! Both refer to physical or chemical changes over time. "Kinetics" focuses on the influence of "forces and moments" on the motion of objects. It is also called "kinematics" in mechanics, and commonly used when discussing the rate of chemical reactions. "Dynamics" emphasizes the influence of energy on changes, a broader category, including kinetics. Since the mid-1920s, physicists have gradually adopted the term "dynamics" in textbooks. Corresponding to "statics" (the study of the properties of a system in an equilibrium state), it is more appropriate to use "dynamics" for studying the effect of thermal energy on the overall and local conformation and motions of macromolecular chains. Therefore, this book translates use "dynamics" instead of "kinetics". However, when discussing the process of the conformational transition of chain, "kinetics" will still be used. When discussing the phase transition of macromolecular solutions, the mesoglobular phase and macromolecular colloids in Chapters 10-12, the "kinetic stability" is still used to describe the non-equilibrium metastable state of a system to distinguish the "dynamic stable" in the thermodynamic equilibrium state.

Macromolecular dynamics discusses the influence of energy on the conformational changes of macromolecular chains. To simplify the discussion, let us first consider the undisturbed state of a chain in the solution. Each Kuhn segment in the macromolecular chain is treated as a small bead and a short stretchable spring. Therefore, a macromolecular linear chain becomes a series of N small beads with no size and no hydrodynamic interaction, [Rouse model, J. Chem. Phys., 1953, 21: 1272]. Each two beads are connected together by a small spring with an average length of b. The energy corresponding to each pair of spring and bead is only one thermal energy ($k_B T$). Therefore, in one dimension, the energy density per unit length (k_s, the force constant) is $k_B T/b$. Every dimension is equal, so that in three dimensions, the force constant of a small spring is three times that of one dimension, that is, $k_s = 3k_B T/b$. Under the agitation of thermal energy, the length and orientation of each small spring are changing randomly, and the position of each small bead is also constantly moving.

In order to deal with such a complex macromolecular chain composed of N units, three important concepts are needed: the degrees of spatial freedom, normal coordinates, and normal mode. Each degree of spatial freedom is one independent direction of motion. A bead can move in three independent (orthogonal) directions in a three-dimensional space, so that it has three degrees of spatial freedom, that is, the commonly used coordinate system with three orthogonal directions of $x, y,$ and z. Its movement towards any direction can always be decomposed into components in these three directions. In other words, a combination of the components in the three directions can express the motion of one bead; two beads have six degrees of spatial freedom; and so on, N beads have $3N$ degrees of spatial freedom. From another perspective, it can also be said that for any bead, a three-dimensional coordinate (x_i, y_i, z_i) with three degrees of spatial freedom is required to express its position. The three-dimensional coordinates ($x_i, y_i, z_i, i = 1, 2, \cdots, N$) with $3N$ degrees of spatial freedoms are required to express the positions of N beads.

In the polar coordinate system, the position of any point in the three-dimensional space to the origin can be expressed as a vector ($\vec{r}_i$) pointing from the origin to the point. The position of N beads requires N such vectors ($\vec{r}_i, i = 1, 2, \cdots, N$). The derivative of each vector with respect to time ($d\vec{r}_i/dt$) is the velocity of the corresponding bead ($\vec{u}_i = d\vec{r}_i/dt$). Each moving object in the liquid receives a resistance ($f\vec{u}_i$) that prevents its movement, where $f = 6\pi\eta b$ is the coefficient of friction, related to the liquid viscosity (η) and the bead size (b). The faster the motion, the greater the resistance. Physically, every bead with a certain size and mass (m) is excessively damped in the liquid, and quickly approaches to a constant velocity, that is, it cannot accelerate for a long time. Therefore, at a frequency of less than 10^6 Hz or when the time is longer than 1 μs, the force related to the acceleration motion ($md\vec{u}_i/dt$) can be ignored. If only considering the interaction between each bead and its two adjacent beads, and the random force of solvent molecules on each bead at each instant ($\vec{f}_i(t)$), the motion equation of N beads is

$$f\vec{u}_i = k_s(\vec{r}_{i-1} - \vec{r}_i) + k_s(\vec{r}_{i+1} - \vec{r}_i) + \vec{f}_i(t) \quad i = 1, 2, \cdots, N$$

Careful readers will notice that the two beads at both ends have only one bead next to each other. For macromolecules, N is much larger than 1. Therefore, it is conceivable to add one bead at each end, not affecting the motion of the entire chain. The above equation looks simple and perfect. However, these N beads are connected together. The motion of a bead will not only affect the two adjacent beads, but also other nearby beads; conversely, the motions of other beads will also affect its motion. The affecting distance depends on how strong the interaction between the beads is. Physically, it is called as the correlation length between the beads. Even if only considering the interaction between two adjacent beads, the three-dimensional coordinates of $3N$ degrees of spatial freedom of N beads are not completely independent, but there is a certain degree of interaction. It is due to the interdependence among variables that there is basically no way to solve the above N equations simultaneously no matter whether in the intuitive space of the ordinary three-dimensional or polar coordinates.

Therefore, an abstract multidimensional space is needed, whose dimensional coordinates are not only independent of each other, but also capable of describing the coupled motions of N beads. This is the so-called normal coordinates, which are a linear combination of the displacement coordinates in the normal three-dimensional space, i.e.,

$$\vec{\mathbf{p}}_n(t) = \frac{1}{N} \sum_{i=1}^{N} \cos \frac{in\pi}{N} \vec{\mathbf{r}}_i(t) \quad n = 1, 2, \cdots, N$$

The motions it depicts are the so-called normal modes. **It is equivalent to the diagonalization of a matrix in linear algebra. In a normal three dimension coordinators, each item (the coefficient in front of each item in the previous equation) in a matrix with $N \times N$ items is non-zero. The multiplications of such a matrix by one matrix results in an abstractive normal coordinator space, making all the items in the matrix become zero except for the diagonal elements (eigenvalues). In such away, the motion equations expressed by $\vec{\mathbf{p}}_n(t)$ can be conveniently solved.**

When $n = 0$, $\vec{\mathbf{p}}_0(t) = \left[\sum_{i=1}^{N} \vec{\mathbf{r}}_i(t) \right]/N$. According to the definition of an object's center of mass, $\vec{\mathbf{p}}_0(t)$ represents the time trajectory of the centroid. When $n > 1$, $\vec{\mathbf{p}}_n(t)$ represents the internal motions of the beads in the chain with respective to its center of mass. $\vec{\mathbf{p}}_1(t)$ is equivalent to a vector along the chain, representing the chain orientation. Due to its continuous changes in length and direction, it can be roughly regarded as the rotation of the chain around its center of mass. $\vec{\mathbf{p}}_2(t)$ is

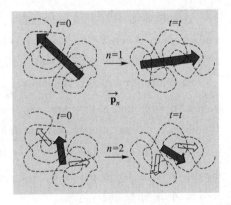

equivalent to two vectors facing away from the centroid, in basically opposite directions, and constantly changing in size and direction. In comparison with $\vec{\mathbf{p}}_1(t)$, it is more sensitive to the details of the change of the chain conformation, as shown above. As n increases, $\vec{\mathbf{p}}_n(t)$ involves smaller and smaller local conformational changes. When $n = N$, $\vec{\mathbf{p}}_N(t)$ corresponds to the smallest change of conformation, the relative displacement of two adjacent beads. A combination of the two equations above leads to the following equation of motion in the normal coordinate.

$$f\frac{\mathrm{d}\vec{\mathbf{p}}_n(t)}{\mathrm{d}t} = \frac{f}{N}\sum_{i=1}^{N}\cos\frac{in\pi}{N}\frac{\mathrm{d}\vec{\mathbf{r}}_i(t)}{\mathrm{d}t} = \frac{f}{N}\sum_{i=1}^{N}\cos\frac{in\pi}{N}\vec{\mathbf{u}}_i \quad (n=1,2,\cdots,N)$$

$$= \frac{k_s}{N}\sum_{i=1}^{N}\cos\frac{in\pi}{N}(\vec{\mathbf{r}}_{i-1}(t) + \vec{\mathbf{r}}_{i+1}(t) - 2\vec{\mathbf{r}}_i(t)) + \frac{1}{N}\sum_{i=1}^{N}\cos\frac{in\pi}{N}\vec{\mathbf{f}}_i(t)$$

$$\cong \frac{k_s}{N}\sum_{i=1}^{N}\left[\cos\frac{n(i-1)\pi}{N} + \cos\frac{n(i+1)\pi}{N} - 2\cos\frac{in\pi}{N}\right]\vec{\mathbf{r}}_i(t) + \frac{1}{N}\sum_{i=1}^{N}\cos\frac{in\pi}{N}\vec{\mathbf{f}}_i(t)$$

$$\cong \frac{k_s}{N}\sum_{i=1}^{N} - \left(\frac{n\pi}{N}\right)^2\cos\frac{in\pi}{N}\vec{\mathbf{r}}_i(t) + \frac{1}{N}\sum_{i=1}^{N}\cos\frac{in\pi}{N}\vec{\mathbf{f}}_i(t)$$

$$\frac{\mathrm{d}\vec{\mathbf{p}}_n(t)}{\mathrm{d}t} = -\frac{k_{s,n}}{f}\vec{\mathbf{p}}_n(t) + \frac{1}{fN}\sum_{i=1}^{N}\cos\frac{in\pi}{N}\vec{\mathbf{f}}_i(t)$$

$$\frac{\mathrm{d}\vec{\mathbf{p}}_n(t)}{\mathrm{d}t} = -\frac{1}{\tau_n}\vec{\mathbf{p}}_n(t) + \vec{\Delta}_n(t) \quad n=1,2,\cdots,N \tag{7.1}$$

where

$$k_{s,n} = \left(\frac{n\pi}{N}\right)^2 k_s; \quad \tau_n = \frac{f}{k_{s,n}} = \frac{2\eta b^3 N^2}{\pi k_B T n^2}; \quad \text{and} \quad \vec{\Delta}_n(t) = \frac{1}{fN}\sum_{i=1}^{N}\cos\frac{in\pi}{N}\vec{\mathbf{f}}_i(t)$$

where $\langle R^2 \rangle = b^2 N$, the mean square end-to-end distance of an undisturbed chain; $\langle R_g^2 \rangle = \langle R^2 \rangle / 6$, the mean square radius of gyration; $D = k_B T/(fN)$, the translational diffusion coefficient; and $t_{R_g} = \langle R_g^2 \rangle/(6D)$, the time required to diffuse a distance of $\sqrt{\langle R_g^2 \rangle}$. Substituting them into τ_n,

$$\tau_n = \frac{12 t_{R_g}}{\pi^2}\frac{1}{n^2} \approx \frac{t_{R_g}}{n^2}$$

The dimension of τ_n is time, the characteristic relaxation time of the nth internal normal mode motion. When $n=1$, $\tau_1 \approx t_{R_g}$. Physically, it is the time required to relax the length of an average radius of gyration of a chain. As n increases, it quickly becomes shorter.

Since every $\vec{\mathbf{r}}_i(t)$ is a random variable, $\vec{\mathbf{p}}_n(t)$ is also a random variable. Both of the time average $\vec{\mathbf{p}}_n(t)$ and $\vec{\Delta}_i(t)$ are zero, respectively, that is, $\langle \vec{\mathbf{p}}_n(t) \rangle = 0$ and $\langle \vec{\Delta}_i(t) \rangle = 0$. Therefore, τ_n is not directly solvable from eq. (7.1). Similar to the aforementioned dynamic LLS, one can first calculate the time correlation function of the normal coordinate. According to the definition, different $\vec{\mathbf{p}}_n(t)$ are not correlated to each other; $\vec{\mathbf{p}}_n(t)$ and $\vec{\Delta}_i(t)$ are also not correlated to each other, so that

$$\frac{\mathrm{d}\langle \vec{\mathbf{p}}_n(0) \cdot \vec{\mathbf{p}}_n(t) \rangle}{\mathrm{d}t} = -\frac{1}{\tau_n}\langle \vec{\mathbf{p}}_n(0) \cdot \vec{\mathbf{p}}_n(t) \rangle + \langle \vec{\Delta}_n(0) \cdot \vec{\Delta}_n(t) \rangle \quad n=1,2,\cdots,N \tag{7.2}$$

where $\langle \vec{\Delta}_n(0) \cdot \vec{\Delta}_n(t) \rangle = \left[\langle \vec{\mathbf{f}}_n(0) \cdot \vec{\mathbf{f}}_n(t) \rangle \sum_{i=1}^{N}\cos\frac{in\pi}{N}\cos\frac{in\pi}{N}\right]/(fN)^2$. Based on the definition of an random interaction force $\vec{\mathbf{f}}_n(t)$, $\langle \vec{\mathbf{f}}_n(0) \cdot \vec{\mathbf{f}}_n(t) \rangle = A\delta(t)$. It can be proven that $A = 6fk_BT$. The sum in the above equation is

$$\sum_{i=1}^{N} \cos\frac{in\pi}{N}\cos\frac{in\pi}{N} = \frac{1}{2}\sum_{i=1}^{N}\left[\cos\frac{(i+i)n\pi}{N} + \cos\frac{(i-i)n\pi}{N}\right] = \frac{N}{2}$$

In the derivation, $2\cos x \cos y = \cos(x+y) + \cos(x-y)$ was used. Since $N \gg 1$, the first item in the sum is equal to 1 only when $N \gg 1$ and the rest items approach 0. The second item is always equal to 1, so that when $= 0$,

$$\langle \vec{\mathbf{p}}_n(0) \cdot \vec{\mathbf{p}}_n(0) \rangle = \langle p_n^2(0) \rangle = \frac{b^2 N}{\pi^2 n^2} = \frac{6\langle R_g^2 \rangle}{\pi^2 n^2} \approx \frac{2}{3}\frac{\langle R_g^2 \rangle}{n^2}$$

which represents the mean square fluctuation amplitude of the nth normal mode. When $n = 1$, $\langle p_1^2(0) \rangle \approx 2\langle R_g^2 \rangle/3$, i.e., the mean square fluctuation amplitude of the first normal mode is smaller than the mean square radius of gyration. As n increases, the mean square fluctuation amplitude of the normal mode sharply decreases. Substituting $\langle p_n^2(0) \rangle$ and τ_n into eq (7.2),

$$\langle \vec{\mathbf{p}}_n(0) \cdot \vec{\mathbf{p}}_n(t) \rangle = \langle p_n^2(0) \rangle \exp\left(-\frac{t}{\tau_n}\right) \approx \frac{2\langle R_g^2 \rangle}{3n^2}\exp\left(-\frac{t}{\tau_n}\right) \quad n = 1, 2, \cdots, N \quad (7.3)$$

Both the average fluctuation amplitude ($\langle q_n^2(0) \rangle$) and characteristic relaxation time (τ_n) of the above-mentioned time correlation function are inversely proportional to n^2. As n increases, $\langle \vec{\mathbf{p}}_n(0) \cdot \vec{\mathbf{p}}_n(t) \rangle$ not only decays faster with time, but also its contribution (average fluctuation) to the normal mode (the internal motion of the chain) decreases rapidly.

The normal coordinate $\vec{\mathbf{p}}_n(t)$ is formed by a linear combination of the usual coordinate $\vec{\mathbf{r}}_i(t)$. Conversely, $\vec{\mathbf{r}}_i(t)$ can also be expressed by a linear combination of the normal coordinate $\vec{\mathbf{p}}_n(t)$. The two constitute two mutually "conjugate" space. Substituting the definition of $\vec{\mathbf{p}}_n(t)$ into the following equation leads to

$$2\sum_{n=1}^{N}\cos\frac{in\pi}{N}\vec{\mathbf{p}}_n(t) = 2\sum_{n=1}^{N}\cos\frac{in\pi}{N}\left[\frac{1}{N}\sum_{i=1}^{N}\cos\frac{in\pi}{N}\vec{\mathbf{r}}_i(t)\right]$$

$$= \frac{1}{N}\sum_{i=1}^{N}\vec{\mathbf{r}}_i(t)\sum_{n=1}^{N}2\cos\frac{in\pi}{N}\cos\frac{in\pi}{N}$$

$$= \frac{1}{N}\sum_{i=1}^{N}\vec{\mathbf{r}}_i(t)\sum_{n=1}^{N}\left[\cos\frac{(i+i)n\pi}{N} + \cos\frac{(i-i)n\pi}{N}\right]$$

where $2\cos x \cos y = \cos(x+y) + \cos(x-y)$ is also used in the derivation. In the first items in the square bracket, $\cos[(i+i)n\pi/N] = 1$ only when $(i+i) = 2N$ and the rest items are zero; while the second items are always 1. Thus, the sum of the right side leads to $\vec{\mathbf{r}}_i(t)(1+N)$, so that

$$\vec{\mathbf{r}}_i(t) = 2\sum_{n=1}^{N}\cos\frac{in\pi}{N}\vec{\mathbf{p}}_n(t) \quad i = 1, 2, \cdots, N \quad (7.4)$$

where $(N+1)/N \cong 1$ was used. Similarly, careful readers can notice that the 1st and Nth beads are different from the beads in the middle. Since $N \gg 1$, one can add one bead before and

after the chain, and then ignore them, not affecting the result. Using the above relationship, some kinetic parameters can be obtained.

The first calculable parameter is the relaxation of the chain end-to-end distance that is randomly and ever-changing with time in Chapter 4.

$$\vec{\mathbf{R}}(t) = \vec{\mathbf{r}}_N(t) - \vec{\mathbf{r}}_1(t) = 2\sum_{n=1}^{N}\left(\cos n\pi - \cos\frac{n\pi}{N}\right)\vec{\mathbf{p}}_n(t)$$

where $\cos n\pi = (-1)^n$. The average fluctuation amplitude of $\vec{\mathbf{p}}_n(t)$ decreases rapidly and $\vec{\mathbf{p}}_n(t)$ itself decays sharply as n increases, so that only the first few items ($n < 10$) are normally considered. When $N \gg 1$, $\cos(n\pi/N) \to 1$. Therefore, only when n is an odd number, the item is nonzero, The difference between the two sums in parentheses is -2, so that the above equation becomes

$$\vec{\mathbf{R}}(t) = \vec{\mathbf{r}}_N(t) - \vec{\mathbf{r}}_1(t) \cong -4\sum_{n=1:\text{ odd}}^{N}\vec{\mathbf{p}}_n(t)$$

The random variation of the chain conformation leads to the random changes of $\vec{\mathbf{p}}_n(t)$ and $\vec{\mathbf{R}}(t)$. According to the definition of the normal coordinates, there is no correlation between different $\vec{\mathbf{p}}_n(t)$. Therefore, the respective autocorrelation functions of $\vec{\mathbf{R}}(t)$ and $\vec{\mathbf{p}}_n(t)$ are related to each other as follows.

$$\langle\vec{\mathbf{R}}(0) \cdot \vec{\mathbf{R}}(t)\rangle \cong 16\sum_{n=1:\text{ odd}}^{N}\langle\vec{\mathbf{p}}_n(0) \cdot \vec{\mathbf{p}}_n(t)\rangle = \sum_{n=1:\text{ odd}}^{N}\frac{8Nb^2}{\pi^2 n^2}\exp\left(-\frac{t}{\tau_n}\right)$$

If only considering the first term ($n = 1$) and omitting the rest items of the summation, one can use τ_n defined in eq. (7.1), the mean square end-to-end distance of the undisturbed chain $\langle R_0^2\rangle = b^2 N \sim M$, and the Contour length ($L_C = bN \sim M$) to rewrite the above equation as

$$\langle\vec{\mathbf{R}}(0) \cdot \vec{\mathbf{R}}(t)\rangle = \frac{8\langle R_0^2\rangle}{\pi^2}\exp\left(-\frac{t}{\tau_1}\right) \approx \langle R_0^2\rangle\exp\left(-\frac{t}{\tau_1}\right) \qquad (7.5)$$

where $\tau_1 \propto N^2 \propto M^2$. According to the definition of τ_1 in eq. (7.1), the length of τ_1 can be roughly estimated. In the cgs system, $N \sim 10^3$ for ordinary polymers ($M \sim 10^5\,\text{g/mol}$); the solvent viscosity, $\eta \sim 10^{-2}$; $k_B T \sim 10^{-14}$, so that $\tau_1 \sim 10^{-3}\,\text{s}$. In other words, the slowest relaxation time related to the rotation and vibration of the entire chain is in the order of milliseconds. The relaxations of other internal modes are faster. On the other hand, the characteristic relaxation time increases with the chain length. Therefore, it is easier to observe and measure a longer macromolecular chain.

The second calculable parameter is the motion of the centroid of the macromolecular chain. As mentioned earlier, the normal coordinate $\vec{\mathbf{p}}_0(t)$ represents the centroid of a chain, and its motion can also be described by the trajectory of the normal coordinate $\vec{\mathbf{p}}_n(t)$. After a given period of time (t), the position of the centroid randomly moves from its initial $\vec{\mathbf{r}}_G(0)$ to a new position $\vec{\mathbf{r}}_G(t)$, so that its mean square distance is

$$\langle[\vec{\mathbf{r}}_G(t) - \vec{\mathbf{r}}_G(0)]^2\rangle = \langle[\vec{\mathbf{p}}_0(t) - \vec{\mathbf{p}}_0(0)]^2\rangle = \langle\int_0^t\int_0^t \vec{\mathbf{\Delta}}_n(t')\vec{\mathbf{\Delta}}_n(t'')\,dt'dt''\rangle$$

$$\langle\,[\,\vec{\mathbf{r}}_G(t)-\vec{\mathbf{r}}_G(0)\,]^2\,\rangle=\frac{1}{(fN)^2}\int_0^t\int_0^t\sum_{i=1}^N\sum_{j=1}^N\langle\,\vec{\mathbf{f}}_i(t')\,\vec{\mathbf{f}}_j(t'')\,\rangle\,\mathrm{d}t'\mathrm{d}t''$$

where $\vec{\mathbf{f}}_i(t)$ is a random force, all cross terms disappear after the time average. There is only N same self-terms left after the double summations: $\langle\,\vec{\mathbf{f}}_i(t')\,\vec{\mathbf{f}}_j(t'')\,\rangle=6fk_BT\delta(t''-t')$; and only when $t''=t'$, nonzero. The above equation becomes

$$\langle\,[\,\vec{\mathbf{r}}_G(t)-\vec{\mathbf{r}}_G(0)\,]^2\,\rangle=\frac{1}{(fN)^2}\int_0^t\sum_{j=1}^N 6fk_BT\mathrm{d}t=\frac{6k_BT}{fN}t$$

On the other hand, in a three-dimensional space, the mean square distance of a randomly walking particle ($\langle\,[\,\vec{\mathbf{r}}(t)-\vec{\mathbf{r}}(0)\,]^2\,\rangle$) is equal to six times of the product of its translational diffusion coefficient (D) and its walking time (t), i.e., $\langle\,[\,\vec{\mathbf{r}}(t)-\vec{\mathbf{r}}(0)\,]^2\,\rangle=6Dt$, so that

$$D=\frac{k_BT}{fN}=\frac{k_BT}{6\pi\eta bN}=\frac{k_BT}{6\pi\eta\,L_c}\sim M^{-1} \tag{7.6}$$

However, in the Θ solvent (quasi-ideal undisturbed state), the scaling exponents between the experimentally measured characteristic relaxation time of the first normal mode (τ_1) as well as the translational diffusion coefficient (D) of the centroid and the molar mass of macromolecules are $3/2$ and $1/2$, respectively, which deviate seriously from 2 and 1 predicted by eqs. (7.5) and (7.6). The failure of the Rouse model in dilute solutions is mainly due to the assumption that the segments are independent of each other, and the hydrodynamic interaction is completely ignored, that is, the motion of a segment can affect solvent molecules around it, and the motions of solvent molecules will affect the motions of other chain segments.

In order to introduce the hydrodynamic interaction and overcome the failure of Rouse model in dilute solutions, about three years later, Zimm [J. Chem. Phys., 1956, 24:269] used a pre-averaged approximation of the Oseen tensor, i.e.,

$$\langle\,\mathbf{H}_{ij}\,\rangle=\frac{\mathbf{I}}{6\pi\eta_s}\left\langle\frac{1}{|\,\vec{\mathbf{r}}_i(t)-\vec{\mathbf{r}}_j(t)\,|}\right\rangle$$

He successfully decoupled the hydrodynamic tensor in the equation of motion of $\vec{\mathbf{r}}_i(t)$, making it solvable in normal coordinates. Physically, it is not difficult to see that each element (H_{ij}) in $\langle\,\mathbf{H}_{ij}\,\rangle$ is equivalent to the reciprocal of the friction coefficient of a thin rod-like object. One end of the thin rod is at $\vec{\mathbf{r}}_i(t)$ and the other end is at $\vec{\mathbf{r}}_j(t)$, $|\,\vec{\mathbf{r}}_n(t)-\vec{\mathbf{r}}_m(t)\,|$ can be regarded as the direct distance between any two chain segments in the chain, or as the rod length ($\vec{\mathbf{b}}_{nm}(t)$), but the rod length and direction change continuously and randomly with the chain conformation, so that a time average is required. The derivation is cumbersome, so it is omitted. The final result is

$$H_{ij}=\frac{1}{(6\pi^3)^{1/2}\eta_s b\,|\,i-j\,|^{\alpha_R}}$$

In the normal coordinates, considering that $N\gg 1$, H_{ij} can be further transformed into (the derivation is also omitted),

$$h_{nm} = \frac{1}{N^2} \sum_{i=1}^{N} \sum_{j=1}^{N} \frac{1}{(6\pi^3)^{\frac{1}{2}} \eta b \, |i-j|^{\alpha_R}} \cos\frac{in\pi}{N} \cos\frac{jm\pi}{N}$$

$$= \frac{\Gamma(1-\alpha_R)}{2(3\pi^3)^{\frac{1}{2}} \pi^{1-\alpha_R} \eta b N^{\alpha_R} n^{1-\alpha_R}} \delta_{nm} \quad (\text{besides } i=j=0)$$

Substituting it into the equation of motion in the normal coordinates, and following the derivation in the Rouse model, one can obtain the same result in eq. (7.1), i.e.,

$$\frac{d\vec{\mathbf{p}}_n(t)}{dt} = -\frac{1}{\tau_n}\vec{\mathbf{p}}_n(t) + \vec{\Delta}_n(t) \quad n=1,2,\cdots,N \qquad (7.7)$$

where the expressions of τ_n and $\vec{\Delta}_n(t)$ are identical to that in the Rouse model. The difference is that in the Rouse model, the friction coefficient $f = 6\pi\eta b$ and $k_{s,n} = \left(\frac{n\pi}{N}\right)^2 k$, while in the Zimm model,

$$f = \frac{1}{h_{nm}} = \frac{2(3\pi^3)^{\frac{1}{2}} \pi^{1-\alpha_R} \eta b}{\Gamma(1-\alpha_R)}\left(\frac{n}{N}\right)^{1-\alpha_R} \cong 2(3\pi^3)^{\frac{1}{2}} \eta b \left(\frac{n}{N}\right)^{1-\alpha_R} \quad \text{and} \quad k_{s,n} = \left(\frac{n\pi}{N}\right)^{2\alpha_R+1} k_s$$

In the above approximation, $\Gamma(1-\alpha_R) \approx \pi^{1-\alpha_R}$; $1/2 \leqslant \alpha_R \leqslant 3/5$ are used. Substituting the above equation and $k_s = 3k_B T/b^2$ into τ_n, one can obtain the characteristic relaxation time of each normal mode in the Zimm model and the normal coordinates

$$\tau_n = \frac{f}{k_{s,n}} = \frac{2(3\pi)^{\frac{1}{2}} \eta b^3 N^{3\alpha_R}}{3\pi^{2\alpha_R} k_B T n^{3\alpha_R}} \approx \frac{44}{\pi^{2\alpha_R+1}} \frac{t_{R_g}}{n^{3\alpha_R}} \approx 4\frac{t_{R_g}}{n^{3\alpha_R}} \sim M^{3\alpha_R} \qquad (7.8)$$

In the first normal mode, $\tau_1 \approx 4\, t_{R_g}$. In comparison with the Rouse model, the relaxation is about 4 times slower, which can be attributed to the excessive hydrodynamics damping. In the Θ solvent, $\alpha_R = 1/2$ and $\tau_1 \propto M^{3/2}$, similar to the experimental observation. According to the same derivation in the Rouse model, the mean square fluctuation amplitude of the nth normal mode can be obtained,

$$\langle \vec{\mathbf{p}}_n(0) \cdot \vec{\mathbf{p}}_n(0) \rangle = \langle p_n^2(0) \rangle = \frac{b^2 N^{2\alpha_R}}{2\pi^2 n^{3\alpha_R}} = \frac{6\langle R_g^2 \rangle}{2\pi^2 n^{3\alpha_R}} \approx \frac{1}{3}\frac{\langle R_g^2 \rangle}{n^{3\alpha_R}}$$

In comparison with the Rouse model, the mean square fluctuation amplitude of the first normal mode is ~ 50% smaller, attributing to the excessive hydrodynamic damping. The nth normal mode $\vec{\mathbf{q}}_n(t)$ dependence of the time autocorrelation function of the chain end-to-end distance $\vec{\mathbf{R}}(t)$ is as follows.

$$\langle \vec{\mathbf{R}}(0) \cdot \vec{\mathbf{R}}(t) \rangle = \sum_{n=1;odd}^{N} \frac{6N^{2\alpha_R} b^2}{\pi^2 n^{3\alpha_R}} \exp\left(-\frac{t}{\tau_n}\right) \approx \frac{2\langle R^2 \rangle}{\pi^2} \exp\left(-\frac{t}{\tau_1}\right) \approx \langle R_g^2 \rangle \exp\left(-\frac{t}{\tau_1}\right) \quad (7.9)$$

where $\langle R^2 \rangle = N^{2\alpha_R} b^2$, the scaling relationship between the mean square end-to-end distance and the segment number; $\langle R^2 \rangle = 6\langle R_g^2 \rangle$; and

$$\tau_1 = \frac{6\eta b^3 N^{3\alpha_R}}{\pi^{2\alpha_R} k_B T} \approx 4t_{R_g} \sim M^{3\alpha_R}$$

Similarly, you can roughly estimate the length of τ_1. In the cgs system, for macromolecules with a molar mass in the range of $10^5 - 10^6$ grams and common solvents, $\eta \sim 10^{-2}$, $b \sim 10^{-7}$, $N \sim 10^3 - 10^4$, and $k_B T \sim 10^{-14}$. Therefore, $\tau_1 \sim 10^{-3}$s. Readers should note that whether in the Rouse model or the Zimm model, if there is no entanglement, the characteristic relaxation time of macromolecular chains is usually in the order of milliseconds. In other words, it takes only one thousandth of a second for a deformed linear macromolecular chain (for example, a stretched chain) to return to its equilibrium conformation (a random coil), extremely fast. According to the same derivation in the Rouse model, one can obtain the translational diffusion coefficient of the centroid of the macromolecular chain,

$$D = \frac{k_B T}{f} \sim \frac{k_B T}{\eta b N^{\alpha_R}} \sim M^{-\alpha_R} \tag{7.10}$$

Readers who are interested in understanding the detailed process of the above derivation can further refer to the relevant monographs from the former USSR Russian Lifshits School [Grosberg, Khokhlov, "Statistical Physics of Macromolecules", originally published in Russia in Moscow, 1990; Later, AIP Press (English Ed.), New York, 1994] and the British Edwards School [Doi, Edwards, "The Theory of Polymer Dynamics", Cambridge, New York, 1989]. In contrast, the French de Gennes School [de Gennes, "Introduction to Polymer Dynamics", 1990] approach is easier for teachers and students of chemistry, polymer and other non-physics departments to understand and accept. The brief introduction is as follows:

If the end-to-end distance of a linear flexible chain is $\vec{R}(0)$ and $\vec{R}(t)$ at time $t = 0$ and $t = t$, respectively, as shown on the right, and taking its undisturbed average conformation state (random coil) as a reference point, the force generated by the change of the chain conformation is $k_{coil} \vec{R}(t)$

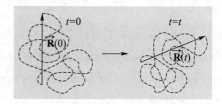

and the frictional resistance expressed by the chain moving in a liquid is $f_{coil} \partial \vec{R}(t)/\partial t$. Both of them must reach equilibrium.

$$\frac{f_{coil} \partial \vec{R}(t)}{\partial t} = k_{coil} \vec{R}(t) \rightarrow \vec{R}(t) = \vec{R}(0) \exp\left(-\frac{t}{\tau_c}\right)$$

where

$$k_{coil} = \frac{3 k_B T}{\langle R^2 \rangle} = \frac{3 k_B T}{b^2 N^{2\alpha_R}}, \quad f_{coil} = 6\pi\eta R_h \approx 5\eta b N^{\alpha_R} \quad \text{and} \quad \tau_c = \frac{f_{coil}}{k_{coil}} \approx \frac{1.7\eta b^3 N^{3\alpha_R}}{k_B T} \approx 5 t_{R_g}$$

In the above derivation, all solvent molecules inside the random coil were assumed to move with the macromolecular chain, i.e., no drainage effect. Each chain is equivalent to a small sphere with a radius of R_h: $R_g / R_h \approx 1.5$. $R = \sqrt{\langle R^2 \rangle} = b N^{\alpha_R} = \sqrt{6} R_g$ (Chapter 4) is also used. The result obtained in this way is very close to the above τ_1. Substituting f_{coil} into the Stokes equation, the translational diffusion coefficient of the centroid of the chain is exactly the same as eq. (7.10).

$$D = \frac{k_B T}{6\pi\eta R_h} \approx \frac{k_B T}{4\pi\eta R_g} \approx 0.0786 \frac{k_B T}{\eta R_g}$$

The coefficient is very close to 0.0829 calculated based on the renormalization group theory. Obviously, de Gennes's approach is relatively simple and easy. However, this method cannot get the motion details of a chain, including its average fluctuation range and characteristic relaxation time of each normal mode. Is the above assumption of "no drainage effect" physically reasonable?

To answer this question, the discussion on page 238 of the book by Grosberg and Khokhlov is borrowed, i.e., the velocity distribution of solvent molecules in a moving random coil chain. First, let us assume that inside a random-coiled chain the number density of the chain segments is ρ_n, the chain segments have the same velocity $(\vec{u}_0)$, and the friction coefficient between every chain segment and solvent molecule is f. If there is a velocity difference between a chain segment and its adjacent solvent molecules $(\Delta\vec{u} = \vec{u}_0 - \vec{u})$, a friction force $(f\Delta\vec{u})$ will be exerted on its adjacent solvent molecules. Ignoring the influence of inertia and assuming that it has little disturbance to the velocity and the liquid volume is incompressible, one can use the Navier–Stokes equation published by Claude–Louis Navier in 1822 and perfected by George Gabriel Stokes in 1842 to find that in the vertical motion direction (y), the solvent molecule follows the following equation

$$\eta \frac{\partial^2 \vec{u}}{\partial y^2} + fC\Delta\vec{u} = 0$$

where $f = 6\pi\eta b$; $C = N/(\pi R^3/6)$; $fC = 36\eta bN/(b^3 N^{3\alpha_R}) = 36\eta/(b^2 N^{3\alpha_R-1})$, here R is the average end-to-end distance of the chain. Physically, it is not difficult to imagine that if the chain center is placed at infinity in the y direction, the edge of the chain is close to $y = 0$. The above equation has two boundary conditions: in the chain center $(y \to \infty)$, the solvent molecules and the chain segment have the same speed, $\Delta\vec{u} = 0$; outside the chain $(y \to 0)$, the solvent molecules do not move, $\vec{u} = 0$. Under these conditions, the above equation is solvable, and the final result is

$$\vec{u}(y) = \vec{u}_0 \left[1 - \exp\left(-\frac{y}{\lambda} \right) \right]$$

where $\lambda = \left(\frac{\eta}{fC} \right)^{1/2} = \frac{bN^{(3\alpha_R-1)/2}}{6}$. λ is the characteristic attenuation (penetration) length, and its ratio to the average radius of gyration of the chain $(R_g = \sqrt{6} bN^{\alpha_R})$ is as follows.

$$\frac{\lambda}{R_g} = \frac{N^{(3\alpha_R-1)/2}}{6\sqrt{6} \text{ or } 7N^{\alpha_R}} = \frac{N^{(\alpha_R-1)/2}}{6\sqrt{6} \text{ or } 7} = 0.068N^{(\alpha_R-1)/2}$$

In the Θ and good solvents, $\lambda/R_g = 0.068N^{-1/4}$ and $\lambda/R_g = 0.063N^{-1/5}$. For ordinary polymers, $N \sim 10^3$; that is, $\lambda/R_g < 1.7\%$. After converted to volume, more than 95% of solvent molecules in a random coiled chain follow the chain moving with the same velocity, especially in good solvents. The assumption of "no draining effect" is correct.

Experimentally, as discussed in Chapter 5, in a dilute solution of macromolecules, and under the premise that the relaxation of the system is only related to the translational diffusion, dynamic LLS can be used to measure the mutual translational diffusion coefficients at different concentrations and different scattering angles, and then the extrapolation of them respectively to the zero scattering angle and concentration can result in the translational diffusion coefficient of the centroid of the macromolecular chain. Methods, such as the forced Rayleigh scattering, the fluorescence bleaching recovery, and the pulsed field gradient nuclear magnetic resonance can also be used to directly measure the translational diffusion coefficient of the labeled chains.

According to the above discussion of chain dynamics, a general expression of experimentally measurable intrinsic viscos ity can also be obtained (the derivation is omitted),

$$[\eta] = \frac{RT}{2M\eta_s} \sum_{n=1}^{N} \tau_n$$

Experimentally, the characteristic relaxation time (τ_n) of macromolecular chains can also be measured. For example, for macromolecules with a permanent dipole moment along the chain direction, the dielectric relaxation spectrometer can be used to measure the characteristic relaxation time of the average end-to-end distance of the chain, which is mainly τ_1.

In order to observe the internal motions of a macromolecular chain, one has to "stand" inside the chain, in other words, use a "ruler" with its marks much smaller than the chain size, and its time resolution much faster than the segment motions. In the discussion of LLS in Chapter 5, the scattering vector is defined as $q = 4\pi n \sin(\theta/2)/\lambda_0$, its reciprocal is physically equivalent to an optical "ruler" of measuring objects. When a HeNe laser ($\lambda_0 \approx 633$ nm) is used, and if $n \approx 1.4$, the observation range is 38 nm $\leqslant 1/q \leqslant 300$ nm.

When the scattering angle is very small, $1/q$ is usually larger than the size of polymer chains ($R_g < 100$ nm). Therefore, LLS is not able to measure the internal motions of a polymer chain. The entire macromolecular chain is equivalent to a small particle. Static LLS can only measure its overall size, while dynamic LLS only measure the diffusion related linewidth (Γ) that is related to the translational diffusion coefficient (D) as $\Gamma = Dq^2$. This is equivalent to standing in a far distance away and watches a fast moving train. Only the overall movement of the train can be seen, but not the sitting, lying, walking, and varied postures of individual passengers inside the carriage. To measure the internal motions of the chain, it is necessary to increase the scattering vector q (i.e., "close to the train") to make $1/q \leqslant R_g$.

When q is sufficiently large, LLS measures the local information whose size is $1/q$. Physically, it is not difficult to imagine that the size of a chain itself is not important anymore, similar to standing inside a carriage to watch passengers, the outer size of the carriage is not important, The measured Γ should have nothing to do with the chain size. According to the scaling theory of de Gennes, in the dynamic LLS, the linewidth due to the internal motions of a chain is scaled to qR_g as $\Gamma \propto (qR_g)^n$. Since the entire chain is randomly diffusing, just like the train is moving, the observed total linewidth should include two parts: the motions of translation and internal motions, their product, namely $\Gamma \propto Dq^2(qR_g)^z$. Note that $D \propto 1/R_g$. Therefore, if Γ is independent of R_g, z can only be 1. Experimental results have confirmed that when $qR_g \geqslant 4$, the scaling relationship between the measured linewidth and the scattering vector is indeed $\Gamma \propto q^3$.

The scaling theory provides no system-related details, but the "devils" are often hidden in the details! For a uniform object, in the three-dimensional space, the scaling relationship between its volume (V) and linearity (l) is always $V \propto l^3$; in the two-dimensional space, the relationship between with its area (S) and l is always $S \propto l^2$; in the one-dimensional space, the relationship between its length (L) and l is always $S \propto l^2$. However, when calculating the volume, area or length of a specific object, one must know the detail of its shape, i.e., the proportional constant.

If hoping to use dynamic LLS to measure the average fluctuation amplitude of the previous normal modes and the corresponding characteristic relaxation time (τ_n), one has to return to more rigorous theoretical calculations because the scaling method becomes helpless. For any two segments i and j in the chain, their positions at time 0 and t are $\vec{r}_i(0)$ and $\vec{r}_j(0)$ as well as $\vec{r}_i(t)$ and $\vec{r}_j(t)$. Define $\vec{r}_{ij}(t) = \vec{r}_i(t) - \vec{r}_j(0)$, whose physical meaning is the distance (vector) of the ith segment from the initial position of the jth segment after an interval of time t. In the normal coordinate system, eq. (7.4) can be used to obtain

$$\vec{r}_{ij}(t) = [\vec{p}_0(t) - \vec{p}_0(0)] + 2\sum_{n=1}^{N}\left[\cos\frac{ni\pi}{N}\vec{p}_n(t) - \cos\frac{nj\pi}{N}\vec{p}_n(0)\right]$$

Different normal modes are independent from each other. Therefore, for any given two segments, the average value of $\exp[i\vec{q} \cdot \vec{r}_{ij}(t)]$ is as follows.

$$\langle\exp[i\vec{q} \cdot \vec{r}_{ij}(t)]\rangle = \langle\exp[i\vec{q} \cdot (\vec{p}_0(t) - \vec{p}_0(0))]\rangle \times$$
$$\prod_{n=1}^{N}\left\langle\exp\left[2i\vec{q} \cdot \left(\cos\frac{ni\pi}{N}\vec{p}_n(t) - \cos\frac{nj\pi}{N}\vec{p}_n(0)\right)\right]\right\rangle$$

In the normal coordinates, $\vec{p}_0(t)$ is the position vector of the centroid after a period of time t. Note that here t is actually $\Delta t = t - 0$. Therefore, the first term of the product on the right (denoted as I) means the average characteristic relaxation time (τ_0) required for the centroid of the chain to move a distance of $1/q$, corresponding to the characteristic line width ($\Gamma_0 = 1/\tau_0$). This relaxation is expressed as

$$I = \exp\left(-\frac{t}{\tau_0}\right) = \exp(-\Gamma_0 t)$$

After a more complicated calculation (the specific process is omitted), the expression of the second term of the product on the right (denoted as II) is also obtainable,

$$II = \langle\exp[i\vec{q} \cdot \vec{r}_{ij}(0)]\rangle\left\langle\exp\left[-4q^2\sum_{n=1}^{N}k_n\cos\frac{ni\pi}{N}\cos\frac{nj\pi}{N}\left[1 - \exp\left(\frac{t}{\tau_n}\right)\right]\right]\right\rangle$$

where k_n is a model-dependent constant. Multiplying the above two equations and substituting the result into the previous equation, one can get

$$\langle \exp[i\vec{\mathbf{q}} \cdot \mathbf{r}_{ij}(t)] \rangle = \text{I} \times \text{II}$$

$$= \exp(-\Gamma_0 t) \left\langle \exp[i\vec{\mathbf{q}} \cdot \mathbf{r}_{ij}(0)] \exp\left[-4q^2 \sum_{n=1}^{N} k_n \cos\frac{ni\pi}{N} \right.\right.$$

$$\left.\left. \cos\frac{nj\pi}{N} \left[1 - \exp\left(\frac{t}{\tau_n}\right) \right] \right] \right\rangle$$

In comparison with eq. (5.40), the above equation is the dynamic structure factor between different segments on a macromolecular chain; the average is the double sum of i and j. After the normalization, it becomes the time autocorrelation function of the electric field of scattered light.

$$g^{(1)}(q,t) = \frac{\langle \exp[i\vec{\mathbf{q}} \cdot \mathbf{r}_{ij}(t)] \rangle}{\langle \exp[i\vec{\mathbf{q}} \cdot \mathbf{r}_{ij}(0)] \rangle} = \exp(-\Gamma_0 t) J(q,t)$$

where $J(q,t)$ represents the relaxations of all normal modes (the internal motions of a chain). Its specific expression is as follows.

$$J(q,t) = \frac{\sum_{i=1}^{N} \sum_{j=1}^{N} \exp[i\vec{\mathbf{q}} \cdot \mathbf{r}_{ij}(0)] \exp\left[-4q^2 \sum_{n=1}^{N} k_n \cos\frac{ni\pi}{N} \cos\frac{nj\pi}{N} \left[1 - \exp\left(-\frac{t}{\tau_n} \right) \right] \right]}{\sum_{i=1}^{N} \sum_{j=1}^{N} \exp[i\vec{\mathbf{q}} \cdot \mathbf{r}_{ij}(0)]}$$

When $t \to 0$, $g^{(1)}(q,t) \to 1$ (the macromolecular chain has not start to move). When $t \to \infty$, II approaches a constant. Therefore, $g^{(1)}(q,t) \propto \exp(-\Gamma_0 t)$, which physically means that all the faster internal motions have are fully relaxed, leaving only the slower relaxation of the centroid. If the relaxation of the centroid is attributed to the translational diffusion motion, $\Gamma_0 = D q^2$. Physically, $\exp(-\Gamma_0 t) J(q,t)$ in the above equation stems from the simultaneous occurrence of the translational diffusion relaxation of the centroid and the relative motions of all the segments to the centroid (various normal modes). In other words, the motions of all normal modes are "carried" on the centroid motion. The numerical solution of each normal mode (internal motion) is obtainable from calculations.

Perico, Piaggio, Cuniberti [J. Chem. Phys., 1975, 62:2690; 1975, 62:4911] pointed out that under the conditions of no draining and $q^2 \langle R_g^2 \rangle > 1$, $g^{(1)}(q,t)$ mainly depends on the first five relaxations ($n = 0, 1, 2, 3,$ and 4); namely, the translational diffusion plus four main normal modes related to the internal motions of the chain. According to the decreasing order of the average fluctuation amplitude (A_n, the contribution to the intensity of scattered light), the above equation can be rewritten as

$$g^{(1)}(q,t) = \frac{\sum_{n=0}^{N} A_n \exp(-\Gamma_n t)}{\sum_{n=0}^{N} A_n} \tag{7.11}$$

They also calculated all the values of A_n in the range $1 \leqslant q^2 \langle R_g^2 \rangle \leqslant 100$. Using the Zimm model, one can obtain

$$\Gamma_n = q^2 D + \frac{\pi^{2\alpha_R} k_{\mathrm{B}} T}{6\eta \, b^3 N^{3\alpha_R}} n^{3\alpha_R} \tag{7.12}$$

where $\langle R^2 \rangle = b^2 N^{2\alpha_R}$ or written as $R = \sqrt{\langle R^2 \rangle} = b\ N^{\alpha_R}$; $\langle R^2 \rangle = 6\langle R_g^2 \rangle$; $R = \sqrt{6}\ R_g$. In the above equation, the linewidth can be normalized by the translational diffusion corresponded linewidth ($q^2 D$), and $D = k_B T/(6\eta\pi R_h)$ can be substituted in, the result is as follows.

$$\frac{\Gamma_n}{q^2 D} = 1 + \frac{\pi^{2\alpha_R+1} R_h}{6\sqrt{6} R_g q^2 \langle R_g^2 \rangle} n^{3\alpha_R} \approx 1 + \frac{0.56}{q^2 \langle R_g^2 \rangle} n^{3\alpha_R} \quad n = 0,1,2,\cdots,N \qquad (7.13)$$

In the above derivation, $\alpha_R = 1/2$ and $R_g/R_h \approx 1.3$ as well as $\alpha_R = 3/5$ and $R_g/R_h \approx 1.5$ were respectively used for linear flexible chains in the Θ and good solvents. To facilitate writing and discussion, $x = q^2 \langle R_g^2 \rangle$ is now defined. The plot of the normalized linewidth versus x is independent of the chain size and the segment number, only related to n. After deducting the translational diffusion of the centroid of a macromolecular chain itself, the internal motions of the segments relative to the centroid is naturally independent of the chain length, so that the above conclusion is physically reasonable.

When de Gennes considered using dynamic LLS to study the internal motions of polymer chains in the 1970s, he pointed out that these internal normal modes could be observed only at $qR_g \sim 2.5$, and asserted that for normal polymers ($R_g \sim 50$ nm), the normal modes cannot be measured with visible light, and a vacuum ultraviolet light source must be used. However, there was no vacuum ultraviolet laser at the time. Even with such a light source and a highly sensitive vacuum ultraviolet single photon detector today, it is still difficult to adjust the entire optical path under vacuum ultraviolet to achieve the LLS required precision.

Careful readers will think that since $qR_g \sim 2.5$ is required, if the wavelength cannot be shortened to increase q, why not use an ultra-long macromolecular chain? For a HeNe laser ($\lambda_0 \sim 633$ nm) and a scattering angle of $150°$, $1/q \sim 40$ nm, i.e., R_g must be larger than 100 nm, which can be realized experimentally. However, based on the above discussion, the normal modes (internal motions) of a macromolecular chain are always coupled with its translational diffusion of the centroid. In order to separate the two, a series of linewidths (Γ) must be measured in the range $qR_g < 1$ first, and then (Γ/q^2) is plotted against q^2. Its extrapolation to $q \rightarrow 0$ leads to D. Only in this way, the relaxation related to the translational diffusion of the centroid can be subtracted in the analysis of the time autocorrelation function of the intensity of scattered light. If $R_g \sim 150$ nm, the LLS experiments must be conducted at an angle smaller than $15°$. However, the low angle limit of most of the LLS instruments is $15°$. This is why the experimental studies have been far behind the theoretical calculations.

It was until the early 1990s that Chu and his collaborators [Macromolecules, 1991, 24: 6832] studied this complicated problem of the internal motions of macromolecular chains for the first time by combining an ordinary laser light scattering instrument with a special small-angle laser light scattering instrument with a prism scattering sample cell. However, they did not explain why the relatively faster and weaker second normal mode instead of the theoretically predicted slowest and strongest first normal mode was observed.

A few years later, after adopting a solid laser light source with a thinner beam and a shorter wavelength, and improving the optical detection system of a commercial instrument, the minimum measurable angle of the LLS instrument in the author's laboratory has reached $5.6°$, so that the internal motions of macromolecules were systematically studied by laser light scattering.

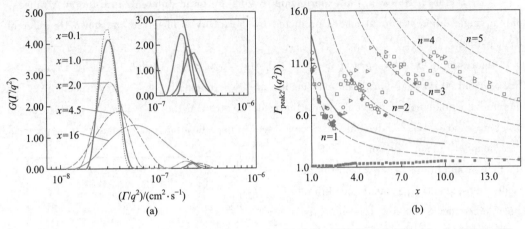

Wu C, Chan K K, Xia K Q. Macromolecules, 1995, 28: 1032
(Figure 4, permission was granted by publisher)

Wu C, Chan K K, Xia K Q. Macromolecules, 1995, 28: 1032
(Figure 5, permission was granted by publisher)

The figure (a) above shows the experimental results of three polystyrene samples with ultra-high molar mass ($M_w \sim 10^7$ g/mol) measured using dynamic LLS. When $x < 1$, the linewidth distribution after normalized by q^2 contains only one single peak, and does not change with x. When $x \geqslant 1$, there appears a second small peak in the range $10^{-7} - 10^{-6}$ cm^2/s, much faster than the translation diffusion of the centroid of the entire chain. The inset in the figure (a) above enlarges and clearly shows how this small peak changes with x. The interested readers can refer to the original literature to learn more about the experimental details.

On one hand, the use of a serie of narrowly distributed, ultra-long polystyrene chains with $R_g \sim 130-310$ nm makes qR_g as high as 4, so that faster and weaker high-order normal modes (internal motions of the chain on a smaller scale) can be observed. On the other hand, since the translational diffusion coefficient of these ultra-long macromolecular chains can be accurately measured in a very small scattering angle range, $q^2 D$ can be used to normalize the linewidth of the second small peak related to the internal motions, as shown in figure (b) above, where different small solid symbols near the bottom abscissa represent the linewidth of the first large peak related to the translational diffusion after the normalization ($\Gamma_{peak1}/q^2 D$). It can be seen that when $x < 2$, $\Gamma_{peak1}/q^2 D \cong 1$, no chain length dependence, because the chain length dependence has been reflected in the translational diffusion coefficient. As x increases, the ratio only slightly increases, indicating a small amount of internal motions slowly coupled into the translational diffusion.

As predicted by the theory, the linewidth of the second peak of three different polystyrene samples after normalization ($\Gamma_{\text{peak2}}/q^2 D$) is basically independent of the molar mass. The solid line in the figure(b) above represents the result of the weighted summation ($n = 1, 2, \cdots, N$) of $\Gamma_n/q^2 D$ in eq. (7.12) based on the numerical solution of A_n in the literature, representing the theoretically predicted dependence of the linewidth of the second peak after the normalization ($\Gamma_{\text{peak2}}/q^2 D$) on x. However, the experimental results are quite different. In different x ranges, only a single different normal modes predicted by the theory is observed, as shown by different dashed lines.

Interestingly, in the ranges of $1 < x < 3$, $3 < x < 6$, $6 < x < 10$ and $10 < x < 15$, only the first, second, third and fourth normal modes are experimentally observed, respectively. The qualitative explanation is that only when the scale involved in the normal mode is close to the observation scale of $1/q$, dynamic LLS can "see" this relaxation mode, similar to a physical "resonance", not dependent of the fluctuation amplitude (energy) of a normal mode. Subsequent experiments on different macromolecular solutions repeatedly showed the same phenomenon. Therefore, the existing theory must be modified.

As mentioned earlier, when $x \gg 1$, both strict derivation and de Gennes's scaling theory predict that Γ/q^3 approaches a constant. The experimental result is consistent with the theoretical prediction, but the experimental constant is only 70% – 80% of the theoretical prediction, as shown in figure(a) on the right. In the theory, the predicted constant is from the sum of all weighted normal modes, while dynamic LLS can only gradually "see" different normal modes one by one according to its different observation scales ($1/q$), which qualitatively explains why the measured constant is smaller. It should be pointed out that when the theory about the internal motions was established, dynamic LLS measurement was not considered. Up to now, there is no method that can measure all weighted normal modes at once or

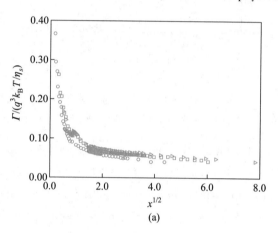

Wu C, Chan K K, Xia K Q. Macromolecules, 1995, 28:1032

(Figure 6, permission was granted by publisher)

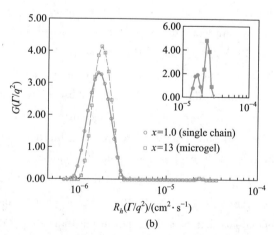

Wu C, Choun S Q. Macromolecules, 1996, 29:1574

(Figure 5, permission was granted by publisher)

measure each normal mode one by one. In short, theoretically, we need to consider why all normal modes cannot be detected by dynamic LLS.

In the 1970s, de Gennes predicted that "$qR_g \sim 2.5$" was a necessary condition for measuring the normal modes of a macromolecular chain based on the status of dynamic LLS instruments at that time. However, a modern dynamic LLS spectrometer uses a higher power laser, a fast single-photon detector, a larger memory and a fast computer developed in the period of 1980-2000. It has a much better measurement accuracy and faster data processing. The old instruments have no comparison. In the past two decades, the experimental results from the measurements of different macromolecular solutions have repeatedly confirmed that as long as "$qR_g \geqslant 1$" is reached, dynamic LLS can detect the second small peak related to the normal modes no matter how small its area is (proportional to the intensity of scattered light), as shown on the above, where the inset shows the enlarged small peaks.

Therefore, by using dynamic LLS to measure the linewidth distribution of a very dilute macromolecular solution at different scattering angles, the scattering angle at which the second small peak appears can be used to estimate the mean square radius of gyration of the chain, i.e., $\langle R_g^2 \rangle \sim 1/q^2$, which can be mutually verified with $\langle R_g^2 \rangle$ measured by static LLS. Besides for macromolecular chains, dynamic LLS can in principle measure any fast relaxation with a length scale of $10^0 - 10^3$ nm. The figure (b) above also shows the experimental results of poly (N-isopropyl acrylamide) (PNIPAM) spherical microgels in the swelling state. In order to compare with the same linear chain, the abscissa ($\Gamma/q^2 = D$) is multiplied by different hydrodynamic radius (180 and 160 nm, regardless of dimension), equivalent to removing the influence of two sizes on D.

The enlarged second small peak shows that only when $x = 13$, i.e., $1/q \sim R_g/3.6 \sim 160/3.6 \approx$ 44 nm, dynamic LLS can detect the thermal energy agitated fluctuation of the swollen microgel network. In other words, the maximum correlation length of the internal density fluctuation of the microgel network is ~ 44 nm. This is the first time to learn from experiments that thermal energy cannot excite the entire microgel network to fluctuate, but only one part of the network. It is known that in the fully swollen state, the average distance between two adjacent cross-linking points of the microgel network is ~ 5 nm, implying that the correlation length spans about 10 cross-linking points, that is, the excessive damping of the liquid limits the fluctuation of a chain link between two adjacent crosslinking points, which can only affect about 5 chain links on each side.

In addition, when $x \gg 1$, Γ/q^3 approaches a constant, only $\sim 6\%$ of that predicted by theory. It further shows that the cross-linking restricts most of the internal motions of the chain. It also indirectly confirms the above discussion that dynamic LLS is not able to observe all the normal modes predicted by theory. This example clearly illustrates that dynamic laser light scattering can not only measure the hydrodynamic radius distribution of macromolecular chains or colloidal particles, but also get microscopic properties that cannot be obtained by other methods. Therefore, one should properly and fully use this powerful method.

Another example is to use dynamic LLS to determine the Θ conditions of a polymer solution. When discussing the conformation of macromolecular chains in Chapter 4, the Θ state was defined as the zero second virial coefficient ($A_2 = 0$); namely, the disappearance of the excluded volume effect that includes the hardcore volume of the segments in a chain and the dispersive attraction among the segments in a chain. Obviously, this is the overall property of individual chains.

However, when discussing how to determine the Θ condition of a macromolecular solution in all macromolecule textbooks, the concentration dependence of a certain physical property (scattered light intensity, reduced viscosity, osmotic pressure difference, etc.) is measured first, and then the conditions (temperature, pressure, solvent composition, etc.) at which the physical property becomes independent of the concentration is found, i.e., the Θ condition of the solution.

Note what is measured in such a well accepted method is the condition when the excluded volume effect including the hardcore volume of the chain and the dispersion attraction (between chains, solvent molecules, and between chains and solvent molecules) disappears. **All textbooks do not explain why the experimentally measured interchain excluded volume effect in the solution can be used to replace the theoretically discussed intrachain excluded volume effect in a chain when determining the Θ conditions of a macromolecular solution.**

When the theory of macromolecular physics was developed in the 1940s and 1950s, it was impossible to measure physical properties and solution behavior of individual chains. This might be why the theatrically discussed excluded volume effect between the segments in a chain was replaced by the experimentally measured excluded volume effect between molecules in the solution, only a stopgap measure. As mentioned above, some physical properties of individual chains can be measured by modern experimental methods, including using dynamic LLS to measure the internal motions (various normal modes) of individual chains.

The question is whether a certain physical property of individual chain can be used to directly determine the Θ condition of a macromolecular solution. It is known qualitatively that in a good solvent, the chain expansion due to the hardcore volume of the segments dominates the chain contraction induced by the dispersion attraction. The chain has a more expanded conformation in a good solvent than in the Θ solvent, so that there is a net repulsion between the segments. While in a poor solvent, the dispersion attraction dominates the chain expansion, so that the macromolecular chain contracts, i.e., there is a net attractive force between the segments. In the Θ state, the two cancel each other, so that there is no net force on the segments. Theoretically, the thermal energy should be able to excite more internal motions (more normal modes) inside the chain. Experimentally, one should observe an increase of the intensity of scattered light from the internal motions. In other words, the area ratio (A_r) of the two peaks obtained in dynamic LLS should reach an inflection point. Based on eq. (7.11), the area ratio is

$$A_r = \frac{\sum_{n=1}^{N} A_n}{\sum_{n=0}^{N} A_n} \ (\text{theoretically}) \quad \text{and} \quad A_r = \frac{A_{peak2}}{A_{peak1} + A_{peak2}} \ (\text{experimentally})$$

Under the condition of $1/q < R_g$, the normal modes of individual chains and microgels were measured using dynamic laser light scattering. An inflection point of A_r is indeed observed near the Θ temperature, as shown below.

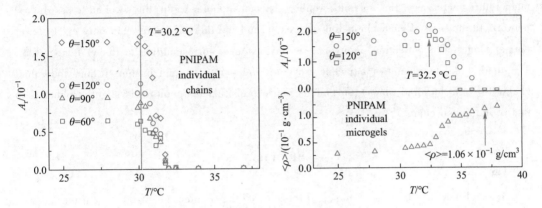

Dai Z J, Wu C. Macromoles, 2010, 43:10064.

(Figures 6 and 11, permission was granted by publisher)

In a good solvent, for a given temperature, dynamic laser light scattering measures more normal modes as the observation scale $(1/q)$ becomes smaller, reflecting in an increase of A_r significantly. For a given scattering angle, as the temperature increases, the thermally sensitive PNIPAM chain gradually shrinks, and R_g decreases. On the one hand, the excluded volume effect weakens as the temperature increases, so that thermal energy can excite more normal modes; on the other hand, the chain shrinkage is equivalent to a decrease of q, leading to a less number of the measurable normal modes in dynamic LLS. The two opposite effects cancel each other out, qualitatively explaining why A_r is nearly independent of the temperature. Near 30℃, A_r changes obviously. The inflection point does not depend on the observation scale, slightly lower than Θ.

The follow-up study of polystyrene in cyclohexane also further proved that the inflection point (~36℃) is close, but slightly higher than the Θ temperature measured by turbidimetry (34.5℃), which is reasonable. The inter-chain interaction between different chains in solution is weaker than the intrachain interaction between different segments in a chain, so that a lower temperature is required to reach the same excluded volume effect. The above discussion of a macromolecular chain becoming the "softest" under the Θ condition was later confirmed by the temperature-dependent stretching experiment of a macromolecular chain.

When the cooling rate is not infinitely slow, for solutions with upper or low critical temperature, the Θ temperature measured by the turbidity method is always lower or higher than the true Θ temperature. It is not difficult to understand that when the chain collapses completely, the internal motions of the chain disappears. In the microgel, the cross−linking between chains has restricted the motions of the chain segments. Therefore, the microgel in the upper right picture has a two orders of magnitude smaller A_r than individual chains. Although A_r is only about one or two thousandth, it is still measurable, showing the sensitivity of modern dynamic LLS. In the upper right picture, the chain density of the microgel increases as the temperature increases (the solvent property becomes poorer). In the swollen cross−linked network, the tension of each chain link is greater, and the thermal energy can only excite fewer normal modes. The inflection point of A_r is obvious, also corresponding to the Θ temperature. The chain links between two adjacent cross−linking points are much shorter than individual chains (several hundred times), so that its phase transition temperature is slightly higher than that of individual chains.

Summary

In addition to the translational diffusion of the center of mass (centroid) of a macromolecular chain in a dilute solution, its N segments agitated by thermal energy also randomly move relative to the centroid (the internal motions). Each segment has its own three−dimensional coordinates (x, y, z) or say, a position vector ($\vec{r}$). Therefore, there are N position vectors ($\vec{r}_i$, $i = 1, 2, \cdots, N$); corresponding to N motion equations. Since the segments are interconnected, N position vectors are also interdependent. In a usual coordinate system, it is a very complicated and difficult to solve these N equations simultaneously. Therefore, N position vectors can be linearly combined to form N independent normal coordinates ($\vec{p}_n$, $n = 0, 1, 2, \cdots, N$), in which $\vec{p}_0$ represents the motion of the centroid. Mathematically, the transformation from one coordinate system $\vec{r}_i$ to another $\vec{p}_n$ is the diagonalization of a matrix in linear algebra. In this new coordinate system, N motion equations are solvable in different special models, resulting in N normal modes ($\vec{p}_n(t)$, $n = 1, 2, \cdots, n$), related to the internal motions of a macromolecular chain. As n increases, the normal mode gradually involves the internal motions with a smaller scale.

The complicated and tedious theoretical derivation has its perfect and beautiful side, but it also makes most experimentlists dizzy. It is not complicated to understand the fluctuation amplitude and characteristic relaxation time of the normal mode relaxation inside a macromolecular chain without mathematical equations and only on the basis of some physical images. If the hydrodynamic interactions between different molecules (segments) are ignored, for a linear flexible macromolecular chain whose average conformation is a random coil, although the segment position relative to the centroid is ever−changing, the maximum amplitude of the internal motion of a chain (i.e., the rotation/vibration of the entire chain) should be close to but not larger than the chain size itself (its average radius of gyration, $R_g = \sqrt{\langle R_g^2 \rangle}$); the slowest relaxation time should be close to the time required for the chain diffuse a distance of its own size. The smallest variation size (i.e., $n = N$) is the length of a chain segment (b), and the relative motions between two adjacent chain segments. If considering the excessive damping of the hydrodynamic interaction, the maximum fluctuation amplitude should be smaller than its average radius of gyration, and the slowest relaxation time should be longer than the time required for the chain to diffuse its one size. Although such deduction based on physical images is not able to give details of how the normal modes depend on the chain length, but it provides a basic concept of magnitude.

In the Rouse model, the hydrodynamic interaction is ignored, so that it is more suitable for a concentrated solution or melts in which the hydrodynamic interaction has been shielded. The average fluctuation amplitude of the time autocorrelation function of $\vec{\mathbf{p}}_n(t)$: $\langle \vec{\mathbf{p}}_n(0) \cdot \vec{\mathbf{p}}_n(0) \rangle = \langle p_0^2(0) \rangle \approx \langle R_g^2 \rangle / n^2$, proportional to $\langle R_g^2 \rangle$, inversely proportional to n^2; the characteristic relaxation time: $\tau_n \approx t_{R_g}/n^2$, proportional to the average time (t_{R_g}) required for the chain to diffuse a distance of R_g, inversely proportional to n^2. When $n = 1$, the strictly deduced result is the same as the deduction based on the physical image. As n increases, the normal modes becomes weaker and faster!

In the Zimm model, the hydrodynamic interaction is considered, more suitable for a dilute solution or a system with strong interaction between macromolecules and solvent molecules. The average fluctuation amplitude of the time autocorrelation function of $\vec{\mathbf{p}}_n(t)$: $\langle p_0^2(0) \rangle \approx 0.3 \langle R_g^2 \rangle / n^{\alpha_{R_g}}$; the characteristic relaxation time: $\tau_n \approx 4 t_{R_g}/n^{\alpha_{R_g}}$, where $1/2 \leqslant \alpha_{R_g} \leqslant 3/5$. In comparison with the Rouse model, due to the excessive damping, the fluctuation amplitude of the first normal mode is smaller, and the characteristic relaxation time becomes longer. However, the changing rate of the normal mode as n increases is slower. Similarly, when $n = 1$, the result matches the deduction based on physical images.

Experimentally, the normal mode inside a macromolecular chain is measurable, especially the first normal mode. The methods include, but not limited to, dielectric relaxation, birefringence relaxation, and dynamic LLS. The first two methods have requirements for properties of the chains to be measured. Dynamic LLS is more general for measuring the normal modes, and also the key content introduced in this chapter.

Basic Principles:

Dynamic LLS uses the light scattered by macromolecules or particles to measure their motions. The resolution of the "ruler" is related to the wavelength and the reciprocal of the scattering vector ($1/q$). When $1/q$ is much larger than the average radius of gyration (R_g), dynamic LLS can only detect the diffusion of the centroid of a macromolecule, but not the internal motions inside. Just like to stand in the distance to watch a high-speed train, one can only see the moving train as a whole, but not activities of passengers inside each carriage. When

$1/q \leqslant R_g$, from the time autocorrelation function of the intensity of scattered light, besides the characteristic relaxation time related to the translational diffusion, one can also obtain the characteristic relaxation time distribution corresponding to the internal motions (normal modes), equivalent to standing inside a carriage to observe different activities of passengers.

Measurable Physical Quantities:

A digital time correlator is used to measure the time autocorrelation function of the intensity of scattered light of macromolecular solutions at different concentrations and at different scattering angles. First, in the range $1/q > R_g$, after extrapolating the apparent translational diffusion coefficient to the infinite dilute concentration and zero scattering angle, one can get the average translational diffusion coefficient of the centroid. Then, in the range $1/q \leqslant R_g$, one can measure the time autocorrelation function to obtain the characteristic relaxation time distribution related to the internal motions. Generally speaking, only when $R_g > 50$ nm or $M \sim 10^6$ g/mol, can the condition of $1/q \leqslant R_g$ be met.

Data Processing:

The Laplace inversion of each measured time autocorrelation function of the intensity of scattered light can lead to a corresponding linewidth (characteristic relaxation time) distribution $G(\Gamma)$. For a narrow distribution of macromolecule samples, in the range $1/q > R_g$, the linewidth or characteristic relaxation time distribution is a single peak. When $1/q > R_g$, there will appear a second small peak related to the internal motions (the normal modes) on the right in the larger linewidth (low−frequency) region. Normalizing its corresponding linewidth (Γ_{peak2}) by the linewidth ($q^2 D$) corresponding to the first large peak, one can obtain the average linewidth of the normal modes. In addition, the area ratio (A_r) of the second small peak to the first large peak represents the contribution of all normal modes to the intensity if scattered light. In addition to measuring the internal motions (normal modes) of macromolecular chains, the measured Γ_{peak2} and A_r can also be used to get microscopic information that is difficult to obtain by other methods. This chapter only lists two examples of using dynamic LLS to find the fluctuation amplitude inside the microgel and the Θ conditions of the solution.

Therefore, what a waste if one only uses an expensive research−type LLS instrument to measure the hydrodynamic radius distribution of macromolecular chains or colloidal particles. A combination of static and dynamic LLS and other methods can obtain many microscopic properties that cannot be obtained by other experimental methods. It is advised for readers to understand the LLS principles, especially the theory of dynamic LLS, and then make good use of it! The figure below summarizes how to use dynamic LLS to measure the normal modes of a chain, compare the average fluctuation amplitude and characteristic relaxation time of the time autocorrelation function of the normal modes in the Rouse and Zimm models, where D is the translational diffusion coefficient. The experimental results in dilute solutions agree well with those predicted by the Zimm model.

$$\langle \overrightarrow{\mathbf{p}_0}(0)\overrightarrow{\mathbf{p}_0}(t)\rangle \begin{cases} \Gamma_0 = q^2 D \\ \tau_0 = 1/\Gamma_0 \\ \langle p_0^2(0)\rangle \end{cases}$$

$$\langle \overrightarrow{\mathbf{p}_n}(0)\overrightarrow{\mathbf{p}_n}(t)\rangle \begin{cases} \Gamma_0 + \Gamma_n \\ \tau_n = 1/\Gamma_n \\ \langle p_n^2(0)\rangle \end{cases}$$

$$q = \frac{4\pi n}{\lambda_0}\sin\frac{\theta}{2}$$

$$\langle R^2\rangle = b^2 N = 6\langle R_g^2\rangle$$

$$R_g/R_h \approx 1.5$$

$$\tau_{R_g} = \langle R_g^2\rangle/(6D)$$

Rouse Model

$$\tau_n \approx \tau_{R_g}/n^2 \sim M^2$$

$$\langle p_n^2(0)\rangle \approx \langle R_g^2\rangle/n^2$$

$$D = \frac{k_\mathrm{B}T}{6\pi\eta bN} \sim M^{-1}$$

Zimm Model

$$\tau_n \approx 4\tau_{R_g}/n^2 \sim M^{3\alpha_{R_g}}$$

$$\langle p_n^2(0)\rangle \approx 0.3\,\langle R_g^2\rangle/n^2$$

$$D = \frac{k_\mathrm{B}T}{6\pi\eta bN^{\alpha_{R_g}}} \sim M^{-\alpha_{R_g}}$$

第八章 大分子溶液中链的构象变化

依据第四章中的讨论已知,在溶液中,一条无特殊相互作用的线型柔性大分子链,因两个对链尺寸有相反依赖性的、均源于熵的作用达至一个扩展的、具有一定尺寸的、无规线团状的平衡构象。随着溶剂性质逐渐地变差,排除体积效应(包括硬核体积和色散吸引作用二者产生的综合效应)逐渐减弱,导致链收缩。当色散吸引产生的负排斥体积效应正好抵消了链本身的排斥体积产生的效应时,一条真实大分子链就具有与一条没有硬核体积和没有相互作用的理想链相似的一些整体性质,称作"准理想状态"。在此状态,链的平均回转半径(R_g)与摩尔质量(长度)之间的标度指数为1/2。注意:一条真实链的硬核体积永不消失,更不可能为负,是色散吸引产生的"负排除体积效应"使得排除体积在一些情形中(如回转半径的链长依赖性)"等效地"消失了,但并非真正地消失了。

相比而言,小分子在一个二元混合物(溶液)中只有溶解和不溶解两种状态,没有链构象随溶剂性质的变化。因此,大分子构象对溶剂性质和链长的依赖性导致了新的物理:大分子特有的物理,大分子物理。无任何外力时,如果以准理想状态(也称作无扰状态)作为一个参考状态(点),随着溶剂性质变好,排除体积效应逐渐地增强,故大分子链就逐渐地扩张至一个最大的平均回转半径(R_g),链尺寸与摩尔质量之间的标度指数为3/5。相反,如果溶剂性质逐渐地变差,排斥体积效应越来越负,链段之间的吸引逐渐地变强,因此,链应该逐渐地蜷缩成一个均匀的单链小球,此时的标度指数理应降至1/3。另一方面,有外力存在时,不难想象,一个足够强的拉力应可克服熵弹性,使得一条处于无规线团构象的大分子链延展和伸直。对一个完全伸直的构象,标度指数显然为1。

上述的"线团至小球"和"线团至伸展"两个大分子构象转变是自20世纪60年代以来大分子物理中两个最重要的命题,它们的理论研究分别始于20世纪60和70年代初。物理上要问的是,第一,这两个变化会不会在均相溶液中发生?第二,如果可以,它们是不是一个类似相变的变化?第三,如是,它们是类似一级还是二级相变的转变?第四,它们与链长之间是何种定量和普适的关系?实验上的研究在20世纪70年代末期方才开始。尽管许多实验室尝试使用了不同的实验方法和大分子溶液,但无人能够证实理论上的预测。直至20世纪90年代中叶和21世纪初,才终于在实验上证实了这两个重要的构象转变,并回答了与这两个命题有关的一系列问题。详情如下。

从"线团至小球"的转变

20 世纪 60 年代初,基于 Flory 平均场理论,Stockmayer[Makromol. Chem.,1960,35:54]提出,随着溶剂性质逐渐变差,链段间的相互吸引逐渐增强,所以,一条线型柔性大分子链的构象将从一个扩展的无规线团变成一个均匀的、蜷缩的单链小球。他还指出,由于这些蜷缩的单链小球的溶解度太低,不易在一相区观察到热力学稳定的、蜷缩的单链小球状态。

Ptitsyn 和合作者[J. Polym. Sci. Part C,1968,16:3509]也利用 Flory 模型,计算了一条大分子链上链段之间的相互作用能。在不考虑链的连接性的前提下,他们得到膨胀因子 $\alpha[=R_g(T)/R_g(\Theta)]$ 与相互作用参数之间的关系,其中 $R_g(T)$ 和 $R_g(\Theta)$ 分别是温度为 T 和 Θ 时的根方均回转半径($\sqrt{R_g^2}$)。他们还利用蒙特卡络法模拟和计算了"线团至小球"的构象转变,发现如果链段之间的相互吸引为主导,这一构象转变发生在一个非常狭窄的区间。

如果这一构象变化发生在形成两相的相变中,链内折叠和链间聚集将几乎同时发生。实验上,将这两个完全不同的过程分开,只观察单链的构象转变,相当困难,几乎不能。后来,Sanche[Macromolecules,1979,12:980]利用平均场理论推算了大分子链的尺寸。在计算过程中近似地考虑了链段间的各种高阶相互作用后发现,当大分子分别处于小球和扩展的无规线团这两种链构象时,$R_g(T)$ 与 Kuhn 链段数 N 之间有着显而易见的标度关系:$R_g(T) \propto N^{1/3}$ 和 $R_g(T) \propto N^{3/5}$。膨胀因子可由下式解出。

$$\frac{7(1-\alpha^2)}{3N} = \left(\frac{\Theta}{T}\right)\frac{\phi}{2} + \frac{\ln(1-\phi)}{\phi} + 1 \tag{8.1}$$

其中,$\phi_0 = \phi\alpha^3$,ϕ_0 和 ϕ 分别是回转半径为 $R_g(T)$ 和 $R_g(\Theta)$ 的真实链和理想链在链的空间中占有的体积分数,均小于 1。在良溶剂中,$\alpha > 1$,所以,$\phi_0 > \phi$。当一条大分子链足够长时($N \to \infty$),$\phi = (19/27)^{1/2}N^{-1/2}$。对一条溶胀的链,$\phi \ll 1$,将 $\ln(1-\phi)$ 展开可重写上式如下:

$$\alpha^6(1-\alpha^2) + 0.102 + \cdots = 0.180\alpha^3\tau N^{1/2} \tag{8.2}$$

其中,$\tau = (T-\Theta)/\Theta$,为约化温度。上式显示了,给定链长时,膨胀因子和约化温度之间的关系;当溶液体系从良溶剂趋近准理想状态时,$\alpha \to 1$,以及 $\tau N^{1/2} \to 0.102/0.180$。

因此,当链长趋向无穷时,即 $N \to \infty$,约化温度必须趋于零。理论上,"线团至小球"的变化应趋于类似一级相变的突变。对给定的有限链长,上式显示了一个连续变化,类似二级相变。在第三章中讨论混合熵变时,作用接触点数(Z)包含在表达全部色散效应的参数 χ_{sp} 中。在推导上式时,Sanche 假定了 χ_{sp} 仅随温度变化,忽略了 Z 也可能随着大分子链的体积分数而变,即接触点数会随着链段浓度增加。除了上述列举的研究者,其他理论学家对"线团至小球"转变的理论也贡献良多,尤其是莫斯科大学物理系的已故 Lifshits 教授和他的许多学生们,包括 Khokhlov 和 Grosberg。然而,源于理论物理的论文涉及艰深的数学,故在本书中略去。

物理上,不难理解,随着链的蜷缩,其体积分数增加,链段之间的距离更近。因此,应有更多的相互作用接触点(机会),即 Z 可随着 ϕ 增加。另外,随着链的体积分数增加,还需考虑多体相互作用。Kholodenko 和 Freed[J. Phys. A:Math. Gen.,1984,17:2703]利用场论研究了转变级数与多体相互作用之间的关系,发现其取决于三体互相作用的强度。在20 世纪 80 年代末,钱人元和其合作者[Macromolecules,1993,26:2950;Macromol. Chem. Rapid Commun.,1993,14:747]也开始在实验的基础上定性地研究与单链粒子有关的问题,包括单链和寡链高分子粒子内的密度和结晶行为。

然而,自 20 世纪 70 年代末起,实验学家们经过大量的实验仅观察到无规线团的有限蜷缩,无法得到热力学稳定的单链小球。即随着溶剂性质逐步变差,在扩展的无规线团蜷缩成一个热力学稳定的单链小球之前,就发生了链间聚集,即相变,溶液从一相分成两相,导致实验失败。Tanaka 和其合作者[Nature,1979,281:208;Phys. Rev. Lett.,1980,44:796]最早发表了有关该转变的实验结果。然而,其所观察到的单链小球位于两相区,并非热力学稳定。Chu 和其合作者[Macromolecules,1987,20:2833;1988,21:1178;1995,28:180;1996,29:1824]经过长达十年的研究,也没观察到热力学稳定的单链蜷缩小球。

鉴于实验上的屡试屡败,Grosberg 和 Kuznetsov[Macromolecules,1993,26:4249]在1993 年甚至著文断言,利用目前的样品制备方法和仪器设备很难观察到这一现象(热力学稳定的单链小球)。他们退而求之地询问,(1)能否在沉淀发生之前,在实验中形成和研究单链小球的状态?(2)能否在非平衡的条件下,获知小球内的链具有一个简单的、褶皱的无规线团还是一个复杂的、缠绕和打结的构象?(3)能否在非平衡的两相区内,观察到大分子链的蜷缩和聚集动态过程?

在自然界,许多蛋白质链在细胞内都可很容易地折叠成一个稳定的球状构象,但是无法解释为何将一条高聚物链折叠成一个蜷缩的小球构象如此之难。研究"线团至小球"转变的主要是物理学家和高分子物理学家。为了在实验上研究该转变,溶液必须尽可能地稀,使得溶剂性质可以在一相区内充分地从良向不良变化。另一方面,如在第三章中所讨论,大分子链的相图是一个链长的函数,即长链发生溶液相变早于短链,故所使用的大分子样品必须具有很窄的摩尔质量(链长)分布。

对不善化学合成的物理学家,商业化的超高摩尔质量和窄分布的聚苯乙烯标准样品就成了他们不二的选择。聚苯乙烯仅可溶解在有机溶剂中,其中链段之间的相互作用主要为色散吸引,远弱于水溶液中的疏水作用和氢键相互作用。因此,可采用水溶性聚合物,以增强链段之间的相互作用。具有低临界溶液温度(LCST ~ 32 ℃)、热敏性的聚(N-异丙基丙烯酰胺)(PNIPAM)就是一个很好的选择。然而,制备窄分布的超长水溶性聚合物链是另一个极其困难的问题。Kubota 和其同事[J. Phys. Chem.,1990,94:5154]就试图利用变化溶液温度来研究 PNIPAM 在水溶液中的构象转变。由于采用了一个摩尔质量分布相对较宽的 PNIPAM 样品($M_w/M_n>1.3$),他们在链间聚集之前仅观察到单链的有限蜷缩。

有了一个确定的方向,制备超高摩尔质量且窄分布的 PNIPAM 就成了不得不越过的第一个障碍。在聚合反应前,先将 N-异丙基丙烯酰胺单体在甲苯和正己烷的混合溶液中重结晶三次,以提高单体纯度;同时,充分地干燥了聚合溶剂苯,以减少链转移和副反应。

反应后,将干燥后的粗产品溶于无水丙酮中,再缓慢地滴入非溶剂无水正己烷中沉出,以除去未反应单体和可溶性杂质。过滤和干燥后得到的 PNIPAM 呈纤维状。将如此纯化后的样品进一步在无水丙酮和无水正己烷的混合溶液中经多次分级后,得到一个高摩尔质量($M_w \sim 6 \times 10^6$ g/mol)和较窄分布($M_w/M_n \sim 1.3$)的级份,其与超高摩尔质量和极窄分布的要求仍有较大差距。

在多次分级失败以后,Wu 和 Zhou[Macromolecules,1995,28:5388;1995,28:8381]意识到研究"线团至小球"的构象转变,并不需要得到宏观量级的超高摩尔质量和极窄分布的干燥固体样品;相反,仅需几微克存在于溶液中的超高摩尔质量和极窄分布的大分子链。突破传统思维后,先将摩尔质量最高的一个样品配成 $\sim 10^{-4}$ g/mL 的水溶液,再过滤一至两滴这样的 PNIPAM 水溶液至含有约 1 mL 无尘水的光散射池中。重复此过程直至得到超长的 PNIPAM 链($M_w > 10^7$ g/mol)和极窄的分布宽度($M_w/M_n < 1.10$)。最终的 PNIPAM 浓度为 10^{-6} g/mL。

在解决了样品制备的问题后,仍有第二个障碍。依据第五章中激光光散射的讨论,实验上的观察长度,即散射矢量(q)的倒数,应大于链的平均回转半径(R_g),即 $qR_g < 1$。在此条件下,利用静态激光光散射测得时间平均散射光强对散射角度和浓度的依赖性,据此可得重均摩尔质量、方均回转半径和二阶维里系数。为了测量超长 PNIPAM 链,不得不改造商品化的激光光散射仪,包括使用光束较细的固体激光、减少杂散光和精心调节入射以及接收光路。因此,最小的测量角度从通常的 15° 降低至 6°,故满足了 $qR_g < 1$。

下图(a)综合地显示了 PNIPAM 单链在"线团"和"小球"两个构象状态下的静态和动态光散射结果。随着温度升高,流体力学半径分布向左(小尺寸)移动,同时瑞利比值的倒数对散射矢量的依赖性显著地降低(较小斜率),但截距不变,显示了无链间聚集,链收缩是一个单链过程。其还显示链尺寸(R_g 和 R_h)显著地变小,PNIPAM 大分子链的尺寸蜷缩了约 8 倍,即体积改变了约 500 倍,这是"线团至小球"转变的第一个实验证据。

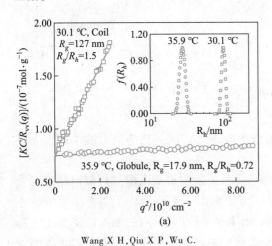

Wang X H,Qiu X P,Wu C.

Macromolecules,1998,31:2972.

(Figure 1,permission was granted by publisher)

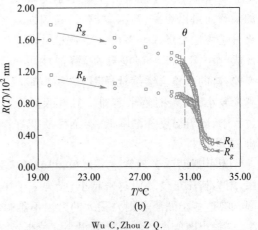

Wu C,Zhou Z Q.

Macromolecules,1995,28:8381.

(Figure 1,permission was granted by publisher)

文献中,研究者常常"忘记"或不知道研究"线团至小球"转变必须在一相区内进行,即他们不得不证明在转变中没有链间聚集。由于平均流体力学半径对少量的链间聚集并不敏感,仅依赖动态光散射所测的平均流体力学半径变化是不够的。检测是否有链间聚集的最好方法就是一个不变的截距,如上图(a)所示。如前所论,"散射光强正比于散射粒子(大分子)摩尔质量的平方",故散射光强对任何聚集都异常敏感。即使仅有二聚体形成,散射光强也会翻倍。

上图(b)总结了两个具有不同分子量的 PNIPAM 样品在水溶液中的链尺寸对温度的依赖性。随着温度升高,平均回转半径从大于变为小于平均流体力学半径。这一变化源于它们不同的物理定义。平均回转半径反映了链段在空间中的分布;而平均流体力学半径则指的是一个和链在溶液中具有相同平动扩散系数的一个均匀小球的半径。在良溶剂中,$T < \Theta$,平均回转半径和平均流体力学半径随温度升高缓慢地变小,但它们的比值几乎为一个常数,接近"线团"构象的理论值(1.504)。在不良溶剂中,$T > \Theta$,链的平均尺寸随温度升高急剧变小,最后趋于一个常数。

成功观察到一条中性高分子链在溶液中从"线团至小球"的构象变化和形成热力学稳定的单链小球使得研究其反过程"小球至线团"的构象转变成为可能。下图总结了在一个加热和冷却的循环中,PNIPAM 单链的平均尺寸如何随温度变化。其中有两点值得注意。第一,插图中平均回转半径与平均流体力学半径之比(R_g/R_h)不是预期的单调下降;第二,无论是在升温还是降温的过程中,在准理想(Θ)状态(31.4 ℃)附近,有明显的滞后现象。

首先,让我们考察第一点。在加热过程中,平均回转半径与平均流体力学半径之比(R_g/R_h)先从 ~1.5 降至 ~0.6,又回升至 ~0.78,接近一个均匀小球的预期值 0.774。R_g/R_h 随温度的变化存在一个明显的小凹坑。其定性的解释如下。一条真实的柔性大分子链不是无限的柔软。可以想象,当其收缩时,其表面无可避免地会形成许多由一定数量链段形成的小圈。随着温度升高,溶剂性质变差,链的构象会继续地蜷缩,这些小圈越来越小,故蜷缩越来越难。在相反的

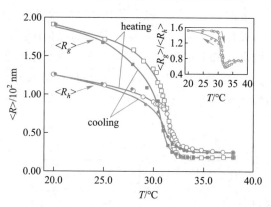

Wu C,Wang X H. Phys. Rev. Lett.,1998,80:4092.

(Figure 2,permission was granted by publisher)

过程中,随着温度逐渐降低,溶剂性质变好,小球构象逐渐扩展。充满张力的小圈首先融出、在表面扩展。

为了帮助读者理解这一现象,可以想象用一只手掌握住和挤压由一根细铜丝组成线团,用力挤压可使线团越来越小,成为一个小球,但表面上一定有许许多多微微突出的小圈。将这些小圈压平则要很大的力。不难想象,因为这些小圈的存在,表面的密度比内部小,这些小圈所占的质量分数很小。依据定义,表面上的这些低密度小圈对回转半径的贡献很小,但是表面上包裹着一些溶剂分子的小圈则会稍稍增大流体力学半径。因此,与一个均匀小球相比,表面带有许多小圈的小球具有较小的回转半径与流体力学半径之比。

文献中先后有两个实验室在研究"线团至小球"构象变化时,也曾经观察到这样的小凹坑。也许由于理论比值是一个单调地从 ~1.5 到 ~0.8 的下降,他们均先入为主地将其归于实验误差。然而,正是这一重复出现的小凹坑(较低的 R_g/R_h 的比值)导致了一个发现:新的大分子链构象,"融化球"[Phys. Rev. Lett.,1996,77:3053],被誉为大分子研究的一个地标。

关于第二点,滞后现象主要出现在准理想(Θ)状态(31.4 ℃)附近。对给定的温度,在"小球至线团"的反过程中,链的平均尺寸均偏小。这一滞后现象更明显地反映在插图中的回转半径与流体力学半径之比上。这一滞后现象完全出乎理论意料。注意,图中的每个数据点都是在温度达到平衡后获得的,完全不同于在温度扫描实验中常见的滞后。合理的猜想如下,在 PNIPAM 单链蜷缩的过程中,链的体积分数和链段浓度逐步增加,增加了形成链内氢键的概率。在冷却的过程中,这些在蜷缩过程中形成的额外链内氢键无法在 Θ 状态附近全部断开,限制了链的扩展,故链的平均尺寸较小。随着温度进一步地下降,溶剂性质变得更好,高温下形成的额外链内氢键全部断开,平均链尺寸回归初始值。

为了验证这一猜想,可采用无法生成链内氢键的柔性中性聚(N,N-二乙基丙烯酰胺)(PDEAM)在实验上去掉那些可能生成的链内氢键。利用 PDEAM 可重复以上"线团至小球"转变的实验。右图分别显示了 PNIAPAM 和 PDEAM 的化学结构。前者含有氢键的受体和给体[C＝O 和 H—N(Pr-i)],故其在较高的温度下蜷缩时可能形成链内氢键,即 C＝O…H—N(Pr-i)—,但后者不能。

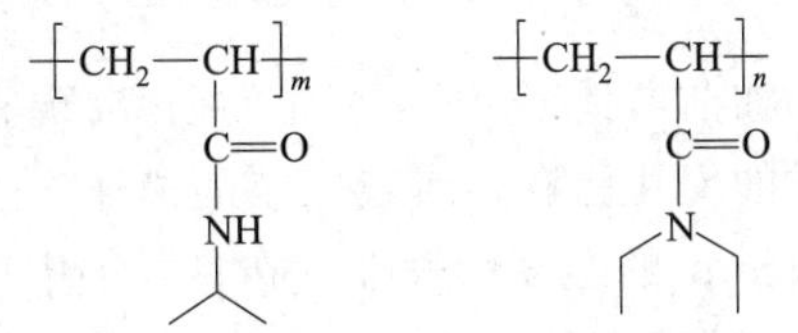

然而,制备水溶性、窄分布的超长大分子链永远都是一个挑战。为了抑制链的转移及终止,最终发展了一个利用低温本体聚合制得重均摩尔质量高达 10^6 g/mol 的 PDEAM 样品的方法。将纯化后的样品在无水丙酮和无水正己烷的混合溶液中进一步经多次分级得到重均分子量为 ~6×10^6 g/mol 且分布较窄的级份。将该级份进一步分级,最终得到所需的 PDEAM 样品(M_w ~1.7×10^7 g/mol 和 M_w/M_n ~1.06)。

利用这一样品,进一步研究了"线团至小球"的构象转变以及其反过程,并得到了平均回转半径和平均流体力学半径随温度的变化。下图(a)显示,在一个升温和降温的循环中,PDEAM 的升温和降温曲线几乎重叠在一起,完全没有滞后现象,与 PNIPAM 完全不同。经过十多年实验研究后,终于证实了当初的猜想,即先前观察到的滞后现象确实可归于链在蜷缩过程中形成的链内氢键。

此实验结果也排除了当初的一个疑惑:热力学稳定的 PNIPAM 单链小球的形成是否因为链内的氢键,以及是否为一个特例? 在温度 31~35 ℃内,同样观察到偏小的 R_g/R_h 比值,证实了"熔化球"构象的存在,且与有无链内氢键无关。在高温下,与 PNIPAM 相比,稳定的 PDEAM 单链小球的链密度稍低,其 R_g/R_h 比值稍大于均匀实心球的预期比值,表明稳定的 PDEAM 单链小球含有更多的溶剂,且部分地"排"溶剂,导致其流体力学半径较小,R_g/R_h 比值稍大。这些结果说明链内氢键对单链最终的蜷缩程度有一定影响。

poly(N-isopropylacryl-amide) (PNIPAM)

poly(N,N-diethylacryl-amide) (PDEAM)

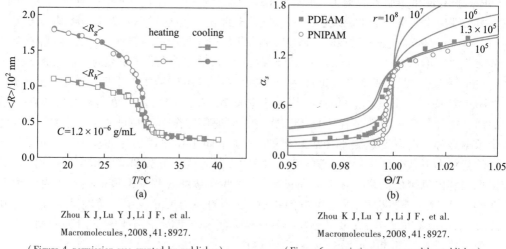

Zhou K J, Lu Y J, Li J F, et al.
Macromolecules, 2008, 41 : 8927.
(Figure 4, permission was granted by publisher)

Zhou K J, Lu Y J, Li J F, et al.
Macromolecules, 2008, 41 : 8927.
(Figure 6, permission was granted by publisher)

　　如前所述,经过一百年的研究,现有的高分子理论对在良溶剂中高分子链构象随溶剂性质的变化已有相对较完整的描述。上图(b)比较了 PDEAM 和 PNIPAM 单链的静态膨胀因子 α_s 的温度依赖性。在良溶剂中(这里 $T < \Theta$),PDEAM 和 PNIPAM 的实验结果和由式(7.2)算出的理论曲线吻合,与聚苯乙烯在环己烷中的结果类似[J. Chem. Phys., 1980, 73 : 5971]。在不良溶剂中($T > \Theta$),随着温度升高,溶剂性质变差。测得的 α_s 数值不仅比理论的预期值低很多,而且随着温度升高,下降得也更快。PNIPAM 的 α_s 又比 PDEAM 的下降得更低、更快。从实验的角度观看,"线团至小球"转变已接近类似一级相变的突变。它们的 α_s 值分别趋于 0.20 和 0.14,进一步显示在蜷缩状态,PDEAM 单链小球较为松散。

　　大分子链随着溶剂性质变差而蜷缩的现象还可用式(7.2)中的折合膨胀因子 $\alpha^3 \tau M_w^{1/2}$ 与 $\tau M_w^{1/2}$ 来描绘。右图显示源自两个不同摩尔质量的 PNIPAM 样品的结果。以 Θ 状态作为参考点,$\alpha^3 \tau M_w^{1/2} = 0$。随着 $|\tau|$ 的增加,溶剂性质变差,链的体积收缩,α^3 递减,但 $\alpha^3 \tau$ 增加。当 $\partial(\alpha^3 \tau M_w^{1/2})/\partial \alpha \propto 6\alpha^5 - 8\alpha^7 = 0$ 时,$\alpha = 0.866$,$\alpha^3 \tau M_w^{1/2}$ 达至极大。由右图的理论曲线可知,当大分子链的链段数为 10^5 时(实验上的极限),当 α_s 从 1.0 降到 0.4 左右时,τ 的对应变化范围小于 0.01。如果 $\tau M_w^{1/2}$ 很大,且 $\alpha < 0.4$,

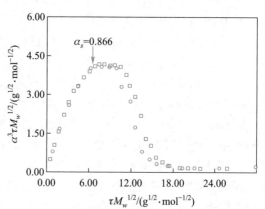

Wu C, Zhou Z Q. Macromolecules, 1995, 28 : 8381.
(Figure 9, permission was granted by publisher)

式(7.2)中的第一项可忽略不计,$\alpha^3 \tau M_w^{1/2}$ 趋于一平台常数(0.102/0.180 = 0.567)。当 $\tau M_w^{1/2} < 10$ 时,实验结果与理论预期基本吻合。确实在 0.866 处达至极大,存在一个热力学稳定的平台。但是,在 $\tau M_w^{1/2} > 10$ 以后,$\alpha^3 \tau M_w^{1/2}$ 迅速下降,偏离理论预测。虽然,$\alpha^3 \tau M_w^{1/2}$ 仍然趋于一常数,但远小于理论值。

起初,这些在不良溶剂中的差异被归于链在蜷缩状态中形成的额外链内氢键,即链内氢键的形成进一步促进了链的蜷缩,一个自加速过程。然而,PDEAM 链无法形成链内氢键,上图中的实验结果推翻了这一最初解释。这些理论和实验之间的差别可能主要归于平均场理论的假定,即链段之间的色散吸引作用和链段浓度无关。在良溶剂中,一条大分子链处于一个扩展的状态,链尺寸 $R(T)$ 和膨胀因子 α 对温度依赖性相对较小,链段在链内的体积分数变化不大。故理论和实验比较吻合。

在不良溶剂中,排斥体积效应为负,即链段之间存在净的吸引作用。目前的理论仍然无法定量地描绘大分子链的构象变化。定性的解释可能至少有如下两种。其一,平均场理论关于链段间的相互作用和链段浓度无关的假定需要修改,引入作用点数目(Z)和浓度的关系,但这可能会动摇平均场理论已有的一些结论,这是一个值得理论学家关注和研究的问题。其二,作为溶剂的水具有一些特殊性质,不仅与高分子链发生水合作用,而且还可通过分子间的氢键作用互相之间自络合,依赖于温度。这要求实验学家不但要找到一个非水溶剂,而且要在一相区测量"线团至小球"构象转变,并达到一个热力学的稳定状态。

在成功地研究了"线团"和"小球"构象之间的每一个稳定状态后,仍需研究的是"线团"和"小球"之间的转变动力学,包括构象变化的内在机理和该过程的特征弛豫时间。与稳态研究相比,尽管也有少量尝试,但没有成功的动力学研究。Chu 和 Ying 以及他们的合作者[Macromolecules,1995,28:180;1996,29:1824]使用 0.01 mm 的薄壁样品池,试图减少溶液温度变化的时间,发现了两个弛豫过程,其特性弛豫时间为分钟量级。其他人也试图利用激光光散射来研究这一动力学问题,但始终无法克服变温和测量速度相对太慢的两大难题。Liu 和其合作者[Phys. Rev. Lett.,2006,96:27802]采用快速的停流法研究了这一构象转变,也发现 PNIPAM 链的蜷缩包括了两个相对较快的弛豫过程,其特征弛豫时间分别约为 12 ms 和 270 ms。

理论上,在 20 世纪 80 年代中叶,de Gennes[J. Phys. Lett.,1985,46:639]就预测,一条大分子链的折叠可能始于一个伸展的无规线团,经过一个"香肠状"的构象,终于一个小球状态。第一个过程的特征弛豫时间约为毫秒量级。后来,Grosberg 和 Kuznetsov[Macromolecules,1993,26:4249]假定大分子链在不良溶剂中的蜷缩分作两步:先快速地皱缩,再续以缓慢地链内缠结(打结),构象最终成了一个稳定的单链小球。但实验上则束手无策,人们不得不借助计算机模拟和唯象模型研究大分子链从"线团至小球"的构象变化动力学。Kuznetsov[J. Chem. Phys.,1995,103:4807;1996,104:3338]和 Halperin 等人[Phys. Rev. E,2001,61:565]都陆续地预测构象变化的首个过程应是链上的局部"成核",其特性弛豫时间应和链长无关。在这些研究中,都假定了链长无限,以及从"线团至小球"的构象变化为一级相变。否则,也就没有所谓的在亚稳态里的"成核"过程了。

要验证理论上的预测,实验上不得不解决两个问题。首先,要有一个远快于毫秒的加热方法,可将水迅速地从 Θ 溶剂变成不良溶剂,因而可在加热过程中,避免链构象的变化。其次,一个时间分辨能力可达微秒级、远快于动态激光光散射的检测方法。先前提及的停流法是一种常见的研究动力学的方法,但其死时间通常是几个毫秒,并不适合研究大分子构象的变化,尤其是初始的动力学过程。为了解决这些技术难题,可利用生物学中测量蛋白质折叠(链更短、过程更快)的方法,其中使用了一个波长为 1540 nm 的红外纳秒脉冲激光加热水溶液,外加一个快速荧光和一个光散射检测器。该方法可在 20 ns 内将照射光斑处的水溶液温度升高约 3 ℃,以及在 90°的散射角实时地检测溶液的散射光强和荧光强度,时间分辨率为微秒量级。

Ye 和其同事[Macromolecules,2007,40:4750]成功地利用该法研究了"线团至小球"构象转变的动力学,其中利用了 8-苯氨基-1-萘磺酸铵盐(ANS)作为荧光探针。当环境由亲水性变为疏水性时,其发射的荧光会极大地增强。右图显示了当溶液迅速地被一个 20 ns 的激光脉冲加热后,ANS 的荧光强度如何随时间变化。没有 PNIPAM 时,ANS 的荧光强度基本为一个常值,不依赖于温度(菱形符号)。当溶液中存在 PNIPAM 时,ANS 荧光强度在 0.5 ms 内迅速增加(圆圈符号),随后,趋向于一个最大值。插图显示,ANS 荧光强度的变化可以

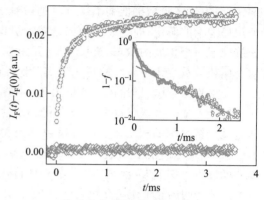

Ye X D,Lu Y J,Shen L,et al.

Macromolecules,2007,40:4750.

(Figure 1,permission was granted by publisher)

分为快、慢两个过程,其对应的特征弛豫时间分别为(0.11±0.01) ms 和(0.83±0.06) ms。

整个蜷缩过程在约 4 ms 内完成。因此,常规的停流法应该无法观察到这两个快速过程。先前利用停流法测得的两个过程很可能源于大分子的链间聚集。为了证明上图中所观察到的荧光光强变化确实源于单链的构象变化,还同时测量了加热后散射光强随时间的变化。如前所述,二聚体的形成也会导致散射光强增加一倍。实验结果显示在荧光光强变化的时间窗口内,散射光强几乎为一个常值,证实观察到的仅是单链的构象变化。

成功地在 PNIPAM 链的水溶液中观察到两个动力学过程后,他们进一步研究了这两个过程对链长的依赖性。利用前述的分级方法,得到重均摩尔质量在 $1.2 \times 10^6 \sim 2.3 \times 10^7$ g/mol 范围内、且摩尔质量分布相对较窄($M_w/M_n \sim 1.3$)的一系列级份。在动力学的研究中,无须达到 $M_w/M_n < 1.0$。下图(a)显示,快过程的特征弛豫时间并不依赖链长,与理论预测吻合,即快弛豫对应的是"成核"过程,仅与链上的局部短链节有关。

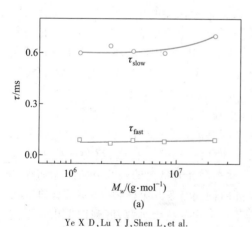

(a)

Ye X D,Lu Y J,Shen L,et al.

Macromolecules,2007,40:4750.

(Figure 2,permission was granted by publisher)

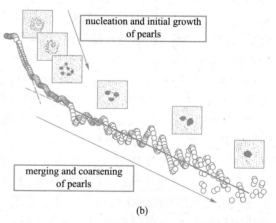

(b)

Ye X D,Lu Y J,Shen L,et al.

Macromolecules,2007,40:4750.

(Figure 4,permission was granted by publisher)

随后,慢过程的特征弛豫时间随着链长缓慢增加。理论上曾预测,其特征弛豫时间仅微弱地依赖链段数,标度关系为 $\tau \sim N^{1/5}$。在这一较慢的过程中,先前形成的大于临界尺寸的"核"会不断地"吞"进与之连接的两边链段、变大。同时,那些小于临界尺寸的"核"则会自动解开(溶解),其组成链段被附近的大核"吞噬"。最终,仅剩下一个单链"核",形成一个单链小球构象。该过程也被称为"粗化",其中,链与溶剂之间的界面逐步减小,故表面能的降低补偿了因聚集而丧失的平动熵和链段间的相互作用能。整体自由能下降,所以该过程自发地进行。

上图(b)形象地描绘了随溶剂性质变差,大分子单链在一相区中从"线团至小球"构象转变中的蜷缩动力学,包括两个过程:"成核"和"粗化"。实验结果显示该"线团至小球"构象转变是典型的一级相变。

小　结

与两种小分子的二元混合物相比,随着溶剂性质变差,溶液中的大分子链在一相区内具有一个独特的构象变化,从无规线团至均匀小球。为了阐明和证实这一小分子物理中没有的新的物理现象,理论和实验学家们自 20 世纪 60 年代起,进行了深入的研究。直至 20 世纪 90 年代中期,方在实验上证实了这一近代大分子物理中的重要命题,即在一相区中,随着溶剂性质变差,大分子链构象可从一个无规线团转变成一个热力学稳定的小球。

在后续的十几年中,揭示了大分子链仅仅皱褶在单链小球中,并无理论上预测的链内互穿和打结;发现了该构象转变过程中,存在着一个全新的"熔化球"构象、被誉为"大分子研究中的一个地标";观察到在反过程("小球至线团")中,存在一个与分子内氢键有关的滞后现象。实验结果证实这一构象变化为一级转变,包含了先后两个过程:与链长无关的链内"成核"和略微依赖于链长的链内"粗化"。实验上,这一近代大分子物理中与知识有关的重要问题已被解决。剩下的仅是如何修饰现存理论,使其可定量地处理大分子链在不良溶剂中的行为,包括定量地引入色散吸引对链内链段体积分数(浓度)的依赖性。

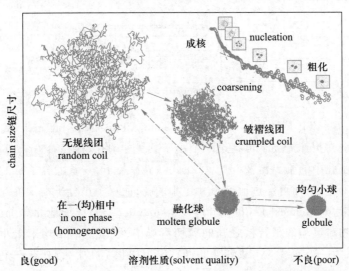

读者应该注意,文献中,"线团至小球"一词不断地被错误地使用,毫无理解。这一近代大分子物理中的重要命题,没有问链构象是否会随溶剂性质变差从无规线团蜷缩成一个小球,而是其可否在一相区中发生,不涉及相变。在一个相变的过程中,大分子的链内蜷缩和链间聚集会同时发生。在相变过程中的构象改变从来不是理论上所讨论的"线团至小球"的构象变化。因此,无论何时谈及和引用"线团至小球"构象转变时,言者必须肯定她/他观察到的发生在均相溶液中,即没有任何链间聚集。希望阅读过本节的详细讨论后,读者在阅读有关文献时,可以去伪存真,不被误导。

本节内容基于笔者和合作者先前的一篇文献综述[Acta Polymerica Sinica,2017,9:1389]。上图形象地总结了有关研究这一问题所得到的主要结论。

从"线团至伸展"的转变

在一个拉伸外力下,一条大分子链的构象也能经历从"线团至伸展"的转变。这是近代大分子物理中另一个重要命题,也始于 20 世纪 60 年代中叶,先由理论物理学家提出,其包含了小分子物理中没有的新物理。这一大分子链构象转变除了具有理论和学术价值以外,还有许多重要的应用。例如,高聚物材料加工中的链的单轴和双轴取向、高聚物纺丝过程中链的牵引拉伸、大分子超滤、蛋白质和核糖核酸链穿越细胞核膜的迁移,甚至涉及春蚕和蜘蛛的吐丝过程。正因如此,这一近代大分子物理研究中的重要命题延伸出了各种相关的重要问题,在过去 50 多年里受到了广泛的关注。为了阐明、证实和解决该命题以及与其相关的问题,许多理论和实验学家投入了大量的精力。

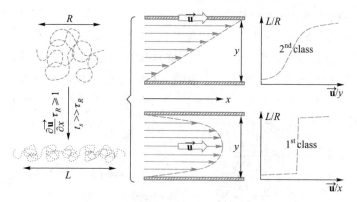

回顾历史,Peterlin[J. Polym. Sci.,Polym. Lett.,1966,4:287]首先指出,一条中性的线型柔性大分子链在拉伸流场下可穿过一个小于其尺寸的孔,存在着一个临界应变速率,其对应着一个在小孔中的临界流量($q_{c,t}$)。在此临界流量下,一条大分子链的构象在流场的强剪切下发生形变,从而可穿过比其尺寸还小的孔,而一个小分子则仅可穿过孔径大于其尺寸的孔。后来,de Gennes[J. Chem. Phys.,1974,60:5030]详细地计算了这一物理现象,并指出在两个相对运动的平行板之间的剪切流场中,线型链的平均末端距(R)随着剪切速率逐渐增大,为一个连续的二级转变。而在一个柱状小孔中,链处在一个沿着小孔纵向的速度梯度拉伸流场中,构象变化是一个一级相变的突变过程,如上图所示。

定性地,无规线团和伸直构象的自由能分别处于相对的最低点,二者之间隔着一个能量较高的亚稳态。当拉伸链的长度短于一个临界尺度(l_c)时,其会自动地回到无规线团状态;只有被拉伸链的长度达到l_c时,链构象将自动地进一步伸展,直至其自由能达到一个相对的最低点。

更有趣的是,de Gennes,以及随后,Pincu[Macromolecules,1976,9:386]计算了线型大分子链穿越小孔的临界流量,发现其与链长和小孔尺寸均无关。后续的理论研究逐步地拓展到具有各种组构的大分子链的超滤。理论进展远远超前于实验研究。既对实验研究提出了挑战,也困惑了实验研究者。这是因为他们经过多年的研究一直没有观察到预测的一级"线团至伸展"的突然转变,以及剪切场强度(临界流量)和大分子链的长度和组构以及小孔的形状和大小之间的定量关系。

实验上,常常错误地认为只要施加足够强的力,就可将一条链拉伸和取向。例如,在将大豆蛋白加工成纤维状结构的行业,这就是一个常见的误区。将宏观的类似肌肉纤维状的结构认作微观上大豆蛋白链拉伸后的取向,而不知施力仅可使一个物体沿着施力的方向加速运动,而不能使之拉伸和取向。日常经验告诉我们,将一根绳子拉直,双手不得不向相反的方向移动,或双手以不同速度向同一个方向移动。十六级超强台风可将一个物体刮向天空、使其在空中旋转和飞奔,但不会将其拉伸成一个线型物体。拉伸一个物体,不同大小的力必须施在其两边或两端,换而言之,其两边或两端必须以不同的速度移动。更精确地说,正是力或速度梯度(单位长度上的力差或速度差)决定了一个物体可否被拉伸和取向。

数学上,力和速度梯度分别记为$\partial F/\partial L$和$\partial u/\partial L$。在拉伸流场中,$\partial F/\partial L$可换算成$\partial u/\partial L$。日常经验还告诉我们,伸直一根卷曲的绳子,不仅需要一个力梯度,还需要力梯度施加一段足够长的时间。可以想象,谁也不能在千分之一秒内用双手将一条卷曲的绳子拉直。物理上,为了拉伸一个物体,不但要有足够大的力或速度梯度,而且该力或速度梯度还必须施加足够长的时间(t_S)。

依据第四章中有关大分子链构象的讨论已知,当一根大分子链被完全拉直后,构象只有一个,构象熵为零。在等温和等压下,一条大分子链总是试图回到其无规线团构象以增加构象熵,朝着自由能减少的方向移动。因此,在撤掉外力之后,一条伸直的大分子链就会迅速地反弹、缩回到其平衡时的无规线团构象。可粗略地估计一条大分子链从一个变形构象反弹(弛豫)回到其平衡构象所需的时间(τ_R)。

由上一章的讨论已知,在大分子稀溶液中链间无缠绕的情况下,如果考虑流体力学相互作用(Zimm模型),该过程对应着一系列依赖于链长的内部简正模式,它们的特征松弛时间如下

$$\tau_n \sim M^{3\alpha_{R_g}}$$

其中,$1/2 \leq \alpha_{R_g} \leq 3/5$。对普通高聚物稀溶液而言,最慢的第一简正模式的特征弛豫时间为$\tau_1 \sim 1$ ms;$\tau_R \approx \tau_1$。上述两个条件在数学上被表达为

$$\left(\frac{\partial u}{\partial L}\right)\tau_R > 1 \quad 和 \quad t_S \gg \tau_R \tag{8.3}$$

如果大分子溶液的浓度较大,且链远长于其缠结长度,即链上的链段数(N)远大于其形成缠结的最小链段数,缠结链段数(N_e),大分子链就会互相穿透、互相缠结;在每一条链占据的空间里,同时还存在许多条其他大分子链。因此,一个变形的大分子构象向其平衡时的平均构象松弛的特征弛豫时间就一定会更慢,具体数值,将在下一章中详细讨论。

在 cgs 系统中,链的尺寸约为 10^{-5} cm;可见链两端的速度差必须达到 10^{-2} cm/s 以上;即两端的速度之差需要达到每秒 10^3 个链的尺寸。可想象链的一端固定不动,另一端必须每秒移动一千个链的尺寸以上! 实验上,不仅要实现这一速度梯度,而且,大分子链在该速度梯度场中还要停留足够长的时间,方可被拉伸成一个雪茄状的构象。以上的讨论和推导基于动态学。

换个角度,也可利用一条线型大分子链伸展时的能量变化讨论同一问题。以无扰的状态下的方均末端距($\langle R_0^2 \rangle = b^2 N$)作为参考点,在良溶剂中的一条扩展链中达到其平衡时的方均末端距为 $\langle R^2 \rangle = b^2 N^{2\alpha_R}$ 或写成 $R = \sqrt{\langle R^2 \rangle} = bN^{\alpha_R}$,其对应的能量仅是一份热能($k_B T$)。如果链构象变成一个直径为 ξ、长度为 L 的雪茄状构象,好似处在一个直径为 d 的"管子"之中。对线型链,$\xi = d$。一条大分子链可被想象地"分割"成(L/ξ)个"链珠",每个"链珠"由一段含有 N_ξ 个链段的链节构成。这样,$\xi^2 \approx b^2 N_\xi^{2\alpha_R}$ 和 $L/\xi = N/N_\xi = (R/\xi)^{1/\alpha_R}$。每一个链段只可在其所属的"链珠"里无规行走。每个"链珠"对应的能量是一份热能 $k_B T$。整条大分子链的能量从 $k_B T$ 增至 $k_B T(L/\xi) = k_B T(R/\xi)^{1/\alpha_R}$。因此,在良溶剂中,一条线型大分子链的自由能因形变而引致的变化为

$$\Delta A = k_B T \left[\left(\frac{R}{\xi} \right)^{\frac{5}{3}} - 1 \right]$$

当 $\xi < R$ 时,一条链的形变导致自由能增加。依据热力学,在等温、等容或者等压的条件下,一个体系不会自动地朝着自由能增加的方向移动,即自动地从无规线团构象变为雪茄状构象的概率很小。依据统计力学,上述链构象转变的概率(ψ)为

$$\psi = \exp\left(-\frac{\Delta A}{k_B T} \right) = \exp\left[-\left(\frac{R}{\xi} \right)^{\frac{5}{3}} + 1 \right] \quad (\xi \leqslant R) \tag{8.4}$$

注意:也可将概率 ψ 看作一条大分子链钻进孔径为 ξ 的柱状小孔的概率,仅和 R/ξ 之比有关。对一条线型链,当 $\xi = R$ 时,概率为 100%;随着 ξ 的递减,ψ 指数地下降;当 $\xi = R/2$ 时,ψ 降至约 1.5%。换而言之,一条线型链自发地、随机地进入孔径等于或大于其平均末端距的柱状小孔的概率为 100%;而当孔径为链尺寸的一半时,该概率则接近零。

一般而言,如果一条链完全进入"管子"中,雪茄状的链的体积为 $d^2 L$。每个"链珠"的体积为 ξ^3;所以,每条链含有($d^2 L/\xi^3$)个"链珠"。链的能量为 $k_B T(d^2 L/\xi^3)$。与进入"管子"之前的能量($k_B T$)相比,形变导致了能量的增加 $\Delta E_c = k_B T [(d^2 L/\xi^3) - 1]$。另外,如果管中雪茄状的链处在一个流量为 q 的拉伸流场中,流速为 $u = q/d^2$。每一个"链珠"的运动阻力为 $f = 3\pi\eta\xi u$;因此,整条链的运动阻力为 $F = f(d^2 L/\xi^3)$。同样以无扰的状态下的链构象(能量为 $k_B T$)作为一个参考点,在拉伸过程中,拉伸流场对大分子链作用,使得其能量增加,

$$\Delta E_h = \int_0^L F\mathrm{d}L - k_{\mathrm{B}}T = \int_0^L \frac{3\pi\eta q}{\xi^2} L\mathrm{d}L - k_{\mathrm{B}}T = \frac{3\pi\eta q}{2\xi^2} L^2 - k_{\mathrm{B}}T$$

因此,一条链在拉伸流场的作用下进入一个直径为 d 的柱状小孔后的总能量变化为

$$\Delta E = \Delta E_h - \Delta E_c = \frac{3\pi\eta q}{2\xi^2} L^2 - \frac{k_{\mathrm{B}}T d^2}{\xi^3} L$$

其对 L 的微分得 $(\partial\Delta E/\partial L)_{T,p} = 3\pi\eta q L/\xi^2 - k_{\mathrm{B}}T\,d^2/\xi^3$。注意:能量随长度的变化率(微分)为力。令 $(\partial\Delta E/\partial L)_{T,p} = 0$,即二力平衡时,可得一个极小值,即

$$L_{\min} = \frac{k_{\mathrm{B}}T d^2}{3\pi\eta q\xi}$$

显然,$L_{\min} = \xi$;物理上,代表着链上仅有第一个(层)"链珠"进入了小孔,即

$$q_{\min} = \frac{k_{\mathrm{B}}T}{3\pi\eta}\left(\frac{d}{\xi}\right)^2 \tag{8.5}$$

值得注意:在推导上式的过程中,并未假定链的组构。所以,对于不同的组构,上述公式是普适的。在其实际的使用中,只要根据具体的链组构算出其相应的 ξ,即可利用上述公式获得一条大分子通过一个柱状小孔的临界流量,以下记为 q_c。一条大分子链能否通过一个柱状小孔完全取决于流量(q)和这一临界流量(q_c)的相对大小。当 $q \geqslant q_c$ 时,通过;反之,挡住。de Gennes 在 20 世纪 70 年代分别推出了星型和支化型大分子链的临界流量,但没有得到一个如同上式的普适公式,而且还犯了一个错误,导致了一个错误的结论。由于一直没有实验上的验证,该错误在很长时间内一直没有得到纠正。

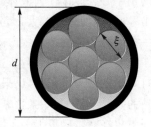

阅读 de Gennes 的综述[Adv. Polym. Sci.,1999,38:92],读者应该注意,在推导支化链的临界流量时,即该文献中的式(30),他错误地假定了一层"链珠"的能量是一份热能 $k_{\mathrm{B}}T$(仅在线型链时成立)。实际上,一层"链珠"的数目为柱状小孔与一个"链珠"的截面积之比;即有 $(d/\xi)^2$ 个"链珠",如右图所示;每个"链珠"的能量为 $k_{\mathrm{B}}T$,故一层"链珠"的能量为 $k_{\mathrm{B}}T$ $(d/\xi)^2$。正是因为此错误,de Gennes 才错误地得到了对支化链的四次方,而不是上式中正确的二次方。

对于线型链,de Gennes 将每个"链珠"当作一个实心小球,$d = \xi$;式(8.5)成为

$$q_{c,\mathrm{linear}} = \frac{k_{\mathrm{B}}T}{3\pi\eta} \tag{8.6}$$

有趣的是,依据上式,线型链的临界流量仅依赖于温度和液体黏度(稀溶液中为溶剂黏度),与链长和孔径毫不相关。直觉上,需要一个更大的流体切变力方可将一条更长的链"挤过"一个给定的小孔。另一方面,对于给定的大分子链,孔越小,所需的流体切变力也应更大。因此,在文献中和申请的研究计划里,有人不时地、错误地提出利用小孔分离不同长度的线型链。可见,直觉有时不对!还是需要理论指导!

定性的解释是,当一条链变成雪茄状后,只要其直径不大于孔径,便可通过小孔,和链长无关。换而言之,如果将一根细的硬棒穿过一个小孔,当硬棒的直径大于孔径,则无法

穿过;反之,可穿过,与棒长无关。依据上式,在 cgs 系统里,估算可得线型链的临界流量 $q_{\min} \approx 10^{-13}$ cm^3/s,极低! 因此,在通常的过滤实验中,流量远大于该临界值,故完全不用担心线型大分子是否会被一个过滤膜挡住。如果使用 200 nm 孔径的膜,临界流速也只有约 10^{-3} cm/s,即约为每秒十个微米。

对于化学、高聚物和其他非物理系的读者而言,用"力"而非"能量"直接推导上述公式也许更易理解。在讨论气体和液体压强时,已知每个"粒子"产生的压强正比于其到达体系中任意一点的概率,正比常数是概率为 100% 时的能量密度。仅有热能时,$p = k_B T/V$,这里 V 是一个粒子可达及的体积。讨论受限一条大分子时,每个"粒子"是链上的一个短链节,其可及体积为其所属的"链珠"体积(ξ^3),$p = k_B T/\xi^3$。当一条大分子链在小孔外时,每个链段的可及体积为 R^3,$p = k_B T/R^3$。注意,$\xi \leq d$。如果 $R \leq d$,大分子链进出小孔自由,如入无人之地;相反,则因孔内压强较大,而留在孔外,不会自发地进入小孔。严格地说,是一个较小的概率,见式(8.4)。

压强乘以一个"链珠"的截面积(ξ^2)就是约束作用力 $f_c = k_B T/\xi$。熟知物理的读者,也可从"能量($k_B T$)除以'链珠'的尺度(ξ)等于作用力"一步得到该结果。另一方面,每个"链珠"在流场中受到的阻力为 $f_h = 3\pi\eta\xi u = 3\pi\eta\xi(q/d^2)$。当作用在每个"链珠"上的二力达到平衡($f_c = f_h$)时,也可得到式(8.5)。殊途同归。更简单? 读者可自行判断。

Frank,Keller and Mackley[Polymer,1971,12:467]首先采用两束撞击射流拉伸线型大分子链。在该实验中,当两束非常靠近、方向相反的大分子溶液高速水平射流互相撞击产生一个与撞击轴线垂直的碟状液面时,流动方向和流速突然改变了 90°。理论上,在撞击的正中心处应无任何液体。随着初始撞击流速增大,液面逐渐地逼近中心。越是靠近撞击中心,速度梯度越大,大分子链在速度梯度场中停留时间越长。因此,当流速足够大时,一条大分子链在撞击中心附近,沿着流场,在纵向被拉伸。

实验上可达到的射流速度是有限的。如前所述,在良溶剂中无缠结时,链构象的松弛时间正比于 $M^{1.8}$。所以,采用长链大分子是一个更实际的做法。他们发现,在当时可以达到的实验条件下,一条聚苯乙烯链被拉伸至断裂的摩尔质量大约是 10^6 g/mol。因此,在制备超高摩尔质量大分子溶液时,无论是搅拌还是过滤,均要避免高速切变场。在射流实验中,如果链的取向造成偏光,则可采用双折射检测法,但对高速流动而言,并非易事。

另一方面,将一个静止的液体推入一个几十纳米大小的柱状小孔,在小孔的入口处造成极大的速度差,产生一个典型的、理想的、流量为 q、流速为 q/d^2 的具有纵向速度梯度的拉伸流场。因为孔外的流速接近零,小孔内的流量(q)实际上代表了一个速度差 $[(q/d^2)-0]$,除以一个"链珠"的长度得速度梯度。然而,在超滤实验中,定量地研究大分子链在柱状小孔中的超滤行为极具挑战,其受到驱动力、大分子链的组构、柱状小孔和过滤膜的结构等诸多因素的影响。在经过了三十多年的极大努力后,直至 21 世纪初,实验学家仍然无法观察到预测的从"线团至拉伸"的一级构象转变,确实令人沮丧。

可以逐项考察影响一条大分子链穿过一个柱状小孔的因素:驱动力主要有拉伸流场,对带电荷的链,也可用高压电场。读者需要注意,通过一个拉伸流场将一条中性的大分子链挤进和推入一个小孔和施加电场将一条带电荷的大分子链拉进一个小孔并非完全一样。在前者中,大量的溶剂分子在小孔入口和内部被迫穿过扩展的大分子链,高度排水,与大分子链在静止的溶液中平动扩散时,几乎无排水的情况完全不同。而在后者中,带电荷的大分子链及其反离子以及其他离子在电场中迁移,大分子链被拉伸和进入小孔。如何制备组构不同、结构明确的大分子样品以及如何获得理想的稳定拉伸流场始终是大分子超滤实验中的两大难题。

直至 2006 年,Jin 和 Wu[Physical Review Letters, 2006, 96:237801]意外地利用一个特殊的双层过滤膜,厚度分别为 1 μm 和 99 μm,分别含有 $d=$ 20 nm 和 200 nm 的柱状小孔,其中一个大孔套着一个小孔,才首次观察到预测的线型大分子链从"线团至伸展"的一级构象变化,如右图中的插图所示。究其成功原因,该双层结构阻止了不同小孔产生的不同拉伸流场之间的互相干扰,确保了拉伸流场中不含有湍流/旋转成分。如果将该双层结构的过滤膜反过来使用,则仅观察到

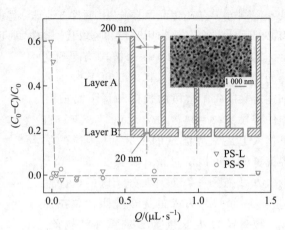

Jin F, Wu C. Acta Polymerica Sinica, 2005, 4:486.
(Figures 2 & 7, permission was granted by publisher)

连续的二级构象变化。另外,通过利用激光光散射和添加一个尺寸小于孔径、总可通过小孔的短链,作为一个内部标定标准,成功地解决了在测量过程中因溶剂挥发造成的浓度变化问题。这样,实现了精确地、定量地检测浓度。即可以精确地、定量地从动态激光光散射测得的长链和短链在线宽分布中对应的峰面积之比测得过滤前后溶液中长链大分子的浓度变化(C_0 和 C)。进而,可算出超滤实验中常用的反映了过滤膜阻挡了大分子链的比例(保留比),$(C_0 - C)/C_0$。有兴趣的读者可查阅原始文献,从而详细地了解二者之间的转换。图中的数据显示,宏观临界流量约为 1.8×10^{-5} cm³/s。膜上小孔数目约为 5×10^8。因此,每个小孔内的平均临界流量约为 10^{-14} mL/s,比理论值小了一个数量级。临界流速为 $u_c = 4 \times 10^{-14}/(2 \times 10^{-6})^2 \approx 10^{-2}$ cm/s。

初始实验结果显示临界流量略微依赖于流量,与理论预测不一致。曾将该不一致归于膜上有几亿个小孔和非均匀孔径。后来,在进一步研究了摩尔质量相差近十倍的大分子链后,实验结果证明理论预测无误,临界流量确实与链长完全无关,如右图所示。为了清楚地比较不同长度的链,特地将横坐标取作对数。尽管临界流量确实不依赖于链长,其值仍然比理论值小了一至两个数量级,取决于溶剂性质。在 Θ 状态时,二者相差最少,也有十倍左右。而

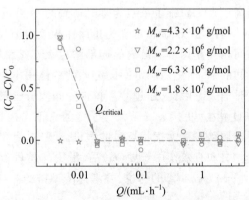

Ge H, Jin F, Li J F, et al. Macromolecules, 2009, 42:4400.
(Figure 1, permission was granted by publisher)

在良溶剂和不良溶剂中,实验和理论值的差别可达 200 倍以上。

检视上述有关从"线团至拉伸"构象转变的理论,不难发现,将每一个受限在孔中的"链珠"处理成一个非"排水"实心小球的假定不妥。在拉伸流场下,链在前行时,大量的溶剂分子被迫穿过大分子链,留在后方,形成强烈的排水效应。"链珠"内部的每个链段均和溶剂分子存在相互摩擦作用,每个"链珠"运动受阻的摩擦阻力远大于 $f = 6\pi\eta\xi$。故有必要引入一个有效作用长度(l_e)取代"链珠"的尺寸 ξ,即 $f = 6\pi\eta l_e$。

如果"链珠"里的每一个链段均受到同样的摩擦阻力,$l_e = bN_\xi$,其中 b 和 N_ξ 分别为前述定义的每个链段的长度和每个"链珠"里的链段数,$\xi = bN_\xi^{\alpha_R}$,故 $l_e = b^{1-1/\alpha_R}\xi^{1/\alpha_R}$。将其代入上述临界流量的公式,从测得的临界流量($q_c$)可以反过来计算最少需要多大的力方可克服一条线型大分子链的熵弹性将其从一个尺寸为 R 的无规线团构象拉伸成一根直径为 d、长度为 L 的雪茄状构象,即 $f = 6\pi\eta b^{1-1/\alpha_R} d^{1/\alpha_R-2} q_c$。对给定的超滤实验,$d$ 为常数;对给定的大分子溶液,η 和 b

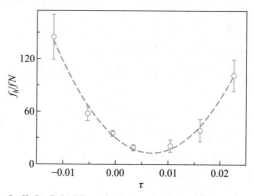

Ge H, Jin F, Li J F, et al. Macromolecules, 2009, 42 : 4400.
(Figure 4, permission was granted by publisher)

为两个常数。上图显示了,随着温度(溶剂性质)变化,拉伸一条在环己烷中的线型聚苯乙烯链所需的最小平均力,其中,τ 为约化温度$(T-\Theta)/\Theta$,代表了溶液偏离 Θ 状态的相对程度。

值得注意:这里克服的是熵弹性,等于热能除以链的尺寸,即 $k_B T/R$,为 $10^{-13} \sim 10^{-14}$ N 比原子力显微镜测量的 $\sim 10^{-11}$ N 至少弱了 100 倍。因此,原子力显微镜测量的不是克服熵弹性将一条大分子链从无规线团拉直的力,而是改变一条伸直和取向的链的化学共价键之间的键角所需的力。现有文献中的卡通图不妥,需要修正。

注意图中的两点:第一,当由排斥体积效应为正(良溶剂)或为负时(不良溶剂),拉伸一条大分子链均需较大的力。图中最小的力对应着一个排斥体积效应消失时的温度。换而言之,在 Θ 状态时,链柔软性最大、最易变形。第二,力的最低点所对应的温度约为 36 ℃,略高于在二阶维里系数为零时得到的 Θ 温度(~ 34.5 ℃)。

理论上,Θ 状态定义为排斥体积效应消失。二阶维里系数源自浓度依赖性,所测的是链间相互作用。而链内的局部浓度大于溶液的平均浓度。因此,链内的色散吸引作用比链间的强,而链内和链间的链段硬核体积本身并无区别,故排除体积效应在一个稍高的温度时消失。因此,通过测量拉伸一条大分子链所需的最小的力,可更好地确定大分子溶液的 Θ 状态、更接近一条大分子链的真正无扰状态。进一步证实了先前利用链的内部简正模式运动测量无扰状态时的假设。现在,有了两个测量大分子链无扰状态的新方法。

引入有效作用长度(l_e)后,沿用前述每个"链珠"的约束作用力 $f_c = k_B T/\xi$ 和摩擦阻力 $f_h = 3\pi\eta l_e (q/d^2)$ 的讨论,并考虑线型链的 $\xi = d$,可以估计临界流量改变了多少,即

$$\frac{q_{c,\text{linear}}}{q_{c,\text{de Gennes}}} = \frac{q_c}{\dfrac{k_B T}{3\pi\eta}} = \frac{d^2}{\xi l_e} = \left(\frac{d}{b}\right)^{1-\frac{1}{\alpha_R}} \tag{8.7}$$

其中，$d \approx 20$ nm；$b \approx 1$ nm。所以，q_c 仅为 $k_B T/(3\pi\eta)$ 的十分之一，与实验值接近。依据上式，临界流量随着孔径的增大递减。标度指数的范围为 $-2/3 \leqslant (1 - 1/\alpha_R) \leqslant -1$。

后来，孔径依赖的实验结果证实了上式预测的临界流量的孔径依赖性。稍后，Freed 和笔者[J. Chem. Phys.，2011，135：144902；Macromolecules，2012，44：9863]利用严格的计算证实了上述结论，详情略去。这一系列实验和理论成功地修改了线型链的"线团至伸展"构象转变的理论和纠正了 de Gennes 当初犯的错误。

当有 f 条臂、每条臂有长度为 b 的 N 个链段的星型大分子链通过一个直径为 d 的柱状小孔时，情况略微复杂。f 条臂将会分成前面和后面两束，假定前面的臂数为 f_f，后面的臂数则为 $f_b = f - f_f$。利用上述"链珠"的描绘，每条前面的臂可以形成 N/N_ξ 个尺寸为 ξ、能量为 $k_B T$ 的"链珠"，$\xi = bN_\xi^{\alpha_R}$，f_f 条前面和 f_b 条后面的臂的总能量为

$$\frac{E}{k_B T} = \frac{Nb^{1/\alpha_R}}{d^{1/\alpha_R}}[f_f^{1/2\alpha_R} + (f - f_f)^{1/2\alpha_R}]$$

将 E 对 f_f 微分，并令 $(\partial E/\partial f_f)_{T,P} = 0$，可得在柱状小孔中一条受限星型链的最小能量，即

$$\frac{Nb^{1/\alpha_R}}{d^{1/\alpha_R}}\left[\frac{1}{2\alpha_R}f_f^{(1/2\alpha_R)-1} - \frac{1}{2\alpha_R}(f - f_f)^{(1/2\alpha_R)-1}\right] = 0 \quad \rightarrow \quad f_f = \frac{f}{2}$$

因此，当一条星型链的臂数一半向前时，其能量最低，最易通过一个柱状小孔。实际情况往往并非如愿。de Gennes 和 Brochard[C. R. Acad. Sci.，Ser. II，1996，323：473]曾分别讨论了 $f_f \leqslant f/2$ 和 $f_f > f/2$ 两种情形，过程较为烦琐。本书提供了一个相当简洁的推导。

一条星型链好似一条八爪鱼，只要 $f_f \neq f/2$，无论是 $f_f \leqslant f/2$，还是 $f_f > f/2$，总是臂的数目较多的一束更难进入柱状小孔，造成更大的阻力。因此，只需寻获该束中"链珠"的尺寸(ξ)与管径(d)之间的关系，然后依据普适的式(8.5)，即可算出一条星型链的临界流量($q_{c,star}$)。如果较多的臂位于前方，$\xi^2 f_f = d^2$；反之，$\xi^2(f - f_f) = d^2$。如果用线型链的临界流量($q_{c,linear}$)归一，可得到 $q_{c,star} = q_{c,linear} f_f$ 或者 $q_{c,star} = q_{c,linear}(f - f_f)$。Li 和合作者[Macromolecules，2012，45：7583]发现不用烦琐地分开讨论，二者可合二为一：

$$q_{c,star} = q_{c,linear}\left(\frac{f}{2} + \left|\frac{f}{2} - f_f\right|\right) \tag{8.8}$$

上式显示，一条星型链通过柱状小孔的临界流量有以下几个特点：第一，临界流量同线型大分子链一样，不依赖于臂(链)长；第二，临界流量与星型链以何种方式进入小孔有关；第三，对于一给定臂长的星型大分子链，临界流量随着臂的数目增加。第一点已被超滤实验证实，如右图所示，其中，将横坐标宏观流量取对数可更清楚地观察临界流量的变化。臂长增加了三倍，但临界流量不变。至于第二点，如果一条星型链以不对称($f_f \neq f/2$)的形式进入小

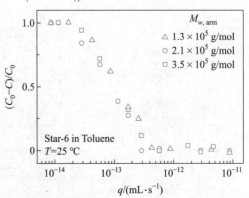

Ge H，Pispas S，Wu C. Polymer Chemistry，2011，2：1071.

（Figure 2，permission was granted by publisher）

孔,其能量较高。不对称和对称方式的能量之差为

$$\Delta E = k_B TN \left(\frac{b}{D}\right)^{1/\alpha_R} \left[\left[\left(\frac{f}{2} + \left|\frac{f}{2} - f_f\right|\right)\right]^{1/2\alpha_R} - \left(\frac{f}{2}\right)^{1/2\alpha_R}\right]$$

因此,星型大分子链以不对称和对称方式进入柱状小孔的概率之比为 $\exp(-\Delta E/k_B T)$。如果溶液处于 Θ 状态 $(\alpha_R = 1/2)$;$N \sim 10^2$;$d/b \sim 10$,对于一个仅有四条臂的星型链,其将一条或三条臂率先伸入柱状小孔的概率降至 $e^{-1} \approx 37\%$;如果臂数增至六,一条或五条臂先伸入柱状小孔的概率更降至 $e^{-2} \approx 14\%$;以此类推,随着臂数的增大,一条星型链以极端非对称的方式穿过一个柱状小孔的概率迅速地减小。

至于第三点,式(8.8)显示,对四条臂的星型链,即使以能量最低的对称方式进入柱状小孔,其临界流量已是线型链的两倍。这提供了一个将流体力学体积相似的线型链和星型链分开的可能性。这一拆分具有重要应用。例如,在制备星型链的"接枝"方法中,先制备窄分布、一端带有反应基团 A 的线型臂,然后让过量的线型臂和一个带有 f 个反应基团 B 的"核"通过 A 和 B 之间的反应形成 f 条臂的星型链。反应完成后,溶液混合物中不可避免地含有大量未反应的线型臂。由第四章中关于星型链回转半径的讨论已知,即使 f 很大时,星型链和其臂的尺寸也只有约 1.7 倍。长期以来,如何分拆它们一直是一个实验难题。

下图显示了在超滤实验后,一个含有线型短链(内标)、三臂星型链和与星型链具有类似流体力学半径的线型链的溶液混合物的流体力学半径分布。先看下左图(a)~(c),除了内标短链峰以外,(a) 当流量很大时,星型链和线型链均通过了柱状小孔,出现在 $20 \sim 30$ nm 的范围处;相反,(c) 当流量很小时,星型链和线型链对应的峰消失,说明它们均无法通过柱状小孔;(b) 当流量处于二者之间时,峰的面积小了大约一半,但无法知道是线型还是星型链被孔挡住了。

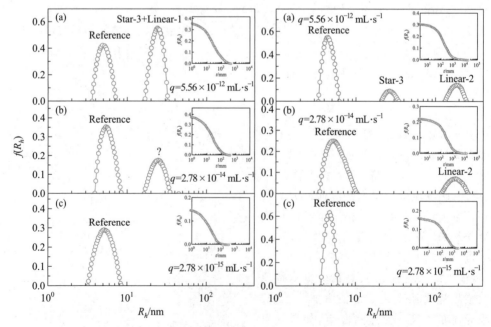

Ge H, Wu C. Macromolecules, 2010, 43: 8711. (Figures 1 and 3, permission was granted by publisher)

上右图(a)~(c)中,线型链换上了很长的线型链($R_h \sim 200$ nm)。(a)当流量很大时,星型链和线型链均可通过小孔,出现了两个峰;相反,(c)当流量很小时,二者均无法通过小孔,对应的双峰消失;(b)当流量介于二者之间时,对应于星型链的峰完全消失,而线型长链对应的峰仍在。正如预测,在此流量下尺寸大的线型长链仍可通过柱状小孔,而尺寸小了近十倍的星型链被挡住了。由此可知,在上左图的(b)中,被挡住的一定也是星型链。这一结果清楚地说明,在超滤实验中,通过选择适当的流量,可以将不同的大分子链按照它们的组构分开,而非像 SEC 那样利用流体力学尺寸进行拆分。

与星型链相比,支化型链的超滤则更为复杂。即使子链中的链段数(N_b)相同,取决于其尺寸与柱状小孔孔径的相对大小,由子链组成的支化母链在一个拉伸流场下,还可被进一步分为可被"强限制"和"弱限制"两类。如果子链的尺寸构象远远大于"链珠"的尺寸,$bN_b^\alpha \gg \xi$,即每个"链珠"里仅含有一个线型链节,故支化母链可被压缩变形,归于"强限制";反之,每个"链珠"里含有一条小的支化链,包括了若干条子链,支化母链就不易被压缩和变形,属于"弱限制"。

Gay,de Gennes,Raphael 和 Brochard[Macromolecules,1996,29:8379]曾经推导出,在良溶剂中,支化链的临界流量($q_{c,\text{branch}}$)与子链和母链中的链段数(N_b 和 N)之间的标度关系为 $q_{c,\text{branch}} \sim N_b^\varphi N^\gamma$。对"强限制"和"弱限制"的支化链,分别有 $\varphi = -1/2$ 和 $\gamma = 1/2$;以及 $\varphi = 2/15$ 和 $\gamma = 2/3$。如前所述,他们错误地将一层(d^2/ξ^2)"链珠"的能量取为一份热能。实际上,每个"链珠"具有一份热能,故他们导出的能量小了(d^2/ξ^2)倍。

正确寻获式(8.5)中受限支化链的一个"链珠"尺寸(ξ)的方法如下:假定每个"链珠"含有 N_ξ 个长度为 b 的链段,以及受限支化链在等于和大于 ξ 的尺寸上的链段密度均匀,那么一个"链珠"的链段密度等于整条具有雪茄状的受限链的链段密度,即

$$\frac{N_\xi}{\xi^3} = \frac{N}{d^2 L_0} \quad \text{或写成} \quad \xi = \left(\frac{N_\xi d^2 L_0}{N}\right)^{1/3}$$

其中,L_0 为一条伸展的链在平衡时的长度。依据第四章的讨论,一条构象变形的链的自由能包含两个部分,一个与二体相互作用有关(也是熵变);另一个和熵弹性有关的能量变化。利用式(4.4),对于变形支化链,链体积(R^3)被 $d^2 L$ 取代;扩展长度的平方(R^2)被 L^2 取代。另外,$R_0^2 \cong b^2 N_b^{1/2} N^{1/2}$。受限和变形的支化链的自由能($A$)为

$$\frac{A}{k_B T} = \frac{b^3 N^2}{d^2 L} + \frac{L^2 - R_0^2}{R_0^2}$$

将自由能对长度(L)微分,再让 $dA/dL = 0$,可得自由能最低时的平衡长度(L_0),

$$\frac{\partial A/\partial L}{k_B T} = -\frac{b^3 N^2}{d^2 L_0^2} + \frac{2 L_0}{R_0^2} = 0 \quad \rightarrow \quad L_0 \approx b\left(\frac{N R_0}{d}\right)^{2/3}$$

对于强受限和弱受限的支化链,ξ 分别由线型链节和支化链节组成。在第四章中,也曾经指出,一条线型链可看作一条特殊的支化链,$N_b = N$,故 $R_0^2 \cong b^2 N_b^{1/2} N^{1/2}$ 也同样适合线型链($R_0^2 \cong b^2 N$)。因此,在良溶剂中,也可用同一个标度关系表达线型链节和支化链节的平均末端距(R)对 N_b 和 N 的依赖,即 $R = \sqrt{R^2} \cong b N_b^\beta N^{\alpha_R}$。$\xi$ 也可被类似地表达成 $\xi \cong b N_b^\beta N^{\alpha_R}$。将上

述各个关系式代入式(8.5),可得

$$q_{c,\text{branch}} = q_{c,\text{linear}} \left(\frac{b}{d} \right)^{\psi} N_b^{\varphi} N^{\gamma} \tag{8.9}$$

其中,

$$\psi = \frac{2(3 - 5\alpha_R)}{3(3\alpha_R - 1)}; \quad \varphi = \frac{6\beta - \alpha_R}{3(3\alpha_R - 1)}; \quad \gamma = \frac{\alpha_R}{3(3\alpha_R - 1)}$$

上式概括了不同的溶剂性质,以及强受限和弱受限支化链的两种情况,其中,分别有 $\beta = 0$ 和 $\alpha_R = 3/5$;以及 $\beta = 1/10$ 和 $\alpha_R = 1/2$。将它们分别代入上式可得

$$\text{强受限}: \frac{q_{c,\text{branch}}}{q_{c,\text{linear}}} = \left(\frac{N}{N_b} \right)^{\frac{1}{4}};$$

$$\text{弱受限}: \frac{q_{c,\text{branch}}}{q_{c,\text{linear}}} = \left(\frac{b}{d} \right)^{2/3} N_b^{1/15} N^{1/3}$$

注意: N/N_b 代表了支化点的数目。对给定的超滤实验条件和大分子,d 和 b 为两个常数,若令上述两式相等,可得到从"强受限"向"弱受限"的条件如下,

$$N_b^* = \left(\frac{d}{b} \right)^{40/19} N^{-5/19}$$

一般而言,$d/b \sim 10$;$N \sim 10^3$。因此,$N_b^* \sim 20$,即约 50 个支化点。对给定的支化链组成,也求出从"强受限"向"弱受限"转变的孔径大小如下:

$$d^* = b N_b^{19/40} N^{1/8}$$

通常,$b \sim 1$ nm;$N_b \sim 10$;因此,$d^* \sim 7$ nm,约为链段长度的十倍左右。实验上可以实现。

上述理论推导显示,在"强受限"情形下,支化链的临界流量随着支化点的数目缓慢地增加。当支化点数目为 10 时,其已接近线型链的两倍。在"弱受限"情形下,支化型链的临界流量随着支化子链的链段数的增加略微地增加,但随着支化链的链段数增加而增加。物理上,上述两式合理。对给定的链段总数,支化链的临界流速,在一种情况下,随支

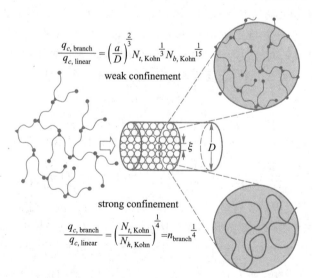

化子链长度的增加而减少(支化点减少,更易压缩)。在另一种情况下,则对支化子链的长度不敏感(已高度支化,不易压缩)。上图形象地总结了超滤中一条支化链"强受限"和"弱受限"的两种不同情况。

实验难度相当大!所以,在早期理论问世以后的三十多年时间里,鲜有可靠的、可检验理论的实验结果。实验上,需要两套样品方可研究临界流量对支化子链长度和支化链的链段总数的依赖性:一套具有类似的链段总数,但是支化子链长度不同;另一套具有相同的支化子链长度,但是链段总数不同。另外,至少需要两个不同孔径的过滤膜,以便研究临界流量对孔径的依赖性。如前所述,在发展了一种"跷跷板型"的大单体以后,利用点击化学反应,将大单体连在一起,形成了所需的两套支化链样品。约在十年前,利用这两套支化链的超滤实验得到了下列方程。

$$q_{c,\text{branch}} \propto d^{-4/5} N_b^{-2/5} N^{1.0\pm0.1}$$

其中,d 的指数与理论值($-2/3$)相差不远,可接受;N_b 的指数仅来源于三个不同子链长度的样品,误差较大,偏离理论值($-1/4$)也不太远。然而,N 的指数却远离理论值($1/4$ 或 $1/3$)。即使将实验测得的 $\alpha_R = 0.46$ 代入 γ 的计算,也仅得到 $2/5$,仍距实验值 1.0 ± 0.1 甚远。为何预测值和实验值之间有如此大的偏差?让我们回去考察早期理论中的假定。

物理上,支化链被压缩在柱状小管中,其构象已变形和扭曲。所以,不可再用原始理论中的假定,即链的尺寸已不再遵循其在溶液中自由时的标度关系。极限是将一条支化链压缩成一个密度均匀的小球,$\alpha_R = 1/3$,$\gamma \to \infty$;显然,也不符合实验结果。因此,可将实验测得的 $\gamma = 1$ 代入理论公式,得到 $\alpha_R = 3/8 = 0.375$,大于 $1/3$,但远小于实验上得到的 0.46,则说明支化链确实被压缩得更加密实,但仍然不是一个密度均匀的小球。

正是基于在实验上确立了不同组构的大分子链的临界流量和链的分子参数之间的一系列定量关系,方有可能利用超滤将不同的大分子链按照组构分开,解决其他实验方法(如尺寸排除色谱)无法解决的问题。例如,在大分子研究中,对如何分离具有类似流体力学体积、不同结构的支化链,一直束手无策。在充分理解了支化链的临界流量在超滤中如何依赖于各种分子性质以后,Li 在笔者的实验室[Macromolecules,2012,45:7583]成功地用超滤将具有相似链段总数、不同支化子链长度的支化大分子分开,终于解决这一实验难题,详情可见相关文献。

在生命科学里,在真核细胞中,蛋白质和核糖核酸链通过核膜上众多核孔的迁移就是一个典型的和大分子链如何穿过小孔相关的有趣问题。在细胞活跃时,每秒钟可有 $\sim10^3$ 条生物大分子链通过核孔。这带来两个问题。第一,为何细胞内的穿过核膜的大分子链全是线型的,特别是 RNA?意外还是自然选择?这仅是因为遗传的需要还是有其他原因?第二,因为生物大分子以如此高的频率通过核孔进出核膜,它们会在核孔里面互相碰撞吗?如是,效率就会很低。因此,它们可能不会在核孔内碰撞,为何?除了生物上已知的依靠生物能量的主动运输机理以外,是否还存在仅由热能驱动的被动运输机理?

有关多孔薄壳中空小球的研究显示,在热能的扰动下,一个充满液体、溶胀的中空小球会出现类似"呼吸"的振动,如右图所示。为了显示这一振动,右图夸张了振动幅度,微小的热能不足以改变整个中空小球的尺寸。其振动频率(f)正比于多孔薄壳的弹性模量,反比于介质黏度和小球的半径(r),$f = m/r$,其中,m为一个含有弹性模量和介质黏度的常数。前述有关微凝胶内部涨落的

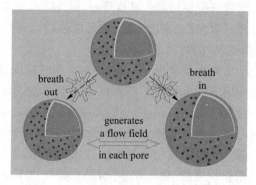

讨论已揭示,热能可以激发凝胶网络中大约 50 nm 的一个内部结构,故热能可激发多孔薄壳的局部振动,好似一个表面波动。

与无孔的密闭中空小球不同,这一波动定会造成中空小球的体积变化。体积变化为 $V = 4\pi r^2 \Delta r$。假定单位表面积上的小孔数目为 σ,小孔总数(n)为 $\sigma 4\pi r^2$。由"呼吸"振动引致的通过每个小孔的流量可由下式求出。

$$q = \frac{\Delta V}{n} f = \frac{m}{\sigma} \frac{\Delta r}{r}$$

显然,"呼吸"引致的通过每个小孔的流量(q)正比于中空小球的相对尺寸变化,而和小球本身的尺寸无关。由于不同细胞的细胞核大小不同,这点对后续讨论至关重要。将典型核膜的 m 和 σ 代入上式,即使假定 $\Delta r / r \sim 1\%$,q 仍然远大于 $q_{c,\text{linear}} \sim 10^{-14}$ mL/s。因此,在膜孔附近的生物大分子链就可被动地被吸入或吐出细胞核。

前述讨论显示,只有线型大分子链的临界流速不依赖于链长。因此,这里的线型组构就极为重要。如果细胞内的生物大分子采用了其他非线型组构,那么它们进出核膜的阻碍不仅较大,而且还会因其大小和组构而变,这不符合进化优势。另外,在每一个瞬间,每一个核孔只能处于"呼"或"吸"的两种状态之一,故这一被动的"吸入"或"吐出"机理还自然地避免了生物大分子链在每个核孔里的碰撞。

这带来一个新的问题:如果一条生物大分子链通过是核孔的时间(t)长于"呼吸"频率的倒数($1/f$),其就会在长度为 l、孔径为 r_p 的膜孔内被反复地吸进和呼出,无法通过核孔。利用流速 $u = q/(\pi r_p^p)$ 可粗略估算 t,即

$$t = \frac{l}{u} = \frac{l\pi r_p^2}{q} = \frac{l\pi r_p^2 n}{\Delta V} \frac{1}{f} = \frac{l\pi r_p^2 n}{4 r^2 \Delta r} \frac{1}{f}$$

在实际的细胞中,$l \sim \Delta r$,$r_p/r \sim 10^{-3}$,以及 $n \sim 10^3$,所以,$t \sim 10^{-3}/f$;可见在"呼出"和"吸进"的过程中,生物大分子链经过核孔的时间远远短于每个核孔"呼吸"一次的时间。因此,完全不用担心一条生物大分子链是否在核孔中振荡、反复来回的问题。

小 结

一个剪切或拉伸流场可改变一条大分子的链构象,驱动链偏离其能量最低的平衡状态。一条偏离平衡状态的大分子链总要试图以其特征弛豫时间($10^{-4} \sim 10^{-3}$ s)回到其平衡

状态。只有同时满足下列两个条件,一条大分子链方才可能被拉伸和形变,直至在流场中达到一个新的平衡构象。第一,剪切速度的梯度与其特征弛豫时间的乘积大于1,意味着形变一个链尺寸所需的时间短于链的特征弛豫时间。换而言之,对一个给定的尺寸(长度),形变快于弛豫。定量而言,形变速度应该达到每秒 $10^3 \sim 10^4$ 个链尺寸。第二,链与切变流场的作用时间长于其特征弛豫时间,即链需要在极快的剪切速度下停留一个足够长的时间。

这一问题的理论研究源于 20 世纪 60 年代中叶。在 70 年代,de Gennes 和他的合作者是理论研究的主要推动者。他们首先发现在一个横向的剪切流场中,一条线型链从"线团至拉伸"构象转变是一个随着剪切速度梯度增加而逐步变化的二级过程;而在一个纵向的拉伸流场中,该过程则为一级突变,称为从"线团至拉伸"构象转变,即存在一个临界流量。其决定了一个流场可否拉伸一条大分子链。随后,他们还在理论上研究了星型和支化型链。然而,经过三十几年的研究,实验上,随着流量增大,仅测得链构象的渐变,没有观察到预测的一级构象突变。

直至 2006 年,通过采用一个特殊结构的双层孔膜和一台小角激光光散射装置,方才在超滤实验中证明这个理论预测的一级"线团至拉伸"构象转变。目前,在大量超滤实验的基础上,已经得到了线型、星型和不同支化型大分子链的临界流量。实验结果修正了de Gennes 早期的一些理论,主要包括:

线型链临界流量的实验值比 de Gennes 预测的小了一到两个数量级,揭示了拉伸流场迫使溶剂分子穿过每个"链珠",造成强烈的"排水"效应,故不可将每个由短链节组成的"链珠"当作一个非"排水"的实心小球。必须用一个有效的流体力学作用长度(l_e)代替一个"链珠"的尺寸(ξ)。因此,$q_{c,\text{linear}} = k_B T/(3\pi\eta)(d/l_e)$;$d/l_e = (d/b)^{1-1/\alpha_R}$,而非 de Gennes 预测的 $k_B T/(3\pi\eta)$,其中,b 是链段长度,$1/2 \leqslant \alpha_R \leqslant 3/5$。该修正已被第一性原理的计算证明。虽然线型大分子链的临界流量虽如理论预计的与链长无关,但却随着孔径增大而减小,并非先前预测的与柱状小孔的孔径无关。

对星型链,两个传统的临界流量与前行的臂数(f_f)和总臂数(f)之间的关系统一成一个数学表达:$q_{c,\text{star}} = q_{c,\text{linear}}(f + |f - f_f|)/2$,与臂长无关,但与臂数成正比。

一条支化链在小孔内的受限状态下,截面上有 d^2/ξ^2 个"链珠",每个"链珠"的能量为$k_B T$,一层"链珠"的能量为 $(d^2/\xi^2)k_B T$,而非 de Gennes 错误地假定的 $k_B T$(将支化链处理成线型链)。临界流量与 (d/ξ) 之间正确的标度指数为 2,而非 4。这一错误得到纠正。

将线型链当作只有一个支化点的特殊支化链,统一地描绘了"强受限"和"弱受限"支化链的临界流量分别与子链和母链的链段数(N_b 和 N)之间的双重标度关系:$q_{c,\text{branch}} = q_{c,\text{linear}}(b/d)^{\psi} N_b^{\varphi} N^{\gamma}$,其中三个标度指数都与溶剂性质($\alpha_R$)有关,$\phi$ 与子链的长度有关。

一条"弱受限"支化链在柱状小孔中受压时,其尺寸与链段数之间的标度指数小于其在溶液中自由时的标度指数,从自由时的 0.46(实验值)和 1/2(理论值)降至 0.38,显示受限支化链更加密实,但仍非密度均匀的小球。因此,支化链的超滤理论仍需完善。

成功地推出了一个临界流量的普适公式:$q_c = q_{c,\text{linear}}(d/\xi)^2$。对不同组构的大分子链,仅需算出 ξ 与 d 之间的关系。线型链:$(d/\xi)^2 = 1$;星型链:$(d/\xi)^2 = (f + |f - f_f|)/2$;不同

的支化链：$(d/\xi)^2 = (b/d)^\psi N_b^\varphi N^\gamma$。

　　利用不同组构的大分子链的临界流量与分子结构参数之间的定量关系,以及选择一个恰到好处的流量,可以利用超滤将一个溶液混合物中流体力学尺寸相近,但组构不同的大分子链分开。从而解决其他分离方法束手无策的问题。下图总结了不同组构的大分子链如何在一个拉伸流场中"爬过"一个孔径为 d、远长于链长的柱状小孔。

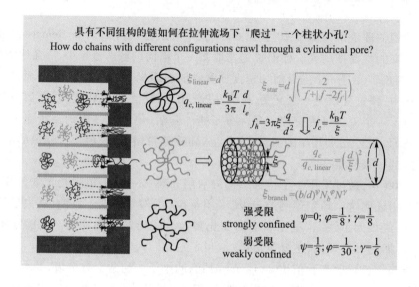

Chapter 8. Transitions of Chain Conformation in Macromolecular Solutions

According to the discussion in Chapter 4, a linear flexible macromolecule chain without special interaction in the solution reaches an expanded, equilibrium random coil conformation with a certain size due to the two entropy-driven, opposite, chain-size dependences. As the solvent quality gradually becomes poorer, the excluded volume effect (including the hardcore volume and the dispersive attraction) gradually weaken, causing the chain shrinking. When the dispersive attraction induced negative excluded volume effect exactly cancels the effect generated by the hardcore volume of the chain itself, a real macromolecular chain has some similar overall properties as an ideal chain with no size and no interaction, called a "quasi-ideal state". In such a state, the scaling exponent between the average radius of gyration (R_g) and the molar mass (length) of the chain is 1/2. Note: **The hardcore volume of a real chain never disappears, let alone negative.** It is the dispersive attraction induced "**negative excluded volume effect**" that makes the excluded volume "**equivalently**" disappears in some cases (e.g., the chain length dependence of the radius of gyration), but not really disappears.

In contrast, small molecules have only two states of soluble and insoluble in a binary mixture (solution), and there is no change in the chain conformation with the solvent quality. Therefore, the solvent quality and chain length dependences of macromolecular conformation leads to new physics: physics specific to macromolecules, macromolecular physics. With no external force, if the quasi-ideal state (also known as the undisturbed state) is taken as a reference state (point), if the solvent quality gradually become better, the excluded volume effect gradually strengthens, so that the macromolecular chain expands to a maximum average radius of gyration (R_g), and the scaling exponent between the size and the molar mass of the chain is 3/5. On the contrary, if the solvent quality gradually deteriorates, the excluded volume effect becomes more negative, and the attraction between the segments gradually enhances, so that a macromolecular chain should gradually collapse into a uniform single-chain globule, and the scaling exponent should decrease to 1/3. On the other hand, in the presence of an external force, it is not difficult to imagine that a sufficiently strong stretching force could overcome the entropy elasticity, making a macromolecular chain with a random coil conformation to be extended and stretched. For a completely stretched chain, the scaling exponent is 1 obviously.

The two above-mentioned "coil-to-globule" and "coil-to-stretch" macromolecular conformational transitions have been two mostly important propositions in macromolecular physics since the 1960s. Their theoretical study began in the early 1960s and 1970s, respectively. Physically, the questions are as follows. 1) Will these two changes occur in a homogeneous solution? 2) If so, are they phase transitions? 3) If yes, are these transitions the first or second-order? 4) How are they quantitatively and universally dependent on the chain length? Experimental research only began in the late 1970s. Although many laboratories had tried by using different experimental methods and macromolecular solutions, no one was able to confirm the theoretical predictions. It was not until the mid-1990s and the beginning of this century that these two important conformational transitions had been finally confirmed experimentally, and a serie of questions related to the two propositions have been answered. The details are as follows.

The "coil-to-globule" Transition

In the early 1960s, based on the Flory mean field theory, Stockmayer[Makromol. Chem., 1960, 35: 54] proposed that as the solvent quality gradually deteriorates, the inter-segment attraction gradually increases, so that the conformation of a linear flexible macromolecular chain changes from an expanded random coil to a uniform collapsed single-chain globule. He also stated that due to the low solubility of these collapsed single-chain globules, it would not be easy to observe the thermodynamically stable collapsed single-chain globular state in the one-phase region.

Ptitsyn and his collaborators[J. Polym. Sci. Part C, 1968, 16: 3509] also used the Flory model, and calculated the interaction energy between the segments on a macromolecular chain. Without considering the connectivity of the chain, they obtained a relationship between the expansion factor $\alpha[= R_g(T)/R_g(\Theta)]$ and the interaction parameter, where $R_g(T)$ and $R_g(\Theta)$ are the root mean square radius of gyration at temperatures T and Θ, respectively. They also used the Monte Carlo method to simulate and calculate the "coil-to-globule" conformational transition, and found that if the attraction between the segments is dominant, this conformational transition occurs in a very narrow interval.

If this conformational transition occurs in the phase transition of forming two phases, the intra-chain contraction and inter-chain aggregation will occur almost simultaneously. It will be rather difficult, if not impossible, to separate these two completely different processes, and only observe the conformational transition of individual chains experimentally. Later, Sanche [Macromolecules, 1979, 12: 980] also used the mean field theory to calculate the chain size by approximately considering various high-order interactions between the segments, and found that when a macromolecule is respectively in the two chain conformations of globule and expanded random coil, there exist two obvious scaling relationship between $R_g(T)$ and the number of Kuhn segments $N: R_g(T) \propto N^{1/3}$ and $R_g(T) \propto N^{3/5}$. The expansion factor is solvable from the following equation.

$$\frac{7(1 - \alpha^2)}{3N} = \left(\frac{\Theta}{T}\right)\frac{\phi}{2} + \frac{\ln(1 - \phi)}{\phi} + 1 \qquad (8.1)$$

where $\phi_0 = \phi\alpha^3$, ϕ_0 and ϕ are the volume fractions in the space occupied by a real and an ideal chain with $R_g(T)$ and $R_g(\Theta)$, respectively, both of which are less than 1. In a good solvent, $\alpha > 1$, so that $\phi_0 > \phi$. When a chain is sufficiently long $(N \rightarrow \infty)$, $\phi = (19/27)^{1/2} N^{-1/2}$. For a swollen chain, $\phi \ll 1$, the above equation can be rewritten as follows by expanding $\ln(1 - \phi)$

$$\alpha^6(1 - \alpha^2) + 0.102 + \cdots = 0.180\alpha^3 \tau N^{1/2} \qquad (8.2)$$

where $\tau = (T - \Theta)/\Theta$, the reduced temperature. The above equation shows the relationship between the expansion factor and the reduction temperature for a given chain length. When the solution approaches the quasi-ideal state from a good solvent, $\alpha \rightarrow 1$, and $\tau N^{1/2} \rightarrow 0.102/0.180$.

Therefore, when the chain length approaches the infinity, i. e., $N \rightarrow \infty$, the reduced temperature must approach zero. Theoretically, the "coil-to-globule" transition should be a sharp transition similar to the first-order phase transition. For a given finite chain length, the above equation shows a continuous transition, similar to the second-order phase transition. In the discussion of the enthalpy change of mixing in Chapter 3, the interaction contacting number (Z) is included in a parameter χ_{sp} that represents the overall dispersion effect. In deriving the above equation, Sanche assumed that χ_{sp} only changes with temperature, ignoring that Z may also change with the volume fraction of a macromolecular chain; namely, the contacting number increases with the segment concentration. Besides the above-mentioned researchers, other theoreticians also contributed a lot in the "coil-to-globule" transition theory, especially the later Professor Lifshits in the Department of Physics, The Moscow State University and his many students, including Khokhlov and Grosberg. However, their papers involve theoretical physics with some difficult mathematics, so that they are omitted in this book.

Physically, it is not difficult to imagine that as the chain shrinks, its volume fraction increases, and the distance between the segments is closer. Therefore, there will be more interaction contacting points (opportunities), i. e., Z can increase with ϕ. In addition, as the volume fraction increases, one has to consider the multi-body interactions. Kholodenko and Freed[J. Phys. A: Math. Gen., 1984, 17:2703] used the field theory to study the relationship between the order of the transition and the multibody interaction, and found that it depends on the strength of the three-body interaction. In the late 1980s, based on the experimental results, Renyuan Qian and his collaborators[Macromolecules, 1993, 26:2950; Macromol. Chem. Rapid Commun., 1993, 14:747] also begun to qualitatively study the problems related to single-chain particles, including the density and the crystallization behavior inside the particles made of single and oligo polymer chains.

However, since the end of the 1970s, experimentalists had only observed the limited shrinking of a random coil chain after many experiments, and not been able to obtain thermodynamically stable single-chain globules. Namely, as the solvent quality gradually deteriorates, the interchain aggregation always appears before an expanded random coil collapses into a thermodynamically stable single-chain globule, i.e., the phase transition, and the solution separates from one phase to two phases, leading to experimental failures. Tanaka and his collaborators [Nature, 1979, 281 : 208; Phys. Rev. Lett., 1980, 44 : 796] are the first to publish an experimental result on this transition. However, his observed single-chain globules were located in the two-phase region, not thermodynamically stable. Subsequently, Chu and his collaborators [Macromolecules, 1987, 20 : 2833; 1988, 21 : 1178; 1995, 28 : 180; 1996, 29 : 1824] carried out a decade-long research, and did not observe the thermodynamically stable single-chain globules.

In view of the repeated experimental failures, Grosberg and Kuznetsov [Macromolecules, 1993, 26 : 4249] even wrote in 1993 that it is difficult to observe the thermodynamically stable single chain globule using the current sample preparation methods and equipment. They pulled back and asked 1) whether the single-chain globule could be formed and studied before the precipitation occurred in an experiment; 2) if yes, whether the chain inside a single-chain globule has a simple crumpled coil or a complicated intrachain entanglement and knotted conformation in a non-equilibrium state; and 3) whether the contracting and aggregating kinetics can be measured in the non-equilibrium two-phase region.

In nature, many protein chains are able to fold into a stable globular conformation in cells easily, but there is no explanation why it is so difficult to fold a polymer chain into a collapsed globular conformation. The "coil-to-globule" transition was mainly studied by physicists or polymer physicists. In order to experimentally study such a transition, the solution has to be as dilute as possible so that the solvent quality can be sufficiently adjusted from good to poor **in the one phase region**. However, in order to have a sufficient intensity of scattered light in such a dilute solution, a ultra-long chain has to be used. On the other hand, the phase diagram of macromolecules is a function of the chain length as discussed in Chapter 3, namely, long chains undergo the phase transition before short ones. So that the chains used have to be narrowly distributed in molar mass (chain length).

For physicists with no skill in synthetic chemistry, commercially available narrowly distributed polystyrene standards with a ultrahigh molar mass are their only choice. Polystyrene is only soluble in organic solvents, in which the intersegment interaction is the dispersion attraction, far weaker than the hydrophobic and hydrogen bonding in aqueous solutions of proteins. Therefore, water soluble polymers can be used to increase the strength of the intersegment interaction. Thermally sensitive poly(N-isopropyl acrylamide) (PNIPAM) with a lower critical solution temperature is a good choice. However, the preparation of narrowly distributed ultralong water-soluble polymer chains is another extremely difficult problem. Kubota and his coworkers [J. Phys. Chem., 1990, 94 : 5154] tried to study the conformation transition of PNIPAN in aqueous solutions by changing the solution temperature. Only a limited shrinking of individual chains was observed before the interchain aggregation because they used a PNIPAM sample with a relatively broad molar mass distribution ($M_w/M_n > 1.3$)

With a defined direction, the preparation of ultra-high molar mass and narrowly distributed water soluble PNIPAM became the first obstacle that has to be overcome. Before the polymerization, N-isopropyl acrylamide monomer was recrystallized three times in a mixed solution of toluene and n-hexane to improve its purity; at the same time, the solvent benzene used in polymerization was sufficiently dried to reduce the chain transfer and negative reaction. After the reaction, the dried crude product was dissolved in anhydrous acetone, and then slowly dropped into and precipitate in non-solvent anhydrous n-hexane to remove unreacted monomers and soluble impurities. The PNIPAM sample obtained after the filtration and drying is fibrous. After further fractionating such a purified sample multi-times in a solution mixture of anhydrous acetone and anhydrous n-hexane, one fraction with a high molar mass ($M_w \sim 6 \times 10^6$ g/mol) and a narrow distribution width ($M_w/M_n \sim 1.3$) was obtained, still far away from the requirement of the ultra-high molar mass and extremely narrow distribution.

After failed in many repeated fractionations, Wu and Zhou[Macromolecules, 1995, 28 : 5388; 1995, 28 : 8381] finally realized that it is not necessary to obtain a macroscopic amount of the required PNIPAM solid sample. Instead, only a few micrograms of PNIPAM with a ultra-high molar mass and an extremely narrow distribution in a dilute solution is required in the study of the "coil-to-globule" conformational transition. After breaking through the traditional thinking, an aqueous solution of the previously obtained PNIPAM sample with a concentration of $\sim 10^{-4}$ g/mL was prepared first, and then, one or two drops of this PNIPAM aqueous solution was filtered into a LLS cell containing ~ 1 mL of dust-free water. The process was repeated until a dust-free solution of PNIPAM with ultra-long PNIPAM chains ($M_w \sim 10^7$ g/mol) and a narrower distribution ($M_w/M_n < 1.10$) was obtained. The final PNIPAM concentration is 10^{-6} g/mL.

After the problem of sample preparation is solved, there is still a second experimental obstacle. According to the discussion of LLS in Chapter 5, the experimental observation length, that is, the reciprocal of the scattering vector (q), should be longer than the average radius of gyration (R_g) of the measured chain, i.e., $qR_g < 1$. Under this condition, by using static LLS, the angular and concentration dependences of the time-average intensity of scattered light were measured, from which the weight average molar mass, the mean square radius of gyration and the second virial coefficient are obtainable. In order to measure the ultra-long PNIPAM chains, a commercial LLS instrument has to be modified, including using a solid-state laser with a thinner beam, reducing the stray light, and carefully aligning the incident and detecting optical paths. The lowest measurable angle was decreases from the usual 15° to 6°, so that $qR_g < 1$ is satisfied.

The figure (a) below comprehensively shows the static and dynamic LLS results of individual PNIPAM chains in the "coil" and "globule" conformations. As the temperature increases, hydrodynamic radius distribution is shifted to the left, a smaller size, and the reciprocal of the Rayleigh ratio becomes significantly less dependent on the scattering vector (a smaller slope), but the intercept remains, indicating that there is no interchain association and the chain shrinking is a single chain process. It also shows that individual PNIPAM chains have shrunk about eight times in size, i.e., a volume of change of about 500 times, which is the first experimental confirmation of the "coil-to-globule" transition.

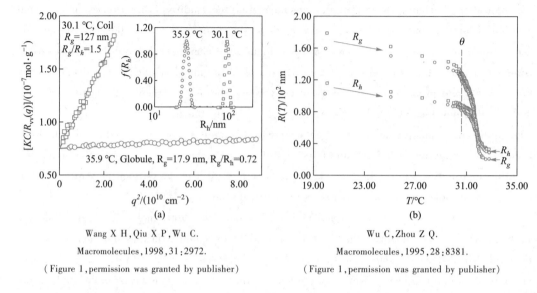

Wang X H, Qiu X P, Wu C.
Macromolecules, 1998, 31:2972.
(Figure 1, permission was granted by publisher)

Wu C, Zhou Z Q.
Macromolecules, 1995, 28:8381.
(Figure 1, permission was granted by publisher)

(a)

(b)

In literature, researchers often "forgot" or did not know that a study of the "coil-to-globule" transition should be conducted in the one phase region; namely, they have to prove that there is no interchain association during the transition. It is insufficient to relay only on a change of the average hydrodynamic radius measured by dynamic LLS since the average hydrodynamic radius is not sensitive to a small amount of interchain aggregates. The best way to check whether there is any interchain association is the constant intercept, as shown in the figure (a) above. As discussed several time before, "the intensity of scattered light is proportional to the square of the molar mass of a scattering particle (macromolecule)", so that it is extremely sensitive to any chain aggregation. Even if only dimers are formed, the intensity of scattered light will double.

The figure (b) above summarizes the temperature dependence of the chain size of two PNIPAM samples with different molar masses in aqueous solutions. As temperature increases, the average radius of gyration changes from larger to smaller than the average hydrodynamic radius. Such a change stems from their different physical definitions. The average radius of gyration is related to the distribution of the spatial distribution of the segments, while the average hydrodynamic radius refers to the radius of a uniform sphere with the same translational diffusion coefficient as the chain in solution. In the good solvent, $T < \Theta$, both of the average radius of gyration and the average hydrodynamic radius slowly decrease as the temperature increases, but their ratio is nearly a constant, close to the theoretical value of the "coil" conformation (1.504). In the poor solvent, $T > \Theta$, the chain size decreases sharply as the temperature increases, and finally approaches a constant value.

The successful observation of the "coil-to-globule" conformational transition and the formation of thermodynamically stable single-chain globules of a neutral polymer chain in the solution made the study of its inverse "globule-to-coil" transition possible. The figure below summarizes how the average sizes of the PNIPAM chains change with the temperature in a heating and cooling cycle. Two points are worth noting. First, the ratio of the average radius of gyration to the average hydrodynamic radius (R_g/R_h) in inset is not as expected to decrease monotonously. Second, no matter whether in the processes of rising and lowering the temperature, there is an obvious hysteresis near the quasi-ideal (Θ) state (31.4 ℃).

First, let us examine the first point. In the heating process, the ratio of the average radius of gyration to the average hydrodynamic radius (R_g/R_h) drops from ~1.5 to ~0.6 first, and then rises to ~0.78, very close to 0.774 predicted for a small uniform ball. The change of R_g/R_h with the temperature has an obvious small pit. Its qualitative explanation is as follows. A real flexible macromolecular chain is not infinitely flexible. It can be imagined that when it contracts, its

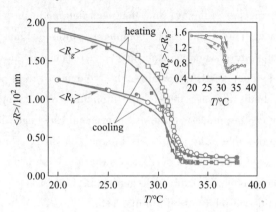

Wu C, Wang X H. Phys. Rev. Lett., 1998, 80: 4092.

(Figure 2, permission was granted by publisher)

surface inevitably has many small loops made of a number of segments. As the temperature increases, the solvent quality deteriorates, the chain conformation continues to contract. These small loops become smaller and smaller, so that the contraction becomes more and more difficult. In the opposite process, the globular conformation gradually expands as the temperature decreases, and the solvent quality becomes better. In this process, , the small loops with a full of tension first "melt out" and expand on the surface.

In order to help readers understand this phenomenon, one can imagine holding and squeezing a coil made of a thin copper wire with one hand. The forceful squeezing can make the coil smaller and smaller, become a small ball, but there must be many small loops protruded slightly on the surface. A much larger force is required to flatten these small loops. It is not difficult to imagine that due to the existence of these small loops, the surface has a lower density than the inside, and the mass fraction of these small loops is small. By definition, the presence of these low-density loops on the surface has little contributions to its radius of gyration, but these small circles on the surfaces together with some entrapped solvent molecules slightly increase the hydrodynamic radius. Therefore, a ball with small loops on its surface has a smaller ratio of the radius of gyration to the hydrodynamic radius than a uniform small ball. In the literature, two laboratories successively observed such a small pit when they studied the "coil-to-globule" conformational transition, too. Maybe, due to the theoretical ratio is a monotonous decrease from ~1.5 to ~0.8, they subjectively attributed such a small pit to the experimental error. However, it was this recurring pit (lower R_g/R_h ratio) that led to a discovery: a new conformation of macromolecular chains, the molten globule [Phys. Rev. Lett., 1996, 77: 3053], called a landmark of macromolecular research.

Regarding the second point: The hysteresis phenomenon mainly occurs near the quasi-ideal (Θ) state (31.4 ℃). For a given temperature, the average size of the chain in the reverse "globule-to-coil" transition is smaller. This hysteresis is more clearly reflected in the ratio of the radius of gyration to the hydrodynamic radius in the inset. This hysteresis is completely unexpected in theory. Note that each data point in the figure was measured after the temperature reached the equilibrium, completely different from the hysteresis observed normally in the temperature scanning experiments. A reasonable guess is as follows. In the collapsing process of individual PNIPAM chains, the volume fraction and segment concentration of the chain gradually increase, thereby the probability of forming intra-chain hydrogen bonds increases. In the cooling process, these extra intrachain hydrogen bonds formed during the collapsing process are not able to be completely broken near the Θ state, limiting the chain expansion, so that the average chain size is smaller. As the temperature further decreases, the solvent quality becomes much better, the extra hydrogen bonds formed at high temperatures are broken, and the average chain size returns to the initial value.

In order to verify this conjecture, those possible intrachain hydrogen bonds can be removed experimentally by using linear flexible neutral poly (N,N-diethyl acrylamide) (PDEAM) chains that are not able to form an intrachain hydrogen bonds. The above "coil-to-globule" transition experiments can be repeated by using PDEAM. The figure on the right respectively shows the chemical structures of PNIPAM and PDEAM. The former

$$\left[CH_2-CH\right]_m \qquad \left[CH_2-CH\right]_n$$
$$\overset{|}{C}=O \qquad\qquad \overset{|}{C}=O$$
$$\overset{|}{NH} \qquad\qquad \overset{|}{N}$$

poly(N-isopropylacryl-amide) (**PNIPAM**) poly(N, N-diethylacryl-amide) (**PDEAM**)

contains hydrogen bond acceptors and donors [$\overset{\diagdown}{\underset{\diagup}{C}}=O$ and H—N(Pr-i)], so that it can form the intrachain hydrogen bonds when it shrinks at higher temperatures, i. e., $\overset{\diagdown}{\underset{\diagup}{C}}=O\cdots H—$ N (Pr-i)—, but not the later.

However, it is always a challenge to prepare water-soluble, narrowly distributed ultra-long macromolecular chains. In order to inhibit the chain transfer and termination, a method was finally developed to prepare PDEAM samples with a weight average molar mass as high as 10^6 g/mol by the low-temperature bulk polymerization. The purified sample was further classified in a solution mixture of anhydrous acetone and anhydrous n-hexane to obtain a fraction with a weight average molar mass of $\sim 6\times 10^6$ g/mol and a narrow distribution. It is further fractionated to obtain a desired PDEAM sample ($M_w \sim 1.7\times 10^7$ g/mol and $M_w/M_n \sim 1.06$).

Using this sample, the "coil-to-ball" conformational transition and its reverse process were studied, and the temperature dependences of the average radius of gyration and the average hydrodynamic radius change were obtained. The figure (a) below shows that in a cycle of heating and cooling, the heating and cooling curves of PDEAM nearly overlap with each other, no hysteresis, very different from that of PNIPAM. After more than 10 years of experimental studies, the original conjecture was finally confirmed, i.e., the previously observed hysteresis is indeed attributed to the intrachain hydrogen bonds formed during the chain shrinking process.

These results also eliminate an initial doubt whether the formation of thermodynamically stable PNIPAM single-chain globule is due to the intrachain hydrogen bonding, and whether it is a special case. In the temperature range of 31-35 ℃, a small R_g/R_h ratio was also observed, further confirming the existence of the "molten globule" conformation, which has nothing to do with whether there are the intrachain hydrogen bonds. At higher temperatures, in comparison with PNIPAM, stable PDEAM single-chain globules have a lower chain density, its R_g/R_h ratio is slightly larger than the ratio expected for a uniform solid ball, indicating that stable PDEAM single-chain globules contain more solvent inside, and partially "draining" of solvent, leading to a smaller hydrodynamic radius and a higher R_g/R_h ratio. These results reveal that the intrachain hydrogen bonding has a certain influence on the final collapsing extend of individual chains.

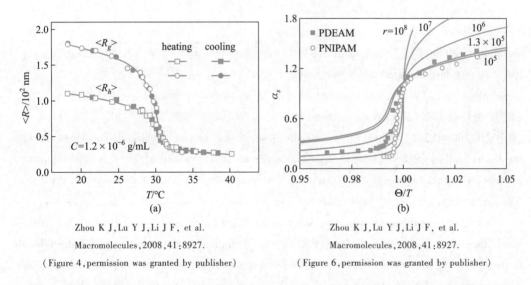

Zhou K J, Lu Y J, Li J F, et al.
Macromolecules, 2008, 41:8927.
(Figure 4, permission was granted by publisher)

Zhou K J, Lu Y J, Li J F, et al.
Macromolecules, 2008, 41:8927.
(Figure 6, permission was granted by publisher)

As mentioned above, after one hundred years of research, the existing polymer theory has a relatively complete description of the conformational change of polymer chains in good solvents as the solvent quality varies. The figure (b) above compares the temperature dependence of static expansion factor α_s of individual PDEAM and PNIPAM chains. In a good solvent (here $T < \Theta$), the experimental results of both PDEAM and PNIPAM are in good agreement with the theoretical curve calculated by eq. (7.2), similar to the results of polystyrene in [J. Chem. Phys., 1980, 73:5971]. In poor solvents ($T > \Theta$), as the temperature increases, the properties of the solvent deteriorate. The measured values of α_s are not only much lower than the theoretical expected ones, but also decreases much faster as the temperature increases. The α_s of PNIPAM drops lower and faster than that of PDEAM. From an experimental point of view, the "coil-to-globule" transition is close to a sudden change similar to the first-order phase transition. Their αs values approach 0.20 and 0.14, respectively, at higher temperatures; further showing that in the collapsed state, individual PDEAM globules is looser.

The "coil-to-globule" conformational transition of a macromolecular chain as the solvent quality deteriorates can also be described by the reduced expansion factor $\alpha^3 \tau M_w^{1/2}$ and $\tau M_w^{1/2}$ in eq. (7.2), which removes the chain length dependence. The figure below shows the results from two PNIPAM samples with different molar masses. Taking the Θ state as the reference point, $\alpha^3 \tau M_w^{1/2} = 0$. As $|\tau|$ increases, the solvent quality becomes worse, the chain volume shrinks, α^3 decreases, but $\alpha^3 \tau$ increases. When $\partial (\alpha^3 \tau M_w^{1/2})/\partial\alpha \propto 6\alpha^5 - 8\alpha^7 = 0$, $\alpha = 0.86$ and $\alpha^3 \tau M_w^{1/2}$ reaches a maximum. The theoretical curve in the figure on the right shows that when the number of segments of a macromolecular chain is 10^5 (a limit in experiment), when α_s decreases from 1.0 to about 0.4, the corresponding variation range of τ is less than 0.01. If $\tau M_w^{1/2}$ is very large, and $\alpha < 0.4$, the first term in eq. (7.2) can be ignored, $\alpha^3 \tau M_w^{1/2}$ approaches a plateau constant

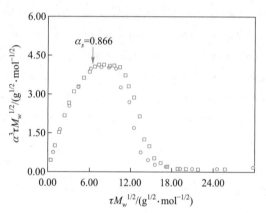

Wu C, Zhou Z Q. Macromolecules, 1995, 28:8381.

(Figure 9, permission was granted by publisher)

(0.102/0.180=0.567). When $\tau M_w^{1/2} < 10$, the experimental results is basically in agreement with the theoretical prediction. There is indeed a maximum at 0.866, and a thermodynamically stable platform. However, after $\tau M_w^{1/2} > 10$, $\alpha^3 \tau M_w^{1/2}$ decreases rapidly, deviating from the theoretical prediction. Although $\alpha^3 \tau M_w^{1/2}$ still approaches a constant, much smaller than the predicted one.

Initially, these differences in poor solvents were attributed toadditional intrachain hydrogen bonds formed in the collapsed state, that is, the formation of the intrachain hydrogen bonds further promotes the chain contraction, a self-accelerating process. However, the PDEAM chain cannot form intrachain hydrogen bonds, and the experimental results in the figure above overturn this initial explanation. The differences between the theories and experiments may be mainly attributed to the assumption of the mean field theory, that is, the dispersion attraction among the segments is indepedent of the segment concentration. In a good solvent, a macromolecular chain is in an extended state, The chain size $R(T)$ and the expansion factor α have relatively weaker temperature dependence, and the volume fraction of the segments inside the chain varies little, so that the theory and experiment are more consistent.

In poor solvents, the excluded volume effect is negative and there exists net attraction among the segments. Current theories are still not able to quantitatively describe the conformational change of a macromolecular chain. There may be at least two qualitative explanations as follows. First, the assumption of interaction among the segments is independent of the segment concentration in the mean field theory needs to be modified, and the concentration dependence of the interaction point number (Z) is introduced, but this may shake some conclusions of the mean field theory, which is a problem worthy of the theoretician's attention and study. Second, water as a solvent has some special properties, it not only hydrates with polymer chains, but also self-complex with each other through intermolecular hydrogen bonding, depending on the temperature. It requires experimentalists not only to find a non-aqueous solvent, but also measure the "coil-to-ball" conformational transition in the one-phase region and reach a thermodynamically stable state.

After each steady state between the "coil" and "ball" conformations was successfully studied, what still needs to be studied is the transition kinetics between the "coil" and "globule" states, including the internal mechanism of the conformational change and the characteristic relaxation time. In comparison with the study of steady states, in spite of few attempts, there was no successful dynamic study. For example, Chu and Ying and their collaborators[Macromolecules, 1995, 28：180; 1996, 29：1824] used a 0.01 mm thin-walled sample cell to reduce the time of changing the solution temperature and found two relaxation processes, whose characteristic relaxation times were in the order of minutes. Others also tried to use LLS to study this kinetic problem, but they were unable to overcome two major problems that the speeds of the temperature change and measurement are relatively too slow. Liu and his collaborators[Phys. Rev. Lett., 2006, 96：027802] used a fast stop-flow method to study this conformational transition, also found that the shrinking of the PNIPAM chain includes two relatively fast relaxation processes, whose characteristic relaxation times are ~12 and 270 ms, respectively.

Theoretically, in the mid-1980s, de Gennes[J. Phys. Lett., 1985, 46：639] predicted that the folding of a macromolecular chain may start from an expanded random coil, passing a "sausage-like" conformation, and end up in a globular state. The characteristic relaxation time of the first process is in the order of ms. Later, Grosberg and Kuznetsov[Macromolecules, 1993, 26：4249] assumed that the collapse of a macromolecular chain in a poor solvent could be separated into two steps. The chain crumpts quickly first, and then followed by a slow intrachain entanglement (knotting) process, finally the conformation becomes a stable single chain globule. However, experimentalists are helpless. People have to use computer simulations and phenomenological models to study the kinetics of the "coil-to-globule" transition. Kuznetsov[J. Chem. Phys., 1995, 103：4807; 1996, 104：3338] and Halperin et al.[Phys. Rev. E, 2000, 61：565] successively predicted that the first process of the conformation transition should be the local "nucleation" in the chain, whose characteristic relaxation time should be independent of the chain length. In these studies, they assumed the infinite chain length, and the "coil-to-globule" conformational transition is similar to the first-order phase transition. Otherwise, there would be no so-called "nucleation" in the metastable state.

To verify the theoretical prediction, two problems have to be solved experimentally: first, a heating method much faster than milliseconds, which can quickly change water from a good solvent into a poor solvent, so that the conformational change can be avoided during the heating process; and next, a detection method with a time resolution of μs, much faster than dynamic LLS. The stopped-flow method mentioned earlier is a common method for studying kinetics, but its dead time is usually a few ms, which is not suitable for studying the conformational changes of macromolecules, especially the initial kinetics. In order to solve these technical problems, the method of measuring protein folding (shorter chain and faster process) in biology was borrowed, in which an infrared nanosecond pulse laser with a wavelength of 1540 nm is used to heat the aqueous solution, plus a rapid fluorescence and a LLS detector. It can heat the temperature of the measured spot up ~ 3 ℃ within 20 ns, and detect the intensities of fluorescence and scattered light from the solution at a scattering angle of 90° in real time with a time resolution of μs.

Ye and his coworkers [Macromolecules, 2007, 40:4750] successfully used this method to study the kinetics of the "coil-to-globule" conformational transition, in which 8-anilino-1-naphthalenesul-fonate ammonium salt (ANS) was used as a fluorescent probe. When its environment changes from hydrophilic to hydrophobic, its emitting fluorescence greatly increases. The figure below shows how the fluorescence intensity of ANS changes with time after the solution is quickly heated up by a 20 ns laser pulse. Without PNIPAM, the fluorescence intensity of ANS is basically a constant, independent of temperature (diamond symbol). In the presence of PNIPAM, the fluorescence intensity of ANS increases rapidly within 0.5 milliseconds (circle symbol), and then approaches to a maximum.

The inset shows that the change of ANS fluorescence intensity can be separated into fast and slow two processes, whose corresponding characteristic relaxation times are (0.11±0.01) ms and (0.83±0.06) ms, respectively. The entire collapsing process was completed within about 4 ms. Therefore, the conventional stop-flow method should not be able to observe these two rapid processes. The two processes measured in the previous stop-flow method are likely to originate from the interchain aggregation of macromolecules. In order to prove that the

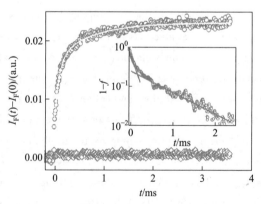

Ye X D, Lu Y J, Shen L, et al.
Macromolecules, 2007, 40:4750.
(Figure 1, permission was granted by publisher)

fluorescence intensity change observed in the figure above is indeed due to the conformational transition of individual chains, the intensity of scattered light with time after the heating was also measured. As mentioned earlier, the formation of dimers will double the intensity of scattered light. Experimental results showed that the intensity of scattered light is nearly a constant within the time window of the fluorescence intensity change, confirming that the observed single-chain conformational change.

After successfully observing two kinetic processes in the conformational transition of PNIPAM chain in aqueous solution, they further studied the chain length dependence of the two processes. Using the fractionation method discussed before, they obtained a series of fractions with a weight-average molar mass in the range of $1.2 \times 10^6 - 2.3 \times 10^7$ g/mol and a relatively narrow molar mass distribution ($M_w/M_n \sim 1.3$). In the kinetic study, it is not necessary to have $M_w/M_n < 1.0$. The figure (a) below shows that the characteristic relaxation time of the faster process has no chain length dependence, agreeing with the theoretical prediction, i.e., the fast process corresponds to the "nucleation" process, only related to short chain links in the chain.

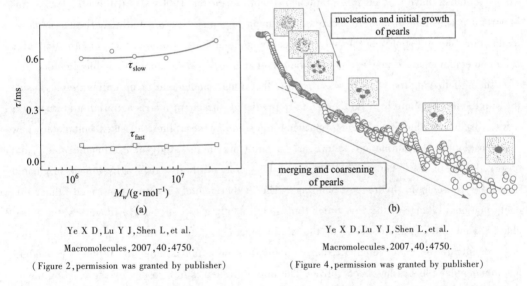

Ye X D, Lu Y J, Shen L, et al.
Macromolecules, 2007, 40:4750.
(Figure 2, permission was granted by publisher)

Ye X D, Lu Y J, Shen L, et al.
Macromolecules, 2007, 40:4750.
(Figure 4, permission was granted by publisher)

The characteristic relaxation time of the subsequent slow process increases slowly with the chain length. It has been predicted theoretically that the characteristic relaxation time of the slower process is weakly dependent on the number of chain segments, and the scaling relationship is $\tau \sim N^{1/5}$. In this slow process, the previously formed "nucleus" larger than the critical size will continue to "swallow" in its connected adjacent segments and become larger. At the same time, those "nucleus" smaller than the critical size will unwound (dissolve) automatically, and its constituent segments will be "swallowed" by nearby nuclei larger than the critical size. In the end, only one single-chain "nucleus" remains, the single-chain globular conformation. This process is also called "coarsening", in which the interface between chain and solvent gradually decreases, so that the decrease in the surface energy compensates for the loss of translational entropy and the energy of interaction between the segments due to the segment aggregation. The overall free energy decreases, so that this process proceeds spontaneously.

The figure above on the right graphically depicts kinetics of the "coil-to-globule" conformational transition of a macromolecular chain in the one phase region as the solvent quality deteriorates, including two processes: "nucleation" and "coarsening". The experimental results reveal that **this "coil-to-globule" conformational transition is a typical first-order phase transition.**

Summary

In comparison with a binary mixture of two small molecules, as the solvent quality deteriorates, macromolecular chains in solution have a unique conformational transition, in the one phase region, from a random coil to a uniform globule. In order to clarify and confirm this new physical phenomenon lacked in the physics of small molecules, theorists and experimentalists have studied this problem in depth since the 1960s. Until the mid-1990s, this important proposition in modern macromolecular physics was experimentally confirmed; namely, in the one phase region, as the solvent quality gradually deteriorates, the conformation of a macromolecular chain transits from a random coil into a thermodynamically stable globule.

In the following ten years, it was revealed that a macromolecular chain only crumples inside the single chain globule with no theoretically predicted intrachain interpenetration and knotting; it was discovered that in this conformational transition process, there exists a completely new "molten globule" conformation, known as "a landmark in the study of macromolecules"; and observed that in the reverse ("globule-to-coil") process, there exists a hysteresis originated from the intrachain hydrogen bonding. The experimental results confirmed that this conformational change is the first order transition, including two successive processes: the intrachain "nucleation", independent of the chain length, and the intrachain "coarsening" that depends slightly on chain length. Experimentally, this important intellectual problem in modern macromolecular physics has been solved. The only thing left is how to modify the existing theory so that it can quantitatively handle the behavior of macromolecular chains in poor solvent, including quantitatively introducing the dependence of dispersion attraction on the intrachain volume fraction (concentration) of the segments, which can be as high as 20%-30%.

Readers should note that in literature, the word "coil-to-globule" have been constantly misused with no understanding at all. This important proposition in modern macromolecular physics did not ask whether the chain conformation will collapse from a random coil into a globule as the solvent quality deteriorates, but **whether it can happen in the one phase region without a phase transition.** During a phase transition, macromolecules will undergo the intrachain contraction and the interchain association at the same time. The conformation change during the phase transition has never been the "coil-to-ball" conformational transition discussed in theory. Therefore, whenever talking and quoting the "coil-to-ball" conformational transition, one has to make sure what she/he observed occurs in a homogeneous solution, i.e., no interchain association. Hopefully, after reading the detailed discussion in this section, readers can remove the falsehood and keep the truth, and not be misled when reading the relevant literature.

This section is based on a previous review by author and collaborators [Acta Polymerica Sinica, 2017, 9: 1389]. The picture below vividly summarizes the main conclusions obtained from the research on this issue.

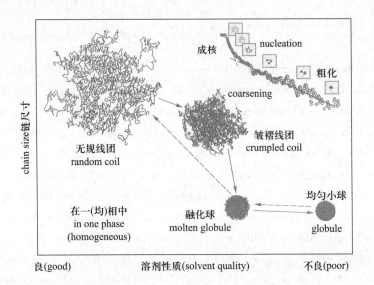

The "coil-to-stretch" transition

Under an external stretching force, the conformation of a macromolecular chain can also undergo a "coil-to-stretch" transition. This is another important proposition in modern macromolecular physics, which also began in the mid-1960s and was first proposed by theoretical physicists, which contains new physics, not existing in small molecular physics. In addition to the theoretical and academic value of this kind of conformational transition, there are many important applications, such as the uniaxial and biaxial orientation of the chain in the processing of polymer materials, and the chain pulling and stretching in the process of polymer spinning, ultrafiltration of macromolecules, migration of protein and ribonucleic acid chains crossing the nuclear membrane, and even the spinning process of spring silkworms and spiders. For this reason, this important proposition in the study of modern macromolecular physics extends various related important issues and has received extensive attention in the past 50 years. Many theorists and experimentalists have spent great efforts in order to clarify, verify and solve the proposition and related problems.

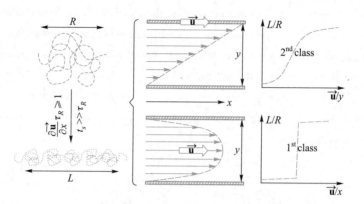

Reviewing history, one can found that Peterlin[J. Polym. Sci., Polym. Lett., 1996, 4:287] first pointed out that a neutral linear flexible macromolecular chain can pass through a pore smaller than its size under an elongation flow field, and there exists a critical strain rate, corresponding to a critical flow rate ($q_{c,1}$) inside the pore. At this critical flow rate, the conformation of a macromolecular chain is deformed under the strong shearing of the flow field, so that it can pass through pores smaller than its size, while a small molecule can only pass through pores with a size larger than its size. Later, de Gennes[J. Chem. Phys., 1974, 60:5030] calculated this physical phenomenon in detail, and pointed out that in the shear flow field between two relatively moving parallel plates, The average end distance (R) of a linear chain gradually increases with the shear rate, a continuous second-order transition. While in a cylindrical pore, the chain is in an elongation flow field with a velocity gradient along the longitudinal direction of the pore. The conformational transition is a sudden first-order transition process, as shown on the figure above.

Qualitatively, free energies of the random coil and the stretched conformation are located at relative minimums, respectively, and a metastable state with a higher energy is between them. When the length of the stretched chain is shorter than a critical length (l_c), it automatically return to the random coil state; only when the length of the stretched length reaches l_c, the chain conformation automatically further extends until its free energy reaches the relative minimum.

More interestingly, de Gennes, and afterwards, Pincu [Macromolecules, 1976, 9:386] calculated the critical flow of linear macromolecular chains passing small pores and found that it has nothing to do with the chain length and the pore size. Subsequent theoretical studies gradually expanded to the ultrafiltration of macromolecular chains with various configurations. The theoretical progress was far ahead of experimental studies. It not only poses a challenge to experimental research, but also confused experimentalists. This is because they had not been able to observe the predicted first order "coil-to-stretch" sudden transition in real experiments, and a quantitative relationship between the shear field strength (critical flow rate) and the chain length, the configuration of macromolecules, the pore size or the pore shape.

Experimentally, one often mistakenly thinks that as long as a sufficiently strong force is applied, a chain would be stretched and oriented. For example, in the industry of processing soybean protein into a fibrous structure, there is a common misunderstanding. The macroscopic muscle-like fibrous structure is regarded as the microscopic stretched orientation of the soybean protein chains, do not know that an imposed force can only make a subject acceleratory moving in the force direction, not able to stretch or orient it. The daily experience tells us that to straighten a rope, two hands have to move in opposite directions or two hands move at different speeds in the same direction. A grade 16 super-typhoon can blow an object into the sky and make it spin and fly in the air, but it will not be able to stretch the object into a linear one. To straighten an object, different forces have to be applied on its two sides or ends, in order words, its two sides or ends have to move in different velocities. More precisely, it is the force or velocity gradient (the force or velocity difference per unit distance) that determines whether an subject can be stretched and orientated.

Mathematically, the force and velocity gradients are written as $\partial F/\partial L$ and $\partial u/\partial L$, respectively. In an elongation flow field, $\partial F/\partial L$ can be converted to $\partial u/\partial L$. The daily experience also tells us that in order to straighten a curled rope, not only a force gradient is required, but also the force gradient has to be applied for a sufficient long period of time. It is conceivable that no one is able to straighten a curly rope with both hands within a thousandth of a second. Physically, in order to stretch a subject, not only must a sufficient force or velocity gradient, but also this force or velocity gradient has to apply for a sufficient long period of time (t_s).

According to the discussion on the conformation of macromolecular chains in Chapter 4, when a macromolecular chain is completely straightened, there is only one conformation, and the conformational entropy is zero. Under isothermal and isostatic conditions, a macromolecular chain always moves to the direction of reducing free energy by trying to return its random coil conformation to increase the conformational entropy. Therefore, after the external force is removed, a stretched macromolecular chain will quickly rebound and retract to its equilibrium random coil conformation. The time (τ_R) required for a macromolecular chain to rebound (relax) from a deformed conformation to its equilibrium one can be roughly estimated.

It is known from the discussion in the previous chapter that in case of no entanglement among the chains in the dilute solution of macromolecules, if considering the hydrodynamic interaction (Zimm model), this process corresponds to a series of the chain length dependent internal normal modes, their characteristic relaxation times are as follows.

$$\tau_n \sim M^{3\alpha_{R_g}}$$

where $1/2 \leqslant \alpha_{R_g} \leqslant 3/5$. For common dilute solutions of macromolecules, the characteristic relaxation time of the slowest first normal mode is $\tau_1 \sim 1$ ms; $\tau_R \approx \tau_1$. The above two conditions can be mathematically expressed as

$$\left(\frac{\partial u}{\partial L}\right)\tau_R > 1 \quad \text{and} \quad t_s \gg \tau_R \tag{8.3}$$

If the concentration of a macromolecular solution is higher and the segment number (N) in the chain is much larger than the minimum segment number (N_e) required forming entanglements, the macromolecular chains will penetrate and entangle each other. In the space occupied by each chain, there are also many other macromolecular chains. Therefore, the characteristic relaxation time of a deformed conformation relaxing to its equilibrium one must be slower. The specific value will be discussed in detail in the next chapter.

In the cgs system, the size of the chain is $\sim 10^{-5}$ cm; it can be seen that the speed difference between the two chain ends must larger than 10^{-2} cm/s; that is, the speed difference has to reach 10^3 chains per second. One can imagine that one end of the chain is fixed; the other end must move more than a thousand chain sizes per second! Experimentally, not only must this velocity gradient be achieved, but also the macromolecular chain must stay in this velocity gradient field for a sufficient long time before it can be stretched into a cigar-like conformation. The above discussion and derivation are based on dynamics.

From another perspective, the same problem can also be discussed using the energy change when a linear macromolecular chain stretches. Taking the mean square end-to-end distance in the undisturbed state ($\langle R_0^2 \rangle = b^2 N$) as the reference point, the mean square end-to-end distance of an expanded chain in a good solvent after reaching its equilibrium is $\langle R^2 \rangle = b^2 N^{2\alpha_R}$, or written as $R = \sqrt{\langle R^2 \rangle} = bN^{\alpha_R}$, whose corresponding energy is only one thermal energy ($k_B T$). If the chain conformation becomes a cigar-like shape with a diameter of ξ and a length of L, similar to be in a "tube" with a diameter of d. For linear chains, $\xi = d$. A macromolecular chain can be imaginatively "divided" into (L/ξ) "blobs", and each "blob" is composed of a chain link with N_ξ segments. Therefore, $\xi^2 \approx b^2 N_\xi^{2\alpha_R}$ and $L/\xi = N/N_\xi = (R/\xi)^{1/\alpha_R}$. Each chain segment can only walk randomly inside the "blob" to which it belongs. The energy corresponding to each "blob" is one thermal energy ($k_B T$). The energy of the entire macromolecular chain increases from $k_B T$ to $k_B T(L/\xi) = k_B T (R/\xi)^{1/\alpha_R}$. Therefore, in a good solvent, the change of free energy of a linear macromolecular chain due to the deformation is

$$\Delta A = k_B T \left[\left(\frac{R}{\xi} \right)^{\frac{5}{3}} - 1 \right]$$

When $\xi < R$, the deformation of a chain leads to an increase in free energy. According to thermodynamics, under isothermal and isometric or isobaric conditions, a system will not automatically move towards the direction of increasing free energy, that is, the probability of changing from a random coil conformation to a cigar-like shape is small. According to statistical mechanics, the probability (ψ) of the above chains conformation change is

$$\psi = \exp\left(-\frac{\Delta A}{k_B T} \right) = \exp\left[-\left(\frac{R}{\xi} \right)^{\frac{5}{3}} + 1 \right] \quad (\xi \leqslant R) \tag{8.4}$$

Note: The above probability ψ can also be regarded as the probability of a macromolecular chain entering a cylindrical pore with a diameter of ξ. It is only related to the ratio of R/ξ. For a linear chain, when $\xi = R$, the probability is 100%. As ξ decreases, ψ decreases exponentially.

When $\xi = R/2$, ψ drops to ~ 1.5%. In other words, the probability of a linear chain spontaneously and randomly entering a cylindrical pore with a diameter equal to or greater than its average end-to-end distance of the chain is 100%; while such a probability approaches zero when the pore diameter is half of the chain size.

Generally speaking, if a chain completely enters a "tube", the cigar-like chain has a volume of d^2L. Each "blob" has a volume of ξ^3, so that each chain contains (d^2L/ξ^3) "blobs". The chain has an energy of $k_BT(d^2L/\xi^3)$. In comparison with he energy (k_BT) before entering the "pore", the deformation confinement leads to an energy increase of $\Delta E_c = k_BT[(d^2L/\xi^3) - 1]$. In addition, if the cigar-like chain in the tube is in an elongation flow field with a flow rate of q, the flow velocity is $u = q/d^2$. The moving resistance of each "blob" is $f = 3\pi\eta\xi u$, so that the entire chain has a resistance of $F = f(d^2L/\xi^3)$. Similarly, taking the undisturbed chain conformation as a reference point, during the stretching process, the elongation flow field acts on the macromolecular chain, leading to an increase of its energy,

$$\Delta E_h = \int_0^L F \mathrm{d}L - k_BT = \int_0^L \frac{3\pi\eta q}{\xi^2} L \mathrm{d}L - k_BT = \frac{3\pi\eta q}{2\xi^2} L^2 - k_BT$$

Therefore, the total energy change of a chain after entering a cylindrical hole with a diameter of d under the action of an elongation flow field is

$$\Delta E = \Delta E_h - \Delta E_c = \frac{3\pi\eta q}{2\xi^2} L^2 - \frac{k_BT d^2}{\xi^3} L$$

Its differentiation with respect to L results in $(\partial\Delta E/\partial L)_{T,p} = 3\pi\eta q L/\xi^2 - k_BT d^2/\xi^3$. Note that the changing rate (differentiation) of energy with respect to length is a force. Let $(\partial\Delta E/\partial L)_{T,p} = 0$, i.e., the two forces are balanced, a minimum value is obtainable; namely,

$$L_{min} = \frac{k_BT d^2}{3\pi\eta q \xi}$$

Obviously, $L_{min} = \xi$. Physically, it means that only the first (layer) of "blob(s)" enters the small pore, that is

$$q_{min} = \frac{k_BT}{3\pi\eta}\left(\frac{d}{\xi}\right)^2 \tag{8.5}$$

It is worth noting that in the process of deriving the above equation, the chain configuration was not assumed. Therefore, the above equation is universal for different configurations. In its actual use, as long as the corresponding ξ is calculated according to a specific chain configuration, the above formula can be used to obtain the critical flow rate of macromolecules passing through a cylindrical pore, denoted as q_c hereafter. Whether a macromolecular chain can pass through a cylindrical pore depends entirely on the relative size of the flow rate (q) and this critical flow rate (q_c). When $q > q_c$, passing; otherwise, blocked. de Gennes in the 1970s derived the critical flow rates of star and branched macromolecular chains, respectively, but did not obtain a universal equation like the above one, and made the following mistakes, leading to a wrong conclusion. As there had been no experimental verification, such an error has not been corrected for a long time.

Reading the review from de Gennes[Adv. Polym. Sci.,1999, 38:92], readers should pay attention that when deriving the critical flow rate of branched chains, i.e., equation (30) in the literature, he incorrectly assumed the energy of a layer of "blobs" is one thermal energy $k_B T$ (only valid for linear chains). In fact, the number of "blobs" in a layer is the ratio of the cross-section areas of the cylindrical pore to one "blob"; namely, there are $(d/\xi)^2$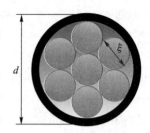
"blobs", as shown on the right. Each "blob" has one $k_B T$, so that the entire energy of one layer of "blobs" is $k_B T (d/\xi)^2$. It was this mistake that led de Gennes erroneously obtained the fourth power for branched chains, instead of the correct quadratic in the above equation.

For linear chains, de Gennes treats each "blob" as a solid ball, $d = \xi$; eq. (8.5) becomes

$$q_{c,\text{linear}} = \frac{k_B T}{3\pi\eta} \qquad (8.6)$$

Interestingly, according to the above equation, the critical flow rate of linear chains only depends on temperature and liquid viscosity (solvent viscosity in dilute solutions), and has nothing to do with the chain length and the pore size. Intuitively, it takes a larger fluid shear force to "squeeze" a longer chain through a given small pore. On the other hand, for a given macromolecular chain, the smaller the pore, the greater the fluid shear force required. Therefore, in the literature and in the research proposals, people mistakenly propose using small pores to separate linear chains of different lengths. It can be seen that intuition is sometimes wrong! Still need theoretical guidance!

The qualitative explanation is that after the chain becomes cigar-shaped, it can pass through a small pore as long as the chain has a diameter not larger than the pore, regardless of the chain length. In other words, if passing a thin hard rod through a small pore, when the rod diameter is larger than the pore diameter, it will not be able to pass through the pore; otherwise, it will pass through, independent of the rod length. According to the above equation, in the cgs system, the critical flow rate of linear chains is estimated to be $q_{\text{min}} \approx 10^{-13}$ cm^3/s, extremely low! Therefore, in a normal filtration experiment, the flow rate is much larger than the critical value, so that there is no need to worry about whether linear macromolecular chains are blocked by the filter membrane. If a membrane with a 200 nm pore size is used, the critical flow rate is only $\sim 10^{-3}$ cm/s, which is about ten micrometer per second.

For readers in chemistry, polymers, and other non-physics departments, it may be easier if the above equation is directly derived from "force" instead of "energy". In the discussion of gas and liquid pressure, it has been known that the pressure generated by each "particle" is proportional to its probability of reaching any point in the system, and the proportional constant is the energy density when the probability is 100%. If there is only thermal energy, $p = k_B T/V$, where V is the reachable volume of one particle. When discussing a restricted macromolecule, each "particle" is a short chain link, and its reachable volume is the "blob" volume to which the segment belongs (ξ^3), $p = k_B T/\xi^3$; when a macromolecular chain is outside the pore, the reachable volume of each segment is R^3, $p = k_B T/R^3$. Note that $\xi \leqslant d$. If $R \leqslant d$, a macromolecular chain enters and exits the pore freely, as in no one's land; on the contrary, it stays outside the pore because of a higher pressure in the pore, and will not spontaneously enter the pore. Strictly speaking, it is lower probability, as shown in equation (8.4).

The pressure multiplied by the cross-section area of a "blob" (ξ^2) is the confining force $f_c = k_B T/\xi$. Readers who are familiar with physics can also get the result from "the energy ($k_B T$) divided by the "blob" size (ξ) equals to the force" in one step. On the other hand, each "blob" in the flow field experiences a resistant force $f_h = 3\pi\eta\xi u = 3\pi\eta\xi(q/d^2)$. When the two forces acting on each "blob" reach an equilibrium ($f_c = f_h$), eq. (8.5) can also be obtained. Different routes lead to the same destination. Is it easier? Readers can judge for themselves.

Frank, Keller and Mackley[Polymer, 1971, 12:467] first used two impinging jets to stretch linear macromolecular chains. In this experiment, when two very close, but opposite high-speed macromolecular solution horizontal jets are impinged each other to produce a liquid dish, perpendicular to the impinging axis, the flow direction and velocity suddenly change 90°. In theory, there is no liquid at the exact impinging center. As the initial impinging velocity increases, the liquid surface gradually approaches the center. The closer to the impact center, the greater the velocity gradient, the longer the macromolecular chain stays in the velocity gradient field. Therefore, when the flow velocity is sufficiently high, a macromolecular chain is stretched in the vertical direction along the flow field near the impinging center.

The experimentally achievable jet velocity is limited. As mentioned earlier, when there is no entanglement in a good solvent, the relaxation time of the chain conformation is proportional to $M^{1.8}$. Therefore, using long macromolecular chains is a more practical approach. They found that under the experimental condition available at the time, the minimum molar mass of a polystyrene chain that could be broken by the stretching was $\sim 10^6$ g/mol. Therefore, when preparing a solution of macromolecules with an ultra-high molar mass, a high shear field should be avoided no matter whether it is stirring or filtering. In the jet experiment, if the chain orientation causes polarization, the birefringence detection method can be used, but not easy for a high-speed flow.

On the other hand, pushing a static liquid into a cylindrical pore with a size of tens of nanometers causes a great speed difference at the pore entrance, resulting in a typical and ideal elongation flow field with a flow rate of q and a longitudinal velocity gradient of q/d^2. Since the flow velocity outside the pore is nearly zero, the flow rate (q) inside the pore actually represents a velocity difference$[[(q/d^2) - 0]$, dividing the "blob" length leads to the velocity gradient. However, in the ultrafiltration experiment, it is rather challenging to quantitatively study the ultrafiltration behavior of macromolecular chains in a cylindrical pore, which is influenced by the driving force, the macromolecular chain configuration, the structures of the cylindrical pore and the membrane, etc. After spending much effort for more than 30 years, experimentalists were still not able to observe the predicted first-order "coil-to-stretch" conformational transition, making people very frustrated.

The factors that affect a macromolecular chain through a cylindrical pore can be investigated one by one. The driving force is mainly an extensional flow field, and a high-voltage electric field can also be used for a charged chain. Readers need to note that pushing a neutral macromolecular chain into a small pore through an extensional flow field is not the same as applying an electric field to pull a charged macromolecular chain into a small pore. In the former, a large number of solvent molecules are forced to pass through the extended macromolecular chains at the entrance and inside the small pore, highly draining, which is completely different from the situation where macromolecular chains diffuse in a solution with nearly no drainage. In the latter, the charged macromolecular chain and its counter ions and other ions migrate in the electric field, and the macromolecular chain is stretched and enters the pores. How to prepare macromolecular samples with different configurations and defined structure and how to obtain an ideal stable extensional flow field are always two major problems in macromolecular ultrafiltration experiments.

Until 2006, Jin and Wu [Physical Review Letters, 2006, 96: 237801] accidentally used a special double-layer filter membrane with a thickness of 1 and 99 μm, containing d = 20 and 200 nm, respectively, in which each smaller cylindrical pore is surrounded by a larger cylindrical pore. The first-order "coil-to-stretch" conformational transition of linear macromolecular chains was observed for the first time, as shown in the inset on the right. The reason for its success is that the double layer structure prevents the

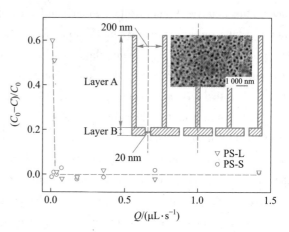

Jin F, Wu C. Acta Polymerica Sinia, 2005, 4: 486.

(Figures 2 & 7, permission was granted by publisher)

interference among different flow fields generated by different pores and ensures that each elongation flow field contains no turbulence and rotational components. If the membrane filter with a double layer structure is used reversibly, only a continuous second-order conformational transition was observed. In addition, by using LLS and additionally added short chains with a size smaller than the pore size, which can always pass through the small pore, as an internal calibration standard, the problem of the concentration variation caused by the solvent volatilization during the measurement was successfully solved. In this way, the accurate and quantitative measurement was achieved.

Namely, the area ratio of two peaks in the linewidth distribution, corresponding to long chains and short reference chains, measured by dynamic LLS can be used to quantitatively and accurately determine the concentration change of long chains in the solution before and after the filtration (C_0 and C). Further, the retention ratio, ($C_0 - C$)$/C_0$ is calculated, which is commonly used in ultrafiltration experiments, reflecting the blocking ratio of macromolecular chains by the filter. Interested readers can refer to the original literature to understand the conversion relationship in detail. The data in the figure above shows that the macroscopic critical flow rate is ~ 1.8×10^{-5} cm^3/s. The number of small pores in the membrane filter is ~ 5×10^8. Therefore, the average critical flow rate in each small pore is ~ 10^{-14} mL/s, It is an order of magnitude smaller than the theoretical value. The critical flow rate is $u_c = 4 \times 10^{-14} / (2 \times 10^{-6})^2 \approx 10^{-2}$ cm/s.

The initial experimental results showed that the critical flow rate slightly depended on the flow rate, inconsistent to the theoretical prediction. The inconsistence was attributed to hundreds of millions of small pores in the membrane and the non-uniform pore size. Later, after studying macromolecular chains with nearly ten-time different molar masses, the experimental results proved that the theoretical prediction was correct, i. e., the critical flow rate was indeed completely independent of the chain length, as shown on

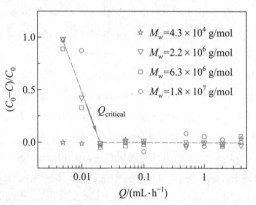

Ge H, Jin F, Li J F, et al. Macromolecules, 2009, 42:4400.

(Figure 1, permission was granted by publisher)

the right. In order to clearly compare the chains with different lengths, the abscissa is specifically taken as a logarithm. In spite that the critical flow rate is indeed independent of the chains, its value is still one to two orders of magnitude smaller than the theoretical value, depending on the solvent quality. In the Θ state, the difference between the two is the least, about ten times; in the good solvent and the poor solvent, the difference between the experimental and theoretical values can be more than 200 times.

Examining the above-mentioned theories about the " coil-to-stretch " conformational transition, it is not difficult to find that the assumption of each "blob" confined inside the pore as a completely "non-draining" solid ball is incorrect. Under the elongation flow field, when the chain is moving forward, a large number of solvent molecules are forced to pass through a macromolecular chain and stay behind, forming a strong drainage effect. There is friction between each chain segment inside the "blob" and solvent molecules, and the friction force of each "blob" hindered from movement is much greater than $f = 6\pi\eta\xi$. Therefore, it is necessary to introduce an effective length (l_e) to replace the "blob" size (ξ), that is, $f = 6\pi\eta l_e$.

If every segment in the "blob" is subjected to the same friction, $l_e = bN_\xi$, where b and N_ξ are, as defined before, the segment length and the number of segments in each "blob", respectively, $\xi = bN_\xi^{\alpha_R}$, so that $l_e = b^{1-1/\alpha_R}\xi^{1/\alpha_R}$. Substituting it into the equation of the critical flow rate, one can use the measured critical flow rate (q_c) to reversely calculate the minimum force (f) required to overcome the entropy elasticity of a linear macromolecular chain and stretch it from a random coil conformation with a size of R to a

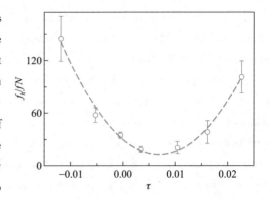

Ge H, Jin F, Li J F, et al. Macromolecules, 2009, 42:4400.

(Figure 4, permission was granted by publisher)

"cigar-like" conformation with a diameter of d and a length of L, i.e., $f = 6\pi\eta\, b^{1-1/\alpha_R}d^{1/\alpha_R-2}q_c$. For a given ultra-filtration experiment, d is a constant; for a given macromolecular solution, η and b are also two constants. The figure above shows that as the temperature (solvent properties) changes, the minimum average force required to stretch a linear polystyrene chain in cyclohexane, where τ is the reduced temperature $(T - \Theta)/\Theta$, representing the relative degree of deviation of the solution from the Θ state.

It is worth noting what is overcome here is the entropy elasticity, which is equal to the thermal energy divided by the chain size, i.e., $k_B T/R$, which is $10^{-13}-10^{-14}$ Newton, at least 100 times weaker than the $\sim 10^{-11}$ Newton measured by the atomic force microscope. The atomic force microscopy does not measure the force of overcoming the entropy elasticity to straighten a macromolecular chain from a random coil, but the force required to alternate the bond angle between chemical covalent bonds of a stretched and oriented chain. Therefore, the cartoons in literature are improper, and modifications are required.

Pay attention to two points in the figure: First, when the excluded volume effect is positive (good solvent) or negative (poor solvent), it takes a stronger force to stretch a macromolecular chain. The minimum force in the figure corresponds to the temperature at which the excluded volume effect disappears. In other words, in the Θ state, the chain is the most flexible and deformable. Second, the temperature corresponding to the minimum force is about 36 ℃, slightly higher than the Θ temperature (~ 34.5 ℃) obtained when the second virial coefficient is zero.

Theoretically, the Θ state is defined as the disappearance of the excluded volume effect. The second virial coefficient is derived from the concentration dependence and measures the interchain interaction. The local concentration within the chain is higher than the average concentration of the solution. Therefore, the intrachain dispersion attraction is stronger than the interchain dispersion attraction, while there is no difference between the intrachain and interchain hardcore volume of the segment, so that the excluded volume effect disappears at a slightly higher temperature. Therefore, by measuring the minimum force required to stretch a macromolecular chain, one can determine the Θ state of a macromolecular solution better, closer to the true undisturbed state of a chain. It further confirms the previous assumption when using the internal normal motions of the chain to measure the undisturbed state. There are now two new methods of determining the undisturbed state of a macromolecular chain.

After introducing the effective length (l_e), following the previous discussion of the confinement force $f_c = k_B T / \xi$ and the friction resistant force $f_h = 3\pi\eta \, l_e (q/d^2)$ of a "blob", and considering that $\xi = d$ for a linear chain, one can estimate how much the critical flow rate changes, namely

$$\frac{q_{c,\text{linear}}}{q_{c,\text{de Gennes}}} = \frac{q_c}{\dfrac{k_B T}{3\pi\eta}} = \frac{d^2}{\xi \, l_e} = \left(\frac{d}{b}\right)^{1-\frac{1}{\alpha_R}} \tag{8.7}$$

where $d \approx 20$ nm; $b \approx 1$ nm. Therefore, q_c is only one tenth of $k_B T / (3\pi\eta)$, which is close to the experimental value. According to the above equation, the critical flow rate decreases as the pore size increases. The range of the scale index is $-2/3 \leqslant (1 - 1/\alpha_R) \leqslant -1$.

Later, the pore size dependence of experimental results confirmed the pore size dependence of the critical flow rate predicted by the above equation. Later, Freed and the author[J. Chem. Phys., 2011, 135: 144902; Macromolecules, 2012, 44: 9863] used rigorous calculations to confirm the above conclusions, the details are omitted. This series of experiments and theories successfully modified the theory of the "coil-to-stretch" conformation transition of linear chains and correct from and corrected the initial mistakes made by de Gennes.

The situation is slightly more complicated when a star-shaped macromolecular chain with f arms and each arm having N segments with a length of b passes through a cylindrical pore with a diameter of d. The f arms will be divided into front and back two bundles. Assuming that the number of front arms is f_f, the number of back arms is $f_b = f - f_f$. Using the above "blob" description, each front arm can form N/N_ξ "blobs" with a size of ξ and an energy of $k_B T$, $\xi = bN_\xi^{\alpha_R}$. The total energy of f_f front and f_b behind arms is

$$\frac{E}{k_B T} = \frac{N \, b^{1/\alpha_R}}{d^{1/\alpha_R}} \left[f_f^{1/2\alpha_R} + (f - f_f)^{1/2\alpha_R} \right]$$

Differentiating E with respect to f_f and letting $(\partial E/\partial f_f)_{T,p} = 0$, one can obtain the minimum energy of a star chain confined in a cylindrical pore, namely

$$\frac{N \, b^{\frac{1}{\alpha_R}}}{d^{\frac{1}{\alpha_R}}} \left[\frac{1}{2\alpha_R} f_f^{\left(\frac{1}{2\alpha_R}\right) - 1} - \frac{1}{2\alpha_R} (f - f_f)^{\left(\frac{1}{2\alpha_R}\right) - 1} \right] = 0 \quad \rightarrow \quad f_f = \frac{f}{2}$$

Therefore, when the arms of a star chain are half forward, its energy is the lowest, most easily passing through a cylindrical pore. The actual situations are often not as one wishes. de Gennes and Brochard[C. R. Acad. Sci., Ser. II, 1996, 323: 473] discussed two cases, $f_f \leqslant f/2$ and $f_f > f/2$, respectively and the process is somehow cumbersome. This book provides a fairly concise derivation as follows.

A star chain is like an octopus. As long as $f_f \neq f/2$, no matter whether it is $f_f < f/2$ or $f_f > f/2$, it is always more difficult for the bundle with more arms to enter the cylindrical pore, causing a higher resistance. Therefore, one only needs to find the relationship between the "blob" size (ξ) in the bundle and the pore diameter (d), and then use the universal eq. (8.5) to calculate the critical flow rate of a star chain ($q_{c,\text{star}}$). If more arms are in the front, $\xi^2 f_f = d^2$; otherwise, $\xi^2 (f - f_f) = d^2$. If the critical flow rate ($q_{c,\text{linear}}$) of linear chains is used for normalization, one gets $q_{c,\text{star}} = q_{c,\text{linear}} f_f$ or $q_{c,\text{star}} = q_{c,\text{linear}} (f - f_f)$. Li and his coworkers [Macromolecules, 2012, 45: 7583] found that there is no need to separate the discussion tediously at all. The two are combined into one as

$$q_{c,\text{star}} = q_{c,\text{linear}} \left(\frac{f}{2} + \left| \frac{f}{2} - f_f \right| \right) \tag{8.8}$$

The above equation shows that the critical flow rate of a star chain through a cylindrical pore has the following characteristics. First, the critical flow rate is the same as a linear macromolecular chain, independent of the arm (chain) length; second, the critical flow rate is related to how the star chain enters the pore; third, for star chains with a given arm length, the critical flow rate increases as the arm number. The first point has been confirmed by the ultrafiltration experiment, as shown on the right, where the macroscopic flow rate on the

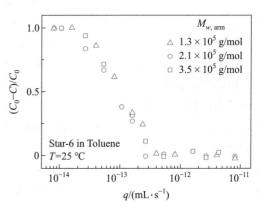

Ge H, Pispas S, Wu C. Polymer Chemistry, 2011, 2: 1071.
(Figure 2, permission was granted by publisher)

abscissa is logarithmically spaced to clearly reveal the variation of critical flow rate. The arm length has been increased three times, but the critical flow rate remains. As for the second point, if a star chain enters the pore asymmetrically ($f_f \neq f/2$), its energy is higher. The energy difference between the asymmetric and symmetric ways is

$$\Delta E = k_{\text{B}} TN \left(\frac{b}{D} \right)^{1/\alpha_R} \left[\left[\left(\frac{f}{2} + \left| \frac{f}{2} - f_f \right| \right) \right]^{1/2\alpha_R} - \left(\frac{f}{2} \right)^{1/2\alpha_R} \right]$$

Therefore, the probability ratio of a star chain entering a cylindrical pore in an asymmetric and symmetric manner is $\exp(-\Delta E / k_{\text{B}} T)$. If the solution is in the Θ state ($\alpha_R = 1/2$); $N \sim 10^2$; $d/b \sim 10$, for a star chain with only four arms, the probability of one or three arms entering the cylindrical pore first is reduced to $e^{-1} \approx 37\%$; if the arm number increases to six, the probability of one or five arms entering the cylindrical pore first will be reduced to $e^{-2} \approx 14\%$; and so on, the probability of a star chain passing through a cylindrical pore in an extremely asymmetrical manner rapidly decreases as the arm number increases.

As for the third point, eq. (8.8) shows that for a four-arm star chain, even if it enters the cylindrical pore in a symmetrical manner with the lowest energy, its critical flow is already twice that of a linear chain. This provides a possibility to separate linear and star chains with a similar hydrodynamic volume. This separation has important applications. For example, in the "grafting-to" method of preparing star chains, narrowly distributed linear arms with one reactive end group A are prepared first, and then, an excess amount of the linear arms react with one molecule with f reactive groups B to form a star chain with f arms through a reaction between A and B. After the reaction is completed, the solution mixture inevitably contains a large amount of unreacted linear arms. From the discussion on the radius of gyration of star chains in Chapter 4, even when f is large, the size of a star chain and its arm is only ~1.7 times. How to separate them has been an experimental problem for a long time.

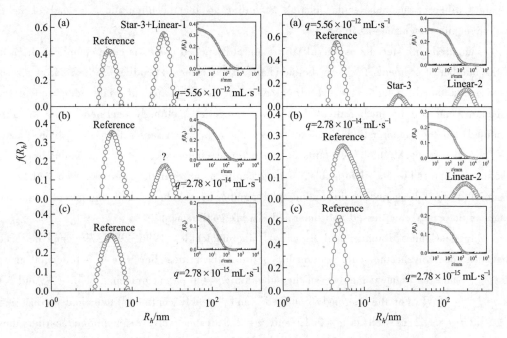

Ge H, Wu C. Macromolecules, 2010, 43: 8711. (Figures 1 and 3, permission was granted by publisher)

The figure above shows hydrodynamic radius distributions of a solution mixture containing linear short chain (internal standard), three-arm star chain, and linear chains with a hydrodynamic radius similar to the star chain after the ultrafiltration. Look at the left side of the figure (a) ~ (c) first, except for the peak related to the internal standard short chains, (a) when the flow rate is very high, both the star and linear long chains pass through the cylindrical pore and appear in the range of 20-30 nm. On the contrary, (c) when the flow rate is low, the corresponding peaks of the star and linear long chains disappear, indicating that they are not able to pass through the cylindrical pore; (b) when the flow rate is in between, the peak area is about half smaller, but it is impossible to know whether star or linear chains are blocked by the pore?

On the right side of the figure (a) ~ (c), the linear chain is replaced with very long linear chain ($R_h \sim 200$ nm). (a) When the flow rate is high, both the star chain and the linear long chain can pass through the small pore, and two peaks appear; on the contrary, (c) when the flow rate is very low, neither of them can pass through the small pore, and the corresponding two peaks disappear; (b) when the flow rate is in between, the peak corresponding to the star chain disappears completely, while the peak corresponding to the linear long chains remains. As predicted, the linear long chain with a larger size can still pass through the cylindrical pore under this flow rate, while the star chain with a size nearly ten times smaller is blocked. Therefore, (b) on the left, the blocked chain must also be the star chain. This result clearly shows that in an ultrafiltration experiment, by selecting the appropriate flow rate, one can separate different macromolecular chains according to their configuration, instead of using hydrodynamic dimensions as in SEC.

Compared with star chains, the ultrafiltration of branched chains is more complicated. Even if the number of segments (N_b) in the subchains is the same, depending on the relative size of its size and the diameter of the cylindrical pore, the branched parent chain composed of the subchains can still be further divided into two categories, "strongly confined" and "weakly confined". If the conformational size of the subchain is much larger than the "blob" size, $bN_b^{\alpha R} \gg \xi$, that is, each "blob" contains only one linear chain link, so that the parent branched chain is compressible and deformable, "strong confinement"; on the contrary, each "blob" is made of a small branched chain, including a number of subchains, and the parent branched chain is not easily compressed and deformed, "weak confinement".

Gay, de Gennes, Raphael and Brochard [Macromolecules, 1996, 29:8379] once deduced that in a good solvent, the scale relationship between the critical flow rate of a branched chain ($q_{c,branch}$) and the segment numbers of the subchain and the parent branched chain (N_b and N) is $q_{c,branch} \sim N_b^{\varphi} N^{\gamma}$. For the "strongly confined" and "weakly confined" branched chains, $\varphi = -1/2$ and $\gamma = 1/2$; as well as $\varphi = 2/15$ and $\gamma = 2/3$, respectively. As mentioned earlier, they mistakenly took the energy of a layer of (d^2/ξ^2) "blobs" as one heat energy. Actually, each "blob" has one heat energy, so that their deduced energy is (d^2/ξ^2) times smaller.

The method to correctly find the "blob" size (ξ) of the confined branched chain in eq. (8.5) is as follows: Assume that each "blob" contains N_ξ segments with a length of b, and the confined branched chain has a uniform segment density over a size equal or larger than ξ, then the segment density of a "blob" is equal to the segment density of the entire confined "cigar-like" chain, namely

$$\frac{N_\xi}{\xi^3} = \frac{N}{d^2 L_0} \quad \text{or} \quad \xi = \left(\frac{N_\xi d^2 L_0}{N} \right)^{1/3}$$

where L_0 is the length of a stretched chain at equilibrium. According to the discussion in Chapter 4, the free energy of a conformational deformed chain consists of two parts, one is related to the two-body interaction (also entropy change); the other is the energy change related to the entropy elasticity. Using eq. (4.4), for the deformed branched chain, the volume of the chain (R^3) is replaced by d^2L; The square of the expanded length (R^2) is replaced by L^2. In addition, $R_0^2 \cong b^2 N_b^{1/2} N^{1/2}$. The free energy ($A$) of the confined and deformed branched chain is

$$\frac{A}{k_B T} = \frac{b^3 N^2}{d^2 L} + \frac{L^2 - R_0^2}{R_0^2}$$

Differentiating the free energy with respect to the length (L), and let $dA/dL = 0$, one can obtain the equilibrium length (L_0) at the minimum free energy,

$$\frac{\partial A/\partial L}{k_B T} = -\frac{b^3 N^2}{d^2 L_0^2} + \frac{2L_0}{R_0^2} = 0 \quad \rightarrow \quad L_0 \approx b \left(\frac{NR_0}{d} \right)^{2/3}$$

For the "strongly confined" and "weakly confined" branched chains, each "blob" is composed of a linear link and a branched link, respectively. In Chapter 4, it has been known that a linear chain can be regarded as a special branch chain, $N_b = N$, so that $R_0^2 \cong b^2 N_b^{1/2} N^{1/2}$ is also suitable for a linear chain ($R_0^2 \cong b^2 N$). Therefore, in a good solvent, the same scaling relationship can also be used to represent the dependence of the average end-to-end distance (R) of the linear link and the branched link on N_b and N, i.e., $R = \sqrt{R^2} \cong b N_b^{\beta} N^{\alpha_R}$. ξ can also be similarly represented as $\xi \cong b N_b^{\beta} N^{\alpha_R}$. Substituting the above relations into eq. (8.5), one gets

$$q_{c,branch} = q_{c,linear} \left(\frac{b}{d} \right)^{\psi} N_b^{\varphi} N^{\gamma} \tag{8.9}$$

where

$$\psi = \frac{2(3 - 5\alpha_R)}{3(3\alpha_R - 1)}; \quad \varphi = \frac{6\beta - \alpha_R}{3(3\alpha_R - 1)}; \quad \gamma = \frac{\alpha_R}{3(3\alpha_R - 1)}$$

The above equation summarizes different solvent qualities, as well as the "strongly confined" and "weakly confined" two categories, where $\beta = 0$ and $\alpha_R = 3/5$; as well as $\beta = 1/10$ and $\alpha_R = 1/2$, respectively. Respectively substituting them into the above equation results in

$$\text{strongly confined}: \frac{q_{c,branch}}{q_{c,linear}} = \left(\frac{N}{N_b} \right)^{\frac{1}{4}};$$

$$\text{weakly confined}: \frac{q_{c,branch}}{q_{c,linear}} = \left(\frac{b}{d} \right)^{2/3} N_b^{1/15} N^{1/3}$$

Note: N/N_b represents the number of branching points. For given ultrafiltration experimental conditions and macromolecules, d and b are two constants. If the above two equations are equal, the conditions for the transition from the "strongly confined" to the "weakly confined" can be obtained as follows:

$$N_b^* = \left(\frac{d}{b} \right)^{40/19} N^{-5/19}$$

Generally speaking, $d/b \sim 10$; $N \sim 10^3$. Therefore, $N_b^* \sim 20$, that is, about 50 branch points. For a given composition of branched chains, the pore size from the "strongly confined" to the "weakly confined" is also be calculable as follows.

$$d^* = b N_b^{19/40} N^{1/8}$$

Usually, $b \sim 1$ nm; $N_b \sim 10$. Therefore, $d^* \sim 7$ nm, ~ 10 times the segment length. It is achievable experimentally.

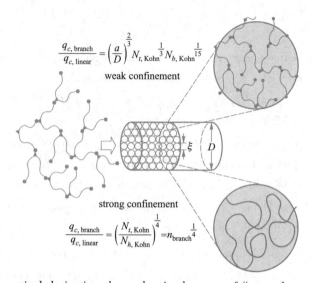

$$\frac{q_{c,\,\text{branch}}}{q_{c,\,\text{linear}}} = \left(\frac{a}{D}\right)^{\frac{2}{3}} N_{t,\,\text{Kohn}}^{\frac{1}{3}} N_{b,\,\text{Kohn}}^{\frac{1}{15}}$$

weak confinement

strong confinement

$$\frac{q_{c,\,\text{branch}}}{q_{c,\,\text{linear}}} = \left(\frac{N_{t,\,\text{Kohn}}}{N_{h,\,\text{Kohn}}}\right)^{\frac{1}{4}} = n_{\text{branch}}^{\frac{1}{4}}$$

The above theoretical derivation shows that in the case of "strongly confined", the critical flow rate of branched chains increases slowly with the number of branch points. When the number of branch points is 10, it is nearly twice that of the linear chain. In the case of "weakly confined", the critical flow rate of branched chains increases very slightly with the segment number of the subchain, but increases with the segment number of the branched chain. Physically, the above two equations are reasonable. For a given total number of segments, in one case, the critical flow rate of the branch chain decreases with the subchain length (the branching point decreases, easier to be compressed), and in another case, the critical flow rate of the branch chain is not sensitive to the subchain length (highly branched, not easy to be compressed). The figure above schematically summarizes the two different cases of "strongly restricted" and "weakly restricted" of a branched chain in ultrafiltration.

The experiment is rather difficult! Therefore, in the first thirty years after the early theory came out, there are few reliable experimental results that can test the theory. Experimentally, two sets of samples are required before the dependences of the critical flow rate on the subchain length and the total number of segments: one with a similar total segment number but different subchain lengths; and another with the same subchain length but different total segment number. In addition, at least two membrane filters with different pore sizes have to be used to study the pore size dependent critical flow rate. As previously mentioned, after a clever "seesaw-type" macromonomer was developed, using the click chemistry, the macromonomers are linked together to form the required two sets of branched chains. The ultrafiltration experiments of using these two sets of branched chains led to the following equation about ten years ago.

$$q_{c,\,\text{branch}} \propto d^{-4/5} N_b^{-2/5} N^{1.0 \pm 0.1}$$

where the exponent on d is not far away from the theoretical value ($-2/3$), acceptable; the exponent on N_b comes from only three samples with different subchain lengths, so that the error is large, deviating from the theoretical value $(-1/4)$ but not too much. However, the exponent on N is far away from the theoretical value $(1/4$ or $1/3)$. Even if the experimentally measured $\alpha_R = 0.46$ is substituted into the above calculation of γ, only $2/5$ is obtained, still far away from the experimental value $(1.0 + 0.1)$. Why is there such a big difference between the predicted value and the experimental result? Let us go back to examine the original assumptions in the earlier theory.

Physically, the branched chain is compressed in the cylindrical pore, and its conformation has been deformed and distorted. Therefore, the assumption in the original theory can no longer be used, i.e., the chain size does not follow the scaling relationship when it is free in solution.

The limit is to compress a branched chain into a small ball with a uniform density, $\alpha_R = 1/3$, $\gamma \to \infty$. Obviously, it does not meet the experimental result. Therefore, the experimentally measured $\gamma = 1$ can be substituted into the theoretical equation to obtain $\alpha_R = 3/8 = 0.375$, larger than $1/3$, but much smaller than the experimentally obtained 0.46, which indicates that the branched chain is indeed compressed and denser, but it is still not a small ball with a uniform density.

It is precisely based on the experimentally determined a series of quantitative relationships between the critical flow rate of macromolecular chains with different configurations and the molecular parameters of the chains that one can possibly use the ultrafiltration to separate different macromolecular chains according to their configurations, and solve problems that cannot be solved by other experimental methods (such as size exclusion chromatography). For example, in macromolecular researches, it had been helpless ton separate branched chains with a similar hydrodynamic volume but different structures. After fully understanding how the critical flow rate of branched chains depends on various molecular properties in the ultrafiltration, Li [Macromolecules, 2012, 45: 7583] successfully used ultrafiltration in the author's laboratory to separate the branched macromolecules with a similar total segment number but different subchain lengths, and finally solved this difficult experimental problem. Details can be found in the relevant literature.

In life sciences, the migration of protein and ribonucleic acid chains through numerous nuclear pores on the nuclear membrane of eukaryotic cells is a typical and interesting problem related to how macromolecular chains pass through a small pore. When the cell is active, $\sim 10^3$ biological macromolecular chains pass through the nuclear pore every second. This brings up two questions. First, why are these membrane-crossing macromolecular chains linear in the cells, especially RNA? an accident or a natural choice? Is it purely due to the requirements of biological inheritance or is it due to other reasons? Second, since biological macromolecules pass through the nuclear pores in to and out of the nuclear membrane at such a high frequency, would they collide with each other inside the nuclear pores? If yes, the efficiency would be very low. Therefore, it is likely that they do not collide with each other inside the nuclear pores, but why? In addition to the biologically known active transport mechanism that relies on biological energy, is there a passive transporting mechanism driven only by thermal energy?

The study on a porous and thin-shell hollow sphere shows that under the agitation of thermal energy, a liquid-filled and swollen hollow sphere exhibits "breathing-like" vibration, as shown on the right. In order to show this vibration, the figure on the right exaggerates the vibration amplitude. The tiny thermal energy is not sufficient to alternate the size of the entire hollow sphere. The vibrating

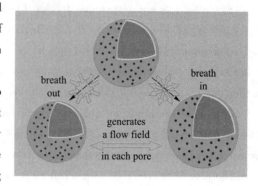

frequency (f) is proportional to the elastic modulus of the porous shell, and inversely proportional to the medium viscosity and the radius of the sphere (r), $f = m/r$, where m is a constant containing the elastic modulus and the medium viscosity. The above discussion of the internal fluctuations of a microgel has revealed that thermal energy can only excite a 50 nm internal structure of a gel network, so that thermal energy can excite a local vibration of the porous shell, like a surface wave.

Unlike a non-porous airtight hollow sphere, this fluctuation will definitely cause a volume change of the hollow sphere. The volume change is $\Delta V = 4\pi r^2 \Delta r$. Assuming the number of the small pores per unit surface area is σ, that is, the total number of the small pores (n) is $\sigma 4\pi r^2$, one can calculate the flow rate through each small pore induced by the "breathing" vibration as follows.

$$q = \frac{\Delta V}{n} f = \frac{m}{\sigma} \frac{\Delta r}{r}$$

Obviously, the "breathing" induced flow rate (q) through each pore is proportional to the relative change of the size of the small hollow sphere, which is very important for subsequent discussion since the nucleus of different cells has different sizes. Substituting the typical m and σ of the nuclear membrane into the above equation and even assuming that $\Delta r/r \sim 1\%$, q is still much larger than $q_{c,\text{linear}} \sim 10^{-14}$ mL/s. Therefore, the biological macromolecule chains near the membrane pores can be passively sucked in or spitting out of the cell nucleus.

The above discussion shows that only for linear macromolecular chains, the critical flow rate is independent of the chain length. Therefore, the linear configuration here is extremely important. If biological macromolecules inside the cell nuclear membrane had adopted other non-linear configurations, the hindrance for them to enter and exit the nuclear membrane would not only be larger, but also vary with their size and configuration. This is not in line with evolutionary advantages. In addition, at every moment, each nuclear pore can only be in one of the two states of "breathing" or "breathing", so that this passive "breathing in" or "spitting out" mechanism naturally avoids the collision of biological macromolecules inside each nuclear pore.

This brings about a new problem: if the time (t) for a biological macromolecule chain to pass through a nuclear pore is longer than the reciprocal of the "breathing" frequency ($1/f$), it would be repeatedly inhaled and exhaled inside a nuclear pore with a length of l and a size of r_p, not passing through the nuclear pore. Using the flow velocity $u = q/(\pi r_p^2)$, one can estimate t, i.e.,

$$t = \frac{l}{u} = \frac{l\pi r_p^2}{q} = \frac{l\pi r_p^2 n}{\Delta V}\frac{1}{f} = \frac{l\pi r_p^2 n}{4 r^2 \Delta r}\frac{1}{f}$$

In actual cells, $l \sim \Delta r$, $r_p/r \sim 10^{-3}$, and $n \sim 10^3$, so that $t \sim 10^{-3}/f$. In the process of "exhaling" and "inhaling", the time for biological macromolecular chains to pass through the nuclear pore is much shorter than that for each nuclear pore to "breath" once. Therefore, there is no need to worry about the problem whether a biological macromolecular chain oscillates back and forth inside the nuclear pore.

Summary

A shearing or stretching flow field alternates the chain conformation of a macromolecule, driving the chain to deviate from its lowest energy equilibrium state. A chain that deviates from its equilibrium state always tries to return to its equilibrium state with its characteristic relaxation time ($10^{-4} - 10^{-3}$ s). Only when the following two conditions are met at the same time, can a macromolecular chain be stretched and deformed until reaching a new equilibrium conformation in the flow field. First, the product of the reciprocal of shear rate gradient and its characteristic relaxation time is larger than one, implying that the time required to deform one chain size is shorter than the characteristic relaxation time of the chain. In other words, for a given size (length), the deformation is faster than the relaxation. Quantitatively speaking, the deformation speed should reach the size of $10^3 - 10^4$ chains per second. Second, the interaction time between the chain and the shear flow field is longer than its characteristic relaxation time, that is, the chain needs to stay at a very fast shear rate for a sufficiently long time.

The theoretical research on this issue originated in the mid-1960s. In the 1970s, de Gennes and his collaborators were the main promoters of theoretical research. They first discovered that in a transverse shear flow field, the "coil-to-stretch" conformational transition of a linear chain is a secondary process, i.e., a gradual stretching as the shear velocity gradient increases; while in a longitudinal elongation flow field, the process is the first order, called the "coil-to-stretch" conformational transition, that is, there exists a critical flow rate. It determines whether a flow filed can stretch a macromolecular chain. Afterwards, they also theoretically studied star and branched chains. However, after more than 30 years of research, experimentally, only a gradual change of the chain conformation as the shear rate increases had been measured. The predicted first-order sudden conformational transition was not observable.

Until 2006, the first-order "coil-to-stretch" conformational transition predicted by this theory was proved in an ultrafiltration experiment by using a double-layered membrane with a special structure and a small-angle LLS device. At present, based on a large number of ultrafiltration experiments, the critical flow rates of linear, star, and different branched macromolecular chains have been observed. The experimental results have led to some revision of de Gennes's early theories, mainly including

The experimental values of the critical flow rate of linear chains are one to two orders of magnitude smaller than that predicted by de Gennes, which reveals that the elongation flow field forces solvent molecules to pass through each "blob", causing a strong "drainage" effect, so that each "blob" consisting of a short chain link cannot be treated as a "non-draining" solid ball. The size (ξ) of a "blob" has to be replaced by an effective hydrodynamic length (l_e). Therefore, $q_{c,\text{linear}} = k_B T / (3\pi\eta) (d/l_e)$; $d/l_e = (d/b)^{1-1/\alpha_R}$, not $k_B T / (3\pi\eta)$ predicted by de Gennes, where b is the segment length, $1/2 \leqslant \alpha_R \leqslant 3/5$. Such a revision has been proved by a first-principle calculation. Although the critical flow rate of linear macromolecular chains has nothing to do with the chain length as predicted by theory, but it decreases with an increase of the pore size, not as previously predicted that it is independent of the diameter of a cylindrical pore.

For star chains, two traditional relationships between the critical flow rate and the number of forward arms (f_f) and the total arm number (f) is united into one mathematic expression: $q_{c,\text{star}} = q_{c,\text{linear}}(f + |f - f_f|)/2$, not related to the arm length, but proportional to the arm number.

In the confined state of a branched chain in the small pore, there are d^2/ξ^2 "blobs" on the cross section; each "blob" has energy of $k_B T$, and a layer of the "blobs" have a total energy of $(d^2/\xi^2) k_B T$, not $k_B T$ mistakenly assumed by de Gennes (treating a branched chain as a linear one). The correct scaling exponent between the critical flow rate and (d/ξ) is 2, not 4. This mistake was corrected.

Regarding the linear chain as a special branched chain with only one branch point, two double-scaling relationships respectively between the critical flow rate of the "strongly confined" and "weakly confined" branched chains and the segment numbers of the subchain and the parent chain (N_b and N): $q_{c,\text{branch}} = q_{c,\text{linear}} (b/d)^{\psi} N_b^{\varphi} N^{\gamma}$, where the three scaling exponents are all related to the solvent quality (α_R), Φ is related to the subchain length.

When a "weakly confined" branched chain is compressed inside a cylindrical pore, the scaling exponent between its size and the segment number is smaller than the scaling exponent when it is free in the solution, decreasing from 0.46 (experimental value) and 1/2 (theoretical value) when it is free to 0.38, revealing that the confined branched chain is more denser, but still not a ball with a uniform density. Therefore, it is still necessary to perfect the ultrafiltration theory of branched chains.

A universal equation of the critical flow rate has been successfully derived: $q_c = q_{c,\text{linear}}$

$(d/\xi)^2$. For macromolecular chains of different configurations, it is only necessary to calculate the relationship between ξ and d. Linear chains: $(d/\xi)^2 = 1$; star chain: $(d/\xi)^2 = (f + |f - f_f|)/2$; and different branched chains: $(d/\xi)^2 = (b/d)^\psi N_b^\varphi N^\gamma$.

Utilizing the quantitative relationship between the critical flow rate and molecular parameters of macromolecular chains with different configurations and selecting an appropriate flow rate, one use the ultrafiltration to separate macromolecular chains with a similar hydrodynamic size but different configurations in a solution mixture, so that the problem that other separation methods are helpless has been solved. The figure below summarizes how macromolecular chains with different configurations "crawl" through a cylindrical pore with a diameter of d and a length much longer than the chain length in an elongation flow field.

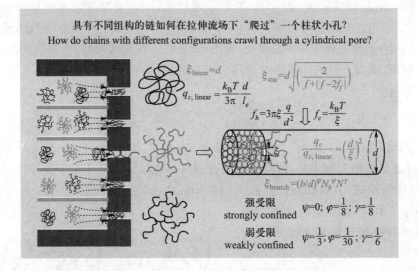

第九章　大分子亚浓溶液

　　沿用第三和第四章中的符号定义,溶液中有 N_p 条长度相同的大分子链,每条链含有 N 个长度为 b、质量为 m_0 的链段。链的质量为 $m = m_0 N$,摩尔质量为 $M = m N_{AV}$。溶液中链的总重量(严格地说,应是质量)为 $W = m N_p$。不难想象,随着大分子浓度的增加,链间距逐渐缩短、互相作用逐渐增强。最终,达到一个链和链之间开始互相接触(交叠)的浓度,记为 C^*。质(重)量浓度的常用单位是克/毫升($\mathrm{g/mL}$)。在此浓度时,所有链的体积之和等于溶液的体积,而每条链的体积约等于其平均末端距($R = b N^{\alpha_R}, 1/2 \leqslant \alpha_R \leqslant 3/5$)的立方,即 $R^3 N_p = V$。因此,

$$C^* = \frac{W}{V} = \frac{m_0 N N_p}{(b N^{\alpha_R})^3 N_p} = \frac{m_0}{b^3 N^{3\alpha_R - 1}} = \frac{M}{R^3 N_{AV}} \tag{9.1}$$

　　其中,对给定的大分子,b, m_0, R 和 M 均为常数。R 和 M 为两个可测的物理量,而且 $R \propto M^{\alpha_R}$。所以,C^* 随着链长增加而递减,递减速度取决于溶剂性质。大分子浓度高于接触浓度时,溶液称作"亚浓溶液"。对聚合物而言,$R \sim 10^{-5}\,\mathrm{cm}, M \sim 10^5\,\mathrm{g/mol}$,故 $C^* \sim 10^{-3}\,\mathrm{g/mL}$。

　　与小分子不同,大分子链在溶液中的构象和溶液结构会随着浓度进一步地增加出现以下两种完全不同的变化。第一,如果链的长度(L)短于缠结长度(L_e),或曰大分子的链段数(N)小于缠结链段数(N_e),链构象蜷缩、尺寸变小、保持触碰状态,即使在熔融状态,浓度达到了 100%,大分子链也不会互相进入对方所占据的空间。其物理性质类似小分子溶液或溶体。第二,当 $L > L_e$ 时,链之间会互相穿插,在每条大分子链占据的空间(体积,$R^3 = b^3 N^{3\alpha_R}$)里面共有 N_{p,R^3} 条大分子链,如右图所示。对一个均匀体系,局部与整体的链段密度相等,所以

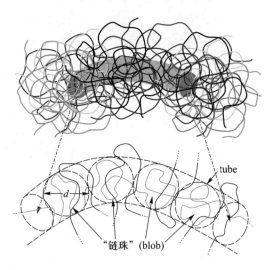

$$\frac{N_{p,R^3}}{R^3} = \frac{N_p}{V} \quad \rightarrow \quad N_{p,R^3} = \frac{R^3 N_{AV}}{M} C = \frac{C}{C^*} \tag{9.2}$$

　　可见,对于可缠结的大分子链,当浓度大于接触浓度时,每条链占据的空间内的链的数目(N_{p,R^3})正比于浓度。在 Θ 状态时,对于给定的浓度,$N_{p,R^3} \propto N^{1/2}$,随链长增加。在稀

溶液中，$C<C^*$，$N_{p,R^3}=1$。对常见高聚物而言，$b\sim10^{-7}\,\mathrm{cm}$；$m_0N_{AV}\sim10^2$；和 $N\sim10^3$。假定浓度约为 $10^{-1}\,\mathrm{g/mL}$，$N_{p,R^3}\sim10-100$，这意味着几十条链互相纠缠在同一个空间内。随着浓度的进一步增加，两个紧邻缠结点之间的链节长度逐渐缩短，当链节长度趋向一个链段长度时，大分子溶液进入另一个称作"浓溶液"的区域，常记作 C^{**}。因此，$C^{**}=m_0/b^3$，故

$$C^{**}=C^*N^{3\alpha_R-1} \tag{9.3}$$

当溶液浓度很高时，即使在良溶剂中，一个链段被其他链上的大量链段围绕，色散吸引增强。理论和中子散射实验均已证明排除体积效应趋向消失，故可认为浓度高的溶液处于 Θ 状态，$\alpha_R=1/2$，即 $C^{**}/C^*=N^{1/2}$。对普通的高聚物，$N\sim10^3$，比值已超过 30 倍。因此，在"亚浓溶液"和"浓溶液"之间存在着一个很宽阔的浓度区间，且范围随着 N 递增。

链的纠缠造成了大分子一些独特的、小分子欠缺的物理性质，如前述的黏弹性。严格地寻获这些性质之间的定量关系不可避免地涉及物理上的多体问题，非常复杂，远远超出本书内容，故略去。下面，仅利用 de Gennes 的"管子"模型、"蛇行"图像和标度理论来描绘大分子链在亚浓溶液中的静态和动态行为。

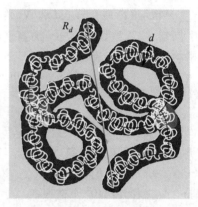

如右图所示，每条大分子链想象地处于一个由其周围互相缠结的链构成的"管子"之中，类似在上一章中已讨论的一条链在柱状小管中的情形。不同的是这里链并没有被拉伸，仍然保持着一个无规线团的构象。因此，困住一条链的"管子"自然也是无规线团状。千万不要将这里的"管子"和前一章中直的柱状小孔混淆。一条链上的每个链段均在"管子"中的局部范围内无规行走，造成有限的局部密度涨落，最大的步长就是管径(d)，长于一个链段(b)。为此，可将大分子链想象地"分割"成一段段的短链节。每个短链节含有 N_d 个链段，形成一个小的无规线团，即前述的"链珠"。每个短链节的方均末端距开方为管径，$d=bN_d^{\alpha_R}$。这样，一条大分子链就可被描绘成由一串 N/N_d 个"链珠"构成。因此，链和"管子"的构象可被描绘成无扰时，一个步长为 d 的、N/N_d 步无规行走后的轨迹。其平均末端距的开方为 $R_d=d\,(N/N_d)^{1/2}$。每个"链珠"内的短链节均可被拉直，故其方均末端距的开方短于其伸直长度：$d=bN_d^{\alpha_R}<bN_d\,(1/2\leqslant\alpha_R\leqslant3/5)$。因此，一条伸直的链长度($L_c=bN$)远远长于伸直的一串"链珠"或"管子"($L_{c,d}=d(N/N_d)$)。

所以，每条线型大分子链的构象可被描绘成两种不同步长(b 和 d)的无规行走轨迹。然而，大分子链的平均末端距是一个真实的物理存在，不因其不同的描绘而改变，故 $d\,(N/N_d)^{\alpha_R}=d\,(L_{c,d}/d)^{\alpha_R}=bN^{\alpha_R}$。对于一个均匀溶液，每个链节和每条链占据的体积内的链段密度相等，

$$\frac{N_d}{d^3}=\frac{NN_{p,R^3}}{R^3}$$

利用 $N_{p,R^3}=C/C^*$，可进一步得到

$$\frac{1}{b^3 N_d^{3\alpha_R - 1}} = \frac{N_{p,R^3}}{b^3 N^{3\alpha_R - 1}} \quad \rightarrow \quad \left(\frac{N}{N_d}\right)^{3\alpha_R - 1} = N_{p,R^3} \quad \rightarrow \quad \frac{N}{N_d} = \left(\frac{C}{C^*}\right)^{\frac{1}{3\alpha_R - 1}} \quad (9.4)$$

在上述的描绘中,没有一条大分子链特殊。每一条大分子链均处于由周围别的大分子链组成的"管子"之中。同时,其本身也是围绕着其他大分子链的"管子"中的一员。如前所述,当浓度很高时,排除体积效应消失,链处在 Θ 状态,$\alpha_R = 1/2$。因此,链的平均末端距与稀溶液中链的平均末端距($R = bN^{\alpha_R}$)之比为

$$\frac{R_d}{R} = \frac{d(N/N_d)^{1/2}}{bN^{\alpha_R}} = \frac{bN_d^{\alpha_R}(N/N_d)^{1/2}}{bN^{\alpha_R}} = \left(\frac{N}{N_d}\right)^{\frac{1}{2} - \alpha_R} = \left(\frac{C}{C^*}\right)^{\frac{1/2 - \alpha_R}{3\alpha_R - 1}}$$

在 Θ 状态,$\alpha_R = 1/2$。依照定义,链的尺寸与浓度无关,因此,$R_d = R$。而在良溶剂中,上式成为 $R_d = R(C/C^*)^{-1/8}$。所以,在良溶剂的亚浓溶液中,链的平均末端距随着浓度的增加而缩短。当 C/C^* 从 10 增至 40 时,链收缩的程度从 75% 变至 63%。注意,链的收缩并没有改变 R 和 N 二者之间在亚浓溶液中的标度关系,$R = d(N/N_d)^{1/2}$。改变的仅是"链段"长度($d/N_d^{1/2} = bN_d^{\alpha_R - 1/2}$)。浓度增大,$N_d$ 变小,导致 R 变小,仿佛总的链段数没变,但每个"链段"逐渐地变短。d 和 N_d 均与链长无关,仅依赖于浓度。

同样,微观参数 d(管径)随着重(质)量浓度(C)增大而变小。依据"一个均匀体系中的平均微观浓度和宏观浓度相等"这一引理可直接得到二者之间的依赖关系,具体如下:

$$\frac{m_0 N_d}{d^3} = C \rightarrow d = b^{\frac{1}{1 - 3\alpha_R}}\left(\frac{C}{m_0}\right)^{\frac{\alpha_R}{1 - 3\alpha_R}} \quad \rightarrow \quad d = R\left(\frac{C}{C^*}\right)^{\frac{\alpha_R}{1 - 3\alpha_R}} \quad (9.5)$$

推导时,利用了 $d = bN_d^{\alpha_R}$ 和式(9.1),$C^* = m_0/(b^3 N^{3\alpha_R - 1})$。亚浓溶液中的管径 d 仅依赖于质(重)量浓度和链段长度,与链长无关。在 Θ 和良溶剂中,上式分别成为 $d = R(C/C^*)^{-1}$ 和 $d = R(C/C^*)^{-3/4}$。对普通高聚物链,$R \sim 10^2$ nm,故当 $C/C^* \sim 10$ 时,$d \sim 10$ nm,对应着大约 10^2 个链段。

链段的无规运动引致密度 $\rho(r)$ 围绕着平均密度 ρ_0 涨落,或高或低,$\Delta\rho(r) = \rho(r) - \rho_0$,或正或负。仅考虑近程相互作用时,密度涨落的链段间距自相关函数为

$$\langle \Delta\rho(r)\Delta\rho(0)\rangle = \langle \rho(r)\rho(0)\rangle - \rho_0^2 \quad \rightarrow \quad \frac{\langle \Delta\rho(r)\Delta\rho(0)\rangle}{\rho_0^2} = \frac{\langle \rho(r)\rho(0)\rangle}{\rho_0^2} - 1$$

可以证明 $\langle \rho(0)\rho(0)\rangle = 2\rho_0^2$;$\langle \Delta\rho(r)\Delta\rho(0)\rangle/\rho_0^2$ 是归一的、密度涨落的链段间距自相关函数。随着 r 的增加,其从 100% 弛豫到 0%。上式由 Leonard Ornstein 和 Frits Zernike [Royal Netherlands Academy of Arts and Sciences(KNAW)Proceedings,1914,17:793] 于一百多年前提出,具体推导略去,数学表达为

$$g_{OZ}(r) = \frac{\langle \Delta\rho(r)\Delta\rho(0)\rangle}{\rho_0^2} = \frac{1}{4\pi\xi^2 r}\exp\left(-\frac{r}{\xi}\right) \quad (9.6)$$

其中,ξ 是特征相关长度,代表了在任意一点处的密度涨落可波及范围。将上式的两边同时乘上 $4\pi r^2 dr$ 后,代表了在半径为 r、厚度为 dr 的球壳里发现另一个链段的概率,

$$4\pi r^2 g_{OZ}(r)dr = \frac{r}{\xi^2}\exp\left(-\frac{r}{\xi}\right)dr = \frac{r}{\xi}\exp\left(-\frac{r}{\xi}\right)d\left(\frac{r}{\xi}\right)$$

当 $r \to 0$ 时,相关程度最高,但球壳体积却趋向零,故该概率为零。显而易见,在 $r \to 0$ 处,只有自己,没有其他链段。另一方面,当 $r \to \infty$ 时,该概率也趋向零。在 $r = \xi$ 处,该概率达至最大,这就是 ξ 的物理意义。当 $r = \xi$ 时,方均密度涨落已从最高衰减了约 37%。当 $r = 3\xi$ 时,上式已几乎衰减为零。换而言之,密度涨落波及的范围限于一个半径为 3ξ 的球体内。上式的 Fourier 变换即为静态结构因子,

$$p(q) = \frac{1}{V\rho_0^2} \int \langle \Delta\rho(r)\Delta\rho(0) \rangle \exp(i\vec{q} \cdot \vec{r})\,\mathrm{d}r = \frac{p(0)}{1 + \xi^2 q^2}$$

与在第五章中讨论静态激光光散射一节时的静态结构因子相比,$\xi^2 = \langle R_g^2 \rangle / 3 \approx R^2/18$,即 $\xi \approx R/4$。利用激光光散射,可先测量一个大分子亚浓溶液在不同角度的时间平均散射光强 $I(q)$;再将 $1/I(q)$ 对 q^2 作图。依据上式可得一条直线,从其截距和斜率分别可得 $1/I(0)$ 和 ξ^2。右图显示了在一个紫外辐射诱导交联大分子链的反应中,静态特征相关长度(斜率)随着交联程度的增加而递减。大分子链之间的交联限制了链的运动和迁移,密度涨落波及的

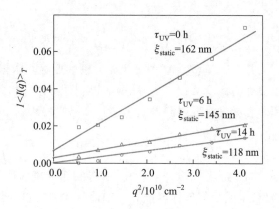

Ngai T, Wu C, Chen Y. Macromolecules, 2004, 37: 987.
(Figure 7, permission was granted by publisher)

范围变小,因此静态特征相关长度变短,详情可参考图中的文献。**注意:ξ 反映的是,在给定温度和压强下,整个溶液的一个集体性质,而 R_g 则是单个大分子链的一个性质。在讨论中,也可将其认作一个位于"链珠"内的"短链节"的性质。密度涨落受限于管径(d),故一定有 $\xi < d$。**

在大分子链缠结的状态下,热能已很难在短时间内搅动整条链的质心移动,仅仅"链珠"中的短链节上的 N_d 链段在"链珠"体积(d^3)内无规行走。如果将每个短链节看作一个"粒子",其质心无规行走可产生压强,$\pi_d = k_B T/d^3$。依据定义,该压强与溶剂分子造成的压强之差为渗透压差。在第五章中讨论渗透压差时已知,既然是压强之差,可将溶剂分子造成的压强取作参考零点。相比之下,在稀溶液中,热能的搅动可使每条链在溶液体积(V)内无规移动,产生一个压强,$k_B T/V$。理想状态下,N_p 条链产生的压强应为 N_p 倍,即 $\pi = k_B T N_p/V$。将 N_d/d^3 和 N_p/V 转为质量浓度,可得

$$\pi_d = RT \left(\frac{N_{AV} R^3 C^{3\alpha_R}}{M^{3\alpha_R}} \right)^{\frac{1}{3\alpha_R - 1}} \quad \text{和} \quad \pi = RT\frac{C}{M}; \quad \to \quad \frac{\pi_d}{\pi} = \left(\frac{C}{C^*} \right)^{\frac{1}{3\alpha_R - 1}} \tag{9.7}$$

其中,利用了改写的式(9.1):$(C^* N_{AV}/M)^{1/(3\alpha_R-1)} R^{3/(3\alpha_R-1)} = 1$。在 Θ 和良溶剂中,上式分别成为 $\pi_d/\pi = (C/C^*)^2$ 和 $\pi_d/\pi = (C/C^*)^{5/4}$。在 cgs 系统里,对亚浓溶液中典型的高聚物链,$R \sim 10^{-5}\,\mathrm{cm}$;$M \sim 10^5\,\mathrm{g/mol}$;$C \sim 10^{-2}\,\mathrm{g/mL}$;$C/C^* \sim 10$。因此,在 Θ 溶剂中,$\pi_d/\pi \sim 10^2$,每条大分子链在亚浓溶液中产生一个很大的压强。

如果将一个孔径为 d 的柱状小孔插入亚浓溶液,只要 $\xi < d$,大分子链就会因压强差而

自发地进入小孔,概率为100%。反之,如果 $\xi>d$,进入小孔的概率就和前述讨论一条大分子链进入小孔的情形完全相同。仅需注意,链尺寸(R)已被特征相关长度(ξ)取代,

$$\psi = \exp\left[-(\xi/d)^{\frac{5}{3}} + 1\right] \quad (\xi \geqslant R) \tag{9.8}$$

物理上,只要记住在缠结状态,一条大分子链的许多物理性质就与其长度无关或者与其长度只有特定的关系,就可以采用 de Gennes 的标度理论。理论上,任何一个物理量(Y)以浓度趋向零时的理想状态作为一个参考基准均可展开成一个浓度(C)的维里多项式,

$$Y = Y_{\text{ideal}}\left[1 + A_2\left(\frac{CN_{\text{AV}}}{M}R^3\right) + A_3\left(\frac{CN_{\text{AV}}}{M}R^3\right)^2 + \cdots\right] = Y_{\text{ideal}}f\left(\frac{CN_{\text{AV}}}{M}R^3\right) \tag{9.9}$$

其中,$CN_{\text{AV}}R^3/M$ 为一个无量纲的变量。依据式(9.1),$CN_{\text{AV}}R^3/M = C/C^*$。$f(C/C^*)$ 也被称为标度函数。R 与平均回转半径相关,和 M 一起是两个可测的物理量(例如,可利用静态激光光散射测得)。注意,$R = bN^{\alpha_R}$;$M = m_0NN_{\text{AV}}$。在浓度趋向两个极端时,

$$f\left(\frac{CN_{\text{AV}}}{M}R^3\right) \to 1 \quad (C \to 0); \qquad f\left(\frac{CN_{\text{AV}}}{M}R^3\right) \to \left(\frac{CN_{\text{AV}}}{M}R^3\right)^z \quad (C \gg 1)$$

与 z 阶项$(C/C^*)^z$ 相比,所有的低阶项均可忽略。先以亚浓溶液中的静态相关长度为例,说明如何作标度推导。如前所述,在极稀溶液中,$\xi \to \sim R/4 = bN^{\alpha_R}/4$;所以

$$\xi \approx \frac{bN^{\alpha_R}}{4}\left(\frac{C}{C^*}\right)^z = \frac{bN^{\alpha_R}}{4}\left(\frac{C}{m_0}\right)^z b^{3z}N^{(3\alpha_R-1)z} = \frac{1}{4}\left(\frac{C}{m_0}\right)^z b^{3z+1}N^{(3\alpha_R-1)z+\alpha_R}$$

推导中利用了略微变化的式(9.1),$1/C^* = b^3N^{3\alpha_R-1}/m_0$。依据定义,$\xi$ 和大分子的链长无关,故$(3\alpha_R - 1)z + \alpha_R = 0$ 或写成 $z = \alpha_R/(1 - 3\alpha_R)$。代入上式,可得

$$\xi \approx \frac{R}{4}\left(\frac{C}{C^*}\right)^{\frac{\alpha_R}{1-3\alpha_R}}$$

与式(9.5)比较,$\xi \approx d/4$。物理上的解释是,每个"链珠"独立,其密度涨落的特征尺寸为 $d/4$。以一个链段为起点,在距离为 4ξ 之处,找到另一个链段的概率已几乎为零。读者应该注意,在文献和其他教科书中,将 ξ 当作或画作管径 d 是极其误导的,物理概念模糊。

再以亚浓溶液中的渗透压差为例,在极稀溶液中$(C \to 0)$,$\pi_d \to \pi = k_{\text{B}}TN_{\text{AV}}C/M$。注意:这里没有将 $k_{\text{B}}N_{\text{AV}}$ 写成摩尔气体常数 R,是为了避免和链的平均末端距(R)混淆。所以,

$$\pi_d = \frac{k_{\text{B}}TN_{\text{AV}}C}{M}\left(\frac{C}{C^*}\right)^z = \frac{k_{\text{B}}T}{R^3}\left(\frac{C}{C^*}\right)^{z+1} = k_{\text{B}}T\left(\frac{C}{m_0}\right)^{z+1}b^{3z}N^{(3\alpha_R-1)(1+z)-3\alpha_R}$$

由于链的缠结,在短时间内,整条大分子链已经无法运动一段距离,故 π_d 应该与链长无关,$(3\alpha_R - 1)(1 + z) - 3\alpha_R = 0$ 或写成 $z = 1/(3\alpha_R - 1)$。代入上式得式(9.7),殊途同归,

$$\frac{\pi_d}{\pi} = \left(\frac{C}{C^*}\right)^{\frac{1}{3\alpha_R-1}}$$

另一个例子是溶液和溶剂化学势之差($\Delta\mu$)随着溶液中链的摩尔数(n_p)从稀溶液到

亚浓溶液的变化，$(\partial \Delta \mu / \partial n_p)_{T,p}$。由热力学已知，该变化与渗透压差($\pi$)相关。对给定的浓度，在等温的条件下，溶剂自由能的变化为 $dG=Vdp$，其随溶剂摩尔数的变化为溶剂化学势的变化($d\mu=Vdp$)。对于一个含有 n_s 摩尔溶剂的极稀大分子溶液，假定体积不变，两边同时积分。左边从溶剂的化学势积分到溶液的化学势得 $\mu_{\text{solution}}-\mu_{\text{solvent}}=\Delta\mu$；右边从 0 积分到 π 得 $V\pi$，故 $\Delta\mu=V\pi$。已知 $C=W/V=(W/M)M/V=n_p M/V$ 和 $\pi=N_{\text{AV}}k_B TC/M$。因此，

$$\left(\frac{\partial \Delta \mu}{\partial n_p}\right)_{T,p}=\left[\frac{\partial(V\pi)}{\partial(CV/M)}\right]_{T,p}=M\left(\frac{\partial \pi}{\partial C}\right)_{T,p}=N_B k_B T$$

其中，$(\partial \pi/\partial C)_{T,p}$ 是渗透压差的压缩系数。当 $C\to0$ 时，$(\partial \pi/\partial C)_{T,p}\to k_B TN_{\text{AV}}/M$；所以

$$\left(\frac{\partial \pi}{\partial C}\right)_{T,p,\rho}=\frac{k_B TN_{\text{AV}}}{M}\left(\frac{C}{C^*}\right)^z=\frac{k_B T}{CR^3}\left(\frac{C}{C^*}\right)^{1+z}=\frac{k_B T}{C}\left(\frac{C}{m_0}\right)^{1+z}b^{3z}N^{(3\alpha_R-1)(1+z)-3\alpha_R}$$

由于链缠结，$(\partial \pi/\partial C)_{T,p,\rho}$ 同样和链长无关，所以 $(3\alpha_R-1)(1+z)-3\alpha_R=0$ 或写成 $z=1/(3\alpha_R-1)$。代入上式，

$$\left(\frac{\partial \pi}{\partial C}\right)_{T,p,\rho}=\frac{k_B TN_{\text{AV}}}{M}\left(\frac{C}{C^*}\right)^{\frac{1}{3\alpha_R-1}}$$

在 Θ 和良溶剂中，分别有 $\alpha_R=1/2$ 和 $3/5$，故有 $(\partial \pi/\partial C)_{T,p,\rho}\sim C^2$ 和 $\sim C^{5/4}$。在静态激光光散射的讨论中，已知大分子溶液和溶剂的散射光强之差(ΔI)正比于 $C/(\partial \pi/\partial C)_{T,p,\rho}$。因此，在 Θ 和良溶剂中，当 $C\leqslant C^*$ 时，$\Delta I\sim C$；而当 $C>C^*$ 时，分别有 $\Delta I\sim C^{-1}$ 和 $\Delta I\sim C^{-1/4}$。因此，利用静态激光光散射，通过测量 ΔI 随着浓度的变化，可以从其斜率改变处获得链在溶液中互相接触的重叠浓度(C^*)。右图显示随着溶剂的蒸发，聚苯乙烯在良溶剂苯中的

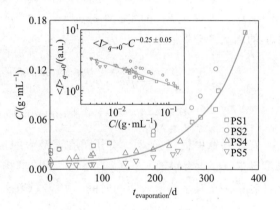

Li J F,Li W,Huo H,et al.Macromolecules,2008,41:901.
(Figure 1,permission was granted by publisher)

浓度逐渐增加，插图中的数据清楚地显示，平均散射光强与浓度的依赖性确实是理论上预测的-1/4。

最后一个例子是前面已经讨论过的大分子链在亚浓溶液中的平均末端距(R_d)。当 $C\to0$ 时，$R_d\to R=bN^{\alpha_R}$；当 $C\gg1$ 时，

$$R_d=bN^{\alpha_R}\left(\frac{C}{C^*}\right)^z=\left(\frac{C}{m_0}\right)^z b^{3z+1}N^{(3\alpha_R-1)z+\alpha_R}$$

物理上，作为大分子链在亚浓溶液中的平均末端距，R_d 一定依赖于链长，假定在亚浓溶液中其与链长的依赖性为 $R_d=bN^{\alpha'_R}$，那么 $(3\alpha_R-1)z+\alpha_R=\alpha'_R$，可得 $z=(\alpha'_R-\alpha_R)/(3\alpha_R-1)$。另一方面，已知在 Θ 状态下($\alpha_R=1/2$)，链的尺寸与浓度无关。因此，α'_R 必须也等于 1/2，即

与在 Θ 无扰状态时相同,在"管子"中的链的整体尺寸或"管子"尺寸(平均末端距)与溶剂性质无关。代入上式,可得链的平均末端距对浓度的依赖,

$$R_d = R \left(\frac{C}{C^*} \right)^{\frac{1/2 - \alpha_R}{3\alpha_R - 1}}$$

与先前推出的完全相同。换成实验上可测的方均末端距,$\langle R_d^2 \rangle / \langle R^2 \rangle = (C/C^*)^{-1/4}$(在良溶剂中),其中,给定大分子后,$R$ 和 C^* 均为常数,链随着浓度的增加而收缩。

在讨论了亚浓溶液的一些常见静态性质后,下面将阐述部分重要的动态行为。在讨论之前,先回顾以下有关"管子"模型中相关参数之间的关系。每条链可被看作一串 N/N_d 个相连的"链珠",每个"链珠"本身是一个含有 N_d 个链段和具有无规线团构象的短链节,其平均末端距 $d = b N_d^{\alpha_R}$。每条链困在一个由围绕着它的链构成的、弯曲的"管子"之中,链和困住它的"管子"具有相同的无规线团构象和平均末端距:$R = bN^{1/2} = d (L_{c,d}/d)^{1/2}$,或写成 $L_{c,d} = bN/N_d^{1/2}$,其中,$L_{c,d}$ 为"管子"的伸直长度。由于 $N_d > 1$,$L_{c,d}$ 远小于链的伸直长度($L_c = bN$)。

一般而言,一条链在稀溶液中无规行走时的摩擦系数和平动扩散系数分别为 $f = 6\pi\eta d$ 和 $D = k_B T/f$。当 $C \leqslant C^*$ 时,有两种极限情形:"排水"(Rouse 模型),每个长度为 b 的链段均受到一个摩擦阻力,$d = bN$;"非排水"(Zimm 模型),每条链均可被当作一个尺寸为 R(严格地说,应该是 R_h)的实心小球,$d = R_h \sim R = bN^{\alpha_R}$。对应的平动扩散系数就是在稀溶液中链的相互平动扩散系数,记为

$$D_0 = \frac{k_B T}{6\pi\eta bN} \qquad \rightarrow \qquad D_0 \sim N^{-1} \qquad (C < C^*, \text{"排水"})$$

$$D_0 = \frac{k_B T}{6\pi\eta R} \qquad \rightarrow \qquad D_0 \sim N^{-\alpha_R} \qquad (C < C^*, \text{"非排水"}) \tag{9.10}$$

在前一章中,已经证明大分子链在溶液中的接近非排水;实验结果也已证明,在 Θ 和良溶剂中,D_0 非常接近 $\sim N^{-1/2}$ 和 $\sim N^{-0.59}$,和"非排水"的预期标度指数相符。链通过平动扩散弛豫一个其平均末端距的时间 $[\langle R^2 \rangle/(6D_0)]$ 分别为

$$\tau_0 = \frac{\pi\eta bNR^2}{k_B T} = \frac{\pi\eta b^3}{k_B T} N^{1+2\alpha_R} \qquad \rightarrow \qquad \tau_0 \sim N^{1+2\alpha_R} (C \rightarrow 0 < C^*, \text{"排水"})$$

$$\tau_0 = \frac{\pi\eta R^3}{k_B T} = \frac{\pi\eta b^3}{k_B T} N^{3\alpha_R} \qquad \rightarrow \qquad \tau_0 \sim N^{3\alpha_R} (C \rightarrow 0 < C^*, \text{"非排水"}) \tag{9.11}$$

对常见的高聚物,在 cgs 体系中,$\eta \sim 10^{-2}$;$b \sim 10^{-7}$;$k_B T \sim 10^{-14}$;$N \sim 10^3$;所以 $\tau_0 \sim 10^{-3}$s。显然,这两种情形分别对应着 Rouse 和 Zimm 模型中的第一简正模式的特征弛豫时间(τ_1),其中,分别有 $\alpha_R = 1/2$(Rouse 模型)和 $1/2 \leqslant \alpha_R \leqslant 3/5$(Zimm 模型)。

当 $C > C^*$ 时,由于缠结,在短时间内,不仅整条大分子链而且每个"链珠"里的链节已不可长距离移动。链段在每个"链珠"的局部范围内无规行走引致的密度涨落的特征相关长度为 ξ。如果在 ξ^3 空间中有 N_ξ 个链段以及这 N_ξ 个链段可被看作一个小球,其尺寸为平均末端距则为 $\xi = b N_\xi^{\alpha_R}$。每个小球的摩擦系数也取决于排水($f = 6\pi\eta b N_\xi$)和非排水($f = 6\pi\eta\xi$)。对应的平动扩散系数为

$$D_\xi = \frac{k_B T}{6\pi\eta b N_\xi} \sim N^0 \qquad \rightarrow \qquad D_\xi = D_0 \left(\frac{C}{C^*}\right)^{\frac{1}{3\alpha_R-1}} \quad (C > C^*, \text{"排水"})$$

$$D_\xi = \frac{k_B T}{6\pi\eta\xi} \sim N^0 \qquad \rightarrow \qquad D_\xi = D_0 \left(\frac{C}{C^*}\right)^{\frac{\alpha_R}{3\alpha_R-1}} \quad (C > C^*, \text{"非排水"}) \qquad (9.12)$$

其中,利用了 $\xi = b N_\xi^{\alpha_R}$ 和 $N/N_\xi = (C/C^*)^{1/(3\alpha_R-1)}$。$N_\xi, \xi$ 和 D_ξ 与链长无关;D_0 和 C^* 与浓度无关。因为,该弛豫源于体积 ξ^3 中所有链段的协同运动,故 D_ξ 也被称为协同扩散系数,记作 D_{coop}。利用动态激光光散射,可以在实验上测量因密度涨落引致的在频率空间里线宽的光强分布 $G(\Gamma)$ 或由该分布算出的 z 均线宽 $\langle\Gamma\rangle$,其中,$\Gamma = D_\xi q^2$。因此,可在不同的散射角度测量 Γ,然后从 Γ 对 q^2 的作图可得 D_ξ。

右图显示了利用动态激光光散射测得的一个窄分布的聚苯乙烯样品在良溶剂苯中的亚浓溶液(C/C^* 高达30)的散射光强时间自相关函数 $G^{(2)}(t)$,其中,原本为延迟时间(t)的横坐标经过了 q^2 的修正,从而消除了散射角度的影响。修正后,来自四个不同散射角度的时间自相关函数曲线重叠在一起。它们的 Laplace 反演对应的都是一个很窄的线宽分布,只有一个较快弛豫峰。其中的插图显示了平均线宽对 q^2 作图是一条过原点的直线,说明该弛豫的物理

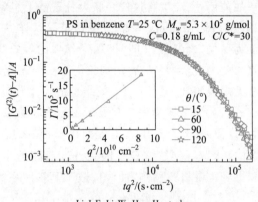

Li J F, Li W, Huo H, et al.

Macromolecules, 2008, 41:901

(Figure 7, permission was granted by publisher)

本质是平动扩散,直线的斜率是 D_ξ;对应的特征弛豫时间为短链节的质心扩散距离 ξ 所需的时间,$\tau_\xi = \xi^2/(6D_\xi)$,即

$$\tau_\xi = \frac{\pi\eta\xi^3 N_\xi^{1-\alpha_R}}{k_B T} \sim N^0 \qquad \rightarrow \qquad \tau_\xi = \tau_0 \left(\frac{C}{C^*}\right)^{-\frac{2\alpha_R+1}{3\alpha_R-1}} \quad (C > C^*, \text{"排水"})$$

$$\tau_\xi = \frac{\pi\eta\xi^3}{k_B T} \sim N^0 \qquad \rightarrow \qquad \tau_\xi = \tau_0 \left(\frac{C}{C^*}\right)^{-\frac{3\alpha_R}{3\alpha_R-1}} \quad (C > C^*, \text{"非排水"}) \qquad (9.13)$$

对常见的聚合物,$\xi = 10^{-6} \sim 10^{-5}$ cm。因此,$\tau_d = 10^{-6} \sim 10^{-5}$ s。随着浓度增大极速地变短,线宽增加,弛豫过程加快。对应线宽的范围为 $10^5 \sim 10^6 s^{-1}$,正如上图所示。

比较式(9.10)和式(9.12)可见,稀溶液中的 R_h 被亚浓溶液中的 ξ 取代,在讨论大分子稀溶液时已知,$1/R_h = \langle 1/r \rangle$。这里,可用类似的证明,

$$D_\xi = \frac{k_B T}{6\pi\eta} \left\langle \frac{1}{r} \right\rangle$$

其中,r 是 N_ξ 个链段中任意两个协同运动的链段之间的距离。其分布遵循式(9.6),故

$$\left\langle \frac{1}{r} \right\rangle = \int_V \frac{1}{r} g_{OZ}(r)\, \mathrm{d}V = \int_0^\infty \frac{1}{r} g_{OZ}(r)\, 4\pi r^2 \mathrm{d}r = \frac{1}{\xi}$$

因此,在获得 D_ξ 后,就可利用上式算出动态相关长度 (ξ)。在 20 世纪七八十年代,许多实验室研究了不同高聚物的亚浓溶液。大量的实验结果已证实,在 Θ 和良溶剂中,存在标度关系 $D_\xi \sim C^{1.0}$ 和 $D_\xi \sim C^{0.75} D_\xi$,与预期的"非排水"的标度指数相符。

以上是在较短的时间内,链的动态学行为。如果给以一个足够长的时间,每条困在由其他链组成的"管子"中的链应该可沿着"管子"缓慢地向前"爬行"(reptation),其头部(一个端基)可以向不同的方向无规行走一个管径的长度$(d,$ 步长),进入一段新的"管子",尾部(另一个端基)则跟随着在"管子"里向前一步。仿佛"管子"的一端在不断地延伸,另一端在不断地消失,但总长度不变,好像一条缓慢地爬行的"蛇"。

如前所述,可以将这样一条大分子链上想象地分割成一串短链节,每个短链节的平均末端距为管径 d,共有 N/N_d 条短链节。这样,亚浓溶液中的每条链就成为一条由 N/N_d 个长度为 d 的"新链段"(即"链珠")组成的等效线型链。文献和书中也将其称为原始链,其伸直长度为 $L_{c,d} = d(N/N_d)$,构象仍为一个无规线团。所以,困住它的"管子"的构象也是一个无规线团。"管子"或等效线型链与原始链有相同的平均末端距。

一条大分子链沿着"管子"无规行走的摩擦系数和平动扩散系数分别为 $f = 6\pi\eta l_d$ (N/N_d) 和 $D = k_B T/f$,其中,l_d 是每个短链节的有效流体力学尺寸,其同样有"排水"$l_d = bN_d$ 和"非排水"$l_d = bN_d^{\alpha_R}$ 之分。因此,链沿着"管子"的运动为一个近似的一维的"蛇行",文献和书中也常将其称作"曲线扩散",其扩散系数为

$$D_c = \frac{k_B T}{6\pi\eta b N_d} \frac{N_d}{N} \sim N^{-1} \qquad \rightarrow \qquad D_c = D_0 \quad (C > C^*, \text{"排水"})$$

$$D_c = \frac{k_B T}{6\pi\eta b N_d^{\alpha_R}} \frac{N_d}{N} \sim N^{-1} \qquad \rightarrow \qquad D_c = D_0 \left(\frac{C}{C^*}\right)^{-\frac{1-\alpha_R}{3\alpha_R - 1}} \quad (C > C^*, \text{"非排水"}) \quad (9.14)$$

其中,利用了 $N/N_d \sim (C/C^*)^{1/(3\alpha_R-1)}$;$N_d$ 不依赖于链长。在"排水"情况下,流体力学相互作用消失,亚浓溶液中的曲线扩散就是稀溶液中 Rouse 模型里的平动扩散,其随链长增长而变慢。"非排水"时,在 Θ 和良溶剂中,分别有 $D_c = D_0 (C/C^*)^{-1}$ 和 $D_c = D_0 (C/C^*)^{-1/2}$。显然,与稀溶液中的平动扩散相比,"曲线扩散"随着浓度增加显著变慢。每条链在"管子"中无规地行走,其扩散一根管子长度$(L_{c,d} = d(N/N_d))$所需的时间为 $\tau_c = L_{c,d}^2/(6D_c)$。在"排水"和"非排水"时,分别为

$$\tau_c = \frac{\pi\eta}{k_B T} b N L_{c,d}^2 = \frac{\pi\eta b N R^2}{k_B T} \frac{d^2}{R^2} \left(\frac{N}{N_d}\right)^2 \sim N^2 \qquad \rightarrow \qquad \tau_c = \tau_0 \left(\frac{C}{C^*}\right)^{\frac{2(1-\alpha_R)}{3\alpha_R-1}}$$

$$\tau_c = \frac{\pi\eta}{k_B T} \frac{b N_d^{\alpha_R} L_{c,d}^2}{N_d/N} = \frac{\pi\eta R^3}{k_B T} \frac{d^3}{R^3} \left(\frac{N}{N_d}\right)^3 \sim N^3 \qquad \rightarrow \qquad \tau_c = \tau_0 \left(\frac{C}{C^*}\right)^{\frac{3(1-\alpha_R)}{3\alpha_R-1}} \qquad (9.15)$$

其中,也利用了 $d = bN_d^{\alpha_R}$,$R = bN^{\alpha_R}$ 和 $N/N_d = (C/C^*)^{1/(3\alpha_R-1)}$。注意:无论是在 Θ 还是在良溶剂中,对给定的浓度,d 和 N_d 为两个常数,不依赖于链长。当 $N \sim 10^3$ 时,每条链在亚浓溶液中沿着"管子"的"曲线扩散"很慢。反过来,对给定的链长,随着浓度增加,"曲

线扩散"也越来越慢。当 $C/C^* \sim 20$ 时，$\tau_c/\tau_0 \sim 10^2$，慢了一百倍。

如果"管子"是直的，链无规行走一个"管子"的拉伸长度后，其质心也一定移动了同样的距离。然而，"管子"是一个卷曲的无规线团，当链沿着"管子"走了一个"管子"的拉伸长度后，其质心实际上仅移动了很短的距离。例如，如果一条大分子链围绕着其质心移动了一个"管子"的伸直长度，质心可以几乎不变。

由前一章已知，在稀溶液中，链之间没有缠结，链的质心无规地行走。链内各个链段相对质心的复杂运动构成各种简正模式弛豫。质心平动扩散一个链的平均末端距的特征弛豫时间为 $\tau_R = \langle R^2 \rangle / 6D_m$，或写成 $\langle R^2 \rangle \approx D_m \tau_R$，略去正比常数。如前所述，在亚浓溶液中，链之间互相缠结，链困于一个由别的链构成的"管子"之中。一条大分子链可被描绘成一条含有 N/N_d 个长度为 d 的"新链段"的想象链，其伸直长度为 $L_{c,d} = d(N/N_d) < L_c = bN$。

在"管子"中，链的质心已经无法自由地在三维空间里扩散，只能沿着"管子"进行近似一维的扩散。最大步长为管径 d。行走每一步所需的时间为 $t_d \approx d^2/D_c$。需要 $L_{c,d}/d$ 步方才可以走完一条"管子"的伸直长度。在无扰的状态下，整条链无规行走 $L_{c,d}/d$ 步后的方均末端距为 $d^2(L_{c,d}/d) = dL_{c,d}$。大分子链的方均末端距是一个物理存在，与如何描绘它无关。因此，$d^2(L_{c,\xi}/d^2) = b^2N$，即 $L_{c,d} = b^2N/d$。依据定义，一条大分子链的质心移动一步之前（$t=0$）和之后（$t=t_d$）的数学表达为

$$\vec{r}_G(0) = \frac{1}{L_{c,d}} \int_0^{L_{c,d}} \vec{r}(l,0)\,\mathrm{d}l \quad \text{和} \quad \vec{r}_G(t_d) = \frac{1}{L_{c,d}} \int_d^{L_{c,\xi}+d} \vec{r}(l,t_d)\,\mathrm{d}l$$

依据上式，在这样的移动后，除了头尾的两小节长度为 d 的"管子"（d）分别移出和移入了"管子"，中间的部分保持不变。所以，在求解上述两个位置矢量之差的计算中，中间所有链段的积分相同，相互抵消；故仅需计算头尾两个移动了长度 d 的部分，即

$$\vec{r}_G(t_d) - \vec{r}_G(0) = \frac{1}{L_{c,d}} \int_0^d \vec{r}(L_{c,d}+l)\,\mathrm{d}l - \frac{1}{L_{c,d}} \int_0^d \vec{r}(0+l)\,\mathrm{d}l$$

由于每个链段的运动无规且互相独立，上式的时间平均为零，即 $\langle \vec{r}(L_{c,d}+d) \rangle = \langle \vec{r}(d) \rangle$ 和 $\langle \vec{r}_G(t_d) \rangle = \langle \vec{r}_G(0) \rangle$。因此，为了比较运动速度，不得不使用统计上的方均值，即

$$\langle (\vec{r}_G(t_d) - \vec{r}_G(0))^2 \rangle = \frac{1}{L_{c,d}^2} \int_0^d \int_0^d \langle [\vec{r}(L_{c,d}+l) - \vec{r}(l)]^2 \rangle \,\mathrm{d}l$$

其中，右边积分里边的平均为大分子链移动了一段短距离（d）后的方均末端距，近似地等于移动前的方均末端距，即

$$\langle [\vec{r}(L_{c,d}+d) - \vec{r}(d)]^2 \rangle \approx \langle [\vec{r}(L_{c,d}) - \vec{r}(0)]^2 \rangle = b^2N$$

对给定的大分子，b^2N 为一常数。这样，前式中右边的积分就相当简单，得 b^2Nd^2。因此，

$$\langle (\vec{r}_G(t_d) - \vec{r}_G(0))^2 \rangle = \frac{b^2Nd^2}{L_{c,\xi}^2} = d^2\frac{N_d}{N}$$

其中，利用了 $L_{c,d} = d(N/N_d)$ 和 $d = bN_d^{\alpha_R}$。由于 $N \gg N_d$，即使在时间 t_d 内，链的两端和整条链向前移动了一段短距离（d），但是质心移动的平均距离远小于 d。物理上，给定

大分子和浓度,质心平动扩散系数(D_{GC})应为一常数,等于质心移动的方均距与所需时间之比。由于每个链段的运动随机且独立,扩散一步的距离和所经时间与扩散一条链的尺寸(平均末端距)和所经时间对应的两个平动扩散系数应该相同。增加移动距离和所经时间并不改变这一比值。因此,

$$D_{GC} \approx \frac{\langle (\vec{r}_G(t_d) - \vec{r}_G(0))^2 \rangle}{t_d} = \frac{d^2 N_d}{N} / \frac{d^2}{D_c} \quad \rightarrow \quad D_{GC} = \frac{k_B T}{6\pi\eta l_d} \frac{N_d^2}{N^2} \quad (9.16)$$

在推导时,利用了 $L_{c,d} = d(N/N_d)$ 和 $f = 6\pi\eta l_d (N/N_d)$。注意,在亚浓溶液中,l_d,b,d 和 N_d 均不依赖于 N;故 $D_{GC} \sim N^{-2}$。在"排水"和"非排水"情况下,分别有 $l_d = bN_d$ 和 $l_d = bN_d^{\alpha_R}$。代入上式后,分别可得质心在"排水"和"非排水"情况下的平动扩散系数,

$$D_{GC} = \frac{k_B T}{6\pi\eta bN_d} \frac{N_d^2}{N^2} = \frac{k_B T}{6\pi\eta bN} \frac{N_d}{N} \sim N^{-2} \quad \rightarrow \quad D_{GC} = D_0 \left(\frac{C}{C^*}\right)^{\frac{1}{1-3\alpha_R}}$$

$$D_{GC} = \frac{k_B T}{6\pi\eta bN_d^{\alpha_R}} \frac{N_d^2}{N^2} = \frac{k_B T}{6\pi\eta bN^{\alpha_R}} \frac{N_d^{2-\alpha_R}}{N^{2-\alpha_R}} \sim N^{\alpha_R-2} \quad \rightarrow \quad D_{GC} = D_0 \left(\frac{C}{C^*}\right)^{\frac{2-\alpha_R}{1-3\alpha_R}} \quad (9.17)$$

它们对应的特征弛豫时间,$\tau_{GC} = L_{c,d}^2/(6D_{GC})$,如下:

$$\tau_{GC} = \frac{\pi\eta}{k_B T} \frac{bN_d L_{c,d}^2}{(N_d/N)^2} \sim N^4 \quad \rightarrow \quad \tau_{GC} = \tau_0 \left(\frac{C}{C^*}\right)^{\frac{3-2\alpha_R}{3\alpha_R-1}} \quad (\text{"排水"})$$

$$\tau_{GC} = \frac{\pi\eta}{k_B T} \frac{bN_d^{\alpha_R} L_{c,d}^2}{(N_d/N)^2} \sim N^4 \quad \rightarrow \quad \tau_{GC} = \tau_0 \left(\frac{C}{C^*}\right)^{\frac{4-3\alpha_R}{3\alpha_R-1}} \quad (\text{"非排水"}) \quad (9.18)$$

在 Θ 和良溶剂中,在排水的情况下,分别有 $D_{GC} = D_0 (C/C^*)^{-2}$ 和 $D_{GC} = D_0 (C/C^*)^{-3/2}$ 以及 $\tau_{GC} = \tau_0 (C/C^*)^4$ 和 $\tau_{GC} = \tau_0 (C/C^*)^{9/4}$;在非排水的情况下,分别有 $D_{GC} = D_0 (C/C^*)^{-3}$ 和 $D_{GC} = D_0 (C/C^*)^{-7/4}$ 以及 $\tau_{GC} = \tau_0 (C/C^*)^5$ 和 $\tau_{GC} = \tau_0 (C/C^*)^{11/4}$。在亚浓溶液中,大分子链质心的平动扩散随着浓度和链长的增加,极速变慢。对典型的聚合物,在 cgs 体系中,$d \sim b \sim 10^{-7}$ cm,$N \sim 10^3$。因此,$\boldsymbol{D_{GC} \sim 10^{-10}}$ cm²/s。链的方均距约为 10^{-10} cm²,所以,链的质心移动一个平均末端距的时间($\boldsymbol{\tau_{GC}}$)约为 $10^{-10}/10^{-10} \sim 10^0$ s。在亚浓溶液中,大分子链质心的平动扩散比在稀溶液中慢了 $10^3 \sim 10^4$ 倍。

当一个外力作用的时间短于大分子链的质心运动一个平均末端距所需的时间时,质心在外部作用下变化很小,仅是链的构象在该外部作用下被拉伸或压缩而变形,偏离其平衡状态。当外部作用撤除后,变形的链迅速地弛豫回到其原来的平衡状态,宏观表现就是一个弹性体;相反,当一个作用的时间长于 τ_{GC} 时,大分子链的质心就可在该作用下移动,从一个位置移动到另一个较远的位置,运动引起摩擦,摩擦消耗能量,体系发热,当外部作用撤除后,链就会在该新的质心位置恢复其平衡构象。宏观表现就是一个黏性的流动液体。其流动可慢可快,无论快慢,终究是一个可以流动的液体。

依照上述讨论,原则上,取决于作用时间和特征弛豫时间的相对长短,任何体系都可呈现弹性或黏性。如果特征弛豫时间远远长于或短于通常的时间尺度,体系则显示弹性或黏性。一个典型的例子是内衣的松紧带,初始穿戴时,弹性很好,较紧。但在穿戴数月或数年后(取决于其质量),就慢慢地失去弹性、变长,好像慢慢地"流动"。物理上,这是

因为构成松紧带的大分子链的特征弛豫时间很长的缘故。因此,当体系的特征弛豫时间与日常的时间尺度相近时,体系在宏观上表现出可见的黏性或弹性,取决于外部作用的强度和时间。由上述两式可知,对通常的聚合物而言,$\tau_{GC} \sim 1$ s。这就是为何长链大分子的浓溶液随着测量时间在秒级附近变化时,可以同时表现出黏性和弹性。

可进一步利用标度理论得到上式,以及在"管子"中无规行走的步长(l_1)对浓度的依赖性。已知当 $C \to 0$ 时,在"排水"和"非排水"情况下,分别有 $D_{GC} \to D_{GC}(C \to 0) = D_0 = k_B T/(6\pi\eta bN)$ 和 $D_0 = k_B T/(6\pi\eta bN^{\alpha_R})$,故

$$D_{GC} = D_0 \left(\frac{C}{C^*}\right)^z = \frac{k_B T}{6\pi\eta}\left(\frac{C}{m_0}\right)^z b^{3z-1} N^{(3\alpha_R-1)z-1} \quad (\text{完全"排水"})$$

$$D_{GC} = D_0 \left(\frac{C}{C^*}\right)^z = \frac{k_B T}{6\pi\eta}\left(\frac{C}{m_0}\right)^z b^{3z-1} N^{(3\alpha_R-1)z-\alpha_R} \quad (\text{完全"非排水"}) \quad (9.19)$$

推导中,利用了式(9.1)。已知 $D_{GC} \sim N^{-2}$,在两种情况下分别有 $(3\alpha_R - 1)z - 1 = -2$ 和 $(3\alpha_R - 1)z - \alpha_R = -2$,即 $z = -1/(3\alpha_R - 1)$ 和 $z = (\alpha_R - 2)/(3\alpha_R - 1)$。将其代入式(9.18)同样可得式(9.16),异功同曲!

$$D_{GC} = D_0 \left(\frac{C}{C^*}\right)^{-\frac{1}{3\alpha_R-1}} (\text{完全"排水"}); \quad D_{GC} = D_0 \left(\frac{C}{C^*}\right)^{-\frac{2-\alpha_R}{3\alpha_R-1}} (\text{完全"非排水"})$$

将式(9.16)等于式(9.17),再利用式(9.1),可得管径对浓度的依赖性。

$$d \sim b^{\frac{8-9\alpha_R}{2(1-3\alpha_R)}} \left(\frac{C}{m_0}\right)^{\frac{1}{2(1-3\alpha_R)}} \quad (C > C^*, \text{"排水"})$$

$$d \sim b^{\frac{8-9\alpha_R}{2(1-3\alpha_R)}} \left(\frac{C}{m_0}\right)^{\frac{2-\alpha_R}{2(1-3\alpha_R)}} \quad (C > C^*, \text{"非排水"}) \quad (9.20)$$

其中,$C/m_0 = \rho_b = NN_p/V$(链段数密度)。在 Θ 和良溶剂中,在"排水"和"非排水"的情况下,分别有 $d \sim b^{-7/2}\rho_b^{-1}$ 和 $d \sim b^{-13/8}\rho_b^{-5/8}$ 以及 $d \sim b^{-7/2}\rho_b^{-3/2}$ 和 $d \sim b^{-13/8}\rho_b^{-7/8}$。对给定的大分子,$b$ 为常数;d 随浓度增加而变短,即,"管子"变细。

在大分子的亚浓溶液中,每条链处在一个由其周围互相缠绕的其他链组成的"管子"之中,同时其本身还组成围绕其他链的"管子"。没有一条链特殊,无法区分。因此,无论是测量 D_c 还是 D_{GC},均需标记少量的链作为探针。例如,在中子散射实验中,采用氘原子代替探针链上的氢原子。在荧光时间相关光谱或者荧光漂白恢复光谱的测量中,将荧光分子接在少量的探针链上。在动态激光光散射中,采用和大分子具有相同折射指数的溶剂,再添加少量长度相近,但折射指数不同的其他大分子作为探针链,通过测量这些添加的探针链,间接地获得原本链的动态行为,前提是这些添加的链和原本的链之间无特别的相互作用。

如果利用激光动态光散射直接测量一个大分子亚浓溶液,理论上仅可测得每个"链珠"上的链段在链珠的小体积内的平动扩散引致的弛豫,与 τ_ξ 相关。然而,自20世纪70年代起,在利用激光动态光散射研究不同的大分子亚浓溶液时,不同的实验室一致地观察到一个额外的弛豫,较慢,具有较大的线宽(较低的频率);即在光强权重的线宽分布中出现了第二个小的峰。这样一个额外的弛豫(峰)在"管子"和"蛇行"理论中一直完全没有

被预计过,在其他测量(如流变学测量)中并没有出现。

因此,这一较慢的弛豫引起了理论和实验学家之间长达四十多年的争论。理论学家也分成了两派。信奉 de Gennes 理论的学者常常将其归于一些不可避免的实验缺陷,包括样品内的杂质、在过滤溶液时没有完全去除的尘粒或微小气泡和可能存在浓度和温度梯度。实验学家虽已尽力地设法避免这些潜在的问题,但永远不能彻底地排除这些可能性。

制备适合动态激光光散射实验的无尘亚浓溶液传统上包括以下步骤:首先,配制黏度适合、可过滤的稀溶液;其次,需将最少 10 mL 的稀溶液除尘,并注入光散射样品池,这很难做;再次,在无尘的条件下,缓慢地挥发 50% ~ 90% 的溶剂至不同浓度。在挥发的过程中,不可避免地形成上浓下稀的大分子浓度梯度。先前的讨论显示,大分子链的质心平动扩散一个 10^{-5}cm 的距离需要 1 ~ 10 s;那么扩散 1 cm 的距离则需要 10^{10} ~ 10^{11} s,这是一个天文数字! 所以,实验学家永远无法排除这些可能的干扰。由于缺乏决定性的实验结果,争论就逐渐地被搁置了,没有结论。成为一个大分子溶液研究中悬而未决的问题。

直至 2008 年,利用高真空阴离子聚合直接在激光光散射样品池中合成窄分布、高摩尔质量的聚苯乙烯和制备高浓度、无尘的大分子溶液,方才解决了一系列前述的问题。首先,过滤低黏度的单体溶液代替了过滤黏稠大分子溶液,在确保无尘的前提下,方才开始聚合反应;其次,在聚合反应中,反应程度逐渐地增加。缠绕大分子链无法运动一个较长的距离,但是单体作为小分子在溶液中仍可来去自如,继续和链的活性末端反应,使得链长继续增加,接触浓度降低,导致 C/C^* 增加,其消除了可能的浓度梯度的疑虑。因此,制备了一系列无尘的、均匀的、高浓度的聚苯乙烯分别在苯和环己烷中的溶液。

在讨论式(9.13)中涉及的图显示了在良溶剂苯中,即使 C/C^* 高达 30,确实只观察到一个与每个"链珠"中链段的平动扩散有关的快速弛豫过程。随后对不同链长、浓度和温度的测量结果都证实了 de Gennes 的理论预测。在短时间内,大分子链之间因互相缠结无法移动,只有"链珠"内的链段在 ξ^3 的小体积内无规自由行走。动态激光光散射测得的线宽为 $\Gamma = D_\xi q^2 = (6\pi\eta\xi/k_B T)q^2$。$\xi$ 随浓度增加变短。在亚浓溶液中,大分子的浓度已经很高,理论上,已不可再用溶剂黏度(η_0)代替 η。依据前述有关黏度的讨论,$\eta \sim C$;浓度为局部质量浓度,$C = m_0 N_\xi/\xi^3$;$\eta\xi \sim C\xi$;$\xi = bN_\xi^{\alpha_R}$。因此,$\Gamma \sim m_0/(b^{1/\alpha_R}\xi^{2-1/\alpha_R})$。因为 $\xi \sim C^{\alpha_R/(1-3\alpha_R)}$,

$$\Gamma \sim b^{\frac{1}{\alpha_R(3\alpha_R-1)}} m_0^{\frac{2\alpha_R-1}{3\alpha_R-1}} C^{\frac{2\alpha_R-1}{3\alpha_R-1}} \tag{9.21}$$

随着浓度增大,ξ 变短,Γ 增加,特征弛豫时间(τ)缩短,动态激光光散射测得的特征弛豫时间分布中的快弛豫峰向更快的方向(左边)移动。

然而,将同样的一系列实验从在非常良溶剂苯中平移到良溶剂环己烷中(温度远高于 Θ 温度,~35℃),动态激光光散射的结果显示,一旦溶液进入亚浓区域,除了一个预计的快速弛豫,会立即出现了一个约慢 10^2 倍的慢弛豫过程。

如下图中散射光强权重的特征弛豫时间分布所示,其中,为了比较,横坐标弛豫时间(t)已经过温度和黏度的矫正。这一系列决定性的平行实验使得该慢弛豫过程的存在不容置疑和不可否认。问题是为何会出现这一不在理论预计之中的慢弛豫过程? 首先,考察图中散射光强权重的特征弛豫时间分布对温度的依赖性。随着温度下降,溶剂逐渐地从一个良的溶剂变成一个 Θ 溶剂,链逐渐地收缩,C^* 变高。由于 C 不变,故 C/C^* 变小。

在 37℃时，$C/C^* < 1$，溶液进入稀溶液区间，故只有一个快弛豫峰。在稀溶液中，该快弛豫对应着链的质心平动扩散一个 R 的距离所需的时间。

在亚浓溶液中，其对应着每个"链珠"内的链节的质心平动扩散一个 ξ 的距离所需的时间。而在其他两个温度，溶液刚刚进入亚浓区间，$R \sim \xi$，故三个快的弛豫峰相近。随着温度升高，弛豫稍微加快。依据上式，温度升高，α_R 增加，链的尺寸更大。对给定的浓度，ξ 更短，特征弛豫时间更短，弛豫加快，对应的弛豫峰左移。相比，慢弛豫峰的左移更明显，意味着局部黏度较大。然而，实验上无法直接测得局部的微观黏度。如果假定快弛豫峰不变，可间接地推出局部微

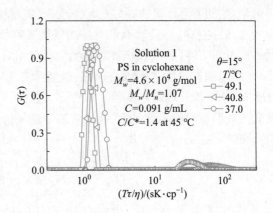

Li J F, Li W, Huo H, et al. Macromolecules, 2008, 41:901.

（Figure 12, permission was granted by publisher）

观黏度，详情可见上图所引的文献。实验结果还显示，与慢弛豫过程对应的线宽 Γ_{slow} 也与 q^2 成正比。注意：溶液浓度略微高于接触浓度。不难想象若干条缠结的链仍可作为一个整体在溶液中平动扩散，但慢很多。所以，此时的慢弛豫过程对应着一群链的平动扩散。

为了更好地探索慢弛豫过程的本质，实验上采用长链聚苯乙烯将浓度比增加到传统溶剂蒸发不可能达至的 $C/C^* \sim 51$。溶液在短时间内不流动，好像一个溶胶。下图（a）显示了 $G^{(2)}(t)$ 对观察尺度（即散射矢量）的依赖性，其中，为了比较角度效应，横坐标上的延迟时间（t）已被 q^2 修正。源自五个不同散射角度的数据在短延迟时间范围内重叠在一起，对应着一个由平动扩散引致的弛豫，其动态相关长度为 ξ。它反映了每个"链珠"内的链节在一个 ξ^3 的空间中无规行走。在下图（a）中，慢弛豫好像随着散射角度增大而变慢，这是因为延迟时间被 q^2 修正后造成的一个错觉而已。实际上，如果用 t 代替 tq^2 作图，五个慢弛豫曲线就会叠在一起，其不依赖于散射矢量。慢弛豫已与平动扩散无关。

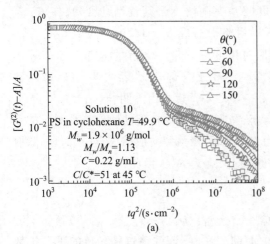

Li J F, Li W, Huo H, et al. Macromolecules, 2008, 41:901.

（Figure 19, permission was granted by publisher）

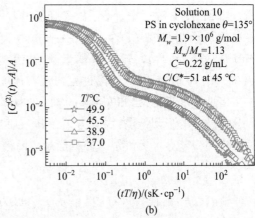

Li J F, Li W, Huo H, et al. Macromolecules, 2008, 41:901.

（Figure 21, permission was granted by publisher）

对于同一个聚苯乙烯在环己烷中的亚浓溶液,在不同温度下测得的散射光强权重的时间相关函数中,两个不同的弛豫(快和慢)清晰可见,如上图(b)所示。依据上式,对给定的浓度,随着温度升高,α_R 变大,ξ 变短,特征弛豫时间也变短,弛豫加快。图中,对应的快弛豫部分左移。另外,不仅对应的慢弛豫部分也随着温度升高而左移、变快,而且其相对高度也在下降,意味着其对散射光强贡献相对地变小。换而言之,其对应的涨落幅度变小,其暗示随着排除体积效应增强,慢弛豫逐渐消失。在亚浓溶液中,链本身处于一个无扰状态,其方均末端距不依赖于温度。依据式(9.19),管径(d)随温度(α_R)变化。散射角度和溶液温度的依赖性显示,亚浓溶液中的慢弛豫不仅和观察尺度无关,而且也和管径无关,故其不是一个局部性质。

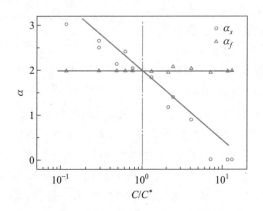

Li J F,Li W,Huo H,et al.Macromolecules,2008,41:901.

(Figure 22,permission was granted by publisher)

快和慢弛豫的线宽(Γ)对散射矢量(q)的依赖性可一般性地写成 $\Gamma_{\text{fast}} \sim q^{\alpha_f}$ 和 $\Gamma_{\text{slow}} \sim q^{\alpha_s}$。上图显示了两个指数的浓度依赖性,其中,为了比较不同的链长,C 已被 C^* 约化。显然,快弛豫过程始终对应的是 $\Gamma_{\text{fast}} \sim q^2$,反映了每个"链珠"内的链段在其体积内无规行走,导致一个源于密度涨落的动态相关长度(ξ),其随浓度升高逐渐地变短,$\xi \sim C^{-3/4}$。另一方面,随着浓度增加,α_s 从 3 降至 0。可以粗略地写成 $\alpha_s = 3/2(1 - \log(C/C^*))$。在稀溶液的范围内,$C/C^* < 10^{-1}$,$\alpha_s = 3$,在大散射角度下测量一条链的内部运动。$\alpha_s$ 逐渐地从 3 降至 2,源于在非零散射角度测量时,部分内部运动混入平动扩散。在 $C/C^* \geqslant \sim 10$ 时,$\alpha_s = 0$,弛豫过程与观察尺度($1/q$)无关。其原因不明。

借助标度理论,已知在 $C \to 0$ 时,在小角动态激光光散射观察到的弛豫过程对应的是大分子质心的平动扩散,线宽为 $\Gamma_0 = Dq^2$;而当 $C \gg 1$ 时,慢弛豫对应的线宽(Γ_{slow})和散射矢量无关,$\Gamma_{\text{slow}} \sim q^0$。假定与慢弛豫对应的密度涨落尺度为 R_{slow},因此,

$$\Gamma_{\text{slow}} = \Gamma_0 (qR_{\text{slow}})^z$$

显然,$z = -2$;代入可得 $\Gamma_{\text{slow}} = 1/\tau_{\text{slow}} = D/\xi_{\text{slow}}^2 = 1/(\xi_{\text{slow}}^2/D)$,故 $\tau_{\text{slow}} = \xi_{\text{slow}}^2/D$,等效于链在稀溶液中扩散一段长度为 ξ_{slow} 的距离所需的时间。注意,该弛豫过程和链的平动扩散无关。从动态激光光散射测得的 $\Gamma_{\text{slow}} = 1/\tau_{\text{slow}}$ 和 D,可估出 ξ_{slow}。结果显示,在 $C/C^* = 51$

时,$\tau_{\text{slow}} \sim 10^{-2}$s 和 $D \sim 10^{-7}$cm^2/s。因此,$\xi_{\text{slow}} = 10^{-4} \sim 10^{-5}$cm。其比散射矢量的倒数($1/q$)以及链的平均末端距($R$)稍大。弛豫很慢,但相关长度较短。注意,人们常误将"运动慢"和"尺寸大"关联。当阻力很大时,小粒子也运动得很慢。

在实验上肯定了慢弛豫的存在,问题是其是否可归于大分子链在"管子"里"蛇行"造成的密度涨落。如是,de Gennes 的"管子"模型和"蛇行"理论稍做修改即可。理论上,"蛇行"的特征弛豫时间和链长的 3 次方成正比(实验上得到的是 3.4 次方)。因此,对给定的质量浓度,改变链长即可回答上述问题。在下图中,短链和长链的 C/C^* 分别为 10 和 35。ξ 只依赖于浓度,与链长无关,故二者的快弛豫过程相近。另一方面,链长增加了约 8 倍,τ_{slow} 仅增加了不到 5 倍。如果慢弛豫与链在"管子"里"蛇行"有关,τ_{slow} 则应增加至少 500 倍。显然,实验结果否定了将慢弛豫归于链在"管子"中的"蛇行"。

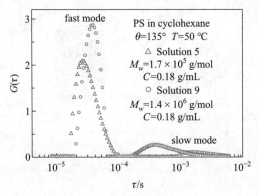

Li J F,Li W,Huo H,et al.Macromolecules,2008,41:901.

(Figure 7,permission was granted by publisher)

至今仍无一个理论可以定量地解释为何当溶剂性质不是极好时,在亚浓溶液中会出现慢弛豫过程,定性的解释为,当排除体积效应还没有达到最大和链还没有充分扩展时,每一个"链珠"并非如理论上假定的完全独立。换而言之,"链珠"相关,其特征相关长度为 ξ_{slow}。这样一个慢弛豫的出现并不仅限于在中性大分子亚浓溶液中,在小分子和大分子的凝胶化和玻璃化的过程中、在浓的胶体中、在无盐的聚电解质稀溶液中均会出现一个类似的慢弛豫。为了证实动态激光光散射中观察到的慢弛豫确实源于不同"链珠"之间的相关,进一步研究了链间相互作用对该慢弛豫的影响。显然,将大分子链逐步地交联在一起是一个最极端的增强链间相互作用的方式。

下图显示了散射光强时间相关函数对浓度的依赖性。链上接了少量的香豆素,所以它们在紫外光的照射下可发生化学交联。光照前,较低浓度溶液的时间相关函数仅有一个快弛豫,而随着浓度升高,出现另一个慢弛豫。随着光交联的进行,其逐步变慢(右移)、其对应的散射光强增加。同时,快弛豫过程的相对光强贡献逐渐减少,在高浓度和长时间辐照后看似消失。插图显示,此时溶液已不可流动。动态激光光散射仍然可观察到这一快弛豫,其特征弛豫时间保持在 10^{-4}s。这一系列结果证实了慢弛豫与链间相互作用有关。

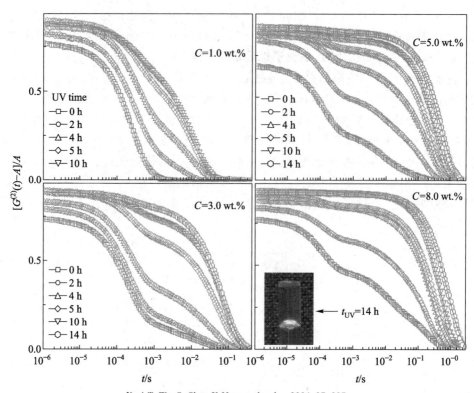

Ngai T,Wu C,Chen Y.Macromolecules,2004,37:987.

(Figure 8,permission was granted by publisher)

用动态光散射测量无盐的聚电解质稀溶液也可观察到快和慢弛豫,其中快弛豫也与链长无关。至于慢弛豫的起源,理论学家和实验学家从 20 世纪 80 年代起就一直争论不休。同样也有接受和否认"慢弛豫"存在的两群人,最后也无定论,不了了之。争论的原因也包括样品的制备,如水溶液中"无盐"的定义;电荷在同一条链和不同链上的分布状况;样品中可能存在的不溶解"聚集团簇"等。

直至 2009 年,在解决了样品制备问题后,才释疑了与溶液配制相关的问题。一个决定性的实验是在中性大分子链上引入可与二氧化碳反应生成带电荷基团的对叠氮基甲基。将二氧化碳气体或者氮气通入装有其溶液的激光光散射样品池后,可在线地使得大分子链在中性状态和带电荷状态之间转变。如下图(b)所示,通入二氧化碳气体后,中性的大分子链成为聚电解质,故溶液的导电性显著增加。通入氮气后,恢复中性状态。

对应着导电性的变化,链为中性时,动态激光光散射测得的弛豫时间分布仅含有一个和链的质心平动扩散相关的峰;通入二氧化碳气体后,出现了两个弛豫峰;再通入氮气后,两个弛豫峰消失,完全回归中性链时的弛豫时间分布,如上图(b)所示。进一步的实验和分析显示,两个弛豫均由平动扩散引致。快弛豫归于单根聚电解质链上的链节与其反离子的耦合扩散,独立于链长和浓度;而慢弛豫则和链间作用相关,等效于一群链的协通平动扩散。在稀溶液中的链间作用源于链上电荷的长程静电作用,或曰较强的排除体积效应。加盐可屏蔽长程静电作用。在盐存在时重复上述实验,无论链是否带电荷,弛豫时间

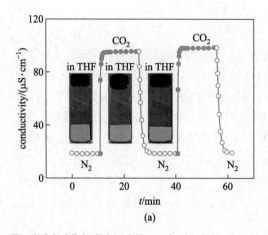

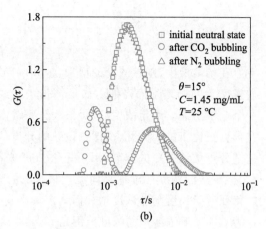

Zhou K J, Li J F, Lu Y J, et al. Macromolecules, 2009, 42:7146.
(Figure 3, permission was granted by publisher)

Zhou K J, Li J F, Lu Y J, et al. Macromolecules, 2009, 42:7146.
(Figure 6, permission was granted by publisher)

分布始终只有一个峰,其位置不变。这些实验结果已经令人信服地说明慢弛豫是一个真实的存在,其与链间作用直接相关。在无盐时,长程静电作用使得每条聚电解质链的等效排除体积变得很大。即使在稀溶液中,体系也变得"拥挤",链的运动受阻。

文献中,曾经常常将慢弛豫的出现归于动态激光光散射实验的可能缺陷和体系的静态不均匀性。面对上述两个决定性实验的系列结果,无人可以再否定慢弛豫的真实存在。可定性地断言,其反映了一个"拥挤"体系虽然在静态上是均匀的,但在动态上并非均匀,存在着不同的时间尺度或曰不同相关长度的弛豫。这一动态不均匀性可归于"粒子"间没有屏蔽的相互作用,造成一个"拥挤"的效应。

在一个"拥挤"的体系或者"聚会"中,每个"粒子"(分子、大分子链、胶体粒子或者人)都处于一个由别的粒子组成的"笼子"之中,也是其周围粒子的"笼子"的一部分,没有任何一个粒子特殊。短时间内,每个粒子只可以在自己的"笼子"里活动,引起一个小的体积内的密度涨落,对应着快弛豫。由于每个粒子的无规运动,"笼子们"应该有一个平均"寿命"。当测量或等待时间长于这一"寿命"时,每个粒子就有机会迁移出原来的旧"笼子",进入另一个新"笼子"。在亚浓溶液中,这意味着虽然整条链沿着"管子"蛇行,无法横向运动。但是,链上的每个"链珠"(短链节)还可和其周围的"链珠"产生一个协同运动,有机会横向跃迁出原有"管子",进入邻近的"管子"。"笼子"的平均特征寿命可以很长,取决于相互作用。

慢弛豫反映了粒子在不同"笼子"之间的迁移,引致一个密度涨落,其特征相关长度为"笼子"间的相互作用可波及的范围,长于"笼子"的线性尺度。其对应的特征弛豫时间是"笼子"的平均"寿命"。给定体积后,"拥挤"可因粒子数增多,如在亚浓溶液和浓的胶体中;也可源于粒子间较强的排斥,造成很强的排除体积效应,等效于每个粒子的排除体积增大,如在无盐稀溶液中带电荷粒子和玻璃化温度附近熔体中的小分子。以上仅是定性的解释。完整的理论处理将不可避免地涉及物理中的多体问题,十分复杂,仍无定量的解释,仍是一个悬而未决的问题。

小　　结

在稀溶液中,随着浓度的增加,链间的距离逐渐缩短,达至一个接触浓度(C^*),其随着链尺寸(长)增加而降低。如果链长于缠结长度,继续增加浓度导致链间的缠结,形成一个亚浓溶液。给定浓度后,每条大分子链占据的空间里还含有多条其他互相贯穿的链,其数目正比于链长的开方,即链段数的开方($N^{1/2}$)。随着浓度的增加,每条链逐渐地被其他链分隔成许多短的链节;每条短链节可被看作一个"链珠",链节们彼此相互独立。因此,整条链的构象逐渐地趋向无扰状态下的无规线团。另一方面,给定链长后,密度涨落的特征弛豫长度随着浓度增加而缩短。当其接近链段长度(b)时,溶液进入浓溶液区间,记为 C^{**}。亚浓溶液的浓度范围是 $C^* \leqslant C \leqslant C^{**}$。

在亚浓溶液中,每条链都在一个由其周围互相缠结的链构成的、直径为"d"的"管子"之中。链和围住其的"管子"的构象均为无规线团。链只可沿着"管子"无规地前行或后退,缓慢地在无规线团状的"管子"里"蛇行",最大步长为"管子"的直径。同时,每一条链也是围住其附近链的"管子"的一部分,没有任何一条链特殊。在短时间内,整条大分子链已经无法长距离迁移,仅有链段在其所属的"链珠"内无规行走,造成密度涨落,其特征相关长度记为 ξ,注意:$\xi \approx d/4$。因此,一条大分子链由一串小的短链节("链珠")构成;每节短链节含有 N_d 个链段,也具有一个无规线团构象,其平均末端距为管径 d("链珠"的直径)。亚浓溶液中的处于"管子"中的大分子链具有以下静态性质。

一条大分子链或一个"管子"在亚浓溶液中的构象可看作是步长为 d 的 N/N_d 步无规行走后的轨迹:一个无规线团。随着浓度逐渐地增大,"链珠"间的排除体积效应慢慢地趋于零,大分子链在亚浓溶液中趋向一个无扰状态,其平均末段距为 $d(N/N_d)^{1/2}$。由于每个"链珠"内的短链节也是一个无规线团,其伸直后的长度为 bN_d,一定大于 $d = bN_d^{\alpha_R}$。所以,"管子"的伸直长度 $L_{c,d}=d(N/N_d)$ 远短于链本身的伸直长度 $L_c=bN$。一条大分子链的构象可看成步长为 b 的 N 步无规行走后的轨迹,平均末端距为 $R=bN^{\alpha_R}$。一条链的平均末端距是一个物理存在,不依赖于其描绘方式,故 $bN^{\alpha_R}=d(N/N_d)^{\alpha_R}$。注意:每个"链珠"内链节的标度指数仍然随着溶剂性质变化,$1/2 \leqslant \alpha_R \leqslant 3/5$。

动态上,现今的理论预计了在亚浓溶液中的三种与大分子链和链段相关的运动:相互扩散、曲线扩散和质心扩散。首先,当观察时间远短于"链珠"质心移动一个 d 的长度所需的时间时,"链珠"的位置基本上保持不变,整条链的位置也保持不变。仅有"链珠"内的链段在一个小体积(d^3)中无规行走,也可看作"链珠"中短链节的质心围绕着其平衡位置作相互平动扩散,其对应的扩散系数和特征弛豫时间分别记为 D_ξ 和 τ_ξ。短链节上链段的无规行走引致小范围内的密度涨落,对应着一个特征相关长度 ξ,约为管径 d 的四分之一。显然,该弛豫与链长无关,但随着浓度增加,ξ 逐渐变短,弛豫加快。由于体系中每个"链珠"和其位相相差 π 的对应的"链珠"的质心运动并非同步,所以动态激光光散射可测量到该快速密度涨落。对典型的高聚物,在散射光强的时间自相关函数中,该快速密度涨落对应着一个较快的弛豫过程,其特征弛豫时间为 $10^{-6} \sim 10^{-5}$ s,取决于浓度的大小。

其次,当观察时间和一条链沿着弯曲的"管子"扩散一个"管子"的伸直长度($L_{c,\xi}$)所需时间(对典型的高聚物,为 $10^{-2} \sim 10^{-1}$ s)相近时,"链珠"内的密度涨落早已趋于平均密度,溶液等效于一个由相同的、均匀分布的"粒子"组成的无密度涨落的体系。源自各个"粒子"的散射光的电场因相消干涉,对测得的散射光强没有贡献。因此,动态激光光散射无法测得这一运动引起的弛豫。这样一个沿着弯曲的"管子"或前或后地无规行走被称作"曲线扩散",其对应的扩散系数和特征弛豫时间分别记为 D_c 和 τ_c。如果链在一条直的管子中行走,其质心将随着链一起运动。但是,"管子"的形状是一个弯曲的无规线团,所以,链沿着管子行走一段 $L_{c,d}$ 距离后,其质心也许移动一点或不动。例如,链围绕着其质心,在弯曲的"管子"行走了一个"管子"的长度,质心不变。

最后,当观察时间接近链的质心移动一个链的平均末端距 10^{-5} cm 所需的时间(对典型的高聚物,1~10 s)时,大分子链在热能的搅动下已沿着"管子"缓慢地、无规地或前或后地"蛇行"了一段很长的时间,其质心也缓慢地移动了一小段距离,称作"质心扩散",其对应的扩散系数和特征弛豫时间分别记为 D_{GC} 和 τ_{GC}。同样,由于干涉相消,"链珠"们对散射光强没有贡献。因此,无法利用动态激光光散射测得这一运动引起的密度涨落。

只有通过标记少数链,方可测量后两种弛豫过程。在动态激光光散射中,先选择折射指数和大分子尽可能相近的溶剂,以减弱或消除它们的散射光强,再选择长度、结构和性质相近,但折射指数不同的大分子作为标记链;在中子散射和核磁共振中,选择加入少量氘代的、长度相近的大分子链;在荧光漂白恢复和强迫瑞利散射中,则选择在少数大分子链上接上适当的荧光基团。读者可参考相关的文献。

实验上,除了上述三种理论上预测的弛豫以外,动态激光光散射还观察到因"链珠"间或链间的相互作用导致的一个额外的慢弛豫,其特征弛豫时间随体系而变。除了在极端良好的溶剂中链已充分扩展以外,在其他溶剂中,亚浓溶液中的"链珠"之间并非如理论上假定的完全孤立,而是存在着一定程度的移动性。每个"链珠"均处于由其邻近的"链珠"构成的一个"笼子"之中。在极端良好的溶剂中(排除体积效应最大),每个"链珠"充分地扩展,因此每个"链珠"被其邻近的"链珠"堵住了,等效于被固定在一个固定的"笼子"之中,在相当长的时间内都不可移动。在其他溶剂中,每个"链珠"均无规地移动一点点,"笼子"并非一个固定的结构,其密度涨落对应着一个特征相关长度和特征弛豫时间("笼子"的平均寿命),反映了"链珠"在邻近"笼子"之间"跳跃"。每个"链珠"本身也是邻近"链珠"的"笼子",没有一个"链珠"特殊,反映了溶液的静态均匀性,但动态并不均匀。

当每个"链珠"都不可移动时,等效地说,"笼子"的特征弛豫时间趋于无穷大,故无"笼子"效应。这就是为何聚苯乙烯在极良溶剂苯中,即使在 C/C^* 高达 51 时,动态激光光散射仍然只测得一个与"链珠"内的链段无规运动相关的快弛豫。当"笼子"因相互作用而具有一个特征相关长度和对应的特征弛豫时间时,溶液中每对散射电场位相差为 π(相消干涉)的"链珠"并不同时"跃出"它们各自的"笼子",导致一个净的散射光强,故在动态激光光散射测得的特征弛豫时间分布中出现一个额外的、理论上完全没有预计的慢弛豫峰,其随着排除体积效应减弱变慢,对散射光强的贡献增加,反映了每个"链珠"均可跃迁更长的距离,涨落幅度更大。

该慢弛豫过程并非大分子亚浓溶液特有,而是普遍地出现在各类"拥挤"的体系中,包括在溶胶至凝胶和玻璃化的转变中,在无盐的聚电解质稀溶液和浓胶体里。如果一个拥挤的体系中的每个"粒子"(短链节、聚电解质链、胶体粒子、小分子等)可以移动,而非完全互相独立,每个"粒子"就处于一个由其他邻近粒子构成的"笼子"之中,其本身也为其他邻近粒子的"笼子"。没有一个"粒子"特殊,故静态上,体系是均匀的。注意:在两种情况下有"完全独立":没有相互作用或静止不动。

由于粒子的无规运动,每个"笼子"随机地涨落,具有一定的平均寿命和大小。每个粒子可在其"笼子"内无规运动,对应着一个与平动扩散相关的快弛豫;粒子还可以在不同的"笼子"中随机地跃迁,对应着一个与"笼子"的平均尺寸和寿命有关的慢弛豫。这可等效于一个含有"粒子"的体系,粒子本身运动很慢(并非一定因为粒子很大),每个粒子内部含有许多快速和无规运动的小单元。在短时间内,仅可看到粒子内部各个单元的涨落和弛豫;当观察时间足够长时,就可看到"粒子"本身的运动。因此,体系在动态上并不均匀。这定性地解释了为何动态激光光散射可观察到一快、一慢两个弛豫。

当观察时间与"笼子"的特征弛豫时间相近时,就可观察到"粒子"在不同的"笼子"之间缓慢地跃迁。所以,动态激光光散射测得的慢弛豫应是"笼子"的松弛过程,对应的特征弛豫时间就是"笼子"的平均寿命。当排除体积效应减弱时,每个"粒子"可跃迁更长的距离,等效于一个更长的特征相关长度或一个更大的"笼子",导致散射光强增加,特征弛豫时间变长。由于 de Gennes 的"管子"模型和"蛇行"理论假定了,除了沿着"管子"很慢地"蛇行"以外,"链珠"横向不动,故没有"笼子"效应。在大多数真实的"拥挤"体系中,"粒子"仍然可以很慢地移动,并非静止,以及相互作用总是存在。然而,至今仍然缺乏一个描绘该"笼子"效应的定量理论。这应该仍然是一个值得理论研究的问题。

下图总结了大分子链在亚浓溶液中的一些静态性质和在非排水条件下的三个动态扩散性质。

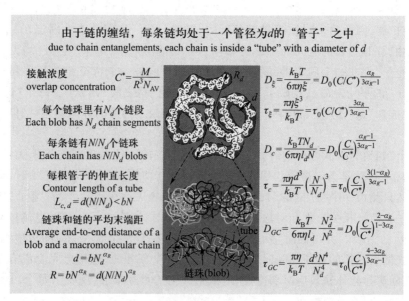

由于链的缠结,每条链均处于一个管径为d的"管子"之中
due to chain entanglements, each chain is inside a "tube" with a diameter of d

接触浓度
overlap concentration
$$C^* = \frac{M}{R^3 N_{AV}}$$

每个链珠里有N_d个链段
Each blob has N_d chain segments

每条链有N/N_d个链珠
Each chain has N/N_d blobs

每根管子的伸直长度
Contour length of a tube
$$L_{c,d} = d(N/N_d) < bN$$

链珠和链的平均末端距
Average end-to-end distance of a blob and a macromolecular chain
$$d = bN_d^{\alpha_R}$$
$$R = bN^{\alpha_R} = d(N/N_d)^{\alpha_R}$$

链珠(blob)

tube

$$D_\xi = \frac{k_B T}{6\pi\eta\xi} = D_0 (C/C^*)^{\frac{\alpha_R}{3\alpha_R - 1}}$$

$$\tau_\xi = \frac{\pi\eta\xi^3}{k_B T} = \tau_0 (C/C^*)^{\frac{3\alpha_R}{3\alpha_R - 1}}$$

$$D_c = \frac{k_B T N_d}{6\pi\eta l_d N} = D_0 \left(\frac{C}{C^*}\right)^{\frac{\alpha_R - 1}{3\alpha_R - 1}}$$

$$\tau_c = \frac{\pi\eta d^3}{k_B T} \left(\frac{N}{N_d}\right)^3 = \tau_0 \left(\frac{C}{C^*}\right)^{\frac{3(1-\alpha_R)}{3\alpha_R - 1}}$$

$$D_{GC} = \frac{k_B T}{6\pi\eta l_d} \frac{N_d^2}{N^2} = D_0 \left(\frac{C}{C^*}\right)^{\frac{2-\alpha_R}{1-3\alpha_R}}$$

$$\tau_{GC} = \frac{\pi\eta}{k_B T} \frac{d^3 N^4}{N_d^4} = \tau_0 \left(\frac{C}{C^*}\right)^{\frac{4-3\alpha_R}{3\alpha_R - 1}}$$

Chapter 9. Macromolecular Semidilute Solutions

Following the definitions of symbols in Chapters 3 and 4, there are N_p macromolecular chains with the same length in the solution, and each chain contains N segments with a length of b and a mass of m_0, so that the mass of each chain is $m = m_0 N$, The molar mass is $M = m N_{AV}$; the total weight of the chains in the solution (strictly speaking, it is mass) is $W = m N_p$. It is not difficult to imagine that as the concentration of macromolecules increases, the interchain distance will gradually shorten and the interaction will gradually become stronger. Eventually, it will reach a contact (overlap) concentration, denoted as C^*, in which the chains start to touch each other. The common unit of the weight (mass) concentration is g/mL. The sum of the volumes of all macromolecular chains is equal to the solution volume, and the volume of each chain is approximately equal to the cube of its average end-to-end distance ($R = b N^{\alpha_R}, 1/2 \leqslant \alpha_R \leqslant 3/5$), that is, $R^3 N_p = V$. Therefore,

$$C^* = \frac{W}{V} = \frac{m_0 N N_p}{(b N^{\alpha_R})^3 N_p} = \frac{m_0}{b^3 N^{3\alpha_R - 1}} = \frac{M}{R^3 N_{AV}} \tag{9.1}$$

where for a given macromolecule, b, m_0, R and M are all constants. R and M are two measurable physical quantities, and $R \propto M^{\alpha_R}$. Therefore, C^* decreases as the chain length increases, and the decreasing rate depends on the solvent quality. When the concentration of macromolecules is higher than the overlap concentration, the solution is called "semidilute solution". For polymers, $R \sim 10^{-5}$ cm, $M \sim 10^5$ g/mol, so that $C^* \sim 10^{-3}$ g/mL.

Unlike small molecules, the conformation of macromolecular chains in solution and the solution structure will undergo two completely different changes as the concentration increases. First, if the chain length (L) is shorter than the entanglement length (L_e), or say, the segment number of macromolecules (N) is less than the number of entangled segments (N_e), the chain conformation shrinks, and the size becomes smaller, keeping the overlap state. Even in the molten state and the concentration reaches 100%, macromolecular

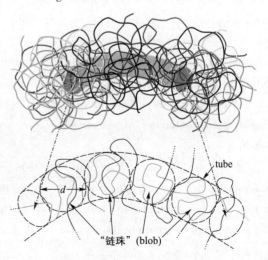

tube

"链珠" (blob)

chains will not enter each other's space. Its physical properties are similar as small molecule solutions or solutions. Second, when $L > L_e$, the chains will interpenetrate each other, and there is a total of N_{p,R^3} macromolecular chains in the space occupied by each macromolecular chain (volume, $R^3 = b^3 N^{3\alpha_R}$), as shown on the right. For a homogeneous system, the local and overall chain segment densities are equal, so that

$$\frac{N_{p,R^3}}{R^3} = \frac{N_p}{V} \qquad \rightarrow \qquad \boldsymbol{N_{p,R^3}} = \frac{R^3 N_{AV}}{M} \boldsymbol{C} = \frac{\boldsymbol{C}}{\boldsymbol{C}^*} \tag{9.2}$$

For the entangled macromolecular chains, when the concentration is higher than the overlap concentration, the number of chains in the space occupied by each chain (N_{p,R^3}) is proportional to the concentration. At the Θ state, for a given concentration, $N_{p,R^3} \propto N^{1/2}$, increasing with the chain length. In a dilute solution, $C < C^*$, $N_{p,R^3} = 1$. For common polymers, $b \sim 10^{-7}$ cm, $m_0 N_{AV} \sim 10^2$, and $N \sim 10^3$. If the concentration is $\sim 10^{-1}$ g/mL, $N_{p,R^3} \sim 10 - 100$, which means that a dozen of chains are entangled in the same space. As the concentration further increases, the length of the chain link between the two adjacent entanglement points gradually decreases. When the length of a chain link approaches the length of a chain segment, the macromolecular solution enters another region, called the "**concentrated solution**", often denoted as C^{**}. Therefore, $C^{**} = m_0/b^3$, so that

$$C^{**} = C^* N^{3\alpha_R - 1} \tag{9.3}$$

When the concentration of a solution is high, even in a good solvent, one segment is surrounded by a large number of segments on other chains, and the dispersion attraction is enhanced. Both theory and neutron scattering experiments have proved that the excluded volume effect tends to disappear, so that the solution with a high concentration can be considered in the Θ state, $\alpha_R = 1/2$, i.e., $C^{**}/C^* = N^{1/2}$. For common polymers, $N \sim 10^3$, the ratio has already exceeded 30 times. Therefore, there is a wide concentration range between "semidilute solution" and "concentrated solution", and the range increases with N.

It is precisely due to the chain entanglement that the macromolecules have some unique physical properties that small molecules lack, such as the aforementioned viscoelasticity. Strictly finding quantitative relationships among these properties inevitably involves many − body problems in physics, which is very complicated and far exceeds the content of this book, so it is omitted. In the following, only de Gennes's "tube" model, "reptation" image and scaling theory are used to describe **static** and **dynamic behavior** of macromolecular chains in semidilute solutions.

As shown in the figure on the right; each macromolecular chain imaginatively lies in a "tube" made up of intertwined chains around it, similar to the situation where a chain is in a cylindrical tube as discussed in the previous chapter. The difference is that the chain is not stretched here and still maintains a random coil conformation. Therefore, **the "tube" with a trapped chain is naturally a random coil, too.** One should not confuse the "tube" here with the straight cylindrical pore in the previous chapter. Each segment in a chain walks randomly within a local range in the "tube", causing a limited local density fluctuation. The largest step is the tube diameter (d), longer than a segment (b). To this end, a macromolecular chain can be imaginatively "divided" into short segments. Each of them contains N_d segments to form a small random coil, the aforementioned "blob". The root mean end–to–end distance of each short chain link is the pipe diameter, $d = bN_d^{\alpha_R}$. In this way, a macromolecular chain can be described as a string of N/N_d "blobs". Therefore, the chain conformation and the "tube" can be described as a trajectory of N/N_d steps undisturbed random walking with a step length of d. The root mean end–to–end distance of the "tube" is $R_d = d \, (N/N_d)^{1/2}$. The short chain link in each "chain bead" can be straightened, so that its root mean end–to–end distance is shorter than its straight length: $d = bN_d^{\alpha_R} < bN_d \, (1/2 \leqslant \alpha_R \leqslant 3/5)$. Therefore, the length of a straight chain ($L_c = bN$) is much longer than the straight string of "blobs" or "tube" ($L_{c,d} = d(N/N_d)$).

Therefore, the conformation of each linear macromolecular chain can be described as two kinds of random walking trajectories with different step lengths (b and d). However, the average end–to–end distance of a macromolecular chain is a real physical existence, and does not change with different descriptions, so that $d \, (N/N_d)^{\alpha_R} = d \, (L_{c,d}/d)^{\alpha_R} = bN^{\alpha_R}$. For a homogeneous solution, the segment densities inside the volumes occupied by each chain link and each chain are equal,

$$\frac{N_d}{d^3} = \frac{NN_{p,R^3}}{R^3}$$

Using $N_{p,R^3} = C/C^*$, one can further get

$$\frac{1}{b^3 N_d^{3\alpha_R-1}} = \frac{N_{p,R^3}}{b^3 N^{3\alpha_R-1}} \quad \rightarrow \quad \left(\frac{N}{N_d}\right)^{3\alpha_R-1} = N_{p,R^3} \quad \rightarrow \quad \frac{N}{N_d} = \left(\frac{C}{C^*}\right)^{\frac{1}{3\alpha_R-1}} \quad (9.4)$$

In the above description, no macromolecular chain is special. Each macromolecular chain is in a "tube" composed of other surrounding macromolecular chains. At the same time, it is also a member of the "pipe" surrounding other macromolecular chains. As mentioned earlier, when the concentration is high, the excluded volume effect disappears, and the chain is in the Θ state, $\alpha_R = 1/2$. Therefore, the ratio to the average end distance ($R_d = d \, (N/N_d)^{1/2}$) of the chain in the "tube" to that in the dilute solution is

$$\frac{R_d}{R} = \frac{d \, (N/N_d)^{1/2}}{bN^{\alpha_R}} = \frac{bN_d^{\alpha_R} \, (N/N_d)^{1/2}}{bN^{\alpha_R}} = \left(\frac{N}{N_d}\right)^{\frac{1}{2}-\alpha_R} = \left(\frac{C}{C^*}\right)^{\frac{1/2-\alpha_R}{3\alpha_R-1}}$$

In the Θ state, $\alpha_R = 1/2$. By definition, the chain size is independent of the concentration, so that $R_d = R$. In a good solvent, the above equation becomes $R_d = R \, (C/C^*)^{-1/8}$. Therefore, in a semidilute solution of a good solvent, the average end–to–end distance of the chain decreases as the concentration increases. When C/C^* increases from 10 to 40, the degree of chain shrinkage changes from 75% to 63%. Note that the chain contraction does not change the scaling relationship between R and N in the semidilute solution, $R = d\ (N/N_d)^{1/2}$. What has changed is only the length of the "chain segment" ($d/N_d^{1/2} = bN_d^{\alpha_R-1/2}$). As the concentration increases, N_d becomes smaller, resulting in a smaller R, as if the total segment number has not changed, but each "segment" gradually becomes shorter. **Both d and N_d are independent of the chain length, and only depend on the concentration.**

Similarly, the microscopic parameter d (the "tube" diameter) becomes smaller as the weight(mass) concentration (C) increases. According to the lemma "the average microscopic concentration and the macroscopic concentration in a uniform system are equal", the dependence between the two can be directly obtained as follows.

$$\frac{m_0 N_d}{d^3} = C \rightarrow d = b^{\frac{1}{1-3\alpha_R}}\left(\frac{C}{m_0}\right)^{\frac{\alpha_R}{1-3\alpha_R}} \qquad \rightarrow \qquad d = R\left(\frac{C}{C^*}\right)^{\frac{\alpha_R}{1-3\alpha_R}} \tag{9.5}$$

In the derivation, $d = bN_d^{\alpha_R}$ and eq. (9.1) are used, $C^* = m_0/(\ b^3 N^{3\alpha_R-1})$. **The tube diameter d in the semidilute solution is dependent of the mass(weight) concentration and the segment length, of macromolecule, and independent of the chain length.** In the Θ and good solvents, the above equations become $d = R\ (C/C^*)^{-1}$ and $d = R\ (C/C^*)^{-3/4}$, respectively. For common polymer chains, $R \sim 10^2$ nm, so that $d \sim 10$ nm when $C/C^* \sim 10$, corresponding to $\sim 10^2$ segments.

The random motions of the segments cause the density $\rho(r)$ to fluctuate around the average density ρ_0, higher or lower, $\Delta\rho(r) = \rho(r) - \rho_0$, positive or negative. When only the short range interaction is considered, the intersegment distance autocorrelation function of the density fluctuation is

$$\langle \Delta\rho(r)\Delta\rho(0)\rangle = \langle \rho(r)\rho(0)\rangle - \rho_0^2 \qquad \rightarrow \qquad \frac{\langle \Delta\rho(r)\Delta\rho(0)\rangle}{\rho_0^2} = \frac{\langle \rho(r)\rho(0)\rangle}{\rho_0^2} - 1$$

It can be proved that $\langle \rho(0)\rho(0)\rangle = 2\rho_0^2$ is the intersegment distance autocorrelation of the density fluctuation. As r increases, it relaxes from 100% to 0%. The above equation was proposed by Leonard Ornstein and Frits Zernike [Royal Netherlands Academy of Arts and Sciences(KNAW) Proceedings, 1914, 17:793] more than one hundred years ago. The specific derivation is omitted, and the mathematical expression is

$$g_{OZ}(r) = \frac{\langle \Delta\rho(r)\Delta\rho(0)\rangle}{\rho_0^2} = \frac{1}{4\pi\xi^2 r}\exp\left(-\frac{r}{\xi}\right) \tag{9.6}$$

where ξ is the characteristic correlation length, representing the reachable range of the density fluctuations at any point. After multiplying both sides of the above equation by $4\pi r^2 \mathrm{d}r$, it represents the probability of finding another segment in a spherical shell with a radius of r and a thickness of $\mathrm{d}r$.

$$4\pi r^2 g_{OZ}(r)\,\mathrm{d}r = \frac{r}{\xi^2}\exp\left(-\frac{r}{\xi}\right)\mathrm{d}r = \frac{r}{\xi}\exp\left(-\frac{r}{\xi}\right)\mathrm{d}\left(\frac{r}{\xi}\right)$$

When $r \to 0$, the degree of correlation is the highest, but the spherical shell volume tends to zero, so the probability is zero. Obviously, at the position of $r \to 0$, there is only one segment itself and no other segments. On the other hand, when $r \to \infty$, the probability also tends to zero. At $r = \xi$, the probability reaches the maximum, which is the physical meaning of ξ. When $r = \xi$, the mean square density fluctuation has decayed from the highest to $\sim 37\%$. When $r = 3\xi$, the above equation has almost decayed to zero. In other words, the range of density fluctuations is limited to a sphere with a radius of 3ξ. The Fourier transformation of the above equation is the static structure factor,

$$p(q) = \frac{1}{V\rho_0^2} \int \langle \Delta\rho(r) \Delta\rho(0) \rangle \exp(i\vec{\mathbf{q}} \cdot \vec{\mathbf{r}}) dr = \frac{p(0)}{1 + \xi^2 q^2}$$

In comparison with the static structure factor discussed in the section of static LLS, Chapter 5, $\xi^2 = \langle R_g^2 \rangle / 3 \approx \langle R^2 \rangle / 18$, i.e., $\xi \approx R/4$. Using LLS, one can measure the time−average intensity of scattered light $I(q)$ of a semidilute solution of macromolecules at different angles first; and then, plot $1/I(q)$ against q^2. According to the above equation, a straight line is obtainable. From its intercept and slope, $1/I(0)$ and ξ^2 can be obtained, respectively. The figure below shows that the static characteristic correlation length(slope) decreases as the crosslinking degree increases in an ultraviolet irradiation induced reaction of crosslinking macromolecular chains. The crosslinking between macromolecular chains restricts the motion and migration of the chain, and the reachable range of density fluctuations becomes smaller, so that the static characteristic correlation length becomes shorter. For details, please refer to the literature in the figure. Note: ξ reflects a collective property of the entire solution at a given temperature and pressure, while R_g is a property of individual macromolecular chains. In the discussion, it can also be regarded as a property of the "short link" located in a "blob". The density fluctuation is limited by the tube diameter(d), so that $\xi < d$.

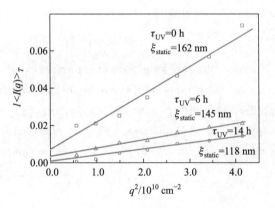

Ngai T, Wu C, Chen Y. Macromolecules, 2004, 37:987.

(Figure 7, permission was granted by publisher)

In the entanglement state of macromolecular chains, it is difficult for thermal energy to agitate the entire chain to move its center of mass in a short period of time, and only N_d segments in the short chain link inside each "blob" to walk randomly within the "blob" volume (d^3). If each short chain link is regarded as a "particle", the random walk of its center of mass can produce a pressure, $\pi_d = k_B T / d^3$. By definition, the difference between this pressure and the pressure generated by solvent molecules is the osmotic pressure difference. It is known in the discussion of the osmotic pressure difference in Chapter 5, since it is a pressure difference, the pressure generated by solvent molecules can be taken as the reference zero point. In contrast, in a dilute solution, thermal energy can agitate each chain to move randomly within the solution volume (V), generating a pressure, $k_B T/V$. In an ideal state, the pressure generated by N_p chains is N_p times, that is, $\pi = k_B T N_p /V$. Converting N_d/d^3 and N_p/V to the mass concentration, one can obtain

$$\pi_d = RT \left(\frac{N_{AV} R^3 C^{3\alpha_R}}{M^{3\alpha_R}} \right)^{\frac{1}{3\alpha_R - 1}} \quad \text{and} \quad \pi = RT \frac{C}{M}; \quad \rightarrow \quad \frac{\pi_d}{\pi} = \left(\frac{C}{C^*} \right)^{\frac{1}{3\alpha_R - 1}} \tag{9.7}$$

where the rewritten eq (9.1) is used: $(C^* N_{AV}/M)^{1/(3\alpha_R - 1)} R^{3/(3\alpha_R - 1)} = 1$. In the Θ and good solvents, the above equation become $\pi_d/\pi = (C/C^*)^2$ and $\pi_d/\pi = (C/C^*)^{5/4}$, respectively. In the cgs system, for typical polymer chains in semidilute solutions, $R \sim 10^{-5}$ cm, $M \sim 10^5$ g/mol, $C \sim 10^{-2}$ g/mL, and $C/C^* \sim 10$. Therefore, in the Θ solvent, $\pi_d/\pi \sim 10^2$. Each macromolecular chain generates a much larger pressure in the semidilute solution.

If a cylindrical pore with a pore diameter of d is inserted into a semidilute solution, as long as $\xi < d$, macromolecular chains will spontaneously enter the pore due to the pressure difference, and the probability is 100%. Conversely, if $\xi > d$, the probability of entering the pore is exactly the same as the situation discussed above for a macromolecular chain entering the pore. Just notice that the chain size (R) is replaced by the characteristic correlation length (ξ),

$$\Psi = \exp \left[-(\xi/d)^{\frac{5}{3}} + 1 \right] \qquad (\xi \geqslant R) \tag{9.8}$$

Physically, as long as remembering that in the entangled state, many physical properties of a macromolecular chain have nothing to do with its length or only have a specific relationship with its length, one can use the scaling theory of de Gennes. In theory, any physical quantity (Y) can be expanded into a virial polynomial of concentration (C) by taking the ideal state at which the concentration tends to zero as a reference benchmark.

$$Y = Y_{\text{ideal}} \left[1 + A_2 \left(\frac{C N_{AV}}{M} R^3 \right) + A_3 \left(\frac{C N_{AV}}{M} R^3 \right)^2 + \cdots \right] = Y_{\text{ideal}} f \left(\frac{C N_{AV}}{M} R^3 \right) \tag{9.9}$$

where $C N_{AV} R^3/M$ is a dimensionless variable. According to eq. (9.1), $C N_{AV} R^3/M = C/C^*$. $f(C/C^*)$ is also called a scaling function. R is related to the average radius of gyration, and together with M are two measurable physical quantities, e. g., they can be measured by static LLS. Note that $R = bN^{\alpha_R}$; $M = m_0 N N_{AV}$. When the concentration approaches the two extremes,

$$f \left(\frac{C N_{AV}}{M} R^3 \right) \rightarrow 1 \quad (C \rightarrow 0); \qquad f \left(\frac{C N_{AV}}{M} R^3 \right) \rightarrow \left(\frac{C N_{AV}}{M} R^3 \right)^z \quad (C \gg 1)$$

In comparison with the z-order term $(C/C^*)^z$, all low-order terms can be ignored. First, let us take the static correlation length in the semidilute solution as an example to illustrate how to do a scaling derivation. As mentioned before, in very dilute solution, $\xi \rightarrow \sim R/4 = bN^{\alpha_R}/4$, so that

$$\xi \approx \frac{bN^{\alpha_R}}{4} \left(\frac{C}{C^*} \right)^z = \frac{bN^{\alpha_R}}{4} \left(\frac{C}{m_0} \right)^z b^{3z} N^{(3\alpha_R - 1)z} = \frac{1}{4} \left(\frac{C}{m_0} \right)^z b^{3z+1} N^{(3\alpha_R - 1)z + \alpha_R}$$

The slightly varied eq. (9.1) is used in the derivation, $1/C^* = b^3 N^{3\alpha_R - 1}/m_0$. According to the definition, ξ is not related to the chain length, so that $(3\alpha_R - 1)z + \alpha_R = 0$, or write as, $z = \alpha_R/(1 - 3\alpha_R)$. Substituting it into the above equation, one can get,

$$\xi \approx \frac{R}{4}\left(\frac{C}{C^*}\right)^{\frac{\alpha_R}{1-3\alpha_R}}$$

In comparison with eq. (9.5), $\xi \approx d/4$. The physical explanation is that each "blob" is independent, and the characteristic size of its density fluctuation is $d/4$. Using one segment as a starting point, the probability of finding another segment at a distance of 4ξ is almost zero. Readers should note that in literature and other textbooks, it is extremely misleading to use ξ as the tube diameter d, and the physical concept is vague.

Taking the osmotic pressure difference in a semidilute solution as an example, in a very dilute solution ($C \to 0$), $\pi_d \to \pi = k_B T N_{AV} C/M$. Note: $k_B N_{AV}$ is not written as the gas constant R here to avoid confusion with the average end-to-end distance (R) of the chain. Therefore,

$$\pi_d = \frac{k_B T N_{AV} C}{M}\left(\frac{C}{C^*}\right)^z = \frac{k_B T}{R^3}\left(\frac{C}{C^*}\right)^{z+1} = k_B T\left(\frac{C}{m_0}\right)^{z+1} b^{3z} N^{(3\alpha_R - 1)(1+z) - 3\alpha_R}$$

Due to the chain entanglement, the entire macromolecular chain cannot move a certain distance in a short time, so that π_d should have nothing to do with the chain length, $(3\alpha_R - 1)(1+z) - 3\alpha_R = 0$ or write as $z = 1/(3\alpha_R - 1)$. The substitution of it into the above equation leads to the same eq. (9.7), but by different ways.

$$\frac{\pi_d}{\pi} = \left(\frac{C}{C^*}\right)^{\frac{1}{3\alpha_R - 1}}$$

Another example is the chemical potential difference between the solution and the solvent ($\Delta\mu$) as the molar number of chains in the solution (n_p) changes from a dilute solution to a semidilute solution, $(\partial\Delta\mu/\partial n_p)_{T,p}$. Known from thermodynamics, this change is related to the osmotic pressure difference (π). For a given concentration, under isothermal conditions, the free energy change of solvent is $dG = Vdp$, and its change with the moles of solvent is the chemical potential change of solvent ($d\mu = Vdp$). For a very dilute macromolecular solution containing n_s moles of solvent, assuming that the volume is a constant, both sides are integrated. The left is integrated from the chemical potential of the solvent to that of the solution, $\mu_{solution} - \mu_{solvent} = \Delta\mu$; the right is integrated from 0 to π, $V\pi$, so that $\Delta\mu = V\pi$. It is known that $C = W/V = (W/M)M/V = n_p M/V$ and $\pi = N_{AV} k_B TC/M$. Therefore,

$$\left(\frac{\partial\Delta\mu}{\partial n_p}\right)_{T,p} = \left[\frac{\partial(V\pi)}{\partial(CV/M)}\right]_{T,p} = M\left(\frac{\partial\pi}{\partial C}\right)_{T,p} = N_{AV} k_B T$$

where $(\partial\pi/\partial C)_{T,p}$ is the compression coefficient of the osmotic pressure difference. When $C \to 0$, $(\partial\pi/\partial C)_{T,p} \to k_B T N_{AV}/M$, so that

$$\left(\frac{\partial\pi}{\partial C}\right)_{T,p,\rho} = \frac{k_B T N_{AV}}{M}\left(\frac{C}{C^*}\right)^z = \frac{k_B T}{CR^3}\left(\frac{C}{C^*}\right)^{1+z} = \frac{k_B T}{C}\left(\frac{C}{m_0}\right)^{1+z} b^{3z} N^{(3\alpha_R - 1)(1+z) - 3\alpha_R}$$

Due to the chain entanglement, $(\partial \pi / \partial C)_{T,p,\rho}$ is also independent of the chain length, so that $(3\alpha_R - 1)(1 + z) - 3\alpha_R = 0$ or write as $z = 1/(3\alpha_R - 1)$. Substituting it in the above equation,

$$\left(\frac{\partial \pi}{\partial C}\right)_{T,p,\rho} = \frac{k_B T N_{AV}}{M} \left(\frac{C}{C^*}\right)^{\frac{1}{3\alpha_R - 1}}$$

In the Θ and good solvents, $\alpha_R = 1/2$ and 3/5 respectively, so that $(\partial \pi / \partial C)_{T,p,\rho}$ $\sim C^2$ and $\sim C^{5/4}$. In the discussion of static LLS, it is known that the difference between the intensities of scattered light (ΔI) from a macromolecular solution and a solvent is proportional to $C/(\partial \pi / \partial C)_{T,p,\rho}$. Therefore, in the Θ and good solvents, when $C \leqslant C^*$, $\Delta I \sim C$; when $C > C^*$, $\Delta I \sim C^{-1}$ and $\Delta I \sim C^{-1/4}$, respectively. Therefore, using static LLS, by measuring the change of ΔI with the concentration, the

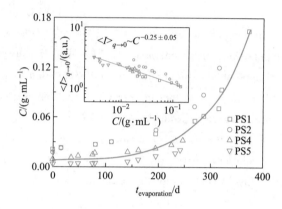

Li J F, Li W, Huo H, et al. Macromolecules, 2008, 41: 901.

(Figure 1, permission was granted by publisher)

overlap concentration (C^*) can be obtained from the slope change, at which the chains start to contact with each other in the solution. The figure above shows that as the solvent evaporates, the concentration in the good solvent benzene gradually increases. The data in the illustration clearly shows that the dependence of the average scattered light intensity on the concentration is indeed theoretically predicted $-1/4$.

The last example is the average end-to-end distance (R_d) of macromolecular chains in semidilute solutions as discussed earlier. When $C \to 0$, $R_d \to R = bN^{\alpha_R}$; when $C \gg 1$,

$$R_d = bN^{\alpha_R} \left(\frac{C}{C^*}\right)^z = \left(\frac{C}{m_0}\right)^z b^{3z+1} N^{(3\alpha_R - 1)z + \alpha_R}$$

Physically, as the average end-to-end distance of macromolecular chains in semidilute solutions, R_d must depend on the chain length. Assuming that its dependence on the chain length in semidilute solutions is $R_d = bN^{\alpha'_R}$, then $(3\alpha_R - 1)z + \alpha_R = \alpha'_R$, one can get $z = (\alpha'_R - \alpha_R)/(3\alpha_R - 1)$. On the other hand, it is known that in the Θ state ($\alpha_R = 1/2$), the chain size is independent of the concentration. Therefore, α'_R must also be equal to $1/2$, i.e., the same as in the Θ undisturbed state, the overall size of the chain in the "tube" or the tube size (the average end-to-end distance) is independent of the solvent quality. Substituting it to the above equation, the concentration dependence of the average end-to-end distance of the chain can be obtained.

$$R_d = R \left(\frac{C}{C^*}\right)^{\frac{1/2 - \alpha_R}{3\alpha_R - 1}}$$

Exactly the same as previously derived. Replaced by the experimentally measurable mean square end-to-end distance, $\langle R_d^2 \rangle / \langle R^2 \rangle = (C/C^*)^{-1/4}$ (in a good solvent), where for a given macromolecule, R and C^* are constants, and the chain shrinks as the concentration increases.

After discussing some common static properties of semidilute solutions, some important dynamic behaviors will be explained as follows. Before discussing, let us review the following relationship between the relevant parameters in the "tube" model. Each chain can be regarded as a string of N/N_d connected "blobs", each "blob" itself is a short chain link with N_d segments and a random coil conformation, whose average end-to-end distance, $d = bN_d^{\alpha_R}$. Each chain is trapped in a curved "tube" formed by other surrounding chains. The chain and the "tube" trapping it have the same random coil conformation and average end-to-end distance : $R = bN^{1/2} = d (L_{c,d}/d)^{1/2}$, or write as $L_{c,d} = bN/N_d^{1/2}$, where $L_{c,d}$ is the straight length of the "tube". Since $N_d > 1$, $L_{c,d}$ is much shorter than the straight length of the chain ($L_c = bN$).

Generally speaking, when a chain randomly walks in a dilute solution, the friction coefficient and the translational diffusion coefficient are $f = 6\pi\eta d$ and $D = k_B T/f$, respectively. When $C \leqslant C^*$, there are two extreme cases : completely "drained" (Rouse model), each segment with a length of b is subjected to a frictional resistant force, $d = bN$; and completely "non-drained" (Zimm model), each chain can be considered as a solid ball with a size of R (strictly speaking, it should be R_h), $d = R_h \sim R = bN^{\alpha_R}$. The corresponding translational diffusion coefficient is the mutual translational diffusion coefficient of the chains in the dilute solution, denoted as

$$D_0 = \frac{k_B T}{6\pi\eta bN} \qquad \rightarrow \qquad D_0 \sim N^{-1} \quad (C < C^*, \text{"draining"})$$

$$D_0 = \frac{k_B T}{6\pi\eta R} \qquad \rightarrow \qquad D_0 \sim N^{-\alpha_R} \quad (C < C^*, \text{"nondraining"}) \qquad (9.10)$$

In the previous chapter, it has been proved that macromolecular chains are nearly non-draining in solution. The experimental results have also proved that in the Θ and good solvents, D_0 is very close to $\sim N^{-1/2}$ and $\sim N^{-0.59}$, consistent with the expected scaling exponents of "non-draining". The times required for a chain to relax its average end-to-end distance by translational diffusion ($\langle R^2 \rangle / (6D_0)$), respectively, are

$$\tau_0 = \frac{\pi\eta bNR^2}{k_B T} = \frac{\pi\eta b^3}{k_B T} N^{1+2\alpha_R} \qquad \rightarrow \qquad \tau_0 \sim N^{1+2\alpha_R} \quad (C \rightarrow 0 < C^*, \text{"draining"})$$

$$\tau_0 = \frac{\pi\eta R^3}{k_B T} = \frac{\pi\eta b^3}{k_B T} N^{3\alpha_R} \qquad \rightarrow \qquad \tau_0 \sim N^{3\alpha_R} \quad (C \rightarrow 0 < C^*, \text{"nondraining"}) \qquad (9.11)$$

For common polymers, in the cgs system, $\eta \sim 10^{-2}$; $b \sim 10^{-7}$; $k_B T \sim 10^{-14}$; $N \sim 10^3$; so that $\tau_0 \sim 10^{-3}$ s. Obviously, these two cases correspond to the characteristic relaxation time (τ_1) of the first normal mode in the Rouse and Zimm models, where $\alpha_R = 1/2$ (Rouse model) and $1/2 \leqslant \alpha_R \leqslant 3/5$ (Zimm model).

When $C > C^*$, due to the entanglement, in a short time, not only the centroid of the entire macromolecular chain but also the chain link inside each "blob" is no long able to move for a long distance. The characteristic correlation length of the density fluctuation caused by random walking of the segment inside a local area of each "blob" is ξ. If there are N_ξ segments inside the space of ξ^3 and these N_ξ segments can be regarded as a small ball, its size is the average end-to-end distance $\xi = bN_\xi^{\alpha_R}$. The friction coefficient of each ball is also dependent of the draining ($f = 6\pi\eta bN_\xi$) and the non-drainage ($f = 6\pi\eta\xi$). The corresponding translational diffusion coefficient is,

$$D_\xi = \frac{k_B T}{6\pi\eta bN_\xi} \sim N^0 \qquad \rightarrow \qquad D_\xi = D_0 \left(\frac{C}{C^*}\right)^{\frac{1}{3\alpha_R - 1}} \quad (C > C^*, \text{"draining"})$$

$$D_\xi = \frac{k_B T}{6\pi\eta\xi} \sim N^0 \qquad \rightarrow \qquad D_\xi = D_0 \left(\frac{C}{C^*}\right)^{\frac{\alpha_R}{3\alpha_R - 1}} \quad (C > C^*, \text{"nondraining"}) \qquad (9.12)$$

where $\xi = bN_\xi^{\alpha_R}$ and $N/N_\xi = (C/C^*)^{1/(3\alpha_R - 1)}$ are used. N_ξ, ξ and D_ξ **are independent of the chain length**; D_0 and C^* **are independent of the concentration.** Since this relaxation comes from the cooperative movement of all the segments inside the volume ξ^3, so that D_ξ is also called the cooperative diffusion coefficient, denoted as D_{coop}. Using dynamic LLS, it is possible to experimentally measure the intensity distribution of the linewidth $G(\Gamma)$ in the frequency space caused by density fluctuation or the z-average linewidth $\langle\Gamma\rangle$ calculated from the distribution, where $\Gamma = D_\xi q^2$. Therefore, Γ can be measured at different scattering angles first, and then, D_ξ can be obtained from the plot of "Γ versus q^2".

The figure on the right shows time auto-correlation functions $G^{(2)}(t)$ of the scattered light intensity of a semidilute solution (C/C^* as high as 30) of a narrowly distributed polystyrene in a good solvent benzene measured by dynamic LLS, where the abscissa, originally the delay time(t), has been corrected by q^2 to eliminate the influence of the scattering angle. After the correction, four time autocorrelation functions from different scattering angles are overlapped together,

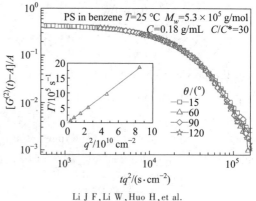

Li J F, Li W, Huo H, et al.

Macromolecules, 2008, 41: 901.

(Figure 7, permission was granted by publisher)

and their Laplace inversions all correspond to a very narrow linewidth distribution with only one faster relaxation peak. The inset shows that the average linewidth plotted against q^2 is a straight line passing through the origin, indicating that the physical nature of the relaxation is translational diffusion. The slope of the straight line is D_ξ; the corresponding characteristic relaxation time is the time required for the short chain link to diffuse a distance of ξ, $\tau_\xi = \xi^2/(6D_\xi)$, i.e.,

$$\tau_\xi = \frac{\pi\eta\xi^3 N_\xi^{1-\alpha_R}}{k_B T} \sim N^0 \qquad \rightarrow \qquad \tau_\xi = \tau_0 \left(\frac{C}{C^*}\right)^{-\frac{2\alpha_R+1}{3\alpha_R-1}} (C > C^*, \text{"draining"})$$

$$\tau_\xi = \frac{\pi\eta\xi^3}{k_B T} \sim N^0 \qquad \rightarrow \qquad \tau_\xi = \tau_0 \left(\frac{C}{C^*}\right)^{-\frac{3\alpha_R}{3\alpha_R-1}} (C > C^*, \text{"nondraining"}) \qquad (9.13)$$

For common polymers, $\xi = 10^{-6} \sim 10^{-5}$ cm, so that $\tau_d = 10^{-6} \sim 10^{-5}$ s. As the concentration increases, it becomes extremely short, the linewidth increases, and the relaxation process speeds up. The range of the corresponding linewidth is $10^5 \sim 10^6$ s^{-1}, as shown in the figure above.

A comparison of eqs. (9.10) and (9.12) shows that R_h in dilute solution is replaced by ξ in semidilute solutions. It is known when discussing dilute macromolecule solutions, $1/R_h = \langle 1/r \rangle$. Here, a similar proof can be used,

$$D_\xi = \frac{k_B T}{6\pi\eta} \left\langle \frac{1}{r} \right\rangle$$

where r is the distance between any two cooperatively moving segments in the N_ξ segments. Its distribution follows eq. (9.6), so that

$$\left\langle \frac{1}{r} \right\rangle = \int_V \frac{1}{r} g_{OZ}(r) \, dV = \int_0^\infty \frac{1}{r} g_{OZ}(r) 4\pi r^2 \, dr = \frac{1}{\xi}$$

Therefore, after having D_ξ, one can calculate the **dynamic correlation length (ξ)** using the above equation. In the 1970s and 1980s, many laboratories studied semidilute solutions of different polymers. A large number of experimental results have confirmed that in the Θ and good solvents, there exist the scaling relationships of $D_\xi \sim C^{1.0}$ and $D_\xi \sim C^{0.75}$, consistent with the expected scaling exponents of "non-draining".

The above is the dynamic behavior of the chain in a relatively short period of time. If given a sufficiently long time, each chain trapped in a "tube" made up of other chains should be able to "reptation" slowly along the "tube", and its head (one end) can walk a length of the "tube" diameter (d, step length) randomly in different directions, enter a new section of the "tube", and the tail (another end) follows one step forward in the "tube". It seems that one end of the "tube" is constantly extending, and the other end is constantly disappearing, but the total length remains the same, like a slowly crawling "snake".

As mentioned before, such a macromolecular chain can be imaginatively divided into a series of short chain links, the average end-to-end distance of each short chain link is the tube diameter d, and there are N/N_d short chain links. In this way, each chain in the semidilute solution becomes an equivalent linear chain composed of N/N_d "new segments" (i.e., "blobs") with a length of d. It is also called the original chain in literature and books, whose straight length is $L_{c,d} = d(N/N_d)$, the conformation is still a random coil. Therefore, the conformation of the "tube" trapped it is also a random coil. The "tube" or equivalent linear chain has the same average end-to-end distance as the original chain.

The friction coefficient and the dynamic diffusion coefficient of a macromolecular chain walking randomly along the "tube" are $f = 6\pi\eta l_d(N/N_d)$ and $D = k_B T/f$, where l_d is the effective hydrodynamics size of each short chain link, which is also divided into "drainage" $l_d = bN_d$ and "non-drainage" $l_d = bN_d^{\alpha_R}$. Therefore, the movement of the chain along the "tube" is approximately one-dimensional "reptation", which is often referred to as "curvature diffusion" in literature and books, and its diffusion coefficient is

$$D_c = \frac{k_B T}{6\pi\eta bN_d}\frac{N_d}{N} \sim N^{-1} \qquad \rightarrow \qquad D_c = D_0 \quad (C > C^*, \text{"draining"})$$

$$D_c = \frac{k_B T}{6\pi\eta bN_d^{\alpha_R}}\frac{N_d}{N} \sim N^{-1} \qquad \rightarrow \qquad D_c = D_0 \left(\frac{C}{C^*}\right)^{-\frac{1-\alpha_R}{3\alpha_R-1}} \quad (C > C^*, \text{"nondraining"}) \quad (9.14)$$

where $N/N_d \sim (C/C^*)^{1/(3\alpha_R-1)}$ is used; N_d is independent of the chain length. In the case of "draining", the hydrodynamic interaction disappears, and the curvature diffusion in a semidilute solution is the translational diffusion in the dilute solution in the Rouse model, which slows down as the chain length increases. When non-draining, in the Θ and good solvents, $D_c = D_0(C/C^*)^{-1}$ and $D_c = D_0 (C/C^*)^{-1/2}$, respectively. Obviously, compared with the translational diffusion in a dilute solution, the "curvature diffusion" becomes significantly slower as the concentration increases. **Each chain walks randomly in the "tube", and the time required for it to diffuse a tube length ($L_{c,d} = d(N/N_d)$) is $\tau_c = L_{c,d}^2/(6D_c)$. In the cases of "draining" and "non-draining", respectively**

$$\tau_c = \frac{\pi\eta}{k_B T}bNL_{c,d}^2 = \frac{\pi\eta bNR^2}{k_B T}\frac{d^2}{R^2}\left(\frac{N}{N_d}\right)^2 \sim N^2 \qquad \rightarrow \qquad \tau_c = \tau_0 \left(\frac{C}{C^*}\right)^{\frac{2(1-\alpha_R)}{3\alpha_R-1}}$$

$$\tau_c = \frac{\pi\eta}{k_B T}\frac{bN_d^{\alpha_R}L_{c,d}^2}{N_d/N} = \frac{\pi\eta R^3}{k_B T}\frac{d^3}{R^3}\left(\frac{N}{N_d}\right)^3 \sim N^3 \qquad \rightarrow \qquad \tau_c = \tau_0 \left(\frac{C}{C^*}\right)^{\frac{3(1-\alpha_R)}{3\alpha_R-1}} \qquad (9.15)$$

where $d = bN_d^{\alpha_R}$, $R = bN^{\alpha_R}$ and $N/N_d = (C/C^*)^{1/(3\alpha_R-1)}$ are also used. Note: **whether in a Θ or a good solvent, d and N_d are two constants for a given concentration, independent of the chain length. When $N \sim 10^3$, the "curvature diffusion" of each chain in the semidilute solution is very slow. Conversely, for a given chain length, as the concentration increases, the "curvature diffusion" is also slower and slower. When $C/C^* \sim 20$, $\tau_c/\tau_0 \sim 10^2$, 10^2 times slower.**

If the "tube" is straight, after the chain randomly walks the "tube" length, its **centroid must have moved the same distance. However, the "tube" is a curled random coil. When the chain travels along the "tube" for a stretched length of the "tube", the centroid actually only moves a short distance. For example, if a macromolecular chain moves around its centroid by the straight length of a "tube", its centroid can be almost unchanged.**

It is known from the previous chapter that in a dilute solution, there is no entanglement among the chains; the centroid of a chain walks randomly. The complicated movements of every segments in the chain relatively to its centroid constitutes various normal mode relaxations. The characteristic relaxation time for the centroid to translationally diffuse the average end-to-end distance of a chain is $\tau_R = \langle R^2 \rangle / 6D_m$, or written as $\langle R^2 \rangle \approx D_m \tau_R$, omitting the proportional constant. As discussed before, in a semidilute solution, the chains are entangled with each other, and the chains are in a "tube" composed of other surrounding and entwining chains. A macromolecular chain can be described as an imaginable chain with N/N_d "new segment" with a length of d, whose straight length is $L_{c,d} = d(N/N_d) < L_c = bN$.

In the "tube", the centroid of the chain can no longer freely diffuse in three-dimensional space, and can undergo only approximately one-dimensional diffusion along the "tube". The maximum step length is the tube diameter d. The time required to walk each step is $t_d \approx d^2/D_c$. It takes $L_{c,d}/d$ steps to complete the straight length of a "tube". In the undisturbed state, the whole chain walks randomly, the mean square end-to-end distance after $L_{c,d}/d$ steps is $d^2(L_{c,d}/d) = dL_{c,d}$. The mean square end-to-end distance of a macromolecular chain is a physical existence, and has nothing to do with how to describe it. Therefore, $d^2(L_{c,\xi}/d^2) = b^2N$; that is, $L_{c,d} = b^2N/d$. According to the definition, the mathematical expression of the centroid of a macromolecular chain before ($t = 0$) and after ($t = t_d$) moving one step is

$$\vec{\mathbf{r}}_G(0) = \frac{1}{L_{c,d}} \int_0^{L_{c,d}} \vec{\mathbf{r}}(l,0)\,\mathrm{d}l \text{ and } \vec{\mathbf{r}}_G(t_d) = \frac{1}{L_{c,d}} \int_d^{L_{c,\xi}+d} \vec{\mathbf{r}}(l,t_d)\,\mathrm{d}l$$

According to the above equation, except the first and last small sector of "tube" with a length of d move out and move in the "tube", respectively, the middle part remains unchanged. Therefore, in the calculation of the difference between the above two position vectors, the integrations of all the segments in the middle are the same, cancelling each other, so that it is only necessary to calculate the first and the last two parts that have moved a length of d, i.e.,

$$\vec{\mathbf{r}}_G(t_d) - \vec{\mathbf{r}}_G(0) = \frac{1}{L_{c,d}} \int_0^d \vec{\mathbf{r}}(L_{c,d} + l)\,\mathrm{d}l - \frac{1}{L_{c,d}} \int_0^d \vec{\mathbf{r}}(0 + l)\,\mathrm{d}l$$

Since the motion of each segment is random and independent with each other, the time average of the above equation is zero, that is, $\langle \vec{\mathbf{r}}(L_{c,d} + d) \rangle = \langle \vec{\mathbf{r}}(d) \rangle$ and $\langle \vec{\mathbf{r}}_G(t_d) \rangle = \langle \vec{\mathbf{r}}_G(0) \rangle$. Therefore, in order to compare the moving speed, one has to use the statistical mean square value, namely

$$\langle (\vec{\mathbf{r}}_G(t_d) - \vec{\mathbf{r}}_G(0))^2 \rangle = \frac{1}{L_{c,d}^2} \int_0^d \int_0^d \langle [\vec{\mathbf{r}}(L_{c,d} + l) - \vec{\mathbf{r}}(l)]^2 \rangle \,\mathrm{d}l$$

where the average on the right side in the integral is the mean square end-to-end distance after the macromolecular chain has moved a short distance (d), which is approximately equal to the mean square end-to-end distance before the movement, namely

$$\langle [\vec{\mathbf{r}}(L_{c,d} + d) - \vec{\mathbf{r}}(d)]^2 \rangle \approx \langle [\vec{\mathbf{r}}(L_{c,d}) - \vec{\mathbf{r}}(0)]^2 \rangle = b^2N$$

For a given macromolecule, b^2N is a constant. In this way, the integral on the right side of the previous equation is rather simple, resulting in b^2Nd^2. Therefore,

$$\langle (\vec{\mathbf{r}}_G(t_d) - \vec{\mathbf{r}}_G(0))^2 \rangle = \frac{b^2Nd^2}{L_{c,d}^2} = d^2 \frac{N_d}{N}$$

where $L_{c,d} = d(N/N_d)$ and $d = bN_d^{\alpha_R}$ are used. Since $N \gg N_d$, even in the time t_d, both ends of the chain and the entire chain move forward a short distance (d), but the average moving distance of the centroid is much smaller than d. Physically, for a given macromolecule and its concentration, the translational diffusion coefficient of the centroid (D_{GC}) should be a constant, equal to the ratio of the mean square distance of the centroid movement to the travel time. Since the movement of each segment is random and independent, the two translational diffusion coefficients related to diffusing one step length and the travel time and diffusing the size of the entire chain (the average end-to-end distance) and the travel time should be the same. Increasing the distance and the travel time does not change this ratio. Therefore,

$$D_{GC} \approx \frac{\langle (\vec{r}_G(t_d) - \vec{r}_G(0))^2 \rangle}{t_d} = \frac{d^2 N_d}{N} / \frac{d^2}{D_c} \qquad \rightarrow \qquad D_{GC} = \frac{k_B T}{6\pi\eta l_d} \frac{N_d^2}{N^2} \qquad (9.16)$$

In the derivation, $L_{c,d} = d(N/N_d)$ and $f = 6\pi\eta l_d(N/N_d)$ are used. Note that in semidilute solutions, l_d, b, d and N_d are independent of N, so that $D_{GC} \sim N^{-2}$. In the cases of "draining" and non-draining", there are $l_d = bN_d$ and $l_d = bN_d^{\alpha_R}$, respectively. Substituting them into the above equation, one can get the translational diffusion coefficients of the centroid in the draining and non-draining cases.

$$D_{GC} = \frac{k_B T}{6\pi\eta bN_d} \frac{N_d^2}{N^2} = \frac{k_B T}{6\pi\eta bN} \frac{N_d}{N} \sim N^{-2} \qquad \rightarrow \qquad D_{GC} = D_0 \left(\frac{C}{C^*} \right)^{\frac{1}{1-3\alpha_R}}$$

$$D_{GC} = \frac{k_B T}{6\pi\eta bN_d^{\alpha_R}} \frac{N_d^2}{N^2} = \frac{k_B T}{6\pi\eta bN^{\alpha_R}} \frac{N_d^{2-\alpha_R}}{N^{2-\alpha_R}} \sim N^{\alpha_R - 2} \qquad \rightarrow \qquad D_{GC} = D_0 \left(\frac{C}{C^*} \right)^{\frac{2-\alpha_R}{1-3\alpha_R}} \qquad (9.17)$$

Their corresponding characteristic relaxation time, $\tau_{GC} = L_{c,d}^2/(6D_{GC})$, as follows,

$$\tau_{GC} = \frac{\pi\eta}{k_B T} \frac{bN_d L_{c,d}^2}{(N_d/N)^2} \sim N^4 \qquad \rightarrow \qquad \tau_{GC} = \tau_0 \left(\frac{C}{C^*} \right)^{\frac{3-2\alpha_R}{3\alpha_R - 1}} \quad (\text{"draining"})$$

$$\tau_{GC} = \frac{\pi\eta}{k_B T} \frac{bN_d^{\alpha_R} L_{c,d}^2}{(N_d/N)^2} \sim N^4 \qquad \rightarrow \qquad \tau_{GC} = \tau_0 \left(\frac{C}{C^*} \right)^{\frac{4-3\alpha_R}{3\alpha_R - 1}} \quad (\text{"nondraining"}) \qquad (9.18)$$

In the Θ and good solvents, in the case of draining, $D_{GC} = D_0 (C/C^*)^{-2}$ and $D_{GC} = D_0(C/C^*)^{-3/2}$ as well as $\tau_{GC} = \tau_0 (C/C^*)^4$ and $\tau_{GC} = \tau_0 (C/C^*)^{9/4}$. In the case of non-draining, $D_{GC} = D_0 (C/C^*)^{-3}$ and $D_{GC} = D_0 (C/C^*)^{-7/4}$ as well as $\tau_{GC} = \tau_0 (C/C^*)^5$ and $\tau_{GC} = \tau_0 (C/C^*)^{11/4}$. In a semidilute solution, the translational diffusion of the centroid of a macromolecular chain becomes extremely slow as the concentration and chain length increase. For typical polymers, in the cgs system, $d \sim b \sim 10^{-7}$ cm, $N \sim 10^3$. Therefore, $D_{GC} \sim 10^{-10}$ cm^2/s. The average square distance of the chain is about 10^{-10} cm^2, so that the time (τ_{GC}) for the centroid of the chain to move one average end-to-end distance is $\sim 10^{-10}/10^{-10} \sim 10^0$s. The translational diffusion of the centroid of a macromolecular chain in a semidilute solution is about $10^3 - 10^4$ times slower than that in a dilute solution.

When an external force acts for a time shorter than the time required for the centroid of a macromolecular chain to move an average end-to-end distance, the centroid changes little under the external action, only the chain conformation is stretched or compressed under the external action to deform, deviating from its equilibrium state. When the external action is removed, the deformed chain quickly relaxes back to its original equilibrium state, and the macroscopic behavior is just an elastomer. On the contrary, when the action time is longer than τ_{GC}, the centroid of a macromolecular chain can be moved under this action from one position to another farther position, the movement causes friction, friction consumes energy, and the system heats up. When the external action is removed, the chain will recover its equilibrium conformation at its new centroid. The macroscopic behavior is a viscous flowing liquid. Its flow can be slow or fast, no matter how fast or slow, it is a flowing liquid after all.

According to the above discussion, in principle, any system can exhibit elasticity or viscosity, depending on the relative length of the action time and the characteristic relaxation time. If the characteristic relaxation time is much longer or shorter than our normal time scale, the system exhibits elasticity or viscosity. A typical example is the elastic band of underwear. When initially worn, it has good elasticity and tightness; but after wearing it for months or years (depending on its quality), it slowly loses its elasticity and becomes longer, as if slowly "flowing". Physically, this is because the characteristic relaxation time of the macromolecular chains that constitutes the elastic band is very long. Therefore, when the characteristic relaxation time of a system is similar to the daily time scale, the system can exhibit visible viscosity or elasticity on a macroscopic scale, depending on the strength and time of the external action. From the above two equations, it can be seen that for ordinary polymer macromolecules, $\tau_{GC} \sim 1\mathrm{s}$. This is why a concentrated solution of macromolecules with long chains can exhibit viscosity and elasticity at the same time as the measurement time changes in the vicinity of seconds.

The scaling theory can be further used to obtain the above equation, and the concentration dependence of the step length of the random walk inside the "tube" (l_1). It has been known that when $C \to 0$, in the cases of "draining" and "non-draining", there are $D_{GC} \to D_{GC}(C \to 0) = D_0 = k_B T / (6\pi\eta b N)$ and $D_0 = k_B T / (6\pi\eta b N^{\alpha_R})$, respectively, so that

$$D_{GC} = D_0 \left(\frac{C}{C^*}\right)^z = \frac{k_B T}{6\pi\eta}\left(\frac{C}{m_0}\right)^z b^{3z-1} N^{(3\alpha_R-1)z-1} \quad (\text{"draining"})$$

$$D_{GC} = D_0 \left(\frac{C}{C^*}\right)^z = \frac{k_B T}{6\pi\eta}\left(\frac{C}{m_0}\right)^z b^{3z-1} N^{(3\alpha_R-1)z-\alpha_R} \quad (\text{"nondraining"}) \qquad (9.19)$$

In the derivation, eq. (9.1) is used. It has been known that $D_{GC} \sim N^{-2}$. In the two cases, $(3\alpha_R - 1)z - 1 = -2$ and $(3\alpha_R - 1)z - \alpha_R = -2$, i.e., $z = -1/(3\alpha_R - 1)$ and $z = (\alpha_R - 2)/(3\alpha_R - 1)$. Substituting them into eq. (9.18) also leads to eq. (9.16), one song played with different ways.

$$D_{GC}=D_0\left(\frac{C}{C^*}\right)^{-\frac{1}{3\alpha_R-1}}(\text{"draining"})\,;\qquad D_{GC}=D_0\left(\frac{C}{C^*}\right)^{-\frac{2-\alpha_R}{3\alpha_R-1}}(\text{"nondraining"})$$

Making that eq. (9.16) equals eq. (9.17), one can use eq (9.1) to obtain the concentration dependence of the tube diameter.

$$d\sim b^{\frac{8-9\alpha_R}{2(1-3\alpha_R)}}\left(\frac{C}{m_0}\right)^{\frac{1}{2(1-3\alpha_R)}}\qquad (C>C^*,\text{"draining"})$$

$$d\sim b^{\frac{8-9\alpha_R}{2(1-3\alpha_R)}}\left(\frac{C}{m_0}\right)^{\frac{2-\alpha_R}{2(1-3\alpha_R)}}\qquad (C>C^*,\text{"nondraining"})\qquad(9.20)$$

where $C/m_0=\rho_b=NN_p/V$ (the number density of segments). In the Θ and good solvents, in the cases of draining and non-draining, there are $d\sim b^{-7/2}\rho_b^{-1}$ and $d\sim b^{-13/8}\rho_b^{-5/8}$ as well as $d\sim b^{-7/2}\rho_b^{-3/2}$ and $d\sim b^{-13/8}\rho_b^{-7/8}$, respectively. For a given macromolecule, b is a constant; d becomes shorter as the concentration increases, that is, the "tube" becomes thinner.

In the semidilute solution of macromolecules, each chain is in a "tube" made up of other chains entwined around it, and at the same time, it is a part of the "tube" surrounding other chains. No chain is special and cannot be distinguished. Therefore, whether to measure D_c or D_{GC}, a small number of chains need to be labeled as probes. For example, in neutron scattering experiments, deuterium atoms are used to replace hydrogen atoms on the probe chains. In the measurement of fluorescence time-correlation spectroscopy or the photo-bleaching recovery spectroscopy, fluorescent molecules are attached to a small number of probe chains. In dynamic LLS, a solvent with the same refractive index is used, and a small amount of other macromolecular chains with similar lengths but a different refractive index as probes, By measuring these added probe chains, indirectly obtains the dynamic behavior of the original macromolecular chain, provided that there is no special interaction between these added chains and the original chains.

If using dynamic LLS to measure a semidilute solution of macromolecules directly, one can only theoretically measure the relaxation caused by the translational diffusion of the segments in each "blob" inside the "blob" volume, relating to τ_ξ. However, since the 1970s, in the dynamic LLS studies of different macromolecular semidilute solutions, different laboratories consistently observe an additional relaxation, which is slower with a larger Γ (a lower frequency), i.e., there is a second small peak in the intensity weighted linewidth distribution ($G(\Gamma)$). Such an additional relaxation (peak) has not been predicted in the "tube" and "reptation" theory at all and does not show up in other measurements (such as rheological measurements).

Therefore, such a slower relaxation has caused more than forty years of disputations between theoreticians and experimentalists. Theoreticians are also divided into two groups. Those who believe in the de Gennes theory often attribute it to some inevitable experimental defects, including impurities in the sample, dust particles or tiny bubbles that are not completely removed when filtering the solution, and a possible concentration or temperature gradient inside the solution. Although experimentalists have tried very hard to avoid these potential problems, they can never completely eliminate these possibilities.

The preparation of a dust–free semidilute solution suitable for dynamic LLS experiments traditionally includes the following steps. First, a dilute solution with a suitable viscosity and filterability is prepared; second, a minimum of 10 mL dilute solution is filtered to remove dust and injected into a dust–free LLS cell, which is difficult to do; third, 50%–90% of solvent is slowly volatilize to different concentrations under a dust–free environment. In the volatilization process, the concentration gradient, decreasing from top to bottom, is inevitable. The previous discussion shows that it takes 1–10 seconds for the centroid of a macromolecular chain to translate a distance of 10^{-5} cm; then it takes 10^{10}–10^{11} seconds to diffuse a distance of 1 cm, which is an astronomical number! Therefore, experimentalists can never rule out these possible interferences. Due to the lack of decisive experimental results, the controversy was gradually shelved with no conclusion. It has become an unresolved problem in the study of macromolecular solutions.

It was not until 2008 that the high–vacuum anion polymerization was used to directly synthesize narrowly distributed polystyrene with a high molar mass and prepare highly concentrated, dust–free solutions in a LLS sample cell, which have solved a number of previous problems. First of all, the problem of filtering a viscose solution of macromolecules is replaced by filtering a monomer solution with a much lower viscosity. The polymerization only starts under the premise of dust–free; secondly, the degree of reaction gradually increases during the polymerization. The entangled macromolecular chains are not able to move a long distance, but monomers as small molecules can still move freely in the solution to continuously react with the active chain ends, making the chain length continue to increase, which lowers the overlap concentration, leading to an increase in C/C^*, thereby any doubt about possible concentration gradients. Therefore, a series of dust–free, uniform, highly concentrated polystyrene solutions, respectively, in benzene and cyclohexane were successfully prepared.

The figure involved in the discussion of eq. (9.13) shows that in a good solvent benzene, even if C/C^* is as high as 30, indeed only one fast relaxation related to the translational diffusion of the segments inside each "blob" is observed. The subsequent measurements of different chain lengths, concentrations and temperatures all confirmed the de Gennes's theoretical prediction: in a short time, macromolecular chains are not able to move due to the entanglement, and only the segments inside each "blob" walk freely in a small volume of ξ^3. The line width measured by dynamic LLS is $\Gamma = D_\xi q^2 = (6\pi\eta\xi/k_B T) q^2$. ξ becomes shorter as the concentration increases. Because the concentration of macromolecules in the semidilute solution is already very high, theoretically, the solvent viscosity (η_0) can no longer be used to replace η. According to the previous discussion about viscosity, $\eta \sim C$; the concentration is the local mass concentration, $. C = m_0 N_\xi/\xi^3$; $\eta\xi \sim C\xi$; $\xi = bN_\xi^{\alpha_R}$. Therefore, $\Gamma \sim m_0/b^{1/\alpha_R}\xi^{2-1/\alpha_R}$. Since $\xi \sim C^{\alpha_R/(1-3\alpha_R)}$,

$$\Gamma \sim b^{-\frac{1}{\alpha_R(3\alpha_R-1)}} m_0^{\frac{2\alpha_R-1}{3\alpha_R-1}} C^{\frac{2\alpha_R-1}{3\alpha_R-1}} \tag{9.21}$$

As the concentration increases, ξ becomes shorter, Γ increases, and the characteristic relaxation time (τ) shortens, and the fast relaxation peak in the characteristic relaxation time distribution measured by dynamic LLS moves towards the faster direction (the left).

However, after translating the same series of experiments from a very good solvent benzene to a good solvent cyclohexane (temperatures much higher than the Θ temperature, ~ 35 °C), the results from dynamic LLS measurements reveal that in addition to the expected fast relaxation, an additional slow relaxation about 10^2 times slower appears immediately as soon as the solution enters the semidilute region.

As shown in the scattered light intensity weighted distribution of the characteristic relaxation time on the right, the abscissa relaxation time (t) has been corrected by temperature and viscosity. This series of decisive parallel experiments has made the existence of the slow relaxation beyond doubt and undeniable. The question is why this slow relaxation is not predicted in the theory. First, let us look at the temperature dependence of the scattered light intensity weighted distribution of the

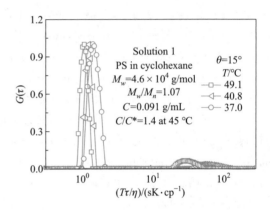

Li J F, Li W, Huo H, et al. Macromolecules, 2008, 41:901.

(Figure 12, permission was granted by publisher)

characteristic relaxation time. As the temperature decreases, the solvent gradually changes from a good solvent to a Θ solvent, the chain gradually shrinks, and C^* becomes higher. Since C remains unchanged, C/C^* becomes smaller. At 37 °C, $C/C^* < 1$, the solution enters the dilute regime, so that there is only one fast relaxation peak. In a dilute solution, the fast relaxation corresponds to the time required for the centroid of the chain to diffuse a distance of ~ R.

In the semidilute solution, it corresponds to the time required for the centroid of the chain link inside each "blob" to diffuse a distance of ξ. While at the other two temperatures, the solution just enters the semidilute regime, $R \sim \xi$, so that the three fast relaxation peaks are similar. As the temperature increases, the relaxation is slightly faster. According to the above equation, when the temperature rises, α_R increases, and the chain size is larger. For a given concentration, ξ is shorter, the characteristic relaxation time is shorter, the relaxation is faster, so that the corresponding relaxation peak shifts to the left. In contrast, the left shift of the slow relaxation peak is more obvious, implying that the local viscosity is higher. However, the local micro-viscosity cannot be directly measured experimentally. If assuming that the fast relaxation peak remains unchanged, the local micro-viscosity can be derived indirectly. For details, readers can refer the literature cited in the figure above. The experimental results also show that the linewidth Γ_{slow} corresponding to the slow relaxation is also proportional to q^2. Note: the solution concentration is slightly higher than the overlap concentration. It is not difficult to imagine that some of the entangled chains can still diffuse as a whole in the solution, but much slower. Therefore, the slow relaxation corresponds to the translational diffusion of a group of chains.

In order to better explore the nature of the slow relaxation process, long-chain polystyrene was used to increase the concentration ratio up to $C/C^* \sim 51$, which was not possibly achieved by the traditional solvent evaporation. The solution is not flowing in a short time, looking like a gel. The figure(a) below on the left shows the observation length(i.e., the scattering vector) dependence of $G^{(2)}(t)$, where in order to compare the effect of angle, the delay time(t) on the abscissa has been corrected by q^2. The data from five different scattering angles overlap with each other in the shorter delay time range, which corresponds to the translational diffusion induced relaxation, whose dynamic correlation length is ξ. It reflects that the short chain link inside each "blob" randomly walk inside a volume of ξ^3. In the figure(a) below, the slow relaxation looks like slowing down as the scattering angle increases. This is due to a illusion from the correction of q^2. In fact, if t instead of tq^2 is plotted, the five slow relaxation curves in the longer delay time range stuck together, which is independent of the scattering vector. The slow relaxation is no longer related to the translational diffusion.

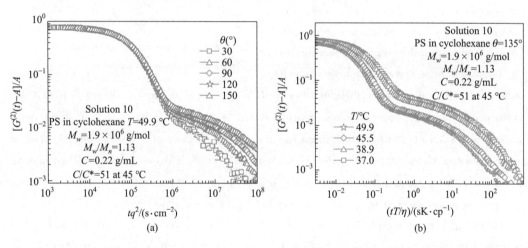

Li J F, Li W, Huo H, et al. Macromolecules, 2008, 41:901.
(Figure 19, permission was granted by publisher)

Li J F, Li W, Huo H, et al. Macromolecules, 2008, 41:901.
(Figure 21, permission was granted by publisher)

For the semidilute solutions of the same polystyrene in cyclohexane, in the scattered light intensity weighted time correlation functions measured at different temperatures, two different relaxation(fast and slow) are clearly visible, as shown in the figure(b) above. According to the above equation, for a given concentration, as the temperature increases, α_R becomes larger, ξ becomes short, the characteristic relaxation time also becomes shorter, and the relaxation is faster. The corresponding part of the relaxation in the figure shifts to the left. In addition, not only the corresponding part of the slow relaxation also shifts to the left and becomes faster as the temperature rises, and its relative height is also decreasing, which means that its contribution to the intensity of the scattered light becomes relatively small. In other words, its corresponding fluctuation amplitude becomes smaller. It implies that as the excluded volume effect enhances, the slow relaxation gradually disappears. In the semidilute solution, the chain itself is in an undisturbed state, and its mean square end-to-end distance is independent of temperature. According to formula (9.19), the tube diameter (d) changes with the temperature (α_R). The scattering angle and the solution temperature dependent results reveal that the slow relaxation in the semidilute solution is not only independent of the observation length, but also the "tube" size, so that it is not a local property.

The scattering vector (q) dependence of the linewidth (Γ) of the fast and slow relaxation can be generally written as $\Gamma_{\text{fast}} \sim q^{\alpha_f}$ and $\Gamma_{\text{slow}} \sim q^{\alpha_s}$. The figure on the right shows the concentration dependence of the two exponents, where C has been reduced by C^* in order to compare different chain lengths. Obviously, the fast relaxation always corresponds to $\Gamma_{\text{fast}} \sim q^2$, which reflects that the random walk of the segments inside each "blob", leading to a dynamic correlation length (ξ) originated from density fluctuation, which

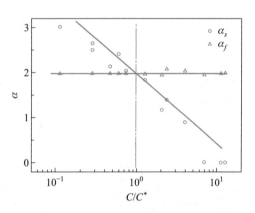

Li J F, Li W, Huo H, et al. Macromolecules, 2018, 41 : 901.

(Figure 22, permission was granted by publisher)

becomes shorter as the concentration increases, $\xi \sim C^{-3/4}$. On the other hand, α_s decreases from 3 to 0 as the concentration increases. It can be roughly written as $\alpha_s = 3/2(1 - \log(C/C^*))$. In the range of dilute solution, $C/C^* < 10^{-1}$, $\alpha_s = 3$, measuring the internal motions of a macromolecular chain at larger scattering angles. α_s gradually decreases from 3 to 2, which is due to a mixture of the internal motions and the translational diffusion at the non-zero angle. When $C/C^* \geqslant \sim 10$, $\alpha_s = 0$, the relaxation is independent of the observation length ($1/q$), The reason is unknown.

With the help of the scaling theory, when $C \to 0$, the relaxation observed in small-angle dynamic LLS corresponds to the translational diffusion of the centroid of macromolecules, and the line width is $\Gamma_0 = Dq^2$; and when $C \gg 1$, the linewidth (Γ_{slow}) corresponding to the slow relaxation is independent of the scattering vector, $\Gamma_{\text{slow}} \sim q^0$. If the correlation length of the density fluctuation corresponding to the slow relaxation is ξ_{slow}. Therefore,

$$\Gamma_{\text{slow}} = \Gamma_0 (q\xi_{\text{slow}})^z$$

Obviously, $z = -2$; substituting it in the above equation leads to $\Gamma_{\text{slow}} = 1/\tau_{\text{slow}} = D/\xi_{\text{slow}}^2 = 1/(\xi_{\text{slow}}^2/D)$, so that $\tau_{\text{slow}} = \xi_{\text{slow}}^2/D$, equivalent to the time required for the chain to diffuse a distance of ξ_{slow} in the dilute solution. Note that the slow relaxation has nothing to do with the translational diffusion of the chain. ξ_{slow} can be estimated from $\Gamma_{\text{slow}} = 1/\tau_{\text{slow}}$ and D measured by dynamic LLS. The result shows that when $C/C^* = 51$, $\tau_{\text{slow}} \sim 10^{-2}$ s and $D \sim$

10^{-7} cm^2/s. Therefore, $\xi_{slow} \sim 10^{-4} - 10^{-5}$ cm, slightly larger than the reciprocal of the scattering vector ($1/q$) and the average end-to-end distance of the chain (R). The relaxation is very slow, but the correlation length is short. Note that **people often mistakenly relates a "slow motion" to a "large size". When the resistance is larger, small particles also move slowly.**

After the existence of the slow relaxation was experimentally confirmed, the question is whether it is attributed to the density fluctuation caused by the "reptation" of the chain inside the "tube". If yes, a slight modification of de Gennes's "tube" model and "reptation" theory would be sufficient. Theoretically, the characteristic relaxation time of the "reptation" is proportional to the third power of the chain length (the experimental result is a power of 3.4). Therefore, for a given mass concentration, a change of the chain length can answer the above question. In the figure below, C/C^* are 10 and 35, respectively, for the short and long chains. ξ only depends on the concentration and not on the chain length, so that their fast relaxations are similar. On the other hand, the chain length increases ~8 times, but τ_{slow} only increases less than 5 times. If the slow relaxation was related to the chain "reptation" in the "tube", τ_{slow} would increase at least 500 times. The experimental results deny the attribution of the slow relaxation to the chain "reptation" in the "tube".

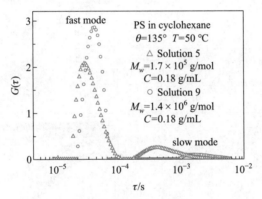

Li J F, Li W, Huo H, et al. Macromolecules, 2008, 41:901.

(Figure 7, permission was granted by publisher)

There is still no quantitative theory to explain why the slow relaxation occurs in the semidilute solutions when the solvent quality is not extremely good. The qualitative explanation is that each "blob" is not completely independent as assumed theoretically when the excluded volume effect has not reached its maximum and the chains are not fully expanded. In other words, the "blobs" are correlated and its characteristic correlation length is ξ_{slow}. The appearance of such a slow relaxation is not limited to in the semidilute solutions of neutral macromolecular chains. In the process of gelation and vitrification of small molecules and macromolecules, in concentrated colloids, and in the dilute solutions of salt-free polyelectrolytes, a similar slow relaxation appears. In order to confirm that the slow relaxation observed in dynamic LLS is indeed from the correlation between different "blobs", the influence of the interaction among the chains on the slow relaxation was further investigated. Obviously, gradually cross-linking macromolecular chains is the most extreme way to enhance the interaction among the chains.

The figure below shows the concentration dependence of the time correlation function of the scattered light intensity. A small amount of coumarin was attached to the chains so that they can be chemically crosslinked together under a ultraviolet irradiation. Before the irradiation, the time correlation function of the solution with a lower concentration has only one fast relaxation. As the concentration increases, another slow relaxation appears. As the photo-crosslinking proceeds, it gradually slows down(shifts to the right), and its corresponding intensity of the scattered light increases. At the same time, the relative intensity contribution of the fast relaxation gradually decreases and seemly disappears at the concentration and after a long time irradiation. The illustration shows that the solution is no longer flowing. Dynamic LLS still observe such a fast relaxation and its characteristic relaxation time remains $\sim 10^{-4}$. This series of results confirm that the slow relaxation is related to the correlation between the chains.

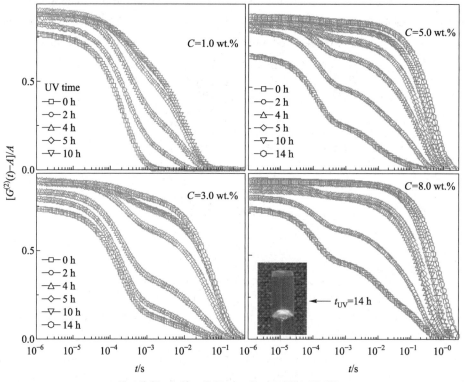

Ngai T, Wu C, Chen Y. Macromolecules, 2004, 37:987.

(Figure 8, permission was granted by publisher)

When using dynamic light scattering to measure the salt-free dilute solutions of polyelectrolytes, two relaxations, fast and slow, can also be observed, and the fast relaxation is also independent of the chain length. As for the origin of the slow relaxation, theorists and experimentalists have been arguing with each other since the 1980s. There are also two groups of people, accepting and denying the existence of the "slow relaxation", and in the end, there is no conclusion. The reason for the controversy also includes sample preparation, for example, the definition of "salt-free" in aqueous solution; the charge distribution on the same and different chains. There are possibly insoluble "clusters" in the sample.

Until 2009, after the problem of sample preparation was solved, the problems related to solution preparation were dispelled. A decisive experiment is the introduction of p-azidomethyl in a neutral macromolecular chain, which can react with carbon dioxide to form charged groups. After carbon dioxide gas or nitrogen gas is passed through its solution in a LLS sample cell, the macromolecular chains can be in situ switched between a neutral state and a charged state. As shown in the figure(b) below, after passing carbon dioxide gas through, the neutral macromolecular chains become polyelectrolytes, so that the solution conductivity significantly increases. After passing nitrogen gas through, the neutral state was restored.

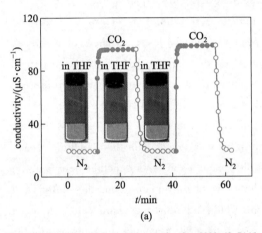 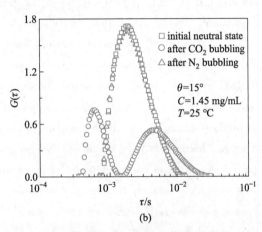

(a)   (b)

Zhou K J,Li J F,Lu Y J,et al.Macromolecules,2009,42:7146.   Zhou K J,Li J F,Lu Y J,et al.Macromolecules,2009,42:7146.

(Figure 3,permission was granted by publisher)   (Figure 6,permission was granted by publisher)

Corresponding to the changes in electrical conductivity, when the chain is neutral, the relaxation time distribution measured by dynamic LLS contains only one peak related to the translational diffusion of the centroid of the chain; after passing carbon dioxide gas through, two relaxation peaks appear; after passing nitrogen gas through again, the two relaxation peaks disappear, completely returning back to the relaxation time distribution when the chains are neutral, as shown in the figure(b) above. Further experiments and analysis show that both the relaxations are caused by translational diffusion. The fast relaxation is attributed to the coupled diffusion of the chain links and their counter ions in individual polyelectrolytes chains, independent of the chain length and concentration; while the slow relaxation is related to the interchain interaction, which is equivalent to the synergistic translational diffusion of a group of chains. The interchain interaction in the dilute solution stems from the long-range electrostatic interaction of the charges on the chain, or the strong excluded volume effect. The addition of salt can shield the long-range electrostatic interaction. After repeating the above experiments in the presence of salt, the relaxation time distribution always has only one peak and its position remains unchanged regardless of whether the chain is charged or not. These experimental results have convincingly demonstrated that the slow relaxation is a real existence, and it is directly related to the interchain interaction. In the absence of salt, the long-range electrostatic interaction makes the equivalent excluded volume of each polyelectrolytes chain much larger. Even in a dilute solution, the system becomes "crowded" and chain movement is affected.

In the literature, the occurrence of the slow relaxation was often attributed to possible defects of dynamic LLS experiment and static inhomogeneity of the system. Faced with series of results of the above two decisive experiments, no one can deny the true existence of the slow relaxation anymore. It can be qualitatively asserted that although a "crowded" system is statically uniform, not dynamically uniform, there exist different relaxations with different time scales, or say, different correlation lengths. This dynamic inhomogeneity can be attributed to the unshielded inter-"particle" interaction, resulting in a "crowding" effect.

In a "crowded" system or "party", each "particle" (molecules, macromolecular chains, colloidal particles, and people) is in a "cage" composed of other particles, and also a part of "cage" for its surrounding particles. No particle is special. In a short time, each particle can only move in its own "cage", causing a density fluctuation within a small volume, corresponding to the fast relaxation. Due to the random movement of each particle, the "cages" should have an average "lifetime". When the measurement or waiting time is longer than this "lifetime", each particle has a chance to migrate out of its original old "cage" and enters another new "cage". In a semidilute solution, this means that although the entire chain reptates along the "tube" and cannot move laterally. However, each "blob" (short chain link) on the chain can generate a coordinated motion with its surrounding "blobs", and there is a chance to jump out of its original "tube" laterally and enter an adjacent "tube". The average characteristic lifetime of the "cage" can be very long, depending on the interaction.

The slow relaxation reflects the migration of particles between different "cages", causing a density fluctuation. Its characteristic correlation length is the reachable range of the inter-cage interaction. longer than the linear dimension of a "cage". Its corresponding characteristic relaxation time is the average "lifetime" of a "cage". For a given volume, the "crowding" can be due to an increase in the particle number, such as in the semidilute solutions and concentrated colloids; it can also come from strong repulsion between particles, resulting in a strong excluded volume effect, which is equivalent to an increase of the excluded volume of each particle, such as charged particles in salt-free dilute solutions and small molecules in the melts near the glass transition temperature. The above is only a qualitative explanation. A complete theoretical treatment will inevitably involve the multi-body problem in physics. It is very complicated, and there is no quantitative explanation. It is still an unresolved problem.

Summary

In a dilute solution, as the concentration increases, the interchain distance gradually decreases, reaching an overlap concentration (C^*), which decreases as the chain size (length) increases. If the chain is longer than the entanglement length, a further increase of the concentration leads to the entanglement between the chains, forming a semidilute solution. For a given concentration, the space occupied by each macromolecular chain also contains many other interpenetrated chains, its number is proportional to the square root of the chain length, that is, the square root of the segment number of each chain ($N^{1/2}$). As the concentration increases, each chain is gradually separated into many short chain links by other chains; each short chain link can be viewed as a "blob"; the chain links are independent from each other. Therefore, the conformation of the entire chain gradually tends to a random coil in an undisturbed state. On the other hand, for a given chain length, the characteristic correlation length of density fluctuation is shortened as the concentration increases. When it is close to the segment length (b), the solution enters the range of concentrated solution, denoted as C^{**}. The concentration range of the semidilute solution is $C^* \leqslant C \leqslant C^{**}$.

In a semidilute solution, each chain is in a "tube" with a diameter of "d" formed by its surrounding entangled chains. The conformation of the chain and the "tube" surrounding it is a random coil. The chain can only move forward or backward randomly along the "tube", slowly "reptates" in the random coil-like "tube", and the maximum step length is the "tube" diameter. At the same time, each chain is also part of the "tubes" surrounding the nearby chains, and no chain is special. In a short period of time, the entire macromolecular chain is unable to migrate a long distance, and only the segments walk randomly within the "blob" to which they belong, causing a density fluctuation. Its characteristic correlation length is recorded as ξ. Note: $\xi \approx d/4$. Therefore, a macromolecular chain is composed of a series of small short chain links ("blobs"); each short chain link contains N_d segments, also has a random coil conformation, and its average end-to-end distance is the tube diameter d (the "blob" diameter). The macromolecular chain in the "tube" in the semidilute solution has the following static properties.

The conformation of the macromolecular chain or "tube" in a semidilute solution can be regarded as the trajectory after a random walk of N/N_d steps with a step length of d: a random coil. As the concentration gradually increases, the excluded volume effect among the "blobs" gradually approaches zero, and the chains tend to a undisturbed state in the semidilute solution, $\alpha_R = 1/2$. Its average end-to-end distance is $d (N/N_d)^{1/2}$. Since the short chain link in each "blob" is also a random coil, its straightened length bN_d must be longer than $d = bN_d^{\alpha_R}$. Therefore, the straightened length of the "tube" $L_{c,d} = d (N/N_d)$ is much shorter than the straight length of the chain itself $L_c = bN$. The conformation of a macromolecular chain can also be regarded as the trajectory after N steps of random walk with a step length of b. The average end–to–end distance is $R = bN^{\alpha_R}$. The average end–to–end distance of a macromolecular chain is a physical existence, independent of its depiction, so that $bN^{\alpha_R} = d\ (N/N_d)^{\alpha_R}$. **Note that the scaling exponent in every "blob" still varies with the solvent quality**, $1/2 \leqslant \alpha_R \leqslant 3/5$.

Dynamically, the current theories predict **three kinds of motions related to macromolecular chains and chain segments in a semidilute solution**: **mutual diffusion, curvature diffusion, and centroid diffusion**. Firstly, when the observation time is much shorter than the time required for the centroid of the "blob" to move a length of d, the position of the "blob" basically remains and so does the entire chain. Only the segments in the "blob" walk randomly in a small volume (d^3). It can also be regarded as the center of mass of the short chain links in the "blob" undergoes a **mutual translational diffusion** around its equilibrium position, its corresponding diffusion coefficient and characteristic relaxation time are denoted as D_ξ and τ_ξ, respectively. The random walk of the segments in the short chain links causes the density fluctuations within a short range, corresponding to a characteristic correlation length ξ, which is about a quarter of the tube diameter d. Obviously, the relaxation is independent of the chain length, but ξ gradually becomes shorter and the relaxation speeds up as the concentration increases. Since the movements of the centroids of each "blob" and its corresponding "blob" with a phase difference of π in the system are not synchronized, dynamic LLS can measure this fast density fluctuation. For typical polymers, in the time autocorrelation function of scattered light intensity, the fast density fluctuation corresponds to a faster relaxation, and its characteristic relaxation time is $\sim 10^{-6} - 10^{-5}$ s, depending on the concentration.

Secondly, when the observation time is close to the time required for a chain to diffuse a straighten length ($L_{c,\xi}$) of a "tube" (for a typical polymer, $\sim 10^{-2} - 10^{-1}$ s) along the curved "tube", the density fluctuations in the "blob" have already tended to the average density, and the solution is equivalent to a system composed of uniformly distributed identical "particles" with no density fluctuation. The electric field of the scattered light from each "particle" does not contribute to the measured scattered light intensity due to destructive interference. Therefore, dynamic LLS is not able to measure such a movement induced relaxation. Such a back–and–forth random walk in the curved "tube" is called the "**curvature diffusion**", and its corresponding diffusion coefficient and characteristic relaxation time are denoted as D_c and τ_c, respectively. If the chain walked in a straight tube, its centroid will move along with the chain. However, the shape of the "tube" is a curved random coil. Therefore, after the chain walks along the tube for a distance of $L_{c,d}$, its centroid may move a little or do not move. For example, if the chain moves a tube length in the tube around its centroid, its centroid remains.

Finally, when the observation time is close to the time required for the centroid of the chain to move the average end–to–end distance ($\sim 10^{-5}$ cm) of a chain (1 – 10 s for a typical polymer), the macromolecular chain has slowly and randomly "reptated" back–and–forth along the "tube" for a long time under the agitation of thermal energy. Its centroid has also slowly moved a short distance, which is called the " **centroid diffusion** ", and its corresponding diffusion coefficient and characteristic relaxation time is denoted as D_{GC} and τ_{GC}, respectively. Similarly, due to the destructive interference, the "blobs" contribute no scattered light intensity. Therefore, it is impossible to use dynamic LLS to measure the density fluctuation caused by this movement.

Only by labelling a small number of chains, can the latter two relaxation processes be measured. In dynamic LLS, first select a solvent with a refractive index that is as close as possible to the macromolecule to reduce or eliminate their scattered light intensity, and then select another macromolecule with a similar length, structure and properties, but a different refractive index as the probe chain; in neutron scattering and nuclear magnetic resonance, a small amount of deuterated macromolecular chains with a similar length are added; in fluorescence photo-bleaching recovery and forced Rayleigh scattering, a few appropriate fluorescent groups are attached to the macromolecular chain. Readers can refer to related literature.

Experimentally, in addition to the above three theoretically predicted relaxations, dynamic LLS also observed an additional slow relaxation due to the inter-"blob" or interchain interaction, and its characteristic relaxation time varies with the system. Except in an extremely good solvent in which the chain is fully expanded, the "blobs" in other solvents in the semidilute solution are not completely isolated from each other as theoretically assumed, but there exists a certain degree of mobility. Each "blob" is in a "cage" formed by its neighboring "blobs". In the extremely good solvent(the maximum excluded volume effect), each "blob" is fully expanded so that every "blob" is jammed by other adjacent "blobs" and equivalently confined inside a fixed "cage" and not able to move even for a fairly long time. In other solvents, each "blob" can randomly move a little. The "cage" is not a fixed structure. Its density fluctuation corresponds to a characteristic correlation length and characteristic relaxation time (the average lifetime of the "cage"), reflecting the "jumping" of "blobs" between adjacent "cages". Each "blob" itself is also a "cage" to its adjacent "blob". No "blob" is special, reflecting the static homogeneous but the dynamic inhomogeneous.

When each "blob" is not be able to move, equivalently say, the characteristic relaxation time of the "cage" approaches infinite, so that there is no "cage" effect. That is why for polystyrene in a very good solvent benzene, even when the C/C^* is as high as 51, dynamic LLS still observe only one fast relaxation related to the random movement of the segment inside the "blob". When the "cage" has a characteristic correlation length and a corresponding characteristic relaxation time due to the interaction, each pair of two "blobs" in the solution with a phase difference of π(destructive interference) do not "jump" out of their respective "cages" at the same time, which leads to a net scattered light intensity, so that there appears an additional slow relaxation in the characteristic relaxation time distribution measured by dynamic LLS, not predicted in theory, which slows down as the excluded volume effect weakens, but its contribution to the scattered light intensity increases, reflecting that each "blob" can jump a longer distance. The amplitude of fluctuation is larger.

This slow relaxation process is not unique to the semidilute solutions of macromolecules, but commonly occurs in various "crowded" systems, including during the sol-to-gel and glass transitions, in salt-free polyelectrolytes dilute solutions and concentrated colloids. If each "particle" (a short chain link, a polyelectrolyte chain, a colloidal particle, a small molecule, etc.) can move inside a crowded system, not completely independent from each other, each "particle" is in a "cage" composed of other adjacent particles, itself is also a "cage" for other adjacent particles. No "particle" is special, so that statically, the system is homogeneous. Note: there are two cases for the "complete independence", no inter-particle interaction or no movement at all.

Due to the random movement of each particle, each "cage" randomly fluctuates with a certain average life and size. Each particle can move randomly in its "cage", corresponding to a fast relaxation related to the translational diffusion; particles can also jump randomly between different "cages", corresponding to a slow relaxation related to the average size and lifetime of the "cage". This can be equivalent to a system containing "particles". The particles themselves move very slowly (not necessarily because the particles are very large), and each particle contains many small fast and randomly moving units. Therefore, the system is dynamically inhomogeneous. This qualitatively explains why dynamic LLS can observe two relaxations, one fast and one slow.

When the observation time is close to the characteristic relaxation time of the "cage", the "particle" can be observed to slowly jump between different "cages". Therefore, the slow relaxation measured by dynamic LLS should be the relaxation of the "cage", and the corresponding characteristic relaxation time is the average lifetime of the "cage". When the excluded volume effect weakens, each "particle" can jump a longer distance, equivalent to a longer characteristic correlation length or a larger "cage", resulting in an increase in the intensity of scattered light and a longer characteristic relaxation time. Since de Gennes's "tube" model and "reptation" theory assumed that "blobs" are laterally stationary besides its slow reptation along the "tube", there is no "cage" effect. In most real "crowded" systems, the "particles" can still slowly moving, not stationary, and there always exists some interaction. However, there still lacks a quantitative theory to describe the "cage" effect. This should still be a problem worthy of theoretical study.

The figure below summarizes some static properties of macromolecular chains in the semidilute solutions and three dynamic diffusion properties under non-drained conditions.

由于链的缠结，每条链均处于一个管径为d的"管子"之中
due to chain entanglements, each chain is inside a "tube" with a diameter of d

接触浓度
overlap concentration
$$C^* = \frac{M}{R^3 N_{AV}}$$

每个链珠里有N_d个链段
Each blob has N_d chain segments

每条链有N/N_d个链珠
Each chain has N/N_d blobs

每根管子的伸直长度
Contour length of a tube
$$L_{c,d} = d(N/N_d) < bN$$

链珠和链的平均末端距
Average end-to-end distance of a
blob and a macromolecular chain
$$d = bN_d^{\alpha_R}$$
$$R = bN^{\alpha_R} = d(N/N_d)^{\alpha_R}$$

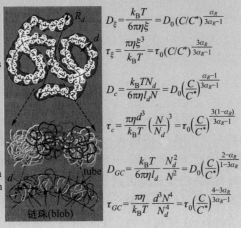

$$D_\xi = \frac{k_B T}{6\pi\eta\xi} = D_0 (C/C^*)^{\frac{\alpha_R}{3\alpha_R - 1}}$$

$$\tau_\xi = \frac{\pi\eta\xi^3}{k_B T} = \tau_0 (C/C^*)^{\frac{3\alpha_R}{3\alpha_R - 1}}$$

$$D_c = \frac{k_B T N_d}{6\pi\eta l_d N} = D_0 \left(\frac{C}{C^*}\right)^{\frac{\alpha_R - 1}{3\alpha_R - 1}}$$

$$\tau_c = \frac{\pi\eta d^3}{k_B T} \left(\frac{N}{N_d}\right)^3 = \tau_0 \left(\frac{C}{C^*}\right)^{\frac{3(1-\alpha_R)}{3\alpha_R - 1}}$$

$$D_{GC} = \frac{k_B T}{6\pi\eta l_d} \frac{N_d^2}{N^2} = D_0 \left(\frac{C}{C^*}\right)^{\frac{2-\alpha_R}{1-3\alpha_R}}$$

$$\tau_{GC} = \frac{\pi\eta}{k_B T} \frac{d^3 N^4}{N_d^4} = \tau_0 \left(\frac{C}{C^*}\right)^{\frac{4-3\alpha_R}{3\alpha_R - 1}}$$

第十章　大分子溶液的相变

　　大分子溶液的相变和日常生活和工业生产息息相关。生活中常常不自觉地利用相变（包含沉淀）将可溶和不可溶的大分子分开；在合成纤维的溶液纺丝中，利用调节溶剂挥发、相变和拉伸速度来控制大分子取向，以便获得所需的力学性质；在形成多孔中空纤维和多孔滤膜时，控制相变速度可调节孔径分布；在沉淀聚合反应中，利用相变控制合成高聚物的摩尔质量；用可结晶的聚合物制造材料时，利用相变速度调节结晶度，获得具有不同力学性能的材料；在结构生物学中利用相变促进蛋白质分子结晶等，不胜枚举。

　　如第三章所述，一个大分子可否溶解在一个溶剂中，取决于混合过程中的焓变和熵变。混合焓变取决于大分子和溶剂分子之间的相互作用，而混合熵变则源于混合中大分子和溶剂分子微观排列方式的增加，即体系总自由度的增加。由热力学已知，对给定的温度和溶液组成，混合自由能与混合焓和混合熵的依赖关系如下，

$$\Delta G_{mix} = \Delta H_{mix} - T\Delta S_{mix}$$

　　只有当 $\Delta G_{mix} < 0$ 时，大分子方可溶于溶剂中形成一个分子水平上的、均匀的、一相二元混合物（溶液）；反之，则形成一个两相二元混合物，具有一个宏观界面的两个共存的饱和溶液（两相）。一个是大分子在溶剂中的饱和溶液（稀相）；另一个是溶剂在大分子中的饱和溶液（浓相）。微观上，大分子和溶剂分子不断地、来回地穿越界面，各自的速率相等，形成一个宏观的动态平衡。

　　由于链的缠结，溶剂分子渗入大分子的缠结网络之中，浓相中溶质的体积分数能够超过 50%。而小分子溶液的浓相中则只有较少的溶质分子。一般的大分子教科书并没有强调这一显著的差别。在一相和两相之间的变化称为相变。依据一些热力学状态变量（如体积和密度）随着温度或压强的变化是否连续，可进一步将该转变分为非连续的一级和连续的二级相变。详情可参阅热力学教科书。

　　严格地说，在分子水平上，两种物质总是互溶，唯一的区别仅是溶解度。当溶解度低于万分之一时，通常称为二者"不溶"。日常生活中，也常说"油水不溶"。当油水混合时，分成上下两层，上面的油层里含有饱和的水，下面的水层中存在饱和的油。在一个二元混合中，熵通常增加，熵变为正。如果焓减少，焓变为负，依据上式，混合可以自发地进行。如果焓变为正，加热溶液一般可以促进溶解。存在一个上临界溶液温度（upper critical solution temperature，UCST），使得混合自由能为负。一般而言，形成二元混合物时自由能的变化可写成

$$dG = \left(\frac{\partial G}{\partial p}\right)_{T,n} dp + \left(\frac{\partial G}{\partial T}\right)_{p,n} dT + \left(\frac{\partial G}{\partial n_A}\right)_{p,T,n_B} dn_A + \left(\frac{\partial G}{\partial n_B}\right)_{p,T,n_A} dn_B$$

其中,$(\partial G/\partial p)_{T,n}=V$ 和 $(\partial G/\partial T)_{p,n}=-S$;$(\partial G/\partial n_A)_{p,T,n_B}$ 和 $(\partial G/\partial n_B)_{p,T,n_A}$ 分别被称为 A 和 B 的化学势,记作 μ_A 和 μ_B。物理本质是 **Gibbs** 自由能随组分 A 或 B 的变化率。对于大分子溶液,A 和 B 可分别代表溶剂小分子和大分子。注意,通常以及下面所说的"混合自由能"指的是"在混合中自由能的变化",切记!

第三章中,已讨论了混合自由能对组成的依赖性。为方便讨论,将式(3.14)重写如下:

$$\frac{\Delta G_{\mathrm{mix}}}{nRT}=X_p(1-X_p)\chi_{sp}+\frac{X_p}{n_v}\ln X_p+(1-X_p)\ln(1-X_p)$$

其中,$X_p(1-X_p)\chi_{sp}$ 和 $(X_p/n_v)\ln X_p+(1-X_p)\ln(1-X_p)$ 分别对应着混合焓变和混合熵变。除以摩尔热能 RT 后,ΔG_{mix} 已无量纲。在第三章中,已将 $\Delta G_{\mathrm{mix}}/(RT)$ 对 X_p 作图,并讨论过其不对称性源于大分子和溶剂分子的体积之差。读者可重新阅读第三章中有关内容。

将上式分别对大分子和溶剂的摩尔数(n_p 和 n_s)求偏微分,可得溶液中大分子和溶剂的化学势(μ_p 和 μ_s)。显然,在计算中,涉及 X_p 分别对 n_p 和 n_s 的偏微分。在第五章讨论渗透压差和超速离心时,已详细地推导过这些微分计算,即式(5.10)和式(5.20),故略去。分别利用纯大分子和纯溶剂的化学势(μ_p^* 和 μ_s^*,实际上就是各自的摩尔自由能)作为参照点,可分别得到溶液中大分子和溶剂化学势对大分子组成(X_p)的依赖性。计算包括两步,先求($\Delta G_{\mathrm{mix}}/nRT$)对 X_p 的偏微分、再乘以 X_p 对 n_p 或 n_s 的偏微分。如果不愿涉及详细的数学计算,读者可以跳过下面两条虚线隔开的部分,仅使用结果:式(10.1)。

$$\left(\frac{\partial \Delta G_{\mathrm{mix}}}{\partial n_s}\right)_{T,p,n_p}=\left(\frac{\partial \Delta G_{\mathrm{mix}}}{\partial X_p}\right)_{T,p,n_p}\left(\frac{\partial X_p}{\partial n_s}\right)_{T,p,n_p}$$

$$\left(\frac{\partial \Delta G_{\mathrm{mix}}}{\partial n_p}\right)_{T,p,n_s}=\left(\frac{\partial \Delta G_{\mathrm{mix}}}{\partial X_p}\right)_{T,p,n_s}\left(\frac{\partial X_p}{\partial n_p}\right)_{T,p,n_s}$$

其中,

$$\left(\frac{\partial X_p}{n_s}\right)_{T,p,n_p}=\frac{\partial\left(\dfrac{n_v n_p}{n_s+n_v n_p}\right)}{\partial n_s}=-\frac{n_v n_p}{(n_s+n_v n_p)^2}=-\frac{X_p(1-X_p)}{n_s}=-\frac{X_p^{\ 2}}{n_v n_p}$$

和

$$\left(\frac{\partial X_p}{n_p}\right)_{T,p,n_s}=\frac{\partial\left(1-\dfrac{n_s}{n_s+n_v n_p}\right)}{\partial n_p}=\frac{n_v n_s}{(n_s+n_v n_p)^2}=\frac{X_p(1-X_p)}{n_p}=\frac{n_v(1-X_p)^2}{n_s}$$

$$\Delta\mu_p=\mu_p-\mu_p^*=\left(\frac{\partial \Delta G_{\mathrm{mix}}}{\partial n_p}\right)_{T,p,n_s}=RT\left[\ln X_p-(N-1)(1-X_p)+\chi n_v(1-X_p)^2\right]$$

和 (10.1)

$$\Delta\mu_s=\mu_s-\mu_s^*=\left(\frac{\partial \Delta G_{\mathrm{mix}}}{\partial n_s}\right)_{T,p,n_p}=RT\left[\ln(1-X_p)+\left(1-\frac{1}{n_v}\right)X_p+\chi X_p^2\right]$$

上式中,变量为大分子的体积分数。可转换成可测的大分子摩尔分数($x_p=n_p/n$)或者更直接的质量浓度(C):$X_p = n_v x_p = C v_p$,其中,v_p是大分子的比容。注意:它不是体积,而是体积与质量之比。对一个溶液而言,它是溶液体积随着组分质量的变化率。为了方便讨论,将约化后的摩尔混合自由能$\Delta G_{mix}/(nRT)$记为f,其一阶偏微分为$f' = \Delta \mu_p/(nRT)$;其二阶和三阶偏微分也可求,分别记为f''和f'''。计算颇为烦琐,故略去。

右图显示了f,f',f''和f'''随大分子体积分数的变化,其中,为了清楚地显示f在低浓度(靠近左边纵轴)时的变化,特意选择了$n_v = 2$和$\chi_{sp} = 2.1$。由图可见即使在$n_v = 2$时,$\Delta G_{mix}/(nRT)$已不对称。从左边纵轴开始,随着浓度的逐渐增加,ΔG_{mix}始于零,先降低,再升高,由负变为正,达至一个峰值。注意,高聚物教科书中一般将N选为100或更大,使得f看上去好像从零点开始就向上升高,完全没有显示,f

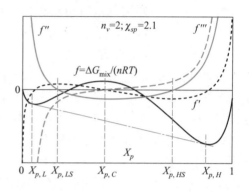

从零开始先逐渐地降低,然后再升高,由负变正这一变化,实属误导。相反,如果从右边的纵轴出发,向左移动,即逐渐地增加溶液中溶剂的含量,可见$\Delta G_{mix}/(nRT)$也是始于零,先降低,再升高,由负变为正,在同一个峰值汇合,变化趋势更加明显。该峰顶位置对应着一个上临界溶液温度(UCST)和一个临界组成($X_{p,C}$)以及$f' = f''' = 0$和f''的最低值。

随着大分子体积分数的逐渐增加,$\Delta G_{mix}/(nRT)$的一阶导数(f',黑虚线)逐渐增加,由负变正,达至一高点,再逐渐地降至零。如果从右边纵轴出发,向左移动,随着溶剂体积分数的增加,f'先由正变负,抵达一低点,再逐渐地升至零。这两个高、低点称为"旋节"点,分别对应着$f'' = 0$的两个点(实蓝线):$X_{p,LS}$和$X_{p,HS}$。从两边出发的汇合点($f' = 0$)对应着前述的大分子的临界组成($X_{p,C}$),其不在$X_p = 0.5$的对称处,归于大分子和溶剂分子的体积不对称。从高点到低点,f'随着X_p的增加而下降。

依据热力学,混合物中任何一种物质的化学势都随着该物质组成的增加而升高,最大值为纯的该物质的化学势。因此,随着添加更多的大分子,其化学势不增反减,这违背了物理规律。所以,体系不可能沿着这一下降路径形成一个均匀混合物。相反,体系在该浓度区间必定会分离成一个大分子在溶剂中的饱和溶液(稀相,体积分数和体积分别为$X_{p,L}$和V_L)和另一个溶剂在大分子里的饱和溶液(浓相,体积分数和体积分别为$X_{p,H}$和V_H)。大分子和溶剂在每一相都有各自的化学势:$f'(X_{p,L})$和$f'(X_{p,H})$;以及$f'(X_{s,L})$和$f'(X_{s,H})$。

两相达到平衡时,不仅$f'(X_{p,L}) + f'(X_{s,L}) = f'(X_{p,H}) + f'(X_{s,H})$,而且$f'(X_{p,L}) = f'(X_{p,H})$和$f'(X_{s,L}) = f'(X_{s,H})$。否则,就在两相之间出现大分子向一相和溶剂分子向另一相的净流动。这就是为何两个"双节"点不是出现在两个$f' = 0$处,虽然在这两点处有$f'(X_{p,L}) + f'(X_{s,L}) = f'(X_{p,H}) + f'(X_{s,H})$,但是$f'(X_{p,L}) \neq f'(X_{p,H})$和$f'(X_{s,L}) \neq f'(X_{s,H})$。两个"双节"点稍稍地移向左边,是$f$上的一条切线通过的两个节点,故曰"双节"线。

自由能是一个状态函数,故其变化不依赖于路径。依据定义,此处的自由能变化可由$f'(X_p)$对X_p的积分获得,即$\Delta G = \int f'(X_p) \, dX_p$。因此,在上图中,沿着$f'$的路径和沿着分相的路径(平行于$X_p$轴)积分所得的$\Delta G$必须相等。如果沿着$f'$的路径,从左边$f' = 0$(自由能左边最低点)处积分到右边 $f' = 0$(自由能右边最低点)处,可见,以 $f = 0$ 为界,以上的积分面积小于以下的积分面积,显然不等于沿着平行路径(始终有 $f = 0$)积分所得的矩形面积。同样,如果从左边的节点处积分到右边的节点处,且用连接两个节点的水平线为界,以上的积分面积等于以下的积分面积,互相填补,正好等于沿着水平线积分的面积,不因路径不同而变,符合热力学原理。

相变后,总混合自由能是每一相的混合自由能乘上其体积分数后的加和,数学表达如下:

$$f_{\text{two phases}} = \frac{V_L}{V_L + V_H} \frac{\Delta G_{\text{mix}}(X_{p,L})}{n_{\text{box}} RT} + \frac{V_H}{V_L + V_H} \frac{\Delta G_{\text{mix}}(X_{p,H})}{n_{\text{box}} RT} \qquad (10.2)$$

它就是图中一条连接混合自由能(f)上两个双节点的直线,偏离 f 上的两个最低点。如前所述,两个节点分别为 $X_{p,L}$ 和 $X_{p,H}$,称为"双节点"。如果溶质和溶剂分子具有相同的体积,切线就正好通过两个最低点,但这仅是特例。在化学势从 $f'(X_{p,L})$ 到其旋切高点以及从 $f'(X_{p,H})$ 到其旋切低点的两个区间内,化学势随着 X_p 的增加而逐渐升高,并不违背热力学原理,且混合自由能仍为负值,应可自发地形成一个均相溶液。但是,正如上图所示,在这两个区间内,黑实线(形成一相的混合自由能)上的任何一点均高于蓝色点线上的对应一点(分成两相后总的混合自由能)。

对一个等温和等压的过程而言,自由能总是自发地降低。因此,在这两个区间内,大分子溶液总是倾向于分离成稀、浓两相以降低混合自由能。因此,一相是一个亚稳态。微观上,相分离涉及分子间的聚集以及聚集体与其介质(溶剂)之间形成界面。聚集降低自由能,正比于聚集体(这里称作引致相变的"核")的体积,即其线度(尺寸)L 的三次方,而形成界面则增加了表面能,正比于聚集体的表面积,即 L 的二次方。对一个等温和等容过程而言,体系中因聚集而导致的自由能变化为

$$\Delta A \sim -\rho_E L^3 + \sigma_E L^2 \rightarrow \left(\frac{\partial \Delta A}{\partial L}\right)_{T,V} \sim -3\rho_E L^2 + 2\sigma_E L \qquad (10.3)$$

其中,ρ_E 和 σ_E 分别是能量的体密度和面密度。与聚集体形状有关的正比常数已被略去。当 L 较小时,$(\partial \Delta A/\partial L)_{T,V} > 0$,聚集非自发,其相反的解聚自发。因此,即使有机会形成一个聚集体,其也会立即解聚,不能继续长大,尽管聚集可以降低总的自由能。另一方面,当核的尺寸 L 足够大时,$(\partial \Delta A/\partial L)_{T,V} < 0$,此时聚集会自发地进行。

$(\partial \Delta A/\partial L)_{T,V} = 0$ 对应着一个临界尺寸:$L_C \sim \sigma/\varepsilon$。聚集是一个随机涨落的过程。给体系一个足够长的诱导时间,总有机会出现一个尺寸大于 L_C 的瞬间聚集体。一旦体系里有了这样一个"核",分子就会自发地在其上继续聚集,最终实现宏观的相分离。统计物理学告诉我们,这一诱导时间 $\tau_{\text{ind}} \sim 1/[\exp(\varepsilon/RT) - 1]$。显然,若无分子间的相互作用,即无相变,即 $\varepsilon \to 0$,$\tau_{\text{ind}} \to \infty$;当 $\varepsilon \gg RT$ 时,$\tau_{\text{ind}} \to 0$。τ_{ind} 取决于与双节线的距离。

现实中,这样的"核"总是存在,其可以是一粒灰尘、一个小气泡、器壁上的一个缺陷等。一般而言,对一个真实的大分子溶液,当浓度抵达 $X_{p,L}$ 或 $X_{p,H}$ 时,就会自动地分成稀相和浓相,而不会继续从左向着 $f'(X_{p,L})$ 的高点或从右向着 $f'(X_{p,H})$ 的低点前进,保持一相,形成过饱和溶液。蒸馏一个大分子溶液时,如果温度低于 UCST,随着溶剂蒸发,浓度逐渐增加而出现相分离。如果利用同一个容器蒸馏大分子溶液,通常每次均需在溶液中加入

少量的沸石以提供小的气泡作为"核"，以避免过热，暴沸而伤及操作者。

在上图中，随着溶液温度升高，X_{sp}变小，$X_{p,LS}$和$X_{p,HS}$以及$X_{p,L}$和$X_{p,H}$相互趋近，最后，四个点汇聚在临界体积分数（$X_{p,c}$）一点上，对应着一个位于$f'=f'''=0$处的临界温度（T_c）。对于一个给定的温度，随着大分子体积分数的增加，体系先从一个溶液变成一个浓度为$X_{B,L}$的饱和溶液；再进入亚稳态。当体积分数达到和大于$X_{p,LS}$时，体系一定分成体积分数分别为$X_{p,L}$和$X_{p,H}$的两相。

同样，如果从纯大分子出发，逐渐添加溶剂，少量的溶剂总是可以溶入大分子样品，使其溶胀，形成溶剂分子在大分子中的均相溶液。当体积分数达到和小于$X_{p,H}$时，体系开始进入另一个亚稳态。继续增大溶剂的体积分数（降低X_p）至$X_{p,HS}$，体系又一定分成两相。因此，对一给定的温度，随着体积分数的变化，共可得到四个点：双节线上的两个双节点：$X_{p,L}$和$X_{p,H}$；和旋节线上的两个旋节点：$X_{p,LS}$和$X_{p,HS}$。

注意：以上的讨论基于常见的"温度对组分"的相图。实际上，也有"压强对组分"的相图，讨论和结论类似。一般而言，压强对大分子溶液的影响较小，恕不赘述。恒压下，改变温度可得到一系列这样的一套四个点。

将所有的双节点和双节点分别连接，可得两条曲线，如右图所示，被称为双节线（实线）和旋节线（虚线）。这两条曲线之间的阴影区域为上述讨论的亚稳态区，体系虽然倾向分成两相，但需要形成一个"核"。诱导时间随着离开双节线的距离变化，越近越长。曲线上最高的一点（T_c）即为上临界溶液温度（UCST）。当溶液温度高于其时，大分子可以任何比例溶解在溶剂中形成一个在分子水平上均匀的混合物。与临界温度对应的就是大分

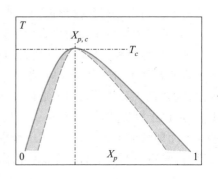

子的临界组成（$X_{p,c}$）。临界温度和临界组成只依赖于溶剂和大分子的物理本质。图中的曲线有时也可上下颠倒（常常发生在水和其他物质的混合物中）。这样，曲线有一个最低点，称作下临界溶液温度（lower critical solution temperature，LCST）。当温度低于其时，大分子和溶剂可以任何比例混合，形成一个的均匀二元混合物。依据$\Delta G_{mix}=\Delta H_{mix}-T\Delta S_{mix}$，混合熵变和混合焓变二者必定均为负值。该反转的热力学解释如下。

混合大分子和溶剂时，总的混合熵变是它们各自熵变之和，$\Delta S_{mix}=\Delta S_{mix,p}+\Delta S_{mix,s}$。两者一般都为正，混合增加它们的熵。如果大分子和溶剂分子之间有很强的吸引相互作用，许多溶剂分子吸附在一条大分子链上可减少溶剂的平动熵。如果溶剂的熵减大于大分子在溶解过程中的熵增，总的混合熵变则可为负。这样，升温将会导致ΔG_{mix}增加，不利于溶解，甚至导致相变。水就是一个典型的例子，水分子可形成较强的氢键作用。在一个水溶液中，如果水分子较强地、优先地吸附在大分子链上，水分子的平动熵就会减少。另外，疏水的大分子主链也会排斥水分子，减少水分子的有效可达及体积和熵，故大分子水溶液常有一个LCST。降温可增大它们的溶解度。

对一给定的混合组成，可利用不同的实验方法，如常用的浊度法，测得上述的双节线，得到表观的"相图"。实验上，先在一相区中配制一个大分子溶液，再用一个固定的速率

逐渐地改变温度,并跟踪测量透光率或溶液的浊度。一旦温度触及双节线,溶液进入两相区,大分子链开始相互聚集。如在静态激光光散射中所述,散射光强度与质量的平方成正比。在大部分光被大聚集体散射之后,透射光强度下降。与透射光强度变化对应的温度被当作双节点线上的一点。

几乎现今每本聚合物教科书中都在使用 Flory 于 1952 年以这种方式测量的不同长度的聚苯乙烯在环己烷中的"相图"。理论上,温变速度越慢,所测的点就越接近双节线。实验上,温变速度总是一个有限的数值。大分子的聚集需要时间,故这样测得的"相变"温度总是落在上图的阴影亚稳态区,而不是真正的双节线上的一点。偏离的程度取决于温变速率,越快偏离越多。读者应该注意,目前大分子教科书中使用的相图只是表观的。

实验上,先在不同的温变速率下测得一系列表观的"相变温度",然后,将表观"相变温度"外推到温变速率无穷慢的相变温度,用其逼近双节线上真正的相变温度。也可外推到温变速率无穷快,得到旋节线上的一点。改变溶液浓度,重复测量,最终可得到一个二元混合物的双节线和旋节线,此法十分烦琐和费时。因此,常常只是在一个固定的温变速率下,测量浊度变化,从而得到一个粗略的相变温度。这样的相图已足以应付和解决大部分应用问题。据笔者所知,文献中尚无任何有关大分子溶液旋节线的测量和数据。

为了得到大分子溶液的一个真正相图,就不得不采用两相平衡的方法。先在一相区配制一个尽量接近临界组成的大分子溶液,然后缓慢地将温度降至两相区内的一个温度,保持恒温,波动需在 ±0.01℃ 以下。当两相完全分离、达到动态平衡时,利用折射或其他光学方法分别测得每相里大分子的浓度,从而得到双节线上的两点($X_{p,L}$ 和 $X_{p,H}$),如右图所示。改变温度,重复以上实验可测

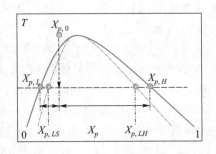

绘相图(双节线)。其中,两相的浓度由温度和双节线的两个相交点决定。两相的体积(V_L 和 V_H)与总体积(V)的关系由"杠杆规则"(类似跷跷板中的力矩相等原理)决定,即

$$V_L = \frac{x_{p,H} - x_{p,0}}{x_{p,H} - x_{p,L}}V \quad \text{和} \quad V_H = \frac{x_{p,0} - x_{p,L}}{x_{p,H} - x_{p,L}}V \tag{10.4}$$

这一平衡方法极端耗时。其原因是溶液分成稀相和浓相两层时,在稀相中,一条典型大分子链的扩散系数约为 $10^{-7}\,\mathrm{cm}^{-2}/\mathrm{s}$。因此,扩散一段宏观距离($\sim 10^{-1}\,\mathrm{cm}$)的时间约为 $10^5\,\mathrm{s}$,$1 \sim 2$ 天;而浓相已为亚浓溶液,由前一章节的讨论已知,在亚浓溶液中,大分子链的质心扩散一个链的尺寸($\sim 10^{-5}\,\mathrm{cm}$)需时约为 $1\,\mathrm{s}$;以此类推,扩散 $\sim 10^{-1}\,\mathrm{cm}$ 约需 $10^8\,\mathrm{s}$,约 3 年,极其缓慢。虽然浓相始于链的聚集,聚集体的尺寸很小,但达到平衡的时间仍需以周或月计算。如要测得一个好的相图,可能需以年作为时间单位。Xia, An 和 Shen[Journal of Chemical Physics, 1996, 105: 6018]利用此法,历时近两年测绘了五条聚甲基丙烯酸甲酯在 3-辛酮中的相图。

超长的平衡时间仅是获得真正的大分子溶液相图的一个障碍。依据上述讨论,大分子溶液相图还有链长(N,链段数)依赖性,反映在 n_v 的数值上。因此,测绘相图所用的大

分子样品必须具有较窄的链长分布($M_w/M_n<1.1$)。研究相图的链长依赖性需要不同链长的窄分布样品,链长最少要相差十倍以上。除了聚苯乙烯和聚甲基丙烯酸甲酯等少数商品化的标准样品以外,获得一套不同长度、窄分布的大分子并不是一件易事。还有,采用传统相分离和平衡的方法,每个样品最少需要 0.5~1.0 g,更是难上加难。这也是为何真正的大分子相图在文献中屈指可数、罕见。

如一开始所述,正是链长依赖性造就了大分子物理。与小分子相比,大分子相图的链长依赖性引入了新的物理。早在 20 世纪 50 年代物理学家就已经发现,对小分子相图而言,在二元混合物的临界点(温度 T_c 和摩尔分数 x_c)附近,两相的浓度之差($|x_H - x_L|$)和约化温度($\varepsilon = |T_c - T|/T_c$)之间存在着一个与二元混合物中具体物质无关的、普适的标度关系:$|x_H - x_L| \sim \varepsilon^\beta$。平均场理论预计 $\beta = 1/2$。但经过多年的理论和实验研究,终于在 20 世纪 60 年代末期,通过实验证实了 $\beta = 1/3$,一个普适常数,称作 Ising 指数。

大分子和溶剂分子的体积相差极大,这一普适的浓度差对约化温度的依赖关系是否仍然成立? 如果成立,标度指数为何? 在 Ising 模型中,大、小分子的空间维度和序参量维度相同,分别为 3 和 1。因此,该标度关系应该不变。有限的实验数据基本上支持这一理论判断。但真正的、高质量大分子相图太少,所以,在很长的一段时间内,并无定论。除了浓度差对约化温度的依赖性以外,另一个问题是,大分子溶液是否还存在着一个浓度差对链长的普适依赖性? 如果肯定,浓度差对约化温度和链长均具有普适依赖性就可写成

$$X_{p,H} - X_{p,L} \sim \varepsilon^\beta N^\xi \tag{10.5}$$

其中,ξ 也是一个与具体的大分子和溶剂性质无关的普适常数。在近四十年的研究中,不同的理论预测了不同的链长依赖性。没有足够多的、令人信服的实验数据可以辨伪存真。

早期,Flory[Principles of Polymer Chemistry,Cornell University Press;Ithaca,NY,1953]依据平均场理论指出,$\beta = 1/2$ 和 $\xi = -1/4$。其缺陷已被后来的理论和实验证实,尤其是 $\beta = 1/3$ 已被肯定地证实和广泛地接受。后来,de Gennes[Scaling Concepts in Polymer Physics;Cornell University Press;Ithaca,NY,1979]指出,对足够长的链,大分子溶液的临界状态接近 Θ 状态,可以忽略二体相互作用,只考虑三体相互作用,并且 $\xi = (\beta - 1)/2$。如将平均场的 $\beta = 1/2$ 代入,$\xi = -1/4$,与 Flory 的推论相符,代入正确的 $\beta = 1/3$,可得 $\xi = -1/3$。稍后,Muthukumar[Journal of Chemical Physics,1986,85:4722]沿着这一思路,考虑到大、小分子溶液的不同之处,即大分子溶液不仅有和小分子相同的因分子质心运动引起的分子间密度涨落,还有小分子溶液没有的,链内各个链段相对运动导致的链内密度涨落。混合熵变与平均场理论中的相同,故只需考虑混合焓变的差别,$\Delta H_{mix}/RT = X_p(1 - X_p)\chi_{sp}$,即

$$\frac{\Delta H_{mix}}{RT} = \left(\frac{1}{2} - A_2\right)X_p(1 - X_p) + \left(A_3 - \frac{1}{6}\right)X_p^3 - (A_2 + A_3 X_p)\alpha^6 X_p^2 + \frac{512\pi^2\alpha^6}{3}X_p^3 \tag{10.6}$$

其中,前两项和后两项分别反映了分子间(质心)和分子内(链段)相对运动引起的密度涨落;α 为扩展因子,见式(8.2)。已知,在大分子溶液中,当 $N \to \infty$ 时,$T_c \to \Theta$,二体相互作用(A_2)接近消失。如果仅考虑三体(A_3)相互作用,上式有解析结果,$\beta = 1/3$ 和 $\xi = -2/9$。有关详细信息,读者可以阅读原始论文。

尽管理论上有不同的预测,在 20 世纪 80 年代后期,这方面的理论探讨逐渐搁置。有

限的几个实验结果显示$-1/3 \leqslant \xi \leqslant -1/10$,没有定论。其主要原因仍是缺乏足够、可信的实验数据(大分子相图)来区分、判断和证实不同的理论。显然,正确的理论只有一个,但也可能所有现存的理论均不正确。直至最近,借助结构生物学中开发的利用微流体在纳升溶液(10^{-9}L)中寻获蛋白质结晶条件的方法,实验上方有突破。由于这是大分子溶液研究中最新和最重要的结果之一,以飨有兴趣的读者,特简介如下。

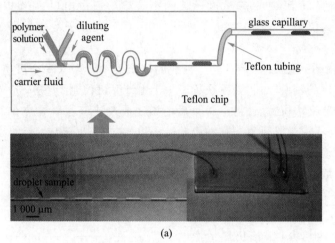

(a)

Shangguan Y G, Guo D M, Feng H, et al. Macromolecules, 2014, 47:2496.

(Figure 2, permission was granted by publisher)

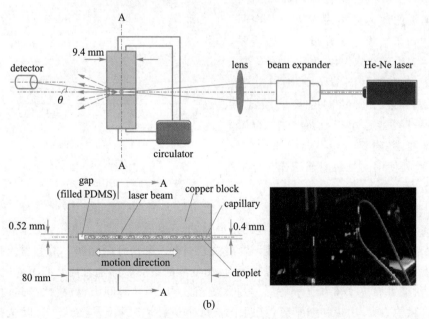

(b)

Shangguan Y G, Guo D M, Feng H, et al. Macromolecules, 2014, 47:2496.

(Figure 3, permission was granted by publisher)

如上图(a)所示,一块用聚四氟乙烯制作的微流体"芯片"上刻有一个横截面为0.02×0.02 cm^{2}的微通道,其共有三个入口,分别对应大分子溶液、溶剂和载体(一个与溶剂不混合的含氟液体)以及一个出口。每一入口的进液量由计算机控制的推进泵精确计量。当大分子溶液和稀释溶剂的推进量之和与载液的推进量接近时,载液就会将大分子溶液断割成一个又一个的小液滴,每两个之间为一小段封装载液,其有效地阻止了不同小液滴之间的混合。每个小液滴的体积约为 10^{-8} L,即约 10 nL。如果逐渐降低大分子溶液的推进量、同时增加稀释溶剂的推进量,那么每个小液滴中的大分子浓度就会逐渐降低,形成一串浓度递减的小液滴。通过出口,将它们注入一根玻璃毛细管。理论上,仅用 1 μg 样品就可制得一串 50 个大分子浓度等距递减的、体积为 10 nL 的液滴。

如上图(b)所示,将这根含有几十个液滴的玻璃毛细管嵌入一个可精密控温、两边各带有一个狭缝窗口的铜块,再将该铜块置于一个由计算机控制、步进马达驱动的精密平行移动平台上。调节铜块位置,使得一扩束后的激光通过狭缝窗口聚焦在毛细管中的液滴上,并使毛细管与入射激光垂直。在铜块的另一边,在与入射光成 ~4° 的小角度位置上安放一个散射光强检测器。恒温后,沿着与激光光束垂直的方向移动平台,就可让激光光束依次扫描每一个液滴,先测得其散射光强(I),再除以其浓度(C)。

由静态激光光散射原理已知,在散射角度趋向零时,该比值(I/C)与重均摩尔质量成正比,即与散射粒子质量的平方和粒子数的乘积成正比,$I/C \sim M^2 N$,意味着即使溶液中的大分子链出现两两相聚(溶液仍然清澈透明),摩尔质量和粒子数也将各自增加一倍和减少一半,导致散射光强增加一倍。小角度的散射光强对大分子链在两相区内的轻微聚集远比浊度法中的透射光强敏感。达到给定温度以后,大分子在一相区保持单链状态,I/C 不随时间变化;而在两相区,由于链间的聚集(注意:聚集体尺寸可能远小于临界尺寸),I/C 将明显增加,并随着时间不断增加,如右图所示,此处使用的是聚醋酸乙烯在异丁醇中的溶液。

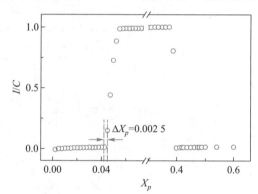

Shangguan Y G,Guo D M,Feng H,et al. Macromolecules,2014,47:2496.

(Figure 6,permission was granted by publisher)

图中的每个小圆圈代表了一个液滴的散射光强与其浓度之比。因此,扫描液滴完全等效于在大分子相图上,固定温度,逐渐变化大分子组成(浓度)。由于链在两相区内聚集,散射光强开始增加的两个浓度标记着溶液分别进入和离开两相区。因此,得到在此给定的温度下,双节线上的两点。在两相区内,相分离的诱导时间随着离开双节线距离的增加而变短、聚集加快、I/C 增加。而一旦进入旋节线内,诱导时间为零,故此区间内各个液滴的 I/C 相差不大,出现一个平台,其两个端点对应着旋节线上的两点。所以,在每一个温度下,一次扫描可得外面两点和里面两点,共四点。改变温度,重复扫描,并将不同温度下获得的外面两点以及里面两点分别连接,可得大分子溶液相图(双节线和旋节线)。此新颖方法的好处之一是无须等待两相达至动态平衡,即可快速和精确地测绘大分子溶液相图。

右图显示了利用该法获得的五条不同长度的聚醋酸乙烯在苯中的溶液相图,其中, N 为聚合度(单体数)。显然,结合微流体技术和小角激光光散射测绘大分子溶液相图具有传统方法无法比拟的优势:第一,利用制备型排除体积色谱柱,很容易获得几毫克不同摩尔质量、窄分布的大分子样品,可供上述实验使用,从而解决了如何制备窄分布大分子样品的问题;第二,无须经年累月地等待两相溶液达至动态平衡,从而可以快速和准确地测绘不同大分子溶液的相图。

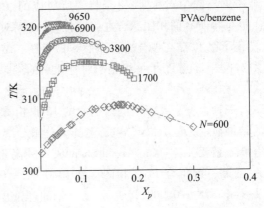

Wu C,Li Y.Macromolecules,2018,51:5863.
(Figure 2,permission was granted by publisher)

理论上,如果有足够量的不同长度的窄分布大分子样品,也许无须采用此法。可直接配制几十个浓度的大分子溶液,将它们置于恒温水槽,然后再利用小角激光散射扫描每一个浓度,测量散射光强,得到同样的结果。但是,除了制备样品的问题以外,在水槽中恒温和测量几十个溶液是另一个实验问题。使用微流体的另一个好处是每个液滴的尺度比常规溶液最少小了十倍,除了可以很快地到达给定温度以外,更重要的是两相平衡达至的时间缩短了一百多倍,从三个月缩短至一天,十分可观,尽管在此法的测量中并不需要达至平衡。注意,液滴的尺寸不可太小,否则界面层(<1 μm)就会影响实验结果。

上图清楚地显示,大分子溶液的相图已严重地不对称。临界体积分数(X_c ,双节线最高点)已经远远偏离0.5。偏离随着链长的增加,临界体积分数左移、临界温度上移。理论上,当 $N \to \infty$ 时, $T_c \to \Theta$ 。 X_c 和 N 存在一个标度关系: $X_c \sim N^{-\gamma}$ 。利用最小二乘法拟合上图中的数据,可得每个相图的 X_c 。综合源自四个不同大分子在不同溶剂中的17张相图得到 $\gamma = 0.37 \pm 0.01$ 。为了方便分析这些严重倾斜的相图,Sanchez[Journal of Applied Physics,1985,58:2871]引入了一个无量纲的序参数(Ψ),使得不对称的相图重新对称。

$$\Psi = \frac{X_p}{X_p + R_c(1 - X_p)} \tag{10.7}$$

其中,对一个给定的大分子溶液, R_c 是数据分析中的一个拟合参数,随着链长增加而递减。Sanchez在上述文献中指出, R_c 的物理意义不明。显然,当 $R_c = 1$ 时, $\Psi = X_p$ 。因为不对称源于一条大分子链和一个溶剂分子的体积之差, R_c 可使相图重新对称,故其一定与链长有关。利用 $X_p + X_s = 1$,可将上式重写成

$$\Psi = \frac{V_p}{V_p + R_c V_s} = \frac{V_p/V_s}{V_p/V_s + R_c}$$

文献中有限的数据显示 R_c 与临界体积分数有关,即 $R_c \approx 2X_{p,c}/(1 - X_{p,c}) = 2(V_{p,c}/V_{s,c})$ 。作对称化操作时,需要先将上图中的每一个温度按照定义换算成约化温度(ε),并

以 $\varepsilon_{max}N^{0.3} \leqslant 0.075$ 作为一个标准,选择每个相图中充分靠近临界点的数据,然后通过拟合得到在 Ψ 空间中对称时,每个相图所对应的 Ψ_c 和 R_c,再分别求出 Ψ_c 和 R_c 各自的平均值。也可将所有的相图综合在一起拟合,得到对应的平均 Ψ_c 和 R_c。借助计算机,现在通常采用后者,以提高拟合的可靠性和稳定性。结果显示 $\Psi_c = 0.325 \pm 0.002 \approx 1/3$,与链长无关。下图(a)显示了一组相图对称化后的结果。进一步利用下式

$$|\Psi - \Psi_c| = \Psi_0 \ (\varepsilon N^b)^{\beta} \tag{10.8}$$

对三种大分子在不同的溶剂中的 12 张相图,通过三变量 Levenberg-Marquardt 拟合得到的结果显示,Ψ_0 不依赖链长,但随大分子和溶剂变化。b 和 β 二者均确实与链长和大分子溶液体系无关,与 Xia,An 和 Shen 从一种高聚物溶液的 5 张相图中得到的结果相似。因此,重新一起拟合 17 张相图得到 $b = 0.47 \pm 0.02$,略微小于平均场理论所预计的 $1/2$;以及 $\beta = 0.328 \pm 0.009$,接近 Ising 模型的理论预期值 $1/3$。

由上式和下图(a)已知,经过如此对称化操作以后,Ψ 以 Ψ_c 为其对称轴,故可将 Ψ_0 左面所有的数据点镜面翻转到右面,εN^b 对 $|\Psi - \Psi_c|/\Psi_0$ 作图是一条近似三次方的曲线,$\varepsilon N^b = (|\Psi - \Psi_c|/\Psi_0)^{1/\beta}$,如下图(b)所示。注意:纵坐标从上至下递增。图中,每种符号代表了一种溶液,每种溶液包括了 4~5 不同的链长,4 种不同的溶液,共计 17 张相图。它们叠加成一条曲线,充分地证实了 b 和 β 的普适性。

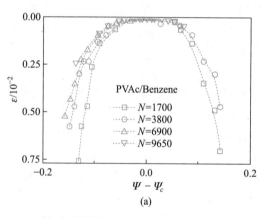

Wu C,Li Y.Macromolecules,2018,51:5863.

(Figure 3,permission was granted by publisher)

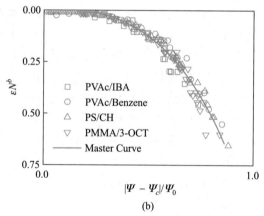

Wu C,Li Y.Macromolecules,2018,51:5863.

(Figure 4,permission was granted by publisher)

实验上,需要将对称空间的 Ψ 换算成体积分数(X_p),方可进一步换成可测的质(重)量浓度。依据上式中 Ψ 的定义,将其在 $X_{p,c}$ 附近展开,可得

$$\Psi = \Psi_c \left[1 + \frac{R_c \Psi_c}{X_{p,c}} \left(\frac{X_p}{X_{p,c}} - 1 \right) + \cdots \right] \approx \Psi_c \left[1 + \frac{\Psi_c}{1 - X_{p,c}} \left(\frac{X_p}{X_{p,c}} - 1 \right) + \cdots \right]$$

对给定的大分子溶液,Ψ_c 和 $X_{p,c}$ 为两个常数。理论上,当链足够长时,在充分靠近临界点的范围内,二元混合相图对称,故 $\Psi_H - \Psi_L = 2|\Psi - \Psi_c|$ 以及 $X_{p,H} - X_{p,L} = 2|X_p - X_{p,c}|$。

$$|\Psi - \Psi_c| = \frac{\Psi_c^2}{1 - X_{p,c}} \frac{|X_p - X_{p,c}|}{X_{p,c}} \sim \frac{|X_p - X_{p,c}|}{X_{p,c}}$$

其中，$X_{p,c} \sim N^{-\gamma}$。将上式结合式（10.7），并将 $\beta = 0.328$，$b = 0.47$ 和 $\gamma = 0.37$ 代入，可得

$$|X_p - X_{p,c}| = X_0 \varepsilon^{\beta} N^{b\beta-\gamma} = X_0 \varepsilon^{0.328} N^{-0.216} \tag{10.9}$$

其中，X_0 是一个随体系而变的正比常数。这就是二元混合物的普适标度关系。尽管二元混合物的种类千变万化，但它们在临界点附近的相图只取决于两个物理不变量（普适标度指数），与组成二元混合物的具体物质无关，可见物理之美！实验结果得到的 $|X_p - X_{p,c}|$ 与链长之间的标度指数（$\xi \approx -0.22$）非常接近 Muthukumar 预测的 $-2/9$。

物理学上，当大分子链足够长时，靠近临界点的二元混合物趋向 Θ 状态，二体相互作用接近消失。在临界点附近，两相的组成非常接近，密度剧烈地涨落，相关长度显著地变长。因此，可合理地假定原本很弱三体相互作用成为主导。

再考察 R_c 的物理意义。平均场理论只考虑二体相互作用。对称化操作后，大分子和溶剂分子的体积之差已被矫正，故 Ψ 应为 $1/2$；即在临界状态时，溶液中，所有大分子的体积（$V_{p,c}$）和所有溶剂分子的体积（$V_{s,c}$）与 R_c 的关系必定为 $R_c V_{s,c} = V_{p,c}$，或写成 $R_c = V_{p,c}/V_{s,c}$。大量实验结果已显示 $\Psi \approx 1/3$；意味着 $R_c = 2(V_{p,c}/V_{s,c})$。显然，R_c 的物理意义是，在处于临界状态的溶液中大分子的总体积与溶剂分子的总体积之比；换算成摩尔数之比，$R_c = 2n_v n_{p,c}/n_{s,c}$，其中，$n_v$ 代表的是一条大分子链在体积上等效于多少溶剂小分子。注意，$V_{p,c}$ 和 $V_{s,c}$ 随着大分子链长的增加分别变小和变大；所以，R_c 随着链长增加变小。实验结果显示，虽然它们都随链长变化，但 $R_c V_{s,c}/V_{p,c}$ 是一个物理不变量。目前，仍无一个理论可以定量地解释这一结果。

除了其理论上的重要性以外，上述的普适标度关系还有广泛的应用。从此以后，绘制一个大分子溶液相图，再已无须配制数十个不同浓度的溶液和测量它们的相变温度。依据上式，仅有三个取决于体系的未知变量：$X_{p,c}$，X_0 和 T_c。因此，测绘任何一个大分子溶液的相图仅需在临界点附近配制三个具有不同浓度（X_p）的溶液，测得它们各自的相变温度，代入上式，联解三个方程可得三个未知数，即可绘制出其相图。

对于可在溶液中结晶的大分子，其相图较为复杂。仅以具有 UCST 的溶液为例，如果结晶温度高于相变温度且浓度足够高时，随着降至结晶温度，大分子溶液因结晶开始分成两相，一相为大分子晶体（纯大分子，相图中的右轴），另一相为大分子在溶剂中的饱和溶液。当温度进一步降低时，结晶相里，结晶量增多，但浓度不变（已经 100%，不可再高），另一相中的饱和浓度递减。

当结晶温度低于相变温度时，溶液则会先进行相变，分成稀相和浓相两个饱和溶液。随着温度降至结晶温度，浓相中出现结晶，溶液中三相共存，两个饱和溶液外加结晶。溶液中存在两个动态平衡，稀相和浓相两个饱和溶液之间，以及浓相和结晶相之间。共聚物、聚电解质和混合溶剂中的溶液相图则更为复杂，超出本书范畴。有兴趣的读者可参阅相关专著和文献。

小　结

溶剂性质随着温度或压强的改变发生变化。溶剂性质逐渐变差使得一个均匀的大分

子溶液(一相)分成两个均匀的溶液(两相,稀相和浓相),称为相变。稀相是大分子在溶剂中的饱和溶液;浓相则为溶剂在溶胀的大分子中的饱和溶液。两相达到一个动态平衡时,微观上,溶剂分子和大分子各自不断地以相同速率在两相之间穿梭来回,宏观浓度保持不变。两个饱和溶液的体积可借助相图,由"杠杆规则"获得。相变与否取决于混合自由能的增减,包含了混合焓变和混合熵变的综合效应。对一个等温和等压过程,体系总是自发地趋向其自由能减少的方向。由于压强对大分子溶液的影响不大,仅以温度变化为例。

在温度对组成的相图中,连接每一组成对应的相变温度,可得一条分割一相区和两相区的曲线,称为双节线。两相区还可被进一步按照分相是否需要诱导时间,分成两个区域。将与诱导时间为零的温度和组成点连接起来,可得另一条曲线,称作旋节线,对应的是混合自由能随大分子摩尔分数的二阶导数为零处。因此,对给定的温度,逐步地增加浓度可导致一个溶液依次经过一个双节点($X_{p,L}$),两个旋节点($X_{p,LS}$ 和 $X_{p,HS}$)和另一个双节点($X_{p,H}$)。

双节线和旋节线之间为亚稳态区。虽然在此区间,自由能仍然减少,但分成两相可进一步降低自由能,故溶液倾向于相变,但需要一个大于临界尺寸的"核"。分相所需的诱导时间,取决于大分子间的相互作用能和界面能二者之间的相对大小,随着偏离双节线的距离的增加而缩短。在双节线上时,为无限长;在旋节线上时,为零。实验上,由于溶液中总是存在可以作为"核"的杂质(灰尘、气泡等),分相总是始于很短的时间,但最终达至两相动态平衡的时间则由大分子链的平动扩散控制。由于大分子在亚浓溶液中的扩散极慢,所以,达到动态平衡所需的时间可为经年累月,取决于链长和浓度。

通常,混合熵变为正。依据热力学原理,在等温过程中,$\Delta G_{mix} = \Delta H_{mix} - T\Delta S_{mix}$,故升温有助溶解,存在一个上临界溶液温度(UCST),高于此温度,ΔG_{mix} 始终为负,大分子和溶剂可自发地以任何比例互相混合,形成一个均匀的一相二元共混物(溶液)。当大分子和溶剂分子之间存在很强的优先吸附时,溶剂的平动熵减少,混合熵变可为负值,导致一个下临界溶液温度(LCST)。低于此温度,ΔG_{mix} 始终为负,大分子和溶剂可以任何比例自发地互相混合,形成一个溶液。一些大分子溶液可具有上、下临界溶液温度,如聚苯乙烯在环己烷中。显然,实验上可观察的上和下临界溶液温度必须分别低于和高于溶剂的沸点和冰点。

当温度逐渐逼近上或下临界溶液温度时,上述的四个点($X_{p,L}$, $X_{p,LS}$, $X_{p,HS}$ 和 $X_{p,H}$)趋于一点:一个具有临界温度(T_c)和临界组成($X_{p,c}$)的临界状态,混合自由能随组成变化的一阶和三阶导数均为零。在临界点附近,稀相和浓相的浓度对称于 $X_{p,c}$,即 $(X_{p,H} - X_{p,L})/2 = X_{p,H} - X_{p,c} = X_{p,c} - X_{p,L}$。现已证实,此浓度之差分别与约化温度($\varepsilon = |T - T_c|/T_c$)和链长(链段数,$N$)之间存在着两个与具体溶液(大分子和溶剂)无关的普适标度关系 $|X_p - X_{p,c}| = X_0\varepsilon^\beta N^\xi = X_0\varepsilon^{0.328}N^{-0.216}$,与理论预测的 $|X_p - X_{p,c}| \sim \varepsilon^{1/3}N^{-2/9}$ 接近。

下图小结了本章所讨论的内容,包括二元混合物相图的来源和相图中普适的标度关系。

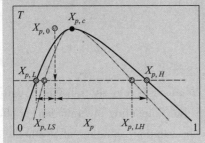

$$f = \frac{\Delta G_{\text{mix}}}{nRT} = X_p(1 - X_p)\chi_{sp} + \frac{X_p}{n_v}\ln X_p + (1 - X_p)\ln(1 - X_p)$$

双节线：切线节点 $f'(X_{p,L}) = f'(X_{p,H})$ 和 $f'(X_{s,L}) = f'(X_{s,H})$

旋节线：$f'' = 0 \rightarrow X_{p,LS}$ 和 $X_{p,LH}$ 临界点：$f'' = f''' = 0 \rightarrow X_{p,c}$

普适标度律（Universal Scaling）

$$|X_p - X_{p,c}| \propto \varepsilon^{1/3} N^{-2/9}$$

杠杆规则（The Level Rule）

$$V_L = \frac{X_{p,H} - X_{p,0}}{X_{p,H} - X_{p,L}} V \qquad V_H = \frac{X_{p,0} - X_{p,L}}{X_{p,H} - X_{p,L}} V$$

Chapter 10. Phase Transition of Macromolecular Solutions

The phase transition of macromolecular solutions is closely related to daily life and industrial production. In daily life, the phase transition (including precipitation) is often used to separate soluble and insoluble macromolecules; in solution spinning of synthetic fibers, the orientation of macromolecules is controlled by adjusting solvent volatilization, phase transition and drawing speed in order to obtain the desired mechanical properties; in the formation of porous hollow fibers and porous membranes, the phase transition rate is controlled to adjust the pore size distribution; in the precipitation polymerization, the phase transition is used to control the average molar mass of the synthesized polymer; When using a crystalline polymer to make materials, the phase transition rate is often used to adjust the crystallinity to produce materials with different mechanical properties; in structural biology, the phase transition is often used to promote the protein crystallization; and so on, countless.

As discussed in Chapter 3, whether a macromolecule can dissolve in a solvent depends on the changes of enthalpy and entropy during the mixing. The mixing enthalpy depends on the interaction between macromolecules and solvent molecules, while the mixing entropy comes from the increase in the number ways of microscopically arranging macromolecules and solvent molecules in the mixing, that is, the increase in the total degree of freedom of the system. Known from thermodynamics, after a given temperature and solution composition, the dependence of the mixing free energy on the mixing enthalpy and the mixing entropy is as follows:

$$\Delta G_{mix} = \Delta H_{mix} - T\Delta S_{mix}$$

Only when $\Delta G_{mix} < 0$, macromolecules can dissolve in a solvent to form a homogeneous, one-phase binary mixture (solution) at the molecular level; otherwise, a two-phase binary mixture is formed, two coexistent saturated solutions (two phases) with a macroscopic interface. One is a saturated solution of macromolecules in solvent (dilute phase); the other is a saturated solution of solvents in macromolecules (concentrated phase). Microscopically, macromolecules and solvent molecules traverse the interface back and forth continuously at the same rate, forming a macroscopic dynamic equilibrium.

Due to the chain entanglement, solvent molecules penetrate into the entangled network of macromolecules, and the volume fraction of solvent in the concentrated phase can exceed 50%. In the concentrated phase of a small molecule solution, there are much less solute molecules. The general textbooks for macromolecules do not emphasize this significant difference. The change between one phase and two phases is called the phase transition. Depending on whether some thermodynamic state variables(such as volume and density)change continuously with temperature or pressure, the transition can be further divided into discontinuous first-order phase transitions and continuous second-order phase transitions. For details, readers can refer to the textbooks of thermodynamics.

Strictly speaking, at the molecular level, two substances are always soluble, and the only difference is the solubility. When the solubility is less than one part in 10,000, it is usually called "insoluble" between the two. We often say that "oil and water are insoluble" in daily life. When oil and water are mixed, it is divided into upper and lower layers. The upper oil layer contains saturated water, and the lower water layer contains saturated oil. In a binary mixing, entropy usually increases and the change of entropy is positive. If enthalpy decreases, the change of enthalpy is negative, according to the above equation, the mixing can proceed spontaneously. If the change of enthalpy is positive in the mixing, heating the solution can generally promote the dissolution. There is an upper critical solution temperature(UCST)that makes the mixing free energy zero. Generally speaking, the change of free energy in the formation of a binary mixture can be written as

$$dG = \left(\frac{\partial G}{\partial p}\right)_{T,n} dp + \left(\frac{\partial G}{\partial T}\right)_{p,n} dT + \left(\frac{\partial G}{\partial n_A}\right)_{p,T,n_B} dn_A + \left(\frac{\partial G}{\partial n_B}\right)_{p,T,n_A} dn_B$$

where $(\partial G/\partial p)_{T,n} = V$ and $(\partial G/\partial T)_{p,n} = -S$, $(\partial G/\partial n_A)_{p,T,n_B}$ and $(\partial G/\partial n_B)_{p,T,n_A}$ are called the chemical potentials of A and B, denoted as μ_A and μ_B. The physical essence is the changing rate of Gibbs free energy with component A or B. For macromolecular solutions, A and B can represent solvent small molecules and macromolecules, respectively. Note that the "mixing free energy" used usually and hereafter refers to the "change of free energy in mixing", remember!

In Chapter 3, the composition dependence of the mixing free energy was discussed. For the convenience of discussion, eq. (3. 14) is rewritten as follows.

$$\frac{\Delta G_{mix}}{nRT} = X_p(1 - X_p)\chi_{sp} + \frac{X_p}{n_v}\ln X_p + (1 - X_p)\ln(1 - X_p)$$

where $X_p(1 - X_p)\chi_{sp}$ and $(X_p/n_v)\ln X_p + (1 - X_p)\ln(1 - X_p)$ correspond to the mixing enthalpy and the mixing entropy, respectively; after dividing by the molar thermal energy RT, ΔG_{mix} is dimensionless. In Chapter 3, $\Delta G_{mix}/(RT)$ has been plotted against X_p, and it has been discussed that its asymmetry stems from the volume difference between macromolecules and solvent molecules. Readers can re-read the relevant content in Chapter 3.

Partially differentiating the above equation with respect to molar numbers of macromolecules or solvent(n_p or n_s)in the solution leads to the chemical potential of macromolecules or solvents (μ_p or μ_s). Obviously, the calculation involves the partial differential of X_p with respect to n_p and n_s, respectively. When discussing the osmotic pressure difference and ultracentrifugation in Chapter 5, these differential calculations have been derived in detail, i. e., eqs. (5. 10) and

(5.20), so that they are omitted. Respectively using the chemical potentials of pure macromolecules and pure solvents (μ_p^* and μ_s^*, in fact, their respective molar free energy) as the reference points, the macromolecular composition (X_p) dependences of the chemical potentials of macromolecule and solvent in the solution are obtainable. The calculation includes two steps, first, partial derivative of ($\Delta G_{mix}/nRT$) to X_p, and then multiply by partial derivative of X_p to n_p or n_s. If not wishing to involve detailed mathematical calculations, readers can skip the part separated by the two dashed lines below, and only use the results: eq. (10.1).

$$\left(\frac{\partial \Delta G_{mix}}{\partial n_s}\right)_{T,p,n_p} = \left(\frac{\partial \Delta G_{mix}}{\partial X_p}\right)_{T,p,n_p} \left(\frac{\partial X_p}{\partial n_s}\right)_{T,p,n_p}$$

$$\left(\frac{\partial \Delta G_{mix}}{\partial n_p}\right)_{T,p,n_s} = \left(\frac{\partial \Delta G_{mix}}{\partial X_p}\right)_{T,p,n_s} \left(\frac{\partial X_p}{\partial n_p}\right)_{T,p,n_s}$$

where

$$\left(\frac{\partial X_p}{\partial n_s}\right)_{T,p,n_p} = \frac{\partial \left(\frac{n_v n_p}{n_s + n_v n_p}\right)}{\partial n_s} = -\frac{n_v n_p}{(n_s + n_v n_p)^2} = -\frac{X_p(1 - X_p)}{n_s} = -\frac{X_p^2}{n_v n_p}$$

and

$$\left(\frac{\partial X_p}{\partial n_p}\right)_{T,p,n_s} = \frac{\partial \left(1 - \frac{n_s}{n_s + n_v n_p}\right)}{\partial n_p} = \frac{n_v n_s}{(n_s + n_v n_p)^2} = \frac{X_p(1 - X_p)}{n_p} = \frac{n_v(1 - X_p)^2}{n_s}$$

$$\Delta \mu_p = \mu_p - \mu_p^* = \left(\frac{\partial \Delta G_{mix}}{\partial n_p}\right)_{T,p,n_s} = RT[\ln X_p - (N - 1)(1 - X_p) + X n_v(1 - X_p)^2]$$

and (10.1)

$$\Delta \mu_s = \mu_s - \mu_s^* = \left(\frac{\partial \Delta G_{mix}}{\partial n_s}\right)_{T,p,n_p} = RT\left[\ln(1 - X_p) + \left(1 - \frac{1}{n_v}\right)X_p + X X_p^2\right]$$

In the above equation, the variable is the volume fraction of macromolecules. It can be converted into a measurable molar fraction of macromolecules ($x_p = n_p/n$) or a more direct mass (weight) concentration (C): $X_p = n_v x_p = C v_p$, where v_p is the specific volume. Note that it is not a volume, but a ratio of volume to mass. In a solution, it is the rate of change of the solution volume with the component mass. For the convenience of discussion, the reduced mixing free energy $\Delta G_{mix}/(nRT)$ per mole is denoted as f, its first order partial differential is $f' = \Delta \mu_p/(nRT)$; its second and third order partial differentials are also obtainable, respectively, denoted as f'' and f'''. The calculations are quite cumbersome, so that it is omitted.

The figure below shows the macromolecular volume fraction dependence of f, f', f'' and f'''. In order to clearly reveal how f varies in the low concentration range, close to the left vertical axis, $n_v = 2$ and $X_{sp} = 2.1$ are specifically selected. It shows that even when $n_v = 2$, $\Delta G_{mix}/(nRT)$ is already asymmetric. Starting from the left vertical axis, as the concentration gradually increases, ΔG_{mix} starts from zero, first decreases, then increases, from negative to positive, and reaches a peak value. Note that in polymer textbooks, N is generally selected as 100 or larger, so that f looks as if it is rising from zero, and there is no display at all. There is no indication at all that f starts from zero, gradually decreases first, and then, increases from negative to positive, which is indeed misleading. On the contrary, if starting

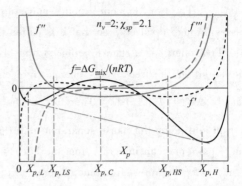

from the right vertical axis, moving to the left, i.e., gradually increasing the solvent fraction in the solution, $\Delta G_{\text{mix}}/(nRT)$ also starts from zero, decreases first, and then increases, from negative to positive, converging at the same peak value, The trend of change is more obvious. **The peak position corresponds to the upper critical solution temperature(UCST) and the critical composition($X_{p,C}$) as well as $f'=f'''=0$ and the lowest value of f''.**

As X_p gradually increases, the first derivative of $\Delta G_{\text{mix}}/(nRT)$ (f', black dashed line) gradually increases from negative to positive, reaches a high point, and then gradually decrease to zero. If starting from the right vertical axis and moving to the left, as the solvent volume fraction increases, f' changes from positive to negative, reaches a low point, and then gradually increases to zero. The high and low two points are called "**spinode**" points, corresponding to two points(yellow blue lines) at which $f''=0$: $X_{p,LS}$ and $X_{p,HS}$. The confluence point($f'=0$) from both sides corresponds to the aforementioned **critical composition of macromolecules** ($X_{p,C}$), which is not the symmetric point of $X_p=0.5$, attributing to the asymmetric volumes of macromolecules and solvent molecules. From the high point to the low point, f' decreases as X_p increases.

According to thermodynamics, chemical potential of any substance in a mixture increases as its composition increases. Therefore, as more macromolecules are added, its chemical potential does not increase but decrease, which violates the physical law. Therefore, the system cannot form a homogeneous mixture along this descending path. On the contrary, the system **must** separate into two phases: one saturated solution of macromolecules in solvent(dilute phase, the volume fraction and volume are $X_{p,L}$ and V_L, respectively) and another saturated solution of solvent in macromolecule(dense phase, the concentration and volume are $X_{p,H}$ and V_H, respectively). Macromolecules and solvent have their own chemical potentials in each phase: $f'(X_{p,L})$ and $f'(X_{p,H})$ as well as $f'(X_{s,L})$ and $f'(X_{s,H})$.

When the two phases reach an equilibrium, not only $f'(X_{p,L})+f'(X_{s,L})=f'(X_{p,H})+f'(X_{s,H})$, but also $f'(X_{p,L})=f'(X_{p,H})$ and $f'(X_{s,L})=f'(X_{s,H})$. Otherwise, there will be a net flow of macromolecules to one phase and solvent molecules to the other phase between the two phases. This is why the two "binode" points does not appear at two points of $f'=0$, although at these two points $f'(X_{p,L}) + f'(X_{s,L}) = f'(X_{p,H}) + f'(X_{s,H})$, but $f'(X_{p,L}) \neq f'(X_{p,H})$ and $f'(X_{s,L}) \neq f'(X_{s,H})$. The two "spinode" points move slightly to the left, which are two nodes. One tangent passes through them on f, so that it is called the "binode" line.

Free energy is a state function, so that its change is independent of the path. According to the definition, the free energy change here can be obtained by the integral of $f'(X_p)$ to X_p; namely, $\Delta G = \int f'(X_p) \, dX_p$. Therefore, in the above figure, ΔG obtained by integrating along the f' path and along the phase separation path (parallel to the X_p axis) must be equal. If following the f' path, one can integrate from the left $f' = 0$ (the lowest free energy point on the left) to the right $f' = 0$ (the lowest free energy point on the right). If taking $f = 0$ as a boundary, the above integral area is smaller, so that the total integral area is obviously not equal to the rectangular area obtained by the integration along the parallel direction (always with $f = 0$). Similarly, if integrating from the left node to the right node, and using the horizontal line connecting them as a boundary, the integrated areas above and below are equal, filling each other up, so that the total integral area is exactly equal to the integral area along the horizontal line (another rectangular area), independent of the path, in line with the principle of thermodynamics.

After the phase transition, the total mixing free energy is a sum of the mixing free energy of each phase multiplied by its volume fraction, which is mathematically expressed as follows

$$f_{\text{two phases}} = \frac{V_L}{V_L + V_H} \frac{\Delta G_{\text{mix}}(X_{p,L})}{n_{\text{box}} RT} + \frac{V_H}{V_L + V_H} \frac{\Delta G_{\text{mix}}(X_{p,H})}{n_{\text{box}} RT} \tag{10.2}$$

It is a straight line connecting the two points on the reduced mixed free energy (f) in the graph, deviating from the two lowest points on f. As mentioned earlier, the two tangent points are $X_{p,L}$ and $X_{p,H}$, which are called "binode" points. If solvent and solute molecules have the same volume, the tangent line will exactly pass through the two lowest points, but this is only a special case. In the two ranges from $f'(X_{p,L})$ to the high binode point and from $f'(X_{p,H})$ to the binode point, the chemical potential gradually increases with X_p, not violating the principle of thermodynamics, and the mixing free energy is still negative, a homogeneous solution should be spontaneously formed. However, as shown in the figure above, in these two ranges, at any point on the black solid line (the mixing free energy) is higher than the point on the blue dotted line (the total mixing free energy after the two phases are formed).

For an isothermal and isostatic process, free energy always decreases spontaneously. Therefore, in these two ranges, a solution of macromolecules always tends to separate into dilute and concentrated two phases to lower the mixing free energy. Therefore, one phase is a metastable state. Microscopically, the phase separation involves the aggregation of molecules and the formation of an interface between the aggregate and its medium (solvent). The aggregation reduces the free energy, which is proportional to the volume of an aggregate (herein referred to as a "nucleus" that induces the phase transition), i. e., the cube of its linear dimension (size), L; while the formation of an interface increases the surface energy, which is proportional to the surface area of an aggregate, i.e., the square of L. For an isothermal and isometric process, the free energy change due to the aggregation in the system is

$$\Delta A \sim -\rho_E L^3 + \sigma_E L^2 \rightarrow \left(\frac{\partial \Delta A}{\partial L}\right)_{T,V} \sim -3\rho_E L^2 + 2\sigma_E L \qquad (10.3)$$

where ρ_E and σ_E are the volume and surface densities of energy, respectively. The proportional constants related to the shape of aggregates are omitted. When L is small, $(\partial \Delta A/\partial L)_{T,V} > 0$, aggregation is not spontaneous, and the opposite disaggregation is spontaneous. Therefore, even if there was a chance to form an aggregate, it would immediately disaggregate and could not continue to grow, although the aggregation can reduce the total free energy. On the other hand, when L is sufficiently large, $(\partial \Delta A/\partial L)_{T,V} < 0$, the aggregation will proceed spontaneously.

$(\partial \Delta A/\partial L)_{T,V} = 0$ corresponds to a critical dimension: $L_C \sim \sigma/\varepsilon$. The aggregation is a process of random fluctuations. Given the system a sufficiently long induction time, there is always a chance that a transient aggregate larger than L_C will appear. Once there is such a "nucleus" in the system, molecules will continue to gather on it spontaneously, and finally reach a macroscopic phase separation. According to statistical physics, this induction time $\tau_{ind} \sim 1/[\exp(\varepsilon/RT) - 1]$. Obviously, if there is no interaction between molecules, there is no phase transition, i.e., $\varepsilon \rightarrow 0$, $\tau_{ind} \rightarrow \infty$; when $\varepsilon \gg RT$, $\tau_{ind} \rightarrow 0$. τ_{ind} depends on the distance from the binode curve.

In reality, such a "nucleus" always exists. It can be a grain of dust, a small bubble, a defect on the wall, etc. Generally speaking, for a real solution of macromolecules, when the concentration reaches $X_{p,L}$ or $X_{p,H}$, it will spontaneously separate into a dilute and a concentrated phase, and will not continue to move from left to $f'(X_{p,L})$ or move from the right to $f'(X_{p,H})$ to maintain one phase, forming a supersaturated solution. In the distillation of a macromolecular solution, if the temperature is lower than USTC, as solvent evaporates, the concentration gradually increases, and the phase separation occurs. If the same container is used to distill solutions, usually a small amount of zeolite must be added to the solution each time to provide small bubbles as "nucleus" to avoid the overheating, bumping to injury the operator.

In the figure above, as the solution temperature increases, χ_{sp} becomes smaller, $X_{p,LS}$ and $X_{p,HS}$ as well as $X_{p,L}$ and $X_{p,H}$ approach each other, and finally the four points converge on the critical volume fraction($X_{p,c}$), corresponding to a critical temperature(T_C) at $f' = f''' = 0$. For a given temperature, as the volume fraction of macromolecules increases, the system changes from a solution to a saturated solution with a concentration of $X_{p,L}$ first, and then, enters a metastable region. When the volume fraction reaches and exceeds $X_{p,LS}$, the system must separate into two phases, whose volume fractions are $X_{p,L}$ and $X_{p,H}$, respectively.

Similarly, if starting from pure macromolecules and gradually adding solvent, a small amount of solvent can always dissolve into the macromolecular sample to swell it to form a homogeneous solution of solvent molecules in a macromolecule, making it swell. When the volume fraction reaches and is less than $X_{b,H}$, the system begins to enter another metastable state. Further increase the volume fraction of the solvent(X_p decreases) to $X_{p,HS}$, the system must separate into two phases. Therefore, for a given temperature, as the volume fraction changes, a total of four points can be obtained: two binode points on the binode line: $X_{p,L}$ and $X_{p,H}$; and two spinode points on the spinode line: $X_{p,LS}$ and $X_{p,HS}$.

Note: The above discussion is based on a common "temperature vs composition" phase diagrams. Actually, there are also phase diagrams of "pressure vs composition", and the discussion and conclusion are similar. Generally speaking, the effect of pressure on macromolecular solutions is much smaller, so that it will not be repeated here. Under a constant pressure, a variation of temperature can result in a series of such a set of four points.

The connection of all the binode points and spinode points separately lead to two curves, as shown on the right, which are called binode lines (solid lines) and spinode lines (dashed lines). The shaded area between the two curves is the aforementioned metastable region. Although the system tends to separate into two phases, it requires the formation of a "nucleus". The induction time decreases as the distance from the binode line, the closer it is, the longer the induction time is. The highest point (T_c) on the curve is the upper critical solution temperature (UCST). For temperatures higher than UCST, macromolecules can dissolve in solvent with any ratio to form a homogeneous mixture at the molecular level. The critic composition ($X_{p,c}$) corresponds to the critical temperature. The critical temperature and critical composition only depend on the physical nature of solvent and macromolecules. The curve in the figure above can sometimes be upside down (it often occurs in mixtures of water and other substances). In this way, the curve has a lowest point, called the lower critical solution temperature (LCST). When temperature is lower than it, macromolecules and solvent can be mixed in any proportion to form a homogeneous binary mixture. According to $\Delta G_{mix} = \Delta H_{mix} - T\Delta S_{mix}$, both the mixing entropy and mixing enthalpy must be negative. The thermodynamic explanation of this reversal is as follows.

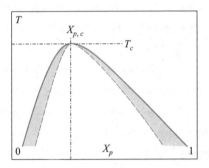

When mixing macromolecule and solvent, the total mixing entropy is the sum of their respective entropies, $\Delta S_{mix} = \Delta S_{mix,p} + \Delta S_{mix,s}$. Generally, both of them are positive, and the mixing increases their entropy. If there is strong attractive interaction between macromolecule and solvent, the adsorption of many solvent molecules on a macromolecular chain reduces the translational entropy of solvent. If the decrease of entropy is more than the increase of entropy of macromolecule during the dissolution, the total mixing entropy can be negative. In this case, an increase of temperature will lead to an increase of ΔG_{mix}, not conducive to the dissolution and even leads to a phase transition. Water is a typical example. Water molecules can form a strong hydrogen bond. In an aqueous solution, if water molecules are strongly preferentially adsorbed on macromolecular chains, the translational entropy of water molecules is reduced. On the other hand, the hydrophobic main chain of a macromolecule also repels water molecules, reducing the effective volume reachable for water molecules and their entropy, so that aqueous solutions of macromolecules often have a LCST. Cooling increases their solubility.

For a given mixture composition, different experimental methods, such as the commonly used turbidity method, are used to measure the aforementioned binode line and obtain an apparent "phase diagram". Experimentally, a macromolecular solution is prepared in one phase region first, and then, the solution temperature is gradually changed at a fixed rate, and the light transmittance, or say, turbidity of the solution, is measured. Once the solution temperature hits the binode curve, and the solution enters the two phase region, macromolecular chains start to aggregate with each other. As discussed in static LLS, the scattered light intensity is proportional to the square of mass. After most of the light scattered by large aggregates, the transmitted light intensity drops. The temperature corresponding to the change of the transmitted light intensity is used as a point on the binode line.

Nearly every current polymer textbook is using the "phase diagrams" of polystyrene with different lengths in cyclohexane measured by Flory in 1952 in such a way. Theoretically, the slower the temperature changing rate, the closer the point to the binode curve. Experimentally, the temperature change rate is always finite. The aggregation of macromolecules takes time, so that the "phase transition" temperature measured in this way always falls in the shaded metastable region in the figure above, rather than a point on the true binode line. The deviation depends on the rate of temperature change, the faster the rate, the more the deviation. Readers should note that the phase diagrams currently used in polymer textbooks are only apparent.

Experimentally, a series of apparent "phase transition temperatures" are measured at different rates of temperature change first, and then, the apparent "phase transition temperatures" are extrapolated to a phase transition temperature with an infinitely slow rate of temperature change, using it to approach the true phase transition temperature on the binode line. It can also be extrapolated to an infinitely fast rate of temperature change to get a point on the spinode line. Changing the solution concentration and repeating the measurement can finally obtain a binode curve and a spinode curve of a binary mixture. This method is very tedious and time-consuming. Therefore, it is often used with only one fixed rate of temperature change to measure the turbidity change to obtain a rough phase transition temperature. Such obtained phase diagrams are sufficient to deal with and solve most application problems. As far as the author knows, there is no measurement and data about spinode curve of a macromolecular solution in the literature.

In order to obtain a true phase diagram of a solution of macromolecule, a two-phase equilibrium method has to be adopted. In the one phase region, a

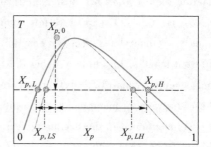

macromolecular solution with its composition as close to the critical composition as possible is prepared first, and then, the temperature is slowly lowered a temperature in the two-phase region, keeping a constant temperature, and the fluctuation must be below ±0.01 ℃. When the two phases are completely separated and a dynamic equilibrium is reached, the refraction or other optical methods is used to measure the concentration of macromolecules in each phase to obtain two points ($X_{p,L}$ and $X_{p,H}$) on the binode line, as shown above. By change the temperature and repeating the above experiment, one can map the phase diagram (the binode line), where the concentrations of the two phases are determined by the two intersection points between the horizontal temperature and binode line. The relationship between the volume of the two phases (V_L and V_H) and the total volume (V) is determined by the "lever rule" (similar to the equal moment principle in a seesaw); namely,

$$V_L = \frac{x_{p,H} - x_{p,0}}{x_{p,H} - x_{p,L}}V \text{ and } V_H = \frac{x_{p,0} - x_{p,L}}{x_{p,H} - x_{p,L}}V \tag{10.4}$$

This equilibrium method is extremely time-consuming. The reason is that when the solution is separated into two layers of a dilute phase and a concentrated phase, the diffusion coefficient of a typical macromolecular chain in the dilute phase is about 10^{-7} cm^{-2}/s. Therefore, the time to diffuse a macroscopic distance ($\sim 10^{-1}$ cm) is about 10^5 s, 1–2 days; while the concentrated phase is already a semidilute solution. Based on the discussion in the previous Chapter, it takes about 1 s for the centroid of a macromolecular chain to diffuse a chain size ($\sim 10^{-5}$ cm); and so on, it takes about 10^8 s to diffuse $\sim 10^{-1}$ cm, about 3 years, extremely slow. Although the concentrated phase begins with the chain aggregation and the size of the aggregates is still small, but the time to reach an equilibrium still needs weeks or months. To obtain a good phase diagram, one may need to use years as the time unit. Xia, An and Shen [Journal of Chemical Physics, 1996, 105:6018] spent about two years to map five phase diagrams of poly(methyl methacrylate) in 3-octanone by using this method.

The long equilibration time is only one obstacle to obtaining a true phase diagram of a macromolecular solution. Based on the above discussion, the phase diagram of a macromolecular solution is also dependent on the chain length (N, the segment number), reflected in the value of n_v. Therefore, the macromolecule sample used to map the phase diagram must have a narrow chain length distribution ($M_w/M_n < 1.1$). The study of the chain length dependence of the phase diagram requires narrowly distributed samples with different chain lengths, which must differ by at least ten times. Except for a few commercial standard samples, such as polystyrene and poly(methyl methacrylate), it is not easy to obtain a set of narrowly distributed macromolecules with different lengths. In addition, using the traditional phase separation and equilibrium method, for each sample, at least 0.5–1.0 grams are required, which is even more difficult. This is why true phase diagrams of macromolecular solutions are rare in the literature.

As mentioned in the beginning, it is the chain length dependence that creates macromolecular physics. Compared with small molecules, the chain length dependence of the phase diagram of macromolecules introduces new physics. As early as the 1950s, physicists discovered that for the phase diagram of small molecules, near the critical point (temperature T_c and mole fraction x_c) of a binary mixture, the difference between concentrations of the two phase ($|x_H - x_L|$) and the reduced temperature ($\varepsilon = |T_c - T|/T_c$), there is a universal scaling relationship that is independent of specific substances in a binary mixture: $|x_H - x_L| \sim \varepsilon^\beta$. The mean field theory predicted that $\beta = 1/2$. However, after years of theoretical and experimental research, at the end of the 1960s, it was finally confirmed through experiments that $\beta = 1/3$, a universal constant, called the Ising exponent.

The volume of macromolecules and solvent molecules are very different. Does this universal dependence of the concentration difference on the reduced temperature still hold true? If so, what is the scaling exponent? In the Ising model, the spatial and order parameter dimensions of small and macromolecules should be the same, which are 3 and 1, respectively. Therefore, such a scaling relationship should remain unchanged. Limited experimental data basically supported this theoretical judgment. However, there are only few real, high-quality phase diagrams of macromolecules in literature, so that no conclusion had been drawn for a long time. In addition to the dependence of the concentration difference on the reduced temperature, another question is whether there is a universal dependence of the concentration difference on the chain length in macromolecular solutions. If yes, a universal dependence of the concentration difference on the reduced temperature and the chain length can be written as

$$X_{p,H} - X_{p,L} \sim \varepsilon^\beta N^\xi \qquad (10.5)$$

where ξ is also a universal constant, independent of the properties of specific macromolecules and solvents. In nearly four decades of research, different theories predicted different chain length dependencies. There is no enough and convincing experimental data to be able to distinguish between false and true.

Earlier days, according to the mean field theory, Flory [Principles of Polymer Chemistry, Cornell University Press: Ithaca, NY, 1953] showed that $\beta = 1/2$ and $\xi = -1/4$. Its defects have been confirmed by later theories and experiments, especially $\beta = 1/3$ has been confirmed and widely accepted. Later, de Gennes [Scaling Concepts in Polymer Physics, Cornell University Press: Ithaca, NY, 1979] pointed out that for a sufficiently long chain, the critical state of a solution of macromolecule is close to the Θ state. The two-body interaction can be ignored, only the three-body interaction should be considered, and $\xi = (\beta - 1)/2$. If substituting $\beta = 1/2$ from the mean field into it, $\xi = -1/4$, consistent with the prediction from Flory. If substituting the correct $\beta = 1/2$ into it, $\xi = -1/3$. Later, Muthukumar [Journal of Chemical Physics, 1986, 85: 4722] followed this line of thinking, taking into account the difference between solutions of small and macromolecules, that is, macromolecules not only have the same intermolecular density fluctuation induced by the motion of centroid as small molecules, but also an intrachain density fluctuation caused by the relative movement of each segment in the chain, lacking in the solution of small molecules. Since it has the same mixing entropy as the mean field theory, one only needs to consider the difference in the mixing enthalpy, $\Delta H_{mix}/RT = X_p(1 - X_p)\chi_{sp}$, i.e.,

$$\frac{\Delta H_{mix}}{RT} = \left(\frac{1}{2} - A_2\right)X_p(1 - X_p) + \left(A_3 - \frac{1}{6}\right)X_p^3 - (A_2 + A_3X_p)\alpha^6X_p^2 + \frac{512\pi^2\alpha^6}{3}X_p^3 \quad (10.6)$$

where the first two items and the last two items respectively reflect the density fluctuations induced by the relative interchain (centroids) and intrachain (segment) movements between different chains and between different segments. α is the expansion factor, see eq. (8.2). In a solution of macromolecule, when $N \to \infty$, $T_c \to \Theta$, the two-body interaction (A_2) nearly disappears. If only the three-body (A_3) interaction is considered, the above equation has an analytical solution with $\beta = 1/3$ and $\xi = -2/9$. For details, readers can read the original paper.

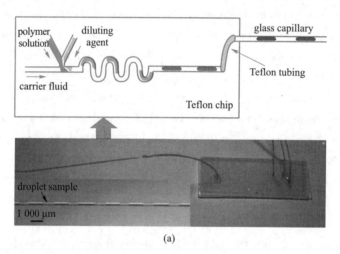

(a)

Shangguan YG, Guo DM, Feng H, et al. Macromolecules, 2014, 47:2496. (Figure 2, permission was granted by publisher)

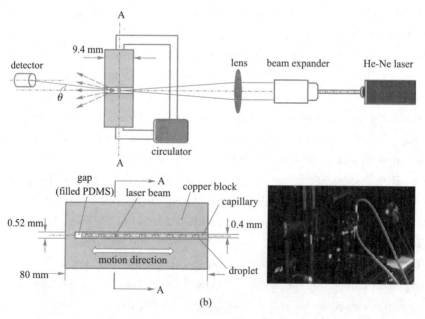

(b)

Shangguan YG, Guo DM, Feng H, et al. Macromolecules, 2014, 47:2496. (Figure 3, permission was granted by publisher)

Although there were different theoretical predictions, in the late 1980s, theoretical discussions in this area were gradually shelved. A limited number of experimental results show that $-1/3 \leqslant \xi \leqslant -1/10$, inconclusive. The main reason is still the lack of sufficient and credible experimental data(phase diagrams of macromolecules) to distinguish, judge and confirm different theories. Obviously, there is only one correct theory, but it is also possible that all existing theories are incorrect. Until recently, using a microfluidics method developed in structural biology to find protein crystallization conditions in a nanoliter solution (10^{-9} mL) , there was a breakthrough in the experiment. Since this is one of the latest and most important results in the study of solutions of macromolecules, for readers who are interested, a special introduction is as follows.

As shown in the figure(a) above, a microfluidic "chip" made of PTFE is engraved with a microchannel with a cross-section of 0.02×0.02 cm^2. It has three inlets, corresponding to macromolecular solution, solvent and a carrier, a fluorine-containing liquid, immiscible with solvent, and one outlet. The liquid intake of each inlet is accurately measured by a computer-controlled propulsion pump. When the sum of the infused amount of the macromolecular solution and the diluent solvent is close to that of the carrier liquid, the carrier liquid breaks the macromolecular solution into one droplet after another, and there is a small segment of encapsulated carrier liquid between each two, effectively preventing the mixing of different droplets. The volume of each droplet is $\sim 10^{-8}$ L, or ~ 10 nL. If the infused amounts of the macromolecular solution and the diluent solvent are gradually reduced and increased at the same time, the concentration of macromolecules in each droplet will gradually decrease to form a series of droplets with decreasing concentration. Through the outlet, they are injected into a glass capillary tube. Theoretically, a string of 50 droplets with a volume of ~ 10 nL and equidistantly decreasing macromolecule concentrations can be produced with only 1 microgram of sample.

As shown in the figure(b) above, this glass capillary tube containing dozens of droplets is inserted into a copper block with a precise temperature control and a slit window on each side. The copper block is mounted on a precision parallel moving platform driven by computer-controlled stepping motor. The position of the copper block is adjusted so that an expanded laser beam is focused on the droplet in the capillary through the slit window, and the capillary is perpendicular to the incident laser. On the other side of the copper block, a light intensity detector is positioned at a small angle ($\sim 4°$) to the incident light. After reaching a constant temperature, the platform is moved along the direction perpendicular to the laser beam, so that the laser bean can scan each droplet in turn, its scattered light intensity(I) is measured first, and then, divided by its concentration(C).

It is known from the principle of static LLS that when the scattering angle approaches zero, the ratio (I/C) is proportional to the weight-average molar mass, that is, proportional to the product of the square of mass of the scattering particle and the number of particles, $I/C \sim M^2 N$ means that even if the macromolecular chains in the solution appear are dimerized(the solution is still clear and transparent), the molar mass and the number of particles will each increase and decrease by a factor of two, resulting in a doubling of the scattered light intensity. For slight aggregation of macromolecular chains in the two-phase region, the scattered light intensity at a small angle is much more sensitive than the transmitted light in the turbidity method. After reaching a given temperature, macromolecules remain a single-chain state in the one-phase region, and I/C does not change with time; while in the two-phase region, due to inter-

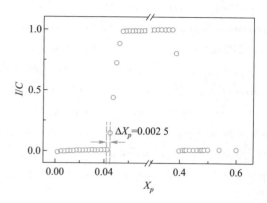

Shangguan Y G, Guo D M, Feng H, et al. Macromolecules, 2014, 47:2496.

(Figure 6, permission was granted by publisher)

chain aggregation(note: the aggregate's size may be much smaller than the critical size), I/C will increase significantly, and it continues to increase over time, as shown above, here is a solution of polyvinyl acetate in isobutanol.

Each small circle in the figure represents the ratio of the scattered light intensity of a droplet and its concentration. Therefore, the droplet scanning is completely equivalent to a variation of the macromolecular concentration in the phase diagram for a given temperature. Since the chains aggregate in the two phase region, two concentrations at which the scattered light intensity starts to increase mark where the solution enters and leaves the two-phase region. Therefore, at this given temperature, two points on the binode line are obtained. In the two-phase region, the induction time of the phase separation becomes shorter with the distance from the binode line increases, the aggregation speeds up, and I/C increases. Once inside the spinode line, the induction time becomes zero, so that the difference of I/C of each droplet in this region is not big, and a platform appears, and its two end points correspond to two points on the spinode line. Therefore, at each temperature, one scan results in two points outside and two points inside, a total of four points. One can change the temperature, repeat the scanning, and connect the outer two points and the inner two points obtained at different temperatures to obtain the phase diagram(binode line and spinode line)of macromolecular solutions. One of the advantages of this novel method is that there is no need to wait for the two phases to reach dynamic equilibrium, and it can quickly and accurately map the phase diagram of a macromolecular solution.

The figure below shows the solution phase diagrams of five poly(vinyl acetates)of different lengths in benzene obtained by this method, where N is the degree of polymerization(number of monomers). Obviously, the combination of microfluidic technology and low-angle LLS to map the phase diagram of macromolecular solutions has advantages that traditional methods are not able to match: Firstly, it is easy to get few milligrams narrowly distributed macromolecular samples with different molar masses for the above experiments by using a preparative size exclusion chromatography column, solving the sample preparation problem; secondly, there is no need to wait months or years for the two phases to reach a dynamic equilibrium, so that the phase diagrams of different macromolecular solutions can be mapped quickly and accurately.

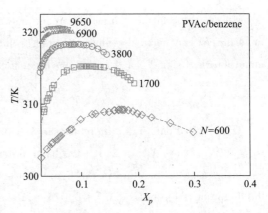

Wu C, Li Y. Macromolecules, 2018, 51:5863.

(Figure 2, permission was granted by publisher)

Theoretically, if there are enough narrowly distributed macromolecule samples with different chain lengths, this method might not be needed. One can prepare dozens of macromolecular solutions with different concentrations directly, place them in a constant temperature water tank, and then use small-angle LLS device to measure their scattered light intensity to obtain the same result. However, in addition to the problem of sample preparation, thermostatting and measuring dozens of solutions in a water tank is another experimental problem. Another advantage of using microfluidics is that the size of each droplet is at least ten times smaller than that of conventional solutions. In addition to reaching a given temperature quickly, the more important is that the time for the two phases to reach an equilibrium is shortened by more than one hundred times, from three months to one day, which is quite impressive, although it is not necessary to reach the equilibrium in the measurement of this method. Note that the droplet size should not be too small, otherwise the interface layer (< 1 μm) will affect the experimental results.

The figure above clearly shows that the phase diagrams of macromolecular solutions have been severely asymmetrical, and the critical volume fraction (X_c, the highest point of the binode line) has deviated far from 0.5. The deviation increases with the chain length. The critical volume fraction shifts to the left and the critical temperature moves up. Theoretically, when $N \rightarrow \infty$, $T_c \rightarrow \Theta$. There is a scaling relationship between X_c and N: $X_c \sim N^{-\gamma}$. Using the least squares fitting method to analyze the data in the figure above, one can get X_c of each phase diagram. A combination of 17 phase diagrams derived from solutions of four different macromolecules in different solvents leads to $\gamma = 0.37 \pm 0.01$. In order to facilitate the analysis of these severely skewed phase diagrams, Sanchez [Journal of Applied Physics, 1985, 58:2871] introduced a dimensionless order parameter (Ψ) to make the asymmetric phase diagram re-symmetric.

$$\Psi = \frac{X_p}{X_p + R_c(1 - X_p)} \tag{10.7}$$

where for a given macromolecular solution, R_c is a fitting parameter in the data analysis, which decreases as the chain length increases. Sanchez pointed out in the above-mentioned literature that the physical meaning of R_c is unknown. Obviously, when $R_c = 1$, $\Psi = X_p$. Since the asymmetry is due to the volume difference between one macromolecular chain and one solvent molecule, R_c can make the phase diagram symmetric again, so that it must be related to the chain length. Using $X_p + X_s = 1$, the above equation can be rewritten as

$$\Psi = \frac{V_p}{V_p + R_c V_s} = \frac{V_p/V_s}{V_p/V_s + R_c}$$

The limited data in the literature shows that R_c is related to the critical volume fraction, that is, $R_c \approx 2X_{p,c}/(1 - X_{p,c}) = 2(V_{p,c}/V_{s,c})$. In order to perform the symmetry operation, one needs to convert each temperature in the figure above to a reduced temperature (ε) first according to the definition, then use $\varepsilon_{max} N^{0.3} \leqslant 0.075$ as a standard to select data that is sufficiently close to the critical point in each phase diagram, and finally get Ψ_c and R_c for each phase diagram when it is made symmetrical in the Ψ space by the fitting. The average values of Ψ_c and R_c can be calculated, respectively. A combination of all the phase diagrams and fit them together to obtain the corresponding average Ψ_c and R_c. With the aid of computers, the latter is now usually used to improve the reliability and stability of the fitting. The results show that $\Psi_c = 0.325 \pm 0.002 \approx 1/3$, independent of the chain length. The left figure below shows the symmetrized results of a set of phase diagrams. Further using the following equation,

$$|\Psi - \Psi_c| = \Psi_0 (\varepsilon N^b)^\beta \tag{10.8}$$

For 12 phase diagrams of three types of macromolecules in different solvents, the three-variable Levenberg–Marquardt fitting results reveal that Ψ_0 is independent of the chain length, but varies with macromolecules and solvents. Both b and β are indeed independent of both the chain length and specific properties of a macromolecular solution, similar to the results obtained by Xia, An and Shen from 5 phase diagrams of one kind of polymer solutions. Therefore, a refitting of 17 different phase diagrams together leads to $b = 0.47 \pm 0.02$, slightly smaller than $1/2$ predicted by the mean field theory; and $\beta = 0.328 \pm 0.009$, close to $1/3$ predicted by the Ising model.

It is known from the above equation and the figure (a) below that after such a symmetry operation, Ψ takes Ψ_c as its axis of symmetry, so that all the data points on the left of Ψ_0 can be mirrored to the right. The plot of εN^6 versus $|\Psi - \Psi_c|/\Psi_0$ is approximately a cubic curve, $\varepsilon N^b = (|\Psi - \Psi_c|/\Psi_0)^{1/\beta}$, as shown in the figure (b) below. Note: the ordinate increases from top to bottom. In the figure, each of the symbols represents one type of solutions; each kind of solutions includes 4–5 different chain lengths, 4 different kinds of solutions, a total of 17 phase diagrams. They are superimposed into one curve, fully confirming the universality of b and β.

Experimentally, it is necessary to convert Ψ in the symmetrical space back to X_p in the volume fraction space before it can be further converted to the measurable mass (weight) concentration. According to the definition of Ψ in the equation above, one can expand it around $X_{p,c}$ to get

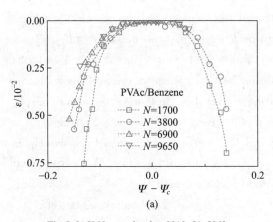

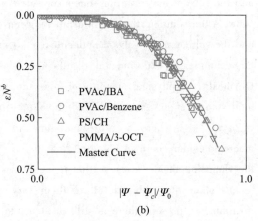

(a)

(b)

Wu C,Li Y.Macromolecules,2018,51:5863.   Wu C,Li Y.Macromolecules,2018,51:5863.

(Figure 3,permission was granted by publisher)   (Figure 4,permission was granted by publisher)

$$\Psi = \Psi_c \left[1 + \frac{R_c \Psi_c}{X_{p,c}} \left(\frac{X_p}{X_{p,c}} - 1 \right) + \cdots \right] \approx \Psi_c \left[1 + \frac{\Psi_c}{1 - X_{p,c}} \left(\frac{X_p}{X_{p,c}} - 1 \right) + \cdots \right]$$

For a given macromolecular solution, Ψ_c and $X_{p,c}$ are two constants. In theory, when the chain is sufficiently long, the phase diagram of a binary mixture is symmetrical in the range of sufficiently close to the critical point, $\Psi_H - \Psi_L = 2 | \Psi - \Psi_c |$ and $X_{p,H} - X_{p,L} = 2 | X_p - X_{p,c} |$.

$$| \Psi - \Psi_c | = \frac{\Psi_c^2}{1 - X_{p,c}} \frac{| X_p - X_{p,c} |}{X_{p,c}} \sim \frac{| X_p - X_{p,c} |}{X_{p,c}}$$

where $X_{p,c} \sim N^{-\gamma}$. Substituting $\beta = 0.328$, $b = 0.47$ and $\gamma = 0.37$ in a combination of the above equation and eq. (10.7) results in

$$| X_p - X_{p,c} | = X_0 \varepsilon^\beta N^{b\beta - \gamma} = X_0 \varepsilon^{0.328} N^{-0.216} \qquad (10.9)$$

where X_0 is a system-dependent proportional constant. **This is the universal scaling relationship of binary mixtures**. Although types of binary mixtures are ever-changing, their phase diagrams near the critical point only depend on two physical invariants(universal scaling exponents), and have nothing to do with specific substances that make up a binary mixture, showing the beauty of physics! The scaling exponent($\xi \approx -0.22$) between $| X_p - X_{p,c} |$ and the chain length obtained from the experimental results is very close to $-2/9$ predicted by Muthukumar.

Physically, when macromolecular chains are sufficiently long, binary mixtures near the critical point approach the Θ state, and the two-body interaction nearly disappears. Near the critical point, the density strongly fluctuates and the correlation length significantly increases. Therefore, it is reasonable to assume that the originally weak three-body interaction becomes dominant.

Let us examine the physical meaning of R_c. The mean field theory only considers the two-body interaction. After the symmetry operation, the volume difference between macromolecule and the solvent molecule has been corrected, so that Ψ_c should be $1/2$; namely, in the critical state, in the solution, the relationship between R_c and the volume of all macromolecules ($V_{p,c}$) and the volume of all solvent molecules ($V_{s,c}$) must be $R_c V_{s,c} = V_{p,c}$, or written as, $R_c = V_{p,c}/V_{s,c}$. A large number of experimental results have shown that $\Psi_c \approx 1/3$; it means $R_c = 2(V_{p,c}/V_{s,c})$. Obviously, the physical meaning of R_c is the ratio of the total volume of macromolecules to the total volume of solvent molecules in a solution in the critical state. In terms of the molar ratio, $R_c = 2n_v n_{p,c}/n_{s,c}$, where n_v represents how many small solvent molecules are equivalent to a macromolecular chain in volume. Note that $V_{p,c}$ and $V_{s,c}$ decrease and increase respectively as the chain length increases, so that R_c decreases as the chain length increases. The experimental results show that although all of them vary with the chain length, $R_c V_{s,c}/V_{p,c}$ is a physical invariant. At present, there is still no theory to be able to explain this result quantitatively.

In addition to its theoretical importance, the above scaling relationship has a wide range of applications. From now on, to map the phase diagram of a macromolecular solution, it is no longer necessary to prepare dozens of solutions with different concentrations and then measure their phase transition temperatures. According to the above equation, there are only three system-dependent unknowns parameter: $X_{p,c}$, X_0 and T_c. Therefore, in order to map the phase diagram of any macromolecular solution, one only needs to prepare three solutions with different concentrations (X_p) near the critical point, measure their respective phase transition temperatures, substitute the results into the above equation, and solve the three equations together to get three unknowns, and draw the phase diagram.

For macromolecules that can be crystallized in solution, the phase diagram is more complicated. Let us only use the solution with UCST as an example. If the crystallization temperature is higher than the phase transition temperature and the concentration is sufficiently high, as the temperature drops to the crystallization temperature, the macromolecular solution will begin to separate into two phases due to crystallization, one phase is macromolecular crystals (pure macromolecules, the right axis in the phase diagram), and another phase is a saturated solution of macromolecules in solvent. When the temperature further decreases, the amount of crystals in the crystalline phase increases, but the concentration remains unchanged (it has been 100% and cannot be higher), and the saturation concentration in the other phase decreases.

When the crystallization temperature is lower than the phase transition temperature, the solution will first undergo a phase transition, separating into two saturated solutions of dilute phase and concentrated phase. As the temperature drops to the crystallization temperature, crystals appear in the concentrated phase. Three phases coexist in the solution, two saturated solutions plus the crystal phase. In the solution, there are two dynamic equilibriums: between the two saturated solutions and between the concentrated phase and the crystalline phase. The phase diagrams of solutions of copolymers, polyelectrolytes, and with a mixture of solvents are more complex and beyond the scope of this book. Interested readers can refer to related monographs and literature.

Summary

The solvent quality changes as the temperature or pressure varies. The gradual deterioration of the solvent quality makes a homogeneous macromolecular solution(one phase)to separate into two homogeneous macromolecular solutions(two phases: dilute and concentrated phases), called **phase transition**. The dilute phase is a saturated solution of macromolecules in solvent; and the concentrated phase is a saturated solution of solvent in swollen macromolecules. When the two phases reach a dynamic equilibrium, microscopically, solvent molecules and macromolecules shuttle back and forth between the two phases at the same rate constantly and the macroscopic concentration remains unchanged. The volume of the two saturated solutions can be obtained by the "lever rule" with the help of the phase diagram. Whether a phase transition can happen depends on the increase or decrease of the mixing free energy, including a combined effect of the mixing enthalpy and the mixing entropy. For an isothermal and isobaric process, a system always tends to the direction of spontaneously decreasing its free energy. Since the pressure effect on the macromolecular solution is small, only the temperature change will be used as an example.

In the phase diagram of temperature versus composition, the connection of the phase transition temperature corresponding to each composition leads to **the binode line** that divides the one-phase region and the two-phase region. The two-phase region is further divided into two regions based on whether the induction time is required to start the phase transition. The connection of the points at which the induction time just approaches zero results in another line, called **the spinode line**, which corresponds to the point at which the second derivative of the mixing free energy with respect to the mole fraction of macromolecule is zero. Therefore, for a given temperature, a gradual increase of the concentration can lead a solution to pass by a binode point($X_{p,L}$), two spinode points($X_{p,LS}$ and $X_{p,HS}$), and another binode point($X_{p,H}$).

There is a metastable region between thebinode line and the spinode line. Although free energy still decreases in this region, free energy can be further lowered if the solution is separated into two phases, so that the solution tends to the phase transition, but requiring a "nucleus" larger than the critical size. The induction time required for the phase separation to start is determined by the relative magnitude of the interaction energy between macromolecules and the interfacial energy, decreasing as the distance away from the binode line increases. When on the binode line, it is infinite; when on the spinode line, it is zero. Experimentally, since there are always impurities(dust, bubbles, etc.) that can act as "nuclei" in the solution, the phase separation always starts in a very short time, but the time to reach the dynamic equilibrium of the two phases is determined by the translational diffusion of the macromolecular chains. Since the diffusion of macromolecules in the semidilute solution is extremely slow, the time required to reach the dynamic equilibrium can be months and years, depending on the chain length and concentration.

Generally, the mixing entropy is positive. According to the principle of thermodynamics, in the isothermal process, $\Delta G_{\mathrm{mix}} = \Delta H_{\mathrm{mix}} - T\Delta S_{\mathrm{mix}}$, so that heating helps the dissolution, and there is an upper critical solution temperature (UCST). Above this temperature, ΔG_{mix} is always negative, so that macromolecules and solvents can spontaneously mix with each other in any ratio, forming a homogeneous one-phase binary mixture (solution). When there is a strong preferential adsorption between macromolecules and solvent molecules, the translational entropy of solvent decreases, and the mixing entropy may become negative, resulting in a lower critical solution temperature (LCST). Below this temperature, ΔG_{mix} is always negative, and macromolecules and solvent spontaneously mix with each other in any ratio to form a solution. Some macromolecular solutions can have both upper and lower critical solution temperatures, e. g., polystyrene in cyclohexane. Obviously, the experimentally observable upper and lower critical solution temperatures must be lower and higher than the boiling point and freezing point of the solvent, respectively.

When the temperature gradually approaches the upper or lower critical solution temperature, the above four points ($X_{p,L}, X_{p,LS}, X_{p,HS}$ and $X_{p,H}$) tend to one point: the critical state with a critical temperature (T_c) and a critical composition ($X_{p,c}$), both the first and third derivatives of the mixing free energy with respect to composition are zero. Near the critical point, the concentrations of the dilute phase and the concentrated phase are symmetrical to $X_{p,c}$, that is ($X_{p,H} - X_{p,L}$)/2 = $X_{p,H} - X_{p,c} = X_{p,c} - X_{p,L}$. It has been confirmed that there is a universal scaling relationship between such a concentration difference and the reduction temperature ($\varepsilon = |T - T_c|/T_c$) and the chain length (the segment number, N), $|X_p - X_{p,c}| = X_0 \varepsilon^\beta N^\xi = X_0 \varepsilon^{0.328} N^{-0.216}$, which is independent of a specific solution (macromolecule and solvent) and close to the theoretically predicted $|X_p - X_{p,c}| \sim \varepsilon^{1/3} N^{-2/9}$.

The following figure summarizes the content discussed in this chapter, including the origin of the phase diagram of a binary mixture and the universal scaling relationship in the phase diagram.

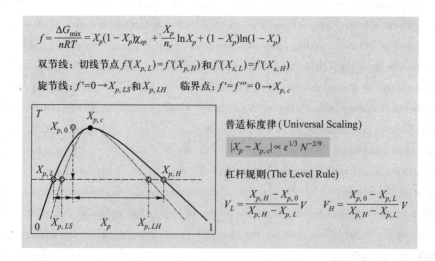

第十一章　大分子溶液中的介观球相

　　如前所述,对均聚物而言,随着温度逐渐下降(UCST)或升高(LCST),溶剂性质逐渐变差,先在一相区内发生从"线团至小球"构象转变,单链蜷缩。进一步改变温度导致在两相区内出现链间聚集,形成宏观的稀、浓两相。两相的浓度差别随着溶剂性质变差而逐渐增加。共聚物在溶剂中的构象也会随溶剂性质变化,但其相变复杂,不仅取决于各个共聚单体的化学性质和它们在链上的具体排列方式,而且还会随着相变"路径"变化,包括变温速率、改变溶剂性质或者浓度的方式等。

　　例如,如果将少量羧化的聚苯乙烯离聚物在四氢呋喃(THF)中的稀溶液迅速地加入大量的水中,由于离聚物不溶于水,混合物立即变成一个浑浊的悬浮液,意味着形成了尺寸大于可见光波长的聚集体。然而,如果将同样的聚苯乙烯离聚物稀溶液,一滴一滴慢慢地加入大量的水中,混合液体仍然清澈透明。问题是不溶于水的聚苯乙烯离聚物链去了何处? 激光光散射的研究揭示,原来它们在水中形成了尺寸远小于可见光波长、几十纳米、仅含有数十根链的稳定聚集体。其平均尺寸和每个聚集体中大分子链的平均数目(聚集数)随着羧化度的增加而递减;但随着离聚物在THF中的起始浓度增加,如右图所示。定量的实验结果显示,随着起始浓度从2.0×10^{-2} g/mL 降到 6.2×10^{-4} g/mL,平均流体力学半径从 18 nm 降至 8 nm;平均聚集数从约为 100 减至约为 18。

　　问题是为何这些不溶于水的离聚物不继续聚集、分离,直至形成宏观的两相(稀相和浓相)。定性的解释是,在相分离中,带负电荷、亲水的羧基倾向于停留在聚集体的表面,而疏水的聚苯乙烯主链则聚集、蜷缩在中间,形成很多个具有核壳结构的粒子。对给定的羧化度,每个聚集体表面的羧基数目(n)正比于聚

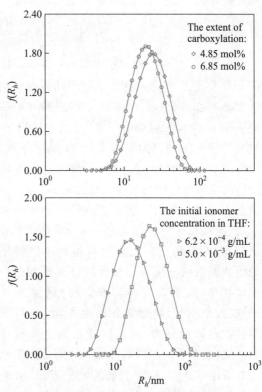

Li M,Jiang M,Zhu L,et al.Macromolecules,2004,37:4989.

(Figures1 and 2,permission was granted by publisher)

集度,也正比于其体积(V)或其尺寸(R)的三次方(密度均匀),而每个聚集体的表面积(S)则正比于其尺寸的平方。因此,每个羧基所占据的表面积(s)为

$$s = \frac{S}{n} \sim \frac{R^2}{R^3} = \frac{1}{R} \tag{11.1}$$

因此,在聚集中,s 随着 R 的逐渐增大而递减。对于每种给定的带电荷基团,s 有一个极小值。达到该值,进一步聚集就变得困难,尽管宏观分相可进一步降低溶液的自由能。实验数据显示,每个离子基团所占的最小平均表面积约为(3.0 ± 0.2) nm^2。另一方面,流体力学半径分布对离聚物起始浓度的依赖也说明上式非控制聚集结构的唯一条件。

理论上,Timoshenko 和 Kuznetsov[Journal of Chemical Physics,2000,112:8163]提出了一个模型,假定溶液的体积是 L^3,有 N_p 条链,每条链有 N 个长度为 b 的链段,并略去所有溶剂分子的自由度,体系有效的哈密尔顿为

$$H = \frac{k_B T}{2L^2} \sum_a (\mathbf{Y}^a - \mathbf{Y})^2 + \frac{k_B T}{2b^2} \sum_a (\mathbf{X}_n^a - \mathbf{X}_{n-1}^a)^2 + \sum_{J > 2} \sum_{\{A\}} A_J^{|M|} \prod_{i=1}^{J-1} \delta(\mathbf{X}_{M_{i+1}} - \mathbf{X}_{M_1})$$

其中,$\mathbf{X}_n^a$ 是第 a 条链上第 n 个链段的坐标,$\mathbf{Y}^a \equiv (1/N) \sum_n \mathbf{X}_n^a$ 和 $\mathbf{Y} \equiv (1/M) \sum_a \mathbf{Y}^a$ 分别是链和体系的质量中心的坐标,$A_J^{|M|}$ 是一套依赖位置的维里系数。

先以仅含有两个不同单体的无规共聚物为例,并利用一个在高斯变分法中通用的二次形式 $H_0 \equiv (1/2) \sum_{M_1, M_2} V_{M_1, M_2} \mathbf{X}_{M_1} \mathbf{X}_{M_2}$ 作为一个尝试哈密尔顿,并且将 V_{M_1, M_2}(有效势能)排除在外,可得到针对二元共聚物的二阶维里系数矩阵的一个特殊参数化结果。

$$A_2^{M_1, M_2} = A_2 + \frac{\sigma_{M_1} + \sigma_{M_2}}{2} \Delta \tag{11.2}$$

其中,A_2 为平均二阶维里系数,σ_{M_1} 和 σ_{M_2} 分别反映了 M_1 和 M_2 两个共聚单体在链上具体的分布方式,二者之和除以 2 是平均分布。给定共聚单体后,M_1 和 M_2 在一个溶剂中的溶解度之差(Δ)可大可小,取决于溶剂和由 M_1 和 M_2 组成的链上单元的相互作用。在典型的例子中,一个单元可溶解于水中(亲水),而另一个则只溶于与水非共混的"油"(有机溶剂)中(亲油),这也是为何将 Δ 称作"两亲性程度"。

如果 $\Delta = 0$,两个单体一样,共聚物就成了均聚物,即下图左轴。当 Δ 较小时,无规共聚物链的溶液行为和均聚物相差无几,即只存在两种状态,一相(一个均匀溶液)或两相(两个饱和溶液)。只有在两亲性程度足够高时,随着溶剂性质逐渐变差(A_2 变得更负),方有可能在单链蜷缩和宏观相分离之间出现一个额外的介观球相区。在介观球相中,有限数目(n_a,聚集度)的链聚在一起,形成大小不一、亚稳态球状聚集体(也可为非球形),出现微相分离,其范围随着两亲性程度增加而扩展,如图中的阴影部分和其局部放大所示。

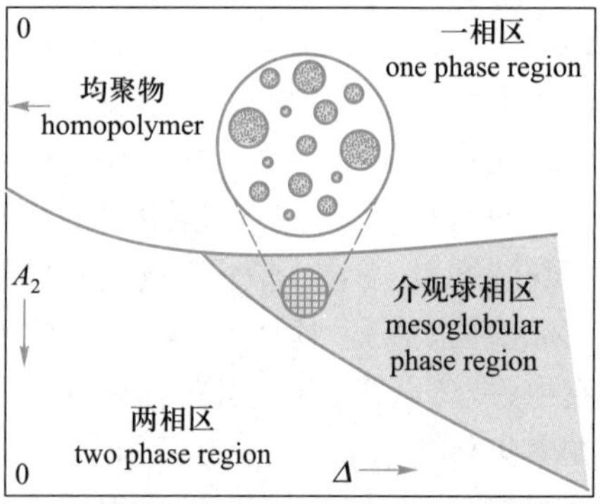

对给定的 Δ,当 A_2 低于该区域上限时,溶液中的共聚物链开始聚集,即微相分离,在溶液中形成许多与蜷缩单链共存的小的微浓相(聚集体)。原理上,这些小的微浓相将会进一步聚集最终形成两个宏观相。但在有限的时间

内(可以很长,经年累月),它们将保留在溶液中。严格地说,共聚物与均聚物分相在本质上并没有区别,只是共聚物聚集体能在亚稳态的时间停留很长而已。现今,为吸引眼球,该过程常被称为"自组装",形成的微浓相(小聚集体)也常被叫作各种"纳米"粒子。

注意:"聚集"是物理过程,而"聚合"则是化学过程,不可将二者混淆。在前述讨论中,有关"线团至小球"的构象转变论及的是在一相区内形成热力学稳定的均聚物单链小球。而在介观球相中形成的共聚物粒子则是处在非热力学稳定的亚稳态,其聚集形成的结构与路径有关。一方面,对于给定的 Δ,聚集数则随着溶剂性质变差而增加;另一方面,对于给定的溶剂性质(温度),聚集数通常随着 Δ 的增加而逐渐降低。在适当的条件下,也可在两相区内生成单链聚集体($n_a \to 1$),但其和"线团至小球"构象转变的问题无关。许多蛋白质大分子就是典型的例子,它们可以在细胞内沿着一定路径折叠成单链结构;也可以在细胞外通过特殊的方法(路径)重新折叠成单链的活性构象。一般而言,形成共聚物单链聚集体的条件相当苛刻,比蛋白质难很多。

Timoshenko 和 Kuznetsov 利用计算机模拟分别计算了 12 条组成相同、各含有 6 个 M_1 链段和 6 个 M_2 链段,但组构不同的"大分子"链:$[M_1 M_2]_6$ 和 $[(M_1)_2 (M_2)_2]_3$。结果显示,如果两个链段一样(均聚物),$\Delta = 0$,在一相中,以单链存在时的混合自由能最低;在两相中,以 12 条链聚集在一起时的混合自由能最低。当两亲性程度 Δ 增大时,在两相区,仍以 12 条链的聚集体具有最低的混合自由能。而在介观球相中,随着溶剂性质变差,聚集度增加,其相应的混合自由能也最低。但在聚集度不同的聚集体之间,混合自由能的差别很小,故可同时存在不同聚集度的聚集体,清楚地显示在介观球相中,聚集体处于亚稳态。

下面,利用两个真实的例子进一步说明为何在介观球相中共聚物的聚集受控于非热力学因素。第一个例子是两亲性共聚物在选择性溶剂中的聚集行为对两亲性程度和链长的依赖性。在计算机模拟中,很容易在同一个共聚组构中调节两亲性程度和链长。实验上,所需的共聚物一共含有四个变量:组分、组构、两亲性程度和链长。在合成实验中,保持三个变量不变,只改变其中一个变量,得到四组不同的共聚物,如非不能,则十分难也。在共聚反应中,仅实现组分和组构相同就已相当不易。原则上,可以固定一种单体,通过采用不同的另一单体实现调节共聚物两亲性程度,但如何保持其他三个变量相似?

为了解决这一问题,实验上,采用聚环氧乙烷大单体(PEO)和 N-异丙基丙烯酰胺分别在纯去离子水中以及水和弱链转移剂(乙醇)的混合溶剂中进行同样条件下的无规共聚。并在单体转化率小于 10% 时,停止反应,以控制共聚单体在链上分布的均匀性。这样,得到不同链长(相差 10 倍),但组分和组构接近的两个共聚物,如右图所示。由于聚 N-异丙

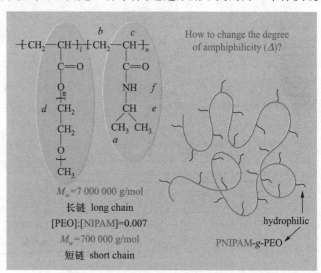

基丙烯酰胺(PNIPAM)是一个热敏性聚合物,通过变温可调节两亲性程度。即,在25~50 ℃的范围内,PEO始终亲水;而PNIPAM具有一个LCST,故其可在~32 ℃从亲水迅速地变成疏水。因此,升高溶液温度可增加两亲性程度和诱导介观球相的生成。

从前一章的讨论中已知,依据热力学原理,对给定种类的大分子溶液和质量浓度,随着溶剂性质变差(温度变化),含有长链的溶液先发生相变,聚集程度更高。因此,同一个温度下,溶液中的聚集体和每个聚集体内的聚集数都应该更大。另外,从直觉出发,在同一条件下,长链应该形成更大的聚集体。然而,实验的结果则完全出乎预料。先看下图(a),平均流体力学半径($\langle R_h \rangle$)随着溶液温度升高变小,在32~33 ℃之间突降,好像是单链蜷缩。但是应该和下图(c)(纵坐标实际上是表观重均摩尔质量,$M_{w,\text{app}}$)放一起考察。在$\langle R_h \rangle$突降之前,散射光强没有增加,故$\langle R_h \rangle$的变化确实可归于单链的收缩,但$\langle R_h \rangle$突降后,$M_{w,\text{app}}$增加了1~5倍,聚集数随着浓度递减,并没有形成单链聚集体。在此过程中,链内收缩主导了链间聚集,所以,$\langle R_h \rangle$单调递减。该结果充分地说明了,不可仅凭动态激光光散射测得的粒径变化来判断是否形成了单链聚集体,必须将静态和动态激光光散射联用方可正确地判断聚集体的微观状态。

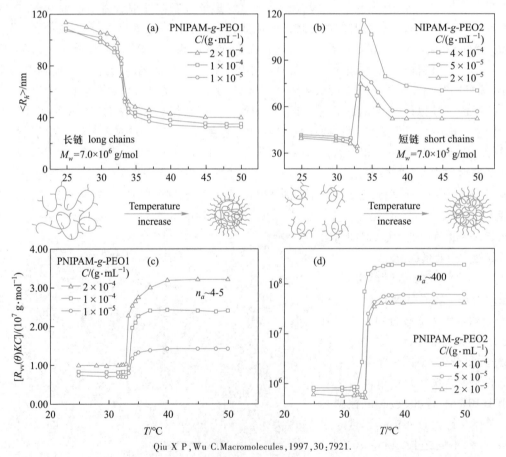

Qiu X P,Wu C.Macromolecules,1997,30:7921.

(Figures 3,4,7,8 and 11,permission was granted by publisher)

再看上图(b)中短链的情况,$\langle R_h \rangle$先随温度升高递减,但$M_{w,app}$不变,反映了单链蜷缩。$\langle R_h \rangle$在32~33 ℃之间突然增加,$M_{w,app}$也突然大了100倍左右,显示链间聚集。随后,$\langle R_h \rangle$随着温度升高而再次下降,但对应的$M_{w,app}$则不变,说明此时聚集已经停止,但链仍在继续收缩。在整个过程中,链间聚集占了主导。比较上图(a)、(b),可知在含有长链的溶液里,不仅聚集体的尺寸(~40 nm)较小,而且聚集数也小了近100倍。即使考虑二者链长的10倍之差,短链聚集体也重了约10倍。另外,在稀溶液中,长链聚集体平均只含有两根链。进一步稀释,可得单链聚集体。如用短链,无论如何稀释,也无法得到单链聚集体。这些出乎预料的结果无法用热力学解释,必须考察动力学的影响。

　　下图(a)显示,如果以几乎接近平衡的方式,将溶液从25 ℃缓慢地一步一步地升至45 ℃时,无论是长链还是短链,聚集体的平均尺寸随着大分子浓度(C)的增加变大。只是在含有长链稀溶液里($<10^{-4}$ g/mL),浓度对$\langle R_h \rangle$的影响不大。随着浓度增大,出现链间聚集的概率增大,形成较大的聚合体,完全合理。下图(b)则显示了加热速率对介观球相中共聚物链聚集的影响。这样形成的聚集体十分稳定,数月内均无明显变化。无论在何浓度,将共聚物溶液从25 ℃一步快速地加热至45 ℃均导致较小的聚集体。在$C = 10^{-5}$ g/mL时,甚至得到了单链聚集体。源自不同升温速率的结果清晰地显示,在介观球相中的聚集依赖于具体路径,并非处于与路径无关的热力学稳定状态。

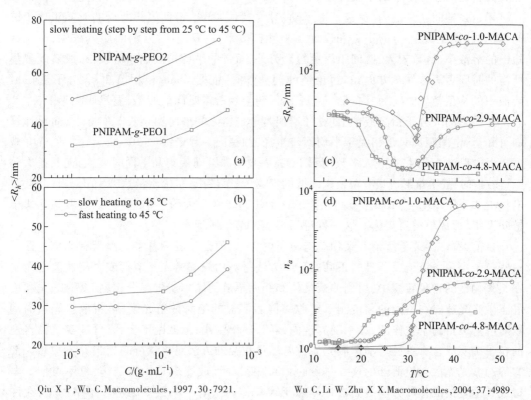

Qiu X P,Wu C.Macromolecules,1997,30:7921.
(Figures 5 and 9,permission was granted by publisher)

Wu C,Li W,Zhu X X.Macromolecules,2004,37:4989.
(Figures 1 and 2,permission was granted by publisher)

　　第二个例子:先在相似长度($M_w = (6.5 \sim 9.4) \times 10^5$ g/mol)的热敏性聚N-异丙基丙烯酰胺上无规地共聚上不同含量(1.0%,2.9%和4.8%)的疏水共聚单体(具有生物活性的疏水 2'-甲基丙烯酰基氨基乙烯-3α,7α,12α-三羟基-5β-胆酰胺,MACA),再利用溶液温度调节共聚物链在水中的溶解性,生成介观球相。依照热力学原理,对给定的温度,疏水单体含量较高的共聚物链在介观球相中更易聚集,形成更大的聚集体,每个聚集体平均含有更多的共聚物链。然而,结果再次完全地出乎预料,如上图(c)和(d)所示。

与前述相同,在溶液温度升至相变温度之前,$\langle R_h \rangle$ 递减,聚集数基本为 1,反映了单链蜷缩,完全正常。在介观球相中,聚集数随着疏水单体含量的增加,不增而减,从几千条链变成了几条链。当疏水单体含量为 4.8% 时,$\langle R_h \rangle$ 随着温度升高一直下降,好像没有任何聚集,但聚集数则显示了少量聚集,反映了高疏水单体含量反而导致了链内收缩的主导作用,与直觉不同,十分异常。

以上两个反常的实验结果充分地显示了在介观球相中的链间聚集并非完全取决于热力学原理,也与动力学过程有关。为何长链和含疏水单体较多的链反而导致了在介观球相中较少的链间聚集?以及为何溶剂性质的变化速率和浓度会影响在介观球相中共聚物链的聚集?微观上,在相分离区域内,小分子溶液中仅会发生分子间的聚集,逐步形成一个稀相和一个浓相,并达至动态平衡。而在介观球相区内,共聚物溶液中,除了像小分子一样出现链间聚集之外,还有一个额外的链内聚集,导致一个蜷缩的链构象。这是小分子和大分子之间在相变中的本质差别。

无论是链间还是链内,聚集都由平动扩散控制。因此,链间聚集所需的时间正比于链间平均"作用距离"(l_I)的平方和反比于平动扩散系数(D)。对给定的质量浓度,在含有短链的溶液中,链数较大,链间距离更短。另外,短链的平动扩散也相对较快,故链间聚集所需的时间更短。另一方面,在讨论链的动力学时,依据 Zimm 模型已知,大分子链在稀溶液中的构象弛豫时间为其扩散一个平均回转半径所需时间的四倍,显然取决于链长。

在通常的大分子稀溶液中,直觉的错误是链与链之间的距离很长。实际上,简单计算即可知道,链间距离只是稍大于平均回转半径,约在同一数量级($\sim 10^2$ nm)。因此,在介观球相中出现相分离时,在绝大多数情形下,链内收缩和链间聚集几乎同时发生,尤其是大分子链较短时,故很难形成只有单根蜷缩链的聚集体。只有当摩尔质量达到 $\sim 10^7$ g/mol,且浓度接近或小于 10^{-5} g/mL 时,链间距离才比平均回转半径大约十倍,使得链内聚集远快于链间聚集,形成单链聚集体。这一推论已被上述结果所证实。

目前,有许多关于所谓共聚物单链粒子的文章,但是作者们基本上都没有严格地证明这些粒子由单条链组成。唯一正确的证明方法是利用静态激光光散射或者超速离心显示在共聚物链的构象转变中,得到的聚集体和起始单链的摩尔质量完全相同。例如,先利用静态激光光散射在不同的角度测量链收缩前后溶液的绝对散射光强,再将散射光强的倒数对散射矢量的平方作图,如果随着链的蜷缩,斜率变小,但截距不变,方可证实这样得到的粒子确实只含有单链。动态激光光散射测量粒径的方法对少量链间聚集并不敏感。带有静态激光光散射检测器的尺寸排除色谱可因稀释而改变聚集体的结构,也不可信。电镜或其他直接成像的方法更无法证明每个聚集体内只有一条共聚物链。不幸的是,在许多现有的文献中,"单链粒子"一词已被乱用和滥用。读者在阅读有关文章或从事这方面研究时要去伪存真。

在第七章中,链蜷缩的动力学研究已经证实,一条大分子链在稀溶液中的构象弛豫时

间快于 10^{-3} s。在如此之短的时间内,链间聚集和链内收缩同时进行,导致局部浓度迅速增大。不难想象,一旦每个聚集体的链内收缩完成了,局部浓度必定高于接触浓度,每个聚集体都成为一小"滴"亚浓溶液,构成数目众多的微浓相悬浮在稀相之中。此后的分相过程,就成了小"液滴"之间或一个"液滴"和单链之间的继续聚集,或建立一个动力学平衡。

可将这些小"液滴"和蜷缩的大分子单链均看成一个个小粒子。在热能的搅动下,它们均在溶液中无规地平动。如果它们能通过无规碰撞逐步地融合成较大的粒子,并最终分离成宏观的两相,每两个碰撞的粒子不仅需要相互作用一段时间(τ_I),而且其一定要远长于粒子内的链的质心移动一条链的末端距所需的时间(τ_R)。只有这样,分别在两个粒子内的链才有足够的时间互相扩散、贯穿、融合。显然,τ_I 与粒子间的平均"作用距离"(l_I)、粒子的平均速度($\sqrt{\langle u^2 \rangle}$)和平动扩散系数($D$)有以下关系:

$$\frac{l_I}{\sqrt{\langle u^2 \rangle}} \leqslant \tau_I \leqslant \frac{l_I^2}{D}$$

上式的物理意义是 τ_I 介于线性碰撞和平动扩散碰撞所需的两个极端时间之间。与 τ_R 相关的质心运动已在第九章中详细讨论,在浓度较高的亚浓溶液中,应该采用 Rouse 模型,即

$$\tau_R \approx \tau_{GC} = \frac{\pi \eta}{k_B T} \frac{b\, N_d L_{c,d}^2}{(N_d/N)^2} = \frac{\pi \eta}{k_B T} \frac{b^3 N^4}{N_d^{3-\alpha_R}} \rightarrow \tau_R = \tau_0 \left(\frac{C}{C^*} \right)^{\frac{3-2\alpha_R}{3\alpha_R - 1}}$$

其中,$\tau_I \sim \tau_0 \sim 10^{-4}$ s 和 $\tau_R \sim \tau_{GC} \sim 10^0$ s。基于前述有关黏弹性的讨论,一个材料是"黏"是"弹"取决于作用时间和其中分子弛豫时间之间的相对长短。在常温和通常的时间尺度上,水是黏性液体,但是,如果用力将一片小石头以尽可能小的角度快速地掠过水面,它会在水面上弹起数次,形成一串水花,仿佛水面具有弹性。两千五百多年前,孙武就在《孙子兵法》中论及:"激水之疾,至于漂石者,势也。",此处的"势":极快之速、极短之时也。

在介观球相中,如果两个发生碰撞的聚集体之间的相互作用时间远远短于粒子内的大分子长链在两个粒子间互相扩散、贯穿、融合所需的时间,每个聚集体均呈弹性,好似两个弹性的玻璃小球,碰撞后互相分开。碰撞将不会(概率极低)使它们成为一个较大的玻璃球。正是每个聚集体里面的链内收缩终止了链间聚集。所以,因为"黏弹"效应,介观球相中的这些小聚集体处于一个动力学(kinetically)稳定状态,不会在较短的时间内进一步聚集、分成宏观的稀、浓两相。

综上所述,完全不同于小分子,大分子链的弛豫较慢,尤其是在亚浓溶液中。一个大分子体系很难达到一个真正的热力学平衡状态,其往往陷在一个能量并非最低的、取决于路径的亚稳态。因此,往往观察到一些无法用热力学解释的现象。利用这一"黏弹性"动力学稳定机理,以及在介观球相的形成中,链间聚集和链内收缩的共存和竞争,就可依次地解释上述"反常的"实验结果。具体如下。

一、"对于给定的共聚物溶液,平均聚集数和平均聚集体尺寸随着浓度降低而递减"。降低浓度增加了链间距离,减慢了链间聚集,但不影响链内的构象变化(链内收缩),故链内收缩相对地快于链间聚集。在同样的时间内,链间聚集相对地减少,形成较小的聚集体。

二、"对于给定的共聚物溶液和浓度,平均聚集数和平均聚集体尺寸随着温变速率增加而递减"。在稀溶液中,链内收缩一般快于链间聚集。故温变速率较快时,链内收缩先于链间聚集,导致链间聚集减少,形成较小的聚集体。

三、"对于给定的大分子质量浓度,平均聚集数和平均聚集体尺寸随着链长增加而递减"。在含有长链的溶液中,链间距离较长,导致链间聚集慢于链内收缩。在聚集体中,链质心的弛豫时间与链长的四次方成正比($\tau_E \sim N^4$)。所以,在竞争中,一旦链内收缩快于链间聚集,长链具有更强的"黏弹性"效果,故聚集体之间的进一步聚集几乎停止,形成较小的粒子。

四、"对于给定的链长和质量浓度,平均聚集数和平均聚集体尺寸随着链上疏溶剂基团的增多而递减"。含有较多疏溶剂基团的共聚物链在相变前的构象就已经处于相对收缩的状态,利于链内收缩。在分相的过程中,为了降低界面能量,疏溶剂的基团更倾向于集中在聚集体内部,它们之间较强的色散吸引进一步延长了链在聚集体中的弛豫时间,有效地减缓了聚集体之间的进一步融合。所以一旦链内收缩完成,聚集停止,形成较小的聚集体。

除了热力学的考量以外,在介观球相中的聚集主要受控于链间聚集和链内收缩速率之间的竞争。如果希望得到较小的亚稳态聚集体,就需要采用长的共聚物链,低的浓度,加快溶剂性质的变速,和适当地增加疏溶剂基团在链上的含量。在生物教科书以及文献中,谈及蛋白质链折叠成一个三维结构时,只是依据热力学原理,考虑了蛋白链将亲水和疏水基团分别移至表面和内部以减少界面能量,形成稳定的、具有生物活性的单链结构,而不知道上述的"黏弹性"动力学稳定机理。在上亿年的生物进化中,除了降低界面能以外,表面的亲水基团缩短了蛋白质链之间的作用时间(τ_I);里面的疏水基团延长了蛋白质链在折叠结构中的弛豫时间(τ_E),使得$\tau_I \ll \tau_E$。蛋白质链"聪明地"一步满足了基于"黏弹性"的动力学稳定条件。上亿年进化的结果使得很多蛋白质链可以轻易地折叠成单链结构。

近三十多年方兴未艾的大分子"自组装"都是研究链在介观球相中的亚稳态行为,包括各种嵌段共聚物在选择性溶剂中的"自组装"、形式多样的反应诱导"自组装"、类蛋白质结构共聚物在水中的"自组装"、甚至无规共聚物在选择性溶剂中的无规聚集也被包装成"自组装"。所形成的大分子聚集体也都被冠以"纳米"粒子,不一而足。希望在该领域里工作的研究者,尤其是年轻的研究者,可以注意到以下一些重要的区别。

聚集和组装:这是两个有着明显不同的内涵和外延的中文单词。"组装"是"聚集"的子集,聚集并非一定是组装。将一堆儿童积木杂乱无章地堆(聚)集在一起不同于将它们有序地组装成一个结构明确的小房子。小分子在不良溶剂中的沉淀通常是聚集,而有序结晶则是某种意义上组装。细胞内的蛋白质常常先折叠成特定的构象、再组装成有序的寡聚体活性结构。这里"无序"和"有序"仅是区分聚集和组装二者的要点之一,不是全部。

柔性嵌段高分子在选择性溶剂中可形成各种形貌的微结构。不溶链段的微聚集与其均聚物在同一不良溶剂中宏观聚集并无区别,本质上都是杂乱的。唯一不同的是,可溶链段阻止了宏观聚集。仅考虑热力学因素时,每个亲溶剂基团稳定界面的能力决定了可稳定的总的界面面积，也即最终的聚集体形貌。仅有热能时，该过程必定遵循热力学的统计无序涨落。因此，这些聚集体的尺寸和聚集数并不相同，也不可能相同。它们的形成过程不是严格意义上的组装（尽管笔者以前也常常将其称为自组装）。那么为何众人将这两个不同的过程混为一谈呢？这里有"无意"和"有意"之分。无意者，人云亦云也；有意者则有浑水摸鱼之嫌。究其原因，无外乎是，"聚集"是一个老的概念，了无新意；而"组装"和"自组装"则看似更"新颖""漂亮"和"吸引眼球"，且符合当今高歌的"创新"潮流！

组装和溶解：无论是聚集还是组装，从概念上讲，都指的是在溶液中，一个一个的单个分子随着溶剂性质的变化"走"到一起的过程和结果。但在文献中，尤其是研究离聚物溶液时，常常看到有人将溶液中大于溶质分子的"粒子"混淆地统称为"聚集体"或"组装体"。然而，他们显然忘了，在制备一个溶液时，起点并不是将一个接一个的分子投入溶剂，而是将一宏观溶质与溶剂混合。在溶解的过程中，溶质通常以单个分子的形式逐渐地分散到溶剂中去。因此，在溶液条件不变时，制备溶液后观察到的"粒子"应是溶解的残留不溶物，而不是溶液中形成的"聚集或组装体"。这一问题，笔者曾经在以前的文章中明确地指出过，但并没引起相关人士的注意，类似的概念错误仍不断地重复出现。因此，在研究溶液中高分子链聚集或组装时，先要确定起始的溶液是否是一个真溶液（即不溶解的"粒子"）。最好的实验方法是静态激光散射或超速离心法。它们可测量绝对平均摩尔质量，其应该接近单链的数值。

沉淀聚合和聚合诱导组装：前者是高分子教科书上典型的合成方法之一，而后者则是一个新创名词，英文曰为 pisa。实际上早在 20 世纪 90 年代，笔者就明确地提出了化学反应诱导自组装的概念（也许还有人更早地提出了这一概念）。"沉淀聚合"和"聚合诱导组装"没有任何本质上的不同，唯一的区别是后者引入了一小段可溶性大分子链作为界面稳定剂，进入了亚稳态的介观球相而已。因此，可否将加入一些稳定剂的微沉淀或微乳甚至乳液聚合也称之为聚合诱导组装呢？笔者认为，"聚合诱导组装"是一个典型的将旧酒装入新瓶吆喝叫卖的例子。需要指出的是，无人否定该方法在应用上的潜在价值，只是这样的标新立异毫无必要，刻意制造新词和标签，造出一个所谓的"新"领域，并以领跑者自居。因此，称这类化学变化为沉淀或微沉淀聚合反应足矣。

被动和主动组装：国际上有人没有深思熟虑就提出了"动态自组装"的新概念，国内外跟风者甚多。在多次的演讲会上，笔者诘问演讲者，何为"静态自组装"？演讲者们无一不语塞。显而易见，他们并没有思考过这一问题或已习惯于遵循名人的导向。组装的过程一定是动态的！因此，根本没有静态自组装。既然没有静态自组装，又何来动态自组装？然而，讨论并未结束。一种争辩是，动态自组装指的是从一个有序状态变为另一个有序状态。非也，物理上对此早就有过描述，即"有序到有序"转变（order-to-order transition）。另一种争辩是，动态自组装指的是过程中需借助除了热能以外的其他能量。错也，物理上也早就有过定义，即仅有热能驱动的称为被动过程，而需要其他额外能量驱动的则是主动过程，因此，正确的说法应为：被动和主动自组装。

焓变和熵变：在等温和等压下，在一个自发的聚集过程中，体系自由能递减，换而言之，焓变（ΔH）和熵变（$-T\Delta S$）之和为负。谁主沉浮？值得指出，这里的主要驱动力往往是

熵变而不是常常误解的焓变(各种相互作用)。带相反电荷大分子链的聚集就是一个典型的例子。文献中往往将其归于静电相互作用(焓变),而忘了在混合两个带相反电荷大分子链之前,每条链都有各自的小分子反电荷,已有静电相互作用。实际上,带相反电荷高分子链的聚集或组装释放了大量的反电荷,从而极大地增加了体系的平动熵,导致自由能降低。因此,该过程的主要驱动力是熵变。与聚电解质有关的物理源于其反离子的物理,严格地说,其不属于大分子物理。与此类似,氢键诱导大分子链的聚集也是由于释放了原本与各自大分子链形成氢键的溶剂分子,导致了体系的熵增加。微沉淀聚合也主要由熵变驱动。反应前,反应单体溶于溶剂,焓和熵同时增加,但熵变项($T|\Delta S|$)大于焓变项(ΔH)。聚合导致体系的平动熵逐步减少,直至 $\Delta H > T|\Delta S|$,发生相变,导致聚集、沉淀。不仅年轻的研究者,一些资深的国内外研究者们也常常犯这样的基本概念错误。

综上所述,尽管诱导大分子链聚集或组装的形式、成因和化学基团千变万化,但均可用现有的热力学和大分子物理定性或定量地解释,没有脱离现有的理论框架和知识体系。"太阳底下无新事"。它们都是介观球相中处于亚稳态的小聚集体,本质上它们是与含有收缩单链的稀相共存的微浓相,实现了亚稳态的微相分离。由于大分子链在聚集体中的弛豫极慢,两个聚集体碰撞作用的时间远短于每个聚集体内部的链的弛豫时间。因此,每个聚集体好似一个玻璃态的小球,两个碰撞的聚集体融合成一个稍大的聚集体的概率极低。聚集体因这一新"黏弹性"稳定机理实现了动力学稳定。

读者需要清醒地认识到,第一,从纯学术和知识创新的角度出发,绝大多数的这类研究都是利用了现有的知识,研究了各种各样的大分子体系,并没有创造新的化学、物理、生物或数学,即无知识上的贡献! 第二,介观球相中共聚物聚集体都处在一个与形成路径有关的亚稳态,而不是处于热力学的稳定状态。因此,这类研究不可能得到新的、普适的物理规律。换而言之,这类研究所得的结论都随体系和形成路径而变。与小分子在胶体里的聚集不同,共聚物在聚集体中的弛豫可以经年累月。一些在介观球相中形成的聚集体往往看似很稳定,但实际上仍然处于亚稳态。正因为如此,此类研究不是严格意义上的基础研究!

正是因为这一足够稳定的亚稳态,在介观球相中生成的各种聚集体粒子可满足许多应用的需求,使得该方面的研究可以产生新的材料,包括环境响应材料、药物输送体系、生物医用辅料等。有志于在该领域继续工作的研究者需将主要的方向和精力聚焦到一些具体的应用问题。对每个研究者而言,研究就是解决问题!"不上书架、便上货架",笔者的这一观点并不新颖,近年来已被许多研究者认同。

研究大分子在介观球相中的聚集要关注解决具体的、源自生产、生活和生物医疗等方面的问题,而不是泛泛地讨论一些潜在和可能的应用。譬如,在药物可控释放的研究中,应该从某一特定的药物以及其在临床上特定的给药方式出发来考虑、选择、设计和合成适当的大分子,寻获适当的路径,实现所要的聚集,解决一个相关的问题,而不是从实验室现有或可合成的大分子出发。不要为了组装而组装。需要牢记,研究的目的是解决问题,而不仅仅是发表文章。"不上书架、便上货架"!

小　结

对小分子的二元共混物,随着二者的相溶性变化,要么在分子水平上的互相溶解,形成一个均匀溶液(一相);要么发生相变,混合物分成两个互为饱和的均匀溶液(两相,稀相和浓相),二者必居其一。两相之间有一个宏观的物理界面。在均聚物与小分子溶剂形成二元共混物中,也有相同的从一相到两相的相变。但在相变之前,在一相区内,还有一个大分子特有的构象蜷缩转变。而在共聚物与小分子溶剂形成二元共混物中,随着共聚单元在溶剂中溶解度之差(两亲性程度,Δ)的增大,还会在原本的两相区出现一个处于亚稳态的介观球相区,其中有限数目的共聚物链聚集在一起,形成一个个分散的微浓相,与一个含有一个个蜷缩单链的饱和溶液共存,即微相分离。在均聚物的相变中,也会出现链间聚集,但聚集体互相之间发生进一步融合,故混合物分成宏观的两相。而在介观球相中形成的聚集体则可在亚稳态停留经年累月。

这一微相分离的过程常被称作各种形式的"自组装"。这些形成微相区的小聚集体粒子也常被冠为"自组装体""纳米粒子"等新颖头衔,但万变不离其宗。读者需要清楚地知道,这些小聚集体都处于亚稳态,体系的自由能处于一个局部范围内的低点,但不是热力学的稳定状态。因此,它们的形成与达至微相分离的路径有关。沿着不同的路径,可落入不同的能量低点,形成不同结构和形态的聚集体,千变万化。因此,共聚物溶液的相变十分复杂。没有普适的定量规律。

如同小分子溶液和均聚物溶液的相变,改变温度、压强、溶剂组成等实验条件,可引致共聚物溶液出现两相分离,进入介观球相区。不同的是,除了溶剂性质以外,共聚物溶液的相变多了一个两亲性程度的变量。而溶剂性质和两亲性程度互相依赖。实验上,无法固定一个,只改变另一个。

依据热力学原理,这些微浓相,称作液滴或聚集体,应该进一步融合,形成宏观的两相。但是在通常的实验时间尺度内,这些聚集体往往较稳定。阻止它们进一步融合分相的正是大分子特有的"黏弹性"提供了动力学稳定的基础。换而言之,在介观球相中,如果两个发生碰撞的聚集体之间的相互作用时间($\tau_I \sim 10^{-4}$ s)远远短于粒子内大分子链在粒子间互相扩散、贯穿、融合所需的时间($\tau_R \sim 10^0$ s),它们就好似两个弹性的玻璃小球,碰撞后分开,而不会因为碰撞而融合成一个较大的聚集体。

微观上,在介观球相中出现微相分离时,链间聚集和链内收缩总是同时发生。对通常的共聚物,链内收缩的平均时间仅略快于链间聚集的时间。因此,当一条共聚物链蜷缩时,总是伴随着链间聚集,很难形成单链聚集体。两亲性共聚物链在收缩时,总是倾向于将亲溶剂的基团移向聚集体的表面以降低疏溶剂单元的表面能。不同的相变途径均会影响链间聚集和链内收缩二者之间的竞争,从而在介观球相中形成含有不同微相结构的亚稳态,包括不同的聚集体尺寸、密度和聚集数。

由于$\tau_I \ll \tau_R$,一旦聚集体中的链内收缩完成了,链间聚集就实际上停止了。剩下的仅是聚集体之间的弹性碰撞,成为一个较大聚集体的概率极低。微相分离被大分子特有的黏弹性"冻"在了介观球相之中,形成了一个动力学(kinetic)稳定的状态。因此,通过降低浓度和加快溶剂性质的变化速率均可减少链间聚集的概率,生成较小的、聚集数较少的聚集体。而增加链长和引入适当的疏溶剂基团则可加快链内收缩和延长聚集体内大分子链的弛豫时间,使得聚集体之间互相扩散、贯穿和融合更加不易,也更利于形成更小的、动力学稳定的聚集体。

　　研究取决于路径的介观球相不可能导致普适的、定量的物理规律和产生新的知识。严格地说,其不是基础研究,但其应用价值不菲。利用其与路径有关的特性,可以通过操纵和调节介观球相的形成条件,得到不同的、具有足够稳定性的各种聚集体,以满足不同应用需求。该方面的研究可以产生新的材料,包括环境响应材料、药物输送体系、生物医用辅料等。有志于在该领域继续工作的研究者需将主要研究方向和精力聚焦到解决一些与生产、生活和医疗有关的具体的、真正的问题。下图小结了本章所讨论的主要内容。

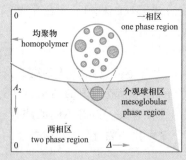

由有限条链聚集而成的微浓相(粒子)与稀相达至动力学平衡,形成一个特殊的处于亚稳态的介观球相。
Micro-concentrated phase (particles) made of a limited number of macromolecular chains is in equilibrium with a dilute phase, forming a special metastable mesoglobular phase.

微浓相的结构(大小和聚集数)和形成路线有关,因此,并无一个普适物理规律来描绘介观球相的生成。
The structure of micro-concentrated phase (aggregation number and size) depends on its formation path, no universal physical law to describe the formation of mesoglobular phase.

动力学:基于"黏弹性"的稳定机理
Kinetics: A stabilization mechanism on the basis of "viscoelasticity"

$$\frac{l_I}{\sqrt{\langle u^2 \rangle}} \leqslant \tau_I \leqslant \frac{l_I^2}{D} \quad \boxed{\tau_I \ll \tau_R} \quad \tau_R \approx \tau_I \left(\frac{C}{C^*}\right)^4$$

快速变化溶剂性质(变温或溶剂速率)、用长链和适当增加疏水成分含量可得小的聚集体粒子。
Fast changing solvent property (rate of changing temperature or solvent), using longer chain and properly increasing hydrophobic content can lead to smaller aggregates (particles).

Chapter 11. Mesoglobular Phase in Macromolecular Solutions

As mentioned earlier, for homopolymers, as the temperature gradually decreases (UCST) or increases (LCST), the solvent properties gradually deteriorate. First, there is a "coil to small ball" conformational transition in the one phase region, individual chains collapses. Further change of the temperature leads to the interchain aggregation the two-phase region, forming macroscopic dilute and concentrated two phases. The concentration difference of the two phases gradually increases as the solvent quality become worse. The conformation of copolymers in the solvent also changes with the solvent quality, but the phase transition is complicated, dependent not only of chemical properties of each comonomer and their specific arrangement on the chain, but also of the "path" of the phase change, including the temperature changing rate, and the way of changing the solvent quality or concentration.

For example, if a dilute solution of a slightly carboxylated polystyrene ionomer in tetrahydrofuran (THF) is quickly added to a large amount of water, the mixture immediately becomes a cloudy suspension because the ionomer is insoluble in water, which means that aggregates with a size larger than the wavelength of visible light are formed. However, if the same dilute solution of polystyrene ionomer is slowly added to a large amount of water drop by drop, the mixed liquid is still clear and transparent. The question is where those water-insoluble polystyrene ionomer chains go. The LLS study revealed that in water they actually formed stable aggregates with a size much smaller than the wavelength of visible light, tens of nanometers, and containing only dozens of chains. The

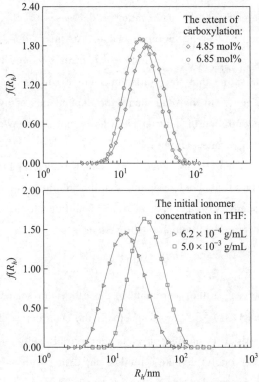

Li M, Jiang M, Zhu L, et al. Macromolecules, 2004, 37: 4989.

(Figures 1 and 2, permission was granted by publisher)

average size of the aggregates and the average number of macromolecular chains inside each aggregate (aggregation number) decrease as the degree of carboxylation increases; but increases with the initial concentration of ionomer in THF, as shown above. The quantitative experimental results showed that as the initial concentration decreased from 2.0×10^{-2} g/mL to 6.2×10^{-4} g/mL, the average hydrodynamic radius decreased from 18 nm to 8 nm; the average aggregation number decreases from ~100 to ~18.

The question is why these water-insoluble ionomers do not continue to aggregate and separate until two macroscopic phases (dilute and concentrated) are formed. The qualitative explanation is that in the phase separation, the negatively charged, hydrophilic carboxyl groups tend to stay on the surface of the aggregate, while the hydrophobic polystyrene backbone aggregates and collapsed in the middle to form individual particles with a core-shell structure. For a given degree of carboxylation, the number of carboxyl groups (n) on the surface of each aggregate is proportional to the degree of aggregation, and also proportional to its volume (V) or the cube of its size (R) (a uniform density), and the surface area (S) of each aggregate is proportional to the square of its size. Therefore, the surface area (s) occupied by each carboxyl group is

$$s = \frac{S}{n} \sim \frac{R^2}{R^3} = \frac{1}{R} \qquad (11.1)$$

Therefore, in the aggregation, s decreases as R gradually increases. For each given charged group, s has a minimum value. At this value, further aggregation becomes difficult, even macroscopic phase separation can further reduce the free energy of the solution. Experimental data shows that the minimum average surface area occupied by each ionic group is ~3.0 ± 0.2 nm^2. On the other hand, the dependence of hydrodynamic radius distribution on the initial concentration of ionomer also shows that the above equation is not the only condition to control the aggregate's structure.

Theoretically, Timoshenko and Kuznetsov [Journal of Chemical Physics, 2000, 112:8163] proposed a model, assuming that the solution volume is L^3, with N_p chains, and each chain has N segments with a length of b, and omitting all the degrees of freedom of solvent molecules, the effective Hamiltonian of the system is

$$H = \frac{k_B T}{2L^2} \sum_a (\mathbf{Y}^a - \mathbf{Y})^2 + \frac{k_B T}{2b^2} \sum_a (\mathbf{X}_n^a - \mathbf{X}_{n-1}^a)^2 + \sum_{J>2} \sum_{|A|} A_J^{|M|} \prod_{i=1}^{J-1} \delta(\mathbf{X}_{M_{i+1}} - \mathbf{X}_{M_1})$$

where X_n^a is the coordinate of the nth chain segment on the ath chain, $Y^a \equiv (1/N) \sum_n X_n^a$ and $Y \equiv (1/M) \sum_a Y^a$ are respectively the coordinates of the center of mass of the chain and system, $A_J^{|M|}$ is a set of position-dependent virial coefficients.

First, let us take a random copolymer with only two different monomers as an example, and use a general quadratic form $H_0 \equiv (1/2) \sum_{M_1, M_2} V_{M_1, M_2} X_{M_1} X_{M_2}$ in the Gaussian variational method as a testing Hamilton, and excluding V_{M_1, M_2} (effective potential energy), a special parameterized result for the second-order virial coefficient matrix of the binary copolymer can be obtained.

$$A_2^{M_1,M_2} = A_2 + \frac{\sigma_{M_1} + \sigma_{M_2}}{2}\Delta \tag{11.2}$$

where A_2 is the average second-order virial coefficient, σ_{M_1} and σ_{M_2} respectively reflect the specific distributions of the two comonomers M_1 and M_2 on the chain, and their sum divided by 2 is the average distribution. For a given comonomer, the solubility difference (Δ) between M_1 and M_2 in a solvent can be large or small, depending on the interaction between the solvent and the chain unit composed of M_1 and M_2. In a typical example, one unit is soluble in water (hydrophilic), while another is only soluble in a water-immiscible organic solvent "oil" (hydrophobic), which is why Δ is called for "the degree of amphiphility".

If $\Delta = 0$, the two monomers are the same, and the copolymer becomes a homopolymer, which is the left axis of the figure on the right. When Δ is small, the solution behavior of random copolymer chains is almost the same as that of homopolymers, i. e., there are only two states, one phase (a homogeneous solution) or two phases (two saturated solutions). Only when the degree of amphiphilicity is high enough, as the solvent quality gradually deteriorates (A_2 becomes more negative), can there be an additional mesoglobular phase between one phase

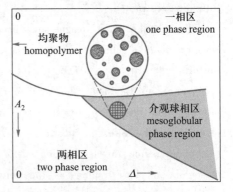

with collapsed single chains and two macroscopic phases. In the mesoglobular phase, a limited number of chains (n_a, the aggregation number) gather together to form metastable spherical aggregates of different sizes (or non-spherical), and a microphase separation occurs, and its range expands as the degree of amphiphilicity increases. As shown in the shaded part and its partial enlargement in the figure.

For a given Δ, when A_2 is lower than the upper limit of the region, the copolymer chains in the solution will start to aggregate, i. e., a microscopic phase separation, forming many small concentrated microphases (aggregates) coexisting with individual collapsed chains in the solution. In principle, these small microphases will further aggregates to form two macroscopic phases eventually; but in a limited time (it can be very long, over months or years), they will remain in the solution. Strictly speaking, there is no essential difference between the phase separation of copolymers and homopolymers, but the copolymer aggregates can stay in the metastable state for a much longer time. Nowadays, in order to attract attention, this process is often called "self-assembly", and the micro-concentrated phases (small aggregates) formed are often called various "nano" particles.

Note: "aggregation" is a physical process, and "polymerization" is a chemical process. The two should not be confused. In the foregoing discussion, the "coil-to-ball" conformational transition is about the formation of thermodynamically stable homopolymer single-chain globules in the one phase region. The copolymer aggregates formed in the mesoglobular phase are in a non-thermodynamically stable metastable state, and its aggregation is related to the path. On the one hand, for a given Δ, the aggregation number increases with the deterioration of the solvent properties; on the other hand, for a given solvent property (temperature), the aggregation number generally decreases with the increase of Δ. Under proper conditions, single-chain aggregates can also be generated ($n_a \rightarrow 1$) in the two phase region, but it has nothing to do with the problem of the "coil-to-globule" conformational transition. Many protein macromolecules are typical examples. They can be folded into single-chain structures along a certain path within the cell; they can also be refolded into a single-chain active conformation through a special method (path) outside the cell. Generally speaking, the conditions for forming copolymer single-chain aggregates are quite harsh and much more difficult than proteins.

Timoshenko and Kuznetsov used computer simulations to calculate 12 linear "macromolecule" chains with the same composition, each containing 6 M_1 and 6 M_2 segments, but different structures: $[M_1 M_2]_6$ and $[(M_1)_2 (M_2)_2]_3$. The results show that if the two kinds of segments are the same (homopolymer), $\Delta = 0$, individual chains have the lowest mixing free energy in the one phase region; while in the two phase region, the aggregation of 12 chains together has the lowest mixing free energy. As the degree of amphiphilicity Δ increases, in the two-phase region, the aggregation of 12 chains still has the lowest mixing free energy. In the mesoglobular phase region, as the solvent quality becomes worse, the aggregation number increases, whose corresponding mixing free energy is also the lowest. However, the difference in the mixing free energy between particles with different aggregation numbers is very small, so that there may exist aggregates with different aggregation numbers at the same time, clearly showing that in the mesospheric phase, the aggregates are in the metastable state.

In the following, using two real examples further illustrate why the aggregation of copolymers in the mesoglobular phase is controlled by non-thermodynamic factors. The first example is the dependence of the aggregation of amphiphilic copolymers in selective solvents on the degree of amphiphilicity and the chain length. In the computer simulation, it is easy to adjust the degree of amphiphilicity and the chain length in the same copolymer configuration. Experimentally, there are four variables in the desired copolymer: composition, configuration, degree of amphiphilicity, and chain length. In the synthetic experiment, it is very difficult to "make sure that the three variables remain unchanged, but only one of the variables is changed to obtain four sets of different copolymers". It is difficult if not impossible. In the copolymerization reaction, it is quite difficult to only achieve the same components and structure. In principle, one monomer can be fixed, and another monomer can be used to adjust the degree of amphiphilicity of the copolymer, but how to keep the other three variables similar?

In order to solve this problem, experimentally, polyethylene oxide macromonomer (PEO) and N-isopropyl acrylamide are randomly copolymerized, respectively, in pure deionized water and in a mixture of water and a weak chain transfer agent (ethanol) under the same conditions. The copolymerization reaction was stopped when the monomer conversion rate is 10% or less in order to control the uniformity of the comonomer distribution on the chain. In such a way, two copolymers with different chain lengths (10 times) but similar composition and configuration were obtained, as shown on the right. Since poly (N-isopropyl acrylamide) (PNIPAM) is a thermal sensitive polymer, the degree of amphiphilicity can be adjusted by a temperature change. Namely, within the range of 25 – 50 ℃, PEO remains hydrophilic; while PNIPAM has a LCST, so that it quickly changes from hydrophilic to hydrophobic at ~ 32 ℃. Thus, an increase of the solution temperature can be used to increase the degree of amphiphilicity and induce the formation of mesoglobular phase.

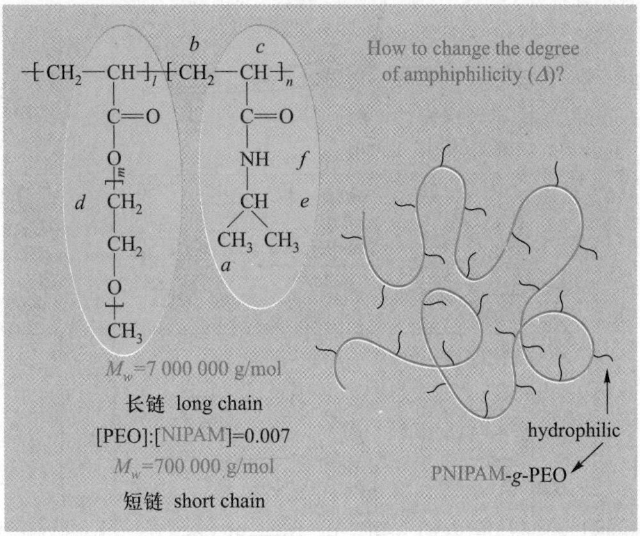

From the discussion in the previous chapter, it is known that according to the principle of thermodynamics, for a given kind of macromolecular solution and mass concentration, as the solvent quality becomes worse (temperature change), the solution containing long chains will undergo the phase change first, and the degree of aggregation will be higher. Therefore, at the same temperature, the aggregates in the solution and the number of aggregates in each aggregate should be larger. In addition, from intuition, under the same conditions, long chains should form larger aggregates. However, the experimental results were completely unexpected, as shown in the figures below. Let us look at the figure (a) below first, the average hydrodynamic radius ($\langle R_h \rangle$) decreases as the solution temperature rises, and drops suddenly between 32–33 ℃, which seems like a single chain collapse. However, the figure (c) below (the ordinate is actually the apparent weight-average molar mass, $M_{w,\mathrm{app}}$) should be examined together. The scattered light intensity did not increase before $\langle R_h \rangle$ drops, so that the change of $\langle R_h \rangle$ can indeed be attributed to the single-chain contraction, but after $\langle R_h \rangle$ drops, $M_{w,\mathrm{app}}$ increases about 1–5 times, and the aggregation number decreases with the concentration. No single-chain aggregates were formed. In this process, the intrachain contraction dominates the interchain aggregation, so that $\langle R_h \rangle$ decreases monotonically. This result fully illustrates that the particle size change measured by dynamic LLS cannot be used to judge whether single-chain aggregates are formed. The static and dynamic LLS must be combined to correctly judge the microscopic state of the aggregates.

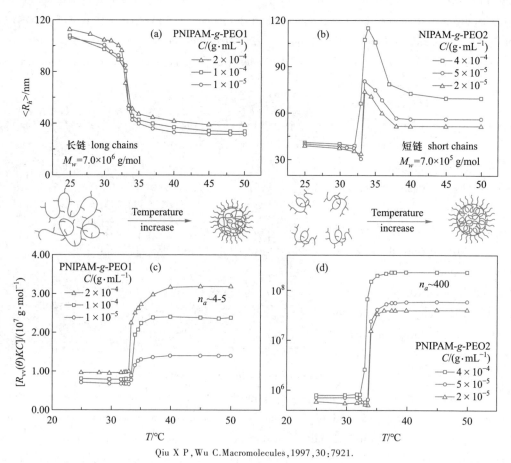

Qiu X P, Wu C.Macromolecules,1997,30:7921.

(Figures 3,4,7,8 and 11,permission was granted by publisher)

Looking at the figure (b) above, $\langle R_h \rangle$ first decreases as the temperature raises, but $M_{w,app}$ remains unchanged, reflecting the single chain collapsing. $\langle R_h \rangle$ suddenly increases between 32–33 ℃, and $M_{w,app}$ suddenly is ~ 100 times larger, indicating the interchain aggregation. Subsequently, $\langle R_h \rangle$ decreases again as the temperature increases, but the corresponding $M_{w,app}$ remains unchanged, indicating that the aggregation has stopped at this time, but the chains are still shrinking. In the whole process, interchain aggregation dominates. A comparison of the figures (a) and (b) above reveals that in the solution containing long chains, not only the size of the aggregates (~40 nm) is smaller, but the aggregation number is also ~ 100 times smaller. Even considering the 10 times difference between their chain length, the short–chain aggregates are ~ 10 times heavier. In addition, in a dilute solution, the long–chain aggregates on average contain only two chains. Further dilution can leads to single–chain aggregates. If using the short chains, the single–chain aggregates were not obtainable no matter how dilute the solution is. These unexpected results are not explainable by thermodynamics, so that the kinetic effect has to be considered.

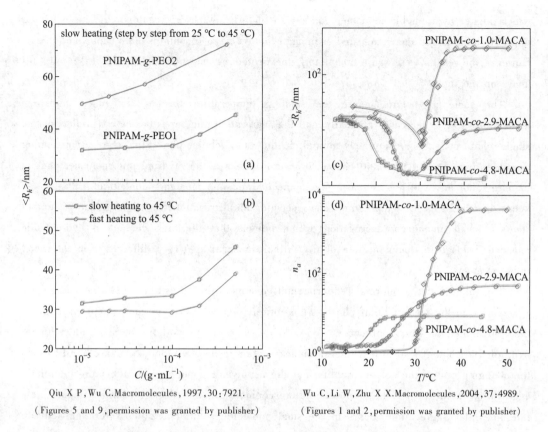

Qiu X P,Wu C.Macromolecules,1997,30:7921.

(Figures 5 and 9,permission was granted by publisher)

Wu C,Li W,Zhu X X.Macromolecules,2004,37:4989.

(Figures 1 and 2,permission was granted by publisher)

The figure(a) above shows that if the solution is slowly raised from 25 ℃ to 45 ℃ step by step in a way that is almost close to equilibrium, whether it is long chain or short chain, the average size of the aggregates increases with the concentration of macromolecules (C). Only in the long chain dilute solution ($<10^{-4}$ g/mL), the concentration has little effect on $\langle R_h \rangle$. As the concentration increases, the probability of interchain aggregation increases, forming larger aggregates, which is completely reasonable. The figure(b) above shows the effect of heating rate on the aggregation of copolymer chains in the mesoglobular phase. Such formed aggregates are very stable, with no significant changes in several months. Regardless of the concentration, heating the copolymer solution up rapidly from 25 ℃ to 45 ℃ in one step leads to smaller aggregates. At $C = 10^{-5}$ g/mL, even single-chain aggregates were obtained. The results from different heating rates clearly show that the aggregation in the mesoglobular phase depends on the specific path, and is not in a thermodynamically stable state that is independent of the path.

The second example:Randomly copolymerize different amounts (1.0%,2.9% and 4.8%) of hydrophobic copolymer monomers (bioactive hydrophobic 2′−methacryloyl−amino−ethylene−3α, 7α, 12α−trihydroxy−5β−cholamide, MACA) on thermally sensitive poly (N−isopropyl acrylamide) with a similar length ($M_w \sim 6.5-9.4\times10^5$ g/mol) first, and then, use the solution temperature to adjust the solubility of the copolymer chains in water, forming a mesoglobular phase. According to the principle of thermodynamics,for a given temperature,copolymer chains with a higher hydrophobic monomer content are more likely to aggregate in the mesoglobular phase, forming larger aggregates, and each aggregate on average contains more copolymer chains. However, the results were again completely unexpected, as shown in the figures (c) and (d) above on the right.

The same as before, before the solution temperature reaches the phase transition temperature, $\langle R_h \rangle$ decreases gradually, and the aggregation number is basically 1, reflecting the single-chain collapsing, completely normal. In the mesoglobular phase, the aggregation number decreases from several thousand chains to several chains as the hydrophobic monomer content increases, not an expected increase. When the hydrophobic monomer content is 4.8%, $\langle R_h \rangle$ keeps decreasing as the temperature raises, seeming no aggregation, but the aggregation number shows a small amount of aggregation, which reflects that the high content of hydrophobic monomer leads to a dominant role of the intrachain shrinkage, very different from intuition, rather strange.

The above two abnormal experimental results fully illustrate that the interchain aggregation in the mesoglobular phase is not entirely dependent on thermodynamic principles, but also related to kinetic processes. Why do the long chains and the chains with a higher hydrophobic content lead to less interchain aggregation in the mesoglobular phase? and why do the changing rate of the solvent quality and the copolymer concentration affect the interchain aggregation in the mesoglobular phase? Microscopically, in the phase separation region, only intermolecular aggregation occurs in the small molecule solution, gradually forming a dilute phase and a concentrated phase, and reaching a dynamic equilibrium. In the mesoglobular phase region, in the copolymer solution, in addition to the interchain aggregation like small molecules, there is an additional intrachain aggregation, resulting in a collapsed chain conformation. This is the essential difference between small molecules and macromolecules in the phase transition.

Whether it is interchain or intrachain, the aggregation is controlled by translational diffusion. Therefore, the time required for interchain aggregation is proportional to the square of the average interchain "interaction distance" (l_I) and inversely proportional to the translational diffusion coefficient (D). For a given mass concentration, in a short-chain solution, the number of chains is larger and the interchain distance is shorter. In addition, the translational diffusion of short chains is relatively fast, so that the time required for the interchain aggregation is short. On the other hand, in the discussion of chain dynamics, according to the Zimm model, it is known that the conformational relaxation time of a macromolecular chain in a dilute solution (the first normal mode) is four times longer than the time required for the chain to diffuse a distance of its average radius of gyration, obviously depending on the chain length.

In the usual dilute solutions of macromolecules, the intuitive error is that the interchain distance is very long. In fact, a simple calculation can tell that the interchain distance is only slightly larger than the average radius of gyration, in the same order of magnitude ($\sim 10^2$ nm). Therefore, when the microphase separation occurs in the mesoglobular phase, in most cases, the intrachain contraction and the interchain aggregation occur almost simultaneously, especially when the macromolecular chains are short, so that it is difficult to form the aggregates with only individual collapsed chains. Only when the molar mass reaches $\sim 10^7$ g/mol and the concentration is close to or less than 10^{-5} g/mL, the interchain distance is about ten times longer than the average radius of gyration, making the intrachain contraction is much faster than the interchain aggregation to form single-chain aggregates. This inference has been confirmed by the above results.

At present, there are many articles about the so-called single-chain copolymer particles, but the authors basically have not rigorously proved that these particles are composed of a single chain. The only correct way to prove it is to use static LLS or ultracentrifugation to show that in the conformational transition of the copolymer chains, the molar masses of the resultant aggregates and initial single chains are exactly the same. For example, one can use static LLS to measure the absolute scattered light intensity of the solution before and after the chain contraction at different angles first, and then, plot the reciprocal of the scattered light intensity against the square of the scattering vector. If the slope decreases with the chain contraction but the intercept remains unchanged, it can be confirmed that such obtained particles indeed contain only a single chain. The method of measuring the particle size by dynamic LLS is not sensitive to a small amount of interchain aggregation. The size exclusion chromatography with a static LLS detector may change the structure of the aggregates due to the dilution, also not reliable. Electron microscopy or other direct imaging methods cannot prove that there is only one copolymer chain inside each aggregate. Unfortunately, in many existing literature, the term "single-chain particle" has been misused and abused. When reading related articles or doing research in this area, readers should get rid of the false and keep the truth.

In Chapter 7, the kinetic study of the chain contraction has confirmed that the conformational relaxation time of a macromolecular chain in dilute solution is faster than 10^{-3} s. In such a short period of time, the interchain aggregation and the intrachain contraction proceed simultaneously, resulting in a rapid increase of the local concentration. It is not difficult to imagine that once the intrachain contraction of each aggregate is completed, the local concentration must be higher than the overlap concentration, and each aggregate becomes a small "droplet" of the semidilute solution, forming a large number of micro-concentrated phases suspended in the dilute phase. The subsequent phase separation process becomes the continuous aggregation between small "droplets" or between a droplet and individual chains, or a kinetic equilibrium is established.

These small "droplets" and the collapsed individual macromolecular chains can be regarded as small particles. Under the agitation of thermal energy, they all move randomly in the solution.

If they can gradually fuse into larger particles through random collisions, and finally separate into two macroscopic phases, not only will the two colliding particles need to interact for a period of time (τ_I), but also it has to be much longer than the time required for the centroid of the chain inside the particle to move one end-to-end distance (τ_R). Only in this way can the chains within the two particles have enough time to diffuse, penetrate, and merge with each other. Obviously, τ_I is related to the average inter-particle "distance" (l_I), the average particle velocity ($\sqrt{\langle u^2 \rangle}$) and the translational diffusion coefficient (D) as

$$\frac{l_I}{\sqrt{\langle u^2 \rangle}} \lesssim \tau_I \lesssim \frac{l_I^2}{D}$$

Its physical meaning is that τ_I is between the two extreme times required for a linear collision and a translational diffusion collision. The centroid movement related to τ_R has been discussed in detail in Chapter 9. In the semidilute solution with a higher concentration, the Rouse model should be used; namely,

$$\tau_R \approx \tau_{GC} = \frac{\pi\eta}{k_B T} \frac{b\, N_d L_{c,d}^2}{(N_d/N)^2} = \frac{\pi\eta}{k_B T} \frac{b^3 N^4}{N_d^{3-\alpha_R}} \qquad \rightarrow \qquad \tau_R = \tau_0 \left(\frac{C}{C^*} \right)^{\frac{3-2\alpha_R}{3\alpha_R-1}}$$

where $\tau_I \sim \tau_0 \sim 10^{-4}$ s and $\tau_R \sim \tau_{GC} \sim 10^{0}$ s. Based on the previous viscoelasticity discussion, a material is "viscous" or "elastic", depending on the relative length between the interaction time and the molecular relaxation time. At a normal temperature and on the usual time scale, water is a viscous liquid. However, if a small rock is forced quickly across the water surface at the smallest possible angle, it will bounce on the water surface several times, forming a series of splashes, as if the water surface is elastic. More than two thousand five hundred years ago, Sun Wu mentioned in "The Art of War by Sun Tzu": "Hitting the water surface so fast that the stone flies is due to the potential". Here, the "potential": extremely fast speed, and extremely short time.

In the mesoglobular phase, if the interaction time between two colliding aggregates is much shorter than the time required for long chains of macromolecules in the particles to diffuse, penetrate, and merge between two particles, each aggregate is elastic, just like two elastic glass balls, separating from each other after the collision. The collision would not make them to become a larger glass ball (very low probability). It is the intrachain contraction within each aggregate that terminates the interchain aggregation. Therefore, due to the "viscoelastic" effect, these small aggregates are in a kinetically stable state, and will not further aggregate and separate into macroscopic dilute and concentrated phases in a short time.

In summary, completely different from small molecules, the relaxation of macromolecular chains is slower, especially in semidilute solutions. It is difficult for a macromolecular system to reach a true thermodynamic equilibrium state, and it is often trapped in a metastable state that is not the lowest in energy and depends on the path. Therefore, some thermodynamically unexplainable phenomena are often observed. Using this "viscoelasticity" kinetic stabilization mechanism as well as the coexistence and competition of interchain aggregation and intrachain contraction in the formation of the mesoglobular phase, the above "abnormal" experimental results can be explained in turn. The details are as follows.

1. "For a given copolymer solution, the average aggregation number and the average size of the aggregates decrease with the concentration." A decrease of the concentration increases the interchain distance, and slows down the interchain aggregation, but has no effect on the conformational change inside the chain, so that the intrachain contraction is relatively faster than the interchain aggregation. Within the same time, the interchain aggregation relatively decreases, forming smaller aggregates.

2. "For a given copolymer solution and concentration, the average aggregation number and the average size of the aggregates decrease as the temperature change rate increases." In dilute solutions, intrachain contraction is generally faster than the interchain aggregation. Therefore, when the temperature change rate is faster, the intrachain contraction precedes the interchain aggregation, which leads to less interchain aggregation, forming smaller aggregates.

3. "For a given macromolecular mass concentration, the average aggregation number and the average size of the aggregates decrease as the chain length increases". In a long chain solution, the interchain distance is longer, making the interchain aggregation slower than thee intrachain contraction. In the aggregate, the relaxation time of the chain centroid is proportional to the fourth power of the chain length ($\tau_E \sim N^4$). Therefore, in the competition, once the intrachain contraction is faster than the interchain aggregation, long chains have a stronger "viscoelastic" effect, so that further inter-aggregate aggregation nearly stops, forming smaller particles.

4. "For a given chain length and mass concentration, the average aggregation number and the average size of the aggregates decrease with the solvophobic groups on the chain increases". The conformation of the copolymer chain with more solvophobic groups is already in a relatively shrinking state before the phase change, which is conducive to the intrachain contraction. In the process of phase separation, in order to reduce the interfacial energy, the solvophobic groups tend to concentrate inside the aggregate, and the strong dispersion attraction between them further prolongs the relaxation time of the chain in the aggregate and effectively slows down further fusion between the aggregates. Therefore, once the intrachain contraction is complete, the aggregation stops and smaller aggregates are formed.

In addition to thermodynamic consideration, the aggregation in the mesoglobular phase is mainly controlled by the competition between interchain aggregation and intrachain contraction. If wishing to get smaller metastable aggregates, one needs to use a long copolymer chain and a lower concentration, speed up the change rate of solvent quality, and appropriately increase the content of solvophobic groups in the chain. In biology textbooks and literature, when talking about the folding of a protein chain into a three-dimensional structure, it is only based on the principle of thermodynamics that the protein chain moves the hydrophilic and hydrophobic groups to the surface and the interior, respectively, with less interface energy to form a stable, biologically active single-chain structure, not knowing the above-discussed "viscoelasticity" kinetic stabilization mechanism. In the biological evolution of hundreds of millions of years, in addition to reducing the interface energy, the hydrophilic group on the surface shortens the interaction time between protein chains (τ_I); the hydrophobic groups inside prolong the relaxation time of the protein chain in the folded structure (τ_E), such that $\tau_I \ll \tau_E$. The protein chain "smartly" satisfies the kinetic stability condition based on "viscoelasticity" in one step. The hundreds of millions of years of evolution enable many protein chains to be easily fold into single-chain structures.

The macromolecular "self-assembly" that has been in the ascendant for more than 30 years is the study of the metastable behavior of chains in the mesoglobular phase, including the "self-assembly" of various block copolymers in selective solvents, and various forms reaction-induced "self-assembly", the "self-assembly" of protein-like copolymers in water, and even the random aggregation of random copolymers in selective solvents are packaged as "self-assembly". All of these macromolecular aggregates are also addressed as "nano" particles, and that's all. Hope that researchers are still working in this field, especially young researchers, can realize some important differences as follows.

Aggregation and assembly: These are two Chinese words with obviously different connotations and extensions. "Assembly" is a subset of "aggregation", and aggregation is not necessarily assembly. Aggregating (gathering) piles of children's building blocks in a disorderly manner is different from assembling them in an orderly manner into a well-defined structured little house. The precipitation of small molecules in poor solvents is usually aggregation, while an ordered crystallization is assembly in a sense. Proteins in cells are often folded into a specific conformation, and then assembled into an ordered oligomeric bioactive structure. Here "disordered" and "ordered" are only one of the points that distinguish between aggregation and assembly, not all of them.

Flexible block polymers can form microstructures with various morphologies in selective solvents. There is no difference between the micro-aggregation of insoluble segments and the macro-aggregation of homopolymers in the same poor solvent, and they are essentially messy. The only difference is that the soluble segments prevent macroscopic aggregation. When only thermodynamic factors are considered, the ability of each solventophile group to stabilize the interface determines the total interface area that can be stabilized, that is, the final morphology of the aggregates. When there is only thermal energy, the process must follow the statistical random fluctuation of thermodynamics. Therefore, the size of these aggregates and the aggregation number are not the same, nor can they be the same. Their formation process is not strictly assembled (although the author also often referred it as self-assembly before). So why do people confuse these two different processes? There are "unintentional" and "intentional" here.

Those who are unintentional are just follower; those who are intentional are doing it with purposes. The reason is that "aggregation" is an old concept without any new ideas; while "assembly" and "self-assembly" seem to be more "novel", "beautiful" and "attractive", and conform to the highly praised "innovative" trend of today!

Assembly and dissolution: Whether it is aggregation or assembly, conceptually speaking, it refers to the process and result of individual molecules in a solution "coming" together with a change of the solvent quality. However, in the literature, especially when studying ionomer solutions, it is often seen that the "particles" in the solution larger than the solute molecules are confusingly collectively referred to as "aggregates" or "assemblies." However, they obviously forgot that when preparing a solution, the starting point is not to put one molecule after another into a solvent, but to mix a macroscopic solute with a solvent. In the process of dissolution, the solute is usually gradually dispersed into the solvent in the form of individual molecules. Therefore, when the solution conditions remain unchanged, the "particles" observed after preparing the solution should be insoluble residuals, rather than "aggregates or assemblies" formed in the solution. This issue was clearly pointed out by the author in previous articles, but it did not arouse the attention of relevant people. Similar conceptual errors continue to recur. Therefore, when studying the aggregation or assembly of polymer chains in a solution, it is necessary to determine whether the starting solution is a true solution (no insoluble "particles"). The best experimental method to check it is static LLS or ultracentrifugation. They can measure the absolute average molar mass, which should be close to the value of individual chains.

Precipitation polymerization and polymerization-induced assembly: the former is one of the typical synthesis methods in polymer textbooks, while the latter is a newly created term, called pisa in English. In fact, as early as the 1990s, the author clearly put forward the concept of chemical reaction-induced self-assembly (perhaps someone put forward this concept earlier). There is no essential difference between "precipitation polymerization" and "polymerization-induced assembly". The only difference is that the latter introduces a short soluble chain as an interface stabilizer, entering the metastable mesoglobular phase. Therefore, can we also call the microprecipitation or microemulsion or even emulsion polymerization with some added stabilizers as polymerization-induced assembly? The author believes that "polymerization-induced assembly" is a typical example of putting old wine into new bottles and shouting and selling. It needs to be pointed out that no one denies the potential value of this method in application, but it is not necessary to deliberately create new words, labels, a so-called "new" field, and claim to be a leader. Therefore, it is sufficient to call this type of chemical change as precipitation or micro-precipitation polymerization.

Passive and active assembly: Some people in the world put forward the new concept of "dynamic self-assembly" without careful consideration. There are many followers at home and abroad. In many lectures, the author questioned speakers, "what is static self-assembly"? The speakers were all speechless. Obviously, they have not thought about this problem or are used to following the guidance of celebrities. The assembly process must be dynamic! Therefore, there is no static self-assembly at all. Since there is no static self-assembly, how comes dynamic self-assembly? However, the discussion did not end here. One argument is that dynamic self-assembly refers to changing from one ordered state to another. No, it has been described in physics for a long time, that is, the "order-to-order transition". Another argument is that dynamic self-assembly refers to the need to rely on other energy other than thermal energy in the process. Wrong, it has been defined in physics for a long time, that is, what is driven by only thermal energy is called a passive process, and what needs additional energy to drive is an active process. Therefore, the correct statement should be: passive and active self-assembly.

Enthalpy change and entropy change: In an isothermal and isobaric spontaneous aggregation process, free energy of a system decreases. In other words, the sum of the enthalpy change (ΔH) and the entropy change ($-T\Delta S$) is negative. Who controls the process? It is worth pointing out that the main driving force here is often the entropy change rather than the often misunderstood enthalpy change (various interactions). The aggregation of oppositely charged macromolecular chains is a typical example. In the literature, it is often attributed to electrostatic interaction (enthalpy change), and it is forgotten that before mixing two oppositely charged macromolecular chains, each chain has its own small molecular counter ions, and there is electrostatic interaction already. In fact, the aggregation or assembly of oppositely charged polymer chains releases a large number of counter ions, which greatly increases the translational entropy of the system and reduces the free energy. Therefore, the main driving force of this process is entropy change. Strictly speaking, physics related to polyelectrolytes originates from physics of its counter ions, not belonging to macromolecular physics. Similarly, hydrogen bonding induced aggregation of macromolecular chains is due to the release of solvent molecules that originally formed hydrogen bonds with the respective macromolecular chains, resulting in an increase in the entropy of the system. Micro-precipitation polymerization is also mainly driven by entropy changes. Before the reaction, the reaction monomer is dissolved in the solvent, and the enthalpy and entropy increase simultaneously, but the entropy variable term ($T|\Delta S|$) is greater than the enthalpy variable term (ΔH). Polymerization causes the translational entropy of the system to gradually decrease until $\Delta H > T|\Delta S|$, a phase transition occurs, leading to the aggregation and precipitation. Not only young researchers, but also senior domestic and foreign researchers often make such basic conceptual errors.

In summary, although the forms, causes, and chemical groups that induce the aggregation or assembly of macromolecular chains are ever-changing, they can all be explained qualitatively or quantitatively with the existing thermodynamics and macromolecular physics, without departing from the existing theoretical framework and knowledge system. "There is nothing new under the sun." They are all metastable small aggregates in the mesoglobular phase. In essence, they are micro-concentrated phases coexisting with dilute phases containing individual collapsed chains, achieving a metastable microphase separation. Since the relaxation of macromolecular chains in the aggregates is extremely slow, the collision and interaction time of two aggregates is much shorter than the relaxation time of the chains inside each aggregate. Therefore, each aggregate is like a glassy ball, and the probability of two colliding aggregates fusing into a slightly larger aggregate is extremely low. The aggregates are kinetically stable due to this novel "viscoelastic" stabilization mechanism.

Readers need to be soberly aware that, first, from the perspective of pure academic and knowledge innovation, most of this type of research makes use of existing knowledge and studies various macromolecular systems without creating new chemistry, physics, biology or mathematics, that is, no intellectual contribution! Second, the copolymer aggregates in the mesoglobular phase are in a metastable state, depending on the formation path, rather than in a thermodynamically stable state. Therefore, it is impossible for this kind of research to obtain new and universal laws of physics. In other words, the conclusions of this type of research vary with the system and formation path. Unlike the aggregation of small molecules in colloids, the relaxation of copolymers in aggregates can last for years. Some aggregates formed in the mesoglobular phase often seem to be very stable, but in fact they are still in a metastable state. It is due to these reasons that this kind of research is not basic research in a strict sense.

It is precisely because of this sufficiently stable metastable state that various aggregates generated in the mesoglobular phase can meet the needs of many applications, so that research in this area can produce new materials, including environmentally responsive materials, drug delivery system, biomedical accessories, etc. Researchers interested in continuing to work in this field need to focus their main directions and energy on some specific application problems. For every researcher, **research is about solving problems**! **"One should put his results either on the books shelf or on the goods shelf."** This point of view from author is not new, but it has been recognized by many researchers in recent years.

The study of the aggregation of macromolecules in the mesoglobular phase should focus on solving specific problems originating from production, daily life and biomedicine, rather than discussing some potential and possible applications in general. For example, in the research of controlled release of drugs, one should consider, select, design and synthesize appropriate macromolecules for a specific drug and its clinically specific delivery method, find a proper path, obtain the desired aggregates, and solve one related problem, instead of starting from the existing or synthesizable macromolecules in laboratory. Do not assemble for the sack of assembly. One needs to keep in mind that the purpose of research is to solve problems, not just to publish papers. **"One should put his results either on the books shelf or on the goods shelf."**

Summary

For binary mixtures of small molecules, as the compatibility of the two components changes, either they mix with each other at the molecular level to form a uniform solution (one phase) or the mixture separates into two mutually saturated uniform solutions (two phases, dilute phase and concentrated phase), one of the two. There is a macroscopic physical interface between the two phases. In the binary mixture of homopolymer and small molecule solvent, there is the same phase change from one phase to two phases. But before the phase transition, there is a special conformational transition of macromolecules in the one phase region. In the binary mixture of copolymer and small molecule solvent, as the solubility difference (amphiphilicity, Δ) of the copolymerized unit in the solvent increases, there will be a mesoglobular phase in the original two-phase region, in which a limited number of copolymer chains aggregate together to form individually dispersed micro-concentrated phases, coexisting with a saturated solution containing individual collapsed copolymer chains, that is, a microphase separation. In the phase transition of homopolymers, interchain aggregation will also occur, but the aggregates further merge with each other, so that the mixture separates into two macroscopic phases. While the aggregates formed in the mesoglobular phase can stay in the metastable state for many months or even years.

This process of microphase separation is often called as various forms of "self-assembly". These small aggregate particles that form the microphase region are often referred to as novel titles such as "self-assembled subjects" and "nanoparticle", but their underline principle never changes. Readers need to know clearly that these small aggregates are in a metastable state, and the free energy of the system is locally at a relatively low point, but it is not a thermodynamically stable state. Therefore, their formation is related to the path of microphase separation. Along different paths, it can fall into different low energy points, forming aggregates with different structures and morphologies, which are ever-changing. Therefore, the phase transition of the copolymer solution is very complicated. There is no universal quantitative law.

Like the phase transitions of small molecular solution and homopolymer solution, a change of the experimental conditions such as temperature, pressure, solvent composition, etc. can lead the copolymer solution to be separated into two phases, and enter the mesoglobular phase region. The difference is that in addition to the solvent quality, the phase transition of the copolymer solution has an additional variable of the degree of amphiphilicity. The solvent quality and the degree of amphiphilicity depend on each other. Experimentally, there is no way to fix one, and only change another.

According to the principle of thermodynamics, these micro-concentrated phase, called droplets or aggregates, should further merge to form macroscopic two phases. However, in the usual experimental time scale, these aggregates are often relatively stable. It is the unique "viscoelasticity" of macromolecules that prevents them from further fusion and phase separation, which provides the basis for kinetic stability. In other words, in the mesoglobular phase, if the interaction time between two colliding aggregates ($\tau_I \sim 10^{-4}$s) is much shorter than the time required for macromolecular chain to diffuse, penetrate and fuse between the two particles ($\tau_R \sim 10^0$s), they are like two elastic glass balls that separate after collision, and will not fuse into a larger aggregate due to the collision.

Microscopically, when the microphase separation appears in the mesoglobular phase, the inter-chain aggregation and intrachain contraction always occur simultaneously. For common copolymers, the average time for intrachain contraction is only slightly faster than the time for interchain aggregation. Therefore, when a copolymer chain shrinks, it is always accompanied by interchain aggregation, difficult to form single-chain aggregates. When the amphiphilic copolymer chain contracts, it always tends to move the solvophilic group to the surface of the aggregate to reduce the surface energy of the solvophobic unit. Different phase transition pathways affect the competition between the interchain aggregation and intrachain contraction, thus forming a metastable state with different microphase structures in the mesoglobular phase, including different aggregate sizes, densities and aggregation numbers.

Due to $\tau_I \ll \tau_R$, once the intrachain contraction in the aggregate have completed, the interchain aggregation actually stops. What is left is the elastic collision between the aggregates, and the probability of becoming a larger aggregate is extremely low. The microphase separation is "frozen" in the mesoglobular phase by the unique viscoelastic properties of macromolecules, forming a kinetic stable state. Therefore, reducing the concentration and accelerating the rate of solvent quality change can reduce the probability of the interchain aggregation, resulting in smaller aggregates with a smaller aggregation number. Increasing the chain length and introducing appropriate solvophobic groups can speed up the intrachain contraction and prolong the relaxation time of the macromolecular chains in the aggregates, making it more difficult to diffuse, penetrate and merge between he aggregates, and is also more conducive to the formation of smaller, kinetically stable aggregates.

The study of the path-dependent mesoglobular phases is not able to result in universal and quantitative physical laws and generate new knowledge. Strictly speaking, it is not a basic research, but its application value is very high. Using its path dependent characteristics, one can manipulate and adjust the formation conditions of the mesoglobular phase to obtain various aggregates with sufficient stability to meet different application requirements. Research in this area can produce new materials, including environmental response materials, drug delivery systems, and biomedical excipients. Researchers who are interested in continuing to work in this field need to focus their main direction and energy on solving some specific and real problems related to production, life and medical treatment. The following figure summarizes the main content discussed in this chapter.

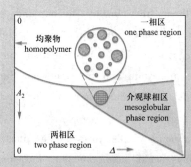

由有限条链聚集而成的微浓相(粒子)与稀相达至动力学平衡,形成一个特殊的处于亚稳态的介观球相。
Micro-concentrated phase (particles) made of a limited number of macromolecular chains is in equilibrium with a dilute phase, forming a special metastable mesoglobular phase.

微浓相的结构(大小和聚集数)和形成路线有关,因此,并无一个普适物理规律来描绘介观球相的生成。
The structure of micro-concentrated phase (aggregation number and size) depends on its formation path, no universal physical law to describe the formation of mesoglobular phase.

动力学:基于"黏弹性"的稳定机理
Kinetics: A stabilization mechanism on the basis of "viscoelasticity"

$$\frac{l_I}{\sqrt{<u^2>}} \leqslant \tau_I \leqslant \frac{l_I^2}{D} \qquad \boxed{\tau_I \ll \tau_R} \qquad \tau_R \approx \tau_I \left(\frac{C}{C^*}\right)^4$$

快速变化溶剂性质(变温或溶剂速率)、用长链和适当增加疏水成分含量可得小的聚集体粒子。
Fast changing solvent property (rate of changing temperature or solvent), using longer chain and properly increasing hydrophobic content can lead to smaller aggregates (particles).

第十二章　大分子胶体

两种物质可以在分子水平上均匀地混合成一个稳定的溶液,为一相二元混合物。一种物质也可以小聚集体的形式均匀地、较稳定地分散在另一种物质之中形成胶状分散体,简曰:胶体,为两相二元混合物。聚集体也常被称作胶体粒子。聚集体的尺寸一般小于一个微米。当聚集体较大时,混合物也常被称作悬浮混合物(液),其与胶体之间并无严格的界限和物理上的区分,可笼统地称作胶体悬浮混合物(液)。以此定义,处于介观球相的两亲性共聚物溶液也可被视为一种特殊的胶体分散液。

一个溶液和一个胶体分散液最本质的物理区别在于,在溶液的一相区,没有大于分子尺度的物理界面。而胶体则至少有两相,聚集体和分散介质之间有着明显的微观物理界面。依据定义,无论是溶液还是胶体,其可为固、液或气态,并非限于液态。例如,细小液滴分散在气体中形成气溶胶和一些金属合金溶液等。简而言之,胶体分散液的种类千变万化;形成的方式也各有千秋。各种文献、专著和专利更是不计其数。下面,主要讨论一些胶体的稳定机理,尤其是大分子胶体特有的"黏弹性"动力学稳定机理,以及接枝在胶体粒子表面上的大分子刷和微凝胶等特殊的胶体分散体系,而略去对传统大分子胶体的介绍。

很早已知,两亲性小分子可在一个选择性的溶剂中聚集形成胶束,分数在体系中。两亲性的共聚物大分子也可在一个选择性的溶剂中聚集形成许多分散在稀相(含一个个蜷缩大分子链的饱和稀溶液)的微浓相(大分子聚集体)。一般而言,将一个给定质量、线度为 L 的物质分割 n 次成小的、线度为 L/n 颗粒时,粒子数为 $L^3/(L/n)^3 = n^3$。所有粒子的总表面积(S)是初始的表面积(S_0)的 n 倍,数学表达为

$$S_0 \sim L^2 \text{ 和 } S \sim \left(\frac{L}{n}\right)^2 n^3 = L^2 n \qquad \rightarrow \qquad \frac{S}{S_0} = n$$

因此,总表面积随着线性分割的次数增加。如果一个不溶的宏观物质可分散到一个介质中形成微小的分子聚集体,其表面积(能量)将极大地增加。依据热力学原理,这将不会自发地发生。相反,在分散液中,不溶的聚集体总是倾向于聚集在一起,降低界面能量。

一个典型的例子是将油和水混合。在正常的情况下,混合物自动地分成两层(相):水相中含有饱和的油,而油相中则含有饱和的水。微观上,油、水分子不断地、来回地穿越界面,各自的速率相同,达到一个热力学稳定的动态平衡。如果没有挥发,两相的体积和组成不变。剧烈地搅拌或摇晃(输入能量)可使油以小的液滴分散到水中。但是,一旦搅拌或摇晃停止,两个碰撞小油滴之间油分子的扩散时间远短于油滴之间的相互作用时间,

每次碰撞都导致一个较大油滴的形成，不断聚集，混合物瞬间恢复至宏观的上下两层（相）。

在热力学的驱动下，大分子聚集体（微浓相）也倾向于继续聚集，降低界面能量，但受制于前述的"黏弹性"效应，即一个动力学（kinetic）稳定机理，依赖于链间聚集和链内收缩之间的竞争。由于两个碰撞聚集体内的链互相扩散、缠结和融合的时间比两个聚集体之间碰撞的相互作用时间长一千倍以上。一旦链内收缩完成了，溶液中聚集体之间的碰撞就好似两个玻璃小球之间的弹性碰撞，融合成一个稍大的小球的概率几乎为零，即聚集体之间的进一步聚集在实际上已停止，形成动力学（kinetic）稳定。前提是聚集体的尺寸很小，密度接近溶剂（分散介质），热能可驱动聚集体无规运动，不会沉淀。

因此，为了将一个不溶的物质分散到一个介质中形成相对稳定的胶体，往往需要加入可分布在聚集体表面上的两亲性稳定剂（也称作分散剂或表面活性剂）。界面稳定剂除了减少界面能以外（热力学考量），还缩短了两个碰撞聚集体之间的作用时间和阻止了二者间的分子交换和融合（动力学机理）。这样形成的各种胶体在生活和生产上都有着极其广泛的应用。例如，在工业化的高聚物合成中，在表面活性剂的帮助下，油性的反应单体（如苯乙烯、丙烯酸酯等）先被分散成小的液滴，形成乳液或悬浮液，再开始聚合反应。这样的方式有利传热控温、降低黏度、改善均匀性和增加聚合速率。在日常生活中，利用清洁剂和洗涤剂中含有的表面活性剂，将锅碗瓢勺上的油脂和衣服上的油污剥离，分散到水中冲走，等等，不胜枚举。

稳定剂的种类繁多、千变万化。按所带电荷，分为阴离子、阳离子和中性稳定剂；按照分子大小，可分为小分子、寡聚物和大分子稳定剂；按照与界面的相互作用，可分为物理吸附和化学键连；不一而足。然而，万变不离其宗！注意：在多数的情况下，除了受制于热力学以外，大分子胶体的形成还受动力学的控制。即使是对完全一样的体系，不同的改变溶剂性质的方法或不同的反应速率会导致不同的胶体结构。因此，除了完全受制于热力学的少数体系以外，并无普适的物理规律。换而言之，这里只是运用，而无法创造知识。

下面，先讨论热力学控制的体系。这里，控制的物理参数是"每个稳定基团在界面上占据的表面积(s)"，对一个给定的体系，s 为一常数。一个典型的例子是受热力学控制的微乳聚合，其中稳定剂与待分散的油性单体的含量比较接近。被稳定剂包裹的"油滴"大小仅有 10 nm 左右。

在相当长的时间内，微乳聚合领域的研究人员仅定性地知道，随着稳定剂与油性单体质量比（投料比，W_s/W_m）的增加，所得聚合物粒子逐渐变小。在传统的"粒径对投料比"的作图中得一条曲线，粒径随着投料比的增加，单调下降。实验上，已知稳定剂和油性单体的投料质量（W_s 和 W_m）；稳定剂分子的摩尔质量（M_s）和单体密度（ρ）也是已知的两个参量。实验上，

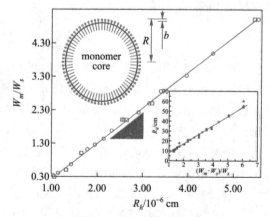

Wu C.Macromolecules,1994,27:298.

（Figures 2,permission was granted by publisher）

利用动态激光光散射测得的"油滴"半径(R)包括了表面上厚度为b的稳定层,见上图中的示意图。假定稳定剂在"油滴"表面上和水中的分配系数为$0 < \gamma < 1$。从每个"单体油滴"的表面积与体系中"单体油滴"的总数之积可得体系中"单体液滴"的总表面积(S)。另外,利用分配系数(γ)得到吸附在界面上的稳定剂总质量,除以稳定剂的摩尔质量可得界面上稳定剂分子的总数(N_s),即

$$S = 4\pi (R - b)^2 \frac{W_m}{\rho \frac{4\pi (R - b)^3}{3}} = \frac{3 W_m}{(R - b)\rho} \quad \text{和} \quad N_s = \frac{\gamma W_s}{M_s} N_{AV}$$

由S和N_s之比,可再算出每个稳定剂分子在界面上平均所占据的面积(s)如下:

$$s = \frac{S}{N_s} = \frac{3 W_m}{(R - b)\rho} \frac{M_s}{\gamma W_s N_{AV}} \quad \rightarrow \quad \frac{W_m}{W_s} = s \frac{\gamma \rho N_{AV}}{3 M_s} R - s \frac{\gamma \rho N_{AV}}{3 M_s} b \quad (12.1)$$

注意,上式与上图参考文献中的公式略有差别,更加简洁;其中,对给定的体系,上式右边R前面的系数里的参量中,除了s以外,均为常数;截距除以斜率可得分散剂分子伸出界面的厚度(b)。

上图源于由十六烷基三甲基溴化铵稳定的苯乙烯微乳。数据清楚地显示,W_m/W_s对R的作图为一条直线,证实s的确为一个常数。由直线的斜率和截距分别可获$s = 0.182 \text{ nm}^2$和$b = 1.0 \text{ nm}$,两个以前无法获得微观参数。上式于1994年发表后,Antonietti和其合作者[Macromolecular Chemistry and Physics, 1995, 196:441]将他们以前的数据重新作图,如上图中的插图所示,完全落在上式所预测的线上。

这样的作图方法看似只是简单地将原本的W_s/W_m换成W_m/W_s,但在本质上却完全不同。前者没有任何物理意义,只是给出了变化趋势的表观曲线,犹如将数据列表。而依据上述推论得到的后者则有明确的物理意义。所得直线的斜率和截距分别对应着两个重要的微观参数:每个分散剂分子在分散相和连续相的界面上占据的平均表面积(s)和露出界面的长度(b)。以正视听,有必要指出,Antonietti和其合作者在他们的文章中将这样作图法称作"The Antonietti-Wu"作图法,其实,这一作图法和Antonietti及其研究合作者毫无关系。如果真的要命名,也应是"吴氏作图法(The Wu Plot)"。

上述结论"每个稳定基团在界面上所占据的表面积为一常数"也可被延伸到其他由热力学主导的过程。例如,由共聚单体A和B组成的线型两亲性嵌段共聚物链在选择性溶剂中形成胶束状的核壳聚集体,如右图中的示意图所示,其中稳定基团不再是带电荷的小分子,而是一条亲溶剂的大分子嵌段。除了可通过选择不同的A和B改变两亲性程度以外,两个嵌段各自的聚合度(m和n,链段长度)也为变量,记为$A_m B_n$。假定核壳聚集体由N_p条两嵌段共聚物链聚集而成,其中核由B

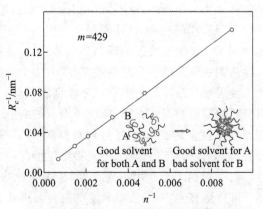

Gao J, Wu C. J. Macromolecules, 2000, 33:645.
(Figures 2, permission was granted by publisher)

嵌段组成,半径为R_c,密度为ρ,那么

$$\rho\,\frac{4\pi\,R_c^3}{3}=\frac{M_0}{N_{AV}}n\,N_p$$

另一方面,每个核的界面面积$(4\pi\,R_c^2)$也正比于每条 A 嵌段所占据的平均界面面积(s)和聚集数,即 $4\pi\,R_c^2=s\,N_p$。与上式结合可得

$$\rho\,\frac{s\,R_c}{3}=\frac{M_0}{N_{AV}}n\qquad\rightarrow\qquad\frac{1}{R_c}=\frac{s\rho\,N_{AV}}{3\,M_0}\frac{1}{n}\qquad(12.2)$$

上式已被写成一个实验可测的形式。如果将 B 嵌段氘代,再利用小角中子散射,可以测得R_c。上图证实了上式,对给定的亲溶剂链段(m),$1/R_c$对 $1/n$ 的作图为一条直线,s 确实为一个常数,并可由斜率获得。图中的数据源于聚环氧乙烷和聚环氧丙烷两嵌段共聚物。在室温下,聚环氧丙烷在水中为不溶嵌段。

另外,对密集地接枝在表面上的短链,接枝层的厚度(δ)和单位面积上的链数$(\sigma,$接枝密度$)$以及可溶解嵌段的长度$(m,$链段数$)$的依赖关系为$\delta\propto m\,\sigma^{1/3}$;而$\sigma=N_p/(4\pi\,R_c^2)$。利用第四章中的定义,$k$个单体组成一个长度为 b 的 Kuhn 链段。在正常情况下,B 嵌段的聚集不能迫使 A 嵌段在界面上伸直。作为柔软线型链,每条接枝 A 嵌段的构象应仍为或稍微变形的无规线团。假定其仍然处于一个无扰状态,其方均末端距为$\langle R^2\rangle=b^2m/k$;而 $s=4\pi\langle R^2\rangle=4\pi\,b^2m/k$。再与 $4\pi\,R_c^2=s\,N_p$结合,得$\sigma=k/(4\pi\,b^2m)$。因此,$\delta\propto m^{2/3}$,如右图所示。

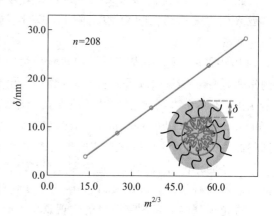

Gao J,Wu C.Macromolecules,2000,33:645.
(Figures 3,permission was granted by publisher)

将 $s=4\pi\langle R^2\rangle=4\pi\,b^2m/k$ 代入上式,还可得$1/R_c\sim m/n$。因此,对给定的 n,$1/R_c$对 m 的作图也为一条直线,可参阅上图文献中的图 2。阅读该文献时,读者会发现,笔者和其学生并没做任何耗时、费钱的聚合反应,也没有借助大型设备做小角中子散射的实验,仅是利用了文献中已发表的现存数据,以及上述不太复杂的推算,得到了两嵌段共聚物在选择性溶剂中聚集的一些基本规律,等效于别人帮他们做了大量的艰辛实验。

读者可从该实例中得到一个教训。永远不要为了文章而急急忙忙地发表没有经过认真分析的实验数据。否则,就成了别人的实验员。请记住"**研究是为了解决问题,而不仅仅是为了发表文章**"。有了好的数据,需要认真分析、建立定量或至少半定量的模型、不仅要理解和解释自己的数据,还要将结论推而广之,可以预测和指导其他类似类型的实验。

另一个例子是利用不同长度的、水溶性的聚环氧乙烷大单体与疏水的苯乙烯或甲基丙烯酸甲酯在 4:1 的乙醇和水的混合物里共聚得到无皂(没有表面活性剂)的、稳定的聚苯乙烯(PS)或聚甲基丙烯酸甲酯(PMMA)粒子。加入大量乙醇是为了增加疏水共聚单

体的溶解度。在共聚反应中,水溶性的聚环氧乙烷接枝在生成的疏水聚苯乙烯或聚甲基丙烯酸甲酯链的一端上。聚合导致平动熵减少,出现微相分离,其中疏水的 PS 或 PMMA 链向内聚集,形成一个"核";亲水的 PEO 留在了表面,形成了一个"壳",好似在界面上接枝了一个"刷子",减少了界面的能量,有利于热力学稳定。

同时,还分别减少和增加了相互作用时间和聚集体内的链在聚集体之间的互相扩散、缠结和融合的时间,使得 $\tau_I \ll \tau_R$,促进了动力学稳定,一石二鸟。只要将稳定基团从带电荷的离子换成中性的 PEO 短链就可再次沿用式(12.1)。右图清晰无误地显示,对两个完全不同的体系和不同长度的聚环氧乙烷短链稳定剂(每一条直线中的不同符号),疏水单体("油")与 PEO 稳定剂的投料质量比(W_m/W_p)对聚集体的流体力学半径(R_h)作图为一直线,进一步佐证了"每个稳定基团在界面

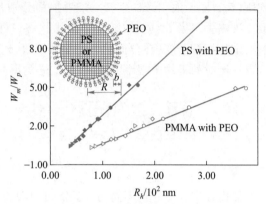

Wu C,Akashi M,Chen M Q.Macromolecules,1997,30:2187.

(Figures 2 and 4,permission was granted by publisher)

上所占据的表面积为一常数"。依据式(12.1),从每条直线的斜率和截距分别可得每条聚环氧乙烷短链在界面上所占据的平均表面积(s)和稳定剂的层厚(b)。对两个体系分别有

$$s = \frac{7.54 \times 10^{-5}}{\gamma} M_P \quad \text{(PS)} \quad \text{和} \quad s = \frac{1.79 \times 10^{-4}}{\gamma} M_P \quad \text{(PMMA)}$$

其中,γ 和 M_P 分别为稳定剂聚环氧乙烷短链在混合溶剂中和界面上的分配系数和平均摩尔质量,完全符合前述的 $s = 4\pi \langle R^2 \rangle \propto M_P$。与 PS 粒子相比,每条同样长度的 PEO 链可稳定更大的 PMMA 表面,可归于 PMMA 在混合溶剂中的溶解度更大,物理上合理。从截距得到的 b 为 2~4 nm,取决于 PEO 的链长,与预期值很近。读者可查阅上图中的文献。

一些具体的、非热力学控制和稳定大分子胶体的例子讨论如下。在胶体制备中,在粒子的表面上接枝一层水溶性链是一种稳定胶体的方法。在生物医药领域,利用一层聚合物刷子阻止蛋白质在表面的吸附是一种常见策略。微观上,每条接枝大分子链可看作处于一根由其周围的链构成的一根长度为 L_0,半径为 R 的"柱状小管"之中,其横截面面积为 $s = \pi R^2$;体积为 $v = sL$,如右图中的插图所示。如在第七章中讨论大分子链在拉伸流场下穿过柱状小孔时所描述的,一条含有 N 个长度为 b 的链段的大分子链在一根"柱状小管"中的自由能包括与熵变有关的两部分:由排除体积效应造成的链扩展和

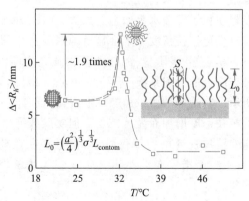

Hu T J,Wu C.Physical Review Letters.1999,83:4105.

(Figures 4,permission was granted by publisher)

由熵弹性引致的扩展链的收缩。用热能归一后的自由能的数学表达如下：

$$\frac{G}{k_B T} \approx \frac{b^3 N^2}{2sL} + \frac{L^2}{R_0^2}$$ (12.3)

其中，第一项是二体相互作用对的密度，其也为熵变。第二项是链的构象偏离无扰方均距（$R_0^2 = b^2 N$）时引致的熵变。二者前面分别与二体排除体积效应和构象熵变有关的常数已被略去。当 $L = R_0$ 时，链处于无扰状态，排除体积效应消失，故第一项为零，$G = k_B T$。对给定的接枝密度（单位面积上的接枝链数，$\sigma = 1/s$），将上式对 L 微分，求平衡时的极值，可得平衡时的平均末端距，即接枝层的平均厚度（L_0）：

$$\left(\frac{\partial G}{\partial L}\right)_{T,p} \approx k_B T\left(-\frac{b^3 N^2}{2s\,L^2} + \frac{2L}{b^2 N}\right) \quad \rightarrow \quad \left(\frac{\partial G}{\partial L}\right)_{T,p} = 0 \rightarrow L_0^3 \approx \frac{b^5 N^3}{4s}$$

依据伸直长度的定义，$L_c = bN$，代入式（12.3），可得平衡时的 L_0：

$$L_0 = \left(\frac{b^2}{4}\right)^{1/3} \sigma^{1/3} L_c \propto \sigma^{1/3} N$$ (12.4)

然而在实验上获得不同的和确定可控的接枝密度，尤其是实现长链高接枝密度，并非易事。文献上有许多达到高接枝密度的报道，多数没有经过严格的证明。唯一确定的方法是将接枝链从表面上用化学方法切割下来，测量其浓度和摩尔质量。可是，在绝大多数情况下，切割下的样品量极少，很难精确地表征。无论是采取从表面引发聚合，还是将一条具有一个活性端基的长链接到表面上，都很难实现高接枝密度。

如果不能有控制地把大量的链接到一个给定的表面上，是否可以在给定接枝链数目后，通过缩小表面积来改变接枝密度？基于这一逆向思维，先在低温下将水溶性线型聚环氧乙烷接枝到球状聚 N-异丙基丙烯酰胺微凝胶上，再通过升温使得微凝胶收缩，故可连续地调节接枝密度。微凝胶表面积与其尺寸的平方成正比，故半径收缩 2.5 倍意味着接枝密度增加 6.25 倍。将它们代入上式可得厚度（$L_0 = \Delta R$）改变 ~1.9 倍，正如上图所示。当接枝密度进一步增加时，厚度反而变小，则源于聚环氧乙烷在高浓度下的团簇化。有兴趣的读者可进一步参阅上图中所列文献，了解详情。

为何在胶体粒子表面接枝大分子，尤其是聚环氧乙烷短链，可以表观上减少蛋白质的吸附？长期以来，其被归于接枝大分子刷的亲水性和立体效应，阻止了蛋白质分子接近和吸附在粒子的疏水界面上。然而，即使在最高的接枝密度，接枝层的含水量也超过 90%，所以变化接枝密度并不会显著地改变蛋白质分子和亲水接枝层的相互作用能（ΔE）。由热力学已知，吸附量（θ）就是蛋白质在水溶液和胶体粒子表面上的分配系数，$\theta_0 = 1 - \exp\left(-\Delta E/k_B T\right)$，如右图所示。对给定的 ΔE，θ_0 为一定值，故蛋白质吸附不应随着接枝密度增加而显著减少。但实验数据显示，吸附量随着接枝密度增大不仅减少，而且还在接枝密度很大时，不减而增。亲水性和立体效应无法解释这一反常现象。

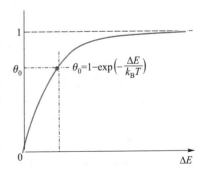

将式（12.4）代入式（12.3），并以无扰状态为参照系，可算出每条接枝链最低的自由能变化为

$$\frac{\Delta G}{k_B T} \approx \left(\frac{b}{2}\right)^{4/3} \sigma^{2/3} N \tag{12.5}$$

一个极端的情形,无扰的接枝链刚刚互相接触,$\sigma_{min} = 1/R_0^2 = 1/(b^2 N)$;另一个极端的情形,接枝链完全伸直,接枝密度最高,$\sigma_{min} = 1/b^2$;其中,$b \sim 1$ nm;因此,

$$\frac{2}{5} N^{1/3} \leqslant \frac{\Delta G}{k_B T} \leqslant \frac{2}{5} N$$

对常用的接枝聚环氧乙烷链($2\,000 \sim 5\,000$ g/mol)而言,$N = 25 \sim 50$。随着接枝密度的变化,

$$k_B T \leqslant \Delta G \leqslant 10 \sim 20\, k_B T$$

接枝链的一端固定,另一端则在热能的激发下,不断地、无规地围绕着平衡位置(L_0)上下波动、振荡。依据统计物理学,当一个热力学体系在给定的实验条件(温度、压强、浓度……)下,达到一个宏观稳定的动态平衡状态后,其自由能必定最低。微观上,在热能的搅动下,组成分子或粒子的各个单元仍在不停地无规运动,使得体系略微地偏离其稳定平衡状态,总是引起自由能增加,造成一个永远为正(增加)的涨落。这里,选择涨落变量为接枝链的平均末端距(L_0),并将自由能围绕着平均末端距处作 Taylor 展开为

$$\delta G = G - G_0 = \left(\frac{\partial G}{\partial L}\right)_{T,p} \delta L + \frac{1}{2}\left(\frac{\partial^2 G}{\partial L^2}\right)_{T,p} (\delta L)^2 + \cdots$$

其中,在 L_0 处的所有自由能的偏微分均为常数。G 无规地涨落,故 δL 的时间平均为零。故经时间平均后,右边的第一项消失。如果不考虑涨落的三次方贡献,上式成为

$$\langle \delta G \rangle = \frac{1}{2}\left(\frac{\partial^2 G}{\partial L^2}\right)_{T,p} \langle (\delta L)^2 \rangle$$

另外,从一般的统计考虑出发,在平衡状态下,每条接枝链自由能的时间平均涨落只有半份热能,即 $\langle \delta G \rangle = k_B T/2$。永远为正,无论接枝链是被压缩还是拉伸。代入上式,可得

$$\langle (\delta L)^2 \rangle = k_B T \Big/ \left(\frac{\partial^2 G}{\partial L^2}\right)_{T,p}$$

上式是式(5.32)的变种。依据式(12.3),可得 $(\partial^2 G/\partial L^2)_{T,p}$ 和 $\langle (\delta L)^2 \rangle$,

$$\left(\frac{\partial^2 G}{\partial L^2}\right)_{T,p} = k_B T\left(\frac{b^3 N^2}{s L^3} + \frac{2}{b^2 N}\right) = k_B T\left(\frac{R_0^2 b N}{s L^3} + \frac{2}{R_0^2}\right) \qquad \rightarrow \qquad \langle (\delta L)^2 \rangle = \frac{R_0^2 L^3}{R_0^4 L_c \sigma + 2 L^3}$$

进而,可求出相对的方均链长涨落 $\langle (\delta L)^2 \rangle / L^2$,其开平方可得 d$L/L$,即

$$\frac{\langle (\delta L)^2 \rangle}{L^2} = \frac{R_0^2 L}{R_0^4 L_c \sigma + 2 L^3} \tag{12.6}$$

其中,R_0^2 和 L_c 与 L 无关。对给定的 σ,当 L 很小或很大时,$\langle (\delta L)^2 \rangle / L^2$ 均很小,故存在一个 $L_{optimal}$,使得 $\langle (\delta L)^2 \rangle / L^2$ 达到一个极大。对上式先求对 L 的偏微分,再令其为零,可得

$$L_{optimal} = \left(\frac{R_0^4 L_c \sigma}{4}\right)^{1/3} = \left(\frac{b^2 \sigma}{4}\right)^{1/3} L_c$$

$L_{optimal}$ 与 $\sigma_{optimal}$ 相关。如同讨论在柱状小孔中的一条线型大分子链,可将一条接枝链按照"管径"(d)分成或 L/d 个"链珠",每个"链珠"含有 N_b 个长度为 b 的链段;"链珠"数

目可由 L/d 或者 N/N_b 求出，二者自然相等，即 $L/d=N/N_b$。注意，$d^2=b^2N_b$ 和 $1/\sigma=s=\pi d^2/4$。利用 $R_0^2=b^2N$，$L=dN/N_b=d^2N/(dN_b)=b^2N_bN/(dN_b)=R_0^2/d=(\pi/4)^{1/2}R_0^2\sigma^{1/2}$，代入上式可得

$$(\pi/4)^{1/2}R_0^2\sigma_{\text{optimal}}^{1/2}=\left(\frac{R_0^4L_c\sigma_{\text{optimal}}}{4}\right)^{1/3} \quad \rightarrow \quad s_{\text{optimal}}=\frac{1}{\sigma_{\text{optimal}}}=\frac{\pi^3b^2}{4} \quad (12.7)$$

将其代入前式，得 $L_{\text{optimal}}\approx L_c/\pi$，约为链的伸直长度的 1/4 至 1/3。上式表达的物理意义很有趣。如要达到接枝链的自由一端有最大的相对涨落，最佳的接枝密度只和链段长度 b 有关，而与总的链长无关，正如预期，因为其是相对涨落。一条链的链段长度则反映了其柔软性，链段越长链越刚性。最佳的接枝密度随着链的刚性降低。这就解释了为何最软的聚环氧乙烷常常被用来接枝表面以减缓体内蛋白质的吸附。将式（12.6）代入式（12.5）和式（12.7），可得如下在最佳的接枝密度时的自由能和接枝链的相对振动幅度。

$$\frac{\Delta G}{k_BT}\approx\left(\frac{b}{2}\right)^{4/3}\left(\frac{4}{\pi^3b^2}\right)^{2/3}N=\frac{N}{\pi^2}\approx\frac{N}{10} \quad \text{和} \quad \frac{\Delta L}{L}=\frac{\sqrt{\langle(\delta L)^2\rangle}}{L}=\frac{\pi}{(6N)^{1/2}}\propto N^{-1/2} \quad (12.8)$$

在最佳的接枝密度时，对典型的聚环氧乙烷接枝链（2 000~5 000 g/mol），接枝链的自由能（G）为 4~6 倍的热能；接枝链相对振动幅度为 26%~18%；绝对涨落振幅：$\Delta L=bN^{1/2}/6^{1/2}=R_g$，这是一个有趣的结果。以上结果的定性理解不难，在接枝密度较低时，$\langle\delta G\rangle$ 会引起自由能的相对变化较大，但接枝链的绝对振动幅度则较小；如果接枝密度接近 σ_{max} 时，接枝链已经伸直，$\langle\delta G\rangle$ 引起的自由能相对变化很小，接枝链的相对振动幅度也很小。因此，存在一个最佳的接枝密度和对应的自由能，使得 $k_BT/2$ 可引致最大的相对振动幅度。因此，每根接枝大分子链的作用仿佛就像一根"分子弹簧"，上下震荡；表面不停地波动起伏，如右图所示。

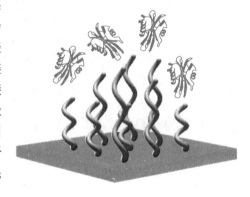

可以想象，与静止的表面相比，蛋白质分子将较不易吸附在这样一个动态的"大分子刷子"的表面。一个具有适当接枝密度的"大分子刷子"可以显著地延缓吸附，而不是阻止吸附。换而言之，吸附多少由热力学控制，取决于蛋白质分子和接枝层表面的相互作用能（ΔE），其随接枝密度的变化并不显著；而吸附快慢则受制于动力学，取决于具体体系的特征吸附时间（τ）。数学上，其表达为 $\theta=\theta_0[1-\exp(-t/\tau)]$，如下图所示。对于给定的大分子分散系，接枝密度的改变实际上影响的是 τ，而不是 ΔE。由于观察和比较吸附量总是在一个给定的吸附时间，而不是等到吸附平衡。右图形象地显示了不同的特征吸附时间可导致表观的，而不是真正的吸附量差异。

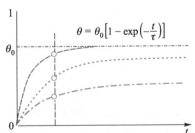

$$\theta=\theta_0\left[1-\exp\left(-\frac{t}{\tau}\right)\right]$$

因此，大分子链接枝在一个表面上可减缓，而不是减少，蛋白质吸附；是基于吸附动力学，而不是源于热

力学吸附平衡。利用这一表观蛋白质吸附的差异已足以解决一些应用问题。例如,在静脉给药的输送体系中,一般而言,只要在 24 小时之内,减缓蛋白吸附,就可避免被体内免疫系统的清理,延长药物在血管中的循环时间,提高药效。另外,在船舶水线下的最后一道油漆的配方中,一个适当的设计可使得油漆表面与海水接触时,可持续地形成不断更新的动态表面,阻止蛋白质和海洋生物的吸附,从而实现海洋防污。

基于上述讨论已知,与小分子胶体不同,一个大分子胶体的形成和稳定,不仅遵循热力学原理,即每个稳定基团在胶体粒子界面上所占据的表面积有一个最小值,对一个给定的体系,其为一常数;而且还受到动力学(kinetics)控制,与形成路径有关,取决于大分子链间聚集和链内收缩二者之间的竞争,形成动力学(kinetics)稳定。

达至动力学稳定一般有以下两个先决条件。第一,聚集体本身不可太大,大约小于一个微米,热能可以搅动每个粒子在分散体系中平动扩散。第二,聚集体和分散介质的密度相近,否则沉降(上浮)力就会克服由热能引致的随机力,导致粒子们的定向运动,沉淀或上浮。在堆集状态,聚集体之间的相互作用时间激增,导致聚集体融合,出现宏观分相。满足这两个先决条件,两个聚集体之间的碰撞一般就不会引致两个碰撞聚集体之间的扩散、缠结和融合,从而实现大分子特有的、基于"黏弹性"的动力学稳定。

显然,将亲溶剂的大分子链化学交联起来形成很小的、溶胀的、密度几乎和分散介质相等的微凝胶,即可满足上述两个先决条件。除了链段的局部运动以外,交联已阻止微凝胶之间的融合。无须任何稳定剂,它们就可长期地、稳定地分散在介质中,形成一类特殊的、有应用前景的大分子胶体,尤其是可在水中溶胀的、对环境(温度、pH、盐浓度等)敏感的微凝胶。

仅以热敏性、球状聚 N-异丙基丙烯酰胺微凝胶为例。在一个无稳定剂的分散体系中,微凝胶的平均流体力学半径在 25 ℃ 和 35 ℃ 时,分别约为 110 nm 和 45 nm。无论是在低温的溶胀状态,还是在高温下的塌缩状态,分散体系仍可长期稳定。因其流体力学体积随着温度降低可变约六倍,可先在 35 ℃ 时配制一个微凝胶体积分数约为 20%(质量分数低于 5%)的胶体分散液,再将温度降至室温,体系中所有溶胀微凝胶的体积分数就超过 100%。它们不得不挤堵在一起,形成一个不能流动的宏观凝胶。每个微凝胶的内部是一个化学交联的凝胶网络,而不同的微凝胶则是物理挤堵在一起,故这是一个化学和物理杂化的宏观凝胶,其多孔结构随着降温速率和方式变化,如右图(c)中的插图所示。

制备天然大分子凝胶的历史比大分子科学悠久。如果大分子链通过物理相互作用(氢键、结晶、强疏水作用等)构成三维网络,也被称作"冻胶"。

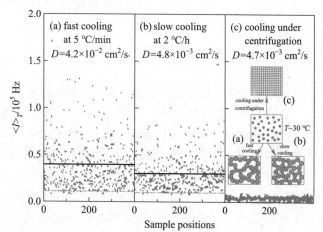

Zhao Y, Zhang G Z, Wu C. Macromolecules, 2001, 34:7804.

(Figures 2 and 4, permission was granted by publisher)

实验上已发现交联的反应速率越快,形成的凝胶就越不均匀,其时间平均散射光强($<I>_T$)随着激光光束的照射位置变化,出现光强跳跃,如上图所示。换而言之,不遵循统计学上无规变量的遍历性,时间平均不等于位置平均,完全不同于大分子溶液。这里有双重原因:一是交联网络的不均匀性;二是处于交联状态的链无法遍历体系里的各点。在很长的时间内,人们将这一不均匀性归于在交联过程中形成了不同大小的支化团簇或微凝胶。

实际上,这一解释忽略了已有的实验结果和实验条件。在紧靠凝胶点之前形成(测得)的支化团簇或微凝胶的尺寸一般也小于微米,与可见光波长接近,远小于通常激光光散射的散射体积的尺寸($\sim 200\ \mu m$)。换而言之,在激光光散射的散射体积内,有上百万个支化团簇或微凝胶。即使大小不同,它们在不同位置的平均散射光强也是一个常数。为何观察到的散射光强跳跃?上图中由不同的降温速率和方式得到的数据解释了这一疑问。

在上述杂化宏观凝胶中,微凝胶的尺寸分布很窄,它们的流体力学半径小于 $500\ \mu m$。显然,在快速冷却[上图(a)]得到的杂化宏观凝胶中,散射光强的跳跃幅度最大,其位置平均散射光强也最高(图中的黑线)。这一冷却速率依赖性首先否定了散射光强的跳跃源自支化团簇或微凝胶的大小不均。在上图(c)中,先通过超速离心将微凝胶密集堆集在光散射池的底部,然后一边离心,一边降温,使得微凝胶在堆集的状态下逐步溶胀。可见,不仅散射光强跳跃消失了,而且由位置平均得到的散射光强最低。

一般而言,在由单体聚合形成化学凝胶的过程中,微观上沿着"线型链—支化—支化团簇—微凝胶—凝胶点(渗滤点)—宏观凝胶"的路径。热能驱动反应混合物中形成的各种大小和组构的大分子通过平动扩散趋向一个均匀密度,而扩散需要时间。如果大分子之间的交联快于扩散,而且交联进一步减慢扩散,那么交联就会在体系内若干个几乎不动的中心附近进行,较小且可动的单体和大分子逐步地向这些交联中心汇集并交联在其上。结果是在这些交联中心之间形成很大的溶剂空洞。

在讨论激光光散射时,已经指出,一块纯净的光学玻璃中的每一个散射单元总有另一个与其对应的、位相相差 $180°$ 的散射单元。二者的散射电场因干涉抵消,故除了零角度(入射方向)以外,纯净的光学玻璃在理论上不产生任何散射光。如果其中有一些气泡,那么就会出现很强的散射。散射光强并非来自气泡里的气体分子,而是源于和每个气泡对应的、具有相同体积的、位相相差 $180°$ 的一小块玻璃。原因是气泡内的分子数远少于其对应的玻璃中的分子数,故气泡的散射光强比对应的一小块玻璃弱很多,它们的散射电场无法干涉抵消。

凝胶散射光强对位置的依赖性也是基于同样的道理。大分子网络相当于"光学玻璃",而每个大的溶剂空洞等效于一个大的"气泡"。一个个溶剂空洞被交联反应固定在不同的位置,大小和分布不遵守统计规律、极不均匀。因此,在不同的位置,散射体积中溶剂空洞的大小和数目不等,造成散射光强跳跃。在上述杂化凝胶的形成中,冷却速率等效于交联速率。迅速冷却时,每个微凝胶均极速膨胀,使得微凝胶没有足够的时间从一个密度较高处向另一个较低处扩散,形成一个个较大的溶剂空洞。而在超速离心下的缓慢冷却中,较大的溶剂空洞无法形成,故得到一个均匀的、散射较弱的、没有散射光强跳跃的杂化凝胶。

这仅是利用微凝胶的独特性质解答科学疑问的一个实例。文献中还有许多关于微凝胶潜在应用的建议和设想,但是绝大多数都是纸上谈兵,近三十年来"只听楼梯响不见人下来"，至今无一个微凝胶的大规模实际应用。不是微凝胶没用，而是研究一直没有聚焦在真正的实际问题，研究者没有"从实践中来，到实践中去"！再举一例微凝胶在生物医疗中的潜在应用。Dai 和其合作者[Journal of Biomaterials Applications，2015，29：1272] 利用材料模量可以影响干细胞分化这一现象，将处于收缩状态的微凝胶与干细胞在约 40 ℃时混合，利用微凝胶在体温下溶胀的特性，在体内对干细胞施加压强，诱导干细胞定向分化成软骨细胞，修复膝盖软骨受损，从而达到"一石二鸟"的作用：提供润滑和生成软骨。

与温度敏感性微凝胶相连的还有一个"体积相变温度"（VPTT）的问题。VPTT 的缩写源于 Dusek 和 Patterson 的一篇文章[Journal of Polymer Sciecne：Part A-2，1968，6：1209]。在这篇理论文章中，他们研究了相邻交联点之间链段数（N_b）均等的理想凝胶，并明确地指出，即使在这样的理想凝胶中，观察"体积相变"还需要同时满足以下三个条件：第一，大分子链和溶剂之间的相互作用（χ_{sp}）越强越好；第二，在干凝胶中，单位体积内的交联点数目（v^*）越大越好，即 N_b 应该越小越好；第三，相邻交联点之间的链节在完全蜷缩的干态和无扰（Θ）状态时的均方末端距之比（$\langle \alpha^2 \rangle_0 = N_b^{2/3}/N_b = N_b^{-1/3}$）越小越好，即 N_b 应该越大越好。

显然，第二和第三个条件是矛盾的。因此，他们在文章的简介中就已开诚公布地、结论性地指出，"在自由溶胀时，要得到这些必要的相变条件是困难的"。然而，绝大多数使用 VPTT 一词或甚至引用这篇文献的人都没有阅读或没有读懂这篇最早的文献，不幸地囫囵吞枣、人云亦云。实验上，除了难以寻获以上可能存在的一个 N_b 的狭窄窗口外，也无法制备均匀的、理想的凝胶网络。在任何一个实际的凝胶中，紧邻交联点之间的链节总是不可避免地存在一个长度分布。正如 Dusek 和 Patterson 指出的，不同长度的链节导致不同的相变温度。如何解释许多宏观凝胶体积在某一温度发生突变的实验结果？在第十章讨论中，已经证明相变温度依赖于链长，无论溶液是具有 UCST 还是 LCST，在给定大分子体积分数后，长链总是比短链先发生相变。

微观上，两个紧邻交联点之间的大分子链节有长有短。因此，随着温度的变化，凝胶内部较长的链节先收缩，形成局部的分相，造成内部应力，但远小于凝胶的宏观切变模量，故其宏观尺寸保持不变。当溶剂性质逐渐变差时，越来越多的链节收缩，内部应力不断地加大，至某一温度时，当内部应力大于宏观切变模量时，凝胶的宏观尺寸（体积）就会突然变小。但是，实验上观察到的这一宏观突变不是 Dusek 和 Patterson 所预测体积相变，而是内部应力和宏观切变模量之间的竞争所致。就好比一根橡皮筋，在拉伸中会突然断裂一样，但那不是一个相变。

右图比较了线型聚 N-异丙基丙烯酰胺大分子单链和微凝胶的相对流体力学体

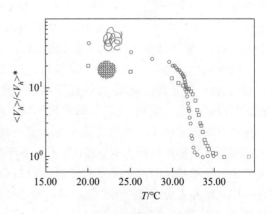

Wu C，Zhou S Q.Macromolecules，1997，30：574.

（Figures 3，permission was granted by publisher）

积对温度的依赖性,其中,$<V_h>*$分别是二者完全收缩时的流体力学体积。注意,线型链不仅比球相微凝胶中两个邻近交联点之间的链节长了近千倍,而且链长分布更窄(M_w/M_n <1.0)。因此,其蜷缩温度比微凝胶更低、温变范围比球相微凝胶更窄。由于内部大分子链之间的互相交联,微凝胶的相对体积变化也更小。对于微凝胶而言,任何的内部应力都会传递到整个微凝胶,反映在其流体力学尺寸的变化上。这也解释了为何在同样的大分子体系中,微凝胶从来没有观察到一个体积突变。

物理上,对一个变化或突然变化是否为"相变"有着严格的定义。有关相变的研究是凝聚态物理一个重要的组成部分。如果是相变,那么紧接着的两个问题就是,几级相变? 在相变(临界)点附近的普适标度律为何? 如同讨论大分子溶液相变时一样。无论是在讨论凝胶还是微凝胶体积变化时,正确的表达是"体积变化温度",而不是"体积相变温度"。在凝胶领域工作的研究者应该清楚为何二者之间有着物理本质的区别,不可乱用!

本章最后一个例子是两种带相反电荷的聚电解质链通过络合在水溶液中形成相对稳定的大分子胶体。在适当的电荷之比、浓度、混合方式等条件下,形成的络合体的表面可带正电荷或负电荷。在生物体内,蛋白质分子本身就由偏碱性和偏酸性的氨基酸组成,在生物环境中,每条蛋白质链可总体带正电荷或负电荷。总体带不同电荷的蛋白质链也常常络合在一起形成一个稳定结构。例如,大豆中的11S球蛋白就含有六条总电荷为正、六条总电荷为负的蛋白质链,组成六对二聚体。近来,这些带不同电荷大分子的络合体也被赋予各种"新奇"的名称,但换汤不换药。恕不一一列举这些"新"的名称,因为实在是了无新意。

微观上,两条分别带有相反电荷的大分子链相遇时,可形成多点络合。每个络合点的作用较弱,故其处于一个不断"开-合"的过程,建立一个动态平衡。但是一旦络合,两条链的分开则不仅要求所有的络合点同时解开,而且还要两条链通过扩散立即分开,概率极低,故一旦络合,则很难分开。教科书和文献中常常将这样的络合归于静电相互作用,但忘了在混合前各自的水溶液中,每条带有电荷的大分子链与其反离子已存在静电相互作用,电中性不可违背。在络合的过程中,一条大分子链上的一个正电荷并不知道其反离子负电荷是一个小分子,还是一个位于另一条大分子链上的负电荷。每形成一个大分子络合点就释放出一对小分子反离子,平动熵增加。正是这一熵增驱动了带有相反电荷的链之间的络合,而非静电相互作用隐含的焓变。

与带有相反电荷的小分子络合相比,当正负电荷的物质的量相近时,大分子的络合很相似,也会出现宏观的沉淀,释放出反离子。然而,大分子链不是无限的柔软,两条链上的相反电荷难以逐一配对,尤其是两种带相反电荷的链的长度相差很大时,不可避免地存在一些没有络合的电荷。它们仍是和混合前一样,由小分子反离子维持着电中性。当正负电荷的物质的量相差很大时,在小分子的络合中,较少的组分就会因络合沉淀,形成一相,而较多组分中剩余的离子和反离子以及因络合释放出的反离子则仍然留在溶液中,形成另一相。而在大分子络合中,带过量电荷的链则会络合在表面,称作过电荷,其既减少了界面能,提供了热力学稳定,又降低了两个碰撞络合体之间的相互作用时间,促进了动力学稳定。

选择一个适当的络合途径可使链内收缩远远快于链间络合,使得络合体的尺寸控制在 500 nm 以下,因此热能足以驱动络合体在混合溶液中无规行走。这样得到的络合体就可在长时间内处于动力学稳定,形成亚稳态的大分子胶体,而不会出现宏观沉淀和分相。近年来,用带阳离子(含氮基团,包括各种氨基)聚电解质络合带阴离子(磷酸)的脱氧核糖核酸(DNA)形成非病毒基因载体(polyplexes,多复合体)就是一个典型的应用的例子。有关的研究在过去三十年中产生了十万多篇文献。然而,绝大部分在这一领域工作的研究者,尤其是学习药学和生物的研究者,并没有注意到大分子络合的复杂性、包括混合路径依赖性、聚合物本身具有链长、组构等特性。

源自不同实验室的大量实验结果显示,在等物质的量电荷(N∶P=1∶1)时,仍有大量的、没有被络合 DNA 链残留在混合溶液中,源于带负电荷的、刚性的 DNA 链远长于聚阳离子链。直至 N∶P=3∶1 时,绝大部分的 DNA 链都被络合,意味着仅有约 1/3 的阳离子参与了络合。少数的没有络合的阳离子位于少量的、未被络合的聚阳离子单链上,自由地在混合溶液中无规扩散;一些裹在复合体内部;更多的都分布在复合体的表面,使其带正电荷,起到了稳定的作用。奇怪的是在后续的、大量的基因转染实验中,不同的实验室也都发现,只有当 N∶P~10∶1 时,基因转染效率才会明显地提高。

有趣的是在长达二十多年的研究中,只有很少的研究者将这两个确定的实验结果合在一起考虑,并问两个显而易见的问题。因为当 N∶P>3 时,混合液中已无自由的、单条的 DNA 链。那么为何只有当 N∶P~10 时,基因转染效率才高?多余的 7 份聚阳离子链去了何处?答案显而易见。其一,多余的聚阳离子链一定是以单链的形式溶于混合液中,自由地扩散,如下图(a)所示;其二、正是那些在混合溶液中自由的聚阳离子链促进了基因转染,图中间的白色荧光点显示了绿色荧光基因在细胞里的表达越多,基因转染效率越高。那些与 DNA 络合的聚阳离子链,仅提供了正电荷,使 DNA 链蜷缩成较小的粒子,便于细胞内吞。

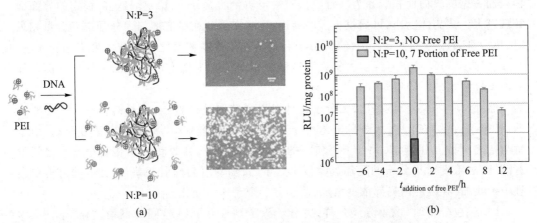

Yue Y N,Jin F,Deng R,et al. Journal of Controlled Release,2011,155:67.

(Graphical Abstract,permission was granted by publisher)

因此,在 N∶P=10∶1 时,加入的聚阳离子链可分为两部分,3 份与 DNA 络合;7 份为混合溶液中的自由聚阳离子链。上图(b)显示了一个决定性实验的结果。纵坐标是经过矫正的基因转染效率,为对数坐标。横坐标为加入 7 份额外的聚阳离子(自由单链)的相对时间,基准 $t=0$,7 份自由聚阳离子链和 DNA/大分子复合物(N∶P=3∶1)同时加入一个基因转染体系(细胞和培养液);$t=-6$ h,代表先加入含有 7 份自由聚阳离子链的溶液,6 个小时后再加入 DNA/大分子复合物(N∶P=3∶1);$t=6$ h,先加入 DNA/大分子复合物(N∶P=3∶1),6 个小时后再加入含有 7 份自由聚阳离子链的溶液。总的氮磷比保持为 N∶P=10∶1。

显然,没有自由聚阳离子链时,即 N∶P=3∶1,基因转染效率很低。无论是先添加,还是后添加自由聚阳离子链,基因转染效率均显著地增加了 10~100 倍。采用不同的细胞和不同的聚阳离子链均得到类似的结果。后续定量的实验进一步证实,自由的聚阳离子链主要有以下四重效应。第一,干扰了内吞体上原本在细胞内壁上的 SNARE 信号蛋白,从而阻断了内吞体与溶酶体的融合,使得基因可在细胞液中释放;第二,弱化了核膜的结构,加速了基因穿过核膜进入细胞核;第三,帮助了基因在细胞核内的转录,生成了更多的核糖核酸(RNA);第四,促进了 RNA 跨过核膜,进入细胞液,最终实现蛋白质合成。详情可见 Cai 和合作者的报告[Journal of Controlled Release,2016,238:71]。

小　结

在稳定剂的协助下,大分子可通过特殊的分相路径,形成许多处于亚稳态的微浓相(聚集体),分散在稀相(大分子的饱和溶液)之中,形成一个处于动力学平衡状态的大分子分散液,也称为大分子胶体,以便有别于通常由小分子聚集而成的胶体。处于亚稳态的大分子聚集体的稳定同时受制于分别基于热力学和动力学两种不同的机理,取决于大分子和分散介质、浓度、稳定剂含量以及形成路径。

依据热力学,对给定的胶体(分散体系),每个稳定基团(离子、水溶性链段等)在聚集体表面占据的界面面积应该为一常数。对给定质量的大分子,稳定剂越多,稳定的界面面积就越大。聚集体变小,可提供更多的界面;反之亦然。在稳定剂与大分子二者含量相近时,热力学效应可占主导。因此,大分子和稳定剂的投料质量之比决定了聚集体的大小。将二者作图,可得一条直线。由斜率和截距分别可得每个稳定剂所占的平均界面面积和界面上稳定剂的平均层厚。

然而,与小分子胶体相比,大分子胶体有着一些明显的差异,均与链长有关。主要有两点:第一,在稀溶液中,大分子链比小分子的平动扩散慢~10^2 倍,而随着浓度升高,由于链与链之间的互相缠结,大分子链质心的运动变得更慢,可变慢~10^4 倍。小分子之间没有缠结,故浓度对小分子的运动影响较小。第二,随着溶剂性质变差,大分子的链构象会逐步地收缩,导致链内局部浓度增加,而小分子则无这一变化。

因此,在分相时,像小分子一样,大分子除了具有分子(链)间的聚集以外,还有链内收缩。在稀溶液中,二者的速率相似,故链间聚集和链内收缩总是相伴而行、互相竞争。一旦构象完成了蜷缩,聚集体内链的运动和两个碰撞聚集体之间链的交换(融合)所需的时间就远远长于两个碰撞聚集体之间相互作用的时间,长一千倍以上。

依据流变学,当相互作用在一个物体上的时间远远短于其中分子在该作用下松弛(运动)一个特征长度的时间,该物体就表现出弹性;反之,黏性。换而言之,当两个大分子聚集体碰撞时,在二者相互作用的时间内,它们各自就像两个弹性的"玻璃"小球,因碰撞而融合成一个稍大聚集体的可能性极低。基于这样的"黏弹性",聚集体被动态(kinetic)地稳定在微相分离状态,一个亚稳态,达至动力学稳定,稳定时间随体系和形成路径而变。

如前所述,达至动力学稳定有两个先决条件:第一,聚集体一般不可太大,约小于 $1\ \mu m$,所以热能可搅动每个聚集体在分散体系中无规行走、平动扩散;第二,聚集体和分散介质的密度相近,否则沉降(上浮)力就会克服由热能引致的无规运动,导致聚集体的定向运动,沉淀或上浮。聚集体堆集在一起,相互作用时间激增,使得聚集体融合,出现宏观分相。如果满足这两个条件,聚集体之间的碰撞就不会引致聚集体之间链的扩散、缠结和融合,从而实现大分子胶体特有的、基于"黏弹性"的动力学稳定。

分相时,在两亲性共聚物链聚集过程中,热力学驱动疏溶剂链段聚集在中心,亲溶剂链段移向表面,以降低界面能。这一过程中也额外地促进了动力学稳定,即分布在界面上的亲溶剂链段减少了聚集体之间相互作用的时间,而疏溶剂链段在内部聚集也进一步减慢了链的松弛,延长了特征弛豫时间。使得两个时间相差更大,更有利于动力学稳定。

上亿年的进化,使得两亲性的蛋白质分子既是典型的代表,又是利用这动力学稳定的佼佼者。因此,许多蛋白质分子可以在细胞内通过核糖体中的特殊途径,一边聚合、一边折叠成动力学稳定的单链聚集体。在细胞外,也可通过特殊的方式,使得一些变性的蛋白质分子重新折叠成具有生物活性的、动力学稳定的单链聚集体。目前,研究蛋白质单链折叠总是利用自由能最低的热力学稳定状态作为参考基点。这一思路可能并不正确,具有生物活性的蛋白质分子很可能是处于一个能量稍高的、动力学稳定的亚稳态。

还有其他特殊的稳定方式。在聚集体的表面接枝亲溶剂的线型链,形成大分子刷,是其中一种。每根接枝链都被其邻近的接枝链环绕。基于降低自由能的热力学原则,它们并不互相穿插和缠绕,而是尽可能地向上伸展,仿佛处于一个由周围接枝链组成的柱状"管子"之中。平衡时,接枝链的平均末端距(刷子的厚度)与接枝密度(单位界面面积上的接枝链数目)的 1/3 次方和链长成正比;接枝链的自由能与接枝密度的 2/3 次方和链长成正比。

热能搅动每条接枝链的自由一端一直无规地上下震动,导致每条接枝链的末端距不断地变化。其平均涨落振幅(ΔL)与平均末端距($\langle L \rangle$)之比,即相对涨落振幅,具有一个极大值,对应着一个最佳的接枝密度,与链长无关,与链段长度(b)的平方成反比。在此最佳接枝密度时,接枝链的自由能(G)等于 $[(N/10) + 1]k_B T$;平均涨落振幅约等于接枝链的平均回转半径($b N^{1/2}/6^{1/2} = \sqrt{\langle R_g^2 \rangle}$);接枝链的平均末端距约为链伸直长度($bN$)的 1/10。因此,$\Delta L/L$ 仅与链长有关。对最佳接枝密度时典型的接枝链($N \sim 30$),$\Delta L/L$ 约为 20% 和 $G \sim 4 \sim 6\ k_B T$。

在生物医药中,在药物载体的表面上接枝大分子链以"阻止"蛋白质吸附是一个常用的策略。显然,大分子链越柔软,最佳接枝密度就越小,每条接枝链占据的界面面积就越大,稳定效果就越好。聚环氧乙烷是目前已知高聚物中最柔软的链,这解释了为何在生物医药的研发中,聚环氧乙烷是最常见的接枝链,且其临床使用已获得批准。

依据热力学原理,蛋白质吸附的百分数仅取决于界面与蛋白质的作用能。接枝密度对作用能的影响甚微,因此,改变接枝密度对蛋白质吸附的影响应该不大。实验上,观察到的吸附量对接枝密度的依赖性源于吸附动力学的变化。每条接枝链的自由端的无规振动构成了一个不断起伏的动态表面,使得蛋白质分子的吸附变得相对困难,延长了吸附动力学中的特征吸附时间。在给定的观察时间,较慢的吸附给出一个相对较低的表观吸附量,造成吸附较少的假象。因此,表面上接枝的大分子刷不是"阻止",而是"延缓"了吸附。采用柔软的链,并选择最佳接枝密度,可使 $\Delta L/L$ 达至最大,减缓蛋白质吸附。文献上的解释需要修正。

　　另一个特殊的稳定方法是交联聚集体内的大分子链,故不同聚集体之间的大分子链的扩散、缠结和融合被阻止。在水中制备的微凝胶就是一个典型的例子。许多聚合物的密度接近水($\sim 1 \text{ g/cm}^3$)。用这些聚合物制备的水溶性凝胶可被分散后成尺寸仅为几百纳米的微凝胶,自然地满足了上述动力学稳定的两个必要条件。在热能的驱动下,这些微凝胶在水中无规地扩散,沉降(上浮)力都不足以使得它们沉降或上浮,分成两层。由于内部的交联,两个相撞的微凝胶也无法融合成一个较大微凝胶。因此,无须任何稳定剂或将制备时添加的稳定剂移去以后,微凝胶仍然可在水中稳定很长时间。通过适当的设计和制备,微凝胶可被用来解决一些科学和应用问题。下图小结了本章中讨论的主要问题。

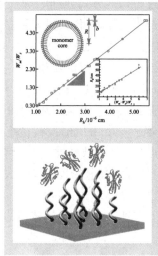

热力学:每个稳定基团所占的表面积为一常数
Thermodynamics: Surface area occupied by each stabilizer is a constant

$$\frac{W_m}{W_s} = s\frac{\gamma \rho N_{AV}}{3M_s}R - s\frac{\gamma \rho N_{AV}}{3M_s}b$$

动力学:基于大分子链"黏弹性"的稳定机理
Kinetics: A stabilization mechanism based on viscoelasticity of polymer

$$\frac{l_I}{\sqrt{\langle u^2 \rangle}} \leqslant \tau_I \leqslant \frac{l_I^2}{D} \qquad \tau_I \ll \tau_R \qquad \tau_R \approx \tau_I \left(\frac{C}{C^*}\right)^4$$

$$\frac{G}{k_B T} \approx \frac{b^3 N^2}{2sL} + \frac{L^2}{R_0^2} \qquad L_0 = \left(\frac{b^2}{4}\right)^{1/3} \sigma^{1/3} L_c$$

最佳接枝密度下:
optimal grafting density $\qquad \dfrac{1}{\sigma_{\text{optimal}}} = s_{\text{optimal}} = \dfrac{\pi^3 b^2}{4}$

$$\frac{\Delta G}{k_B T} = \frac{N}{\pi^2} \;;\; L_{\text{optimal}} = \frac{L_c}{\pi} \;;\; \frac{\Delta L}{L} = \frac{\pi}{(6N)^{1/2}}$$

Chapter 12. Macromolecular Colloids

The two substances can be uniformly mixed into a stable **solution** at the molecular level, which is a one-phase binary mixture. A substance can also uniformly and stably disperse in another substance in the form of small aggregates to form a colloidal dispersion, briefly, **colloids**, a two-phase binary mixture. The aggregates are often called colloidal particles. The size of the aggregates is generally less than one micron. When the aggregates are large, the mixture is often referred to as a suspension mixture (liquid). There is no strict boundary or physical distinction between it and the colloid, which can be generally called a colloidal suspension mixture (liquid). With this definition, the amphiphilic copolymer solution in the mesoglobular phase can also be regarded as a special colloidal dispersion.

The most essential physical difference betweena solution and a colloidal dispersion is that there is no physical interface larger than the molecular scale in the one-phase region of the solution. The colloid has at least two phases, and there is an obvious microscopic physical interface between the aggregate and the dispersion medium. According to the definition, whether it is a solution or a colloid, it can be in the form of a solid, liquid or gas, not limited to liquid. For example, fine liquid droplets are dispersed in the gas to form aerosols and some metal alloy solutions, etc. In short, types of colloids are ever-changing; there are also different ways of forming a colloidal dispersion. There are countless documents, monographs and patents. Below, we mainly discuss the stabilization mechanism of some colloids, especially the unique "viscoelastic" kinetic stabilization mechanism of macromolecular colloids, and special colloidal dispersion systems such as macromolecular brushes grafted on the surface of colloidal particles and microgels. The introduction of traditional macromolecular colloids is omitted.

It has been known for a long time that amphiphilic small molecules can aggregate in a selective solvent to form micelles, dispersed inside a medium. Amphiphilic copolymer macromolecules can also aggregate in a selective solvent to form many micro-concentrated phase (aggregates of macromolecules) dispersed in a dilute phase (saturated dilute solution of individual collapsed macromolecular chains). Generally speaking, when a substance with a given mass and linearity L is divided n times into small particles with linearity L/n, the number of particles is $L^3/(L/n)^3 = n^3$. The total surface area (S) of all the particles will be much larger than the initial surface area (S_0), which is mathematically expressed as,

$$S_0 \sim L^2 \quad \text{and} \quad S \sim \left(\frac{L}{n}\right)^2 n^3 = L^2 n \qquad \rightarrow \qquad \frac{S}{S_0} = n$$

Therefore, the surface area increases with the number of linear divisions. If an insoluble macroscopic substance is dispersed in a medium to form tiny molecular aggregates, its surface area (energy) will greatly increases. According to thermodynamics, this will not spontaneously happen. On the contrary, in dispersion, insoluble aggregates always tend to further aggregate in order to reduce the interface energy.

A typical example is to mix oil with water. Under normal conditions, the mixture spontaneously separates into two coexisting layers (phases): the water phase contains saturated oil and the oil phase contains saturated water. Microscopically, oil and water molecules continue to traverse the interface back and forth, each respectably with the same rate, reaching a thermodynamically stable dynamic equilibrium. If there is no volatilization, the volume and composition of the two phases remain unchanged. Vigorous stirring or shaking (input energy) can disperse the oil into water as small droplets. However, once the stirring or shaking stops, the diffusion time of the oil molecules between the two colliding small oil droplets is much shorter than the interaction time between the oil droplets. Each collision results in the formation of a larger oil droplet, keeping gathering, and the mixture instantly restore to the upper and lower macroscopic levels (phase).

Driven by thermodynamics, the macromolecular aggregates (micro-concentrated phase) also tend to further aggregate and reduce the interface energy, but they are subject to the aforementioned "viscoelastic" effect, i. e., a kinetic stability mechanism, depending on the competition between the interchain aggregation and the intrachain contraction. Since the time required for the inter-diffusion, entanglement and fusion between the chains of two colliding aggregates is more than one thousand times longer than the interaction time of two colliding aggregates. Once the intrachain contraction is completed, the collision between two aggregates in the solution is like an elastic collision between two small glass balls. The probability of fusing into a slightly larger ball is almost zero, i. e., further aggregation between the aggregates is actually impossible, forming a kinetic stability. The premise is that the size of the aggregate is small, the density is close to the solvent (dispersion medium), and the thermal energy can agitate the aggregates to move randomly with no precipitation.

Therefore, in order to disperse an insoluble substance into a medium to form a relatively stable colloid, it is often necessary to add an amphiphilic stabilizer (also called a dispersant or surfactant) that can be distributed on the aggregate surface. In addition to reducing the interface energy (thermodynamic considerations), the interfacial stabilizer also shortens the interaction time between two colliding aggregates and prevents the molecular exchange and fusion between them (kinetic mechanism). The various colloids formed in this way are widely used in life and production. For example, in the industrial synthesis of polymers, the polymerization is initiated after oily reactive monomers (such as styrene, acrylate, etc.) are dispersed into small droplets to form an emulsion or suspension with the help of surfactants. Such a way facilitates the heat transfer anf temperature control, reduces viscosity, improves uniformity and increases the polymerization rate. In daily life, surfactants in cleaning/detergents are used to peel off and disperse the grease on the pots and scoops on the surface and the oily dirt on clothes into the water, so that the can be washed away, etc., there are countless examples.

There are many types of stabilizersin various form. According to their charges, there are anionic, cationic and neutral ones; according to their size, there are small molecular, oligomer and macromolecular stabilizers; and according to their interaction with the interface, there are physical adsorption and chemical bonding, too many to be counted. However, the fundamental principle remains. Note: in most of the cases, besides the restriction of thermodynamics, the formation of macromolecular colloids is also kinetically controlled. Even for the same system, different ways of changing the solvent quality or different reaction rates will lead to different colloidal structures. Therefore, except a few systems completely controlled thermodynamics, there exist no general physical laws. In other words, knowledge is used, not created here.

In the following, let us first discuss thermodynamically controlled systems. Here, the controlling physical parameter is "**the surface area per stabilizer (s)**". For a given system, s is a constant. A typical example is thermodynamically controlled micro-emulsion polymerization, in which the content of stabilizer and oily monomer to be dispersed is relatively close. The size of the "oil droplets" wrapped by the stabilizer is only about 10 nanometers.

For a long time, researchers in the field of microemulsion polymerization only knew qualitatively that as mass ratio of stabilizer to oily monomer (feed ratio, W_s/W_m) increases, the resultant polymer particles gradually become smaller. In the traditional "particle size versus feed ratio" plot, a curve is obtained. The particle size decreases monotonously as the feed ratio increases. Experimentally, the feed weights of stabilizers and oily monomers (W_s and W_m) are known; the molar mass of stabilizer molecules (M_s) and the monomer density (ρ) are also two known parameters. Experimentally, the "oil droplet" radius (R) measured by dynamic LLS includes a stabilizing layer with a thickness of b on the surface, as schematically shown on the right. Assuming that the partition coefficient of the stabilizer on the surface of the "oil droplet" is $0 < \gamma < 1$. The product of surface area of each "monomer oil droplet" and the total number of "monomer oil droplets" is equal to the total surface area (S) of the "monomer oil droplets" in the dispersion system. In addition, the partition coefficient (γ) is used to obtain the total mass of stabilizers on the surface, divided by

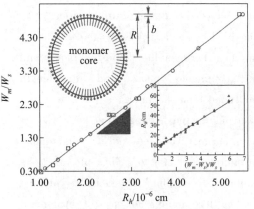

Wu C.Macromolecules,1994,27:298.

(Figures 2, permission was granted by publisher)

the molar mass of the stabilizer, to obtain the total number of stabilizer molecules on the interface (N_s), i.e.,

$$S = 4\pi (R-b)^2 \frac{W_m}{\rho \dfrac{4\pi (R-b)^3}{3}} = \frac{3 W_m}{(R-b)\rho} \quad \text{and} \quad N_s = \frac{\gamma}{M_s} W_s N_{AV}$$

From the ratio of S and N_s, the average area (s) occupied by each stabilizer molecule on the interface can be calculated as follows.

$$s = \frac{S}{N_s} = \frac{3 W_m}{(R-b)\rho} \frac{M_s}{\gamma W_s N_{AV}} \quad \rightarrow \quad \frac{W_m}{W_s} = s \frac{\gamma \rho N_{AV}}{3 M_s} R - s \frac{\gamma \rho N_{AV}}{3 M_s} b \qquad (12.1)$$

Note that the above equation is slightly different that from the reference listed in the figure above, which is more concise; where for a given system, parameters in the coefficient in front of R on the right side are constant except for s; the intercept divided by the slope leads to the thickness (b) of the dispersant molecules protruding from the interface.

The figure above is derived from a styrene micro-emulsion stabilized by cetyltrimethylammo-nium bromide. The data clearly shows that the plot of W_m/W_s versus R is a straight line, confirming that s is indeed a constant. The slope and intercept of the straight line lead to $s = 0.182$ nm^2 and $b = 1.0$ nm, two previously unobtainable microscopic parameters. After the above equation was published in 1994, Antonietti and his collaborators [Macromolecular Chemistry and Physics, 1995, 196:441] remapped their previous data, as shown in the lower right inset in the figure above, completely falling into the line predicted by the above equation.

This kind of plotting method seems to simply replace the original W_s/W_m with W_m/W_s, but it is completely different in essence. The former has no physical meaning, but just giving an apparent curve of the change trend, just like a data list. According to the above inference, the latter has a clear physical meaning. The slope and intercept of the straight line obtained correspond to two important microscopic parameters: the average surface area (s) occupied by each dispersant molecule at the interface between the dispersed phase and the continuous phase and the length protruded from the interface (b). In order to avoid the misleading, it is necessary to point out that Antonietti and its collaborators in their article refer to such a plot as "The Antonietti-Wu Plot". In fact, this plotting method has nothing to do with Antonietti and his research collaborators. If one really

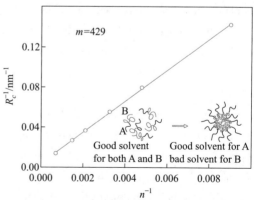

Gao J, Wu C.Macromolecules, 2000, 33:645.

(Figures 2, permission was granted by publisher)

wishes to name it, it should also be "The Wu Plot".

The aforementioned conclusion that "**the surface area occupied by each stabilizing group on the interface is a constant**" can also be extended to other thermodynamically-controlled aggregation processes. For example, linear amphiphilic block copolymer chains made of comonomer A and B form micelle-like core-shell aggregates in a selective solvent, as shown in the schematic inset in the figure above, where the stabilizing group is no longer a small charged molecule, but a macromolecular block that is solvophilic. In addition to changing the degree of amphiphilicity by selecting different A and B or the temperature, the segment numbers (m and n) of the two blocks are also variables, denoted as $A_m B_n$. Assuming that the core-shell aggregate is composed of N_p diblock copolymer chains, the core is made of the B block, with a radius of R_c and a density of ρ, then

$$\rho \frac{4\pi R_c^3}{3} = \frac{M_0}{N_{AV}} n N_p$$

On the other hand, the interface area of each core is also proportional to the interface area occupied by each A block and the aggregation number, i.e., $4\pi R_c^2 = s N_p$. Its combination with the above equation leads to

$$\rho \frac{s R_c}{3} = \frac{M_0}{N_{AV}} n \qquad \rightarrow \qquad \frac{1}{R_c} = \frac{s\rho N_{AV}}{3 M_0} \frac{1}{n} \tag{12.2}$$

The above equation has been written in an experimentally measurable form. If the B block is deuterated, and small angle neutron scattering is used, R_c can be measured. The above equation has been experimentally confirmed (the figure above). For a given solvophilic segment (m), the plot of $1/R_c$ versus $1/n$ is a straight line, and s is indeed a constant, and can be obtained from the slope. The data in the figure is derived from polyethylene oxide and polypropylene oxide diblock copolymers. At room temperature, polypropylene oxide is an insoluble block in water.

In addition, for densely grafted short chains on a surface, the dependence of the thickness (δ) of the grafted layer of the core and the number of macromolecular chains per unit area (σ, the grafting density) and the soluble block length (m, the segment number) is $\delta \propto m \sigma^{1/3}$, and $\sigma = N_p/(4\pi R_c^2)$. Using to the definition in Chapter 4, k monomers form a Kuhn segment with a length of b. Under normal circumstances, the aggregation of the B block is not able to force the A block to straighten at the interface. As a soft linear chain, the conformation of each grafted A block should

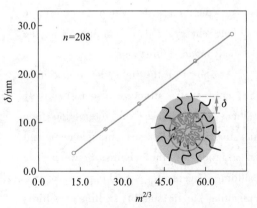

Gao J, Wu C. Macromolecules, 2000, 33: 645.
(Figures 3, permission was granted by publisher)

still be a random coil or a slightly deformed random coil. Assuming that it is still in an undisturbed state, its square mean end-to-end distance is $\langle R^2 \rangle = b^2 m/k$, while $s = 4\pi \langle R^2 \rangle = 4\pi b^2 m/k$. Its combination with $4\pi R_c^2 = s N_p$ leads to $\sigma = k/(4\pi b^2 m)$, so that $\delta \propto m^{2/3}$, as shown above.

A substitution of $s = 4\pi \langle R^2 \rangle = 4\pi b^2 m/k$ into eq. (12.2) leads to $1/R_c \sim m/n$. Therefore, for a given n, the plot of $1/R_c$ versus m is also a straight line. One can refer to Figure 2 in the above reference listed in the figure above. When reading this reference, readers will find that the author and his students did not perform any time-consuming and expensive polymerization reactions, nor did they use a large-scale equipment to conduct small-angle neutron scattering experiments. They only used the existing data published in the literature, and the aforementioned simple derivation to obtain some basic laws of the aggregation of diblock copolymers in a selective solvent, equivalent to that others did a lot of hard experiments for them.

Readers should learn a lesson from this example. One should never rush to publish experimental data that has not been carefully analyzed for the sake of the article. Otherwise, one will become someone else's technician. Remember that "research is about solving problems, not just for publishing papers." With good data, one needs to carefully analyze and establish a quantitative or at least semi-quantitative model, not only to understand and explain one's own data, but also to extend the conclusions to predict and guide other similar types of experiments.

Another example is to use the copolymerization of water-soluble polyethylene oxide (PEO) macromonomers with different lengths and hydrophobic styrene or methyl methacrylate in a 4 : 1 ethanol and water mixture to obtain soap-free (no surfactant) stable polystyrene (PS) or poly(methyl methacrylate) (PMMA) particles. The addition of a large amount of ethanol is for increasing the solubility of the hydrophobic comonomer. In the copolymerization, water-soluble polyethylene oxide is grafted at one end of the resulting hydrophobic PS or PMMA chain. The polymerization leads to a decrease in the translational entropy and a microphase separation in which the hydrophobic PS or PMMA chains aggregate inward to form a "core"; and hydrophilic PEO remains on the surface to form a "shell", as if a "brush" is grafted on the interface, which reduces the interface energy, and is conducive to the thermodynamic stabilization.

As shown on the right, the formation of the "core-shell" structure also respectively reduces and increases the interaction time and the inter-diffusion, entanglement and fusion time of the chains between two colliding aggregates, making $\tau_I \ll \tau_R$, promoting kinetic stability, killing two birds with one stone. As long as changing the stabilizing group from a charged ion to a

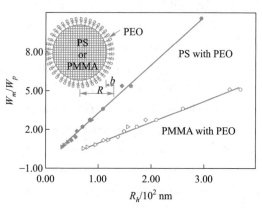

Wu C, Akashi M, Chen M Q. Macromolecules, 1997, 30: 2187.

(Figures 2 and 4, permission was granted by publisher)

neutral hydrophilic PEO short chain, one can once again use eq. (12.1). The figure on the right clearly shows that for two completely different systems and different lengths of PEO stabilizers (different symbols in each straight line), the plot of the weight ratio of hydrophobic monomer ("oil") to PEO stabilizer (W_m/W_p) versus the hydrodynamic radius of the core–shell aggregates (R_h) is a straight line, further confirming that **"the surface area occupied by each stabling group on the interface is a constant."** According to eq. (12.1), the average surface area (s) occupied by each polyethylene oxide short chain at the interface and the thickness of the stabilizing layer (b) are obtainable from the slope and intercept of each straight line. For the two systems, there are

$$s = \frac{7.54 \times 10^{-5}}{\gamma} M_P(\text{PS}) \quad \text{and} \quad s = \frac{1.79 \times 10^{-4}}{\gamma} M_P(\text{PMMA})$$

where γ and M_P are the partition coefficient in the mixed solvent and on the interface and the average molar mass of the PEO stabilizer, respectively, in full compliance with the aforementioned $s = 4\pi \langle R^2 \rangle \propto M_P$. In comparison with the PS particles, each PEO chain with the same length can stabilize a larger PMMA surface, which can be attributed to a higher solubility of PMMA in the mixed solvent, physically reasonable. b obtained from the intercept is 2–4 nm, which depends on the chain length of PEO, and is very close to the expected value. readers can consult the reference in the picture above.

Some specific examples of non–thermodynamic controlled and stabilized macromolecular colloids will be discussed as follows. In the preparation of colloids, grafting a layer of water-soluble chains on the particle surface is a method of stabilizing colloids. In the biomedical field, it is a common strategy to use a layer of polymer brush to prevent the protein adsorption on the surface. Microscopically, each grafted chain can be regarded as being in a "cylindrical tube" with a length of L_0 and a radius of R, formed by its surrounding chains,

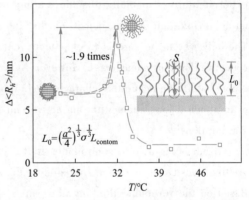

Hu T J, Wu C. Physical Review Letters. 1999, 83 : 4105.

(Figures 4, permission was granted by publisher)

whose cross–section area is $s = \pi R^2$; volume is $v = sL$, as shown in the schematic inset above. As described in Chapter 7 of discussing how a macromolecular chain passes through cylindrical pore under an elongation flow field, the free energy of a macromolecular chain with N segments of length b inside a "cylindrical tube" includes two parts related to the change of entropy: the chain expansion due to the excluded volume effect and the chain contraction due to the entropic elasticity. The mathematical expression of free energy normalized by thermal energy is as follows,

$$\frac{G}{k_B T} \approx \frac{b^3 N^2}{2sL} + \frac{L^2}{R_0^2} \qquad (12.3)$$

where the first term is the density of the two-body interaction pair, which is also the change of entropy. The second term is the entropy change caused when the chain conformation deviates from the undisturbed square end-to-end distance ($R_0^2 = b^2 N$). The constants related to the two-body excluded volume effect and conformational entropy change, respectively, in front of the two terms have been omitted. When $L = R_0$, the chain is in the undisturbed state, the excluded volume effect disappears, so that the first term is zero, $G = k_B T$. For a given graft density (the number of grafted chains per unit area, $\sigma = 1/s$), the above equation is differentiated with respect to L, and the minimum energy at equilibrium is calculated to obtain the average end-to-end distance at equilibrium, that is, the average thickness of the grafted layer (L_0).

$$\left(\frac{\partial G}{\partial L}\right)_{T,p} \approx k_B T\left(-\frac{b^3 N^2}{2s\,L^2} + \frac{2L}{b^2 N}\right) \qquad \rightarrow \qquad \left(\frac{\partial G}{\partial L}\right)_{T,p} = 0 \rightarrow L_0^3 \approx \frac{b^5 N^3}{4s}$$

According to the definition of conture length, $L_c = bN$. Substituting it to eq. (12.3) results in L_0,

$$L_0 = \left(\frac{b^2}{4}\right)^{1/3} \sigma^{1/3} L_c \propto \sigma^{1/3} N \qquad (12.4)$$

However, it is not easy to obtain different definite, controllable grafting densities experimentally, especially to achieve a high graft density of long chains. There are many reports on achieving high grafting density in the literature, most of which have not been strictly proven. The only way to be sure is to chemically cut the graft chain from the surface and measure its concentration and molar mass. However, in most cases, the amount of sample cut from a surface is too small tobe accurately characterized. Whether it is to initiate polymerization from a surface or to attach a long chain with one active end group to a surface, it is difficult to achieve a high grafting density.

If not being able to controllably graft a large number of chains on a given surface, can the grafting density be increased by reducing the surface area for a given number of grafted chains? Based on the reversible thinking, the water-soluble linear polyethylene oxide was grafted onto the spherical poly(N-isopropyl acrylamide) microgel at lower temperatures first, and then, as the temperature increases, the microgel shrinks, so that the grafting density can be continuously adjusted. The surface area of the microgel is proportional to the square of its radius, so that a shrinkage of 2.5 times in size means a 6.25 times increase in the graft density. Substituting them into the above equation leads to a thickness ($L_0 = \Delta R$) change of ~ 1.9 times, exactly as shown in the figure above; When the grafting density further increases, the thickness decreases instead, which is due to the clustering of polyethylene oxide at a high

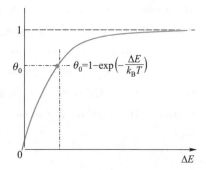

concentration. Interested readers can further refer to the reference listed in the figure above for details.

Why can grafting macromolecules, especially short polyethylene oxide chains, on the surface of colloidal particles apparently reduce the adsorption of proteins? For a long time, it has been attributed to the hydrophilicity and steric effect of the grafted macromolecular brush, which prevents protein molecules from approaching and adsorbing on the hydrophobic interface of the particles. However, even at the highest grafting density, the water content of the grafted layer is still more than 90%, so that a change of the grafting density should not alternate the interaction energy (ΔE) between the protein molecules and the hydrophilic grafted layer significantly. Known from thermodynamics, the adsorption capacity (θ) is the partition coefficient of protein in the aqueous solution and on the surface of colloidal particles, $\theta_0 = 1 - \exp(-\Delta E/k_B T)$, as shown above. For a given ΔE, θ_0 is a constant, so that the protein adsorption should not significantly decrease as the grafting density increases. But experimental data shows that the adsorption not only decreases as the grafting density increases, but also increases when the grafting density is too high. The hydrophilicity and steric effects cannot explain this anomaly.

Substituting eq. (12.4) into eq. (12.3), and using the undisturbed state as a reference state, one can calculate the minimum change of free energy of each grafted chain as

$$\frac{\Delta G}{k_B T} \approx \left(\frac{b}{2}\right)^{4/3} \sigma^{2/3} N \tag{12.5}$$

At one extreme, the undisturbed grafted chain just touched each other, $\sigma_{min} = 1/R_0^2 = 1/(b^2 N)$; at the other extreme, the grafted chain was completely straightened with the highest graft density, $\sigma_{min} = 1/b^2$, where $b \sim 1$ nm. Therefore,

$$\frac{2}{5}N^{1/3} \leqslant \frac{\Delta G}{k_B T} \leqslant \frac{2}{5}N$$

For commonly used polyethylene oxide grafted chains (2,000–5,000 g/mol), $N \sim 25$–50. As the grafting density changes,

$$k_B T \leqslant \Delta G \leqslant 10 \sim 20 \, k_B T$$

One end of the grafted chain is fixed, and the other end is constantly and randomly fluctuating and oscillating around its equilibrium position (L_0) under the agitation of thermal energy. According to statistical physics, when a thermodynamic system reaches a macroscopically stable dynamic equilibrium state under given experimental conditions (temperature, pressure, concentration,...), its free energy must be the lowest. Microscopically, under the agitation of thermal energy, the centroid of molecules or particles are still moving randomly, causing the system to slightly deviate from its stable equilibrium state, always causing an increase in free energy, always resulting in a positive (increase) fluctuations. Here, the fluctuation variable is selected as the average end–to–end distance (L_0) of the grafted chain, and the free energy is Taylor expanded around the average end–to–end distance as

$$\delta G = G - G_0 = \left(\frac{\partial G}{\partial L}\right)_{T,p} \delta L + \frac{1}{2}\left(\frac{\partial^2 G}{\partial L^2}\right)_{T,p} (\delta L)^2 + \cdots$$

where all the partial differentials of free energy at L_0 are constants. G is randomly fluctuating, so that the time average of δL is zero. After the time average, the first term on the right disappears. If the cubic contribution of the fluctuation is ignored, the above equation becomes

$$\langle \delta G \rangle = \frac{1}{2}\left(\frac{\partial^2 G}{\partial L^2}\right)_{T,p} \langle (\delta L)^2 \rangle$$

In addition, from general statistical considerations, in the equilibrium state, the time average fluctuation of free energy of each grafted chain is only half of the thermal energy, i.e., $\langle \delta G \rangle = k_B T/2$, always positive, regardless of whether the grafted chain is contracted or stretched. The substitution of it into the above equation results in

$$\langle (\delta L)^2 \rangle = k_B T \Big/ \left(\frac{\partial^2 G}{\partial L^2}\right)_{T,p}$$

The above equation is a variant of eq. (5.32). According to eq. (12.3), $(\partial^2 G/\partial L^2)_{T,p}$ and $\langle (\delta L)^2 \rangle$ are obtainable.

$$\left(\frac{\partial^2 G}{\partial L^2}\right)_{T,p} = k_B T \left(\frac{b^3 N^2}{s L^3} + \frac{2}{b^2 N}\right) = k_B T \left(\frac{R_0^2 bN}{s L^3} + \frac{2}{R_0^2}\right) \qquad \rightarrow \qquad \langle (\delta L)^2 \rangle = \frac{R_0^2 L^3}{R_0^4 L_c \sigma + 2 L^3}$$

Further, the relative mean square chain length fluctuation $\langle (\delta L)^2 \rangle / L^2$ can be calculated, and its square root leads to dL/L; namely,

$$\frac{\langle (\delta L)^2 \rangle}{L^2} = \frac{R_0^2 L}{R_0^4 L_c \sigma + 2 L^3} \tag{12.6}$$

where R_0^2 is independent of L_c and L. For a given σ, when L is very small or very large, $\langle (\delta L)^2 \rangle / L^2$ is small, so that there is a L_{optimal}, making $\langle (\delta L)^2 \rangle / L^2$ to reach a maximum. The partial differential of the above equation with respect to L first, and then letting it equal to zero lead to

$$L_{\text{optimal}} = \left(\frac{R_0^4 L_c \sigma}{4}\right)^{1/3} = \left(\frac{b^2 \sigma}{4}\right)^{1/3} L_c$$

L_{optimal} and σ_{optimal} are related. Just like discussing a linear macromolecular chain inside a cylindrical pore, one divides a grafted chain into L/d "blobs" according to the "tube diameter" (d), and each "blob" contains N_b segments with a length is b. The number of "blobs" can be calculated by L/d or N/N_b, the two are naturally equal, that is, $L/d = N/N_b$. Note that $d^2 = b^2 N_b$ and $1/\sigma = s = \pi d^2/4$. Using $R_0^2 = b^2 N$, $L = dN/N_b = d^2 N/(dN_b) = b^2 N_b N/(dN_b) = R_0^2/d = (\pi/4)^{1/2} R_0^2 \sigma^{1/2}$. Substituting it into the above equation leads to

$$(\pi/4)^{1/2} R_0^2 \sigma_{\text{optimal}}^{1/2} = \left(\frac{R_0^4 L_c \sigma_{\text{optimal}}}{4}\right)^{1/3} \qquad \rightarrow \qquad s_{\text{optimal}} = \frac{1}{\sigma_{\text{optimal}}} = \frac{\pi^3 b^2}{4} \tag{12.7}$$

The substitution of it into the previous equation leads to $L_{\text{optimal}} \approx L_c/\pi$, which is about 1/3 to 1/4 of the Conture length of the chain. The physical meaning of the above expression is very interesting. To achieve the maximum relative fluctuation of the free end of the grafted chain, the optimal grafting density is only related to the segment length b, not to the total chain length, which is expected, because it is the relative fluctuation. The segment length of a chain reflects its flexibility, and the longer the segment, the more rigid the chain is. The optimal grafting density decreases as the chain rigidity increases. This explains why the most flexible polyethylene oxide is often used to graft a surface to delay the protein adsorption inside the body. Substituting eq. (12.6) into eqs. (12.5) and (12.6), the free energy at the optimal grafting density and the relative fluctuation amplitude of the grafted chain can be obtained respectively as follows.

$$\frac{\Delta G}{k_B T} \approx \left(\frac{b}{2}\right)^{4/3} \left(\frac{4}{\pi^3 b^2}\right)^{2/3} N = \frac{N}{\pi^2} \approx \frac{N}{10} \quad \text{and} \quad \frac{\Delta L}{L} = \frac{\sqrt{\langle (\delta L)^2 \rangle}}{L} = \frac{\pi}{(6N)^{1/2}} \propto N^{-1/2} \quad (12.8)$$

At the optimal grafting density, for a typical polyethylene oxide grafted chain (2,000–5,000 g/mol), the free energy (G) of a grafted chain is 4–6 times the thermal energy; the relative fluctuation amplitude of a grafted chain is 26% – 18%; the absolute fluctuation amplitude: $\Delta L = b \, N^{1/2}/6^{1/2} = R_g$, which is an interesting result. The qualitative understanding of the above results is not difficult. When the grafting density is low, $\langle \delta G \rangle$ will cause a large relative change in free energy, but the absolute

fluctuation amplitude of the grafted chain is smaller; if the grafting density is close to σ_{max}, the grafted chain is fully straightened, the relative change in free energy caused by $\langle \delta G \rangle$ is small, and the relative fluctuation amplitude of the grafted chain is also small. Therefore, there is an optimal grafting density and a corresponding free energy, so that the energy fluctuation $k_B T/2$ can induce the largest relative fluctuation amplitude. Therefore, each grafted macromolecular chain is acting like a "molecular spring", oscillating up and down; the surface is constantly undulating, as shown above

It is conceivable that protein molecules will be less likely to be adsorbed on the surface of such a dynamic "macromolecular brush" than a static surface. A "macromolecule brush" with an appropriate grafting density can significantly slow down the adsorption instead of preventing the adsorption. In other words, the amount of adsorption is controlled by thermodynamics and depends on the interaction energy (ΔE) between the protein molecule and the surface of the grafted layer, and its change with the grafting density is not significant; while the adsorption rate is subject to kinetics, which depends on the characteristic adsorption time (τ) of a specific system.

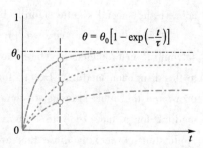

Mathematically, it is expressed as $\theta = \theta_0 [1 - \exp(-t/\tau)]$, as shown on the right. For a given system, a change of the grafting density actually affects τ, not ΔE. Since the observation and comparison of adsorption capacity is always at a given adsorption time, instead of waiting until the adsorption equilibrium. The figure above vividly shows that different characteristic adsorption times can lead to apparent, rather than true, adsorption differences.

Therefore, macromolecular chains grafted on a surface can slow down, not reduce, the protein adsorption; it is based on the adsorption kinetics, not originated from thermodynamic adsorption equilibrium. This difference in the apparent protein adsorption is sufficient to be used to solve some application problems. For example, in the drug intravenous delivery system, generally speaking, as long as the protein adsorption is slowed down within 24 hours, it can avoid being cleaned by the body's immune system, extending the circulation time of a drug in the blood vessel to improve its efficacy. In addition, in the formulation of the last paint under the waterline of a ship, a proper design can make the paint surface continuously form a dynamic surface that is constantly renewed when it comes in contact with seawater, preventing the adsorption of proteins and marine organisms, so as to realize the prevention of marine pollutions.

Based on the above discussion, it is known that, unlike small-molecule colloids, the formation and stability of a macromolecular colloid not only follow the principle of thermodynamics, that is, the surface area occupied by each stabilizing group on the interface of colloidal particles has a minimum value, a constant for a given system; it is also controlled by kinetics, which is related to the formation path and depends on the competition between the interchain aggregation and intrachain contraction of macromolecule chains, forming a kinetic stability.

There are generally two prerequisites for achieving kinetic stability. First, the aggregate itself is not too large, less than about one micrometer, and thermal energy can agitate each particle to translational diffusion in the dispersion system. Second, the densities of the aggregate and the dispersion medium are similar. Otherwise, the sedimentation (buoyant) force overcomes the random force generated by thermal energy, resulting in the directional movement of the particles, precipitation or floating. In the piled state, the interaction time between the aggregates increases sharply, leading to the fusion of aggregates and macroscopic phase separation. If the two prerequisites are met, the collision between two aggregates will generally not lead to the diffusion, entanglement and fusion of macromolecular chains between the two colliding aggregates, so as to achieve the special kinetic stability based on the unique "viscoelasticity" of macromolecules.

Obviously, the above two prerequisites can be met by chemically cross-linking the solvophilic macromolecular chains to form a small, swollen microgel with a density almost equal to the dispersion medium. Except for the local movement of the segments, the crosslinking has prevented the inter-microgel fusion. Without any stabilizers, they can be dispersed stably in the medium for a long time to form a special kind of macromolecular colloids with promising applications, especially those that are able to swell in water and sensitive to the environment

(temperature, pH, salt concentration, etc.).

Let us take only thermally sensitive, spherical poly(N-isopropyl acrylamide) microgel as an example. In a dispersion system without stabilizers, the average hydrodynamic radius of the microgel is ~110 nm and ~45 nm at 25 °C and 35 °C, respectively. Whether it is swollen at lower temperatures or collapsed at higher temperatures, the dispersion system can still be stable for a long time. Since its hydrodynamic volume can change about six times as the temperature decreases, one can prepare a colloidal dispersion with a microgel volume fraction of ~20% (the weight fraction is less than 5%) at 35 °C first, and then, lower the temperature to the room temperature. The volume fraction of all the swollen microgels exceeds 100%. They have to jam together, forming a macroscopic gel that is not able to flow. The inside of each microgel is a chemically cross-linked gel network, but different microgels are jammed together physically, so that this is a chemically and physically hybridized macroscopic gel. Its porous structure changes with the cooling rate and method, as shown below in the schematic inset in the figure(c).

The preparation of natural macromolecular gels has a longer history than the macromolecular science. If the macromolecular chain forms a three-dimensional network through physical interactions (hydrogen bonding, crystallization, strong hydrophobic interaction, etc.), it is also called "physical gels". It has been found in experiments that the faster the cross-linking reaction rate, the more inhomogeneous the gel formed, its time-average scattered light intensity

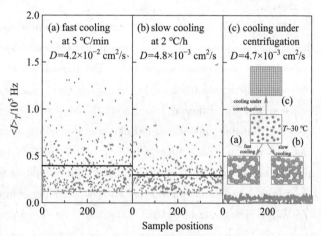

Zhao Y, Zhang G Z, Wu C. Macromolecules, 2001, 34: 7804.

(Figures 2 and 4, permission was granted by publisher)

($<I>_T$) changes with the hitting position of the laser beam, and the light intensity jumps, as shown above. In other words, it does not follow the ergodicity of statistically random variables, and the time average is not equal to the position average, which is completely different from macromolecular solutions. There are two reasons for this. One is the inhomogeneity of the cross-linked network; the other is that the chain in the cross-linked state cannot traverse all points in the system. For a long time, people attributed this inhomogeneity to the formation of branching clusters or microgels of different sizes during the crosslinking process.

In fact, this explanation ignores the existing experimental results and experimental conditions. The size of the branched clusters or microgels formed (measured) just before the gelation point is generally smaller than 1 μm, close to the wavelength of visible light, and much smaller than the size of usual LLS scattering volume (~200 μm). In other words, in the LLS volume, there are millions of branched clusters or microgels. Even with different sizes, their average scattered light intensity at different positions is also a constant. Why is the observed intensity of scattered light jumping? The data obtained from different cooling rates and methods in the figure above answers this question.

In the aforementioned macroscopic hybrid gel, the size distribution of the microgels is narrow, and their hydrodynamic radii are less than 500 μm. Obviously, in the macroscopic hybrid gel obtained during the rapid cooling (the above figure (a)), the jumping amplitude of the scattered light intensity is the largest, and the position average scattered light intensity is also the highest (the black line in the above figure). This cooling rate dependence first negates that the jumping of the scattered light intensity is from the uneven size of the branched clusters or microgels. In the above figure (c), the microgels are densely packed on the bottom of the LLS cell by ultracentrifugation, and then the temperature is lowered while centrifuging, so that the microgels gradually swell in the piled state. It shows that not only the jumping in the scattered light intensity disappears, but also the position average scattered light intensity is the lowest.

Generally speaking, in the process of polymerizing monomers to form a chemical gel, it follows the microscopic route of the "linear chain–branching–branched clusters–microgels–gelation (percolation) point–macroscopic gel". The thermal energy drives the macromolecules of various sizes and structures formed in the reaction mixture to approach a uniform density by translational diffusion, but the diffusion takes time. If the cross–linking between macromolecules is faster than the diffusion, and the crosslinking further slows down the diffusion, then the crosslinking proceeds near several almost stationary centers in the system. Smaller and movable monomers and macromolecules will gradually move towards and crosslink on these cross–linking centers. The result is the formation of large solvent voids between these crosslinking centers.

When discussing laser light scattering, it has been pointed out that each scattering unit in a pure optical glass always has another corresponding scattering unit with a phase difference of 180°. Their scattering electric fields are canceled by interference, so that except for the zero angle (incident direction), a pure optical glass does not produce any scattered light in theory. If there are some bubbles, there will be strong scattering. The scattered light intensity does not come from the gas molecules in the bubbles, but from a small piece of glass with the same volume and 180° phase difference corresponding to each bubble. The reason is that the number of molecules in the bubble is far less than the number of molecules in the corresponding small piece of glass; so that the bubble scatters much less light than corresponding glass and their scattered electric fields are not able to cancel each other by the interference.

The position dependent scattered light intensity is also on the basis of the same principle. A macromolecular network is equivalent to an "optical glass", and each large solvent cavity is equivalent to a large "bubble". The solvent cavities are fixed in different positions by the cross–linking reaction, and their size and distribution do not follow the statistical law and are extremely uneven. Therefore, at different locations, the size and number of solvent holes in the scattering volume are different, leading to the scattered light intensity to jump. In the formation of the aforementioned hybrid gel, the cooling rate is equivalent to the crosslinking rate. When it is rapidly cooled down, each microgel swells extremely fast, so that the microgel does not have enough time to diffuse from one position with a higher density to another position with a lower density, forming many large solvent cavities. In the slow cooling under the ultracentrifugation, larger solvent cavities are not able to formed, so that a uniform, weakly scattered hybrid gel without the jumping of scattered light intensity is obtained.

This is just an example of using the unique properties of the microgels to answer scientific questions. There are many suggestions and ideas about the potential applications of microgels in the literature, but most of them are on paper. In the past 30 years, "only listen to the sound of stairs but not people come down". So far there still lacks a large-scale practical application of microgels. It is not that the microgels are useless, but the research has not been focused on real practical problems, and researchers have not "come from practice to practice"! Let us take one potential application of microgels in biomedicine as an example. Dai and his collaborators [Journal of Biomaterials Applications, 2015, 29:1272] used the phenomenon that the material modulus influences the stem cell differentiation. They mixed the shrinking microgel with stem cells at ~40 ℃, and used the swelling property of the microgels at the body temperature to exert a pressure on stem cells inside the body, induce stem cells to differentiate into chondrocytes, repairs the damaged knee cartilage, and achieves the effect of " one stone kills two birds": providing lubrication and generating cartilage.

There is also a "Volume Phase Transition Temperature" (VPTT) problem connected with temperature-sensitive microgels. The term VPTT originally from a paper by Dusek and Patterson [Journal of Polymer Science: Part A-2, 1968, 6:1209]. In this theoretical article, they studied an ideal gel with a uniform segment number (N_b) between two adjacent crosslinking points, and clearly stated that even in such an ideal gel, the observation of the "volume phase transition" has to meet the following three conditions simultaneously: first, the interaction between the macromolecular chain and the solvent (χ_{sp}) should be as strong as possible; second, in the dried gel, the number of cross-linking points per unit volume (v^*) should be as large as possible, i. e., N_b should be as small as possible; third, the ratio of the mean square end-to-end distances of the chain link between two adjacent crosslinking points in the fully collapsed dried state and the undisturbed (Θ) state ($\langle \alpha^2 \rangle_0 = N_b^{2/3}/N_b = N_b^{-1/3}$) should be as small as possible, i.e., N_b should be as large as possible.

Obviously, the second and third conditions are contradictory. Therefore, in the introduction of the article, they have made an announcement and conclusively pointed out that "it is difficult to obtain these necessary phase transition conditions when swelling freely". However, the vast majority of people who use the term VPTT or even cite this document did not read or did not understand this earliest document, unfortunately, swallowing a date without tasting it, just following others. In experiments, it is difficult to find the above possible narrow window of N_b, and it is also impossible to prepare a uniform and ideal gel network. In any actual gel, there is always a length distribution of the chain links between two adjacent crosslinks. As Dusek and Patterson pointed out, the chain links with different lengths result in different phase transition temperatures. How to explain many experimental observations that the macroscopic gel volume has abrupt changes at a certain temperature? In the discussion of Chapter 10, it has been shown that the phase transition temperature depends on the chain length. Regardless of whether the solution has UCST or LCST, after a given volume fraction of macromolecules, long chains always precede short chains to undergo a phase transition.

Microscopically, the macromolecular chain links between two adjacent crosslinking points have different lengths. Therefore, as the temperature changes, longer chain links inside the gel shrink first, forming a local phase separation, causing an internal stress, but much smaller that the macroscopic shear modulus of the gel, so that its macroscopic size remains unchanged. When the solvent properties gradually deteriorate, more and more chain links shrink and the internal stress continues to increase. At a certain temperature, when the internal stress is larger than the macroscopic shear modulus, the macroscopic size (volume) will suddenly decrease. However, the macroscopic abrupt change observed experimentally is not due to the volume phase transition predicted by Dusek and Patterson, but due to the competition between the internal stress and the macroscopic shear modulus. It is like a rubber band that suddenly breaks during its stretching, but it's not a phase transition.

The figure below compares the temperature dependences of relative hydrodynamic volume of poly(N-isopropyl acrylamide) linear chains and spherical microgels, where $<V_h>*$ is the hydrodynamic volume when they are fully collapsed. Note that the linear chain is not only nearly a thousand times longer than the chain link between two adjacent crosslinks in the microgel, but also has a narrower chain length distribution ($M_w/M_n < 1.0$). Therefore, the linear chain has a lower shrinking temperature and a narrower range of the temperature change than the spherical microgel. Due to the crosslinking, the relative volume change of the microgel is also smaller. For microgels, any internal stress will transmit to the entire microgel, which reflects in the change of its hydrodynamic size. This also explains why in the same macromolecular system, microgels have never observed a sudden volume change.

Physically, there is a strict definition of whether a physical change or sudden change is a "phase transition". The study of phase transition is an important part of condensed matter physics. If it is a phase transition, then the next two questions are, what is the order of the phase transition? And what is the universal scaling law near the phase transition

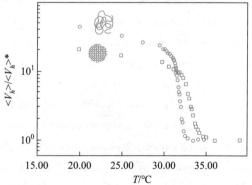

Wu C, Zhou S Q. Macromolecules, 1997, 30: 574.

(Figures 3, permission was granted by publisher)

(critical) point? It is the same as when discussing the phase transition of macromolecular solution. Whether it is discussing the volume change of a microgel or a gel, the correct expression is the "volume change temperature", not "volume phase transition temperature". Researchers working in the field of gels should know why there is a physical difference between the two and should not be used indiscriminately!

The last example in this chapter is the formation of relatively stable macromolecular colloids in an aqueous solution via the complexation of two kinds of polyelectrolytes chains with opposite charges. Under the conditions of a proper charge ratio, concentration, and mixing method, the surface of the complex formed can be positively or negatively charged. In organisms, protein molecules themselves are composed of alkaline and acidic amino acids. In the biological environment, each protein chain can be positively or negatively charged overall. Protein chains with different charges are often complexed together to form a stable structure. For example, the 11S globulin in soy contains six protein chains with a total charge of positive and six protein chains with a total charge of negative, forming six pairs of dimers. Recently, these complexes of differently charged macromolecules have also been given various "novel" names, but "more water is added and cooked without changing the medicine inside". It is not necessary to list these "new" names one by one, because there is really nothing new.

Microscopically, when two macromolecular chains with opposite charges meet, they can form multi-point complexation. The interaction of each complexation point is weak, so that it is constantly in an "open-close" process, establishing a dynamic equilibrium. But once the two chains are complexed, their separation requires all the complex points not only to be unraveled at the same time, but also the two chains are separated immediately by diffusion, the probability is extremely low, so that once complexed, the separation is difficult. Textbooks and literature often attribute such complexation to electrostatic interactions, but forget that in the respective aqueous solutions before the mixing, each charged macromolecular chain and its counter ions already have electrostatic interactions, and the charge neutrality can never be violated. During the complexation, a positive charge on a macromolecular chain does not know whether its counter ion is a small molecule or a negative charge on another macromolecular chain. The formation of a macromolecular complex point releases a pair of small molecular counter ions, and the translational entropy increases. It is this increase in entropy that drives the complexation between oppositely charged chains, not the enthalpy implied by the electrostatic interaction.

In comparison with the complexation of small molecules with opposite charges, when the equivalents of positive and negative charges are similar, the complexation of macromolecules is very similar, leading to macroscopic precipitation, releasing counter ions. However, the macromolecular chain is not infinitely flexible, and the opposite charges on the two chains are difficult to match one by one, especially when the lengths of the two oppositely charged chains are very different, there will inevitably be some uncomplexed charges. As before the mixing, their electrical neutrality is maintained by small molecular counter ions. When the equivalents of positive and negative charges differ greatly, in the complexation of small molecules, fewer components will precipitate due to the complexation to form one phase, while the remaining ions and counter ions in the more components and the counter ions released due to the complexation remain in the solution, forming another phase. In the macromolecular complexation, the chains with an excessive amount of charges will complex on the surface, called the overcharge, which not only reduces the interface energy and provides the thermodynamic stability, but also reduces the time of interaction between the two collision complexes and promotes the dynamic stability.

A proper choice of the complexation pass can make the intrachain contraction much faster than the interchain complexation, and the size of the complexes can be controlled smaller than 500 nm, so that the thermal energy is sufficient to drive the complex to randomly walk in the solution mixture. Such obtained complexes can be kinetically stable for a long time, forming a metastable macromolecular colloid, without the macroscopic precipitation and phase separation. In recent years, polyelectrolytes with cations (nitrogen-containing groups, including various amino groups) are used to complex deoxyribonucleic acid (DNA) with anions (phosphate) to form non-viral gene vectors (polyplexes), which is a typical example of applications. Related research has produced more than 100,000 articles in the past thirty years. However, the vast majority of researchers working in this field, especially those studying pharmacy and biology, have not noticed the complexity of macromolecule complexation, including the mixing path dependence, and polymers themselves have the chain length, configuration and other special properties.

A large number of experimental results from different laboratories show that at the equivalent charge (N : P = 1 : 1), there are still a large number of uncomplexed DNA chains remaining in the solution mixture, since the negatively charged and rigid The DNA chain is much longer than the polycationic chain. Until N : P = 3 : 1, most of the DNA chains are complexed, which means that only about 1/3 of the cations are involved in the complexation. A small number of uncomplexed cations are located on a small number of individual uncomplexed polycationic chains, freely and randomly diffusing in the solution mixture; some are wrapped in the interior of the complex; more are distributed on the surface of the complexes, making it positively charged, playing a stabilizing role. The strange thing is that in subsequent and large numbers of gene transfection experiments, different laboratories also found that it is only when N : P ~ 10 : 1 that the gene transfection efficiency can be significantly improved.

What is interesting is that in more than two decades of research, only few researchers have considered these two definite experimental results together and asked two obvious questions. Since when N : P>3 : 1, there are no individual DNA chains free in the solution mixture. Why is the gene transfection efficiency high only when N : P ~ 10 : 1? Where do those extra 7 portion of polycationic chains go? The answer is obvious. First, the excess polycationic chains must be dissolved as individual chains free in the solution mixture, as shown in the figure (a) below. Second, it is those polycationic chains free in the solution mixture that promote the gene transfection. The white dots in the middle of the figure show the expression of the white fluorescent gene in the cell, the more, the higher the gene transfection efficiency. Those polycation chains that are complexed with DNA only provide positive charges, making the DNA chains to collapse into smaller particles, convenient for endocytosis.

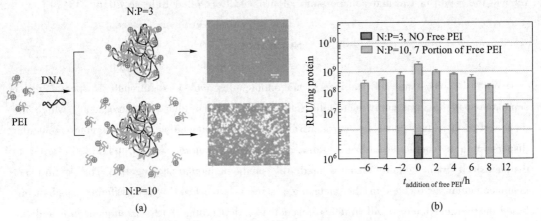

Yue Y N, Jin F, Deng R, et al. Journal of Controlled Release, 2011, 155:67.
(Graphical Abstract, permission was granted by publisher)

Therefore, when N : P = 10 : 1, the added polycationic chains can be divided into two parts, 3 portions are complexed with DNA; 7 portions are free individual chains in the mixed solution. The upper figure (b) shows the results of a decisive experiment. The ordinate is the corrected gene transfection efficiency, which is a logarithmic scale. The abscissa is the relative time of adding 7 portions of polycationic chains (free single chains). The benchmark is $t = 0$ hr, the DNA/macromolecule polyplexes (N : P = 3 : 1) and the solution with 7 portion of free polycationic chains are simultaneously added into the gene transfection system (cells and culture medium); $t = -6$ hrs, the solution with 7 portions of free polycationic chains was added first, and then, after 6 hours, the DNA/macromolecule polyplexes (N : P = 3 : 1) were added; $t = 6$ hrs, the DNA/macromolecule polyplexes (N : P = 3 : 1) were first, and then, 7 portion of free polycationic chains 6 hours later. The total nitrogen to phosphorus ratio is maintained at N : P = 10 : 1.

Obviously, when there is no free polycationic chains, i. e., N : P = 3 : 1, the gene transfection efficiency is very low. Whether free polycationic chains are added before or after the addition of the DNA/macromolecule polyplexes, the gene transfection efficiency is significantly increased by 10 – 100 times. Similar results were obtained with different cells and different polycationic chains. Subsequent quantitative experiments further confirmed that free polycationic chains mainly have the following four-fold effects. First, it interferes with the SNARE signaling protein on the endosome, which is originally on the inner surface of the cell wall, thereby blocking the fusion of the endosome and lysosome, so that the gene can be released in the cytosol; second, it weakens the structure of the nuclear membrane, and accelerates the passage of genes through the nuclear membrane into the nucleus; third, it helps the transcription of genes in the nucleus to generate more ribonucleic acid (RNA); fourth, it promotes RNA to pass through the nuclear membrane into the cytosol, finally realizing the protein synthesis. For details, please refer to the report of Cai and collaborators [Journal of Controlled Release, 2016, 238:71].

Summary

With the assistance of stabilizers, macromolecules can pass through a special phase separation path to form many metastable micro-concentrated phases (aggregates), which are dispersed in the dilute phase (a saturated solution of macromolecules) to form a macromolecular dispersion in a kinetic equilibrium state, also called a macromolecular colloid, in order to distinguish it from colloids usually made of small molecular aggregates. The stability of macromolecular aggregates in the metastable state is governed by two different mechanisms based on thermodynamics and kinetics, respectively, depending on the macromolecule and the dispersion medium, the concentration, the content of stabilizer, and the path of formation.

According to thermodynamics, for a given colloid (dispersion system), the interface area occupied by each stabilizer (ion, water-soluble segment, etc.) on the surface of the aggregate should be a constant. For a macromolecule of a given mass, the more stabilizer, the larger the stabilized interface area. The aggregates become smaller and provide more interfaces; vice versa. When the contents of stabilizer and macromolecule are similar, the thermodynamic effect can be dominant. Therefore, the ratio of the feeding mass of macromolecule to stabilizer determines the size of the aggregates. A plot of the two leads to a straight line. The average interface area occupied per stabilizer and the average thickness of the stabilizer layer on the interface can be obtained from the slope and intercept, respectively.

However, in comparison with small molecular colloids, macromolecular colloids have some obvious differences, which are all related to the chain length. There are two main points. First, in a dilute solution, the translational diffusion of macromolecular chains is $\sim 10^2$ times slower than that of small molecules. As the concentration increases, due to the entanglement between the chains, the movement of the centroid of macromolecular chains becomes slower, up to $\sim 10^4$ times slower. There is no entanglement between small molecules, so that the concentration has little effect on the movement of small molecules. Second, as the solvent quality becomes worse, the chain conformation of macromolecule will gradually contract, resulting in an increase in the local concentration of the chain, while there is no such change for small molecules.

Therefore, during the phase separation, like small molecules, macromolecules not only have the inter-molecule (chain) aggregation, but also the intrachain contraction. In a dilute solution, the rates of the two are similar, so that the interchain aggregation and intrachain contraction always accompany and compete with each other. Once the conformation is collapsed, the time required for the chain to move inside the aggregate and the chain exchange (fusion) between two colliding aggregates is much longer than the interaction time between two collision aggregates, one thousand times longer or more.

According to rheology, when the time ofinteraction on an object is much shorter than the time required for the molecules inside to relax (move) an characteristic length under the interaction, the object exhibits elasticity; otherwise, viscous. In other words, when two macromolecular aggregates collide, each of them behaves like two tiny elastic " glass " ball during the interaction time. The possibility of forming a large aggregate is very low. Based on this " viscoelasticity ", the aggregates are dynamically stabilized in micro-phase separation state, a metastable state, and reaching dynamic stability. The stability time varies with the system and the formation path.

As mentioned earlier, there are two prerequisites for kinetic stability: First, the aggregates are generally not too large, less than 1 μm, so that the thermal energy can agitate each particle to walk randomly and translational diffusion in the dispersed system. Second, the densities of the aggregate and the dispersion medium are similar. Otherwise, the sedimentation (buoyant) force will overcome the random movement caused by the thermal energy, which will lead to the directional movement of the aggregates, precipitation or floating upwards. The aggregates are piled up together The interaction time rapidly increases, making the aggregates to fuse, and a macroscopic phase separation occurs. If these two conditions are met, the collision between the aggregates will not lead to the inter-aggregate diffusion, entanglement and fusion of the chains, so as to achieve the unique kinetic stability of macromolecular colloids based on " viscoelasticity ".

In the phase separation, during the aggregation of amphiphilic copolymer chains, driven by thermodynamics, the solvophobic segments gather in the center and the solvophilic segments move to the surface to reduce the interfacial energy. This process also promotes kinetic stability additionally, that is, the solvophilic segments distributed on the interface reduce the interaction time between the aggregates, and the internal aggregation of the solvophobic segments further slows down the chain relaxation, which prolongs the characteristic relaxation time, making the difference between the two times larger, and more conducive to the kinetic stability.

Hundreds of millions of years of evolution have made amphipathic protein molecules not only a typical representative, but also a leader in using this kinetic stability. Therefore, many protein molecules can pass through special pathways in the ribosome within the cell, polymerizing and folding into kinetic stable single-chain aggregates. Outside the cell, some denatured protein molecules can also be refolded into biologically active and kinetic stable single-chain aggregates by using special methods. At present, the study of protein single-chain folding always uses the thermodynamically stable state with the lowest free energy as the reference point. This idea may not be correct. A protein molecule with biological activity is likely to be in a metastable state with slightly higher energy and kinetically metastable state.

There are other special stabilization methods. One of which is to graft linear solvophilic chains on the surface of aggregates to form macromolecular brushes. Each grafted chain is surrounded by its adjacent grafted chains. Based on the thermodynamic principle of reducing free energy, they do not interpenetrate and entangle with each other, but stretch upwards as much as possible, as if each of them is in a cylindrical "tube" composed of surrounding grafted chains. At equilibrium, the average end–to–end distance of the grafted chains (the brush thickness) is proportional to 1/3 power of the grafting density (the number of grafted chains per unit interface area) and the chain length. The average free energy of a grafted chain is proportional to 2/3 power of the grafting density and the chain length.

The thermal energy agitates the free end of each grafted chain to vibrate constantly and randomly, leading to a constantly change of the end–to–end distance of every grafted chain. The ratio of its average fluctuation amplitude (ΔL) to the average end–to–end distance ($\langle L \rangle$), i.e., the relative fluctuation amplitude, has a maximum value, corresponding to an optimal grafting density, independent of the chain length, but inversely proportional to the square of the segment length (b). At the optimum grafting density, the average free energy (G) of the grafted chains is equal to $[(N/10)+1]k_B T$; the average fluctuation amplitude is approximately equal to the average radius of gyration of the grafted chain ($b N^{1/2}/6^{1/2} = \sqrt{\langle R_g^2 \rangle}$); the average end–to–end distance of the grafted chain is about 1/10 of its conture length (bN). Therefore, $\Delta L/L$ is only related to the chain length. For a typical graft chain ($N \sim 30$) at the optimum graft density, $\Delta L/L$ is about 20% and $G \sim 4$–$6 \, k_B T$.

In biomedicine, it is a common strategy to graft macromolecular chains on the surface of drug carriers to "prevent" the proteins adsorption. Obviously, the flexible the macromolecular chain, the smaller the optimal grafting density, the larger the interface area occupied by each grafted chain, and the best stabilization effect. Polyethylene oxide is currently known as the most flexible chain in polymers, which explains why in the research and development of biomedicine, polyethylene oxide is the most common grafted chain, and its clinical applications have been approved by the Drug Administration Office.

According to the principle of thermodynamics, the percentage of protein adsorption depends only on the interaction energy between the interface and the protein. The grafting density has little effect on the interaction energy, so that a variation of the grafting density should not affect the protein adsorption much. Experimentally, the observation of the grafting density dependence of the protein adsorption is originated from the change of the adsorption kinetics. The random vibration of the free end of each grafted chain constitutes a constantly undulating dynamic surface, making the adsorption of protein molecules relatively difficult and prolonging the characteristic adsorption time in the adsorption kinetics. For a given observation time, the slower adsorption gives a relatively lower apparent amount of the adsorption, creating a false impression of less adsorption. Therefore, **the macromolecular brush grafted on the surface does not "prevent" but "delay" the adsorption**. Using a flexible chain and choosing the optimal grafting density can maximize the relative fluctuation amplitude ($\Delta L/L$), and slow down the protein adsorption. The interpretations in the literature need to be revised.

Another special stabilization method is to crosslink the macromolecular chains in the aggregates, so that the inter-aggregate diffusion, entanglement and fusion are prevented. Microgels prepared in water are a typical example. The density of many polymers is close to water ($\sim 1 \text{ g/cm}^3$). The water-soluble gels prepared from these polymers can be dispersed into microgels with a size of hundreds of nanometers, which naturally meets the above two necessary kinetically stable conditions. Driven by the thermal energy, these microgels randomly diffuse in water. The sedimentation (buoyant) force is not sufficient to make them sink or float, forming two layers. Due to the internal crosslinking, two colliding microgels cannot fuse into a larger microgel. Therefore, without any stabilizer or after the removal of the stabilizer added during preparation, the swollen microgel are still stable in water for a long time. With proper designs and preparation, microgels can be used to solve some scientific and application problems. The figure below summarizes the main issues discussed in this chapter.

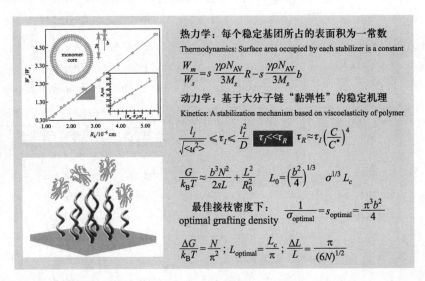